FOR ALL PRACTICAL PURPOSES

PROJECT DIRECTOR

Solomon Garfunkel, *Consortium for Mathematics and Its Applications*

COORDINATING EDITOR, FIRST EDITION

Lynn A. Steen, *St. Olaf College*

CONTRIBUTING AUTHORS

PART I   MANAGEMENT SCIENCE

Joseph Malkevitch, *York College, CUNY*
Rochelle Meyer, *Nassau Community College*
Walter Meyer, *Adelphi University*

PART II   STATISTICS: THE SCIENCE OF DATA

David S. Moore, *Purdue University*

PART III   CODING INFORMATION

Joseph A. Gallian, *University of Minnesota - Duluth*

PART IV   SOCIAL CHOICE AND DECISION MAKING

Steven J. Brams, *New York University*
Bruce P. Conrad, *Temple University*
William F. Lucas, *Claremont Graduate School*
Alan D. Taylor, *Union College*

PART V   ON SIZE AND SHAPE

Donald Albers, *Menlo College*
Paul J. Campbell, *Beloit College*
Donald Crowe, *University of Wisconsin*
Seymour Schuster, *Carleton College*
Maynard Thompson, *Indiana University*

Joseph Blatt, *Chedd-Angier Production Company*

FOR ALL PRACTICAL PURPOSES

INTRODUCTION TO CONTEMPORARY MATHEMATICS

THIRD EDITION

W. H. FREEMAN AND COMPANY
NEW YORK

Library of Congress Cataloging-in-Publication Data

For all practical purposes : introduction to contemporary mathematics.
— 3rd ed.
p.   cm.
Includes index.
ISBN 0-7167-2378-6
1. Mathematics.
QA7.F68   1994
510 — dc20

Front Cover: Wassily Kandinsky, *Swinging*, 1925. Tate Gallery, London/Art Resource, NY
(Copyright 1993 ARS, NY.)

93-30027
CIP

Printed in the United States of America.

Third printing 1995, RRD

Contents

▶▶▶▶▶▶▶▶▶▶▶▶▶▶▶ Preface

We began the preface to the first edition of *For All Practical Purposes* with a basic statement of our philosophy:

Every mathematician at some time has been called upon to answer the innocent question, "Just what is mathematics used for?" With understandable frequency, usually at social gatherings, the question is raised in similar ways: "What do mathematicians do, practice, or believe in?" At a time when success in our society depends heavily on satisfying the need for developing quantitative skills and reasoning ability, the mystique surrounding mathematics persists. *For All Practical Purposes: Introduction to Contemporary Mathematics* is our response to these questions and our attempt to fill this need.

For All Practical Purposes represents our effort to bring the excitement of contemporary mathematical thinking to the nonspecialist, as well as help him or her develop the capacity to engage in logical thinking and to read critically the technical information with which our contemporary society abounds. We attempt to implement for the study of mathematics Thomas Jefferson's notion of an "enlightened citizenry," in which individuals having acquired a broad knowledge of topics exercise sound judgment in making personal and political decisions. Environmental and economic issues dominate modern life, and behind these issues are complex matters of science, technology, and mathematics that call for an awareness of fundamental principles.

To encourage achievement of these goals, *For All Practical Purposes* stresses the connections between contemporary mathematics and modern society. Since the technological explosion that followed World War II, mathematics has become a cluster of mathematical sciences encompassing statistics, computer science, operations research, and decision science, as well as the more traditional areas. In science and industry mathematical models are the tools par excellence for solving complex problems. In this book our goal is to convey the power of mathematics as illustrated by the great variety of problems that can be modeled and solved by quantitative means.

This philosophy has guided our efforts through the second edition and now into our third. Moreover, the success of previous editions of *For All Practical Purposes* and the emergence of new texts that adopt much of our content yield proof that this philosophy has truly changed the nature of Mathematics for Liberal Arts courses across the country. We are justifiably proud of these results and grateful to those faculty who worked with us to make this revolution possible. We have only one concern. It was our expressed desire, through the publication of FAPP, to demonstrate to students and faculty the broad range of modern mathematics and its applications. We did not intend to substitute one orthodoxy for another—to set in stone a more modern table of contents.

As a consequence, in this edition we have added a good deal of new material, including some results (on envy-free fair division) discovered as recently as 1992. We were fortunate indeed to be able to secure the efforts of Professor Joseph A. Gallian to write two new chapters on Coding Theory. This is exciting and accessible material. The point is a simple one—mathematics is dynamic; new mathematics is being invented and applied in new ways every day. No one set syllabus can capture this continuous sense of invention. Our goal, and our text, must be flexible enough to demonstrate to students the contemporary nature of our subject and its myriad of applications to our daily lives.

To that end, we have clarified, reorganized, and revised topics in the third edition of *For All Practical Purposes* in the following ways:

▶ Part I, "Management Science," includes an improved version of the edge walker algorithm, further explication of the sorted edges algorithm, and an expanded discussion on linear programming.

▶ Part II, "Statistics, The Science of Data," includes a discussion on calculating the standard deviation (Chapter 6), more on counting (Chapter 7), and further explication of confidence intervals (Chapter 8). Setting off key principles and definitions within text columns has improved text readability.

▶ Part III, "Coding Information," features two chapters that explore the role of mathematics in coding information in the modern world, from bar codes on milk cartons to identification numbers on your driver's license to television services. Unavailable in any other textbook, this is accessible material to which students can easily relate.

▶ Part IV, "Social Choice and Decision Making," now includes separate chapters on Fair Division (Chapter 13) and on Apportionment (Chapter 14) to more fully develop these topics. A discussion of envy-free fair division makes Chapter 13 the most current and unique treatment of fair division available. Weighted Voting (Chapter 12) has been rewritten and includes a discussion on the Shapley-Shubik power index. Other improvements include further explication of topics such as the Hill-Huntington method (Chapter 14), mixed strategies in Game Theory (Chapter 15), Arrow's theorem, and the Balinski-Young theorem (Chapter 11).

▶ Part V, "On Size and Shape," now devotes individual, shorter chapters to telescopes, Euclidean and non-Euclidean geometry, symmetry and patterns, and tilings.

▶ Two special 8-page color inserts are featured in this edition. The first, a photo essay on fractals by Richard F. Voss, presents a stunningly beautiful and coherent view of fractal geometry, with captions in ordinary language that students can understand. The second, a collection of tilings and Escher sketches and drawings, complements discussions in Chapter 21 (Symmetry and Patterns) and Chapter 22 (Tilings).

▶ Writing projects for every chapter encourage students to explore mathematical concepts in a nontraditional way. Students are encouraged to extend the techniques and ideas introduced in text chapters to the discussion of current social and political events, controversies, and other real life situations.

▶ All chapters include new examples, exercises, and spotlights.

▶ SUPPLEMENTS

We are pleased to provide the following outstanding supplements to accompany the third edition of *For All Practical Purposes:*

▶ An *Instructor's Guide,* prepared by Chris Leary, St. Bonaventure University, offers detailed chapter outlines and summaries, skill objectives, teaching tips, work sheets for in-class or homework assignments, answers to the even-numbered exercises in the text, as well as an additional exercise section (includes answers). Also featured is a helpful digest of the videotape programs, including running times of each segment and suggestions for coordinating the Annenberg videotape programs with lectures.

▶ A set of 100 black-and-white *Transparency Masters* with enlarged illustrations and tables from the text, designed to produce excellent overhead transparencies for use in the classroom or lecture hall. This set includes 40 new images from the text.

▶ A *Telecourse Guide* to accompany the Annenberg video series of 26 half-hour programs. This is a valuable study guide for students enrolled in the telecourse, as well as for students who are enrolled in courses that use the videos in the classroom. Updated to reflect changes within chapters, it includes a cross-reference chart between the videos and the second and third edition texts, program overviews, skill objectives, self-tests, and sample problems. Answers to both self-test questions and the sample problems are included.

▶ A one-hour video, *Geometry in New Technologies*, with five 12-minute presentations that illustrate some of the exciting modern uses of geometry. It covers vertex coloring to resolve conflict situations; motion planning for robots; creating error-correcting codes of photo transmission from outer space; developing soft-tissue x-rays for medical diagnosis, and the use of Euler circuits to solve cost problems. A printed viewer's guide with exercises and solutions accompanies the video.

▶ *Election Theory Software*, with two hands-on exercises that will enhance student understanding of the concepts introduced in "Social Choice and Decision Making." Available in Macintosh and IBM formats.

For more information and to request copies of these supplements, please contact Sales Support, W. H. Freeman and Company, 41 Madison Avenue, New York, NY 10010.

For more information on Telecourse preview, purchase, or rental, please call 1-800-LEARNER, or write The Annenberg CPB Project, P.O. Box 2345, South Burlington, VT 05407-2345.

▶ ACKNOWLEDGMENTS

Since the inception of *For All Practical Purposes*, we have enjoyed the interest and unsolicited contributions of many people, and this time around was no exception. We wish to thank our friends and colleagues, in particular, Kenneth A. Ross at University of Oregon, who offered suggestions, comments, and corrections to the third edition.

We are grateful to the authors of the third edition, several of whom are new to the project:

Part I   Management Science

Joseph Malkevitch, York College, CUNY
Rochelle Meyer, Nassau Community College
Walter Meyer, Adelphi University

Part II   Statistics: The Science of Data

David S. Moore, Purdue University

Part III   Coding Information

Joseph A. Gallian, University of Minnesota-
   Duluth

Part IV   Social Choice and Decision Making

Steven J. Brams, New York University
Bruce P. Conrad, Temple University
William F. Lucas, Claremont Graduate
   School
Alan D. Taylor, Union College

Part V   On Size and Shape

Paul J. Campbell, Beloit College

We are also grateful to the following people, who either evaluated the second edition or carefully reviewed the manuscript for the third:

Mark S. Anderson, Rollins College
Jerry W. Bradford, Wright State University
Helen Burrier, Kirkwood Community College
Bruce P. Conrad, Temple University
Lothar A. Dohse, University of North Carolina-Asheville
John Emert, Ball State University
Sandra Fillebrown, Saint Joseph's
William Gratzer, Iona College
Rodger Hammons, Morehead State University
Robert W. Hunt, Humboldt State University
Sherman Hunt, Community College of Finger Lakes
Phillip E. Johnson, University of North Carolina-Charlotte
Carmelita R. Keyes, Broome Community College
Bennett Manvel, Colorado State University
Christopher McCord, University of Cincinnati
Kay Meeks, Ball State University
John G. Michaels, SUNY-Brockport
John Oprea, Cleveland State University
James Osterburg, University of Cincinnati
Margaret A. Owens, California State University, Chico
Sandra H. Savage, Orange Coast College
Richard Schwartz, College of Staten Island
Joanne R. Snow, Saint Mary's College

Edward L. Thome, Murray State University
David W. Truscott, Delhi State University, College of Technology

We owe our appreciation to the people at W. H. Freeman and Company who participated in the preparation of this book. We wish especially to thank the editorial staff for their tireless efforts and support. Among them are Jerry Lyons, Publisher; Kay Ueno, Development Editor; Christine Hastings, Project Editor; Patrick Shriner, Supplements Editor; Larry Marcus, Photo Researcher; and Scott Zeman, Editorial Assistant.

We also wish to thank the production staff: Ellen Cash, Production Coordinator; Nancy Singer, Designer; Lisa Ginns, Illustration Coordinator; and Maura Fadden Rosenthal, Layout Artist.

Finally, the efforts of the COMAP staff must be recognized. To the production and administrative staff—Roger Slade, Philip McGaw, and Roland Cheyney—go all our thanks. And finally, we recognize the contribution of Laurie Aragon, the COMAP business, development, personnel, etc., manager, who kept this project, as she does all of COMAP, running smoothly and efficiently. To everyone who helped make our purposes practical, we offer our appreciation for an exciting, exhausting, and exhilarating time.

Solomon Garfunkel
COMAP

To The Student

▶ ▶ ▶ ▶ ▶ ▶ ▶ ▶ ▶ ▶ ▶ ▶ ▶ ▶

Have fun. Enjoy. That is what this book is all about. You may at one time or another have wondered why people go into mathematics. Do they have different genes, that they like this stuff? This text is one attempt to show some of the payoffs for working with mathematics. The areas we discuss and the problems we tackle are real. Most of the people you will meet in these pages are alive and much of their work has been done in the last twenty or thirty years. We have tried to show you mathematics as we see it—solving problems we care about and have fun doing.

We do not expect you to become a mathematician. We understand that you won't remember much of the techniques presented here in a year or two. That is not the point. What we hope is that you will do some mathematics, solve some problems, and gain an appreciation for what we do and why. To a large extent mathematics is not about formulas and equations; it is about CDs and CAT-scans, and parking meters, and presidential polls, and computer graphics. Mathematics is about looking at our world and creating representations we can work with to solve problems that count, and we hope it is in this spirit that you use this book. Welcome to the world of contemporary mathematics.

For All Practical Purposes

Part I

Management
Science

Those who were watching live television one night in July 1969 will never forget the spectacle of seeing the first person walk on the moon. The element of danger and uncertainty, heightened by a history of trial and error, added to the suspense of the lunar landing. Before the 1960s, no one really knew if rockets would ever be able to carry humans into space.

Neil Armstrong's first step onto the moon's surface was a triumph for American science and technology and the culmination of a national quest that had begun in the office of President John F. Kennedy. It was Kennedy's goal to put a man on the moon before the decade was out, a goal realized in the Nixon administration.

In the eight years from 1961 to 1969, we moved from a president's vision to the reality of a lunar landing. Most of us think of this achievement in terms of the tremendous scientific advances it represented—in physics, engineering, chemistry, and associated technologies. But there was another side to this far-reaching project. Someone had to set the objectives, commission the work, suffer the setbacks, overcome unforeseen obstacles, and tie together a project with thousands of disparate components. The kind of science responsible for such details is a branch of mathematics called **management science**.

NASA administrators faced many new problems in the Apollo moon project: they had to choose the best design for the

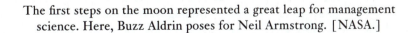

The first steps on the moon represented a great leap for management science. Here, Buzz Aldrin poses for Neil Armstrong. [NASA.]

SPOTLIGHT

Apollo 11 Launch Owes Success to Management Science

Captain Robert F. Freitag, who later became director of Policy and Plans for NASA's space station, in 1969 headed the team responsible for landing the *Apollo 11* safely on the moon. The success of the lunar mission can be traced to management-science techniques that ensured that thousands of small tasks would come together to meet a single giant objective. Freitag shares his observations about that historic event:

> I think the feeling most of us in NASA shared was, "My gosh, now we really have to do it." When you think that the enterprise we were about to undertake was ten times larger than any that had ever been undertaken, including the Manhattan Project, it was a pretty awesome event. But we knew it was the kind of thing that could be broken down into manageable pieces and that if we could get the right people and the right arrangement of these people, it would be possible.
>
> In the case of the Apollo program, it was very important that we take a comprehensive

Captain Robert F. Freitag, NASA.

system-engineering approach. We had to analyze in a very strict sense exactly what the mission was going to be, what each piece of equipment needed was and how it would perform, and all the elements of the system from the concept on through to the execution of the mission, to its recovery back on earth.

spacecraft, design realistic ground simulations, and weigh the priorities of conducting experiments with immediate returns against carrying out tests that would serve long-term goals. When NASA commissioned the Apollo module, it was asking several hundred companies to design, build, test, and deliver components and systems that had never been built before.

Supporting these space age goals, however, were the nuts-and-bolts issues that make up the major concerns of management science, namely, finding ways to make the operations as productive and economical as possible; details of this support are in the Spotlight on the *Apollo 11* launch.

The *Apollo 11* launch is certainly not the only project in which efficiency and timeliness are im-

We started out, in a very logical way, by having a space station in earth orbit. We would then take the lunar spacecraft and build it in orbit, and then send it off to the moon and bring it back. It turned out that this approach was probably a little more risky and took a lot longer, so with the analyses we made, we shifted our whole operation to building a rocket that would go all the way to the moon after it took off from Cape Canaveral. It would then go into orbit around the moon rather than landing on the moon, and from orbit around the moon would descend to the surface of the moon and perform its exploration. Then it would return to its orbit around the moon and come back home.

Well, that was a very comprehensive analysis job. It was probably more deep-seated than the kind of job one would do for building an airplane or a dam because there were so many variables involved. What you do is break it down into pieces: the launch site, the launch vehicles, the spacecraft, the lunar module, and worldwide tracking networks, for example. Then, once these pieces are broken down, you assign them to one organization or another. They, in turn, take those small pieces, like the rocket, and break it down into engines or structures or guidance equipment. And this breakdown, or "tree," is the really tough part about managing.

In the Apollo program, it was decided that three NASA centers would do the work. One

was Huntsville, where Dr. von Braun and his team built the rocket. The other was Houston, where Dr. Gerous and his team built the spacecraft and controlled the flight operations. The third was Cape Canaveral, were Dr. Debries and his team did the launching and the preparation of the rocket.

Those three centers were pieces, and they could break their pieces down into about 10 or 20 major industrial contractors who would build pieces of the rocket. And then each of those industrial contractors would break them down into maybe 20 to 30 or 50 subcontractors — and they, in turn, would break them down into perhaps 300,000 or 400,000 pieces, each of which would end up being the job of one person. But you need to be sure that the pieces come together at the right time, and that they work when put together. Management science helps with that. The total number of people who worked on the *Apollo* was about 400,000 to 500,000, all working toward a single objective. But that objective was clear when President Kennedy said, "I want to land a man on the moon and have him safely returned to the earth, and to do so within the decade." Of course, Congress set aside $20 billion. So you had cost, performance, and schedule, and you knew what the job was in one simple sentence. It took a lot of effort to make that happen.

portant. In a variety of modern projects, ranging from the making of a Hollywood movie to the building of the "Chunnel" (a tunnel under the English Channel), the difference between success and failure depends on whether the project is on time and on budget. Chapters 2 and 3 present some project planning and scheduling techniques that were pioneered in the Apollo program.

But projects are not the only area where organization and efficiency are valuable. Day-to-day operations, such as the delivery of city services discussed in Chapter 1, and the production problems faced by manufacturing firms discussed in Chapter 4, offer big opportunities for cost savings using management science.

Naturally, efficiency is not a modern invention. The Romans were masters of it (for their time), and it is doubtful that their empire would have been as successful as it was without this remarkable flair for organization. Until management science was founded in the modern era, however, efficiency was pursued by trial-and-error— "seat-of-the-pants"—approaches. The distinctive contribution of management science is to add mathematical analysis to the tools available for organizing and decision making.

The value of having scientific principles for operations management was first perceived in World War II. The founders of management science were mathematicians and industrial technicians associated with the armed services who worked together to improve military operations. In applying quantitative techniques to project planning, these pioneers found a new science whose impact reaches beyond the military to many corners of our lives. We will explore some of this new science in the chapters ahead.

1

Street Networks

The underlying theme of management science is finding the best method for solving some problem — what mathematicians call the **optimal solution.** In some cases, it may be to finish a job as quickly as possible. In other situations, the goal might be to maximize profit or minimize cost. In this chapter, our goal is to save time in traversing a street network while checking parking meters, delivering mail, or carrying out some similar task.

Let's begin by concentrating on the parking department of a city government. Most cities and many small towns have parking meters that must be regularly checked for parking violations or emptied of coins. We will use an imaginary town to show how management-science techniques can help to make parking control more efficient.

▶ EULER CIRCUITS

The street map in Figure 1.1 is typical of many towns across the United States, with streets, residential blocks, and a village green. Our job, or that of the commissioner of parking, is to find the most efficient route for the parking-control officer, who travels on foot, to check the meters in an area.

Our map shows only a small area, allowing us to start with an easy problem. But the problem occurs on a larger scale in all cities and towns and, for larger areas, there are almost unlimited possibilities for parking-control routes.

The commissioner has two goals in mind: (1) the parking-control officer must cover all the sidewalks that have parking meters without retracing any more steps than are necessary; and (2) the route should end at the same point from which it began, perhaps where the officer's patrol car is parked. To be specific, suppose there are only two blocks that have parking meters, the two blue-shaded blocks that are side by side toward the top of Figure 1.1. Suppose further that the parking-control officer must start and end at the upper left corner of the left-hand block. You might enjoy working out some routes by trial and error and evaluating their good and bad points. We are going to leave this problem for the moment and establish some concepts that will give us a better method to deal with this problem than trial and error.

We can think of this problem in terms of a structure called a **graph,** one of the many mathematical models that can help to simplify complex problems. A graph is a finite set of dots and con-

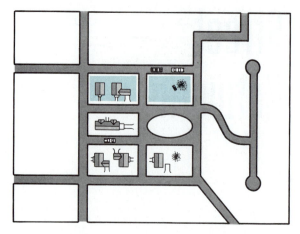

Figure 1.1 A street map for part of a town.

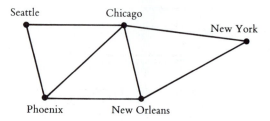

Figure 1.2 The edges of this graph show nonstop routes that an airline might offer.

necting links. The dots are called **vertices** (a single dot is called a **vertex**), and the links are called **edges.** Each edge must connect two different vertices. A **path** is a connected sequence of edges showing a route on the graph that starts at a vertex and ends at a vertex; a path is usually described by naming in turn the vertices visited in traversing it. A path that starts and ends at the same vertex is called a **circuit.** A graph can represent our city map, a communications network, or even a system of air routes.

EXAMPLE: Parts of a Graph

In Figure 1.2, the vertices represent cities and the edges represent nonstop airline routes between them. We see that there is a nonstop flight between New Orleans and Phoenix, but no such flight between Seattle and New York. There are several paths that describe how a person might travel with this airline from Seattle to New York: the path that seems most direct is Seattle, Chicago, New York, but Seattle, Phoenix, New Orleans, New York is also such a path. An example of a circuit is Phoenix, New Orleans, Chicago, Phoenix. It is a circuit be-

cause the path starts and ends at the same vertex. In this chapter we are especially interested in circuits, just as we are in real life; most of us end our day in the same location where we start it — at home!

Returning to the case of parking control in Figure 1.1, we can use a graph to represent the whole territory to be patrolled: think of each street intersection as a vertex and each sidewalk that contains a meter as an edge, as in Figure 1.3. Notice in Figure 1.3b that the street separating the blocks is not explicitly represented; it has been shrunk to nothing. In effect, we are simplifying our problem by ignoring any distance traveled in crossing streets.

The numbered sequence of edges in Figure 1.4a shows one circuit that covers all the meters (note that it is a circuit because its path returns to its starting point). But Figure 1.4b shows another solution that is better because its circuit covers every edge (sidewalk) exactly once. In Figure 1.4b no edge is covered more than once, or *deadheaded* (a term borrowed from shipping, which means making a return trip without a load). Circuits that cover every edge only once are called **Euler circuits,** after the great eighteenth-century mathematician Leonhard Euler (pronounced oy′lur), who first studied them (see Spotlight 1.1, p. 10). Figure 1.4b shows an Euler circuit.

Euler was the founder of the theory of graphs. One of his first discoveries was that some graphs have no Euler circuits at all. For

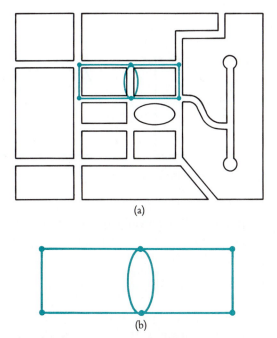

(a)

(b)

Figure 1.3 (a) A graph superimposed upon a street map. The edges show which sidewalks have parking meters. (b) The same graph enlarged.

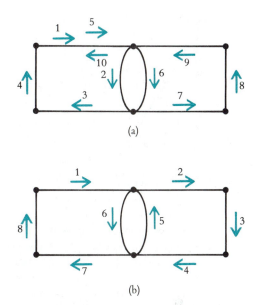

(a)

(b)

Figure 1.4 (a) A circuit and (b) an Euler circuit.

example, in the graph in Figure 1.5b, it would be impossible to start at one point and cover all the edges without retracing some steps: if we try to start a circuit at the leftmost vertex, we discover that once we have left the vertex, we have "used up" the only edge meeting it. We have no way to return to our starting point except to reuse that edge. But this is not allowed in an Euler circuit. If we try to start a circuit at one of the other two vertices, we likewise can't complete it to form an Euler circuit. ▲

Since we are interested in finding circuits, and Euler circuits are the most efficient ones, we will want to know how to find them. If a graph has no Euler circuit, we will want to develop the next best circuits, those having minimum deadheading. These topics make up the rest of this chapter.

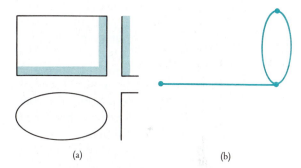

(a)   (b)

Figure 1.5 (a) The three shaded sidewalks cannot be covered by an Euler circuit. (b) The graph of the shaded sidewalks in part (a).

▶ FINDING EULER CIRCUITS

Now that we know what an Euler circuit is, we are faced with two obvious questions:

1. Is there a way to tell by calculation, not by trial and error, if a graph has an Euler circuit?

2. Is there a method, other than trial and error, for finding an Euler circuit when one exists?

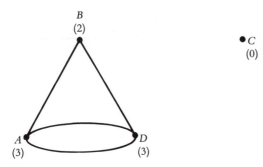

Figure 1.6 Valences of vertices.

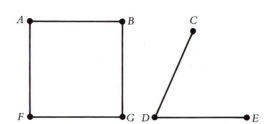

Figure 1.7 A nonconnected graph.

Euler answered these questions in 1735 by using the concepts of valence and connectedness. The **valence** of a vertex in a graph is the number of edges meeting at the point. Figure 1.6 illustrates the concept of valence, with vertices A and D having valence 3, vertex B having valence 2, and vertex C having valence 0. (Isolated vertices such as vertex C are an annoyance in Euler circuit theory. Because they don't occur in typical applications, we henceforth assume that our graphs have no vertices of valence 0.)

Figure 1.3b has four vertices of valence 2, namely, the outer corners of the graph. This graph also has two vertices of valence 4. Notice that each vertex has a valence that is an even number. We'll soon see that this is very significant.

A graph is said to be **connected** if for every pair of its vertices there is at least one path of one or more edges connecting the two vertices. Given a graph, if we can find even one pair of vertices not connected by a path, then we say that the graph is not connected. For example, the graph in Figure 1.7 is not connected because we are unable to join A to D with a path of edges. However, the graph does consist of two "pieces" or connected components, one containing the vertices A, B, F, and G, the other containing C, D, and E. A connected graph will contain a single connected component. Notice that the parking-control graph of Figure 1.3b is connected.

We can now state Euler's theorem, his simple answer to the problem of detecting when a graph G has an Euler circuit:

1. If G is connected and has all valences even, then G has an Euler circuit.

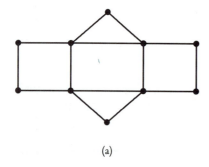

(a)

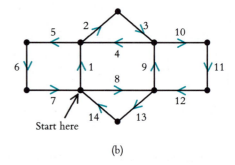

(b)

Figure 1.8 (a) A graph having (b) an Euler circuit.

2. Conversely, if G has an Euler circuit, then G must be connected and all its valences must be even numbers.

In the optional section entitled "Proving Euler's Theorem," you will find an outline of a proof of this theorem.

Since the parking-control graph of Figure 1.3b conforms to the connectedness and even-valence conditions, Euler's Theorem tells us that it has an Euler circuit. We already have found an Euler circuit for Figure 1.4b by trial and error. For a very long graph, however, trial and error may take a long time. It is usually quicker to check whether the graph is connected and even-valent to find out if it has an Euler circuit.

Once we know there is an Euler circuit in a certain graph, how do we find it? Many people find that, after a little practice, they can find Euler circuits by trial and error and they don't need detailed instructions on how to proceed. At this point, you should see if you can develop this skill by trying to find Euler circuits in Figure 1.8a, Figure 1.9a, and Figure 1.10. In doing your experiments, draw your graph in ink and the circuit in pencil so you can erase. Make your graph big and clear so you won't get confused.

If you would like more guidance on how to find an Euler circuit without trial and error, here is a method that works: *never use an edge that is the only link between two parts of the graph that still need to be covered.* Figure 1.9b illustrates this. Here we have started the circuit at A and gotten to D via B and C, and we want to know what to do next.

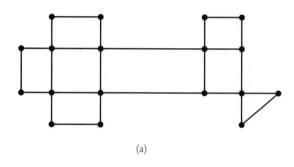

(a)

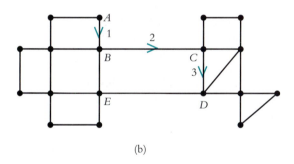

(b)

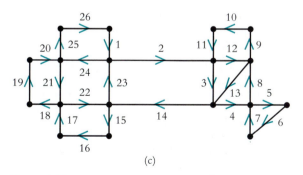

(c)

Figure 1.9 A crucial decision point in finding an Euler circuit.

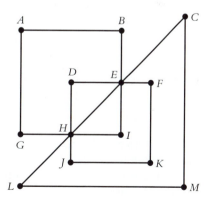

Figure 1.10 A graph with an Euler circuit.

SPOTLIGHT 1.1   Leonhard Euler

Although born in Switzerland, Leonhard Euler (1707 – 1783) spent a large part of his life in St. Petersburg, Russia. He was extremely prolific, publishing over 500 works in his lifetime. His collected works fill 70 volumes. Euler made major contributions to many areas of mathematics, including algebra, geometry, and calculus. A contemporary claimed that Euler could calculate effortlessly, "just as men breathe, as eagles sustain themselves in the air."

Human interest stories about Euler have been handed down through two centuries. He was extremely fond of children and had thirteen of his own, of whom only five survived childhood. It is said that he often wrote difficult mathematical works with a child or two in his lap. He was a prodigy at doing complex mathematical calculations under less than ideal conditions and continued to do them even after he became totally blind later in life. His blindness diminished neither the quantity nor the quality of his output. Throughout his life, he was able to mentally calculate in a short time what would have taken ordinary mathematicians hours of pencil-and-paper work.

Euler's mathematical mind found new mathematics in everyday life. In the old German

Leonhard Euler. (Portrait by Emanuel Handmann, *Bildnis des Mathematikers,* 1753. Oeffentiliche Kunstsammlung Basel, Kunstmuseum.)

town of Konigsberg, people frequently tried to take a Sunday stroll whose route crossed each of the seven bridges in the town exactly once. Euler analyzed this local pastime using what are now known as Euler circuits.

Euler divided his working life between the St. Petersburg Academy in Russia and the Berlin Academy. He moved to Berlin largely because of difficult political conditions in Russia. When asked in Berlin why he spoke so little, Euler replied, "Madam, I come from a country where if you speak, you are hanged."

Going to E would be a bad idea because the uncovered part of the graph would then be disconnected into left and right portions. You will never be able to get from the left part back to the right part

because you have just used the last remaining link between these parts. Therefore, you should stay on the right side and finish that before using the edge from D to E. This kind of thinking needs to be applied every time you need to choose a new edge. Let's see how this works, starting at the beginning at A. From vertex A there are two possible edges and neither of them disconnects the unused portion of the graph. Thus, we could have gone either to the left or down. Having gone down to B, we now have three choices and none of them disconnects the unused part of the graph. After choosing to go from B to C, we find that any of the three choices at C is acceptable. Can you complete the Euler circuit? Figure 1.9c shows one of many ways to do this.

The method just described leaves many edge choices up to you. When there are many acceptable edges for your next step, you can pick one at random. You might even flip a coin. When computers carry out algorithms of this sort, they use *random number generators*. If you don't like the element of chance, you can begin by numbering all the edges of the graph, and, whenever you have a choice, pick the one with the lowest number.

EXAMPLE: Finding an Euler Circuit

Check the valences of the vertices and the connectivity of the graph in Figure 1.8a to verify that the graph does have an Euler circuit. Now try to find an Euler circuit for that graph. You can start at any vertex. When you are done, compare your solution with the Euler circuit given in Figure 1.8b. If your path covers each edge exactly once and returns to its original vertex (is a circuit), then it is an Euler circuit, even if it is not the same as the one we give. ▲

OPTIONAL ▶ **PROVING EULER'S THEOREM**

We'll start by proving that if a graph has an Euler circuit, then it must have only even valences and it must be connected. Let X be any vertex of the

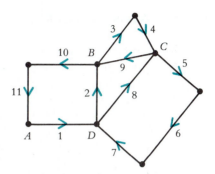

Figure 1.11 An Euler circuit starting and ending at A.

graph. We will show that the edges at X can be paired up and this will prove that the valence is even. Every edge at X is used by the Euler circuit as an outgoing edge (leaving from X) or an incoming edge (arriving at X). As we follow the Euler circuit, each time we arrive at X, the next step involves an outgoing edge which we pair up with the previous incoming one. Since all edges are used by the Euler circuit, none more than once, this pairs up the edges. For example, in Figure 1.11 at vertex B we would pair up edges 2 and 3 and edges 9 and 10. At vertex C we would pair up edges 4 and 5 and edges 8 and 9. Can you see how the pairings would work at D? How about vertex A?

Now to see that a graph with an Euler circuit is connected, note that by following the Euler circuit around we can get from one edge to any other edge (it covers them all) using a portion of the Euler circuit. Since every vertex is on an edge (there are no vertices of 0 valence), we can get from any vertex to any other using a portion of the Euler circuit.

So far, this is not a complete proof of Euler's Theorem. It is also necessary to prove that if a graph has all vertices even-valent and is connected, then an Euler circuit can be found for it. The book by Malkevitch and Meyer in the Suggested Readings section contains an elementary proof of this. ◀

▶ CIRCUITS WITH REUSED EDGES

Now let's see what Euler's theorem tells us about the three-block neighborhood with parking meters, represented by dots in Figure 1.12a. Figure 1.12b shows the corresponding graph. (Since we only use edges to represent sidewalks along which the officer must walk, the sidewalk with no meters is not represented by any edge in the graph.) This graph has two odd valences, so Euler's theorem tells us that there is no Euler circuit for this graph.

Since we must reuse some edges in this graph in order to cover all edges in a circuit, for efficiency we need to keep the total length of reused edges to a minimum. This type of problem, in which we want to minimize the length of a circuit by carefully choosing which edges to retrace, is often called the **Chinese postman problem** (like parking-control routes, mail routes need to be efficient). The problem was first studied by the Chinese mathematician Meigu Guan in 1962; hence the name. Although the Euler circuit theory doesn't deal directly with reused edges or edges of different lengths, we can extend the theory to help solve the Chinese postman problem. The remainder of this chapter is dedicated to solving the Chinese postman problem and discussing applications besides parking control.

In a realistic Chinese postman problem, we need to consider the lengths of the sidewalks, streets, or whatever the edges represent, since we want to minimize the total length of the reused edges. However, to simplify things at the start, we can suppose that all edges represent the same length. (This is often called the *simplified* Chinese postman problem.) In this case, we need only count reused edges and need not measure their lengths. To solve the problem, we want to find a circuit that covers each edge and that has the minimal number of reuses of edges already covered.

To follow the procedure we are going to develop, look at the graph of Figure 1.13a, which is the same graph as in Figure 1.12b, but with labeled vertices. This graph has no Euler circuit, but there is a circuit that has only one reuse of an edge *(CG)*, namely, *ABCDHGCGFBFEA*. Let's draw this circuit so that when edge *CG* is about to be reused, we install a new, extra edge in the graph for the circuit to use. By duplicating edge *CG*, we can avoid reusing the edge. To duplicate an edge, we must add an edge that joins the two vertices that are already joined by the edge we want to duplicate. (It is not a good idea to join vertices that are not already connected by an edge; see Figure 1.15.) We have now created the graph of Figure 1.13b. In this graph the original circuit can be traced as an Euler circuit, using the new edge when needed. The circuit is shown in Figure 1.13c. Our theory will be based on using this idea in reverse, as follows:

1. Take the given graph and add edges, duplicating existing ones until you arrive at a graph that is connected and even-valent. We call this process **eulerizing** a graph, because the graph we produce will have an Euler circuit. (In our

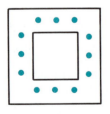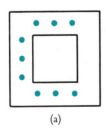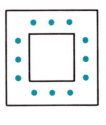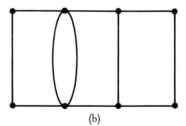

(a)   (b)

Figure 1.12 (a) A street network and (b) its graphic representation.

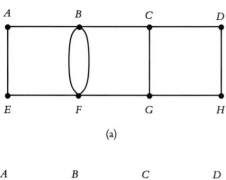

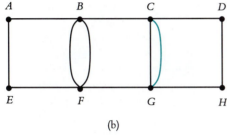

(a)

(b)

(c)

Figure 1.13 Making a circuit by reusing an edge.

graphs, the edges we add are in color, and thus can be distinguished from the original edges, which are black. You may want to create a system to help you remember which edges are original and which are duplicates.)

2. Find an Euler circuit on the eulerized graph.

3. "Squeeze" this Euler circuit from the eulerized graph onto the original graph by reusing an edge of the original graph each time the circuit on the eulerized graph uses an added edge.

EXAMPLE: Eulerizing a Graph

When we eulerize a graph, we first locate the vertices with odd valence. The graph in Figure 1.14a has two, B and C. Next, we add one end of an edge at each such vertex, matching the new edge up with an existing edge in the original graph. Figure 1.14b shows one way to eulerize the graph. Note that B and C have even valence in the second graph. After eulerization, each vertex has even valence. To see an Euler circuit on the eulerized graph in Figure 1.14c, simply follow the edges in numerical order and in the direction of the arrows, beginning and ending at vertex A. The final step, shown in Figure 1.14d, is to squeeze our Euler circuit into the original graph. There are two reuses of

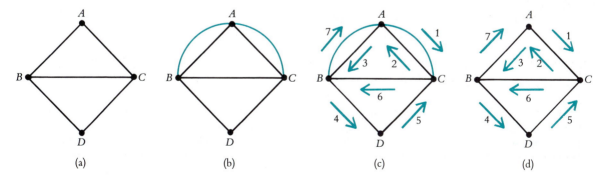

Figure 1.14 Eulerizing a graph.

SP⬤TLIGHT 1.2   The Human Aspect of Problem Solving

▸▸▸ ▸ ▸ ▸ ▸ ▸ ▸ ▸ ▸ ▸ ▸

Thomas Magnanti, professor of operations research and management, heads the Department of Management Science at MIT's Sloan School of Management. Here are some of his observations:

Thomas Magnanti, Department of Management Science, Sloan School of Management, MIT.

Typically, a management-science approach has several different ingredients. One is just structuring the problem — understanding that the problem is an Euler-circuit problem or a related management-science problem. After that, one has to develop the solution methods.

But one should also recognize that you don't just push a button and get the answer. In using these underlying mathematical tools, we never want to lose sight of our common sense, of understanding, intuition, and judgment. The computer provides certain kinds of insights. It deals with some of the combinatorial complexities of these problems very nicely. But a model such as an Euler circuit can never capture the full essence of a decision-making problem.

Typically, when we solve the mathematical problem, we see that it doesn't quite correspond to the real problem we want to solve. So we make modifications in the underlying model. It is an interactive approach, using the best of what computers and mathematics have to offer and the best of what we, as human beings, with our own decision-making capabilities, have to offer.

previously covered edges. Notice that each reuse of an edge corresponds to an added edge. This is generally true in this type of problem: *if you add the new edges correctly, the number of reuses of edges equals the number of edges added during eulerization.* ▲

When adding new edges to eulerize a graph, it is desirable to add only edges that are duplicates of existing edges. Doing so makes the rule, just stated in italics, always true.

In Figure 1.15a, we need to make X and Y even. Adding one long edge from X to Y (Figure

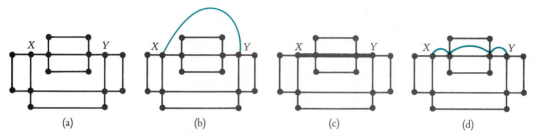

Figure 1.15 Eulerizing when the vertices are more than one edge apart.

1.15b) might seem like an attractive idea, but doing so would duplicate more than one existing edge. (Remember, to duplicate an edge means to add an edge that joins two vertices that are already joined by the edge to be duplicated.) To see why, simply add this long edge and then try to squeeze the Euler circuit on this new graph back into the old graph. This long edge won't count for just one reused edge, since the alternative to using the long edge is not one reused edge, but a whole series stretching from X to Y (shown by the heavy line in Figure 1.15c). If you don't add the long edge, but instead follow the rule that an added edge must duplicate an existing edge, you'll be able to count (in Figure 1.15d) the edges that will be reused by counting the number of duplicate edges that need to be added (three, in this case).

Now that we have learned to eulerize, the next step is to try to get a best eulerization we can— one with the fewest added edges. It turns out that there are many ways to eulerize a graph. It is even possible that the smallest number of added edges can be achieved with two different eulerizations. This is the reason we use the phrase "*a* best eulerization" rather than "*the* best eulerization." Remember, we want a best eulerization because this enables us to find the circuit for the original graph that has the minimum number of reuses of edges.

EXAMPLE: A Better Eulerization

In Figure 1.16a, we begin with the same graph as in Figure 1.14, but we eulerize it in a differ-

ent way — by adding only one edge (see Figure 1.16b). Figure 1.16c shows an Euler circuit on the eulerized graph, and in Figure 1.14d we see how it is squeezed onto the original graph. There is only one reuse of an edge, because we added one edge during eulerization.   ▲

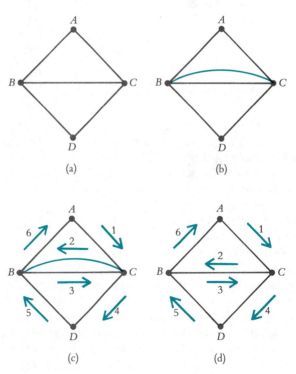

Figure 1.16 A better eulerization of Figure 1.14.

The solution in Figure 1.16 is better than the solution in Figure 1.14 because one reuse is better than two. These examples suggest the following addition to our solution procedure: *try to find the eulerization with the smallest number of added edges.* This extra requirement makes the problem both more interesting and more difficult. For large graphs, a best eulerization may not be obvious. We can try out a few and pick the best among the ones we find, but there may be an even better one that our haphazard search does not turn up.

A systematic procedure for finding a best eulerization does exist, but the process is complicated. There is an especially easy technique for eulerizing rectangular street networks in a best possible way. We will look at rectangular networks first, then we will comment on how to approach nonrectangular ones.

Many street networks are composed of a series of rectangular blocks that form a large rectangle m blocks high by n blocks wide. Such networks are called *rectangular street networks*. Examples of rectangular street networks are shown in Figure 1.17. The graph on the right in each pair shows a best eulerization for the rectangular street network on the left. There appear to be three different eulerization patterns, depending upon whether the rectangle height (m blocks) and width (n blocks) in the original graph are odd or even numbers. In Figure 1.17a, both lengths are 3, both odd; in Figure 1.17b, one length is odd (3) and one is even (4); in Figure 1.17c, both lengths are 4, an even number.

Although the patterns appear different, one technique can be used to create all of them. This technique can be thought of as involving an "edge walker" who walks around the outer boundary of the large rectangle in some direction, say, clockwise. He starts at any corner, say, the upper left corner. As he goes around, he adds edges by the following rules. When he comes to an odd-valent vertex, he links it to the next vertex with an added edge. This next vertex now becomes either even or odd. If it became even, he skips it and continues

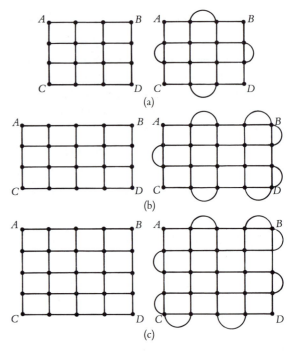

Figure 1.17 Eulerizations of three rectangular networks.

around, looking for an odd vertex. If it became odd, the edge walker links it to the next vertex and then checks this vertex to see whether it is even or odd. Each of the three parts of Figure 1.17 has been eulerized by this method.

In a street network that is not rectangular, the eulerization process is started by locating all the vertices with odd valence and then pairing these vertices with each other and finding the length of the shortest path between each pair. We look for the shortest paths, since each edge on the connecting paths will be duplicated. The idea is to make the pairings cleverly so that the sum of the lengths of those paths is the smallest it can be. With a little practice, most people can find a best or nearly best eulerization using only this idea, trial and error, and a little ingenuity. Those interested in a further discussion can read the following optional section titled "Finding Good Eulerizations."

Suppose we want a perfect procedure for euleriz-ing a graph. What theoretical ideas and methods could we use to build such a tool?

One building block we could use is a method for finding the shortest path between two given vertices of a graph. For example, let us focus on vertices X and Y in Figure 1.18a; both have odd valence. We can connect them with a pattern of duplicate edges, as in Figure 1.18b. The cost of this is the length of the path we duplicated from X to Y. A shorter path from X to Y, such as the one

shown in Figure 1.18c, would be better. Fortu-nately, the *shortest-path problem* has been well studied, and we have many good procedures for solving it exactly, even in large, complex graphs. These procedures are discussed in some of the suggested readings given at the end of this chapter, but are beyond the scope of this text.

But there is more to eulerizing the graph in Figure 1.18a than dealing with X and Y: notice that we have odd valences at Z and W. Should we con-nect X and Y with a path, and then connect Z and W, as in Figure 1.18d? Or should we connect X to Z and Y to W, as in Figure 1.18e? Another alterna-

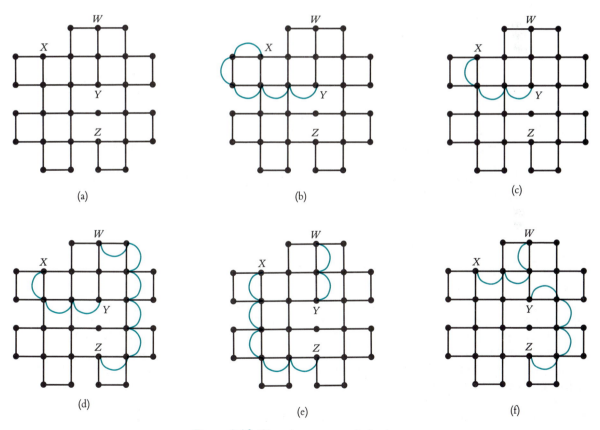

Figure 1.18 Choosing among eulerizations.

tive is to use connections X to W and Y to Z, as in Figure 1.18f. It turns out that the alternatives in both Figures 1.18e and 1.18f are preferable to the one in Figure 1.18d.

At the start, it is often not clear which alternatives are best. The problem is how to pair up vertices for connection to get a set of paths whose total length is minimal. This problem, called the *matching problem*, has also been studied and solved.

As with the shortest-path problem, refer to the Suggested Readings for further details.   ◀

▶ CIRCUITS WITH MORE COMPLICATIONS

Euler circuits and eulerizing have many more practical applications than just checking parking

---

SP⬤TLIGHT 1.3   Israel Electric Company Reduces Meter-Reading Task

▶ ▶ ▶   ▶ ▶ ▶ ▶ ▶ ▶ ▶ ▶ ▶

The Beersheba branch of Israel's major electric company wanted to make the job of meter reading more efficient. When the branch managers decided to minimize the number of people required to read the electric meters in the houses of one particular neighborhood, they set a precedent by applying management science. Formerly, each person's route had been worked out by trial and error and intuition, with no help from mathematics. The whole job required 24 people, each doing a part of the neighborhood in a five-hour shift.

At first, it looks as though one would find a more efficient way of doing the work the same way as in the Chinese postman problem, but there are two important differences. First, the neighborhood was big enough to negate any possibility of having only one route assigned to one person. Instead, it was necessary to find a number of routes that, taken together, covered all the edges (sidewalks). Second, a meter reader who was done with a route was allowed

to return home directly. Thus, there was no reason for the individual routes to return to their starting points; therefore, routes could be paths instead of circuits.

The Beersheba researchers started by doing a partial eulerization in which each odd valence was converted to an even valence, except for two vertices that remained odd. In such a partially eulerized graph, it is possible to find a path, not a circuit, that covers every edge and that starts at one odd-valent vertex and ends at the other.

Next, they chopped the path into five-hour chunks by stepping through the path and marking each time five hours' worth of edges were covered. By following this procedure, researchers managed to cover the neighborhood with 15 five-hour routes, a 40% reduction of the original 24 five-hour routes. Altogether, these routes involve a total of 4338 minutes of walking time, of which 41 minutes (less than 1%) is deadheading.

meters. Almost any time services must be delivered along streets or roads, our theory can make the job more efficient. Examples include collecting garbage, salting icy roads, plowing snow, inspecting railroad tracks, and reading electric meters (see Spotlight 1.3).

Each of these problems has its own special requirements that may call for modifications in the theory. For example, in the case of garbage collection, the edges of our graph will represent streets, not sidewalks. If some of the streets are one-way, we need to put arrows on the corresponding edges, resulting in a directed graph, or **digraph**. The circuits we seek will have to obey these arrows. In the case of salt spreaders and snowplows, each lane of a street needs to be modeled as a directed edge, as shown in Figure 1.19. Note that the arrows on the map and digraph are not in color because these arrows denote restrictions in traversal possibilities, not parts of circuits.

Like salt spreaders, street-sweeping trucks can travel in only one lane at a time and need to obey the direction of traffic. Street sweepers, however, have an additional complication: parked cars. It is very difficult to clean the street if cars are parked along the curb. Yet for overall efficiency, those who are responsible for routing street sweepers want to interfere with parking as little as possible. The common solution is to post signs specifying times when parking is prohibited, such as Thursday between 8 A.M. and 2 P.M. Because the parking-time factor is a constraint on street sweeping, it is important not only to find an Euler circuit, or a circuit with very few duplications, but a circuit that can be completed in the time available. Once again, the theory can be modified to handle this constraint.

Finally, because towns and cities of any size will have more than one street sweeper, parking officer, or garbage truck, a single best route will not suffice. Instead, they will have to divide the territory into multiple routes. The general goal is to find optimal solutions while taking into account traffic direction, number of lanes, time restrictions, and divided routes (see Figure 1.20).

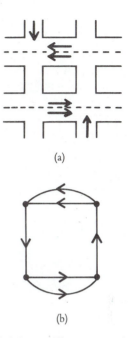

(a)

(b)

Figure 1.19 (a) Salt-spreading route, where each east–west street has two traffic lanes in the same direction, and (b) an appropriate digraph model.

Management science makes all this possible. For example, a pilot study done in the 1970s in New York City showed that applying these techniques to street sweepers in just one district could save about $30,000 per year. With 57 sanitation districts in New York, this would amount to a savings of more than $1.5 million in a single year. In addition, the same principles could be extended to garbage collection, parking control, and other services carried out on street networks.

This plan was not adopted when first proposed. Because city services take place in a political context, several other factors come into play. For example, union leaders try to protect the jobs of city workers, bureaucrats might try to keep their departmental budgets high, and elected politicians rarely want to be accused of cutting the jobs of their constituents. Thus political obstacles can overrule management science.

Despite the complications of real-world problems, management-science principles provide

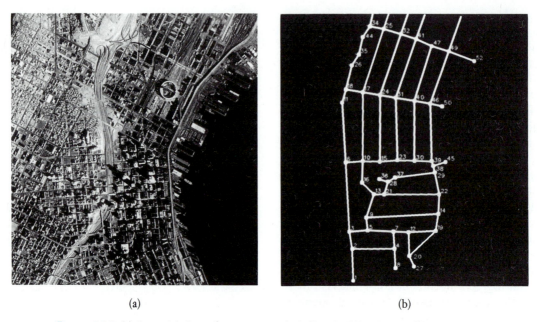

(a)   (b)

Figure 1.20 (a) An aerial view of street networks in Seattle, Washington. [Peter Arnold, Inc., NASA] Today, finding optimal routes within complex street networks is often done with sophisticated computer-based color graphics systems. (b) A computer-generated street network. [Courtesy of Sidney B. Bowne & Son, Consulting Engineers.]

ways to understand these problems by using graphs as models. We can reason about the graphs and then return to the real-world problem with a workable solution. The results we get can have a lasting effect on the efficiency and economic well-being of any organization or community.

▶ REVIEW VOCABULARY

Chinese postman problem   The problem of finding a circuit on a graph that covers every edge of the graph at least once and that has the shortest possible length.

Circuit   A path that starts and ends at the same vertex.

Connected graph   A graph is connected if it is possible to reach any vertex from any specified starting vertex by traversing edges.

Digraph   A graph in which each edge has an arrow indicating the direction of the edge. Such directed edges are appropriate when the relationship is "one-sided" rather than symmetric (e.g., one-way streets as opposed to regular streets).

Edge   A link joining two vertices in a graph.

Euler circuit   A circuit that traverses each edge of a graph exactly once.

Eulerizing   Adding new edges to a graph so as to make a graph that possesses an Euler circuit.

Graph   A mathematical structure in which points (called vertices) are used to represent things of interest, and in which links (called edges) are used to connect vertices, denoting that the connected vertices have a certain relationship.

Management science   A discipline in which mathematical methods are applied to management problems in pursuit of optimal solutions that cannot readily be obtained by common sense.

Operations research   Another name for management science.

Optimal solution   When a problem has various solutions that can be ranked in preference order (perhaps according to some numerical measure of "goodness"), the optimal solution is the best-ranking solution.

Path   A connected sequence of edges in a graph.

Simple circuit   A circuit in which every vertex has a valence equal to two.

Valence (of a vertex)   The number of edges touching that vertex.

Vertex   A point in a graph where one or more edges end.

▶ SUGGESTED READINGS

BELTRAMI, EDWARD J.: *Models for Public Systems Analysis*, Academic Press, New York, 1977. Section 5.4 deals with material similar to that in this chapter. The rest of the book gives a nice selection of applications of mathematics to plant location, manpower scheduling, providing emergency services, and other public service areas. The mathematics is somewhat more advanced than in this chapter.

COZZENS, MARGARET B., AND RICHARD P. PORTER: *Mathematics and Its Applications*, Heath, Lexington, Mass., 1987. Includes a nice discussion of Euler circuit ideas applied to DNA fragments.

MALKEVITCH, JOSEPH, AND WALTER MEYER: *Graphs, Models, and Finite Mathematics*, Prentice-Hall, Englewood Cliffs, N.J., 1974. An introductory text, which includes much the same material as in this chapter, but with a little more detail. A different algorithm for finding Euler circuits is given.

The following two references discuss the shortest-path and matching problems in depth; they are suitable for advanced students and for faculty.

ROBERTS, FRED S.: *Applied Combinatorics*, Prentice-Hall, Englewood Cliffs, N.J., 1984.

TUCKER, ALAN: *Applied Combinatorics*, 2nd ed., Wiley, New York, 1984.

▶ EXERCISES

1. In the graph below, the vertices represent cities and the edges represent roads connecting them. What are the valences of the vertices in this graph? (Keep in mind that E is part of the graph.) What might the valence of city E be showing about the geography?

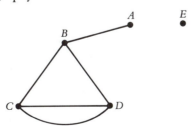

2. What are the valences of the vertices in this graph?

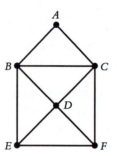

3. Which of these graphs are connected?

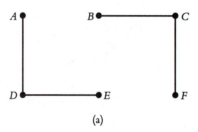   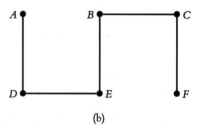

4. In the graphs below, the vertices represent cities and the edges represent roads connecting them. In which graphs could a person located in city A choose any other city and then find a sequence of roads to get from A to that other city?

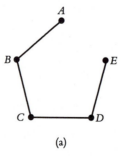   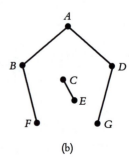

5. Draw a graph to represent five cities A, B, C, D, and E having one road connecting each pair of cities in the following list (when a pair of cities appears twice in the list, there are two roads): (A, B), (A, C), (B, C), (B, C), (C, E), (D, E). What is the valence of each vertex in your graph? What is a real-world consequence of whether or not the graph is connected?

6. Draw a graph to represent six cities A, B, C, D, E, and F having one road connecting each pair of cities in the following list: (A, B), (B, C), (D, E), (D, F). What is the valence of each vertex in your graph? What is a real-world consequence of whether or not the graph is connected?

■ **7.** A postal worker is supposed to deliver mail on all streets represented by edges in the graph below by traversing each edge exactly once. The first day the worker traverses the numbered edges in the order shown in **(a)**, but the supervisor is not satisfied—why? The second day the worker follows the path indicated in **(b)**, and the worker is unhappy—why? Is the original job description realistic? Why?

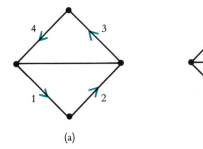

(a)   (b)

8. Examine the paths represented by the numbered sequences of edges in both parts of the figure below. Determine whether each path is a circuit. If it is a circuit, determine if it is an Euler circuit.

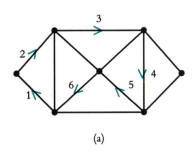   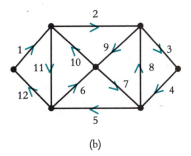

(a)   (b)

■ Discussion exercise.

9. Which graphs in the figure below have Euler circuits? In the ones that do, find the Euler circuits by numbering the edges in the order the Euler circuit uses them. For the ones that don't, explain why no Euler circuit is possible.

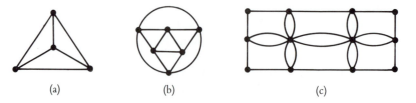

(a) (b) (c)

10. In Figure 1.8, suppose we started an Euler circuit using this sequence of edges: 14, 13, 8, 1, 4 (ignore existing arrows on the edges). What does our guideline for finding Euler circuits tell you not to do next?

11. Find an Euler circuit on Figure 1.10.

12. Find an Euler circuit on the right-hand graph in Figure 1.17a.

13. Find an Euler circuit on the graph of Figure 1.15d (including the blue edges).

14. Each graph below represents the sidewalks to be cleaned in a fancy garden (one pass over a sidewalk will clean it). Can the cleaning be done using an Euler circuit? If so, show the circuit by numbering the edges in the order the Euler circuit uses them. If not, explain why no Euler circuit is possible.

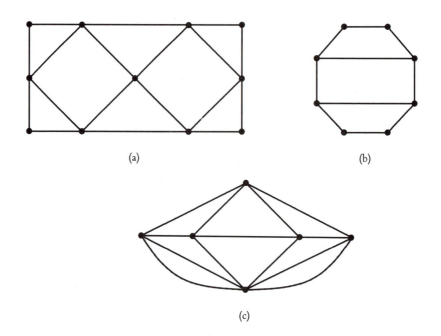

(a) (b)

(c)

15. In the graph below, we see a territory for a parking-control officer that has no Euler circuit. Which sidewalk (edge) could be dropped in order to enable us to find an Euler circuit?

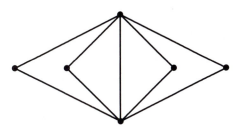

16. In the graph below, we see a territory for a parking-control officer that has no Euler circuit. Which sidewalk (edge) could be dropped in order to enable us to find an Euler circuit?

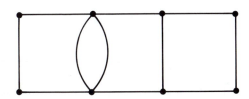

17. A college campus has a central square with sidewalks arranged like those in the graph below. Show how all the sidewalks can be traversed in one circuit; your circuit will have to repeat some edges.

18. In the graph of the figure below, add one or more edges to produce a graph that has an Euler circuit.

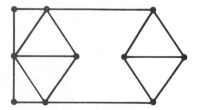

19. For the street network below, draw the graph that would be useful for finding an efficient route for checking parking meters. (*Hint:* Notice that not every sidewalk has a meter; see Figure 1.12 in the text.)

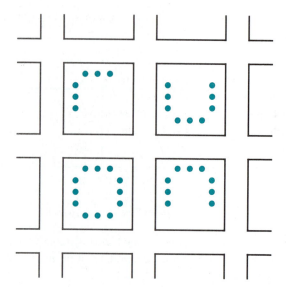

● 20. For the street network in the previous exercise, draw the graph that would be useful for routing a garbage truck. Assume that all streets are two-way and that passing once down a street suffices to collect from both sides.

21. For the street network at the top of the following page, draw the graph that would be useful for finding an efficient route for checking parking meters. (*Hint:* Notice that not every sidewalk has a meter; see Figure 1.12 in the text.)

● Optional exercise.

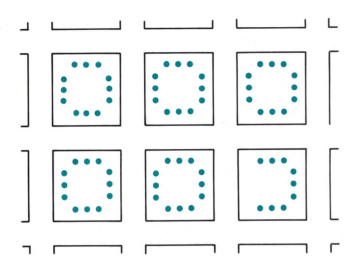

● 22. For the same street network, draw a graph that would be useful for routing a garbage truck. Assume that all streets are two-way and that passing once down a street suffices to collect from both sides.

23. Draw the graph for the parking-control territory shown in the figure below. Label each vertex with its valence and determine if the graph is connected.

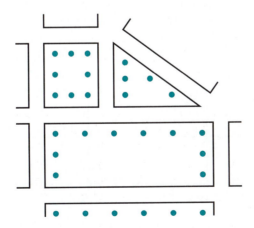

24. Squeeze the circuit shown in graph **(a)** onto graph **(b)**. Show your answers by numbered arrows on the edges.

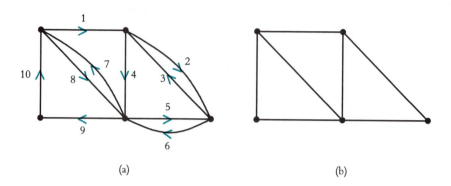

(a)   (b)

Then squeeze the circuit shown in graph **(a)** onto graph **(b)**. Show your answers by numbered arrows on the edges.

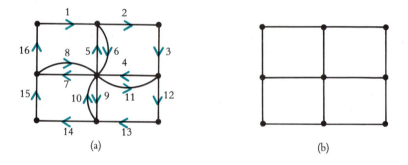

(a)   (b)

25. Find an Euler circuit on the eulerized graph **(b)** of the following figure. Use it to find a circuit on the original graph **(a)** that covers all edges and only reuses edges five times.

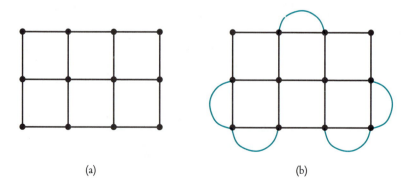

(a)   (b)

26. Find an Euler circuit on the eulerized graph **(b)** of the following figure. Use it to find a circuit on the original graph **(a)** that covers all edges and only reuses edges four times.

(a)

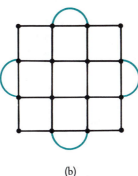

(b)

27. Can you find an eulerization with 7 added edges for a 2-by-5-block rectangular street network? Can you do better than 7?

28. Can you find an eulerization with 6 added edges for a 3-by-5-block rectangular street network? Can you do better than 6?

29. Can you find an eulerization with 9 added edges for a 3-by-6-block rectangular street network? Can you do better than 9?

30. Can you find an eulerization with 10 added edges for a 4-by-6-block rectangular street network? Can you do better than 10?

● 31. Find a circuit in the graph below that covers every edge and has as few reuses as possible. See optional section "Finding Good Eulerizations" for hints.

● 32. Find a circuit in the graph below that covers every edge and has as few reuses as possible. See optional section "Finding Good Eulerizations" for hints.

● Optional exercise.

● 33. In the figure below, all blocks are 1000 by 1000 feet, except for the middle column of blocks, which are 1000 by 4000 feet. Can you find an eulerization in which the total length of all duplicated edges is 8000?

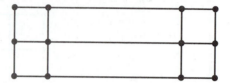

● 34. In the figure below, all blocks are 1000 by 1000 feet, except for the middle column of blocks, which are 1000 by 4000 feet. Can you find an eulerization in which the total length of all duplicated edges is 8000?

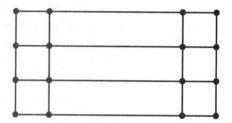

35. Eulerize these rectangular street networks using the same patterns that would be used by the "edge walker" described in the text.
    a. A 5 × 5 rectangle
    b. A 5 × 4 rectangle
    c. A 6 × 6 rectangle

36. Eulerize these rectangular street networks using the same patterns that would be used by the "edge walker" described in the text.
    a. A 6 × 5 rectangle
    b. A 6 × 6 rectangle
    c. A 5 × 3 rectangle

● 37. Find good eulerizations for these graphs, using as few duplicated edges as you can. See the optional section "Finding Good Eulerizations" for hints.

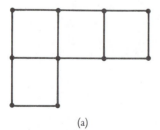

(a)

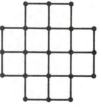

(b)

(c)

● Optional exercise.

● 38. Find good eulerizations for these graphs, using as few duplicated edges as you can. See the optional section "Finding Good Eulerizations" for hints.

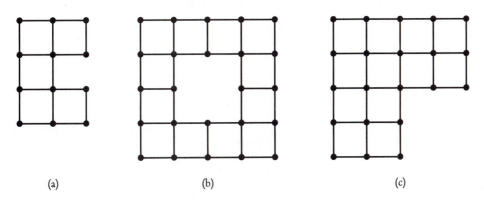

(a)   (b)   (c)

● 39. Find the best eulerizations you can for the two graphs below.

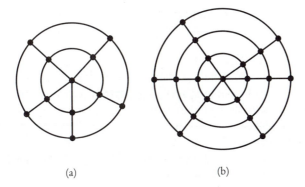

(a)   (b)

● 40. Suppose you had to find circuits on the graph below that have the minimal number of reuses of edges.
    a. Say what that minimum number would be.
    b. Show a circuit with the minimum number of reuses.

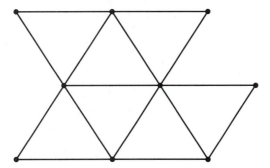

▲ 41. A graph G represents a street network to be traveled by a postal worker who must traverse every street twice, once for each side of the street. In graph G, the edges represent sidewalks. Does such a graph always have an Euler circuit? Explain your answer.

▲ 42. Suppose for a certain graph that it is possible to disconnect it by removing one edge. Explain why such a graph (before the edge is removed) must have at least one odd valence. (*Hint:* Show that it cannot have an Euler circuit.)

■ 43. The word *valence* is also used in chemistry. Find out what it means in chemistry and explain how this usage is similar to the use we make of this term.

■ 44. Is it possible that a street network gives rise to a disconnected graph? If it is possible, draw such a network of blocks and streets and parking meters (in the style of Figure 1.12a). Then draw the disconnected graph it gives rise to.

■ 45. What are the practical reasons why we need to come back to the starting place in our routes? Answer separately for the following cases. (However, if you think returning to the start might not really be necessary, explain why not.)
  a. Parking control
  b. Salt spreading
  c. Delivery of mail by letter carriers

▶WRITING PROJECTS

1. Write a memo of three double-spaced typewritten pages to your local department of parking control (or police department) in which you suggest that management-science techniques like the ones in this chapter be used to plan routes. Assume that the person to whom you are writing is not extensively trained in mathematics, but is willing to read through some technical material, provided you make it seem worth the trouble.

2. Do the same as in writing exercise 1, but to the department in charge of spreading salt on roads after snowstorms.

3. If you were making a recommendation to the mayor of New York City concerning the proposed new street-sweeping routes, designed using the theory of this chapter, would you recommend that the changes be adopted or not? Write a memo (three double-spaced typewritten pages) that outlines the pros and cons as fairly as you can, and then conclude with your recommendation.

▲ Advanced exercise.   ■ Discussion exercise.

2

Visiting Vertices

In the last chapter, we saw that it is relatively easy to determine if there is a circuit traversing the edges of a graph exactly once—for example, a route for street sweepers that covers the streets in a section of a city exactly once. However, the situation changes radically if we make an apparently innocuous change in the problem: When is it possible to find a route along distinct edges of a graph that visits each *vertex* once and only once in a simple circuit? For example, the wiggly line in Figure 2.1a shows a circuit we can take to tour that graph, visiting each vertex once and only once. This tour can be written *ABDGIHFECA*. Note that another way of writing the same circuit would be *EFHIGDBACE*. A different circuit visiting each vertex once and only once would be *CDBIG-FEHAC* (Figure 2.1b). Do not be confused because *C* is written twice when we write down this list of vertices. We can think of the circuit as starting at any of its vertices, but we do start and end at the same vertex.

▶ HAMILTONIAN CIRCUITS

A tour, like the ones marked by wiggly edges in Figure 2.1, that starts at a vertex of a graph and visits each vertex once and only once, returning to where it started, is called a **Hamiltonian circuit.** The concept is named for the Irish mathematician William Rowan Hamilton (1805–1865), who was one of the first to study it. [We now know that the concept was discovered somewhat earlier by Thomas Kirkman (1806–1895), a British minister with a penchant for mathematics.] This problem is typical of the many new problems mathematicians create as a consequence of their professional training. In the situation here, motivated by our success in solving the problem of traversing all the *edges* of a graph, we will investigate visiting all the *vertices* of a graph.

The concepts of Euler and Hamiltonian circuits are similar in that both forbid reuse: the Euler circuit of edges, the Hamiltonian circuit of vertices. However, it is far more difficult to determine which connected graphs admit a Hamiltonian circuit than to determine which connected graphs have Euler circuits. As we saw in Chapter 1, looking at the valences of vertices tells us if a connected graph has an Euler circuit, but we have no such simple method for telling whether or not a graph has a Hamiltonian circuit. Some special classes of graphs are known to have Hamiltonian circuits, and some special classes of graphs are

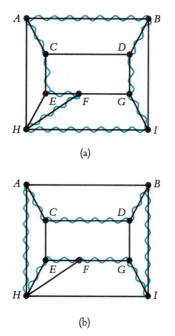

(a)

(b)

Figure 2.1 Wiggled edges illustrate Hamiltonian circuits.

Figure 2.2 An example of one graph from a family of graphs that has no Hamiltonian circuit. The number of vertices m on the left is chosen to be greater than the number of vertices n on the right. The case $m = 4$ and $n = 2$ is shown.

known to lack them. For example, here is a method to construct an infinite family of graphs where each graph in the family cannot have a Hamiltonian circuit. Construct a vertical column of m vertices and a parallel column of n vertices, where m is bigger than n, as shown in Figure 2.2. The figure illustrates the typical case where $m = 4$ and $n = 2$. Now join each vertex on the left in the figure to every vertex on the right. As m and n vary we get a family of different graphs. Any graph obtained in this manner cannot have a Hamiltonian circuit. If a Hamiltonian circuit existed, it would have to include alternately vertices on the left and right of the figure. This is not possible, since the number of vertices on the left and right, m and n, respectively, are not the same. Unfortunately, it is unlikely that a method will ever be found to easily

determine whether or not an arbitrarily chosen graph has a Hamiltonian circuit.

The Hamiltonian circuit problem and the Euler circuit problem are both examples of graph theory problems. Although we posed the Hamiltonian circuit problem merely as a variant of another graph theory problem with many applications (i.e., the Euler circuit problem), the Hamiltonian circuit problem itself has many applications. This is not unusual in mathematics. Often mathematics used to solve a particular real-world problem leads to new mathematics that suggests applications to other real-world situations.

Suppose inspections or deliveries need to be made at each vertex (rather than along each edge) of a graph. An "efficient" tour of the graph would be a route that started and ended at the same vertex and passed through all the vertices without reuse, or repetition; that is, the route would be a Hamiltonian circuit. Such routes would be useful for inspecting traffic signals, or for delivering mail to drop-off boxes, which hold heavy loads of mail so urban postal carriers do not have to carry it long distances. There are many similar examples, but rather than pursue problems involving Hamiltonian circuits in general graphs, we will study instead a more important class of related problems.

EXAMPLE: Vacation Planning

Let's imagine that you are a college student studying in Chicago. During spring break you and a group of friends have decided to take a car trip to visit other friends in Minneapolis, Cleveland, and St. Louis. There are many choices as to the order of visiting the cities and returning to Chicago, but you want to design a route that minimizes the distance you have to travel. Presumably, you also want a route that cuts costs, and you know that minimizing distance will minimize the cost of gasoline for the trip. (Similar problems with different complications would arise for railroad or airplane trips.)

Imagine now that the local automobile club has provided you with the intercity driving distances between Chicago, Minneapolis, Cleveland, and St. Louis. We can construct a graph model with this information, representing each city by a vertex and the legs of the journey between the cities by edges joining the vertices. To complete the model, we add a number called a **weight** to each graph edge, as in Figure 2.3. In this example, the weights represent the distances between the cities, each of which corresponds to one of the endpoints of the edges in the graph. (In other examples the weight might represent a cost, time, or profit.) We want to find a minimal-cost tour that starts and ends in Chicago and visits each other city once. Using our earlier terminology, what we wish to find is a **minimum-cost Hamiltonian circuit**—a Hamiltonian circuit with the lowest possible sum of the weights of its edges.

How can we determine which Hamiltonian circuit has minimum cost? There is a conceptually easy **algorithm,** or mechanical step-by-step process, for solving this problem:

1. Generate all possible Hamiltonian tours (starting from Chicago).

2. Add up the distances on the edges of each tour.

3. Choose the tour of minimum distance.

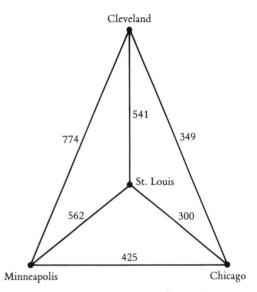

Figure 2.3 Road mileages between four cities.

Steps 2 and 3 of the algorithm are straightforward. Thus, we need worry only about step 1, generating all the possible Hamiltonian circuits in a systematic way. To find the Hamiltonian tours, we will use the **method of trees,** as follows. Starting from Chicago, we can choose any of the three cities to visit after leaving Chicago. The first stage of the enumeration tree is shown in Figure 2.4. If Minneapolis is chosen as the first city to visit, then there are two possible cities to visit next, namely, Cleveland and St. Louis. The possible branchings of

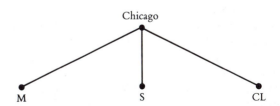

Figure 2.4 First stage in finding vacation-planning routes.

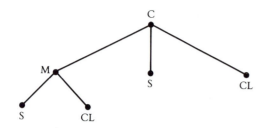

Figure 2.5 Part of the second stage in finding vacation-planning routes.

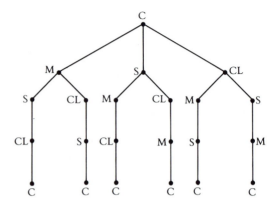

Figure 2.7 Completed tree enumeration of routes for vacation-planning problem.

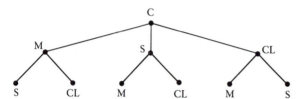

Figure 2.6 Complete second stage in finding vacation-planning routes.

the tree at this stage are shown in Figure 2.5. In this second stage, however, for each choice of first city we visited, there are two choices from this city to the second city visited. This would lead to the diagram in Figure 2.6.

Having chosen the order of the first two cities to visit, and knowing that no revisits (reuses) can occur in a Hamiltonian circuit, there is only one choice left for the next city. From this city we return to Chicago. The complete tree diagram showing the third and fourth stages for these routes is given in Figure 2.7. Notice, however, that because we can traverse a circular tour in either of two directions, the paths enumerated in the tree diagram of Figure 2.7 do *not* correspond to different Hamiltonian circuits. For example, the leftmost path (C–M–S–CL–C) and the rightmost path (C–CL–S–M–C) represent the same Hamiltonian circuit. Thus, among what appear to be six different paths in the tree diagram, there correspond, in fact, only three different Hamiltonian circuits.

These three distinct Hamiltonian circuits are shown in Figure 2.8.

Note that in generating the Hamiltonian circuits we disregard the distances involved. We are concerned only with the different patterns of carrying out the visits. To find the *optimal* route, however, we must add up the distances on the edges to get each tour's length. Figure 2.8 shows that the optimal tour is Chicago, Minneapolis, St. Louis, Cleveland, Chicago. The length of this tour is 1877 miles. ▲

The method of trees is not always as easy to use as our example suggests. Instead of doing our analysis for four cities, consider the general case of *n* cities. The graph model similar to that in Figure 2.3 would consist of a weighted graph with *n* vertices, with every pair of vertices joined by an edge. Such a graph is called **complete** because the edge between any pair of vertices is present in the graph. A complete graph with five vertices is illustrated in Figure 2.9.

Fundamental Principle of Counting

How many Hamiltonian circuits are in a complete graph of *n* vertices? We can solve this problem by

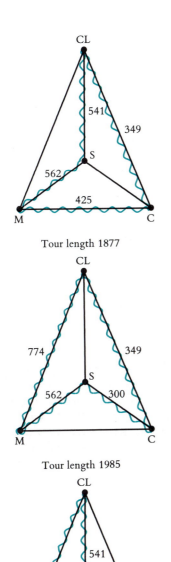

Tour length 1877

Tour length 1985

Tour length 2040

Figure 2.8 The three Hamiltonian circuits for the vacation-planning problem of Figure 2.3.

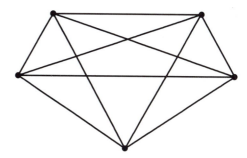

Figure 2.9 A complete graph with five vertices. Every pair of vertices is joined by an edge.

using the same type of analysis that emerged in the enumeration tree. The method of trees is a visual application of the **fundamental principle of counting,** a procedure for counting outcomes in multistage processes. Using this procedure we can count how many patterns occur in a situation by looking at the number of ways the component parts can occur. For example, if Jack has 9 shirts and 4 pair of trousers, he can wear $9 \times 4 = 36$ shirt–pants outfits. Each shirt can be worn with any of the pants. (This can be verified by drawing a tree diagram, but such a diagram is cumbersome for big numbers.)

In general, the fundamental principle of counting can be stated this way: if there are a ways of choosing one thing, b ways of choosing a second after the first is chosen, . . . , and z ways of choosing the last item after the earlier choices, then the total number of choice patterns is $a \times b \times c \times \cdots \times z$.

EXAMPLES: Counting

Here are some other examples of how to use the fundamental principle of counting:

1. In a restaurant there are 4 kinds of soup, 12 entrees, 6 desserts, and 3 drinks. How many different meals can a patron chose from? The

four choices can be made in 4, 12, 6, and 3 ways, respectively. Hence, applying the fundamental principle of counting, there are $4 \times 12 \times 6 \times 3 = 864$ possible meals.

2. In a state lottery a contestant gets to pick a four-digit number that contains no zero followed by an uppercase or lowercase letter. How many such sequences of digits and a letter are there? Each of the four digits can be chosen in 9 ways (that is, 1, 2, . . . , 9), and the letter can be chosen in 52 ways (that is, $A, B, . . . , Z, a, b, . . . , z$). Hence there are $9 \times 9 \times 9 \times 9 \times 52 = 341,172$ possible patterns.

3. A corporation is planning a musical logo consisting of four different notes from the scale C, D, E, F, G, A, and B. How many logos are there to chose from? The first note can be chosen in 7 ways, but since reuse is not allowed, the next note can be chosen in only 6 ways. The remaining two notes can be chosen in 5 and 4 ways, respectively. Using the fundamental principle of counting, $7 \times 6 \times 5 \times 4 = 840$ musical logos are possible. ▲

Returning to the problem of enumerating Hamiltonian circuits for the complete graph with n vertices, the city visited first after the home city can be chosen in $n - 1$ ways, the next city in $n - 2$ ways, and so on, until only one choice remains. Using the fundamental principle of counting, there are $(n - 1)! = (n - 1)(n - 2) \cdots \times 3 \times 2 \times 1$ routes. [The exclamation mark in "$(n - 1)!$" is read "factorial" and is a shorthand notation for the product $(n - 1)(n - 2) \times \cdots \times 3 \times 2 \times 1$. For example, $5! = 5 \times 4 \times 3 \times 2 \times 1 = 120$.]

As we saw in Figure 2.7, pairs of routes correspond to the same Hamiltonian circuit because one route can be obtained from the other by traversing the cities in reverse order. Thus, although there are $(n - 1)!$ possible routes, there are only half as many, or $(n - 1)!/2$, Hamiltonian circuits. Now, if we have only a few cities to visit, $(n - 1)!/2$ Ham-

iltonian circuits can be listed and examined in a reasonable amount of time. To analyze a 6 city problem it would require generation of $(6 - 1)!/2 = 5!/2 = 120/2 = 60$ tours. But for, say, 25 cities, $24!/2$ is approximately 3×10^{23}. Even if these tours could be generated at the rate of 1 million a second, it would take 10 billion years to generate them all. Since large vacation-planning problems would take so long to solve using the method we have described, this method, despite its conceptual ease, is sometimes referred to as the **brute force method**.

▶ TRAVELING SALESMAN PROBLEM (TSP)

If the only benefit were saving money and time in vacation planning, the difficulty of finding a minimum-cost Hamiltonian circuit in a complete graph with n vertices for large values of n would not be of great concern. However, the problem we are discussing is one of the most common problems in operations research. It is usually called the **traveling salesman problem (TSP)** because of its early formulation: determine the trip of minimum cost that a salesperson can make to visit the cities in a sales territory, starting and ending the trip in the same city.

Many situations require solving a TSP:

1. A lobster fisherman has set out traps at various locations and wishes to pick up his catch.

2. The telephone company wishes to pick up the coins from its pay telephone booths.

3. The electric (or gas) company needs to design a route for its meter readers.

4. A minibus must pick up six day campers and deliver them to camp, and later in the day return them home.

5. In drilling holes in a series of plates, the drill press operator (perhaps a robot!) must drill the holes in a predetermined order.

The meaning of cost can vary from one formulation of TSP to another. We may measure cost as distance, airplane ticket prices, time, or any other factor that is to be optimized.

In many situations, the TSP arises as a subproblem of a more complicated problem. For example, a supermarket chain may have a very large number of stores to be served from a single large warehouse. If there are fewer trucks than stores, the stores must be grouped into clusters so that one truck serves each cluster. If we then solve the TSP for every truck, we can minimize total costs for the supermarket chain. Similar vehicle-routing problems for dial-a-ride services and for delivering children to their schools or camps often involve solving the TSP as a subproblem.

▶ STRATEGIES FOR SOLVING THE TRAVELING SALESMAN PROBLEM

Because the traveling salesman problem arises so often in situations where the associated complete graphs would be very large, we must find a faster method than the brute force method we have described. We need to look at our original problem in Figure 2.3 and try to find an alternative algorithm for solving it. Recall that our goal is to find the minimum-cost Hamiltonian circuit.

Nearest-Neighbor Algorithm

Let's try a new approach: starting from Chicago, first visit the nearest city, then visit the nearest city that has not already been visited. We return to the start city when no other choice is available. This approach is called the **nearest-neighbor algorithm.** Applying it to the TSP in Figure 2.3 quickly leads to the tour of Chicago, St. Louis, Cleveland, Minneapolis, and Chicago, with a length of 2040 miles. Here is how this tour is determined. Since we are starting in Chicago, there is a choice of going to a city that is 425, 300, or 349 miles away. Since the smallest of these numbers is 300, we visit next St. Louis, which is

the nearest neighbor of Chicago not already visited. At St. Louis, we have a choice of visiting next cities that are 541 or 562 miles away. Hence, Cleveland, which is nearer (541), is visited. To complete the tour, we visit Minneapolis and return to Chicago, thereby adding 774 and 425 miles to the length of the tour. The nearest-neighbor algorithm is an example of a **greedy algorithm,** because at each stage a best (greedy) choice, based on an appropriate criterion, is made. Unfortunately, this is not the optimal tour. Making the best choice at each stage may not yield the best "global" solution. Even for a large TSP, one can always find a nearest-neighbor route quickly.

Figure 2.10 again illustrates the ease of applying the nearest-neighbor algorithm, this time

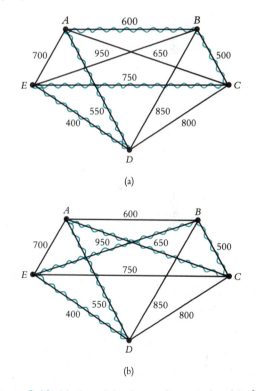

Figure 2.10 (a) A weighted complete graph with five vertices that illustrates the use of the nearest-neighbor algorithm (starting at *A*). (b) TSP tour generated by the nearest-neighbor algorithm (starting at *B*).

to a weighted complete graph with five vertices. Starting at vertex A, we get the tour $ADECBA$ (cost 2800) (Figure 2.10a). Note that the nearest-neighbor algorithm starting at vertex B yields the tour $BCADEB$ (cost 3050) (Figure 2.10b).

This example illustrates that the nearest-neighbor tour can be computed for each vertex of the complete graph being considered, and that different nearest-neighbor tours can be gotten starting at different vertices. Note that even though we may seek a tour starting at a particular vertex, say, A in Figure 2.10, since a Hamiltonian circuit can be thought of as starting at any of its vertices, we can apply the nearest-neighbor procedure, if we wish, starting at vertex B (rather than at A). The Hamiltonian circuit we get can still be thought of as beginning at vertex A rather than B. Even for complete graphs with a large number of vertices, it would still be faster to apply nearest neighbor for each vertex and pick the cheapest of the tours generated (though such a tour might not be optimal) than to apply the brute force method.

Sorted-Edges Algorithm

Perhaps some other easy method would yield optimal solutions. We might start by sorting or arranging the edges of the complete graph in order of increasing cost (or equivalently, arrange the intercity distances in order of increasing distance). Then we can select at each stage that edge of least cost that (1) never requires that three edges meet at a vertex (since a Hamiltonian circuit uses up exactly two edges at each vertex), and that (2) never closes up a circular tour that doesn't include all the vertices. This algorithm will be called the **sorted-edges algorithm.**

Applying the sorted-edges algorithm to the TSP in Figure 2.3 yields the tour Chicago, St. Louis, Minneapolis, Cleveland, and Chicago, since the edges chosen would be 300, 349, 562, 774. Here are the details of how the sorted-edges method works in this example. First, the six weights on the edges listed in increasing order

would be 300, 349, 425, 541, 562, and 774. Since the cheapest edge in this sorted list is 300, this is the first edge that we place into the tour we are building up. Next we add the edge with weight 349 to the tour. The next cheapest edge would be 425, but using this edge together with those already selected would result in having three edges at a vertex (see Figure 2.11a), which is not consistent

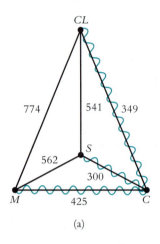

(a)

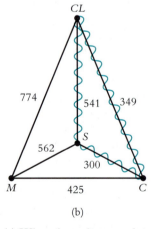

(b)

Figure 2.11 (a) When three shortest edges are added in order of increasing distance, three edges at a vertex are selected, which is not allowed as part of a Hamiltonian circuit. (b) When the edges of distances 300, 349, and 541 are selected, a circuit that does not include all vertices results.

SP TLIGHT 2.1   NP-Complete Problems

▶ ▶ ▶ ▶ ▶ ▶ ▶ ▶ ▶ ▶ ▶ ▶ ▶

Stephen Cooke, a computer scientist at the University of Toronto, showed in 1971 that certain hard, frustrating problems are equivalently difficult. This class of problems, now referred to as **NP-complete problems**, has the following characteristic: if a "fast" algorithm for solving *one* of these problems could be found, then a fast method would exist for *all* these problems.

In this context, "fast" means that as the size n of the problem grows (the number of cities gives the problem size in the traveling salesman problem), the amount of time needed to solve the problem grows no more rapidly than a polynomial function in n. (A polynomial function has the form $a_k n^k + a_{k-1} n^{k-1} +$ $\cdots + a_1 n + a_0$.) On the other hand, if it could be shown that any problem in the class of NP-complete problems required an exponentially growing (3^n is an example of an exponential function) amount of time to solve it as the problem size increased, then all problems in the NP-complete class would share this characteristic. If some NP-complete problems had fast solutions, it seems likely that at least one such fast solution would have been found by now. It has been known for some time that the traveling salesman problem is an NP-complete problem: for this reason it is generally thought that no "fast" algorithm for an optimal solution for the TSP will ever be found.

with having a Hamiltonian circuit. Hence, we do not use this edge. The next cheapest edge, 541, used together with the edges already selected, would create a circuit (see Figure 2.11b) that does not include all the vertices. Thus, this edge, too, would be skipped over. However, we are able to add the edges 562 and 774 without either creating a circuit shorter than one including all the vertices or having three edges at a vertex. Hence, the tour we arrive at is Chicago, St. Louis, Minneapolis, Cleveland, and Chicago. Again, this solution is not optimal because its length is 1985. Note that this algorithm, like the nearest neighbor, is greedy.

Although the edges selected by applying the sorted-edges method to the example in Figure 2.3 are connected to each other at every stage, this does not always happen. For example, if we apply

the sorted-edges algorithm to the graph in Figure 2.10a, we build up the tour first with edge *ED* (400) and then edge *BC* (500), which do not touch. The edges that are then selected are *AD*, *AB*, and *EC*, giving the circuit *EDABCE*, which is the same as the nearest-neighbor circuit starting at vertex *A*.

Although many "quick and dirty" methods for solving the TSP have been suggested and although some methods give an optimal solution in some cases, none of these methods *guarantees* an optimal solution. Surprisingly, most experts believe that no efficient method that guarantees an optimal solution will ever be found (see Spotlight 2.1).

Recently, mathematical researchers have adopted a somewhat different strategy for dealing with TSP problems. If finding a fast algorithm to generate optimal solutions for large problems is

SPOTLIGHT 2.2 — Solving the Elusive Traveling Salesman Problem

More than 20 years ago at Bell Labs, Shen Lin set out to solve the traveling salesman problem. Although the TSP defies any quick mathematical solution, Shen Lin has discovered a method that works fairly quickly on many practical problems. Today, he applies this knowledge to designing private telephone networks for AT&T's corporate customers (see Spotlight 2.5). Here, Shen Lin recalls his earlier work on the TSP:

I started trying to solve the traveling salesman problem back in 1965 and ended in 1972, collaborating with a colleague, Brian Kernahan, in an algorithm that up to this day is considered to be the most efficient, practical method of solving this problem.

In 1965, we knew very little about which kinds of problems were hard and which were easy. So naively, I went in and thought I could solve this problem. Of course, I couldn't. I published my first paper on computer approximations by using the so-called iterative techniques [repeating some operation over and over] to solve them.

Today, I essentially use a *heuristic method.* The difference between this and the so-called classical method is that exact-solution methods usually take too long when you are trying to solve a hard problem like the traveling salesman problem. Also, they are formulated for problems that are very well defined. But in real life, we frequently find that no problem is so well defined, so we need some method of solving problems that is more flexible. We want something that can get at approximate solutions, which may turn out to be optimum solutions. The method must be powerful enough to solve a variety of problems. We cannot guarantee that the solution is the absolute and best one, because to find the best solution would take billions of years of computation, even by the world's fastest computer.

Within a few seconds, however, I could give you the solution to the traveling salesman problem, say, for a few hundred [geographical] points. And I'm sure no human being could look at the map, connect those points, and achieve the same result. I can't *prove* that it's the best solution that could ever be found, but it's usually quite close — within 1 to 2% of the optimum.

unlikely, perhaps one can show that the "quick and dirty" methods, usually called **heuristic algorithms,** come close enough to giving optimal solutions. For example, suppose one could prove that the nearest-neighbor heuristic was never off by more than 25% in the worst case or by more than 15% in the average case. For a medium-size TSP, one would then have to choose whether to spend a lot of time (or money) to find an optimal solution or to use instead a heuristic algorithm to obtain a fairly good solution. Researchers at AT&T Bell Laboratories have developed many remarkably good heuristic algorithms (see Spotlight 2.2). The best-known guarantee for a heuristic algorithm for a TSP is that it yields a cost that is no worse than 1.5 times the optimal cost. Interestingly, this heuristic algorithm involves solving a Chinese postman problem, for which a "fast" algorithm is known to exist.

Throughout our discussions of the TSP we have concentrated on the goal of minimizing the cost (or time) of a tour that visited each of a variety of sites once and only once. One of the things that makes mathematical modeling exciting, however, is the subtle issues that arise in specific real-world situations (or that provide contrast between seemingly similar situations). For example, suppose the TSP situation is that of picking up day campers and taking them to and from the camp. From the point of view of the camp, it may wish to minimize the total length of time that the bus needs to pick up the campers. From the point of view of the parents of the campers, however, they may like the time their children spend on the bus to be as little as possible. For some problems, the tour that minimizes the mean (average) time that a child spends on the bus may not be the same tour that minimizes the total time of the tour. (Specifically, if the bus goes first to pick up the child the farthest from the camp, and then picks up the other children, this may yield a relatively short time on the bus for the kids, but a relatively long time for the tour itself.) It is these subtleties between problems that mathematicians go back to at a later time to examine, after the basic structure of the main problem itself is well understood. It is in this way that mathematics continues to grow, explore new ideas, and find new applications.

▶Minimum-Cost Spanning Trees

The traveling salesman problem is but one of many graph theory optimization problems that have grown out of real-world problems in both government and industry. Here is another.

Example: Pictaphone Service

Imagine that Pictaphone service is to be set up on an experimental basis between five cities. The graph in Figure 2.12 shows the possible links that might be included in the Pictaphone network, with each edge showing the cost in millions of dollars to create that particular link. To send a Pictaphone message between two cities, a direct communication link is not necessary because it is possible to send a message *indirectly* via another city. Thus, in Figure 2.12, sending a message from A to C could be achieved by sending the message from A to B, from B to E, and from E to C, provided the links AB, BE, and EC are part of the network. We assume that the cost of relaying a message, compared with the direct communication link cost, is so small that we can neglect this amount. The problem that concerns us, therefore, is to provide service between any pair of cities in a way that minimizes the total cost of the links.

Our first guess at a solution is to put in the cheapest possible links between cities first, until all cities could send messages to any other city. Such an approach would be analogous to the sorted-edges method that was used to study the traveling salesman problem. In our example, if the cheapest links are added until all cities are joined, we obtain the connections shown in Figure 2.13a.

The links were added in the order ED, AD, AE, AB, DC. However, because this graph contains the circuit $ADEA$ (wiggly edges in Figure 2.13b), it has redundant edges: we can still send messages between any pair of cities using relays after omitting the most expensive edge in the circuit—AE. After deleting an edge of a circuit, a message can still be relayed among the cities of the circuit by sending signals the long way around. After AE is deleted, messages from A to E can be sent via D (Figure 2.13c). This suggests a modified algorithm for our problem,

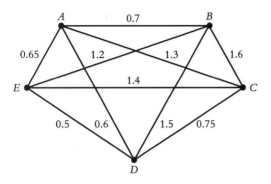

Figure 2.12 Costs of installing Pictaphone service between five cities in millions of dollars.

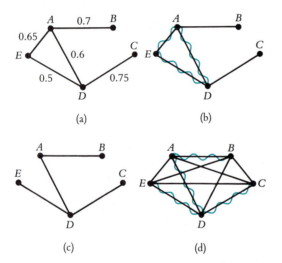

Figure 2.13 (a) Cities are linked in order of increasing cost until all cities are connected. (b) Circuit in (a) highlighted. (c) Most expensive link in circuit in (a) deleted. (d) Highlighted edges show, as a subgraph of the original graph, those links connecting the cities with minimum cost obtained using Kruskal's algorithm.

Kruskal's algorithm: add the links in order of cheapest cost so that no circuits form and so that every vertex belongs to some link added (Figure 2.13d). As in the sorted-edges method for the TSP, the edges that are added need not be connected to each other until the end.

A subgraph formed in this way will be a **tree;** that is, it will consist of one piece and contain no circuits. It will also include all the vertices of the original graph. A subgraph that is a tree and that contains all the vertices of the original graph is called a **spanning tree** of the original graph.

To understand these concepts better, consider the graph G in Figure 2.14a. The wiggled edges in Figure 2.14b would constitute a subgraph of G that is a tree (since it is connected and has no circuit), but this tree would not be a spanning tree of G since the vertices D and E would not be included. On the other hand, the wiggled edges in Figure 2.14c, d show subgraphs of G that include all the vertices of G, but are not trees because the first is not connected and the second contains a circuit. Figure 2.14e shows a spanning tree of G; the wiggled edges are connected, contain no circuit, and every vertex of the original graph is an endpoint of some wiggled edge. ▲

Finding a **minimum-cost spanning tree,** that is, a spanning tree whose edge weights sum to a minimum value, solves the Pictaphone problem. Note that having a different goal in the Pictaphone problem led to a different mathematical question from that of finding a Chinese Postman tour or TSP tour. In this application, the graph theory problem that we need to formulate to solve the applied problem was that of finding a minimum-cost spanning tree. In Figure 2.15a we have a graph model showing the costs of putting in roads to connect new houses in a suburban land development project. Applying the algorithm we have developed, adding the edges in the order of increasing cost, but avoiding the creation of a circuit, yields as a minimum-cost spanning tree the tree indicated by wiggled edges in Figure 2.15b. This tree is the cheapest one that makes it possible to drive between any pair of homes, though the driving distance between some of the homes will be relatively large, since only roads corresponding to

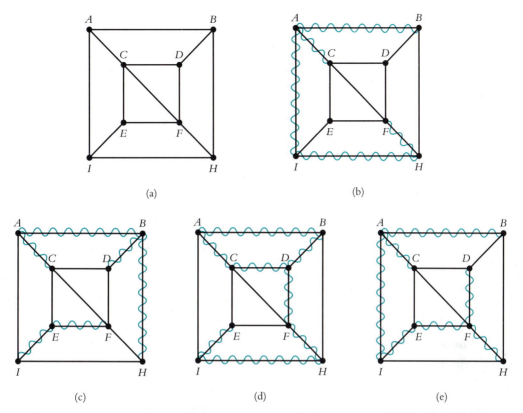

Figure 2.14 (a) A graph to help illustrate the concept of a spanning tree. (b) The wiggled edges are a tree, but not a spanning tree since vertices D and E are not part of the tree. (c) The wiggled edges are not a tree since they do not form a connected collection of edges. All the vertices of the graph are, however, endpoints of wiggled edges. (d) The wiggled edges are not a tree since they contain the edges of the circuit $BDCAB$. All the vertices of the graph are, however, endpoints of wiggled edges. (e) The wiggled edges form a tree and include all of the vertices of the graph as endpoints of wiggled edges.

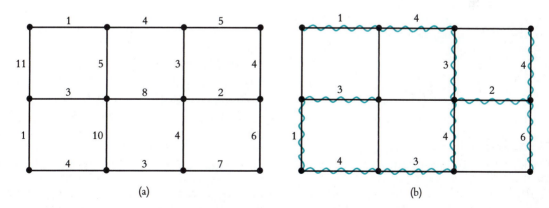

Figure 2.15 (a) A graph showing costs for construction roads between houses. (b) Wiggled edges show a minimum-cost spanning tree for the graph in (a).

SPOTLIGHT 2.3   Some Reminiscences on Shortest Spanning Trees

▶ ▶ ▶ ▶ ▶ ▶ ▶ ▶ ▶ ▶ ▶ ▶ ▶

Dr. Kruskal has written the following thoughts on the minimum-cost spanning algorithm, which he refers to as shortest spanning trees.

In 1951, I entered the graduate mathematics program at Princeton. Among those active in combinatorial mathematics at Princeton then were Al Tucker, who soon became chairman of the department, Harold Kuhn, and Roger Lyndon. For a while, there was a joint Princeton–Bell Laboratories combinatorics seminar, which met alternately at two locations. My first introduction to Bell Laboratories, where I subsequently have spent several decades, came in this connection.

One day, someone in the combinatorics community handed me a blurry, carbon-copy typescript on flimsy lightweight paper about shortest spanning trees. It was in German, and appeared to be the foreign-language summary of a paper written in some other language. I had the typescript only for a limited period, and then passed it on to someone else. The typescript contained no date, and at the time I had no idea of whether it had been published or where. To this day, I have no idea where the typescript had come from, and why it was circulating. While writing my own paper stimulated by this material, I must have been told that the typescript was

wiggled edges will be built. Remember that the weights on the edges of the graph in Figure 2.15a represent the costs of building roads, *not* the driving distance between the houses. Note that Figure 2.15a is not a complete graph, one in which all possible edges are included. Edges that correspond to roads that would be economically prohibitive to build have not been shown in the graph model. Also, in Figure 2.15b, the two edges of weight 5 (shown in Figure 2.15a) do not become part of the minimum-cost spanning tree, because they would create circuits with edges already chosen.

How do we know that the spanning tree found by the algorithm we developed achieves the minimum possible cost? Although this sounds very plausible, our experience with the TSP should suggest caution. Remember that for the TSP, the

sorted-edges algorithm, also a greedy algorithm, did not necessarily give an optimal solution! On what basis should we have more faith in our present algorithm?

Kruskal's Algorithm

The algorithm described here was first suggested by Joseph Kruskal (AT&T Bell Laboratories) in the mid-1950s to solve a problem in pure mathematics proposed by a Czechoslovakian mathematician. In mathematics it is a surprising but not uncommon finding that ideas used to solve problems with no apparent application often turn out to have many real-world uses. Kruskal's solution to the problem of finding a minimum-cost spanning

(part of?) a translation of a 1926 paper by Otakar Boruvka, for I give an exact citation in my paper. However, I have no memory of who told me this, nor do I remember ever seeing the full paper.

By the way, I regret that the phrase "minimum-spanning tree" has become dominant. What is actually meant is the minimum *length* spanning tree (MST), which is more concisely rendered by *shortest* spanning tree (SST). That is the phrase I used in 1954, and continue to use. If you wish to stick with MST rather than SST, however, at least do not make the common error of misconstruing MST as minimal, rather than minimum, spanning tree. The thing being minimized, length, has a unique minimum under very general conditions. [In this book, the term "minimum-cost spanning tree" is used rather than MST, as is common in the technical mathematics literature.]

The chief point of Boruvka's paper, I believe, was the uniqueness of the SST of a graph if the edges of the graph all have distinct lengths. His proof was based on a construction of the SST that I had difficulty in fully grasping. Although the construction has a certain spare elegance, I decided that it was difficult to follow because it was unnecessarily complicated. That led me to devise some much simpler methods, which are described in my 1956 paper. In 1985 Graham and Hell wrote a full history of the SST and closely related ideas. Well over 100 papers dating back to 1909 are cited, though the earliest paper specifically on the SST is Boruvka's work in 1926.

It is interesting to note that I was very uncertain about whether the material in my simple paper, which occupies only two and a half printed pages and requires few equations, was worth writing up for publication. I was a graduate student in my second year, and did not have a good sense of what was publishable. I asked a friend for advice (I don't remember whom), and fortunately received encouragement. It was my second published paper.

tree in a graph with weights is a good example of this phenomenon (see Spotlight 2.3). Kruskal showed that the greedy algorithm described does yield the minimum answer, and his work led to applications of these and related ideas in designing minimum-cost computer networks, phone connections, and road and railway systems. For additional discussion of operations research in the communications industry (see Spotlights 2.4 and 2.5).

Although we have mentioned many routing problems in graphs, we have not discussed one of the most obvious: finding the path between two specified distinct vertices with the sum of the weights of the edges in the path as small as possible. (Here there is no need to cover all vertices or to cover all edges.) We have seen that the weights on the edges have many possible interpretations, including time, distance, and cost. Here are some of the many possible applications:

1. Design routes to be used by a fire engine to get to a fire as quickly as possible.

2. Design delivery routes that minimize gasoline use.

3. Design routes to bring soldiers to the front as quickly as possible.

4. Design a route for a truck carrying nuclear waste.

The need to find shortest paths seems natural. Next we investigate a situation where finding a longest path is the right tool.

SP TLIGHT 2.4

AT&T Manager Explains How Long-distance Calls Run Smoothly

▶▶▶ ▶ ▶ ▶ ▶ ▶ ▶ ▶ ▶ ▶

Although long-distance calls are now routine, it takes great expertise and careful planning for a company like AT&T to handle its vast amounts of telephone traffic. Rich Wetmore was district manager of AT&T's Communications Network Operations Center in Bedminster, New Jersey, in 1988. Here are his responses to questions about how AT&T handles its huge volumes of long-distance traffic and how it tracks its operations to keep things running smoothly.

How do you make sure that a customer doesn't run into a delayed signal when attempting a long-distance call?

We monitor the performance of our AT&T network by displaying data collected from all over the country on a special wallboard. The wallboard is configured to tell us if a customer's call is not going through because the network doesn't have enough capacity to handle it.

That's when we step in and take control to correct the problem. The typical control we use is to reroute the call. Instead of sending the customer's call directly to its destination, we'll route it via a third city — to someplace else in the country that has the capacity to complete the call.

It would seem that routing via another city would take longer. Is the customer aware of this process?

Routing a call via a third city is entirely transparent [imperceptible] to the customer. I'm an ex-pert about the network, and even when I make a phone call, I have no idea how that individual call was routed. It's transparent both in terms of how far away the other person sounds and in how quickly the telephone call gets set up. With the signaling network we use, it takes milliseconds for switching systems to "talk" to each other to set up a call. So the fact that you are involved in a third switch in some distant city is something you would never know.

You want to be sure to keep costs down while supplying enough service to customers. So how do you balance company benefits with customer benefits?

In terms of making the network efficient, we want to do two things. First, we want our customers to be happy with our service and for all their calls to go through, which means we must build enough capacity in the network to allow that to happen. Second, we want to be efficient for our stockholders and not spend more money than we need to for the network to be at the optimum size.

There are basically two costs in terms of building the network. There is the cost of switching systems and the cost of the circuits that connect the switching systems. Basically, you can use operations-research techniques and mathematics to determine cost trade-offs. It may make sense to build direct routing between two switching systems and use a lot of circuits, or maybe to involve three switching systems, with fewer circuits between the main two, and so on.

Made-to-Order Phone Networks Benefit Large-scale Users

▶ ▶ ▶ ▶ ▶ ▶ ▶ ▶ ▶ ▶ ▶ ▶

Shen Lin was executive consultant and head of AT&T Bell Laboratories Network Configuration and Planning Department, Murray Hill, New Jersey. A scientist-turned-salesman for AT&T, he devised a method for solving the traveling salesman problem in 1972 (see Spotlight 2.2). More recently, he developed the "interactive network optimization system" (INOS) for AT&T. Here are his comments on the network:

Before our customers will sign a contract to buy a multimillion-dollar network, they usually want someone to tell them about the methodology used for optimizing their networks. So I am frequently asked to present the network design results to our customers, not only to explain how it serves their needs, but also to give them an idea of how the design was worked out.

Here is how it works: Perhaps hundreds of locations in a large organization need to "talk" to each other. The people may need to talk to one another, or their terminals may need to communicate with computers.

For example, an individual may pick up his phone, make a call, and get charged for a regular long-distance call. If the calling volume is rather high, a company can use the Wide Area Telecommunications Service, commonly called WATS, and reduce the cost by 20% or 30%. Another option is to connect locations via private lines. So the problem is how to configure all communications needs into an optimal network that will minimize overall costs.

Let's consider the first step that must be taken: because not every location can be connected to every other location, a set of locations that are close together will come into what is called a "switch," or concentrator. Their calls will come into the switch and be completed thereafter.

So one of the first questions that has to be answered is, Where do you put the switch? This problem is a mathematically difficult problem in itself. Before we know where to put the concentrators, we have to estimate the cost of calls reaching their destination via the switch, and since we don't know where the switches will be located, we must use many different iterations to get at our optimal answer.

In general, a large corporation may generate about 5 million calls per month, and they may be calling from several hundred — or even 2,000 — locations, to tens of thousands of destinations. The job of optimization is to group all these together in such a way that the total bill is minimized.

INOS does all the investigation and configuration. It's very computation-intensive. To do a customer network, we run something over 300 to 500 million arithmetical instructions, and that's after months of work collecting the traffic data.

Right now, INOS is moving in the direction of configuring data networks. For example, my terminal has to talk to a computer in Massachusetts at this moment, so I have a data line. People may use a dial-up or data line, and in some cases many, many terminals may be connected to use a common line called a multi drop line. All of this can amount to a great deal of savings to the customer simply by solving the mathematical problem of optimum configuration.

►CRITICAL-PATH ANALYSIS

One of the delights of mathematics is its ability to confirm the obvious in certain situations while showing that our intuition is wrong in other circumstances. Our next group of mathematical applications will illustrate this point.

One of the characteristics of recent American life seems to be its fast pace. People are interested in getting things done quickly and efficiently. This means that when you take your car in to be repaired before work, you want to know for sure that the repairs will be done when you get home. You want the trains and the bus that take you to your doctor's appointment to run on time, and when you arrive at the doctor's office, you want there to be a nurse free to take a blood sample and a throat culture. You want your outpatient appointment for an x-ray at the local hospital to occur on schedule. You want the x-ray to be interpreted quickly and the results reported back to your internist. Scheduling machines and people is a big part of modern life. It is important in your own personal daily activities as well as to businesses and governments. Scheduling is involved in running a school, a hospital, an airline, or in landing a person on Mars. Perhaps surprisingly, modern mathematics is a big part of what is involved in solving scheduling problems.

Part of what makes scheduling complicated is that when one performs the tasks that make up a job one must complete, the tasks usually cannot be done in a random order. For example, to make Thanksgiving dinner one must buy and prepare the turkey before putting it in the oven, and one must set the table before serving the food.

If the tasks cannot be performed in arbitrary sequence or order, we can specify the order in an **order-requirement digraph.** Digraph is short for directed graph. Each (directed) edge in a digraph has an arrow on it. Digraphs can be used to model the fact that traffic on a street must go in one direction, or that certain tasks in a job must be

completed prior to other tasks. A typical example of an order-requirement digraph is shown in Figure 2.16. There is a vertex in this digraph for each task. If one task must be done immediately before another, we draw a directed edge, or arrow, from the prerequisite task to the subsequent task. The numbers within the circles representing vertices are the times it takes to complete the tasks. In Figure 2.16 there is no arrow from T_1 to T_5, because task T_2 intervenes. Also, T_1, T_7, and T_8 have no tasks that must precede them. Hence, if there are at least three processors (i.e., people or machines) available, tasks T_1, T_7, and T_8 can be worked on simultaneously at the start of the job.

Let us investigate a typical scheduling problem faced by a business.

EXAMPLE: Turning a Plane Around

Consider an airplane that carries both freight and passengers. The plane must have its passengers and freight unloaded and new passengers and cargo loaded before it can take off again. Also, the cabin must be cleaned before departure can occur. Thus, the job of "turning the plane around" requires the completion of five tasks:

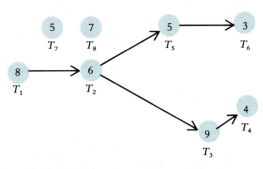

Figure 2.16 Typical order-requirement digraph.

Task *A*	Unload passengers	13 minutes
Task *B*	Unload cargo	25 minutes
Task *C*	Clean cabin	15 minutes
Task *D*	Load new cargo	22 minutes
Task *E*	Load new	
	passengers	27 minutes

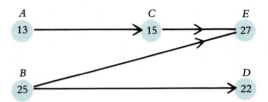

Figure 2.17 Order-requirement digraph for turning an airplane around after landing.

The order-requirement digraph for the problem of turning an airplane around is shown in Figure 2.17. The presence or absence of an edge in the order-requirement digraph depends on the analysis made as part of the modeling process for the problem. It seems natural that one needs to have an arrow between task *A* and task *C*, since before one can clean the cabin, the passengers on the plane should have been unloaded. However, the presence of some arrows may not seem natural, say, perhaps the arrow from task *B* (unload the cargo) to task *E* (load new passengers). This arrow may be due to government rules or requirements. What matters is that the mathematics of solving the problem does not depend on the reason that the order-requirement digraph looks the way that it does. The person solving the problem constructs the order-requirement digraph and then the mathematical techniques we will develop can be applied, regardless of whether or not some business faced with a similar problem might model the problem in a different way.

Because we want to find the earliest completion time, it might seem that finding the shortest path through the digraph (i.e., path *BD* with time length $25 + 22 = 47$) would solve the problem. But this approach shows the danger of ignoring the relationship between the mathematical model (the digraph) and the original problem.

The time required to complete all the tasks, *A* through *E*, must be at least as long as the time necessary to do the tasks on any particular path. Consider the path *BD*, which has length $25 +$

$22 = 47$. Recall that here the term "length" of a path refers to the sum of the times of the tasks that lie along the path. Since task *B* must be done before task *D* can begin, the two tasks *B* and *D* cannot be completed before time 47. Hence, even if work on other tasks (such as *A*, *C*, and *E*) is proceeding during this period, all the tasks cannot be finished before the tasks on path *BD* are finished. The same statement is true for every other path in the order-requirement digraph. Thus, the earliest completion time actually corresponds to the length of the *longest* path. In the airplane example, this earliest completion time is $55 (= 13 + 15 + 27)$ minutes, corresponding to the path *ACE*. We call *ACE* the **critical path** because the times of the tasks on this path determine the earliest completion time. Note that if none of these tasks could go on simultaneously, the time to complete all the tasks would be $13 + 25 + 15 + 22 + 27 = 102$ minutes. However, even though tasks may go on simultaneously, the length of the critical path being 55 shows that completion of the tasks in less than 55 minutes is not possible. Only by speeding up the times to complete the critical-path tasks themselves can a completion time earlier than 55 time units be achieved.

Suppose it were desirable to speed the turnaround of the plane to below 55 minutes. One way to do this might be to build a second jet-

SP TLIGHT 2.6   Every Moment Counts in Rigorous Airline Scheduling

▶ ▶ ▶   ▶ ▶ ▶ ▶   ▶ ▶ ▶ ▶   ▶

When people think of airline scheduling, the first thing that comes to mind is how quickly a particular plane can safely reach its destination. But using ground time efficiently is just as important to an airline's timetable as the time spent in flight. Bill Rodenhizer, who was the manager of control operations for Eastern Airlines in Boston in 1988, is considered to be an expert on airplane turnaround time, the process by which an airplane is prepared for almost immediate takeoff once it has landed. He tells us how this well-orchestrated effort works:

Scheduling, to the airline, is just about the whole ballgame. Everything is scheduled right to the minute. The whole fleet operates on a strict schedule. Each of the departments responsible for turning around an aircraft has an allotted period of time in which to perform its function. Manpower is geared to the amount of ground time scheduled for that aircraft. This would be adjusted during off-weather or bad-weather days or during heavy air-traffic delays.

Most of our aircraft in Boston are scheduled for a 42- to 65-minute ground time. Boston is the end of the line, so it is a "terminating and originating station." In plain talk, that means almost every aircraft that comes in must be fully unloaded, refueled, serviced, and dispatched within roughly an hour's time.

This is how the process works: in the larger aircraft, it takes passengers roughly 20 minutes to load and 20 minutes to unload. During this period, we will have completely cleaned the aircraft and unloaded the cargo, and the caterers will have taken care of the food. The ramp service may take 20 to 30 minutes to unload the baggage, mail, and cargo from underneath the plane, and it will take the same amount of time to load it up again. We double-crew those aircraft with heavier weights so that the work load will fit the time it takes passengers to load and unload upstairs.

While this has been going on, the fueler has fueled the aircraft. As to repairs, most major maintenance is done during the midnight shift, when all but 20 of Eastern's several hundred aircraft are inactive.

We all work under a very strict time frame. There are four functional departments. If any of the four cannot fit its work into its time frame, then it advises us at the control center, and we adjust the departure time or whatever, so that the other departments can coordinate their activities accordingly.

way to help unload passengers. For example, we could unload passengers (task A) in 7 minutes instead of 13. However, reducing task A to 7 minutes does not reduce the completion time by 6 minutes because in the new digraph (Figure 2.18) ACE is no longer the critical (i.e., longest) path. The longest path is now BE, which has a length of 52 minutes. Thus, shortening task A by 6 minutes only results in a 3-minute saving in completion time. This may mean that building a new jetway is uneconomical. Note also that shortening the time to complete tasks that are not on the original critical path ACE will not shorten the completion time at all. Speeding tasks on the critical path will shorten completion time of the job only up to the point where a new critical path is created. ▲

Not all order-requirement digraphs are as simple as the one shown in Figure 2.17. The order-requirement digraph in Figure 2.19 has 12 paths, which can be found by exhaustive search. Examples of such paths are $T_1T_2T_3$, $T_1T_5T_9$, $T_4T_5T_9$, and $T_7T_5T_3$. (Although we have not discussed them here, fast algorithms for finding longest and shortest paths in graphs are known.) The critical path is $T_7T_8T_6$ (length 21), and the earliest completion time for all nine tasks is time 21.

These examples are typical of many scheduling problems that occur in practice (see Spotlight 2.6, p. 52). Perhaps the most dramatic use of criti-

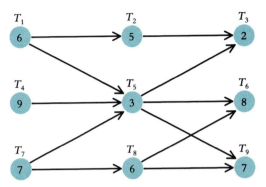

Figure 2.19 An order-requirement digraph with 12 paths, to examine how to find the length of the longest path.

cal-path analysis is in the construction trades. No major new building project is now carried out without first performing a critical-path analysis to ensure that the proper personnel and materials are available at the right times in order to have the project finished as quickly as possible. Many such problems are too large and complicated to be solved without the aid of computers.

The critical-path method was popularized and came into wider use as a consequence of the Apollo project. As we saw in the introduction, this project, which aimed at landing a man on the moon within 10 years of 1960, was one of the most sophisticated projects in planning and scheduling ever attempted. The dramatic success of the project can be attributed partly to the use of critical-path ideas and the related program evaluation and review technique (PERT), which helped keep the project on schedule.

▶ REVIEW VOCABULARY

Algorithm A step-by-step description of how to solve a problem.

Brute force method The method that solves the traveling salesman problem (TSP) by enumerating all the Hamiltonian circuits and then selecting the one with minimum cost.

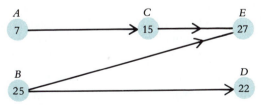

Figure 2.18 Order-requirement digraph for turning an airplane around with reduced times due to construction of new jetway.

Complete graph A graph in which every pair of vertices is joined by an edge.

Critical path The longest path in an order-requirement digraph. The length of this path gives the earliest completion time for all the tasks making up the job consisting of the tasks in the digraph.

Fundamental principle of counting A method for counting outcomes of multistage processes.

Greedy algorithm An approach for solving an optimization problem, where at each stage of the algorithm the best (or cheapest) action is taken. Unfortunately, greedy algorithms do not always lead to optimal solutions.

Hamiltonian circuit A circuit using distinct edges of a graph that starts and ends at a particular vertex of the graph and visits each vertex once and only once. A Hamiltonian circuit can be thought of as starting at any one of its vertices.

Heuristic algorithm A method of solving an optimization problem that is "fast," but that does not guarantee an optimal answer to the problem.

Kruskal's algorithm An algorithm developed by Joseph Kruskal (AT&T Bell Laboratories) that solves the minimum-cost spanning-tree problem by selecting edges in order of increasing cost, but so that no edge forms a circuit with edges chosen earlier. It can be proved that this algorithm always produces an optimal solution.

Method of trees A visual method of carrying out the fundamental principle of counting.

Minimum-cost Hamiltonian circuit A Hamiltonian circuit in a graph with weights on the edges, for which the sum of the weights of the edges of the Hamiltonian circuit is as small as possible.

Minimum-cost spanning tree A spanning tree of a weighted connected graph having minimum cost. The cost of a tree is the sum of the weights on the edges of the tree.

Nearest-neighbor algorithm An algorithm for attempting to solve the TSP that begins at a "home" vertex and visits next that vertex not already visited that can be reached most cheaply. When all other vertices have been visited, the tour returns to home. This method may not give an optimal answer.

NP-complete problems A collection of problems, which includes the TSP, that appear to be very hard to solve quickly for an optimal solution.

Order-requirement digraph A directed graph that shows which tasks precede other tasks among the collection of tasks making up a job.

Sorted-edges algorithm An algorithm for attempting to solve the TSP where the edges added to the circuit being built up are selected in order of increasing cost, but no edge is added that would prevent a Hamiltonian circuit's being formed. These edges must all be connected at the end, but not necessarily at earlier stages. The tour obtained may not have lowest possible cost.

Spanning tree A subgraph of a connected graph that is a tree and includes all the vertices of the original graph.

Traveling salesman problem (TSP) The problem of finding a minimum-cost Hamiltonian circuit in a complete graph where each edge has been assigned a cost (or weight).

Tree A connected graph with no circuits.

Weight A number assigned to an edge of a graph that can be thought of as a cost, distance, or time associated with that edge.

▶ SUGGESTED READINGS

BURR, STEFAN: *The Mathematics of Networks*, American Mathematical Society, Providence, R.I., 1982. A collection of articles dealing with applications of networks.

LAWLER, EUGENE, J. LENSTRA, RINNOY KAN, AND D. SHMOYS (EDS.): *The Traveling Salesman Problem*, Prentice-Hall, Englewood Cliffs, N.J., 1985. This book includes survey and technical articles on all aspects of the TSP.

LUCAS, WILLIAM, FRED ROBERTS, AND ROBERT THRALL (EDS.): *Discrete and Systems Models,* vol. 3: *Modules in Applied Mathematics,* Springer-Verlag, New York, 1983. Chapter 6, "A Model for Municipal Street Sweeping Operations," by A. Tucker and L. Bodin, describes street sweeping and related models in detail. Other models described in this book detail many recent applications of mathematics.

ROBERTS, FRED: *Applied Combinatorics,* Prentice-Hall, Englewood Cliffs, N.J., 1984. The chapters in this book on graphs and related network optimization problems are excellent.

——— : *Graph Theory and Its Applications to Problems of Society,* Society for Industrial and Applied Mathematics, Philadelphia, 1978. A very readable account of how graph theory is finding a wide variety of applications.

▶ EXERCISES

1. For each of the graphs below, write down a Hamiltonian circuit starting at X_1.

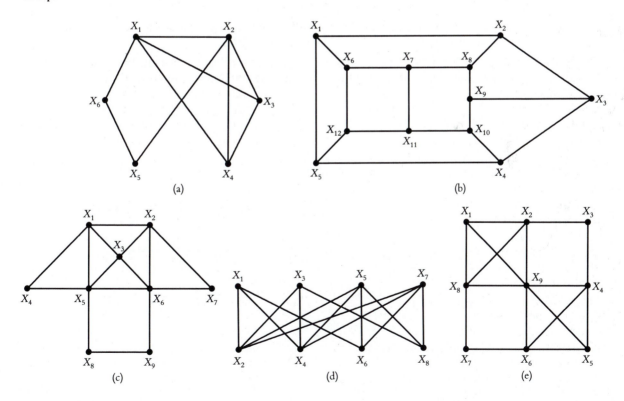

2. For each of the graphs below, add wiggly edges to indicate a Hamiltonian circuit.

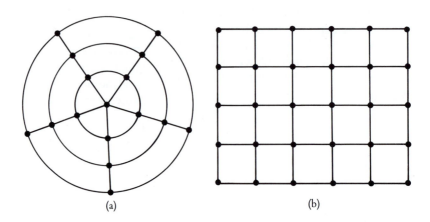

(a)   (b)

3. Suppose two Hamiltonian circuits are considered different if the collections of edges that they use are different. How many other Hamiltonian circuits can you find in the graph in Figure 2.1 different from the two discussed?

4. If the edge X_1X_2 is erased from each of the graphs in Exercise 1, does the resulting graph still have a Hamiltonian circuit?

5. Each of the graphs below has no Hamiltonian circuit. Is it possible to add a single new edge to these graphs and obtain a new graph that has a Hamiltonian circuit?

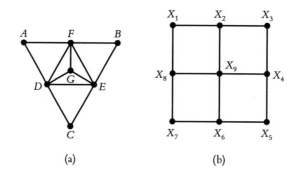

(a)   (b)

▲   6. Explain why the tour *ACEDCBA* is not a Hamiltonian circuit for the following graph. Does this graph have a Hamiltonian circuit?

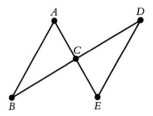

▲   7. Do the following graphs have Hamiltonian circuits? If not, can you demonstrate why not?

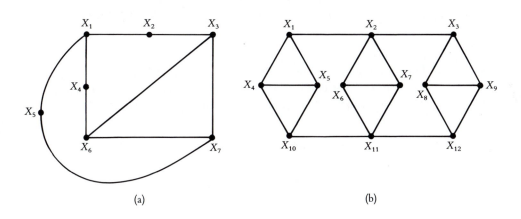

(a)   (b)

▲   Advanced exercise.

8. For each of the graphs below, determine if there is a Hamiltonian circuit.

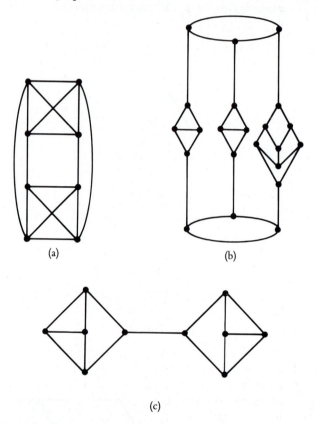

(a)   (b)

(c)

▲   9. a. The graph below is known as a four spokes and three concentric circles graph. What conditions on m and n guarantee that an m spokes and n concentric circles graph has a Hamiltonian circuit? (Assume $m \geq 2, n \geq 1$.)

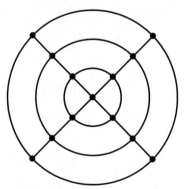

▲   Advanced exercise.

b. The graph below is known as a 3 × 4 grid graph. What conditions on
   m and n guarantee that an $m \times n$ grid graph has a Hamiltonian circuit?

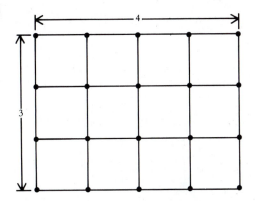

Can you think of a real-world situation in which finding a Hamiltonian circuit in an $m \times n$ grid graph would represent a solution to the problem? If an $m \times n$ grid graph has no Hamiltonian circuit, can you find a tour that repeats a minimum number of vertices and starts and ends at the same vertex?

▲ 10. The n-dimensional cube is obtained from two copies of an $(n-1)$-dimensional cube by joining corresponding vertices. (The process is illustrated for the 3-cube and the 4-cube in the following figure.)

Find formulas for the number of vertices and the number of edges of an n-cube. Can you show that every n-cube has a Hamiltonian circuit? (Hint: Show that if you know how to find a Hamiltonian circuit on an $(n-1)$-cube, then you can use two copies of this to build a Hamiltonian circuit on an n-cube.)

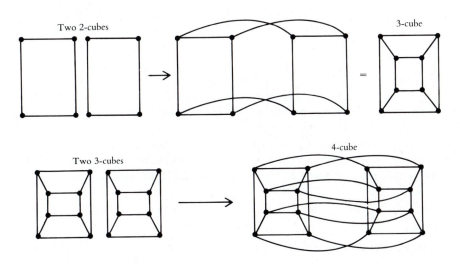

11. To practice your understanding of the concepts of Euler circuits and Hamiltonian circuits, determine for each graph below if there is an Euler circuit and/or a Hamiltonian circuit. If so, write it down.

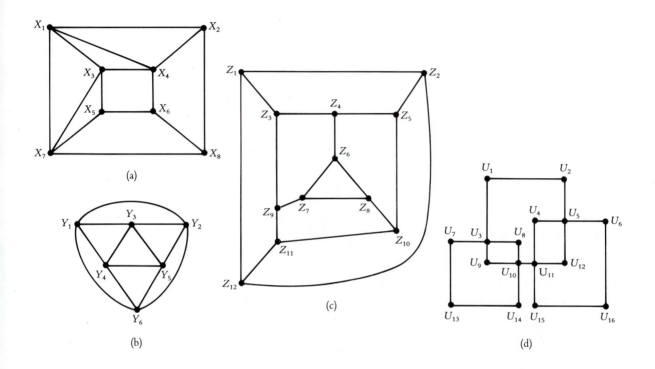

(a)

(b)

(c)

(d)

12. The figure at the top of the following page represents a town where there is a sewer located at each corner (where two or more streets meet). After every thunderstorm, the department of public works wishes to have a truck start at its headquarters (at vertex H) and make an inspection of sewer drains to be sure that leaves are not clogging them. Can a route start and end at H that visits each corner exactly once? (Assume that all the streets are two-way streets.) Does this problem involve finding an Euler circuit or a Hamiltonian circuit?

■ Discussion exercise.

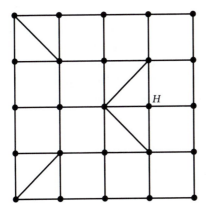

Assume that at equally spaced intervals along the blocks in this graph there are storm sewers that must be inspected after each thunderstorm to see if they are clogged. Is this a Hamiltonian circuit problem, an Euler circuit problem, or a Chinese Postman problem? Can you find an optimal tour to do this inspection?

■ 13. Give examples of real-world situations that can be modeled using a graph and for which finding a Hamiltonian circuit in the graph would be of interest.

14. In planning a camping vacation, Sally can fly to Duluth and then back home, choosing from four different airlines for each direction of her journey. On arriving at Duluth, she can choose from nine different outfitters. How many choices does she have for arrangements?

15. In designing a security system for its accounts, a bank asks each customer to choose a five-digit decimal number, all the digits to be distinct and nonzero. How many choices can the customer make?

16. a. For going outside on a cold winter day, Jill can choose from three winter coats, three wool scarfs, four pairs of boots, and three ski hats. How many outfits might her friends see her in?
   b. If Jill insists on always wearing her red ski hat, how many outfits might her friends see her in?

■ 17. a. In New York State one type of license plate has three letters followed by a three-digit number. Suppose the digits can be chosen from 0, 1, . . . , 9, except that all three digits being zero is not allowed and any letter from A to Z (repeats allowed) can be chosen. How many plates are possible?
   b. Investigate what schemes for license plates are used in your state and determine how many different plates are possible.

18. In the last several years regions that contain large cities that have had telephone service provided via only one area code have had to be divided into service areas with more than one area code. What is the largest number of different phone numbers that can be served using one area code? If an area code cannot begin with a zero, how many different area codes are possible?

19. A restaurant offers 5 appetizers, 10 entrees, and 8 desserts. How many different choices for a meal can a customer make if one selection is made from each category? If three of the desserts are pies and the customer will never order pie, how many meals can the customer chose?

20. Draw complete graphs with four, five, and six vertices. How many edges do these graphs have? Can you generalize to n vertices? How many TSP tours would these graphs have? (Tours yielding the same Hamiltonian circuit are considered the same.)

21. Calculate the values of 5!, 6!, 7!, 8!, 9!, and 10!. Then find the number of TSP tours in the complete graph with ten vertices.

22. The table below shows the mileage between four cities: Springfield, Ill. (S), Urbana, Ill. (U), Effingham, Ill. (E), and Indianapolis, Ind. (I).

	E	I	S	U
E	—	147	92	79
I	147	—	190	119
S	92	190	—	88
U	79	119	88	—

a. Represent this information by drawing a weighted complete graph on four vertices.
b. Use the weighted graph in part **a** to find the cost of the three distinct Hamiltonian circuits in the graph. (List them starting at U.)
c. Which circuit gives the minimum cost?
d. Would there be any difference in parts **b** and **c** if the start vertex were at I?
e. If one applies the nearest-neighbor method starting at U, what circuit would be obtained? Does the answer change if one applies the nearest-neighbor algorithm starting at S? At E? At I?
f. If one applies the sorted-edges method, what circuit would be obtained? Does one get the optimal answer?

23. After a party at her house, Francine (F) has agreed to drive home Mary (M), Rachel (R), and Constance (C). If the times (in minutes) to drive between

her friends' homes are shown below, what route gets Francine back home the quickest?

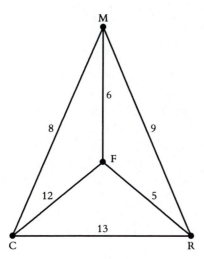

24. A fisherwoman wishes to visit three areas A, B, and C where she has set nets starting from the location where she moors her boat (M). If the times (in minutes) between the locales are given in the figure below, what route to visit the three sites and return to the mooring place would be optimal?

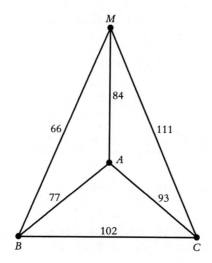

25. For each of the graphs with weights below, apply the nearest-neighbor method (starting at vertex A) and the sorted-edges method to find (it is hoped) a cheap tour.

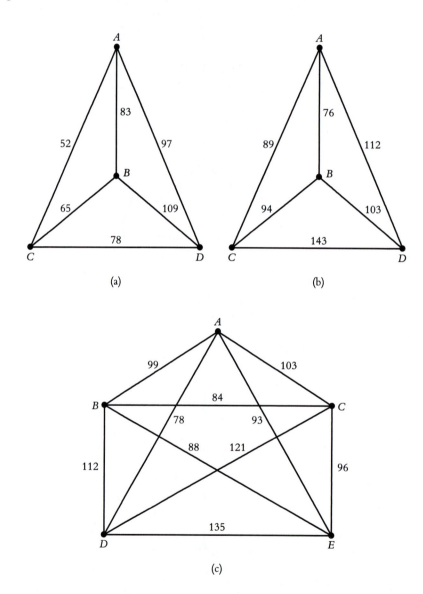

(a)

(b)

(c)

26. a. For each of the complete graphs that follow, find the costs of the nearest-neighbor tour starting at A and of the tour generated using the sorted-edges algorithm.

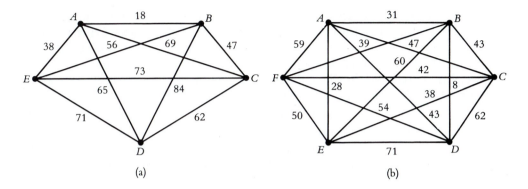

(a)   (b)

▲   b. How many Hamiltonian circuits would have to be examined to find a shortest route for part **a** by the brute force method?

c. Can you invent an algorithm different from the sorted-edges and nearest-neighbor algorithms that is easy to apply for finding TSP solutions? (See Lawler et al., in the Suggested Readings.)

27. An airport limo must take its six passengers to different downtown hotels from the airport. Is this a traveling salesman problem or an Euler circuit problem?

28. a. Solve the six-city TSP shown in the accompanying diagram using the nearest-neighbor algorithm starting at vertex A; starting at vertex B.

b. Apply the sorted-edges method.

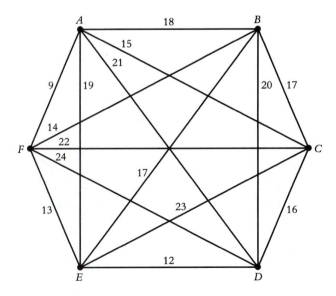

▲   Advanced exercise.

▲ 29. Construct an example of a complete graph on five vertices, with distinct weights on the edges for which the nearest-neighbor algorithm starting at a particular vertex and the sorted-edges algorithm yield different solutions for the traveling salesman problem. Can you find a five-vertex complete graph with weights on the edges in which the optimal solution, the nearest-neighbor solution, and the sorted-edges algorithm solution are all different?

▲ 30. If the brute force method of solving a 20-city TSP is employed, use a calculator to determine how many Hamiltonian circuits must be examined. How long would it take to determine the minimum-cost tour if the cost of tours could be computed at the rate of 1 billion per second? (Convert your answer to years by seeing how many years are equivalent to a billion seconds!)

31. Which of the graphs below are trees?

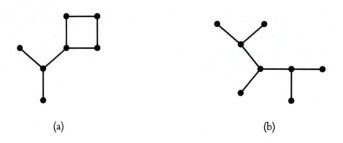

(a)   (b)

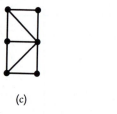   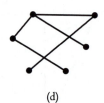

(c)   (d)

▲ Advanced exercise.

32. For each of the diagrams below explain why the wiggled edges are not:
    a. A spanning tree
    b. A Hamiltonian circuit

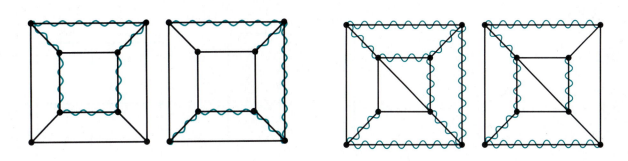

33. Find all the spanning trees in the graphs below.

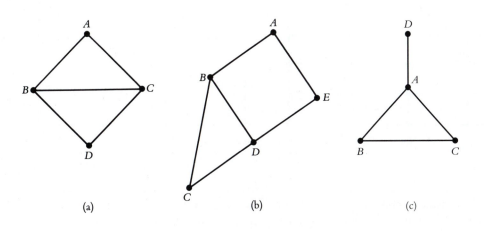

(a)   (b)   (c)

34. Use Kruskal's algorithm to find a minimum-cost spanning tree for graphs (a), (b), (c), and (d).

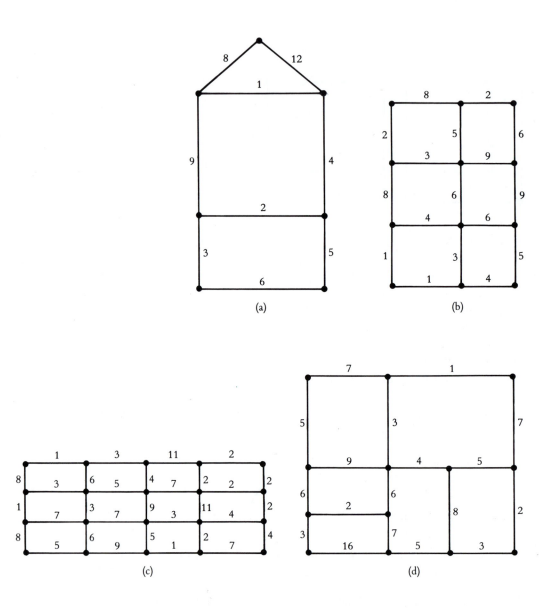

(a)

(b)

(c)

(d)

35. A large company wishes to install a pneumatic tube system that would enable small items to be sent between any of 10 locales, possibly by relay. If the

nonprohibitive costs (in $100) are shown in the graph model below, between which sites should the tube be installed to minimize the total cost?

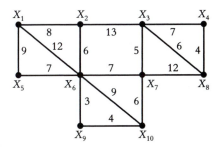

36. If the weight of each edge in Exercise 35 is increased by 1, will the tree that achieves minimum cost for the new collection of weights be the same as the one that achieves minimum cost for the original set of weights?

37. Suppose that Pictaphone service is not possible between two towns where there is a hill of more than 800 feet along a straight line that runs between them. In constructing a model for solving a relay network problem, how would you handle the question of putting a link between two cities with a 1200-foot hill between them?

■ 38. Give examples of real-world situations that can be modeled using a weighted graph and for which finding a minimum-cost spanning tree for the graph would be of interest.

▲ 39. Can Kruskal's algorithm be modified to find a maximum-weight spanning tree? Can you think of an application for finding a maximum-weight spanning tree?

■ 40. Find the cost of providing a relay network between the six cities with the largest population in your home state, using the road distances between the cities as costs. Does it follow that the same solution would be obtained if air distances are used instead?

▲ 41. Would there ever be a reason to find a minimum-cost spanning tree for a weighted graph in which the weights on some of the edges were negative? Would Kruskal's algorithm still apply?

▲ 42. Let G be a graph with weights assigned to each edge. Consider the following algorithm:
  a. Pick any vertex V of G.
  b. Select that edge E with a vertex at V that has a minimum weight. Let the other endpoint of E be W.

▲ Advanced exercise.   ■ Discussion exercise.

c. Contract the edge VW so that edge VW disappears and vertices V and W coincide (see the following figures).

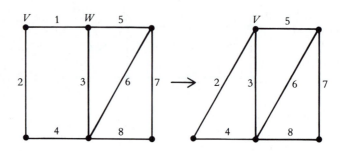

If in the new graph two or more edges join a pair of vertices, delete all but the cheapest. Continue to call the new vertex V.

d. Repeat steps **b** and **c** until a single point is obtained. The edges selected in the course of this algorithm (called Prim's algorithm) form a minimum-cost spanning tree. Apply this algorithm to the following graphs.

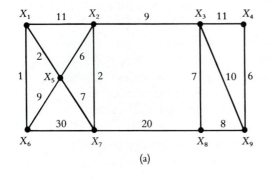

(a)

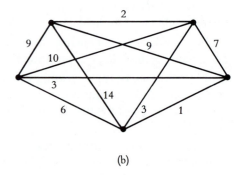

(b)

▲ 43. Determine whether each of the following statements is true or false for a minimum-cost spanning tree T for a weighted connected graph G:

▲ Advanced exercise.

a. T contains a cheapest edge in the graph.

b. T cannot contain a most expensive edge in the graph.

c. T contains one fewer edge than there are vertices in G.

d. There is some vertex in T to which all others are joined by edges.

▲ 44. In the following graphs, the number in the circle for each vertex is the cost of installing equipment at the vertex if *relaying* must be done at the vertex, while the number on an edge indicates the cost of providing service between the endpoints of the edge.

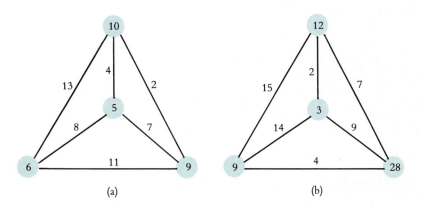

(a)   (b)

In each case, find the minimum cost (allowing relays) for sending messages between any pair of vertices, taking vertex relay costs into account. Would your answer be different if vertex relay costs are neglected? (Warning: Kruskal's algorithm cannot be used to answer the first question. This problem illustrates the value of having an algorithm over relying on "brute force.")

45. Two spanning trees of a (weighted) graph are considered different if they use different edges. Show that the graph below has different minimum-cost spanning trees, though all these different trees have the same cost.

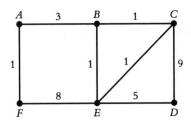

▲ 46. Suppose G is a graph such that all the weights on its edges are different numbers. Show that there is a unique minimum-cost spanning tree.

47. Find a minimum-cost spanning tree for the complete graphs in Exercise 26.

▲ 48. Suppose that a letter requires postage of p (positive integer) and that stamps of various denominations are available, say, $d_1, \ldots, d_v$ (positive integers). We are interested in finding the minimum number of stamps to choose with the available denominations to obtain the postage p exactly.

    a. Give an example to show that unless other conditions are put on p and $d_1, \ldots, d_v$, it might happen that no selection of stamps will achieve the desired postage.

    b. Show that even if the desired postage can be obtained for some choice of stamps, it does not follow that a minimum number of stamps can be achieved by using a greedy algorithm, that is, by selecting at each stage the largest denomination possible.

    c. Show that for at least one choice of denominations, a greedy algorithm will produce the fewest stamps for any given postage!

49. Find the earliest completion time and critical paths for the three order-requirement digraphs below.

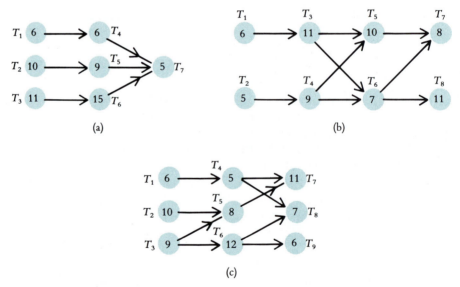

(a)

(b)

(c)

▲ 50. Construct an example of an order-requirement digraph with three different critical paths.

51. In the following order-requirement digraph, determine which tasks, if shortened, would reduce the earliest completion time and which would not.

▲ Advanced exercise.

Then find the earliest completion time if task T_5 is reduced to time length 7. What is the new critical path?

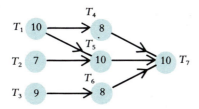

52. To build a new addition on a house, the following tasks must be completed. Construct reasonable time estimates for these tasks and a reasonable order-requirement digraph. What is the fastest time in which these tasks can be completed?

a. Lay foundation.
b. Erect sidewalls.
c. Erect roof.
d. Install plumbing.
e. Install electric wiring.
f. Lay tile flooring.
g. Obtain building permits.
h. Put in door that adjoins new room to existing house.
i. Install track lighting on ceiling.

53. At a large toy store, scooters arrive unassembled in boxes. To assemble a scooter the following tasks must be performed:

TASK 1.   Remove parts from the box.
TASK 2.   Attach wheels to the footboard.
TASK 3.   Attach vertical housing.
TASK 4.   Attach handlebars to vertical housing.
TASK 5.   Put on reflector tape.
TASK 6.   Attach bell to handlebars.
TASK 7.   Attach decals.
TASK 8.   Attach kickstand.
TASK 9.   Attach safety instructions to handlebars.

Give reasonable time estimates for these tasks and construct a reasonable order-requirement digraph. What is the earliest time by which these tasks can be completed.

54. Draw an order-requirement digraph for the following set of tasks, giving a reasonable estimate for the times to do the tasks involved. Find a critical path for the order-requirement digraph that you obtain.

Kitchen remodeling project.

Tasks: Clean the kitchen; scrape walls to remove old paint; prime walls; install wallpaper on walls; scrape paint on ceiling; paint ceiling; replace old floor with new floor tiles; install new stove; install new sink.

55. For the order-requirement digraph below, find the critical path and the task(s) in the critical path whose time, when reduced the *least*, creates a new critical path.

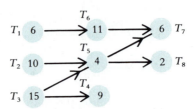

56. Use the fundamental principle of counting to predict that there are 12 paths in Figure 2.19. (Hint: All paths have two edges and must go through one of the three middle vertices. Vertex T_2 has one edge coming in and one edge going out, vertex T_5 has three edges coming in and three edges going out, while vertex T_8 has one edge coming in and two edges going out.)

57. Construct an order-requirement digraph with six tasks that has two critical paths of length 24.

▶WRITING PROJECTS

1. Write an essay about a variety of situations in which you are personally involved for which a solution of the TSP is (perhaps implicitly) required. Explain under what circumstances it might be valuable to carry out a formal mathematical solution to such TSPs rather than use an ad hoc solution.

2. Pick a situation that involves the traveling salesman model and discuss how closely the mathematics describes this situation and what features of the problem are likely to be important but have been neglected in the TSP model.

3. Construct an example, of the kind suggested on page 43, that shows that in a situation where three day campers must be picked up and brought to camp, it may make a difference if the optimization criterion is minimizing distance traveled by the camp bus versus minimizing average time that the children spend on the bus.

4. Determine the six largest cities in the state in which you live. By consulting a road atlas (or by some other means) construct the graph that represents the road distances between your hometown and these six other cities. Now apply (a) the nearest-neighbor method, (b) the sorted-edges method, and (c) the nearest neighbor from each city, and pick the minimum tour method to solve the associated TSP. Do you have reason to believe that the answers you get might include an optimum solution among them?

Chapter 3

Planning and Scheduling

In a society as complex as ours, everyday problems such as providing services efficiently and on time require accurate planning of both people and machines. Take the example of a hospital in a major city. Around-the-clock scheduling of nurses and doctors must be provided to guarantee that people with particular expertise are available during each shift. The operating rooms must be scheduled in a manner flexible enough to deal with sudden emergencies. Equipment used for x-ray, CAT, or NMR scans must be scheduled for maximal efficiency.

Although many scheduling problems are often solved on an ad hoc basis, we can also use mathematical ideas to gain insight into the complications that arise in scheduling. The ideas we develop in this chapter have practical value in a relatively narrow range of applications, but they throw light on many characteristics of more realistic and hence more complex scheduling problems.

▶ SCHEDULING TASKS

Assume that a certain number of identical **processors** (machines, humans, or robots) work on a series of tasks that make up a job. Associated with

each task is a specified amount of time required to complete the task. For simplicity, we assume that any of the processors can work on any of the tasks. Our problem, known as the **machine-scheduling** problem, is to decide how the tasks should be scheduled so that the completion time for the tasks collectively is as early as possible.

Even with these simplifying assumptions, complications in scheduling will arise. Some tasks may be more important than others and perhaps should be scheduled first. When "ties" occur, they must be resolved by special rules. As an example, suppose we are scheduling patients to be seen in a hospital emergency room staffed by one doctor. If two patients arrive simultaneously, one with a bleeding foot, the other with a bleeding arm, which patient should be processed first? Suppose the doctor treats the arm patient first, and while treatment is going on, a person in cardiac arrest arrives. Scheduling rules must establish appropriate priorities for cases such as these.

Another common complication arises with jobs consisting of several tasks that cannot be done in an arbitrary order. For example, if the job of putting up a new house is treated as a scheduling problem, the task of laying the foundation must precede the task of putting up the walls, which in turn must be completed before work on the roof can begin.

Assumptions and Goals

To simplify our analysis, we need to make clear and explicit assumptions:

1. If a processor starts work on a task, the work on that task will continue without interruption until the task is completed.

2. No processor stays voluntarily idle. In other words, if there is a processor free and a task available to be worked on, then that processor will immediately begin work on that task.

3. The requirements for ordering the tasks are given by an order-requirement digraph. (A typical example is shown in Figure 3.1, with task times circled within each vertex. The ordering of the tasks imposed by the order-requirement digraph represents constraints of physical reality. For example, you cannot fly a plane until it has taken fuel on board.)

4. The tasks are arranged in a priority list that is independent of the order requirements. (The

priority list is an ordering of the tasks according to some criterion of "importance," which may in no way reflect physical reality. For example, imagine a construction job with several tasks. Task *A* may have to be done before task *B*, but when task *B* is done a monetary payment will be made. Thus, *B* may be given a higher priority than *A*. The priority list represents, from some point of view, an ordering of the tasks. Another such point of view is to order the tasks in a manner that will help the algorithm being used construct schedules with early completion times. Usually, different points of view for giving priority to tasks are not consistent. Mathematical analysis may sometimes assist in clarifying tradeoffs implicit in these different points of view.)

When considering a scheduling problem, there are various goals one might wish to achieve. Among these are

1. minimizing the completion time of the job

2. minimizing the total time that processors are idle

3. finding the minimum number of processors necessary to finish the job by a specified time

For the moment, we will concentrate on goal 1, finishing all the tasks at the earliest possible time. Note, however, that optimizing with respect to one criterion or goal may not optimize with respect to another.

List-Processing Algorithm

The scheduling problem we have described sounds more complicated than the traveling salesman problem (TSP). Indeed, like the TSP, it is known to be NP-complete. This means that it is unlikely that anyone will ever find a computationally fast algorithm that can find an optimal solu-

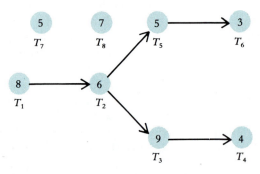

Figure 3.1 A typical order-requirement digraph.

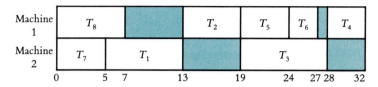

Figure 3.2 The schedule produced by applying the list-processing algorithm to the order-requirement digraph in Figure 3.1 using the list $T_8, T_7, \ldots, T_1$.

tion. Thus, we will be content to seek a solution method that is computationally fast and gives only approximately optimal answers.

The algorithm we use to schedule tasks is the **list-processing algorithm.** In describing it, we will call a task **ready** at a particular time if all its predecessors as indicated in the order-requirement digraph have been completed at that time. For example, in Figure 3.1 at time 0 the ready tasks are T_1, T_7, and T_8, while task T_2 cannot be ready until 8 time units after T_1 is started. The algorithm works as follows: at a given time, assign to the lowest-numbered free processor the first task on the priority list that is ready at that time and that hasn't already been assigned to a processor.

In applying this algorithm, we will need to develop skill at coordinating the use of the information in the order-requirement digraph and the priority list. It will be helpful to cross out the tasks in the priority list as they are assigned to a processor to keep track of which tasks remain to be scheduled. Let's apply this algorithm to one possible priority list, $T_8, T_7, T_6, \ldots, T_1$, using two processors and the order-requirement digraph in Figure 3.1. The result is the schedule shown in Figure 3.2, where idle processor time is indicated by color shading. How does the list-processing algorithm generate this schedule?

Because T_8 (task 8) is first on the priority list and ready at time 0, it is assigned to the lowest-numbered free processor, processor 1. Task 7, next on the priority list, is also ready at time 0 and thus is assigned to processor 2. The first processor

to become free is processor 2 at time 5. Recall that by assumption 1, once a processor starts work on a task, its work cannot be interrupted until the task is complete. Task T_6, the next unassigned task on the list, is not ready at time 5, as can be seen by consulting Figure 3.1. In fact, at time 5, the only ready task on the list is T_1, so that task is assigned to processor 2. At time 7, processor 1 becomes free, but no task becomes ready until time 13. Thus, processor 1 stays idle from time 7 to time 13. At this time, because T_2 is the first ready task on the list not already scheduled, it is assigned to processor 1. Processor 2, however, stays idle because no other ready task is available at this time.

As the priority list is scanned from left to right to assign a processor at a particular time, we pass over tasks that are not ready to find ones that are ready. If no task can be assigned in this manner, we keep one or more processors idle until such time that, reading the priority list from the left, there is a ready task not already assigned. After a task is assigned to a processor, we resume scanning the priority list, starting over at the far left, for unassigned tasks. The remainder of the scheduling shown in Figure 3.2 is completed in this manner.

When Is a Schedule Optimal?

The schedule in Figure 3.2 has a lot of idle time, so it may not be optimal. Indeed, if we apply the list-processing algorithm for two processors to another possible priority list $T_1, \ldots, T_8$, using the digraph in Figure 3.1, the resulting schedule is

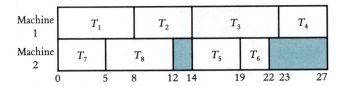

Figure 3.3 The schedule produced by applying the list-processing algorithm to the order-requirement digraph in Figure 3.1 using the list $T_1, T_2, \ldots, T_8$.

shown in Figure 3.3. Here are the details of how this schedule was arrived at. Remember we must coordinate the list $T_1, T_2, \ldots, T_8$ with the information in the order-requirement digraph shown in Figure 3.1. At time 0 task T_1 is ready, so this task is assigned to processor 1. However, at time 0 tasks $T_2, T_3, \ldots, T_6$ are not ready since their predecessors are not done. For example, T_2 is not ready at time 0 since T_1, which precedes it, is not done at time 0. The first ready task on the list, reading from left to right, that is not already assigned is T_7, so task T_7 gets assigned to processor 2. Both processors are now busy until time 5, at which time processor 2 becomes idle (Figure 3.3). Tasks T_1 and T_7 have been assigned; reading from left to right along the list, the first task not already assigned whose predecessors are done by time 5 is T_8, so this task is started at time 5 on processor 2; processor 2 will continue to work on this task until time 12, since the task time for this task is 7 time units. At time 8, processor 1 becomes free, and reading the list from left to right we find that T_2 is ready (since T_1 has just been completed). Thus, T_2 is assigned processor 1, which will stay busy on this task until time 14. At time 12, processor 2 becomes free, but the tasks that have not already been assigned from the list T_3, T_4, T_5, T_6 are not ready, since they depend on T_2 being completed before these tasks can start. Thus, processor 2 stays idle involuntarily until time 14. At this time, T_3 and T_5 become ready. Since both processors 1 and 2 are idle at time 14, the lower numbered of the two, processor 1, gets to start on T_3 because it is the first ready task left to be assigned on the list scanned from left to right. Task T_5 gets assigned to

processor 2 at time 14. The remaining tasks are assigned in a similar manner.

The schedule shown in Figure 3.3 is optimal because the path T_1, T_2, T_3, T_4, with length 27, is the critical path in the order-requirement digraph. As we saw in Chapter 2, the earliest completion time for the job made up of all the tasks is the length of the longest path in the order-requirement digraph.

There is another way of relating optimal completion time for a scheduling problem to the completion time that is yielded by the list-processing algorithm. Suppose that we add all the task times given in the order-requirement digraph and divide by the number of processors. The completion time using the list-processing algorithm must be at least as large as this number. For example, the task times for the order-requirement digraph in Figure 3.1 sum to 47. Thus, if these tasks are scheduled on two processors, the completion time is at least $47/2 = 23.5$ (in fact, 24, since the list-processing algorithm applied to integer task times yields an integer answer), while for three processors the completion time is at least $47/3$ (in fact, 16).

Why is it helpful to take the total time to do all the tasks in a job and divide this number by the number of processors? Think of each task that must be scheduled as a rectangle that is 1 unit high and t units wide, where t is the time allotted for the task. Think of the scheduling diagram with m processors as a rectangle that is m units high, and whose width W is the completion time for the tasks. How small can W be? The area of the rectangle that represents the scheduling diagram must be at least as large as the sum of all the rectangles representing tasks that are "packed" into it. The area of the scheduling diagram rectangle is mW. The combined areas of all the tasks, plus the area of rectangles corresponding to idle time, will equal mW. Width W is smallest when the idle time is zero. Thus, W must be at least as big as the sum of all the task times divided by m. Sometimes the estimate for completion time given by the list-processing algorithm from the length of the critical path gives a more useful value than the approach based on adding task times, and sometimes the opposite is true. For the order-requirement digraph in Figure 3.1, except for a schedule involving one processor, the critical-path estimate is superior. For some scheduling problems, both these estimates may be poor.

The number of priority lists that can be constructed if there are n tasks is $n!$, as can be computed using the fundamental principle of counting. For example, if there are eight tasks: $T_1, \ldots, T_8$, there are $8 \times 7 \times 6 \cdots \times 1 = 40{,}320$ possible priority lists. For different choices of the priority list, the list-processing algorithm we are using will schedule the tasks, subject to the constraints of the order-requirement digraph, in different ways. Different lists may yield schedules with different completion times or different schedules with the same completion time. Different priority lists may yield identical scheduling of tasks (and hence identical completion times). A little later we will see a method that can be used to select a list that, if we are lucky, will give a schedule with a relatively good completion time. In fact, no method is known, except for very specialized cases, of how to choose a list that can be guaranteed to give rise to an optimal schedule when the list algorithm is applied to it.

Strange Happenings

The list-processing algorithm involves four factors that affect the final schedule. The answer we get depends on

1. the times of the tasks

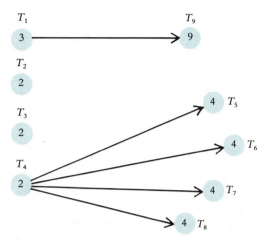

Figure 3.4 An order-requirement digraph designed to help illustrate some paradoxical behavior produced by the list-processing algorithm.

2. the number of processors

3. the order-requirement digraph

4. the ordering of the tasks on the list

To see the interplay of these four factors, consider another scheduling problem, this one associated with the order-requirement digraph shown in Figure 3.4 (the circled numbers are task time lengths).

The schedule generated by the list-processing algorithm applied to the list $T_1, T_2, \ldots, T_9$, using three processors, is given in Figure 3.5.

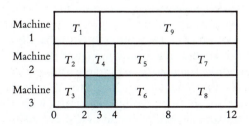

Figure 3.5 The schedule produced by applying the list-processing algorithm to the order-requirement digraph in Figure 3.4 using the list $T_1, T_2, \ldots, T_9$ with three processors.

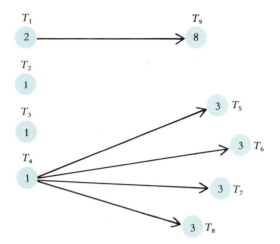

Figure 3.6 The order-requirement digraph obtained from the one in Figure 3.4 by reducing by one unit each of the task times shown there.

Treating the list $T_1, \ldots, T_9$ as fixed, how might we make the completion time earlier? Our alternatives are to pursue one or more of these strategies:

1. Reduce task times.

2. Use more processors.

3. "Loosen" the constraints of the order-requirement digraph.

Let's consider each alternative in turn, changing one feature of the original problem at a time, and see what happens to the resulting schedule. If we adopt strategy 1, reducing the time of each task by one unit, it seems intuitively clear that the completion time would go down. Figure 3.6 shows the new order-requirement digraph, and Figure 3.7 shows the schedule produced for this problem, using the list-processing algorithm with three processors applied to the list $T_1, \ldots, T_9$. The completion time is now 13, longer than the completion time of 12 for the case (Figure 3.5) with *longer* task times. Here is something unexpected! Let's explore further and see what happens.

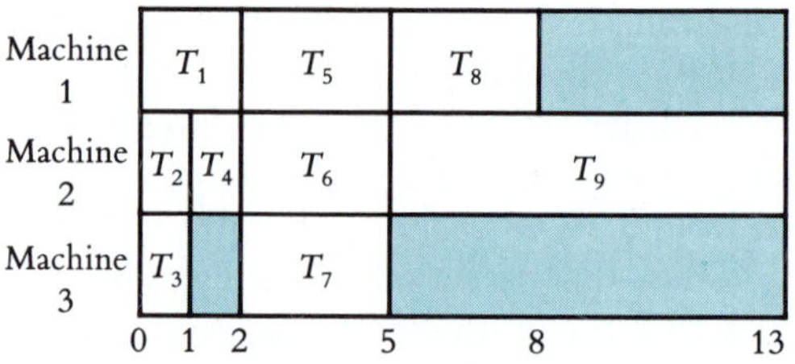

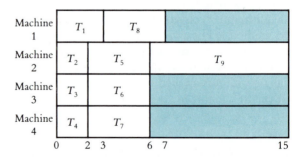

Figure 3.7 The schedule produced by applying the list-processing algorithm to the order-requirement digraph in Figure 3.6 using the list $T_1, T_2, \ldots, T_9$ with three processors.

Next we consider strategy 2, increasing the number of machines. Surely this should speed matters up. When we apply the list-processing algorithm to the original graph in Figure 3.4, using the list $T_1, \ldots, T_9$ and four machines, we get the schedule shown in Figure 3.8. The completion time is now 15! Here is a shock, an even later completion time than for the previous alteration.

Finally, we consider strategy 3, trying to shorten completion time by erasing *all* constraints (edges with arrows) in the order-requirement digraph shown in Figure 3.4. By increasing flexibility of the ordering of the tasks, we might guess we could finish our tasks more quickly. Figure 3.9 shows the schedule using the list $T_1, \ldots, T_9$;

Figure 3.8 The schedule produced by applying the list-processing algorithm to the order-requirement digraph in Figure 3.4 using the list $T_1, T_2, \ldots, T_9$ with four processors.

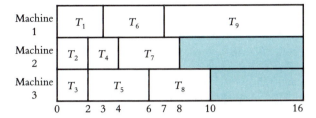

Figure 3.9 The schedule produced by applying the list-processing algorithm to the order-requirement digraph in Figure 3.4, modified by erasing all its directed edges, using the list $T_1, T_2, \ldots, T_9$ with three processors.

now it takes 16 units! This is the worst of our three strategies to reduce completion time.

The failures we have seen here appear paradoxical at first glance, but they are typical of what can happen when a situation is too complex to analyze with naive intuition. Sometimes our common sense leads us astray. The value of using mathematics rather than intuition or trial and error to study scheduling and other problems is that it points up flaws that can occur in unguarded intuitive reasoning.

The paradoxical behavior we see here is a consequence of the rules we set up for generating schedules. Such paradoxical behavior for the list-processing algorithm will not occur for every example you try. In fact, one has to be quite clever to design such examples. The list-processing algorithm has many nice features, including the fact that it is easy to understand and fast to implement. However, the results of the model in some cases can appear strange. Since we have been explicit about our assumptions, we could go back and make changes in these assumptions in hopes of eliminating the strange behavior. But the price we may pay is more time spent in constructing schedules and perhaps even new types of strange behavior. Unfortunately for modern society with its increasing concern with economical and efficient scheduling, recent mathematical research suggests that scheduling is an intrinsically hard problem.

▶ CRITICAL-PATH SCHEDULES

In our discussion so far, we have acted as though the priority list used in applying the list-processing algorithm was given to us in advance based on external considerations. We might, however, consider the question of whether there is a systematic method of *choosing* a priority list that yields optimal or nearly optimal schedules. We will show how to construct a specific priority list based on this principle, to which the list-processing algorithm can then be applied.

Recall from our discussion of critical-path analysis in Chapter 2 that no matter how a schedule is constructed, the finish time cannot be earlier than the length of the longest path in the order-requirement digraph. This suggests that we should try to schedule first those tasks that occur early in long paths, because they can bottleneck the flow of tasks to be completed.

EXAMPLE: Scheduling Two Processors

To illustrate this method, consider the order-requirement digraph in Figure 3.10a. Suppose we wish to schedule these tasks on two processors. Initially, there are two critical paths of length 64: T_1, T_2, T_3 and T_1, T_4, T_3. Thus, we place T_1 first on the priority list. With T_1 "gone," there is a new critical path of length 60, which starts with T_5, so T_5 is placed second on the priority list. At this stage, with T_1 and T_5 removed, we have the residual order-requirement digraph shown in Figure 3.10b. In this diagram there are paths of length 50 (T_2, T_3), 56 (T_6, T_4, T_3), 36 (T_6, T_4, T_7), and 24 (T_8, T_4, T_{10}). Since T_6 heads the path that is currently longest in length, it gets placed third in the priority list. Once T_6 is removed from Figure 3.10b, there is a tie for which is the longest path remaining, since both T_2, T_3 and T_4, T_3 are paths of length 50. When there is a tie between two longest paths, we place next on the priority

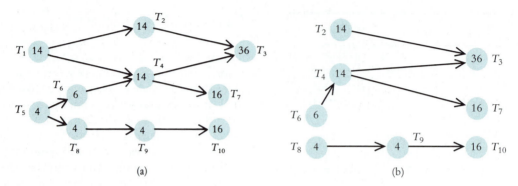

Figure 3.10 (a) An order-requirement digraph used to illustrate the critical-path scheduling method. (b) Residual order-requirement digraph after tasks T_1 and T_5 have been removed.

list the lowest-numbered task heading a longest path. In the example shown here, this means that T_2 is placed next into the priority list, to be followed by T_4. Continuing in this fashion we obtain the priority list $T_1, T_5, T_6, T_2, T_4, T_3,$ T_8, T_9, T_7, T_{10}. Note that the order of T_7 and T_{10} was decided using the rule for breaking ties. The list-processing algorithm is now applied using this priority list and the order-requirement digraph in Figure 3.10a. We obtain the schedule in Figure 3.11. ▲

This example shows that **critical-path scheduling,** as this method is called, can sometimes yield optimal solutions. Unfortunately, this algorithm does not always perform well. For example, the critical-path method employing four processors applied to the order-requirement digraph shown

in Figure 3.12 yields the list $T_1, T_8, T_9, T_{10}, T_{11},$ $T_5, T_6, T_7, T_{12}, T_2, T_3, T_4$ and then the schedule in Figure 3.13. (Note that T_5, T_6, T_7 are thought of as heading paths of length 10.) In fact, there can be no *worse* schedule than this one. An optimal schedule is shown in Figure 3.14.

Many of the results we have examined so far are negative because we are dealing with a general class of problems that defy our using computationally efficient algorithms to find an optimal schedule. But we can close on a more positive note. Consider an arbitrary order-requirement digraph, but assume all the tasks take equal time. It turns out that we can always construct an optimal schedule using two processors in this situation. Ironically, we can choose among many algorithms to produce these optimal schedules. The algorithms are easy to understand (though not easy to prove optimal) and have all been discovered since 1969! Many

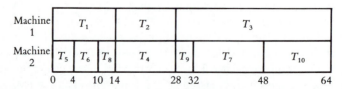

Figure 3.11 The optimal schedule produced by applying the critical-path scheduling method to the order-requirement digraph in Figure 3.10. The list used was $T_1, T_5, T_6, T_2, T_4, T_3, T_8, T_9, T_7,$ T_{10}.

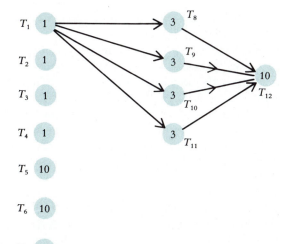

Figure 3.12 An order-requirement digraph used to illustrate how poorly the critical-path scheduling method can sometimes behave.

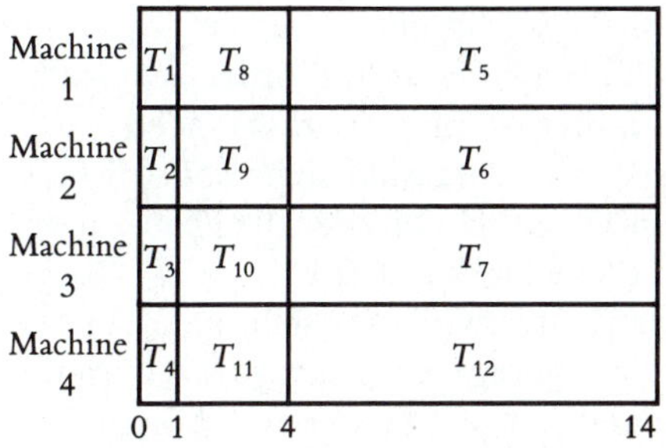

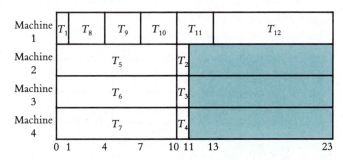

Figure 3.14 An optimal schedule for the order-requirement digraph in Figure 3.12 using four processors.

people think that mathematics is a subject that is no longer alive, and that all its ideas and methods were discovered hundreds of years ago. They assume mathematics consists of nothing more than the arithmetic, algebra, and trigonometry taught in high school. This stereotype is easily broken down, as we have just seen, and it is well to consider that scheduling problems account for only a narrow portion of mathematics in general. In fact, more new mathematics has been discovered and published in the last 30 years than during any previous 30-year period.

▶ INDEPENDENT TASKS

Mathematicians suspect that no computationally efficient algorithm for solving general scheduling problems optimally will ever be found. Due to our limited success in designing algorithms for finding optimal schedules for general order-requirement digraphs, we will consider a *special class* of scheduling problems for which the order-requirement digraph has no edges with arrows. In this case we say that the tasks are **independent** of each other,

Figure 3.13 The schedule produced by applying the critical-path scheduling method to the order-requirement digraph in Figure 3.12 using four processors. The list used was $T_1, T_8, T_9, T_{10}, T_{11}, T_5, T_6, T_7, T_{12}, T_2, T_3, T_4$.

since they can be performed in any order. (No edges with arrows in the order-requirement digraph indicates that no tasks need to precede others; that is, the tasks can be done in any order.) In this section we consider the problem of scheduling independent tasks.

There are two approaches we can consider. To study **average-case analysis** we might ask if the average (mean) of the completion times arrived at by using the list-processing algorithm with all the possible different lists is close to the optimal possible completion time. To study **worst-case analysis** we can ask how far from optimal a schedule obtained using the list-processing algorithm with one particular priority list can be. Average-case analysis is amenable to mathematical solution, but requires methods of great sophistication. For independent tasks, the worst-case analysis can be answered using a surprisingly simple argument developed by Ronald Graham of AT&T Bell Laboratories (see Spotlight 3.1, pp. 86–87). The idea is that if the tasks are independent, no processor can be idle at a given time and then busy on a task at a later time. We will return to Graham's worst-

case analysis after exploring the problem of independent tasks in more detail.

Geometrically, we can think of the independent tasks as rectangles of height 1 whose lengths are equal to the time length of the task. Finding an optimal schedule amounts to packing the task rectangles into a longer rectangle whose height equals the number of machines. For example, Figure 3.15 shows two different ways to schedule tasks of length 10, 4, 5, 9, 7, 7 on two machines. (For convenience, the rectangles in the case of independent tasks are labeled with their task times rather than their task numbers.) Scheduling basically means efficiently packing the task rectangles into the machine rectangle. Finding the optimal answer among all possible ways to pack these rectangles is like looking for a needle in a haystack. The list-processing algorithm produces a packing, but it may not be a good one.

What Graham's worst-case analysis showed for independent tasks is that no matter which list L one uses, if the optimal schedule requires time T, then the completion time for the schedule produced by the list-processing algorithm applied to

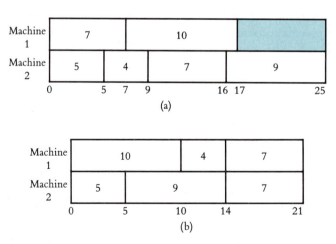

Figure 3.15 (a) A nonoptimal way to schedule independent tasks of time lengths 10, 4, 5, 9, 7, 7 using two processors. (b) An optimal way to schedule independent tasks of time lengths 10, 4, 5, 9, 7, 7 using two processors.

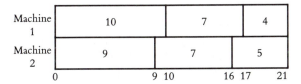

Figure 3.16   The optimal schedule resulting from applying the decreasing-time-list algorithm to a collection of independent tasks. The list used, written down in terms of task times only, is 10, 9, 7, 7, 5, 4.

list L with m processors is less than or equal to $(2 - 1/m)T$. For example, for two machines ($m = 2$), if an optimal schedule yields completion at time 30, then no list would ever yield a completion later than $(2 - \frac{1}{2})(30) = 45$. Although it is of great theoretical interest, Graham's result does not provide much comfort to those who are trying to find good schedules for independent tasks.

Decreasing-Time Lists

Is there some way of choosing a priority list that consistently yields relatively good schedules? The surprising answer is yes! The idea is that when long tasks appear toward the end of the list, they often seem to "stick out" on the right end, as in Figure 3.15a. This suggests that before trying to schedule a collection of tasks, the tasks should be placed in a list so that longest tasks are listed first. The list-processing algorithm applied to a list arranged in this fashion is called the **decreasing-time-list algorithm**. If we apply it to the set of

tasks listed previously (10, 4, 5, 9, 7, 7), we obtain the times 10, 9, 7, 7, 5, 4 and the schedule (packing) shown in Figure 3.16. This packing is again optimal, but it is different from the optimal scheduling in Figure 3.15b. It is worth noting that the decreasing-time list and the list obtained by the critical-path method discussed earlier will coincide in the case of independent tasks. The decreasing-time list can also be constructed for the case where the tasks are not independent, but for general order-requirement digraphs, the decreasing-time list does not produce particularly good schedules.

It is important to remember that the decreasing-time-list algorithm *does not guarantee* optimal solutions. This can be seen by scheduling the tasks with times 11, 10, 9, 6, 4 (Figure 3.17). The schedule has completion time 21. However, the rearranged list 9, 4, 6, 11, 10 yields the schedule in Figure 3.18, which finishes at time 20. This solution is obviously optimal, since the machines finish at the same time and there is no idle time. Note that when tasks are independent, if there are m machines available, the completion time cannot be less than the sum of the task times divided by m.

The problems we have encountered in scheduling independent tasks seem to have taken us a bit far from our goal of *applying* the mathematics we have developed. Sometimes mathematicians will pursue their mathematical ideas even though they have reached a point where there appear to be no applications. Fortunately, it is very common to be

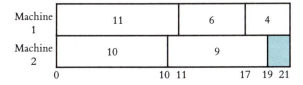

Figure 3.17   The nonoptimal schedule resulting from applying the decreasing-time-list algorithm to a collection of independent tasks. The list used, written down in terms of task times only, is 11, 10, 9, 6, 4.

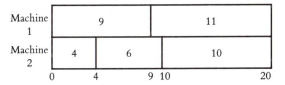

Figure 3.18   The nonoptimal schedule resulting from applying the list-processing algorithm to a collection of independent tasks. The list used, written down in terms of task times only, is 9, 4, 6, 11, 10.

SPOTLIGHT 3.1 Ronald Graham on Mathematics and Mathematicians

Ronald Graham is director of the mathematics sciences research department of AT&T Bell Labs, Murray Hill, New Jersey. In addition to his own research, he supervises some of the country's most distinguished mathematicians as they investigate the puzzles of management science. Here are some of his comments:

On Scheduling

Scheduling is a very interesting area. You have some number of processors that you can think of as computers or as people, and you have a number of tasks or jobs that you would like to get done. One might think that adding more processors to the system would guarantee that you would be done sooner. In fact, just the opposite can happen. It's an example of the old adage, "Too many cooks can spoil the broth." You can add more machines or decrease the amount of time it takes for each job, or relax the constraints between the various jobs, but all of these can actually end up taking longer. To try to understand why that is and how to avoid it is the arena of mathematics.

One of the earliest applications of scheduling came in looking at the Apollo moon shot. In that case, there were three different processors, namely, the three astronauts, who had a large collection of tasks that they were supposed to do — various experiments to run on the way to and from the moon. And of course, there were various constraints. You want the astronauts to

sleep a little every day. Other constraints you might not think of. For example, you have to rotate the capsule so that the same side isn't facing the sun all the time, because of heating problems. Then, within these constraints, you would next like to know: If you sequence the experiments in different orders, how much time could you save over sequencing them in orders that aren't so good?

It turns out that in fact NASA had developed very good sequencing already. All I could show them was that they couldn't improve on it very much, no matter what they might try. And of course this is one of the problems in management science where it's just impossible to enumerate the possibilities. So that's where the mathematics comes in: it enables you to say that no matter what you do, you'll never be able to "do any better than this" or "do it any faster."

You might well ask, how can you know about every one of the possibilities even though you didn't try them all? That's an interesting question, and it really goes to the root of how mathematics is used to analyze the real world. There are several well-known steps in going from a real-world problem to a model of the problem — a model in which you try to capture the essential aspects of the problem, but in a way that can be dealt with mathematically. Once you convert the problem into the world of mathematics, you can say something about all the mathematical possibilities. Then you translate it back to the real-world situation. How well it works depends crucially on how accurate the translation of the model was.

Now it's always good to check a few of the things that you've predicted are going to happen. If the translation is good and you've captured the essence of the problem, then there is a good chance that you can say something sensible about the thing you've studied. If it isn't, you can get some pretty bizarre conclusions. There are famous examples of this. One that comes to mind is the case where the people who first analyzed how bees fly were able to prove mathematically that bees *couldn't* fly. That didn't bother the bees, of course, and the model was eventually modified to show that bees could fly after all.

On the Spirit of Investigation

I think in order to do the best work in any area, you have to enjoy it and be intrigued by it. You have to wonder why this is happening. There are many examples where something slightly out of the ordinary occurs and 99 people notice it yet go on to something else. But one in a hundred — or fewer — is fascinated by that slightly anomalous or even bizarre behavior in one mathematical area or another. They start probing it more deeply and soon find that it's really the tip of a giant iceberg. Once you start to melt it, you have a much deeper understanding about what is really going on. There are those who feel mathematics is a giant game, albeit one that is amazingly relevant to everything that is going on in the scientific world.

Often, half the battle in trying to solve a difficult problem is knowing the right question to ask. If you know what it is you are trying to look for, you can often be very far along in finding the solution. It's useful for me, and for many others who are working on a difficult problem, to look at a special case first. You can work your way up to the full problem by trying easier special cases that you still can't do. You may be bouncing the idea around a bit and not forcing it, and it's amazing how often it happens that the next day or next week something will seem obvious that was very hard to imagine even a week earlier.

On Managing the Mathematicians

One of the rules or axioms that we have here [at Bell Labs] is that each person is best equipped to know how he or she functions optimally. Some people are night people, for example, and they just don't function before 12 o'clock.

In our mathematics group, there are roughly 65 professionals, and quite a few are leading-edge, world-class researchers. People of this caliber you don't so much manage as try to keep up with what they are doing. In many cases, you act as a sounding board, because one of the most useful things you can do with a colleague who has some mathematical ideas is to act as a sympathetic listener. As the person explains the ideas or where he is stuck, he will more often than not gain a much clearer insight into what he's doing and where he's going.

Many times there is much more interactive collaboration. There is quite a lot of joint work that goes on here, not just from within the mathematics area, but among mathematicians, computer scientists, physicists, and chemists. It is a very interactive environment which, I think, works to everyone's benefit.

able to find applications for the "abstract" extensions. This is the case in the current instance.

EXAMPLE: Photocopy Shop and Typing Pool Problems

Imagine a photocopy shop with three photocopiers. Photocopying tasks that must be completed overnight are accepted until 5 P.M. The tasks are to be done in any manner that minimizes the finish time for all the work. Because this problem involves scheduling machines for independent tasks, the decreasing-time-list algorithm would be a good heuristic to apply.

For another example, consider a typing pool at a large corporation or college, where individual typing tasks can be assigned to any typist. In this setting, however, the assumption that the processors (typists) are identical in skill is less likely to be true. Hence, the tasks might have different times with different processors. This phenomenon, which occurs in real-world scheduling problems, violates one of the assumptions of our mathematical model.

Graham's result for the list-processing algorithm — that the finishing time is never more than $(2 - 1/m)T$ (where T represents optimal completion time and m the number of processors) — offers us the small comfort of knowing that even the worst choice of priority list will not yield a completion time worse than twice the optimal time. Compared with the list-processing algorithm, the decreasing-time-list algorithm seems to improve completion times. Thus, it is not surprising that Graham was able to provide an improved *bound*, or time estimate, for this case: the decreasing-list algorithm gives a completion time of no more than $[4/3 - 1/(3m)]T$, where m is the number of processors and T is the optimal time in which the tasks can be completed. In particular, when the number of processors is 2, the schedule produced by the decreasing-time-list algorithm is

never off by more than 17%! Usually, the error is much less. This result is a remarkable instance of the value of mathematical research into applied problems. Note that the optimal completion time T depends on m and that Graham's theoretical analysis is necessary precisely because there is no known algorithm to compute T easily. ▲

▶ BIN PACKING

Suppose you plan to build a wall system for your books, records, and stereo set. It requires 24 wooden shelves of various lengths: 6, 6, 5, 5, 5, 4, 4, 4, 4, 2, 2, 2, 2, 3, 3, 7, 7, 5, 5, 8, 8, 4, 4, and 5 feet. The lumberyard, however, sells wood only in boards of length 9 feet. If each board costs $8, what is the minimum cost to buy sufficient wood for this wall system?

Because all shelves required for the wall system are shorter than the boards sold at the lumberyard, the largest number of boards needed is 24, the precise number of shelves needed for the wall system. Buying 24 boards would of course be a waste of wood and money because several of the shelves you need could be cut from one board. For example, pieces of length 2, 2, 2, and 3 feet can be cut from one 9-foot board.

To be more efficient, we think of the boards as bins of capacity W (9 feet in this case) into which we will pack (without overlap) weights (in this case, lengths) $w_1, \ldots, w_n$, where each $w_i \le W$. We wish to find the minimum number of bins into which the weights can be packed. In this formulation, the problem is known as the **bin-packing problem**. At first glance, it may appear unrelated to the machine-scheduling problems we have been studying; however, there is a connection.

Let's suppose we want to schedule independent tasks so that each machine working on the tasks finishes its work by time W. Instead of fixing the number of machines and trying to find the earliest completion time, we must find the minimum number of machines that will guarantee completion by the fixed completion time *(W)*. Despite this similarity between the machine-scheduling problem and the bin-packing problem, the discussion that follows will use the traditional terminology of bin packing.

By now, it should come as no surprise to learn that no one knows a fast algorithm that always picks the optimal (smallest) number of bins (boards). In fact, the bin-packing problem belongs to the class of NP-complete problems (see Spotlight 2.1), which means that most experts think it unlikely that any fast optimal algorithm will ever be found.

Bin-Packing Heuristics

We will think of the items to be packed, in any particular order, as constituting a list. In what follows we will use the list of 24 shelf lengths given for the wall system. We will consider various **heuristic** methods. Probably the easiest approach is simply to put the weights into the first bin until the next weight won't fit, and then start a new bin. (Once you open a new bin, don't use leftover space in an earlier, partially filled bin.) Continue in the same way until as many bins as necessary are used. The resulting solution is shown in Figure 3.19. This algorithm, called **next fit (NF)**, has the advantage of not requiring knowledge of all the weights in advance; only the remaining space in

the bin currently being packed must be remembered. The disadvantage of this heuristic is that a bin packed early on may have had room for small items that come later in the list.

Our wish to avoid permanently closing a bin too early suggests a different heuristic — **first fit (FF)**: put the next weight into the *first* bin already opened that has room for this weight; if no such bin exists, start a new bin. Note that a computer program to carry out first fit would have to keep track of how much room was left in *all* the previously opened bins. For the 24 wall-system shelves the first-fit algorithm would generate a solution that uses only 14 bins (see Figure 3.20) instead of the 17 bins generated by the next-fit algorithm.

If we are keeping track of how much room remains in each unfilled bin, we can put the next item to be packed into the bin that currently has the most room available. This heuristic will be called **worst fit (WF)**. The name worst fit refers to the fact that an item is packed into a bin with the most room available, that is, into which it fits "worst," rather than into a bin that will leave little room left over after it is placed in that bin (i.e., "best fit"). The solution generated by this approach looks the same as that shown in Figure 3.20. Although this heuristic also leads to 14 bins, the items are *packed* in a different order. For example, the first item of size 2, the tenth item in the list, is put into bin 6 in worst fit, but into bin 1 in first fit.

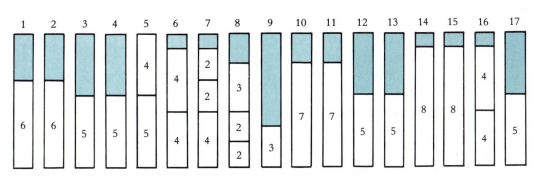

Figure 3.19 The list 6, 6, 5, 5, 5, 4, 4, 4, 4, 2, 2, 2, 2, 3, 3, 7, 7, 5, 5, 8, 8, 4, 4, 5 packed in bins using next fit.

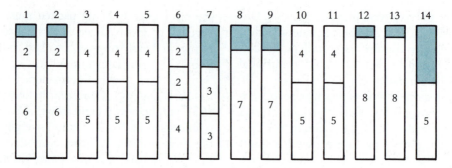

Figure 3.20 The list 6, 6, 5, 5, 5, 4, 4, 4, 4, 2, 2, 2, 2, 3, 3, 7, 7, 5, 5, 8, 8, 4, 4, 5 packed in bins using first fit. Worst fit would yield a packing that would look identical.

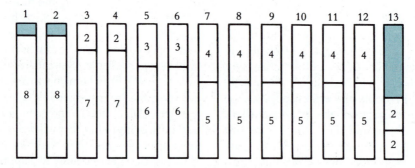

Figure 3.21 The bin packing resulting from applying first-fit decreasing to the wall-system numbers. The list involved, which uses the original list sorted in decreasing order, is 8, 8, 7, 7, 6, 6, 5, 5, 5, 5, 5, 5, 4, 4, 4, 4, 4, 4, 3, 3, 2, 2, 2, 2.

Decreasing-Time Heuristics

One difficulty with all three of these heuristics is that large weights that appear late in the list can't be packed efficiently. Therefore, we should first sort the items to be packed in order of decreasing size, assuming that all items are known in advance. We can then pack large items first and then the smaller items into leftover spaces. This approach yields three new heuristics: **next-fit decreasing (NFD)**, **first-fit decreasing (FFD)**, and **worst-fit decreasing (WFD)**. Here is the original list sorted

by decreasing size: 8, 8, 7, 7, 6, 6, 5, 5, 5, 5, 5, 5, 4, 4, 4, 4, 4, 4, 3, 3, 2, 2, 2, 2. Packing using first-fit-decreasing order yields the solution in Figure 3.21. This solution uses only 13 bins.

Is there any packing that uses only 12 bins? No. In Figure 3.21, there are only 2 free units (1 unit each in bins 1 and 2) of space in the first 12 bins, but 4 occupied units (two 2s) in bin 13. We could have predicted this by dividing the total length of the shelves (110) by the capacity of each bin (board): $\frac{110}{9} = 12\frac{2}{9}$. Thus, no packing could squeeze these shelves into 12 bins — there would always be at least 2 units left over for the thirteenth bin. (In Figure 3.21, there are 4 units in bin 13 because of the 2 wasted empty spaces in bins 1 and 2.) Even if this division had created a zero remainder, there would still be no guarantee that the items could be packed to fill each bin without wasted space. For example, if the bin capacity is 10 and there are weights of 6, 6, 6, 6, and 6, the total weight is 30; dividing by 10, we get 3 bins as the minimum requirement. Clearly, however, 5 bins are needed to pack the five 6s.

None of the six heuristic methods shown will necessarily find the optimal number of bins for an arbitrary problem. How can we decide which heuristic to use? One approach is to see how far from the optimal solution each method might stray. Various formulas have been discovered to calculate the maximum discrepancy between what a bin-packing algorithm actually produces and the best possible result. For example, in situations where a large number of bins are to be packed, first fit (FF) can be off as much as 70%, but first-fit decreasing (FFD) is never off by more than 22%. Of course, FFD doesn't give an answer as quickly as FF, because extra time for sorting a large collection of weights may be considerable. Also, FFD requires knowing the whole list of weights in advance, whereas FF does not. It is important to emphasize that 22% is a worst-case figure. In many cases, FFD will perform much better. Recent results obtained by computer simulation indicate excellent average-case performance for this algorithm.

When modeling real-world problems, we always have to look at the relationship between mathematics and the real world. Thus, first-fit decreasing usually results in fewer bins than next fit, but next fit can be used even when all the weights are not known in advance. Next fit also requires much less computer storage than first fit, because once a bin is packed, it need never be looked at again. Fine-tuning of the conditions of the actual problem often results in better practical solutions and in interesting new mathematics as well. (See Spotlight 3.2, p. 92, for a discussion of some of the tools mathematicians use to verify and even extend mathematical truths.)

► CRYPTOGRAPHY

Not so many years ago secret codes existed primarily for the purpose of allowing diplomats and military personnel to exchange information without their messages being read by people from other countries. Due to changes in telecommunications, banking, and lifestyles, secret codes are now widely used to protect computer files, electronic transfers of funds, and electronic mail.

We conclude with a variant of the bin-packing problem that is of interest to those who make and break secret codes. Given numbers (weights) $w_1, \ldots, w_n$ and a number S, find an algorithm that selects a subcollection of the weights whose sum is exactly S. This problem, also known to be NP-complete, is a special case of what is called the **knapsack problem** because S can be thought of as the capacity of a knapsack to be filled by the items selected from the list of weights. This particular knapsack problem is called the **subset sum problem**. As tame as it may sound, it is related to a revolutionary new proposal for ensuring the security of data stored in computers, such as bank records and classified military information.

The security of data has become an important new branch of **cryptography,** the study of codes and ciphers and how to break them. Traditional cryptography is based on a single key that is used to transform or encrypt a message (plaintext) into garbled form (ciphertext). The same key used to encrypt the message is used to decipher the message. However, in the new scheme, called *public-key cryptography*, there are two keys. One key, made public, is used by anyone wishing to send a secure message to X. The other key is known only to X, the receiver of the message. The keys are designed so that knowledge of the public key does not compromise the private key.

SP**O**TLIGHT 3.2   Using Mathematical Tools

The tools of a carpenter include the saw, T square, level, and hammer. A mathematician also requires tools of the trade. Some of these tools are the proof techniques that enable verification of mathematical truths. Another set of tools consists of strategies to sharpen or extend the mathematical truths already known. For example, suppose that if A and B hold, then C is true. What happens if only A holds? Will C still be true? Similarly, if only B holds, will C still be true?

This type of thinking is of value because such questions will result either in more general cases where C holds or in examples showing that B alone and/or A alone can't imply C. For example, we saw that if a graph G is connected (hypothesis A) and even-valent (hypothesis B), then G has a tour of its edges using each edge only once (conclusion C). If either hypothesis is omitted, the conclusion fails to hold. The figures on the opposite page illustrate this point. On the left you see an even-valent but nonconnected graph, and on the right, a connected graph with two odd-valent vertices; neither graph has an Euler circuit.

Here is another way that a mathematician might approach extending mathematical knowledge. If A and B imply C, will A and B imply both C and D, where D extends the conclusion of C? For example, not only can we prove that a connected, even-valent (hypotheses A and B) graph has an Euler circuit, but we can also show that the first edge of the Euler circuit can be chosen arbitrarily (conclusions C and D). It turns out that being able to specify the first two edges of the Euler circuit may not always be

---

The idea behind public-key cryptography is that certain processes are very easy to carry out in one direction, but very hard to reverse. For example, one can easily add $708 + 259 + 871 + 1836 + 82$ to get 3756, but it is not so easy to find a subcollection of the numbers 1086, 708, 82, 259, 589, 871, and 1836 that add to $S = 3756$. As another example, multiply 2993 by 1362 using pencil and paper or a calculator. Now try to find two numbers whose product is 2,414,203. In case you ran out of patience, the answer is 1111×2173. If you carried out these two problems, you might guess that factoring is "harder" than multiplying.

Because knapsack problems are thought to be computationally hard to solve, schemes have been suggested for applying these ideas to public-key systems. Although recent work has shown that knapsack systems are faulty as a base for public-key systems, other public-key systems based on factoring large numbers appear to be viable. Furthermore, the idea of NP-completeness (see Spotlight 2.1), which arose as a concept in the abstract analysis of the complexity of algorithms, turned out to be related to the mundane issue of how to protect money being transferred between banks! We can see once more how the use of clever

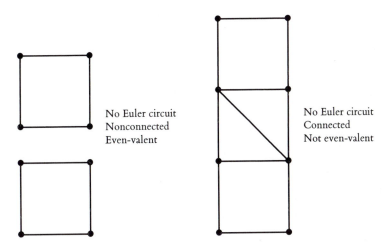

No Euler circuit
Nonconnected
Even-valent

No Euler circuit
Connected
Not even-valent

possible. Mathematicians are trained to vary the hypotheses and conclusions of results they prove in an attempt to clarify and sharpen the range of applicability of the results.

We have seen that machine scheduling and bin packing are probably computationally difficult to solve because they are NP-complete. A mathematician could then try to find the sim-

plest version of a bin-packing problem that would still be NP-complete: What if the items to be packed can have only eight weights? What if the weights are only one and two? Asking questions like these is part of the mathematician's craft. Such questions help to extend the domain of mathematics and hence the applications of mathematics.

(though not necessarily impenetrable) mathematics can affect and enrich our lives. If you are interested in learning more about cryptography and codes used for purposes other than maintaining secrecy, you will find more information about this important subject in Chapter 10.

▶ REVIEW VOCABULARY

Average-case analysis   The study of the list-processing algorithm (more generally, any algorithm) from the point of view of how well it

performs on all the types of problems it may be used on and seeing on average how well it does. (See also Worst-case analysis.)

Bin-packing problem   The problem of determining the minimum number of containers of capacity W into which objects of size $w_1, \ldots, w_n$ $(w_i \leq W)$ can be packed.

Critical-path scheduling   A heuristic algorithm for solving scheduling problems where the list-processing algorithm is applied to the priority list obtained by listing next in the priority list a task that heads a longest path in the order-requirement digraph. This task is then deleted from the order-requirement digraph, and the next task placed in the priority list is obtained by repeating the process.

Cryptography The study of how to make and break codes. These codes are now used primarily for data and computer security rather than for national security.

Decreasing-time-list algorithm The heuristic algorithm that applies the list-processing algorithm to the priority list obtained by listing the tasks in decreasing order of their time length.

First fit (FF) A heuristic algorithm for bin packing in which the next weight to be packed is placed in the lowest-numbered bin already opened into which it will fit. If it fits in no open bin, a new bin is opened.

First-fit decreasing (FFD) A heuristic algorithm for bin packing where the first-fit algorithm is applied to the list of weights sorted so that they appear in decreasing order.

Heuristic algorithm An algorithm that is fast to carry out but that doesn't necessarily give an optimal solution to an optimization problem.

Independent tasks Tasks are independent when there are no edges with arrows between them in the order-requirement digraph.

Knapsack problem Given a knapsack of size W and a collection of weights $w_1, \ldots w_n$, find a largest collection of weights that will fit in the knapsack.

List-processing algorithm A heuristic algorithm for assigning tasks to processors: assign the first ready task on the priority list that has not already been assigned to the lowest-numbered processor that is not working on a task.

Machine scheduling The problem of assigning tasks to processors so as to complete the tasks by the earliest time possible.

Next fit (NF) A heuristic algorithm for bin packing in which a new bin is opened if the weight to be packed next will not fit in the bin that is currently being filled; the current bin is then closed.

Next-fit decreasing (NFD) A heuristic algorithm for bin packing where the next-fit algorithm is applied to the list of weights sorted so that they appear in decreasing order.

Processor A person, machine, robot, operating room, or runway whose time must be scheduled.

Ready task A task is called ready at a particular time if its predecessors as given by the order-requirement digraph have been completed by that time.

Subset sum problem Given an integer W and integer numbers $w_1, \ldots, w_n$, find a subcollection of the numbers whose sum is W.

Worst-case analysis The study of the list-processing algorithm (more generally, any algorithm) from the point of view of how well it performs on the hardest problems it may be used on. (See also Average-case analysis.)

Worst fit (WF) A heuristic algorithm for bin packing in which the next weight to be packed is placed into the open bin with the largest amount of room remaining. If the weight fits in no open bin, a new bin is opened.

Worst-fit decreasing (WFD) A heuristic algorithm for bin packing where the worst-fit algorithm is applied to the list of weights sorted so that they appear in decreasing order.

▶SUGGESTED READINGS

DEMILLO, R., ET AL. (EDS.): *Applied Cryptology,* American Mathematical Society, Providence, R.I., 1983. A relatively recent survey of research work.

FRENCH, SIMON: *Sequencing and Scheduling,* Wiley, New York, 1982. A detailed account of a wide variety of scheduling models, most of them different from the ones treated in this chapter.

GRAHAM, RONALD: "Combinatorial Scheduling Theory," in Lynn Steen (ed.), *Mathematics Today,* Springer-Verlag, New York, 1978. This essay on scheduling is one of many excellent accounts of recent developments in mathematics in this book.

———: "The Combinatorial Mathematics of Scheduling," *Scientific American*, 238(3):124–132 (March 1978). A very readable introduction to scheduling and bin packing.

SIMMONS, G.: "Cryptology, the Mathematics of Secure Communications," *The Mathematical Intelligencer*, 14:233–246 (1979). A survey of cryptology, including early public-key cryptography.

▶ EXERCISES

1. In order to get to a ski resort for a weekend vacation Jocelyn must accomplish a variety of things. She will leave work early at 1 P.M. and must get to the airport to be on a 5 o'clock shuttle to Boston. She then hopes to take a bus to get to the resort. Discuss some of the tasks that must be accomplished to get Jocelyn to the resort by 10 P.M. What are the different types of processors that are involved in getting these tasks done? Can any of these tasks be done simultaneously?

2. For the situation where a family with three children are preparing a Thanksgiving meal for 10 guests, list tasks that must be completed and the types of processors that are involved. Can any of these tasks be done simultaneously?

3. List as many scheduling situations as you can for these environments:
   a. Hospital
   b. Railroad station
   c. Airport
   d. Automobile repair garage
   e. Machine shop
   f. Your home
   g. Your school
   h. Police station
   i. Firehouse

■ 4. Discuss scheduling problems for which it is *not* reasonable to assume that once a processor starts a task, it would always complete that task before it works on any other task. Give examples for which this approach would be reasonable.

5. a. Use the list-processing algorithm to schedule the tasks in the following order-requirement digraph on three processors, using the list $T_1, \ldots, T_{11}$. From the schedule so constructed, for each task list the start and finish time for that task.
   b. Compare your answers in part a with scheduling these tasks on two processors with the same list.

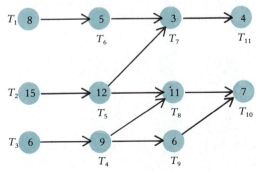

■ Discussion exercise.

6. For the accompanying order-requirement digraph, apply the list-processing algorithm, using three processors for lists a and b. How do the completion times obtained compare with the length of the critical path?

   a. $T_1, T_2, T_3, T_4, T_5, T_6, T_7, T_8$

   b. $T_2, T_4, T_6, T_8, T_1, T_3, T_7, T_5$

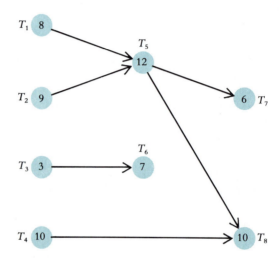

7. Consider the following order-requirement digraph:

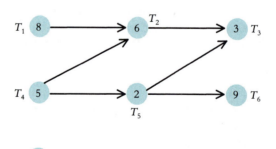

   a. Find the critical path(s).
   b. Schedule these tasks on one processor using the critical-scheduling method.
   c. Schedule these tasks on one processor using the priority list obtained by listing the tasks in order of decreasing time.

d. Do either of these schedules have idle time? How do their completion times compare?

e. If two different schedules have the same completion time, what criteria can be used to say one schedule is superior to the other?

f. Schedule these tasks on two processors using the order-requirement digraph shown and the priority list from step b.

g. Does the schedule produced in step f finish in half the time that the schedule in step a did, which might be expected, since the number of processors has doubled?

8. To prepare a meal quickly involves carrying out the tasks shown (time lengths in minutes) in the following order-requirement digraph:

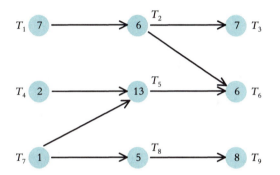

a. If Mike prepares the meal alone, how long will it take?

b. If Mike can talk Mary into helping him prepare the meal, how long will it take if the tasks are scheduled using the list T_5, T_9, T_1, T_3, T_2, T_6, T_8, T_4, T_7 and the list-processing algorithm?

c. If Mike can talk Mary and Jack into helping him prepare the meal, how long will it take if the tasks are scheduled using the same list as in part **b**?

d. What would be a reasonable set of criteria for choosing a priority list in this situation?

9. Discuss different criteria that might be used to construct a priority list for a scheduling problem.

10. If two schedules have the same completion time, can one schedule have more idle time than the other?

11. Give examples of scheduling problems in which
a. processors available can be treated as if they are identical.
b. processors available cannot be treated as if they are identical.

12. Consider the accompanying order-requirement digraph:

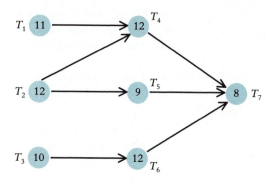

a. Find the length of the critical path.
b. Schedule these seven tasks on two processors using the list algorithm and the lists

(1) $T_1, T_2, T_3, T_4, T_5, T_6, T_7$
(2) $T_2, T_1, T_3, T_6, T_5, T_4, T_7$

c. Does either list lead to a completion time that equals the length of the critical path?
d. Show that no list can ever lead to a completion time equal to the length of the critical path (providing the schedule uses two processors).

▲ 13. For the following schedules, can you produce a list so that the list-processing algorithm produces the schedule shown when the tasks are independent? What are the task times for each task?

Machine 1	T_1		T_4		T_8		T_9	
Machine 2	T_3		T_5			T_7		T_{10}
Machine 3	T_2			T_6		T_{11}	T_{12}	

   0   1   2   3   4   5   6   7   8   9   10   11

(a)

Machine 1: T_1, T_{11}, T_{10}
Machine 2: T_5, T_3, T_{12}
Machine 3: T_7, T_8, T_2, T_{13}
Machine 4: T_4, T_6, T_9

0   2   4   6   8   10   12   14

(b)

■ 14. Once an optimal schedule has been found for independent tasks (e.g., see diagrams in Exercise 13), usually the scheduling of the tasks can be rearranged and the same optimal time achieved (i.e., one can, among other things, reorder the tasks done by a particular processor). Discuss criteria that might be used in implementing the rearrangement process.

15. At a large toy store, scooters arrive unassembled in boxes. To assemble a scooter the following tasks must be performed:

TASK 1.   Remove parts from the box
TASK 2.   Attach wheels to the footboard
TASK 3.   Attach vertical housing
TASK 4.   Attach handlebars to vertical housing
TASK 5.   Put on reflector tape
TASK 6.   Attach bell to handlebars
TASK 7.   Attach decals
TASK 8.   Attach kickstand
TASK 9.   Attach safety instructions to handlebars

   a. Give reasonable time estimates for these tasks and construct a reasonable order-requirement digraph. What is the earliest time by which these tasks can be completed?
   b. Schedule this job on two processors (humans) using the decreasing-time-list algorithm.

■ 16. Some scheduling projects have due dates for tasks (i.e., times by which a given task should be completed). Give examples of circumstances where this situation might arise.

▲ Advanced exercise.   ■ Discussion exercise.

▲ 17. Can you find a schedule (use the order-requirement digraph in Figure 3.4) with a completion time earlier than 12 as shown in the following figure, using a list other than $T_1, \ldots, T_9$? If not, why not?

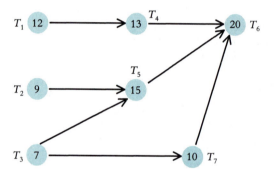

18. Given the accompanying order-requirement digraph:

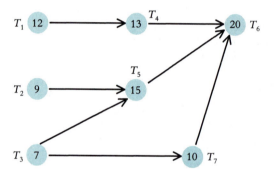

a. Use the list-processing algorithm to schedule these seven tasks on two processors using these lists:

(1) $T_1, T_3, T_7, T_2, T_4, T_5, T_6$

(2) $T_1, T_3, T_2, T_4, T_5, T_6, T_7$

(3) The list obtained by listing the tasks in order of decreasing time

b. Try to determine if any of the resulting schedules are optimal.

c. Schedule the tasks using the critical-path scheduling method. Try to determine if this schedule is optimal.

▲ 19. Describe possible modifications for the list-processing algorithm that would allow for

a. a machine to be "voluntarily" idle.

b. a machine to interrupt work on a task once it has begun work on the task.

▲ Advanced exercise.

20. a. Find the completion time for independent tasks of length 8, 11, 17, 14, 16, 9, 2, 1, 18, 5, 3, 7, 6, 2, 1 on three processors, using the list-processing algorithm.
    b. Find the completion time for the tasks in part **a** on three processors, using the decreasing-time-list algorithm.
    c. Does either algorithm give rise to an optimal schedule?
    d. Repeat for tasks of lengths 19, 19, 20, 20, 1, 1, 2, 2, 3, 3, 5, 5, 11, 11, 17, 18, 18, 17, 2, 16, 16, 2.

21. A photocopy shop must schedule independent batches of documents to be copied. The times for the different sets of documents are (in minutes): 12, 23, 32, 13, 24, 45, 23, 23, 14, 21, 34, 53, 18, 63, 47, 25, 74, 23, 43, 43, 16, 16, 76.
    a. Construct a schedule using the list-processing algorithm on three machines.
    b. Construct a schedule using the list-processing algorithm on four machines.
    c. Repeat parts a and b, but use the decreasing-time-list algorithm.
    d. Suppose union regulations require that an 8-minute rest period be allowed for any photocopy task over 45 minutes. Use the decreasing-time-list algorithm, with the preceding times modified to take into account the union requirement, to schedule the tasks on three human-operated machines.

▲ 22. Could the following schedule have arisen from the list-processing algorithm? Could it have arisen from the application of the list-processing algorithm to a collection of independent tasks?

Machine 1	T_1	T_5	T_8
Machine 2	T_2		T_9
Machine 3	T_3	T_6	T_{10}
Machine 4	T_4	T_7	T_{11}

23. Can you give examples of scheduling problems for which it seems reasonable to assume that all the task times are the same?

24. Find a list that produces the following optimal schedule when the list-processing algorithm is applied to this list. (Assume the tasks are independent.)

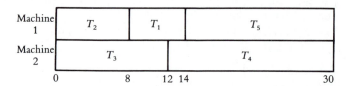

What completion time and schedule are obtained when the decreasing-time-list algorithm is applied to this list?

25. Can you think of situations other than those mentioned in the text where scheduling independent tasks on processors occurs?

26. Can you think of real-world scheduling situations in which all the tasks have the same time and are independent? Can you find an algorithm for solving this problem optimally? (If there are n independent tasks of time length k, when will all the tasks be finished?)

27. a. Show that when tasks to be scheduled are independent, the critical-path method and the decreasing-time-list method are identical.
    b. The (usually unknown) optimal time to complete a specific collection of independent tasks on three machines turns out to be 450 minutes.

    (1) Estimate the worst possible completion time when the list-processing algorithm is used with the worst choice of priority list.
    (2) Estimate the longest possible completion time using the list-processing algorithm and the decreasing-time list.

28. A radio station's policy allows advertising breaks of no longer than 2 minutes, 15 seconds. Using algorithms a and b below, determine the minimum number of breaks into which the following ads will fit (lengths given in seconds): 80, 90, 130, 50, 60, 20, 90, 30, 30, 40. Can you find the optimum solution? Do the same for these ads: 60, 50, 40, 40, 60, 90, 90, 50, 20, 30, 30, 50.
    a. First fit
    b. First-fit decreasing

29. It takes 4 seconds to photocopy one page. Manuscripts of lengths 10, 8, 15, 24, 22, 24, 20, 14, 19, 12, 16, 30, 15, and 16 pages are to be photocopied. How many photocopy machines would be required, using the first-fit-decreasing algorithm, to guarantee that all manuscripts are photocopied in 2 minutes or less? Would the solution differ if worst-fit decreasing were used?

▲ Advanced exercise.

▲ 30. Two wooden wall systems are to be made with pieces of wood with lengths shown in the accompanying diagram. If wood is sold in 10-foot planks and can be cut with no waste, what number of boards would be purchased if one uses the first-fit-decreasing, next-fit-decreasing, and worst-fit-decreasing heuristics, respectively?

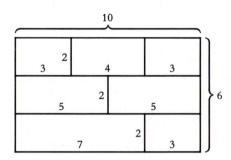

In solving this problem, does it make a difference if the 10-foot horizontal shelves and 6-foot vertical boards employ single-length pieces as compared with using pieces of boards that add up to 10- and 6-foot lengths?

31. A typing pool gets in 30 (independent) tasks that will take the following amounts of time (in minutes) to type: 25, 18, 13, 19, 30, 32, 12, 36, 25, 17, 18, 26, 12, 15, 31, 18, 15, 18, 16, 19, 30, 12, 16, 15, 24, 16, 27, 18, 9, 14.

   a. Using these times as a priority list:

     (1) Use the list-processing algorithm to find the completion time for scheduling these tasks with four secretaries; with five secretaries.

     (2) Repeat the scheduling using the decreasing-time-list algorithm.

     (3) Can you show that any of the schedules that you get are optimal?

   b. If one needs to finish the typing in one hour:

     (1) Use the FFD heuristic to find how many typists would be needed.

     (2) Repeat for the NFD and WFD heuristics.

     (3) Can you show that any of the solutions you get are optimal?

32. Find the minimum number of bins necessary to pack items of size 8, 5, 3, 4, 3, 7, 8, 8, 6, 5, 3, 2, 1, 2, 1, 2, 1, 3, 5, 2, 4, 2, 6, 5, 3, 4, 2, 6, 7, 7, 8, 6, 5, 4, 6, 1, 4, 7, 5, 1, 2, 4 in bins of capacity a through d using the first-fit and first-fit-decreasing algorithms. Can you determine if any of the packings you get are optimal?

   a. 9   c. 11
   b. 10   d. 12

33. Fiberglass insulation comes in 36-inch precut sections. A plumber must install insulation in a basement on piping that is interrupted often by joints. The distances between the joints on the stretches of pipe that must be insulated are 12, 15, 16, 12, 9, 11, 15, 17, 12, 14, 17, 18, 19, 21, 31, 7, 21, 9, 23, 24, 15, 16, 12, 9, 8, 27, 22, 18 inches. How many precut sections would he have to use to provide the insulation if he bases his decision on
   a. next fit?
   b. next-fit decreasing?
   c. worst fit?
   d. worst-fit decreasing?

34. The files that a company has for its employees dealing with utilities occupy 100, 120, 60, 90, 110, 45, 30, 70, 60, 50, 40, 25, 65, 25, 55, 35, 45, 60, 75, 30, 120, 100, 60, 90, 85 sectors. If, after operating systems are installed, a disk can store up to 480 sectors, determine the number of disks to store the utilities if each of these heuristics is used to pack the disk with files:
   a. NF
   b. NFD
   c. FF
   d. FFD

35. Advertisements for the TV show Q are permitted to last up to a total of 8 minutes, and each group of ads can last up to 2 minutes. If the ads slated for Q last 63, 32, 11, 19, 24, 87, 64, 36, 27, 42, 63 seconds, determine if FF and FFD yield acceptable configurations for the ads.

36. Consider the following heuristic for packing bins, which is different from the ones discussed so far and is known as *best fit:* Keep track of how much room remains in each unfilled bin and put the next item to be packed into that bin that would leave the least room left over after the item is put into the bin. (For example, suppose that bin 4 had 6 units left, bin 7 had 5 units, and bin 9 had 8 units left. If the next item in the list had size 5, then first fit would place this item in bin 4, worst fit would place the item in bin 9, while best fit would place the item in bin 7.) If there is a tie, place the item into the bin with the lowest number. Apply this heuristic to the list: 8, 7, 1, 9, 2, 5, 7, 3, 6, 4, where the bins have capacity 10.

■ 37. We have described (in the text and Exercise 36) two algorithms for bin packing called "worst fit" and "best fit." The words "best" and "worst" have connotations in English. However, the performance of algorithms depends on their merits as algorithms, not on the names we give them.
   a. On the basis of experiments you perform with the best-fit and worst-fit algorithms, which one do you think is the "better" of the two?
   b. Can you construct an example where worst fit uses fewer bins than best fit?

38. The best-fit heuristic (see Exercise 36) also has a "decreasing" version, where the list is first sorted in decreasing order. Using bins of capacity 10, apply the best-fit heuristic and its decreasing version to the following list: 6, 9, 5, 8, 3, 2, 1, 9, 2, 7, 2, 5, 4, 3, 7, 6, 2, 8, 3, 7, 1, 6, 4, 2, 5, 3, 7, 2, 5, 2, 3, 6, 2, 7, 1, 3, 5, 4, 2, 6.

▲ 39. How many bins are necessary to pack the numbers 1, 1, 2, 2, . . . , 20, 20 into bins of size 25? Repeat for 1, 1, 1, 2, 2, 2, . . . , 20, 20, 20. Generalize!

▲ 40. In the wall-system example in the text (see Figure 3.20), first fit and worst fit required equal numbers of bins. Can you find an example where first fit and worst fit yield different numbers of bins? Can you find an example where first fit, worst fit, and next fit yield answers with different numbers of bins?

■ 41. A common suggestion for heuristics for the bin-packing problem with bins of capacity W involves finding weights that sum to exactly W. Discuss the pros and cons of a heuristic of this type.

▲ 42. A record company wishes to record all the Beethoven string quartets (16 quartets, each consisting of several consecutive parts called movements) on LPs. It wishes to complete the project on as few records as possible. Recording can be done on two sides as long as the movements are consecutive. Is this an example of a bin-packing problem? (Defend your answer.) If the project were to record the quartets on (standard) tape cassettes or compact discs, would your answer be different?

▲ 43. Two-dimensional bin packing refers to the problem of packing rectangles of various sizes into a minimum number of $m \times n$ rectangles, with the sides of the packed rectangles parallel to those of the containing rectangle.
   a. Suggest some possible real-world applications of this problem.
   b. Devise a heuristic algorithm for this problem.
   c. Give an argument to show that the problem is at least as hard to solve as the usual bin-packing problem.
   d. If you have $1 \times m$ rectangles with total area W to be packed into a single rectangle of area $p \times q = W$, can the packing always be accomplished?

■ 44. In what situations would packing bins of different capacities be the appropriate model for real-world situations? Suggest some possible algorithms for this type of problem.

▲ 45. Can you find an example of weights that, when packed into bins using first fit, use fewer bins than the number of bins used when the first-fit algorithm is applied with the first weight on the list removed?

■ 46. Can you formulate "paradoxical" situations for bin packing that are analogous to those we found for scheduling processors?

▲ Advanced exercise.   ■ Discussion exercise.

▶ WRITING PROJECTS

1. Scheduling is important for hospitals, schools, transportation systems, police services, and fire services. Pick one of these areas and write an essay about the different scheduling situations that come up, types of processors, and extent to which the assumptions of the list-processing model hold for the area you pick.

2. Write an essay that compares and contrasts the basic scheduling problem we investigated with the scheduling version of the bin-packing problem.

3. One of the oversimplifications made in our discussion of scheduling was that there were no "due dates" involved for the tasks making up a job. Develop an algorithm for solving a scheduling problem under the assumption that each task has a due date as well as a time length. You will probably want to decide on a penalty amount that will occur when a due date is exceeded.

4. Consider the problem of scheduling tasks on a single machine. Design different algorithms for achieving different goals. You will probably wish to assume that each task has a due date, such that if the task is not finished by this date, some penalty payment must be made.

Chapter 4

Linear Programming

A corporate manager's job often involves very complicated decisions. One set of decisions involves planning what products the business is to make and determining what resources are needed. In the modern business world, diversification of products provides a company with stability in a climate of changing tastes and needs. So it is not surprising that companies would produce many products, some of which share resource needs. For example, any bakery uses many resources, among which are butter, sugar, eggs, and flour to make its products: cookies, cakes, pies, and breads.

Resources include more than just raw materials. A labor force with appropriate skills, farmland, time, and machinery are also resources. Typically, resources are limited: a farmer owns only so much land, there are only so many hours in a day, in a year of drought the wheat crop is very small. The amount of resources available to a particular company is further limited by the location of the resource and the competition for, and thus the cost of, the resource. This leads to these dilemmas: In order to maximize profit, how much of each product should be produced? How should the available resources be shared among the possible products?

Deciding how much of each product to make is a major management decision. The objective of this branch of management science is to maximize profit. All too often there are so many alternative product mixes that it is impossible to evaluate them all individually. Despite this, millions of dollars may ride on the manager's decision.

In this chapter we learn about **linear programming,** a management-science technique that helps a business allocate the resources it has on hand to make a particular mix of products that will maximize profit. The technique is so powerful that linear programming is said to account for over 50% and perhaps as much as 90% of all computing time used for management decisions in business.

Linear programming is an example of "new" mathematics. It did not originate with the ancient Egyptians or Greeks, nor was it developed in Europe during the time the Western Hemisphere was being explored and settled. Linear programming came into being, along with many other management-science techniques, during and shortly after World War II, in the 1940s; it is quite young as intellectual ideas go. Yet during its short

SP TLIGHT 4.1   Case Studies in Linear Programming

▶ ▶ ▶ ▶ ▶ ▶ ▶ ▶ ▶ ▶ ▶ ▶ ▶

Linear programming is not limited to mixture problems. Here are two case studies that do not involve mixture problems, yet where applying linear programming techniques produced impressive savings:

▶ The Exxon Corporation spends several million dollars per day running refineries in the United States. Because running a refinery takes a lot of energy, energy-saving measures can have a large effect. Managers at Exxon's Baton Rouge plant had over 600 energy-saving projects under consideration. They couldn't implement them all because some conflicted with others, and there were so many ways of making a selection from the 600 that it was impossible to evaluate all selections individually.

Exxon used linear programming to select an optimal configuration of about 200 projects. The savings are expected to be about $100 million over a period of years.

▶ American Edwards Laboratories uses heart valves from pigs to produce artificial heart valves for human beings. Pig heart valves come in different sizes. Shipments of pig heart valves often contain too many of some sizes and too few of others; however, each supplier tends to ship roughly the same imbalance of valve sizes on every order, so the company can expect consistently different imbalances from the different suppliers. Thus, if they order shipments from all the suppliers, the imbalances could cancel each other out in a fairly predictable way. The amount of cancellation will depend on the sizes of the individual shipments. Unfortunately, there are too many combinations of shipment sizes to consider all combinations individually.

American Edwards used linear programming to figure out which combination of shipment sizes would give the best cancellation effect. This reduced their annual cost by $1.5 million.

history, linear programming has changed the way businesses make decisions, from "seat-of-the-pants" methods based on guesswork and intuition to using an algorithm based on available data and guaranteed to produce an optimal decision.

Linear programming has saved businesses millions of dollars. Of all the management-science techniques presented in this book, linear programming is far and away the most frequently used. It can be used in a variety of situations in

addition to the one we study in this chapter. Some of the problems studied in Chapters 1, 2, and 3 can be viewed as linear programming problems. Examples of other uses are in Spotlight 4.1. Linear programming is thus an excellent example of a mathematical technique useful for solving many different kinds of problems that at first do not seem to be similar problems at all. It has been suggested that without linear programming, management science would not exist.

▶Mixture Problems

In this chapter we study one of the easier applications of linear programming, using linear programming to solve **mixture problems.** In mixture problems, limited resources need to be combined into products in a way that maximizes the profits on those products. Mixture problems are widespread because nearly every product in our economy is created by combining resources. For example, various kinds of steel alloys can be produced by combining iron with different amounts of other elements, such as chromium, carbon, and tungsten. How much of each alloy should be made if one has limited quantities of raw materials on hand? Or consider another example: cereal manufacturers use oats, corn, wheat, and rice to produce breakfast cereals that contain either one type of grain or a mix of grains. How many boxes of each product should be made? The specific examples we consider in this chapter are juice mixtures and plywoods. We begin by looking at the typical mixture problems that might confront a juice manufacturer.

The juice manufacturer can sell many different juices and juice combinations; these are the products on which the manufacturer makes profits. Each product is either a pure juice, which would be identical to a resource for the manufacturer, or a mixture of two or more juices. There could be dozens of possible products and many varieties of fruit juice to use as resources. The manufacturer must periodically look at the quantities and prices of unmixed juices and then determine which juice products should be produced in which quantities in order to gain the greatest, or optimum, profit. This is an enormous task, usually requiring the aid of a computer to solve. We examine a much simpler case.

Let's examine the problem for a fruit juice manufacturer that produces and sells just two kinds of juice: appleberry and cranapple. These products are made by combining pure cranberry juice and pure apple juice in the following proportions: 3 quarts of cranberry juice and 1 quart of apple juice make 1 gallon of cranapple; 2 quarts of apple juice and 2 quarts of cranberry juice are combined to make 1 gallon of appleberry. Today, there are 200 quarts of cranberry juice and 100 quarts of apple juice available. The manufacturer makes a profit of 3 cents on a gallon of cranapple and 4 cents on a gallon of appleberry. The manufacturer needs to determine how many gallons of cranapple and how many gallons of appleberry should be produced to obtain the highest profit without exceeding available supplies. (Tomorrow, the available juice supplies may differ, so the manufacturer would then need to solve a similar problem, but with different numbers.)

We list the features or conditions of this problem that make it a mixture problem:

▶ *Resources.* Definite resources are available in limited, known quantities for the time period in question. The resources available here are cranberry juice and apple juice.

▶ *Products.* Definite products can be made by combining, or mixing, the resources. In this example, the products are appleberry and cranapple.

▶ *Recipes.* A recipe for each product specifies how many units of each resource are needed to make one unit of that product.

▶ *Profits.* Each product earns a known profit per unit. (We assumed that every unit produced can be sold.)

▶ *Objective.* The objective in a mixture problem is to find how much of each product to make so as to maximize the profit without exceeding any of the resource limitations.

As is done when creating any mathematical model, here we are beginning by looking at a very simple model. For example, we are assuming that every unit produced will be sold. But if demand for our product changes, we may not be able to sell all of our units. We also have not considered contracts that would require us to supply some of our customers with a minimum number of units, regardless of how much it would cost us to do so. To accurately model such situations we would need more complex models than the one we are learning in this chapter.

In order to maintain a maximum profit, the company needs to rework the numbers in its linear programming model every time circumstances change, and then solve the new problem. When there are fluctuations in the cost or availability of a resource, the amount of that resource that we are willing or able to purchase can vary. This means that we would need to rewrite our resource constraints. Or competition might force us to cut our selling price. Both changes in selling price and in cost of resources affect our profit per unit, and thus require us to rewrite the profit function. And these events do happen. Consider gasolines as our products, crude oil of different grades being the resources. Events in the Middle East, such as wars and crude oil embargoes, threaten to reduce the supply of crude oil in the United States, and as a consequence the price of crude oil increases. Price wars among filling stations in a region occur on occasion, forcing selling prices down.

▶ MIXTURE CHARTS

The most important skill required when solving a mixture problem is the ability to understand its underlying structure. This is as important as being able to do the subsequent algebra and arithmetic. In fact, since linear programming problems can be solved using readily available computer software, extracting the important data from the underlying structure may be the only part of the problem-solving process that must be done by a human being. Understanding the underlying structure means being able to answer these questions:

1. What are the *resources?*

2. What *quantity* of each resource is available?

3. What are the *products?*

4. What are the *recipes* for creating the products from the resources?

5. What are the *unknown quantities?*

6. What is the *profit formula?*

We will display the answers to these questions in a diagram called a *mixture chart*. Then we will translate information in our mixture chart into mathematical statements that we can use to solve the mixture problem.

For the juice manufacturer, our answers to questions 1 through 6 are shown on the mixture chart in Figure 4.1. The resources (question 1) are cranberry juice and apple juice, and the quantities available (question 2) are 200 quarts of cranberry juice and 100 quarts of apple juice, as shown in the labels at the top of the mixture chart. Questions 3 and 4, on products and recipes, are answered in the two rows of boxes in the chart.

The labels at the left side of the chart show the unknowns (question 5): how many units (gallons) of cranapple juice and how many units of appleberry to make. There is one unknown quantity for each product, each mixture. We use x and y to represent the unknown quantities:

$x =$ the (currently unknown) number of
gallons of cranapple to make

$y =$ the (currently unknown) number of
gallons of appleberry to make

The profit for each product (question 6) is shown in the rightmost box for that product.

Now that we have put the given data into the mixture chart, we are ready to translate it into mathematical statements. First, we note that we have one row of information for each product and one column for each resource. We also have an extra column for the profit. We will formulate a mathematical statement corresponding to each of the columns.

For each resource, we develop a mathematical expression reflecting the fact that the manufac-

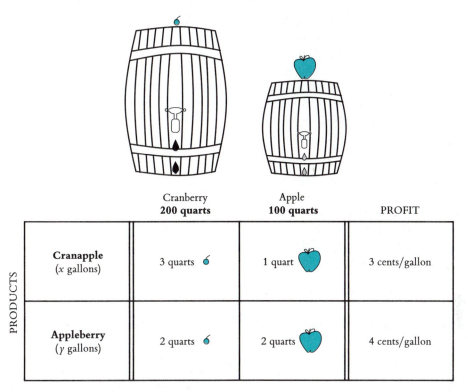

PRODUCTS	Cranberry **200 quarts**	Apple **100 quarts**	PROFIT
Cranapple (*x* gallons)	3 quarts	1 quart	3 cents/gallon
Appleberry (*y* gallons)	2 quarts	2 quarts	4 cents/gallon

Figure 4.1 A mixture chart for fruit juice production.

turer cannot use more of that resource than is available. The number of quarts of cranberry juice needed for x gallons of cranapple is $3x$: 3 quarts of cranberry per gallon of cranapple times x gallons of cranapple. Similarly, $2y$ quarts of cranberry are needed for making y gallons of appleberry. So if the manufacturer makes x gallons of cranapple and y gallons of appleberry, then $3x + 2y$ quarts of cranberry juice will be used. There are only 200 quarts of cranberry available. If all 200 quarts are used, then we would have the equation $3x + 2y = 200$. The situation $3x + 2y > 200$ is impossible, because that inequality states that the amount of cranberry juice used by the manufacturer, $(3x + 2y)$, is *greater than* the amount available (200). On the other hand, $3x + 2y < 200$ is possible, because that inequality states that the amount of cranberry juice used by the manufacturer,

$(3x + 2y)$, is *less than* the amount available (200). When we put together the two possibilities, that the amount of cranberry juice used by the manufacturer could be *less than* or *equal to* the amount of cranberry juice available, we get the inequality

$$3x + 2y \leq 200$$

Note that the numbers 3, 2, and 200 are all in the "cranberry" column. The inequality $3x + 2y \leq 200$ is called a **resource constraint** of the mixture problem; it expresses a limitation, based on availability of a resource, that the solution must respect.

We get another resource constraint from the column for the apple juice resource, namely,

$$1x + 2y \leq 100$$

Most mixture problems also have **minimum constraints**: these are constraints that reflect the real world and tell us that the manufacturer will never make negative quantities of any mixture product. Minimum constraints may seem obvious, but their importance should not be overlooked. In our example we have these minimum constraints:

$$x \geq 0 \quad \text{and} \quad y \geq 0$$

Finally, we express the profit information mathematically. As with the constraints, we again deal with the currently unknown quantities x and y. Since $3x$ is the profit from making x units of cranapple and $4y$ is the profit from making y units of appleberry, we get the **profit formula:**

$$\text{Profit} = 3x + 4y$$

We summarize our analysis of the juice mixture problem:

Maximize the
profit formula: $3x + 4y$
given these
constraints: $3x + 2y \leq 200$
                 $1x + 2y \leq 100$
                 $x \geq 0 \quad \text{and} \quad y \geq 0$

In a mixture problem, our job is to find the values of x and y that make all the constraints true and maximize the profit. For example, if $x = 40$ and $y = 10$, we find that

$3x + 2y = 3(40) + 2(10) = 120 + 20$
   $= 140 \leq 200$ (cranberry constraint satisfied)
$1x + 2y = 1(40) + 2(10) = 40 + 20$
   $= 60 \leq 100$ (apple juice constraint is satisfied)
$40 \geq 0$ and $10 \geq 0$ (both minimum
                          constraints satisfied)
Profit would be $3x + 4y = 3(40) + 4(10)$
                          $= 120 + 40 = 160.$

But perhaps some other values for x and y satisfy all the constraints and give us a bigger profit. We want to find the values of x and y giving us the biggest profit. After we develop a second mixture chart, we discuss how to find these optimal values of x and y.

Here is an actual example of a mixture problem that confronted a lumber company: Plywood Ponderosa of Mexico produces various plywood products by gluing together thin sheets (veneers) of lumber. Some of these products are more profitable than others, yet it is not sensible to produce only the highest profit product because much of the resource lumber would be left over. How much of each product—that is, which product mix—should be made in order to maximize total profit? Because Plywood Ponderosa had too many possible product mixes to consider individually, the company used linear programming to find the product mix that would give the highest overall profit. This turned out to be different from the mix they had been making. In fact, changing to the new mix increased profits by 20%. Real-world mixture problems, like Plywood Ponderosa's problem, deal with very large numbers of resources and products—too many for hand calculation. We will therefore simplify the plywood mixture problem for our example.

EXAMPLE: Making and Using a Mixture Chart

A company produces plywood by using a press to glue veneers together. The veneers come in two grades, A and B, and two kinds of plywood are made, exterior and interior. One panel of exterior plywood requires two panels of grade A veneer, two panels of grade B veneer, and 10 minutes on a press. One panel of interior plywood requires four panels of grade B veneer, no grade A veneers, and 5 minutes on the press. On a certain day there are 1000 panels of grade

A veneer available and 3000 panels of grade B. There are 12 presses, each of which can press four veneer panels into a finished product. Each press can be run for 500 minutes per day, yielding a total of 6000 machine-minutes. The profit per panel is $5 for interior plywood and $6 for exterior plywood. How much of each product should be produced that day to maximize profit?

To create the mixture chart for this problem we need to identify the resources used by the business and the resulting products. Here the resources are the two grades of veneer and the time available on the presses. The products are the exterior and interior plywoods. Once the resources and products have been identified, we set up the columns and rows of the chart and enter the given data. We also assign one un-

known to each product: x for exterior plywood and y for interior plywood. Figure 4.2 shows the mixture chart for the plywood problem.

Now we use the mixture chart to develop the mathematical statements we will need. For the constraints in the plywood problem we get:

▶ *Resource constraints:* These inequalities tell us that the business must use less than or equal to the amount of a resource it has on hand, never more than that amount:

Grade A: $2x + 0y \leq 1000$
(We could omit the y term since it is $0y$)
Grade B: $2x + 4y \leq 3000$
Time: $10x + 5y \leq 6000$

RESOURCES

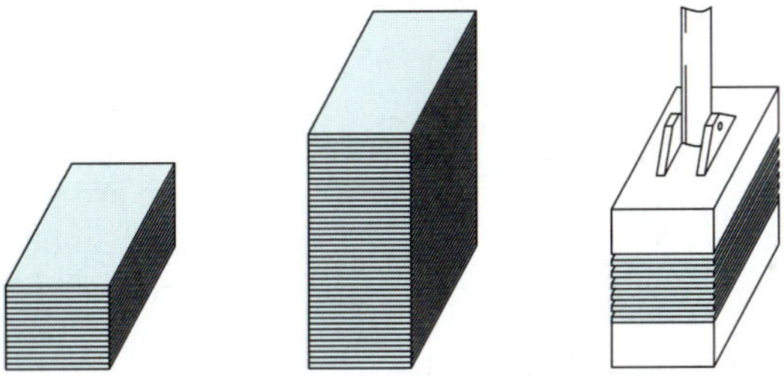

	Grade A **1000 panels**	Grade B **3000 panels**	Presses **6000 minutes**	PROFIT
Exterior (*x* panels)	2 of A	2 of B	10 minutes	$6/panel
Interior (*y* panels)	0 of A	4 of B	5 minutes	$5/panel

Figure 4.2 A mixture chart for plywood production.

▶ *Minimum constraints:* These inequalities tell us that we cannot make a negative number of any product:

$$\text{Exterior plywood: } x \geq 0$$
$$\text{Interior plywood: } y \geq 0$$

▶ *Profit formula:* The profit formula tells us how much profit we make if our production policy is to make x units of one product and y units of the other product.

$$\text{Profit} = 6x + 5y$$

We summarize our analysis of the plywood problem:

Maximize the
profit formula: $6x + 5y$
given these
constraints: $2x + 0y \leq 1000$
                 $2x + 4y \leq 3000$
                 $10x + 5y \leq 6000$
                 $x \geq 0$ and $y \geq 0$ ▲

In the sections that follow, we use our mixture charts to solve the fruit juice and plywood mixture problems.

▶ABOUT LINEAR PROGRAMMING ALGORITHMS

The fruit juice and plywood problems deal with different industries and numbers, but to a mathematician they are similar because both fit into the framework of a mixture problem and can be represented by mixture charts. The task of analyzing a mixture problem is mostly a matter of careful reading and logic. Some people might be fooled into thinking that it is not mathematics at all. But actually, it is an extremely important skill in the practice of applied mathematics. Without the kind of analysis needed to create a mixture chart, we cannot solve any linear programming problem.

What does it mean to find a solution to a linear programming mixture problem? A solution is a **production policy** that tells us how many units of each product to make. Since our examples have just two products, for us a production policy is a pair of values for the unknown quantities x and y. We usually write the pair as (x, y). Thus, every production policy corresponds to some pair (x, y). The optimal production policy has two properties: first, it is possible, that is, it does not violate any of the constraints. Second, the production policy gives the maximum value for the profit formula.

The reader, having studied the previous chapters on management science, may sense that there must be some algorithm that will give us the optimal x and y from the various inequalities and the profit formula, which we get from the mixture chart. There are indeed such algorithms. Most of the algorithms are very algebraic. The algebraic details of these algorithms are easily forgotten. Furthermore, there are computer programs available to solve linear programming problems once the mixture chart has been constructed.

At the heart of every algorithm for linear programming are geometric ideas. This is somewhat surprising, since there seem to be no geometric ideas used to describe a mixture problem. We can see these geometric ideas clearly if we solve small linear programming problems, involving just two products and two or three resources, using a pictorial approach. This pictorial view of linear programming has been valuable to theorists and practitioners alike, and it is readily understandable by nonmathematicians. The pictorial solutions to small problems show us all the essential features of linear programming, but avoid drowning us in the details that come with the larger problems typical of most business applications, problems with dozens or hundreds of products and equally large numbers of resources.

Every algorithm for solving a linear programming problem with two products has these three characteristics. Similar characteristics hold when we apply these algorithms to problems having more than two products.

1. The algorithm can distinguish between "good" production policies, pairs (x, y) that satisfy all the constraints, and those that violate some constraint(s). There are usually many "good" pairs (x, y), each of which corresponds to some production policy: "Make x units of product 1 and y units of product 2."

2. The algorithm makes use of a geometric principle — we shall learn one shortly — to select a special subset of the production policies. Every production policy in the subset will be "good," and the geometric principle guarantees that the maximum profit will occur at one of the production policies in the special subset.

3. The algorithm evaluates the profit formula for (x, y) pairs associated with "good" production policies in the special subset to find which one of them actually gives the maximum profit.

The various algorithms for linear programming differ in the order in which the algorithm evaluates the profit formula for the pairs (x, y) in the special subset and in how quickly the algorithm finds the production policy that gives the optimal profit. The better algorithms are faster at finding the policy giving the optimal profit, because they don't need to evaluate the profit formula at every pair.

We now return to the two problems presented at the start of the chapter, and use the pictorial approach to solve each of them.

▶Graphing the Feasible Region

With just two products, we have two unknown quantities, x and y, representing the production policy. If $x = 7$ and $y = 13$, then the production policy is "Make 7 units of product 1 and 13 units of product 2." The basic pictorial notion we use is that every production policy is a pair of numerical values, such as $(7, 13)$ or $(40, 10)$, or more generally (x, y), which can be interpreted as a point in the plane. In a mixture problem involving two products, we are looking for a pair of values x and y that gives the most profit. This can be interpreted as a search for the best point in the plane.

We can't locate this best point directly, so we begin with another question: Which points are the reasonable candidates for this paragon of points? A candidate for best point must satisfy the constraints we developed from the mixture chart.

We recall that there were two types of constraints, those based on *minimums* and those based on availability of *resources*. Returning to our juice manufacturer, the minimum constraints simply say, for example, that you can't make -6 gallons of appleberry. Since x represents the number of gallons of cranapple and y represents the number of gallons of appleberry, the minimum constraint for appleberry is $y \geq 0$. If we substitute -6 for y, we get the false statement $-6 \geq 0$, telling us that we cannot make -6 gallons of appleberry.

To illustrate the effect of the resource constraints, suppose the fruit juice manufacturer decides to make 70 gallons of cranapple and 20 gallons of appleberry; that is, $x = 70$ and $y = 20$. If we substitute $(70, 20)$ for (x, y) in the inequality for cranberry juice, $3x + 2y \leq 200$, we get $3(70) + 2(20)$ on the left-hand side. Evaluating this expression, we get $210 + 40$, or 250. But 250 is bigger than the 200 on the right-hand side; $250 \leq 200$ is a false statement. Thus the point $(70, 20)$ does not represent a manufacturing decision that can be carried out; it is not feasible to do so. We are not permitted to exceed the supply of even a single resource.

Points that do not violate either type of constraint are called **feasible points**. All others are infeasible points. The collection of feasible points is called the **feasible set**. It is helpful to visualize the feasible set as a portion of the plane or space. We call the visualization the **feasible region**.

What does a feasible region look like? Are the feasible points interspersed with infeasible points in a salt-and-pepper pattern? Or do the feasible points clump together in some type of connected shape? Techniques of algebra and analytic geometry allow us to use the constraint inequalities to determine the precise shape of the feasible region. We will do so for both the juice and the plywood manufacturers.

In the example of the juice manufacturer, we developed these constraint inequalities:

Cranberry juice: $3x + 2y \leq 200$

Apple juice: $\quad 1x + 2y \leq 100$

Minimums: $\quad x \geq 0 \quad$ and $\quad y \geq 0$

Each of these inequalities can be broken down into two parts. Since we read the symbol $\leq$ as "less than or equal to," we have "less than" and "equal to." To graph the part of the plane that satisfies "less than or equal to," we will put together, form the union of, the portion of the plane satisfying "less than" and the portion satisfying "equal to."

Focusing first on the "equal to," we find that associated with the cranberry juice constraint, $3x + 2y \leq 200$, is the equation $3x + 2y = 200$. This is an equation for a straight line in the plane. We recognize that any equation having either an x term, a y term, or both, and some numerical constant, but no other kinds of terms, always represents a line. (The equation cannot have any squared, square root, or other kinds of algebraic combinations of x or y.)

In fact, constraint inequalities always can be associated with equations for lines, hence the term *linear* programming. The *programming* does not refer to a computer, but to a well-defined sequence of steps, or program of action, that solves the kinds of problems we are exploring. Here we are using *program* as a synonym for algorithm. In recent years, with the introduction of computers into business settings, it is true that linear programming is usually carried out by running a computer program.

EXAMPLE: Graphing a Linear Inequality

The graph of the line $3x + 2y = 200$ is shown in Figure 4.3a. There are several ways to plot a line. For many of the lines in our mixture problems, using the intercepts, the points where the line crosses the axes, is a convenient method. We remember that when the line crosses, or intersects, the x axis, the value of y is 0. Substituting $y = 0$ into the equation $3x + 2y = 200$ gives $3x = 200$, or $x = \frac{200}{3}$, which we approximate by 66.7. So one point on the line $3x + 2y = 200$ is (66.7, 0). On the y axis, $x = 0$. Substituting $x = 0$ into our equation $3x + 2y = 200$, we get $2y = 200$, or $y = \frac{200}{2}$, giving us $y = 100$. So we have a second point on our line, namely, (0, 100). In Figure 4.3a, these two points are shown and the line was drawn by connecting them. (Remember, two points determine a line.)

Every point (x, y) on the line $3x + 2y = 200$ is a pair of values for x and y that makes the equality a true statement. The line also divides the whole plane into two "halves," often called *half planes*. It turns out that all the points in one half plane make the inequality $3x + 2y < 200$ a true statement, and all the points in the other half plane make the other inequality $3x + 2y > 200$ a true statement. To find out which half plane corresponds to the inequality $3x + 2y < 200$, we need only test one point of the plane that is not on the line $3x + 2y = 200$. Let's see how that is done.

Suppose we choose the point (50, 50). That point is not on the line because $3(50) + 2(50) = 200$ is not a true statement. In fact $3(50) + 2(50)$ evaluates to $150 + 100$, or 250, which is greater than 200. The point (50, 50) is in the half plane corresponding to the inequality $3x + 2y > 200$. In mixture problems, many constraint inequalities correspond to lines that do not contain the point (0, 0), so that point is frequently used. Using that point we see that it lies in the half plane for which $3x + 2y < 200$.

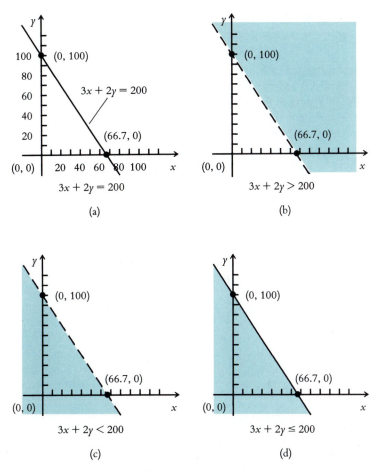

Figure 4.3 Graphing $3x + 2y \leq 200$.

Points on opposite sides of the line satisfy opposite inequalities; we could use either point, or some other one, to determine which half plane corresponds to which inequality. Figure 4.3b shows the half plane corresponding to $3x + 2y > 200$, and Figure 4.3c shows the half plane corresponding to $3x + 2y < 200$.

Returning to our original constraint, $3x + 2y \leq 200$, we wish to show on a graph all the points that make it a true statement. Those are the points that make the equality $3x + 2y = 200$ true and also the points that make the inequality $3x + 2y < 200$ true. That is, all the points on the line in Figure 4.3a and all the points in the half plane in Figure 4.3c are points that make our constraint a true statement. The set of points making $3x + 2y \leq 200$ true is shown in Figure 4.3d. ▲

EXAMPLE: Graphing a Feasible Region

In Figure 4.4a we repeat the graph in Figure 4.3d of the cranberry resource constraint inequality $3x + 2y \leq 200$. In Figure 4.4b we show the graph of the apple constraint inequality $1x + 2y \leq 100$. Note that the two points used to draw that line are $(0, 50)$ and $(100, 0)$, the two intercepts. (Can you find those two intercepts by first substituting $x = 0$ in the inequality $1x + 2y \leq 100$ and solving for y, then repeating the procedure, but this time solving for x when $y = 0$?) You can check that the correct half plane, the one for which $1x + 2y < 100$, has been shaded by substituting the point $(0, 0)$ into that inequality and seeing that the result you get is a true statement.

Now we know that a point in the feasible region must satisfy, or make true, *every* constraint inequality in the problem. Which points in the plane make *both* the cranberry and the apple constraint inequalities true? They are the points that are in not just one but *both* of the shaded regions in Figure 4.4a, b; the set we want is the *intersection* of the two half planes. The region satisfying both resource constraints is shown in Figure 4.4c. [In the next example we show how coordinates of a corner point like $(50, 25)$ can be determined.]

In the juice manufacturer mixture problem we also have two minimum constraint inequalities. The graphs for these inequalities, $x \geq 0$ and $y \geq 0$, are shown in Figure 4.4d, e, respectively. As we just did with the resource constraint inequalities, we need to find the intersection of the two half planes. That intersection is the upper right quarter of the plane, called the first quadrant of the plane. When we intersect the first quadrant of the plane with the graph in Figure 4.4c, we get the final graph, Figure 4.4f, of the feasible region for the juice mixture problem. ▲

We note here that the use of the two intercepts to plot a line will not work in the special cases of horizontal and vertical lines. Equations for horizontal lines have no x term; for example, $3y = 90$ is a horizontal line. It can be graphed by solving for y, giving $y = 30$, and then drawing the horizontal line all of whose y-coordinates are 30. Similarly $5x = 60$ becomes $x = 12$, whose graph is the vertical line all of whose x-coordinates are 12. The x and y axes are special cases of this general case: the x axis corresponds to $y = 0$ and the y axis to $x = 0$.

▶ CORNER POINTS

In the next section we will learn that the corner points of a feasible region are important in finding where in the feasible region we can find the *maximum* profit. We need to be able to calculate the coordinates of those corner points.

Each corner point of a feasible region is the intersection of two lines; the lines, of course, each correspond to one of the constraint inequalities. It is important to keep track of which two lines are intersecting at each corner point, because in general there are intersections of lines that are not corner points of the feasible region. For example, in Figure 4.4b, the point $(100, 0)$ is the intersection of the x axis, that is, the line $y = 0$, and the line $1x + 2y = 100$. But $(100, 0)$ is not a corner point of the feasible region in Figure 4.4f.

EXAMPLE: Finding Corner Points

We look at the graph in Figure 4.4f. We start at some corner and work our way clockwise around the edge of the feasible region. A convenient place to start, if it is a corner, is the origin $(0, 0)$. Although we know the coordinates of the origin, it is useful to note that it is the intersection of two lines having equations $x = 0$ and

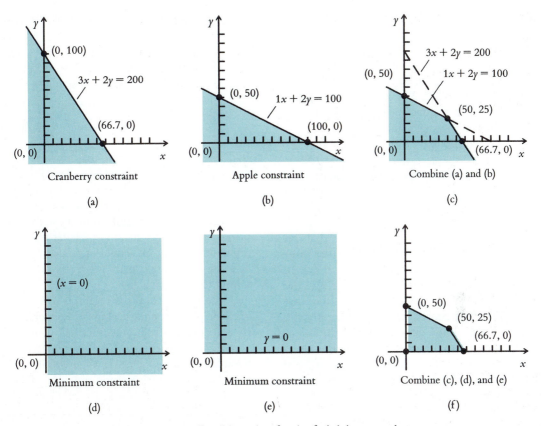

Figure 4.4 Feasible region for the fruit juice example.

$y = 0$ and corresponding to minimum constraints. These equations "solve" the problem of finding the coordinates. In general, we are trying to solve for values of x and y that satisfy both linear equations; then we will have the coordinates of the intersection.

The next corner is the intersection of the lines $x = 0$ and $1x + 2y = 100$. We already found this point when we plotted the line $1x + 2y = 100$. The point is $(0, 50)$.

Continuing clockwise, we come to the intersection of lines $1x + 2y = 100$ and $3x + 2y = 200$. This more general type of intersection can be solved by using multiplication and addition to eliminate one unknown, solving for

the remaining unknown, and then substituting that value into an original equation to get the value of the eliminated unknown. First, we multiply one equation by a positive value and the other by a negative value so that when the two equations are added together, one unknown gets a coefficient of zero:

$$(-3)(1x + 2y = 100) = -3x - 6y = -300$$
$$(1)(3x + 2y = 200) = 3x + 2y = 200$$

It does not matter what numbers we use to multiply as long as we get plus and minus the same coefficient on either both x terms or both y terms. Now we add the new equations together:

$$0x - 4y = -100 \quad \text{or} \quad -4y = -100$$

Solving this, we get $y = 25$. Substituting this y value into $1x + 2y = 100$, we get $1x + 2(25) = 100$, which gives $x = 50$. So the point of intersection seems to be $(50, 25)$. We can check our work in the other original equation: $3(50) + 2(25) = 200$ is a true statement.

The last corner point of this feasible region comes from the intersection of $3x + 2y = 200$ and $y = 0$. As with the other intersection of a resource constraint line and an axis, we already know the coordinates: $(66.7, 0)$. ▲

In some mixture problems, a resource constraint may give us a horizontal or vertical line in our graph. Thus one of the lines already tells us the value of one coordinate, and we can substitute that value into the other equation to get the value of the other coordinate. For example, to find the intersection of the vertical line $x = 5$ and the line $4x + 3y = 38$, we substitute 5 for x, getting $4(5) + 3y = 38$, or $3y = 18$, which gives $y = 6$, so the point of intersection is $(5, 6)$.

EXAMPLE: The Feasible Region for the Plywood Problem

We now return to the Ponderosa Plywood problem, construct its feasible region, and de-termine the corner points. We recall the constraint inequalities:

Grade A: $2x \leq 1000$
Grade B: $2x + 4y \leq 3000$
Time: $10x + 5y \leq 6000$
Minimums: $x \geq 0$ and $y \geq 0$

In Table 4.1 we summarize the information needed to draw the correct half plane for each resource constraint and list the graph in Figure 4.5 in which that half plane is drawn. Forming the correct feasible region from those three half planes plus the first quadrant, arising from the minimum constraints, could be tricky. It will be necessary to know exactly which two lines intersect to form each corner point.

One way to form the region is to start with one half plane, such as in Figure 4.5a, add to it the information from one other half plane, such as in Figure 4.5b, getting part way toward the feasible region, as we see in Figure 4.5d. Then take the new picture and add to it the information from the next half plane, here Figure 4.5c, getting Figure 4.5e. In our case, we are essentially done; all that is left is to restrict the picture to the first quadrant and we have the correct feasible region shown in Figure 4.5f. After you are familiar with constructing feasible regions, you may want to carefully draw all the relevant lines on one graph and then shade in the region.

TABLE 4.1   Using Mixture Chart Data to Determine a Feasible Region

Resource	Equation	Intersections x axis	Intersections y axis	Value at origin	Figure
Grade A	$2x = 1000$	$(500, 0)$	None	< 1000	4.5a
Grade B	$2x + 4y = 3000$	$(1500, 0)$	$(0, 750)$	< 3000	4.5b
Time	$10x + 5y = 6000$	$(600, 0)$	$(0, 1200)$	< 6000	4.5c

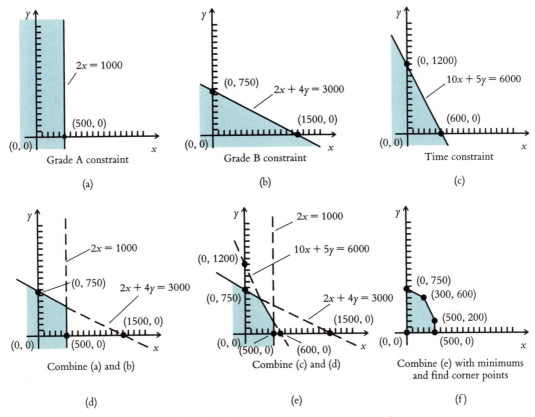

Figure 4.5 Feasible region for the plywood example.

Note that we kept labels on all the line segments, reminding us of which lines intersect to form each corner point. When we have more than two resource constraints, we usually do not need to find the coordinates of all the intersections of pairs of lines, because some of those intersections result in points that are not in the feasible region. For example, the intersection of the lines $2x = 1000$ and $2x + 4y = 3000$ is not a corner point of the feasible region, so determining its coordinates is not necessary. We will calculate the coordinates of the intersection of lines $2x + 4y = 3000$ (Grade B constraint) and $10x + 5y = 6000$ (Time constraint) and let you check that the coordinates of the other corner points have been properly calculated.

First we multiply one equation by a negative number and the other by a positive one so that when we add the two new equations together, one of the unknowns "drops out," that is, has a zero coefficient:

$(-5)(2x + 4y = 3000)$ becomes
$$-10x - 20y = -15{,}000$$
$(4)(10x + 5y = 6000)$ becomes
$$40x + 20y = 24{,}000$$

Adding the equations on the right, we get $(-10 + 40)x + (-20 + 20)y = -15{,}000 + 24{,}000$. Simplifying and "dropping out" the unknown y, we get $30x + 0y = 9000$ or $30x = 9000$. This gives $x = \frac{9000}{30}$, or $x = 300$. Substi-

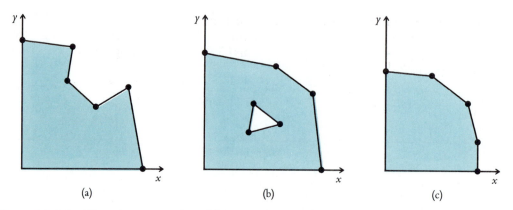

Figure 4.6 A feasible region may not have (a) dents or (b) holes. Graph (c) shows a typical feasible region.

tuting $x = 300$ into one of the two original equations we get $2(300) + 4y = 3000$ which is $600 + 4y = 3000$, or $4y = 2400$, so we get $y = 600$. Thus the point of intersection is $(300, 600)$, as shown in Figure 4.5f. ▲

In general, we can say the following about the shape of a feasible region for a linear programming problem:

1. For typical mixture problems, the feasible region is a polygon in the first quadrant. This is because the minimum constraints require that both x and y be nonnegative. The feasible region has one corner at the origin and one on each coordinate axis. There are usually additional corners off the axes.

2. The region is a polygon that has neither dents (as in Figure 4.6a) nor holes (as in Figure 4.6b). Figure 4.6c is a typical general example.

▶ THE CORNER POINT PRINCIPLE

To evaluate the profit that goes with a given point (x, y), we use the profit formula recorded on the mixture chart. In the juice mixture problem, for

example, to find the profit associated with the point $(40, 10)$, we substitute $x = 40$ and $y = 10$ into the formula for profit:

$$\text{Profit} = 3x + 4y = 3(40) + 4(10) = 160$$

If we had some way of applying this formula to each point of the feasible region, we could then pick out the point with the highest profit. Unfortunately, the feasible region has infinitely many points, making it impossible to compute the profit for each point. Luckily, we can narrow our search because of the

corner point principle: The highest profit value on a polygonal feasible region can always be found at a corner point.

The corner point principle is probably the most important insight in the theory of linear programming. It is the geometric nature of this principle that explains the value of creating a geometric model from the data in a mixture chart.

The corner point principle gives us the following method to solve a mixture problem:

1. Calculate the corner points of the feasible region.

TABLE 4.2   **Evaluating Profit Formula at Corner Points**

Corner	Profit $(= 3x + 4y)$
$(0, 0)$	$3(0) + 4(0) = 0$
$(0, 50)$	$3(0) + 4(50) = 200$
$(50, 25)$	$3(50) + 4(25) = 250$
$(66.7, 0)$	$3(66.7) + 4(0) = 200$ (rounded)

2. Evaluate the profit at each corner point of the feasible region.

3. Choose the corner point with the highest profit as the production policy.

The corners of the feasible region for the juice mixture problem are shown in Figure 4.4f, and Table 4.2 shows the profit at each corner. The highest profit is 250 and occurs at the corner point (50, 25). Therefore, to achieve this profit, we should make 50 gallons of cranapple and 25 gallons of appleberry. (We leave it to the reader to verify that the highest profit in the plywood problem is 4800, and occurs at the corner point (300, 600). This point corresponds to the production policy of making 300 panels of exterior plywood and 600 panels of interior.)

This situation is reminiscent of Alexander the Great's approach to the problem of the Gordian knot, a legendary knot so large and tight and tangled that no one had been able to untie it. Alexander's solution was to slice the knot open with his sword. Mixture problems and other linear-programming problems are like Gordian knots because there are infinitely many feasible points — we can't calculate the profit for all of them. The corner point principle functions like Alexander's sword and cuts through the problem.

You can visualize a mathematical proof of the corner point principle by imagining that each point of the plane is a tiny light bulb that is capable of lighting up. For the juice mixture example, whose feasible region is shown in Figure 4.4f,

imagine what would happen if we ask this question: Will all points with profit = 360 please light up? What geometric figure do these lit-up points form?

In algebraic terms, we can restate the profit question in this way: Will all points (x, y) with $3x + 4y = 360$ please light up? As it happens, this version of the profit question is one mathematicians learned to answer hundreds of years before linear programming was born. The points that light up make a straight line. Furthermore, it is a routine matter to determine the exact position of the line. We call this line the **profit line** for 360; it is shown in Figure 4.7. For numbers other than 360, we would get different profit lines. Unfortunately, there are no points on the profit line for 360 that are feasible, that is, which lie in the feasible region. Therefore, the profit of 360 is unachievable.

If the profit line corresponding to a certain profit doesn't touch the feasible region, then that profit can't be achieved.

Because 360 is too big, perhaps we should ask the profit line for a more modest amount, say, 160,

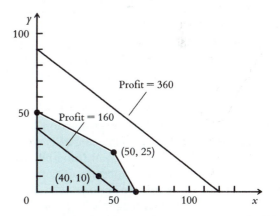

Figure 4.7 The profit line for 360 lies outside the feasible region, whereas the profit line for 160 passes through the region.

to light up. You can see that the new profit line of 160 in Figure 4.7 is parallel to the first profit line and closer to the origin. This is no accident: well-known basic theorems about the slope and y-intercept of straight lines prove it must be so. All profit lines for the profit formula $3x + 4y$ have the same coefficients for x and y, 3 for x and 4 for y, and since the slope is determined by those coefficients, they have the same slope. Changing the constant term from 360 to 160 to some other value has the effect of changing where the line intersects the y axis, but does not affect the slope.

The most important feature of the profit line for 160 is that it has points in common with the feasible region. For example, (40, 10) is on that profit line because $3(40) + 4(10) = 160$, and in addition (40, 10) is a feasible point. This means that it is possible to make 40 gallons of cranapple and 10 gallons of appleberry and that if we do so, we will have a profit of 160.

Can we do better than a 160 profit? As we slowly increase our desired profit from 160 toward 360, the location of the profit line that lights up shifts smoothly upward away from the origin. As long as the line continues to cross the feasible region, we are happy to see it move away from the origin, because the more it moves, the higher the profit represented by the line. We would like to stop the movement of the line at the last possible instant, while the line still has one or more points in common with the feasible region. It should be obvious that this will occur when the line is just touching the feasible region either at a corner point (Figure 4.8a) or along a line segment joining two corners (Figure 4.8b). That point or line segment corresponds to the production policy or policies with the maximum achievable profit. This is just what the corner point principle says: The maximum profit always occurs at a corner or along an edge of the feasible region.

Suppose now that each light bulb (point of the plane) has a color determined by the profit associated with that point. All points with the same profit (i.e., points on a profit line) have the same

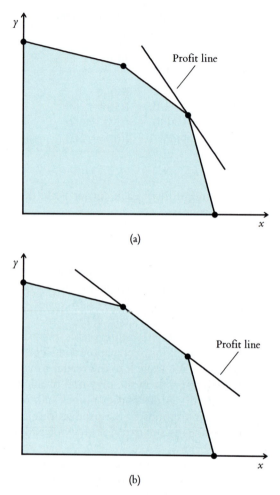

(a)

(b)

Figure 4.8 The highest profit will occur when the profit line is just touching the feasible region, either (a) at a corner point or (b) along a line segment.

color. Furthermore, suppose the colors range continuously from violet to blue to green to yellow to red, just as they do in a rainbow. The cool colors represent low profits, the hot colors, higher ones: the higher the profit, the hotter the color. In effect, we are superimposing a straightened-out rainbow on the picture containing our feasible region. Finding the highest profit point can be thought of as finding the hottest-colored point in the feasible region (Figure 4.9).

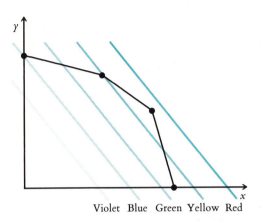

Violet Blue Green Yellow Red

Figure 4.9 If profit lines are colored according to the hues of the rainbow, the highest profit point will be the hottest-colored point in the feasible region.

▶ SUMMARY OF THE PICTORIAL METHOD

Here are the steps we used to solve our linear programming problem to find the optimal production policy in a mixture problem:

1. Read the problem carefully. Identify the resources and the products.

2. Make a mixture chart showing the resources, the products, the recipes for creating the products from the resources, and the profit from each product. Also show the amount of each resource on hand.

3. Assign an unknown quantity, x or y, to each product. Use the mixture chart to write down the resource constraints and the profit formula.

4. Graph the line corresponding to each resource constraint and determine which side of the line is in the feasible region. Sketch the feasible region by putting together, that is, intersecting, the half planes from all the resource constraints plus the first quadrant, which we get from the minimum constraints.

5. Find the coordinates of all the corner points of the feasible region. Some of these have been calculated in order to graph the individual lines.

6. Evaluate the profit formula for each of the corner points. The production policy that maximizes profit is the one that gives the biggest value to the profit formula.

It would be convenient if all linear programming problems yielded simple feasible regions in two-dimensional space. However, two complications arise for practical problems:

1. Sometimes, as in Figure 4.10, we have a great many corners. Naturally, the more corners there are, the more calculations we need to do to determine the coordinates of all of them and the profit at each one.

2. It is not possible to visualize the feasible region as a part of two-dimensional space when there are more than two products. Each product is represented by an unknown, and each unknown is represented by a dimension of space. If we had 50 products, we would need 50 dimensions and couldn't visualize the feasible region.

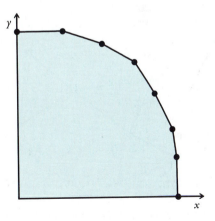

Figure 4.10 A feasible region with many corners.

Most practical linear programming problems present us with both cases, that is, many corners and more than two products. The number of corners literally can exceed the number of grains of sand on the earth. Even with the fastest computer, computing the profit of every corner is impossible.

▶ THE SIMPLEX METHOD

Several methods have been proposed to get around this difficulty. The oldest method is the **simplex method,** which is still the most commonly used. Devised by the American mathematician George Dantzig (see Spotlight 4.2, p. 128), this ingenious mathematical invention makes it possible to find the best corner by evaluating only a tiny fraction of all corners. Using the simplex method, a problem that might be impossible to solve if each corner had to be checked can be solved in a few minutes or even a few seconds on a typical business computer.

The operation of the simplex method may be likened to the behavior of an ant crawling on the edges of a polyhedron (a solid with flat sides), looking for one particular target vertex (Figure 4.11). The ant cannot see where the target vertex is. As a result, if it were to wander along the edges randomly, it might take a long time to reach the target. The ant will do much better if it has a temperature clue to let it know it is getting warmer (closer to the target vertex) or colder (farther from the target vertex).

Think of the simplex method as a way of calculating these temperature hints. We begin at any randomly chosen vertex. All neighboring vertices are evaluated to see which ones are warmer and which are colder. A new vertex is chosen from among the warmer ones, and the evaluation of neighbors is repeated—this time checking neighbors of the new vertex. The process ends when we arrive at the target vertex. (It turns out there are fairly simple algebraic manipulations to do this. The algebra for the simplex method is related to solving several equations, each having several variables. For big problems with many products and resources, there are many equations and variables, but the kind of algebra is the same as for smaller problems—just more of it.)

Part of what the simplex method has going for it in the speed sweepstakes is that it works faster in practice than its worst-case behavior would lead us to believe. Although mathematicians have devised artificial cases for which the simplex method bogs down in unacceptable amounts of arithmetic, the examples arising from real applications are never like that. This may be the world's most impressive counterexample to Murphy's law, which says that if something can go wrong, it will.

Although the simplex method usually avoids visiting every vertex, it may require visiting many intermediate vertices as it moves from the starting vertex to the optimal one. The simplex method has to search along edges on the boundary of the polyhedron. If it happens that there are a great many small edges lying between the starting point and the optimal vertex, the simplex method must operate like a slow-moving bus that stops at every street corner.

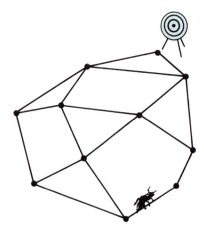

Figure 4.11 The simplex method can be compared to an ant crawling along the edges of a polyhedron, looking for one particular target vertex.

In the introduction to this chapter, we noted that linear programming accounts for over 50% and possibly as much as 90% of nonroutine computer time used for management decisions. Although there are alternatives to the simplex method, as we discuss in a later section, much of that computer time is spent using the simplex method.

Using a Simplex Method Program

There are many programs available that will use the simplex method to give you an optimal production policy if you just give the computer the very same data that we have in our mixture charts. Simplex method programs can be found in a variety of places, among which are spreadsheets and packages of mathematics programs designed for business applications or for finite mathematics courses.

Perhaps you have access to a simplex method program. If you want to use it, before you give the program the numbers in the mixture chart, you will have to tell the program some details about your particular problem.

Typically, a simplex method program asks for the number of resources and the number of products in the problem. These correspond to the number of columns and rows in our mixture chart, with the caution that we do not count the profit column in among the resource columns.

For example, in the plywood problem, we had this mixture chart in Figure 4.2 (note that we have simplified the chart somewhat, but retained the important information):

	Grade A 1000	Grade B 3000	Press time 6000	Profit
Exterior, x	2	2	10	6
Interior, y	0	4	5	5

This problem has 3 resources and 2 products.

Since a simplex method program is designed to solve all sorts of linear programming problems, not just mixture problems, it will use slightly different terminology than we have heard, and will offer you options that you do not need.

We have used the variables x and y for the quantities of the two products. Since typical applications have many products, a simplex method program uses subscripted variables such as x_1, x_2, x_3, x_4, and so on, instead of using consecutive alphabet letters. So the resource constraint for Grade B, which we wrote as $2x + 4y \leq 3000$, would be written as $2x_1 + 4x_2 \leq 3000$.

The "profit formula" is often referred to as the "objective function" by a simplex method program. Using the subscripted variables, the plywood problem has the objective function: $6x_1 + 5x_2$.

The simplex method program will ask you about the resource constraints: it needs to know the number of maximum resource constraints and the number of minimum resource constraints. Our problems have maximum resource constraints; we have a maximum amount of each resource that can be used. So you would respond that the number of maximum resource constraints is the same as the number of resources. For the plywood problem, there are 3 maximum constraints. Our problems did not have minimum *resource* constraints—we were not required to use a certain minimum amount of any resource. So in all our problems, there are 0 minimum resource constraints. We did have minimums, but those were to ensure that we did not reach a solution asking us to make a negative quantity of any *product*.

When you are entering the objective function and individual resource constraints into the computer, you probably will need to enter only the numbers, not the symbols ($+$ and $\leq$) and not the subscripted variable names. The numbers that multiply the variables are called *coefficients*. In the resource constraint $2x_1 + 4x_2 \leq 3000$, the 2 and the 4 are coefficients. The number that stands alone on the other side of the inequality symbol is called the *constant*. In the same resource constraint, the 3000 is the constant. Typically, you would enter the values 2, 4, and 3000 in exactly that order when you describe this resource constraint to the computer. Note that the objective function does not have a constant, so the program will not request one from you.

A simplex method program often gives you a choice of seeing just the solution or the algebraic steps as well. The solution will present the values for the subscripted variables at the optimal corner point. So for the plywood problem, the computer would present our solution of (300, 600) as $x_1 = 300$ and $x_2 = 600$. Some programs also give the maximum value of the objective function, in this case 4800. If a subscripted variable is not listed in the final solution, its value is 0, meaning that none of that particular product should be made.

SPOTLIGHT 4.2   Father of Linear Programming Recalls Its Origins

▶ ▶ ▶ ▶ ▶ ▶ ▶ ▶ ▶ ▶ ▶ ▶ ▶

George Dantzig is professor of operations research and computer science at Stanford University. He is credited with inventing the linear programming technique called the simplex method. Since its invention in the 1940s, the simplex method has provided solutions to linear programming problems that have saved both industry and the military time and money. Dantzig talks about the background of his famous technique:

> Initially, all the work we did had to do with military planning. During World War II, we were planning on a very extensive scale. The civilian population and the military were all performing scheduling and planning tasks, perhaps on a larger scale than at any time in history. And this was the case up until about 1950. From 1950 on, the whole emphasis shifted from military planning to practical planning for the civilian population, and industry picked it up.

The first areas of industry to use linear programming were the petroleum refineries. They used it for blending gasoline. Nowadays, they run all of the refineries in the world (except for one) using linear programming methods. They are one of the biggest users of it, and it's been picked up by every other industry you can think of — the forestry industry, the steel industry — you could fill up a book with all the different places it's used.

The question of why linear programming wasn't invented before World War II is an interesting one. In the postwar period, various technologies just evolved that had never been there before. Computers were one example. These technologies were talked about before; you can go back in history and you'll find isolated papers on them. For example, the very famous French mathematician Joseph Fourier, who invented Fourier series, had a paper on problems similar to linear programming, as did the Belgian mathe-

In a typical solution, other information is given which includes how much of each of the resources is left over after the optimal production policy is carried out. These leftovers are called *slacks* in the standard terminology and are often presented in the solution as values for subscripted variables, such as s_1, s_2, s_3, and s_4. In the plywood problem, the optimal production policy uses up all the time available on the presses and all the Grade B veneer, but leaves 400 panels of Grade A unused. Assuming that we had entered the resource constraint for the Grade A first, the leftover Grade A would correspond to the variable s_1, and the program would tell us that $s_1 = 400$. As with the variables representing products, any slack variable not mentioned in the solution has the value zero.

Earlier, we emphasized that the analysis needed to construct the mixture chart was really

matician de la Valee Pousson. But these were isolated cases that never went anywhere.

In the immediate postwar period, everything just fermented and began to happen, and one of the things that began to happen was linear programming. Mathematicians as well as economists and others who do practical planning and scheduling saw the possibilities. Things then began to happen very rapidly.

In the immediate postwar period, the whole idea of using computers to mechanize the process was an obvious one. The question was then asked, How could you formulate the process as a sort of mathematical system, and how could computers be used to make this happen?

The problems we solve nowadays have thousands of equations, sometimes a million variables. One of the things that still amazes me is to see a program run on the computer — and to see the answer come out. If we think of the number of combinations of different solutions that we're trying to choose the best of, it's akin to the stars in the heavens. Yet we solve them in a matter of moments. This, to me, is staggering. Not that we can solve them — but that we can solve them so rapidly and efficiently.

The simplex method has been used now for close to 40 years. There has been steady work going on trying to use different versions of the simplex method, nonlinear methods, and interior methods. It has been recognized that certain classes of problems can be solved much more rapidly by special algorithms than by using the simplex method. If I were to say what my field of specialty is, it is in looking at these different methods and seeing which are more promising than others.

For example, when the system is too big, we use something called the "decomposition principle" to break it into parts and solve the parts. This has been very efficient for certain classes of problems. Certain methods that have been reported in the news that are quite well known in nonlinear programming, called "interior techniques," have been experimented on very successfully with some kinds of problems, and not so successfully with others.

In my classes at Stanford, we review these different methods, and it's surprising how good some of them are. There's a lot of promise in this — there's always something new to be looked at.

the heart of solving a linear programming problem. Now you see that once you have a mixture chart, you can easily use it to find the data needed for a simplex method program. We hope you have access to such a program and try to use it. For this purpose, we have included some exercises (39 through 42) in this chapter that are larger and more realistic than the ones we have been solving graphically. A graphical solution is only possible for problems limited to two products; these special exercises involve more than two products.   ◄

► AN ALTERNATIVE TO THE SIMPLEX METHOD

In 1984, Narendra Karmarkar (see Figure 4.12), a mathematician working at Bell Laboratories, devised an alternative method for linear programming that avoids such slowdowns by making use of search routes through the interior of the feasible region (see Spotlight 4.3, p. 131). The potential applications of Karmarkar's algorithm are impor-

tant to a lot of industries, including telephone communications and the airlines (see Spotlight 4.4, p. 132). Routing millions of long-distance calls, for example, means deciding how to use the resources of long-distance landlines, repeater amplifiers, and satellite terminals to best advantage. The problem is similar to the juice company's need to find the best use of its stocks of juice to create the most profitable mix of products. We find another example in the airline business. American Airlines worked with Dr. Karmarkar to see if his algorithm could cut fuel costs. According to Thomas Cook, director of operations research for American Airlines, "It's big dollars. We're hoping we can solve harder problems faster, and we think there's definite potential."

In the 1980s, scientists at Bell Labs applied Karmarkar's algorithm to a problem of unprecedented complexity: deciding how to economically build telephone links between cities so that calls can get from any city to any other, possibly being relayed through intermediate cities. Figure 4.13 shows one such linking. The number of possible

Figure 4.12 Narendra Karmarkar, a researcher at AT&T Bell Laboratories, has invented a powerful new linear programming algorithm that solves many complex linear programming problems faster and more efficiently than any previous method. [Courtesy of AT&T Bell Laboratories.]

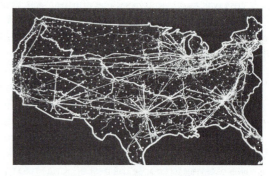

Figure 4.13 A map of the United States showing one conceivable network of major communication lines connecting major cities. Routing millions of calls over this immense network requires sophisticated linear programming techniques and high-speed computers. [Courtesy of AT&T Bell Laboratories.]

linkings is unimaginably large, so picking the most economical one is difficult. For any given linking that one has built, or contemplates building, there is also the problem of deciding how to economically route calls through the network to reach their destinations.

Although difficult, these problems are definitely worth solving. Nat Levine, director of the transmission facilities planning center at Bell Labs, speculated that if one found the best solu-

tion, "the savings could be in the hundreds of millions of dollars."

Work on these problems at Bell Labs involved a linear programming problem with about 800,000 variables, which Karmarkar's algorithm solved in 10 hours of computer time. Scientists involved believe that the problem might have taken weeks to solve if the simplex method had been used.

Karmarkar's algorithm has been incorporated into a software product. The extensive data col-

---

SP**TLIGHT** 4.3   Linear Programming and the Cold War

In 1979, mathematics found itself on the front page of the *New York Times* as a result of a linear programming technique invented by the Russian mathematician L. G. Khachian. This algorithm, called the ellipsoid method, seemed at first to be much better than the simplex method for large problems. Perhaps because so much is at stake in this area of applied mathematics, the story spread like wildfire, with little regard for accuracy. In the popular press it appeared as though the Russians had achieved a breakthrough that might imperil our national security. In particular, it was suggested that the method might be applied to break cryptographic codes that had previously been considered secure.

As it turned out, there was no real reason for alarm. The ellipsoid method doesn't apply to the variant of linear programming that occurs in cryptography. Furthermore, the kinds

of linear programming problems for which it does work better than the simplex method are those that are far larger than any that will ever occur in practice. Best of all, following a tradition of international communication in the mathematics community that dates back thousands of years, the Russian mathematicians made no attempt to keep the ellipsoid algorithm a secret.

In fact, the impact of the ellipsoid algorithm on American mathematics was undoubtedly positive. Although it does not seem to be a practical algorithm, it did break a certain theoretical barrier, thus creating an incentive for other researchers. Narendra Karmarkar, an American mathematician working at Bell Laboratories, broke through this barrier in another way, finding what seems to be a truly practical method for linear programming.

SPOTLIGHT 4.4   Finding Fast Algorithms Means Better Airline Service

▶ ▶ ▶ ▶ ▶ ▶ ▶ ▶ ▶ ▶ ▶ ▶ ▶

Linear programming techniques have a direct impact on the efficiency and profitability of major airlines. Thomas Cook, director of operations research at American Airlines, was interviewed in 1985 concerning his ideas on why optimal solutions are essential to his business:

Finding an optimal solution means finding the best solution. Let's say you are trying to minimize a cost function of some kind. For example, we may want to minimize the excess costs related to scheduling crews, hotels, and other costs that are not associated with flight time. So we try to minimize that excess cost, subject to a lot of constraints, such as the amount of time a pilot can fly, how much rest time is needed, and so forth.

An optimal solution, then, is either a minimum-cost solution or a maximizing solution. For example, we might want to maximize the profit associated with assigning aircrafts to the schedule; so we assign large aircraft to high-need segments and small aircraft to low-load segments. Whether it's a minimum or maximum solution depends on what function we are trying to optimize.

Finding fast solutions to linear programming problems is also essential. If we can get an algorithm that's 50 to 100 times faster, we could do a lot of things that we can't do today. For example, some applications could be real-time applications, as opposed to batch applications. So instead of running a job overnight and getting an answer the next morning, we could actually key

in the data or access the data base, generate the matrix, and come up with a solution that could be implemented a few minutes after keying in the data.

A good example of this kind of application is what we call a major weather disruption. If we get a major weather disruption at one of the hubs, such as Dallas or Chicago, then a lot of flights may get canceled, which means we have a lot of crews and airplanes in the wrong places. What we need is a way to put that whole operation back together again so that the crews and airplanes are in the right places. That way, we minimize the cost of the disruption and minimize passenger inconvenience.

We're working on that problem today, but in order to solve it in an optimal fashion, we need something as fast as Narendra Karmarkar's algorithm. In the absence of that, we'll have to come up with some heuristic ways of solving it that won't be optimal.

The simplex method, which was developed some 40 years ago by George Dantzig, has been very useful at American Airlines and, indeed, at a lot of large businesses. The difference between his solution and Karmarkar's is that if we can get an algorithm that comes up with basically the same optimal answer 50 to 100 times faster, then we can apply that technology to new problems, and even to problems that we wouldn't have tried using the simplex method. I think that's the primary reason for the excitement.

lected from worldwide use of the software support the results at Bell Labs. It appears that for some kinds of linear programming problems, Karmarkar's algorithm is a big improvement over the simplex method.

▶REVIEW VOCABULARY

Corner point principle   The principle that states that there is a corner point of the feasible region that yields the optimal solution.

Feasible point   A possible solution (but not necessarily the best) to a linear programming problem. With just two products, we can think of a feasible point as a point on the plane.

Feasible region   A representation of the feasible set as a portion of n-dimensional space. For problems with just two products, the feasible region is a part of the plane.

Feasible set   The set of all feasible points, that is, possible solutions to a linear programming problem.

Linear programming   A set of organized methods of management science used to solve problems of finding optimal solutions, while at the same time respecting certain important constraints. The mathematical formulation of the constraints in linear programming problems are *linear* equations and inequalities. Mixture problems are usually solved by some type of linear programming.

Minimum constraint   An inequality in a mixture problem that reflects the fact that negative quantities of products cannot be produced.

Mixture problem   A problem in which a variety of resources available in limited quantities can be combined in different ways to make different products. It is usually desired to find the way of combining the resources that produces the most profit.

Production policy   A point in the feasible set, interpreted as specifying how many units of each product are to be made.

Profit formula   The expression, involving the unknown quantities such as x and y, that tells how much profit results from a particular production policy.

Profit line   The set of all feasible points that yields the same profit (for two-dimensional linear programming problems).

Resource constraint   An inequality in a mixture problem that reflects the fact that no more of a resource can be used than what is available.

Simplex method   One of a number of algorithms for solving linear programming problems.

▶SUGGESTED READINGS

ANDERSON, DAVID R., DENNIS J. SWEENEY, AND THOMAS A. WILLIAMS: *An Introduction to Management Science: Quantitative Approaches to Decision Making,* West, St. Paul, Minn., 1985. A business management text with seven chapters covering many applications and aspects of linear programming.

GASS, SAUL I.: *An Illustrated Guide to Linear Programming,* McGraw-Hill, New York, 1970. An engagingly written beginner's approach, which introduces the algebra gently and emphasizes the formulation of problems more than algebraic technique.

GLICKSMAN, ABRAHAM: *Linear Programming and the Theory of Games,* Wiley, New York, 1963. The best elementary approach to a full understanding of the simplex method, including numerous numerical examples.

MALKEVITCH, JOSEPH, AND WALTER MEYER: *Graphs, Models, and Finite Mathematics,* Prentice-Hall, Englewood Cliffs, N.J., 1974. This book has a little more emphasis on "how to" than is given in the present book.

MEYER, WALTER: *Concepts of Mathematical Modeling*, McGraw-Hill, New York, 1984. Chapter 4 has discussion of several other types of linear programming problems, including minimization problems, and the "transportation problem," for which there is a special algorithm, and linear programming problems in which the corner points must have coordinates that are whole numbers, not fractions.

ROLF, HOWARD L., AND GARETH WILLIAMS: *Finite Mathematics*, W. C. Brown, Dubuque, Iowa, 1988. Chapter 3 presents the graphical solution of linear programming problems. Chapter 4 presents the algebra of simplex method, and in section 4.3, the relationship of the steps in that algebra to the realities of the problem are nicely explained.

▶ EXERCISES

Note: Exercises 1 to 8 provide practice in graphing and finding points of intersection. They are here for those students who may need practice in these skills.

1. Graph each of these lines:
   a. $2x + 3y = 12$
   b. $3x + 5y = 30$
   c. $4x + 3y = 24$
   d. $7x + 4y = 42$

2. Graph each of these lines:
   a. $5x + 4y = 20$
   b. $7x + 6y = 84$
   c. $4x + 5y = 60$
   d. $6x + 5y = 15$

In Exercises 3 to 8, graph both lines on the same graph, showing by a dot any intersection of the lines. Use algebra to determine the x- and y-coordinates of any point of intersection. For some pairs of lines, there is no point of intersection; what do you notice about the coefficients of the equations of those lines?

3. $5x + 4y = 22$ and $2x + 4y = 16$
4. $x + 2y = 10$ and $5x + y = 14$
5. $3x + 4y = 18$ and $3x + 4y = 12$
6. $2x + 2y = 14$ and $3x + 4y = 24$
7. $3x + 5y = 30$ and $6x + 15y = 75$
8. $5x + 2y = 20$ and $5x + 2y = 30$

In Exercises 9 to 12, graph the half plane corresponding to the inequality.

9. $5x + 3y \leq 15$
10. $3x + 2y \leq 18$
11. $4x + 5y \leq 30$
12. $7x + 2y \leq 42$

In Exercises 13 to 16, graph the feasible region corresponding to the set of inequalities. Label each line segment bounding the region with the appropriate inequality and give the coordinates of every corner point.

13. $x \geq 0; y \geq 0; 3x + y \leq 9; x + 2y \leq 8$
14. $x \geq 0; y \geq 0; 2x + y \leq 4; 3x + 3y \leq 9$
15. $x \geq 0; y \geq 0; x + 2y \leq 10; 5x + y \leq 14; 4x + 5y \leq 30$
16. $x \geq 0; y \geq 0; 2x + y \leq 8; x + y \leq 5; x + 2y \leq 8$

In Exercises 17 to 20, for each of the following points, determine whether it is a point of the feasible region given: **a.** $(5, 5)$; **b.** $(2, 1)$; **c.** $(1, 3)$; **d.** $(-1, 4)$; **e.** $(2, 4)$.

17. The feasible region of Exercise 13
18. The feasible region of Exercise 14
19. The feasible region of Exercise 15
20. The feasible region of Exercise 16

21. For each of the following points, all feasible for the juice mixture example from the text whose feasible region is graphed in Figure 4.4f, compute the profit using the formula Profit $= 3x + 4y$: $(20, 25)$; $(0, 50)$; and $(30, 10)$.

22. For each of the following points, all feasible for the plywood example from the text whose feasible region is graphed in Figure 4.5f, compute the profit using the formula Profit $= 6x + 5y$: $(200, 250)$; $(0, 350)$; and $(350, 100)$.

In Exercises 23 to 26, for each description, write an appropriate resource-constraint inequality. The unknown to use for each product is given in parentheses.

23. One cake (x) requires 4 cups of flour, and one pie (y) requires 2 cups. There are 28 cups of flour available.

24. Mowing one lawn (x) takes 2 hours of gardening time, and weeding one garden (y) takes 1 hour. There are 40 hours of gardening time available.

25. Manufacturing one package of hot dogs (x) requires 6 ounces of beef, and manufacturing one package of bologna (y) requires 4 ounces of beef. There are 240 ounces of beef available.

26. It takes 30 feet of 12-inch board to make one bookcase (x); it takes 72 feet of 12-inch board to make one table (y). There are 420 feet of 12-inch board available.

27. In the juice mixture problem, x represents gallons of cranapple, y, gallons of appleberry; the feasible region is presented in Figure 4.4f.
 a. What production policy is represented by each corner point? In other words, how many gallons of each juice mixture does each corner point represent?
 b. Which corner point, that is, production policy, represents the maximum profit for each of these profit formulas?

Profit $= 3x + 4y$ (the formula in the text)

Profit $= 2x + 5y$

Profit $= 5x + 3y$

28. For the plywood problem, x represents sheets of exterior plywood, y, sheets of interior plywood; the feasible region is presented in Figure 4.5f.

    a. What production policy is represented by each corner point? In other words, how many panels of each type does each corner point represent?

    b. Which corner point, that is, production policy, represents the maximum profit for each of these profit formulas?

$$\text{Profit} = 6x + 4y \text{ (the formula in the text)}$$

$$\text{Profit} = 3x + 7y$$

$$\text{Profit} = 8x + 3y$$

Exercises 29 to 38 each have several steps. Each entire exercise requires a complete pictorial solution to a mixture problem. Practice in a specific step of the pictorial solution algorithm can be obtained by working out just that step for several problems. The steps are

(a) Make a mixture chart for the problem.

(b) Using the mixture chart, write the resource- and minimum-constraint inequalities. Also write the profit formula.

(c) Draw the feasible region for those constraints and find the coordinates of the corner points.

(d) Evaluate the profit information at the corner points to determine the production policy that best answers the question.

● (e) Compare your answer with the one you get from running the same problem on a simplex algorithm computer program.

29. A refinery mixes high-octane and low-octane fuels to produce regular and premium gasoline. The profits per gallon on the two gasolines are $0.30 and $0.40, respectively. One gallon of premium gasoline is produced by mixing 0.5 gallon of each of the fuels. One gallon of regular gasoline is produced by mixing 0.25 gallon of high octane with 0.75 gallon of low octane. If there are 500 gallons of high octane and 600 gallons of low octane available, how many gallons of each gasoline should the refinery make?

30. A paper recycling company uses scrap paper and scrap cloth to make two different grades of recycled paper. A single batch of grade A recycled paper is made from 40 pounds of scrap cloth and 180 pounds of scrap paper, whereas one batch of grade B recycled paper is made from 10 pounds of scrap cloth and 150 pounds of scrap paper. The company has 100 pounds of scrap cloth and 660 pounds of scrap paper on hand. A batch of grade A paper brings a profit of $500, whereas a batch of grade B paper brings a profit of $250. What amounts of each grade should be made?

● Optional exercise.

31. A bakery produces bread and cake. To produce a cake requires 1 hour of oven time and 2 hours of preparation/decoration time. To produce a loaf of bread requires 1.5 hours of oven time and 1 hour of preparation/decoration time. In any one day there are 12 hours of oven time and 16 hours of preparation/decoration time available. Since the bakery makes only $0.50 per loaf of bread, but clears a profit of $2.50 per cake, should it produce just cakes? What should its production policy be?

32. A car maintenance shop must decide how many oil changes and how many tune-ups can be scheduled in a typical week. The oil change requires 10 minutes of junior mechanics' time plus 5 minutes of senior mechanics' time. The tune-up requires 15 minutes of junior mechanics' time plus 25 minutes of senior mechanics' time. The maintenance shop makes a profit of $20 on an oil change and $30 on a tune-up. What mix of services should the shop schedule if the typical week has available 4000 minutes of junior mechanics' time and 2000 minutes of senior mechanics' time?

33. A manufacturer of hot dogs uses three ingredients, of which these amounts are in stock: beef, 500 pounds; pork, 300 pounds; and grain filler, 400 pounds. The recipe for all-beef hot dogs calls for just beef, 1 pound per package. The recipe for regular hot dogs calls for 0.5 pound of pork and 0.25 pound each of beef and grain filler. If the profit on regular hot dogs is $0.70 and on all-beef hot dogs is $0.80, how many packages of each should be made?

34. An appliance store has 90 square feet for displaying refrigerators and stoves, and can afford $2000 to lease appliances for the display. Suppose that each refrigerator requires 10 square feet of floor and costs $400 to lease. (For simplicity, we ignore the differences that might occur among different models.) Suppose that each stove requires 15 square feet of floor and costs $200 to lease. The store has found that it clears a daily profit of $30 for each refrigerator on display and $20 for each stove. How many of each of these appliances should the store display?

35. In a certain medical office, a routine office visit requires 10 minutes of nurses' time, 5 minutes of doctors' time, and 5 minutes of lab time. The comprehensive office visit requires 5 minutes of nurses' time, 25 minutes of doctors' time, and 10 minutes of lab time. In a typical week, there are 6250 minutes of nurses' time, 11,000 minutes of doctors' time, and 5000 minutes of lab time available. If the medical office clears $30 from a routine visit and $50 from a comprehensive visit, how many of each should be scheduled per week?

36. A bottler uses three pure juices—pineapple, orange, and grapefruit—to make two juice mixtures, pineapple-orange and pineapple-grapefruit, sold in 1-quart bottles. The profit is $0.50 per bottle of pineapple-orange and $0.40 per bottle of pineapple-grapefruit. Each mixture is made by using equal proportions of the two juices in its name. The amounts of juice on hand are 100 quarts of

pineapple juice, 70 quarts of grapefruit juice, and 40 quarts of orange juice. How many of each juice mixture should be produced?

▲ 37. Wild Things raises pheasants and partridges to restock the woodlands and has room to raise 100 birds during the season. The cost of raising one bird is $20 per pheasant and $30 per partridge. The Wildlife Foundation pays Wild Things for the birds; they clear a profit of $14 per pheasant and $16 per partridge. Wild Things has $2400 available to cover costs. How many of each type of bird should they raise?

38. A student has decided that a prospective employer looking at a transcript will value passing a tough course twice as much as passing an easy course. (Assume that the student and the prospective employer rate the course difficulty in the same way.) The student estimates that to pass a typical tough course $75 will be needed in texts and supplies and 150 hours will be needed to study and do homework. The student estimates that the typical easy course demands $50 and 50 hours. This year the student has available 600 hours and $450. How many of each kind of course should the student take? (Hint: The profit could be viewed as 2 "toughness points" for passing a tough course and 1 "toughness point" for passing an easy course.) Would the optimal point be the same if a tough course were valued at three times as much as an easy one?

In Exercises 39 to 42, there are more than two products in the problem, so you cannot solve these problems using the pictorial technique. You can do these steps:

▲ (a) Make a mixture chart for each problem.

▲ (b) Using the mixture chart, write the resource- and minimum-constraint inequalities. Also write the profit formula.

● (c) If you have a simplex method program available, run the program to obtain the optimal production policy.

39. A toy company makes three types of toy, each of which must be processed by three machines: a shaper, a smoother, and a painter. Each Toy A requires 1 hour in the shaper, 2 hours in the smoother, and 1 hour in the painter, and brings in a $4 profit. Each Toy B requires 2 hours in the shaper, 1 hour in the smoother, and 3 hours in the painter, and brings in a $5 profit. Each Toy C requires 3 hours in the shaper, 2 hours in the smoother, and 1 hour in the painter, and brings in a $9 profit. The shaper can work at most 50 hours per week, the smoother 40 hours, and the painter at most 60 hours. What production policy should the toy company follow in order to maximize profit?

40. A rustic furniture company handcrafts chairs, tables, and beds. It has three workers, Chris, Sue, and Juan. Chris can only work 80 hours per month, but Sue and Juan can each put in 200 hours. Each of these artisans has special skills. To make a chair takes 1 hour of Chris's time, 3 hours from Sue, and 2

▲ Advanced exercise.   ● Optional exercise.

from Juan. A table needs 3 hours from Chris, 5 from Sue, and 4 from Juan. A bed requires 5 hours from Chris, 4 from Sue, and 8 from Juan. Even artisans are concerned about maximizing their profit, so what product mix should they stick with if they get $100 profit per chair, $250 per table, and $350 per bed?

41. Taper pins are processed in a certain factory using a lathe, a grinder, and a polisher. These machines are available, respectively, 50, 36, and 81 hours per week. There are four types of taper pins, identified as #1, #2, #6, and #8. Each batch of Pin #1 requires 10 hours of lathe time, 6 hours of grinder time, and 4.5 hours of polisher time, and brings in a profit of $9. Each batch of Pin #2 requires 5 hours of lathe time, 6 hours of grinder time, and 18 hours of polisher time, and brings in a profit of $7. Each batch of Pin #6 requires 2 hours of lathe time, 2 hours of grinder time, and 1.5 hours of polisher time, and brings in a profit of $6. Each batch of Pin #8 requires 1 hour of lathe time, 2 hours of grinder time, and 6 hours of polisher time, and brings in a profit of $4. What production policy produces the maximum profit from batches of taper pins?

42. A gourmet coffee distributor has on hand 17,600 ounces of African coffee, 21,120 ounces of Brazilian coffee, and 12,320 ounces of Colombian coffee. It sells four blends: Excellent, Southern, World, and Special on which it makes these per-pound profits, respectively: $1.80, $1.40, $1.20, and $1.00. One pound of Excellent is 16 ounces of Colombian; it is not a blend at all. One pound of Southern consists of 12 ounces of Brazilian and 4 ounces of Colombian. One pound of World requires 6 ounces of African, 8 of Brazilian, and 2 of Colombian. One pound of Special is made up of 10 ounces of African and 6 ounces of Brazilian. What product mix should the gourmet coffee distributor prepare in order to maximize profit?

▲ 43. How many variables and resource constraints would be associated with a mixture chart for a firm that uses six resources to make four products?

▲ 44. In our colored-light-bulb visualization (see Figure 4.9), we superimposed a straightened-out rainbow on the feasible region. Suppose it were not straightened out, but were curved as in the figure below, and we still wanted to find the hottest-colored point. Would the corner point principle still hold? (This is an example of nonlinear programming.)

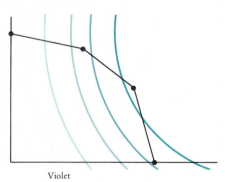

Violet

■ 45. Explain why finding a point of the feasible region which gives the maximum profit would be time-consuming and nearly impossible if we did not have the corner point principle.

● 46. Which steps of the pictorial method are also required of a person who is solving a linear programming problem by using a simplex method computer program?

Exercises 47 to 51 are designed to indicate areas in which the underlying assumptions of the linear programming model may not always fit the reality of a specific situation. These exercises are meant to stimulate discussion and do not have "right answers." Exercise 47 gives a concrete situation that could be used in the other questions.

■ 47. You are in charge of a business that produces sandwiches for snack bars. Make a list of your products (kinds of sandwiches), the resources you would need (sandwich ingredients), and the profit you might expect to get for each product. You need not specify the recipes numerically or the amounts of the resources available.

■ 48. Discuss the validity of the corner point principle if the solution (x, y) to the mixture problem is required not only to lie in the feasible region but also to have integer coordinates. In the sandwich problem (Exercise 47), would there be a useful meaning to a fractional number of some kind of sandwich?

■ 49. In a mixture problem we view recipes as fixed, but in practice this is not always true. For example, bolognas having only slightly different percentages of beef and pork all taste the same to the customers. What might prompt the manufacturer to vary a recipe? What effect might the varying have on the profit function, the feasible region, and the optimal product mix?

■ 50. In the real world, firms often give discounts for large-volume purchases. Does this necessarily contradict the assumption of a fixed (constant) profit on each unit sold? In examining this situation, you should note that discounts for large-volume purchases might not only apply to the products sold by a manufacturer but also to the prices paid by the manufacturer for resources.

■ 51. We learn in economics that prices are determined by the interplay of supply and demand. For example, the price of a product may fall if a large quantity of it is available. In mixture problems, however, we assume a fixed (constant) profit regardless of how much is produced. Is there a contradiction here? Could the model be adjusted to incorporate this economic fact of life?

▶ WRITING PROJECTS

1. Interview a local businessperson who is in charge of deciding the product mix for that business. Must this business take into consideration situations other than minimum and resource constraints? If so, what are these considerations? Find

■ Discussion exercise.

out what methods the person uses to make production policy decisions. Is linear programming used? Are other methods used? If so, what are they? Write a report of your findings, and add some of your own conclusions about the usefulness of linear programming for this business.

2. In economics, it is often useful to distinguish between a firm that has a monopoly (for example, is the only supplier of a product) and firms that supply only a small share of the market. How would the presence of a monopoly affect the relation between production and price? Would the presence of a monopoly tend to ensure the fixed-profit assumption of linear programming, or would it make it more likely that the interplay of supply and demand would have to be considered in order to have a truly realistic model?

Part II

Statistics: The Science of Data

Numerical facts, or data, make up an increasing part of the information we need in order to understand our world. Business executives base decisions on data about the national economy, financial markets, and their firm's own costs, sales, and profits. Engineers gather data on the performance, quality, and reliability of their products. The medical professions watch data on costs as well as data from medical research. Advertisers fine-tune their messages in response to market research data, and politicians use polls of public opinion to shape their campaigns. Citizens and consumers are surrounded by data from all these and other sources. We must be able to understand and communicate with data just as we understand and communicate with words.

The information conveyed by data may be as vital as the fact that 7.2% of the American labor force is out of work, or as trivial as the fact that 57% of American adults think they look younger than their true age. But whether numerical facts are vital or trivial, we must understand where they come from and whether the information they convey is trustworthy. *Statistics* is the science of data — of producing data, of putting them into clear and usable form, and of interpreting them to draw conclusions about the world around us. Just as literacy enables us to use and understand words, a basic knowledge of statistics allows us to use data honestly and skillfully.

◀ ◀ ◀

Olga Markova, women's winner of the 1993 Boston Marathon, crossing the finish line. (© 1993 Rick Friedman/Black Star.)

5

Producing Data

The news media often present information in the form of numbers. Headlines announce that the unemployment rate has dropped to 7.2%. The Gallup poll claims that 45% of Americans are afraid to go out at night because of crime. Where do these numbers come from? Most people aren't personally interviewed to determine their employment status. And the Gallup poll asks only a few of us if fear of being mugged keeps us indoors at night.

The unemployment rate and other data about the labor force are estimated by the Bureau of Labor Statistics from data gathered by the Current Population Survey. The Census Bureau conducts this survey, interviewing 60,000 households each month. Considering that the country has over 95 million households, it seems remarkable that data about such a small group—less than 1 in 1000 households—could represent the unemployment rate of the whole nation. How can information collected from a small number of people justify conclusions about a much larger group?

Even when numbers are not in the headlines, they often affect our lives. When a popular anti-

arthritis drug is recalled because of doubts about its safety, or when the Food and Drug Administration approves a new medication for use in treating AIDS, these decisions are based on clinical trials using relatively few patients. Public confidence about the safety and effectiveness of new drugs relies on statistical conclusions, or *inferences*, drawn from clinical data. Once again, we must draw trustworthy conclusions from relatively little data.

Any use of numbers as evidence depends on the proper production of data. The straightforward way to calculate the nation's unemployment rate would be to interview each and every person living in the United States. The cost of such a large-scale operation would be enormous, however, and the effort required is obviously impractical. Instead, we must rely on information about only a part of the population—in statistical language, a *sample*.

We often draw conclusions about a whole on the basis of a sample. Everyone has sipped a spoonful of soup and judged the entire bowl on the basis of that taste. But a bowl of soup is homogeneous; the taste of a single spoonful represents the whole. Other kinds of observations, such as fear of crime or reaction to a drug, are not so easily gauged. They vary from person to person or from region to region.

Dealing with this kind of variability is the central task of statistics. The first step is to produce data that represent a large group of individuals. To do this, we first state carefully which group we want information about. Statisticians call this group the **population.**

For example, if we want to measure unemployment, we must first define the population we wish to describe. Which age groups will we include? Will we include illegal aliens or people in prisons? What about full-time students? The Bureau of Labor Statistics must answer these questions in order to collect monthly unemployment information (see Spotlight 5.1). In fact, the Current Population Survey defines its population as all U.S. residents (whether citizens or not) 16 years of age and over who are civilians and are not institutionalized. The civilian unemployment rate published in newspapers and in other reports always refers to this specific population.

The clinical testing of a new drug for treating ulcers would also have a particular population in mind. This population could consist of all individuals who have ulcers, or it might be limited to patients with specific types of ulcers or to those in a specific age group. The detailed plan of the clinical trial must specify exactly what eligible population will be studied.

►SAMPLING

A population need not consist of people; it may instead consist of animals or objects. Let's say we want to test a population of newly produced fuses to see if they blow under excessive current, as they should. In this case, gathering information about every member of the population would destroy the entire population. In other cases, such as taking inventory of all items in a large warehouse, boredom and fatigue could prevent an accurate accounting. And the cost of producing data often requires that we not examine the entire population—not even the U.S. government can afford to interview every individual every month to measure unemployment. For all these reasons, we usually gather information about only a few items selected from the population. The part of the population used to draw conclusions about the whole is called a **sample.**

How can we choose a sample that is truly representative of the population? The easiest—but not the best—way to select a sample is to choose individuals close at hand. If we are interested in finding out how many people have jobs, for example, we might go to a shopping mall and ask people passing by if they are employed. A sample selected by taking the members of the population that are easiest to reach is called a *convenience sample*.

Convenience samples often produce unrepresentative data. People at shopping malls tend to be more prosperous than typical Americans; they are also more likely to be teenagers or retired. And when we decide which people to question, we will tend to choose well-dressed, middle-income subjects, and we will tend to avoid poorly dressed, unfriendly, or tough-looking individuals. In short, our shopping mall interviews will not contact a sample that is representative of the entire population, and so will not accurately reflect the nation's rate of unemployment.

Whenever we allow our own convenience to choose a sample, we will find ourselves—whether consciously or unconsciously—sampling individuals on the basis of friendliness, appearance, or income level, thus leaving some parts of the population underrepresented. Our shopping mall sample, for example, will probably overrepresent middle-class and retired people and underrepresent workers and the poor. This misrepresentation will happen every time we take such a sample. That is, it is a systematic error due to a bad sampling method, not just bad luck on one sample. Such a systematic difference between the results obtained

SP⬤TLIGHT 5.1   The Government's Statistician

▶ ▶ ▶ ▶ ▶ ▶ ▶ ▶ ▶ ▶ ▶ ▶ ▶

The commissioner of labor statistics is one of the nation's most influential statisticians. As head of the Bureau of Labor Statistics, the commissioner supervises the collection and interpretation of data on employment, earnings, and many other economic and social trends. These data have a large impact on the U.S. economy. For example, retail price data collected by the bureau are used to adjust many federal and private payments, including social security payments and union wage scales, for the effect of inflation.

The data collected by the Bureau of Labor Statistics are often politically sensitive, as when a report released just before an election shows rising unemployment. For this reason, the bureau must remain objective and independent of political influence. To safeguard the bureau's independence, the commissioner is appointed by the president and confirmed by the Senate for a fixed term of four years. The commissioner must have statistical skill, administrative ability, and a facility for working with both Congress and the president.

Dr. Janet Norwood served three terms as commissioner, from 1979 to 1991, under three

Dr. Janet Norwood, U.S. Commissioner of Labor Statistics, 1979–1991.

presidents. When she retired, the *New York Times* said (December 31, 1991) that she left with "a near-legendary reputation for nonpartisanship and plaudits that include one senator's designation of her as a 'national treasure.'" Norwood says, "There have been times in the past when commissioners have been in open disagreement with the Secretary of Labor or, in some cases, with the President. We have guarded our professionalism with great care."

by sampling and the truth about the whole population is called **bias.** To produce accurate data, we must take specific steps to eliminate bias. In particular, statisticians go to great lengths to eliminate the role of personal choice in the selection of the sample to be measured.

EXAMPLE: Call-in Polls

Television makes heavy use of call-in polls, in which viewers are invited to register their opinions by telephone. Some television stations poll the public daily, asking a question on the

6 o'clock news and reporting the responses on the 11 o'clock news. Viewers are urged to call special 900-prefix telephone numbers with their responses. Dialing one number indicates a "yes" reply; dialing the other, "no." The system works so that talking is not necessary. Completing the telephone call registers a viewer's answer. ▲

Call-in polls have several sources of bias. Households without telephones are automatically excluded. (Although about 94% of U.S. households have telephones, about one-quarter of those in Alaska and one-fifth of those in Mississippi do not.) Because dialing the 900-prefix number incurs a small charge, people in lower income households may be reluctant to pay a fee in order to telephone the station.

The most serious source of bias in call-in surveys is **voluntary response** — that is, the respondents select themselves. Only those who go to the time, trouble, and expense of calling are counted. Any sample chosen by voluntary response draws people with strong feelings, most often negative feelings. So when the newscaster asks the audience if they are afraid to go out at night because of crime, people angry about crime are more likely to call in than those who are not. Voluntary response is a common and serious source of bias in polls.

It is also possible to manipulate the call-in poll. If crime is a local issue with political impact, one political party could arrange for its workers to spend the evening dialing in. Or, because talking isn't needed for registering a response, a computer could be programmed to dial the 900-prefix number repeatedly.

Personal choice is a common source of bias in sampling. Whether it is expressed in the way an interviewer or study team selects its subjects or in the respondents' voluntary response, personal choice can distort the results. To reduce bias, it is vital to reduce the influence of personal choice. The way statisticians eliminate personal choice is to select the sample by *chance*.

▶ RANDOM SAMPLING

Imagine that we have a glass box containing thousands of beads. The beads are all identical except that most of them are light and some of them are dark. The beads form a population. Perhaps the beads represent the American labor force and dark beads represent those who are unemployed (see Figure 5.1).

Our task is to estimate what percentage of the population of beads is dark without examining each bead individually. Suppose that we thoroughly mix the beads in the box and then draw out a sample, using a scoop with 50 recesses in it. Each bead then has the same chance as every other bead of being selected, and each group of 50 beads also has the same chance as any other group of 50 to be selected. This is a simple random sample of size 50.

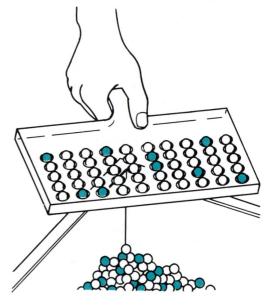

Figure 5.1 A random-sampling demonstrator.

A **simple random sample** of size n is a sample drawn in such a way that every possible sample of n members of the population has the same chance of being in the sample actually chosen.

Because every bead has an equal chance of being chosen, the simple random sample has eliminated bias in selecting the sample. In a survey of employment, a simple random sample gives everyone—rich or poor, male or female, employed or unemployed—an equal chance of being selected.

Suppose we draw a sample from our box of beads and get 12 dark beads. Because 12 out of 50 is the fraction $\frac{12}{50}$ or 0.24, 24% of this sample are dark. Results of simple random sampling are free of bias, so we can use the sample percentage to estimate the truth about the entire population. If 24% of the beads in the sample are dark, we estimate that 24% of the beads in the box are dark. Here is an important statistical technique: *to estimate a characteristic of a population, take a simple random sample and use the sample characteristic as the estimate.*

In principle, we now know how to estimate the nation's rate of unemployment. Put the names of all workers in a hat, mix the names well, and draw out 60,000 names. Because this is a simple random sample, the percent unemployed among those 60,000 persons is an unbiased estimate of the percent unemployed in the entire labor force.

In practice, when a population is too large to fit into a hat, we need an easier way to select a simple random sample. For example, each member of the population can be tagged with a numerical label, and then a sample can be chosen from the set of numerical labels in a way that gives every label an equal chance of being selected. If the population is relatively small, tagging and drawing samples can be done easily with a table of random digits such as Table 5.1. For larger populations, a computer can be programmed to generate random numbers and then to print out the sample chosen.

A **table of random digits** is a string of the digits 0, 1, 2, 3, 4, 5, 6, 7, 8, and 9 chosen in such a way that each entry is equally likely to be any of the 10 possibilities, and each entry is independent of all the other entries.

You can think of writing each of the 10 digits on a tag, putting the tags in a hat, mixing thoroughly, and drawing one. That's the first entry in the table. Put back the tag you drew, mix again, and draw another. That's the second entry. Continue drawing and mixing for hours and you will have a table of random digits like Table 5.1. The digits in Table 5.1 appear in groups of 5 to make the table easier to read and the rows are numbered so we can refer to them, but the groups and row numbers are just for convenience. The entire table is one long string of randomly chosen digits.

There are two steps in using the random digit table to choose a simple random sample.

STEP 1.   Give each member of the population a numerical label of the *same length*. Up to 100 items can be labeled with two digits, up to 1000 items can be labeled with three digits, and so on.

STEP 2.   To choose a simple random sample, go through Table 5.1 looking at successive groups of digits of the length used as labels. This works because, for example, any two-digit group in the table is equally likely to be any of the 100 possible labels 00, 01, . . . , 99. Any group of digits that was not used as a label or that duplicates a label already in the sample is simply ignored.

Here are two examples that will illustrate the technique.

EXAMPLE: Sampling Autos

An auto manufacturer wants to select 5 of the last 50 cars produced on an assembly line for a very detailed quality inspection. To avoid bias, a simple random sample will be chosen.

TABLE 5.1 Random Digits

101	03918	86495	47372	21870	28522	99445	38783	83307
102	10041	35095	66357	64569	08993	20429	28569	63809
103	43537	58268	80237	17407	89680	04655	24678	61932
104	64301	47201	31905	60410	80101	33382	95255	10353
105	43857	42186	77011	93839	28380	49296	63311	49713
106	91823	39794	47046	78563	89328	39478	04123	19287
107	34017	87878	35674	39212	98246	29735	09924	27893
108	49105	00755	39242	50472	39581	44036	54518	46865
109	72479	02741	75732	99808	02382	77201	44932	88978
110	84281	45650	28016	77753	39495	41847	19634	82681
111	61589	35486	59500	20060	89769	54870	75586	07853
112	25318	01995	87789	41212	74907	90734	31946	24921
113	40113	37395	51406	98099	43023	70195	07013	72306
114	58420	43526	15539	24845	15582	16780	95286	69021
115	18075	45894	09875	42869	20618	07699	80671	54287
116	52754	73124	93276	71521	59618	44966	37502	15570
117	05255	53579	08239	99174	75548	95776	42314	13093
118	76032	35569	28738	38092	74669	00749	17832	64855
119	97050	31553	32350	51491	53659	89336	36912	05292
120	29030	43074	84602	95131	22769	44680	68492	33987
121	28124	29686	63745	12313	15745	11570	20953	17149
122	97469	41277	90524	36459	22178	63785	20466	67130
123	91754	40784	38916	12949	76104	20556	34001	59133
124	84599	29798	57707	57392	91757	76994	43827	69089
125	06490	42228	94940	10668	62072	58983	10263	08832
126	30666	02218	89355	76117	75167	69005	42479	79865
127	87228	15736	08506	29759	74257	85594	75154	48664
128	45133	49229	32502	99698	68202	44704	39191	73740
129	55713	98670	57794	64795	27102	83420	26630	95009
130	20390	38266	30138	61250	07527	02014	43972	49370
131	13400	68249	32459	41627	56194	93075	50520	96784
132	08900	87788	73717	19287	69954	45917	80026	55598
133	86757	47905	16890	99047	78249	73739	97076	00525
134	19862	54700	18777	22218	25414	13151	54954	80615
135	96282	11576	59837	27429	60015	40338	39435	94021
136	17463	26715	71680	04853	55725	87792	99907	67156
137	44880	55285	95472	57551	24602	98311	63293	58110
138	61911	78152	96341	31473	58398	61602	38143	93833
139	07769	22819	58373	88466	71341	32772	93643	92855
140	73063	63623	29388	89507	78553	62792	89343	27401
141	24187	60720	74055	36902	22047	09091	79368	35408
142	06875	53335	91274	87824	04137	77579	54266	38762
143	23393	37710	46457	03553	58275	11138	18521	59667
144	00980	73632	88008	10060	48563	31874	90785	78923
145	46611	39359	98036	25351	88031	72020	13837	03121
146	56644	79453	49072	30594	73185	81691	29225	70495
147	98350	36891	04873	71321	29929	37145	95906	41005
148	17444	61728	86112	76261	92519	61569	65672	95772
149	45785	21301	89563	23018	60423	50801	70564	45398
150	54369	08513	36838	19805	67827	74938	66946	01206

00	01	02	03	04	05	06	07	08	09
10	11	12	13	14	15	16	17	18	19
20	21	22	23	24	25	26	27	28	29
30	31	32	33	34	35	36	37	38	39
40	41	42	43	44	45	46	47	48	49

Figure 5.2 The first step in random sampling: assigning labels to 50 cars.

STEP 1. Give each car a numerical label. Because two digits are needed to label 50 cars, all labels will have two digits. Let's begin with 00. Then the labels are as shown in Figure 5.2. It is also correct to use labels 01 to 50 rather than 00 to 49. Be sure to say how you labeled the members of the population.

STEP 2. Look at successive two-digit groups in Table 5.1. Starting at line 140 (any line will do), we read

$$73063 \qquad 63623 \qquad 29388$$

The cars labeled 06, 36, 23, 29, and 38 are chosen. The initial 73 is ignored because it is not used as a label, and the second 36 is ignored as a duplicate. ▲

EXAMPLE: Sampling Households

A town has 4756 households. As part of a national sample, you must choose a simple random sample of 3 of these households to be interviewed. Four digits are needed to label 4756 households. Assign the labels 0000 to 4755 (0001 to 4756 is also OK) to the households, then read four-digit groups from Table 5.1. If we enter at line 130, the households chosen are those labeled 2039, 0382, and 1386. ▲

Clearly, the Bureau of Labor Statistics doesn't use beads in a box or names in a hat to determine who will be interviewed for the Current Population Survey. In fact, the bureau doesn't even use simple random samples. National opinion polls and the Bureau of Labor Statistics employ more complex versions of random sampling. Chance still determines the sample, but the process of selecting from the entire nation is done in stages instead of all at once. The Census Bureau, under contract with the Bureau of Labor Statistics, does the job. The entire country is first divided into about 1900 primary sampling units, or PSUs (see Figure 5.3). Each PSU is a group of neighboring counties. The Census Bureau selects a random sample of PSUs. Within each PSU selected, smaller areas of about 500 inhabitants, called census enumeration districts, are chosen at random. Finally, the Census Bureau selects, also at random, individual households within each of the chosen districts.

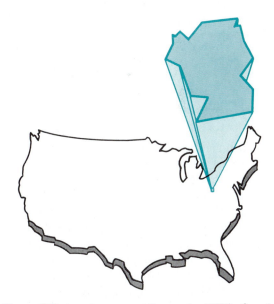

Figure 5.3 A primary sampling unit (PSU) for the Current Population Survey.

SP TLIGHT 5.2   Sampling and Surveys: Taking the National Pulse

▶ ▶ ▶   ▶ ▶ ▶ ▶ ▶ ▶ ▶ ▶ ▶

Dr. Janet L. Norwood, Commissioner of Labor Statistics from 1979 to 1991, shares a number of observations on her agency and on the task of sampling the nation at large through the Current Population Survey.

How We Go about the Business of Sampling

The household survey is a basic labor-force survey that provides us with an enormous amount of data. Specifically, it provides information on the demographic characteristics of people and about their labor-force status: their employment, the length of time they are employed, whether they are job seekers, whether they are job losers, whether they are employed part-time or full-

time, and so on. That survey, which we call the Current Population Survey, is conducted for us on contract by the Bureau of the Census. Interviewers go out to a sample of 60,000 households spread throughout the United States in more than 400 areas of the country. The reason that the sample is so large — and it is a very large sample for a household survey — is that so much data come out of it. We do not set out simply to collect data on how many people are employed or unemployed. We also must have information on men versus women, on blacks, on whites, on Hispanics, and on trends in the youth population. We need to have some information by region and by state. And as you get down to smaller and smaller groups, it is necessary to have a sample size that is sufficient to provide adequate data.

Such a *multistage random sample* offers several practical advantages over a simple random sample. For one, drawing the sample does not require listing every household in the nation. A list of PSUs is used at the first stage and a list of enumeration districts at the second stage. We only need a list of the households in the relatively few enumeration districts chosen; if need be, we can make up that list by walking around those districts. Moreover, the households to be interviewed are clustered in a few locations, so that travel costs for the interviewers are reduced. The price paid for practical-

ity, however, is complexity in actually choosing the sample and in interpreting the results. Because simple random sampling is the essential principle behind all random sampling and because it is also the main building block for more complex designs, we focus our study on simple random sampling.

▶ SAMPLING VARIABILITY

To illustrate the idea of random sampling, we drew 50 beads from a large box of beads (the pop-

On the Impact of Our Agency

Our price data are probably among our best-known indicators. This is because the consumer price index (CPI) — the best measure we have in this country of inflation — is used in a number of government programs as an escalator to keep income up with inflation. In fact, if you take dependents into account, we estimate that more than half the population of this country has income in some way affected by the CPI. And in 1985, income tax brackets will have begun to be adjusted by the CPI for purposes of calculating income tax.

Dealing with Uncertainty

Let's face it. Everyone wants to live in a world of certainty. Even the statistician wants to live in a world of certainty. Our job is to explain to people that there is no such thing as absolute certainty. Every word we use is looked at to be sure that we understand it and that someone looking at it isn't going to read more into it than we intended.

What we *can* do is to provide a set of statistics that have some error surrounding them, something that we often call sampling error or variance, which will tell what the basic tolerances are. If the unemployment rate, for example, moves by one-tenth of a point, we in the Bureau of Labor Statistics say that it was about the same. If it goes up or down two-tenths of a point, we will say that there has been a change, because the sampling error surrounding that number is such that a two-tenths change is outside of that limit.

If somebody wants to know something about BLS data, they pick up a phone and they call me or they call someone on my staff who is an expert in the area. We have made it our business to provide information as rapidly as we can and as openly as we can. I happen to believe that a statistical agency can only be a good one if it is completely open.

ulation). Our first sample contained 12 dark beads. If we draw another sample, we will probably not get 12 dark beads again. For example, when we remix the beads and take a second sample, we may find that only 8 out of 50, or 16%, are dark, a figure quite different from the 24% we got earlier. The results of repeated random sampling vary.

Suppose that the Gallup poll conducted its weekly survey of public opinion twice, selecting two random samples, telephoning those chosen, and asking the same question of the two samples. Two random samples, each selecting 1500 of the more than 190 million U.S. residents age 18 and over, will certainly contain different people. And two distinct groups of people will have somewhat different opinions.

Gallup's report that 45% of Americans hesitate to go out at night because they are afraid of crime refers to the 1500 specific individuals in a specific sample. A second sample would no doubt produce a different result. If the Current Population Survey were conducted twice this month, the two samples would differ, producing two different officially announced unemployment rates.

Random sampling eliminates bias, but it does not eliminate *variability*. The variation from sample to sample in repeated random samples is an inevitable consequence of variability in the population.

How then can we trust the result of a random sample, knowing that a second sample would give a different result? How can we base economic and political decisions on the unemployment rate, knowing that the rate would vary if the Bureau of Labor Statistics took a second sample? In fact, random sampling *can* be trusted. To see why, we need to look more closely at sampling variability.

There are different kinds of variability. The answers obtained by sending an interviewer to a shopping mall vary in a haphazard and unanalyzable way. However, repeated random samples vary in a regular manner because a specific chance mechanism is used in random selection. The long-run results are not haphazard. We see such regular variation in the results of repeated coin tosses or of successive spins of a roulette wheel. Tossing a balanced coin 1500 times is much like choosing a simple random sample of 1500 from a large population, if we imagine that opinion in this population is evenly divided so that heads represents "yes" and tails represents "no."

Let's look more closely at the outcomes of coin tossing. It is unlikely that a person tossing a coin 1500 times will obtain exactly 750 heads (a 50% "yes" response). The first sample might produce 732 heads (49% in favor), a second, 781 heads (52% in favor), and so on. But coin tossing is subject to the laws of *probability*, which dictate how frequently each outcome will occur in the long run. We will introduce probability in Chapter 7, and show in Chapter 8 how the language of probability describes the trustworthiness of statistical conclusions. In particular, probability theory tells us that 1500 tosses of a balanced coin will almost never (only 1 chance in 1000) produce less than 46% or more than 54% heads.

Therefore, 1500 tosses of a balanced coin can be trusted to give a result close to 50% heads.

Similarly, a Gallup poll of 1500 Americans can be trusted to give a result close to the result obtained by polling all 190 million adult Americans. Probability deals with phenomena that are variable, but that nonetheless show a regular pattern in the long run. Tossing a coin and choosing a random sample are such phenomena.

In the case of choosing a simple random sample of 50 beads from a box, we can easily take repeated samples and observe the pattern of outcomes.

EXAMPLE: A Bead-Sampling Experiment

We took 100 simple random samples of size 50 from the box of beads. Our first sample had 12 dark beads out of 50, or 24%. The second sample had 16%, the third 26%, and so on. Table 5.2 displays the results from our 100 samples. We can display the outcomes of all 100 samples in a **histogram** (Figure 5.4). The height of each bar in the histogram shows how often the outcome marked at the base of the bar occurred. For example, the height of the bar marked 16%

TABLE 5.2 Bead-Sampling Results

Percent of dark beads	Number of samples	Percent of dark beads	Number of samples
8	1	22	14
10	1	24	14
12	5	26	8
14	8	28	3
16	11	30	3
18	19	32	0
20	12	34	1

is 11 because 11 of our samples had 16% dark beads. More detail about histograms appears in the next chapter.   ▲

The histogram for our 100 samples in Figure 5.4 shows a lot of scatter, but most of the samples cluster around the 20% mark. In fact, 20% is the actual proportion of dark beads in the box from which these samples were drawn. Only 12 of the samples we took contained exactly 20%, but 70 of the 100 samples had between 16% and 24% dark beads.

The results of random sampling usually do come quite close to the truth about a population. As we'll discover in Chapter 7, the laws of probability make possible exact statements about the accuracy of simple random sampling. We will see, for example, that the outcome of a sample of size 1500 is much more likely than that of a sample of size 50 to come close to the truth. That's why opinion polls interview 1500 people rather than just 50.

The deliberate use of chance in producing data is one of the fundamental ideas of statistics. It is important as a way of removing bias. More important, the proper use of random samples enables us to draw confident conclusions from data because the mathematics of probability provides support for these conclusions.

As we saw earlier, simple random sampling is the building block of all sampling designs. Thus, even though the Current Population Survey and the Gallup poll use more complex random-sampling procedures, the results of their multistage random samples are still described by the laws of probability. Consequently, these samples usually give results close to the truth about the population. A careful statement of a sample result expresses "usually" in terms of probability and "close to the truth" in terms of a *margin of error*.

For example, the Gallup poll takes a sample of about 1500 people each week. To describe the accuracy of the poll results, we ask "What would happen if Gallup took many samples of 1500 people, asking the same question each time?" This is similar in principle to our many samples from the same box of beads. The results are also similar: there is a regular pattern of outcomes that allows

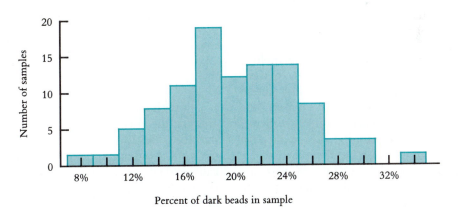

Figure 5.4 Histogram showing sampling variability of the outcomes of a bead-sampling demonstration.

us to state how often the sample result will come close to the truth about the population. Specifically, the probability is 0.95 that a sample percentage from a Gallup poll is within ±3% of the true population value.

Probability describes the regular behavior that appears when random sampling is repeated many times. The probability 0.95 means that in the long run 95% of Gallup's many sample results fall within a margin of error of ±3 percentage points about the truth for the entire population. If the poll reveals that 45% of the sample fear to go out at night, we can be quite confident that between 42% and 48% of the entire population share that fear. The variability in sampling has not vanished, but probability allows us to describe the variability by announcing a margin of error that most samples will meet. Gallup and other polling organizations report these facts in their press releases, but news editors often cut that part of the story.

What about variability in the unemployment rate? Here are the facts in this case. The Census Bureau prefers to announce the margin of error that 90% of Current Population Survey samples will meet: it is about ±0.2%, or two-tenths of one percent. Thus, when the unemployment rate drops from 7.4% to 7.3%, sampling variability may well account for the change. But a change from 7.4% to 7.0% very probably reflects a real drop in the percent of the American labor force who are without a job.

The smaller margin of error in the unemployment rate is due to the larger sample size. A margin of error of ±3% is acceptable in polling public opinion, but not in measuring unemployment. Because data on employment and unemployment are important for economic planning, the federal government is willing to undertake the expense of a large monthly sample. The margin of error attained with any desired probability (such as 0.90 or 0.95) depends on the size of the sample and on the exact sampling design. The mathematics of probability describes these relationships exactly.

► EXPERIMENTATION

Sample surveys gather information on part of the population in order to draw conclusions about the whole. When the goal is to describe a population, as the Current Population Survey's aim is to describe employment and unemployment in the United States, statistical sampling is the right tool to use.

Suppose, however, that we want to study the response to a stimulus, to see how one variable affects another when we change existing conditions. Will a new mathematics curriculum improve the scores of sixth graders on a standard test of mathematics achievement? Will taking small amounts of aspirin daily reduce the risk of suffering a heart attack? Does smoking increase the risk of lung cancer? Observational studies, such as sample surveys, are ineffective tools for answering these questions. Instead, we prefer to carry out experiments.

An **experiment** differs from observation in that the experimenter intervenes actively by imposing a *treatment* on the subjects. A treatment can be any condition that the experimenter is interested in, such as a new math curriculum or an aspirin tablet every day. In sampling, on the other hand, we observe or measure the state of the subjects without trying to change that state by a treatment.

Experiments are the preferred method for examining the effect of one variable on another. By imposing the specific treatment of interest and controlling other influences, we can pin down cause and effect. A sample survey, in contrast, may show that two variables are related, but it cannot demonstrate that one causes the other. Statistics has something to say about how to arrange experiments, just as it suggests methods for sampling.

SP TLIGHT 5.3   Sir Ronald A. Fisher, 1890 – 1962

▷ ▷ ▷ ▷ ▷ ▷ ▷ ▷ ▷ ▷ ▷ ▷

While employed at the Rothamsted agricultural experiment station in the 1920s, British statistician and geneticist R. A. Fisher revolutionized the strategy of experimentation. Experimenters there were comparing the effects of several treatments, such as different fertilizers, on field crops. Because fertility and other variables can change as we move in any direction across the planted field, the experimenters used elaborate checkerboard planting arrangements to avoid bias. Fisher realized that random assignment of treatments to growing plots was simpler and better. He introduced randomization, described more complex random arrangements, such as blocks and Latin squares, and worked out the mathematics of the *analysis of variance* to analyze data from randomized comparative experiments.

Fisher contributed many other ideas, both mathematical and practical, to the new science of statistics. His influential books organized the field. Fisher was both opinionated and

Sir Ronald A. Fisher.

combative. From the 1930s until his death, he was engaged in sometimes vitriolic debates over the appropriate use of statistical reasoning in scientific inference.

EXAMPLE: An Uncontrolled Experiment

The Bigfoot Mountain School District, concerned about the poor mathematics preparation of American children, adopts an ambitious new mathematics curriculum. After three years of the new curriculum, students completing sixth grade have an average achievement score 10% higher than they had before the treatment. Bigfoot Mountain pronounces the curriculum a success, and other systems adopt it.

This experiment has a very simple design. A group of subjects (the students) were exposed to a treatment (the new curriculum), and the outcome (achievement test scores) was observed. Here is the design:

New curriculum $\longrightarrow$ observe test scores

or, in general form,

Treatment $\longrightarrow$ observe response   ▲

Most laboratory experiments use a design like that in the example: apply a treatment and measure the response. In the controlled environment of the laboratory, simple designs are often adequate. But field experiments and experiments with human subjects are exposed to more variable conditions and deal with more variable subjects. They require control of outside factors that can influence the outcome. With greater variability comes a greater need for statistical design.

In Bigfoot Mountain, an atmosphere of concern for education brought about a number of simultaneous changes that could account for the students' achievement-test scores. Elementary teachers were given additional training in mathematics. A parent group began to provide classroom tutors to give children individual help with mathematics. Public concern led parents to pay more attention to their children's progress and teachers to assign more homework.

In these circumstances, mathematics achievement would have increased without a new curriculum. In fact, the new curriculum could even be *less* effective than the old. The Bigfoot Mountain experiment cannot distinguish the effects of the changes in parents and teachers from the effects of the new curriculum. We say that the new curriculum is confounded with the other changes that took place at the same time.

Variables, whether part of a study or not, are said to be **confounded** when their effects on the outcome cannot be distinguished from each other.

The remedy for confounding is to do a *comparative experiment* in which some children are taught from the new curriculum and others from the old. Changes in parents' attitudes and involvement, teacher retraining, and other such variables now operate equally on both groups of students, so that direct comparison of the two curricula is possible. Most well-designed experiments compare two or more treatments.

Once we decide to do a comparative experiment, we need to find a way to assign the students to the two groups. If the groups differ markedly when the experiment begins, bias will result. For example, if we allow students to volunteer for the new curriculum, only adventurous children who are interested in math are likely to sign up for our experimental treatment, and these students are likely to perform well. Personal choice will bias our results in the same way that volunteers bias the results of call-in opinion polls. The solution to the problem of bias is the same for experiments and for samples: use impersonal chance to select the groups.

Let's say the Bigfoot Mountain School District decides to compare the progress of 100 students taught under the new mathematics curriculum with that of 100 students taught under the old curriculum. We select the students who will be taught the new curriculum by taking a simple random sample of size 100 from the 200 available subjects. The remaining 100 students form the **control group;** they will continue in the old curriculum.

The selection procedure is exactly the same as it is for sampling: all 200 members of the population are tagged with numerical labels, beginning with 000 and ending with 199. Next, we consult a table of random digits, inspecting successive three-digit groups. The first 100 labels encountered select the group that will be taught from the new curriculum. Repeated labels and groups of digits not used as labels are ignored. For example, if we begin at line 125 in Table 5.1, the first students chosen are those labeled 064, 106, 102, 022, 188. The remaining 100 students form the control group, to be taught from the old curriculum.

The result is a **randomized comparative experiment** with two groups. Figure 5.5 outlines the design in graphical form. The experiment is comparative because two treatments (the two math curricula) are compared; it is randomized because the subjects are assigned to the treatments by chance.

Randomized comparative experiments are used whenever environmental variables, such as changes in the behavior of Bigfoot Mountain parents and teachers, threaten to confound the results. The results of randomized comparative experiments are as reliable as the results of random samples, and for the same reasons: random selection is governed by the laws of probability. Here is an example with three treatments.

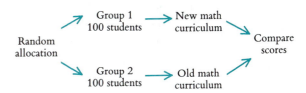

Figure 5.5 Outline of the design of a randomized comparative experiment to evaluate a new mathematics curriculum.

that each label has two digits). Read two-digit groups starting in line 115 of Table 5.1 until 10 turkeys are chosen to make up the first group. Those chosen have labels 18, 07, 09, 28, 20, 15, 24, 27, 21, and 05. Then continue in the table to choose 10 more birds for the second group. The 10 that remain form the third group. ▲

Randomized comparative experiments are common tools of industrial and academic research. They are also widely used in medical research. For example, federal regulations require that the safety and effectiveness of new drugs be demonstrated by randomized comparative experiments. Let's look at a typical medical experiment.

EXAMPLE: Raising Turkeys

Turkeys raised commercially for food are often fed the antibiotic salinomycin to prevent infections from spreading among the birds. Salinomycin can damage the birds' internal organs, especially the pancreas. A researcher believes that adding vitamin E to the diet may prevent injury. He wants to explore the effects of three levels of vitamin E added to the diet of turkeys along with the usual dose of salinomycin. There are 30 turkeys available for the study. At the end of the study, the birds will be killed and each pancreas examined under a microscope.

The researcher decides on a randomized comparative design that allocates 10 birds chosen at random to each of the three levels of vitamin E. See Figure 5.6 for an outline of the design. The turkeys are labeled with tags marked 00 to 29 (01 to 30 is also acceptable, but be sure

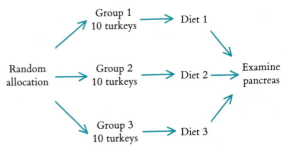

Figure 5.6 The design of a randomized comparative experiment to compare three diets for turkeys.

EXAMPLE: The Physicians' Health Study

There is some evidence that taking low, regular doses of aspirin will reduce the risk of heart attacks. Some physicians also suspect that regular doses of beta carotene (which the body converts into vitamin A) will help prevent some types of cancer. The Physicians' Health Study was a large experiment designed to test these claims (see Spotlight 5.4). The subjects of this study were 22,000 male physicians over 40 years of age. Each physician took a pill every other day over a period of several years. There were four treatments: aspirin alone, beta carotene alone, both, and neither. The subjects were randomly assigned to one of these treatments at the beginning of the experiment.   ▲

The Physicians' Health Study introduces several new ideas important to the proper design of experiments. The first is the importance of counteracting the **placebo effect,** a special kind of confounding. A placebo is a fake treatment, a dummy pill that contains no active ingredient but looks and tastes like the real thing. The placebo effect is the tendency of subjects to respond favorably to any treatment, even a placebo. If subjects given aspirin, for example, are compared with subjects who receive no treatment, the first group gets the benefit of both aspirin and the placebo effect. Any beneficial effect that aspirin may have is confounded with the placebo effect. To prevent confounding, it is therefore important that some treatment be given to *all* subjects in any medical experiment.

In the Physicians' Health Study, all subjects took pills that were identical in appearance, but some pills contained aspirin or beta carotene and some contained a placebo. The study was designed as a **double-blind experiment:** subjects did not know which treatment they were receiving, because this knowledge might influence their reaction. In addition, knowing the subjects' treatments might influence researchers who interview and examine them. Therefore, experimental workers were also kept "blind." Only the study's statistician knew which treatment each subject received.

The Physicians' Health Study is a more elaborate experiment than our earlier examples. Not only are four treatments compared, but two distinct experimental variables are present: aspirin or not, and beta carotene or not. A two-variable experiment, usually called a *two-factor experiment*, allows us to study the interaction, or joint effect, of the two drugs as well as the separate effects of each. For example, beta carotene may reinforce (or counteract) the effect of aspirin on future heart attacks. By comparing these four groups, we can study all these possible interactions. The outline of the design (Figure 5.7) is similar to our earlier

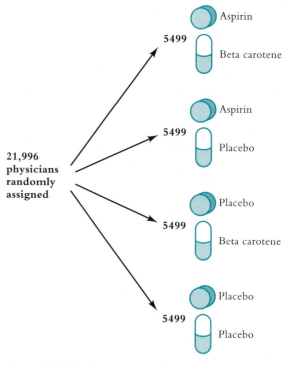

Figure 5.7 The design of the Physicians' Health Study, an experiment with two factors.

SPOTLIGHT 5.4   Using Statistics to Study Disease

▶ ▶ ▶   ▶ ▶ ▶ ▶ ▶ ▶ ▶ ▶ ▶

Dr. Julie Buring, associate director of the Physicians' Health Study, discusses the ways in which epidemiologists gather data on disease.

As an epidemiologist, I look at factors that are involved in the distribution and disease frequency in human populations. What is it about what we do, what we eat, what our environment is, what our occupations are, our history — our medical history, our family history — that leads one group of people to be more or less likely to develop a disease than another group of people? It is these factors that we are trying to identify.

We go at it from a couple of different angles. One is called *descriptive epidemiology,* or looking at the trends of diseases over time, trends of diseases in one population relative to another population.

Another is called *observational epidemiology,* in which we observe what people do. We take a group of people who have a disease and a group of people who don't have a disease. We look at their patterns of eating or drinking, medical history, and what their exposures may have been.

The other way to go about it is to take a group of people who have been exposed to something such as smoking and a group of people who haven't and follow them up over time to see whether they develop the disease or not. Whether we do one design or another, both are observational. That means we don't interfere in the process. We just observe it.

A second type of epidemiology, of which the Physicians' Health Study is an example, is *experimental epidemiology,* sometimes called an intervention study. We take a group of people who have the treatment and a group of people who do not. The difference between the experimental and observational approaches is that in experimental epidemiology, the investigators determine who will be in what treatment group, who will receive the treatment and the placebo, who will be exposed, and who won't be exposed.

From these different approaches — descriptive epidemiology and observational epidemiology — we can judge whether a particular factor causes or prevents the disease that we are looking at.

examples because the basic ideas of randomization and comparison of several treatments remain.

▶ STATISTICAL EVIDENCE

A properly designed experiment, in the eyes of a statistician, is an experiment employing the principles of *comparison* and *randomization:* comparison of several treatments and randomization in assigning subjects to the treatments. As we saw in the math curriculum example, comparison eliminates confounding by environmental variables. The immediate appeal of randomization is the elimination of bias in forming the groups of subjects that will receive the different treatments.

SP TLIGHT 5.5   Experiments and Ethics

Dr. Charles Hennekens.

Dr. Charles Hennekens, director of the Physicians' Health Study, had to concern himself with the goals, design, and implementation of his large-scale study. But other questions also arise in the course of such an experiment. Dr. Hennekens was asked about the ethics of experimenting on human health:

Much has been made of the ethical concerns about randomized trials. There are instances where it would not be ethical to do a randomized trial. When penicillin was introduced for the treatment of pneumococcal pneumonia, which was virtually 100% fatal, the mortality rate plummeted significantly. Certainly it would have been unethical to do a randomized trial, to withhold effective treatment from people who need it.

There's a delicate balance between when to do or not to do a randomized trial. On the one hand, there must be sufficient belief in the agent's potential to justify exposing half the subjects to it.

The future health of the subjects of the Physicians' Health Study may depend on age, past medical history, emotional status, smoking habits, and many other variables known and unknown. Randomization will, on the average, balance the groups simultaneously in all such variables. Because the groups are exposed to exactly the same environmental variables, except for the actual content of the pills, differences among the groups can be attributed to the effect of the medication. That is the logic of randomized comparative experiments.

On the other hand, there must be sufficient doubt about its efficacy to justify withholding it from the other half of subjects who might be assigned to the placebos, the pills with inert ingredients. It was just these circumstances that we felt existed with regard to the aspirin and the beta carotene hypotheses.

A randomized trial done on a newer therapy or drug is best done when the procedure is first introduced. It becomes very difficult, with regard to both feasibility and ethics, to do such trials after a long period of time has elapsed. Treatments begin to be so accepted by the population that it is difficult to find people willing to have the treatment withheld. Others might feel ethically that it would be difficult to withhold treatment.

One example in contemporary times regards the treatment of breast cancer. William Halsted of Johns Hopkins, the father of American surgery, invented the radical mastectomy in the early 1900s as a therapy for breast cancer. It's been used widely for more than half a century. However, after all this time, it became apparent to some investigators that less extensive procedures might accomplish the same results, that is, to keep the age-specific mortality rate from breast cancer in affected women at a low level.

It's only been in recent years that randomized trials of less extensive forms of treatment for breast cancer have been done. It is important that such research be done. It is optimal to do it early so that we get a clear answer at the beginning of the development of new procedures and new drugs.

Dr. Julie Buring, associate director of the Physicians' Health Study, adds:

Sometimes, in the case where certain procedures are not tested in a trial immediately, the reason is that the procedures seem to make sense intuitively. For example, it would make sense that if you remove more of the breast tissue, it might reduce your risk of having a recurrence of the disease. Yet most clinicians, or most medical professionals, only see a small number of patients. It is very hard to get a pattern, to see that there really is no difference between those who receive a radical mastectomy and a less invasive procedure. It is only when you are able to do these trials in thousands of patients that you are able to see that there really is no difference, or just a slight difference. You must have large numbers to be able to pick up those patterns.

Let's be a bit more specific: any difference among the groups is due *either* to the medication *or* to the accident of chance in the random assignment of subjects. It could happen, for example, that men about to have a heart attack were, just by chance, overrepresented in one of the groups. Once again, statistics calls on probability. Because chance was deliberately used in making assignments, the laws of probability tell us how large the differences among the four groups are likely to be if nothing but chance were operating. If we observe differences so large that they would almost never occur just by chance, we are confident that we are seeing the effects of the treatments. Differences among the treatment groups that are so large that they would almost never occur just by chance are called **statistically significant.** As in sampling, larger numbers of subjects increase our confidence in the results. The Physicians' Health Study followed 22,000 subjects in order to be quite certain that any medically important differences among the groups would be detected and that these differences could be attributed to aspirin or to vitamin A. In fact, there were significantly fewer heart attacks among the men who took aspirin than among men who took the placebo. As a result of the Physicians' Health Study, doctors often recommend that men over 50 take small amounts of aspirin regularly.

The logic of experimentation, the statistical design of experiments, and the mathematics of probability combine to give compelling evidence of cause and effect. Only experimentation can produce fully convincing evidence of causation.

EXAMPLE: Smoking and Health

By way of contrast, consider the statistical evidence linking cigarette smoking to lung cancer. This evidence is based on observation rather than experiment. The most careful studies have selected samples of smokers and nonsmokers, then followed them for many years, eventually recording the cause of death. These are called *prospective studies* because they follow the subjects forward in time. Prospective studies are comparative, but they are not experiments because the subjects themselves choose whether or not to smoke. Remember that an experiment must actually impose treatments on the subjects. A large prospective study of British doctors found that the lung-cancer death rate among cigarette smokers was 20 times that of non-smokers; another study of American men aged 40 to 79 found that the death rate from lung

cancer was 11 times higher among smokers than among nonsmokers. Thus, the observed connection between smoking and lung cancer is strong. ▲

This connection is statistically significant; that is, it is far stronger than would occur by chance. We can be confident that something other than chance links smoking to cancer. But observation of samples cannot tell us *what* factors other than chance are at work. Perhaps there is something in the genetic makeup of some people that predisposes them both to nicotine addiction and to lung cancer. In that case, a strong link would be observed even if smoking itself had no effect on the lungs.

The statistical evidence that points to cigarette smoking as a cause of lung cancer is about as strong as nonexperimental evidence can be. First, the connection has been observed in many studies in many countries. This eliminates factors peculiar to one group of people, or to one specific study design. Second, specific ways in which smoking could cause cancer have been identified. Cigarette smoke contains tars that can be shown by experiment to cause tumors in animals. Third, no really plausible alternative explanation is available. For example, the genetic hypothesis cannot explain the rise in lung-cancer rates among women that occurred as more and more women became smokers. Lung cancer, which has long been the leading cause of cancer deaths in men, is now challenging breast cancer as the most fatal cancer for women. Moreover, genetics cannot explain why lung-cancer death rates increase among nonsmokers who are exposed to cigarette smoke from other people.

This evidence is convincing to most people, and almost all physicians accept it. But it is not quite as strong as the conclusive statistical evidence we get from randomized comparative experiments.

MORE ELABORATE
EXPERIMENTS

Many experiments are more complex than a simple comparative randomized design, just as many samples are not simple random samples. If we anticipate that male and female patients will respond differently to aspirin and beta carotene, we can first divide the pool of patients into two **blocks** (males and females) and then randomly assign patients to treatments separately within each block. The division into blocks controls one influential variable, the patient's sex, by including it in the design. If aspirin, for example, has different effects for women and men, these effects are separated in the two blocks. Other influences are still averaged out by randomization. (The Physicians' Health Study was restricted to male subjects because there were not enough female physicians over 40 to form a second block. A Nurses' Health Study is exploring women's response to aspirin and other treatments.) The use of blocks is common. Suppose, to give another example, that agronomists want to compare several varieties of soybeans. Because soil type and local climate may affect the crop, they carry out the experiment in three locations. Each location is a block, and the varieties are randomly assigned to small plots of ground separately in each location.

Even more common than blocking is the simultaneous study of several *factors*, or experimental variables. We saw earlier that the Physicians' Health Study was a two-factor experiment. Industry provides us with another example.

EXAMPLE: An Industrial Experiment

A chemical engineer is trying to determine the most efficient temperature–pressure setting for a production process. She would be foolish to rely on experiments taking into account only

		Factor 1: Temperature (°C)		
		95°	110°	120°
Factor 2: Pressure	100 psi	1	2	3
	150 psi	4	5	6

Figure 5.8 The treatments in a two-factor experiment.

one variable at a time, because the most productive temperature will change according to the pressure and vice versa. An effective experiment must change both temperature and pressure from treatment to treatment. There are therefore two factors. The engineer chooses three temperatures and two pressures that cover the range of interest. Combinations of these temperatures and pressures form six treatments, which are displayed in Figure 5.8. ▲

The combination of several experimental variables with the use of blocks can make experiments prohibitively large and expensive. However, *combinatorics,* the mathematical study of arrangements, can sometimes provide clever arrangements of treatments that help hold down the size and cost of experiments. We will illustrate this by a comparison of motor oils.

EXAMPLE: Comparing Motor Oils

Some makers of motor oil claim that using their product improves gasoline mileage in cars. To test this claim, we want to compare the effect of four different oils on mileage. However, because the car model and the driver's habits greatly influence the mileage obtained, the effect of an oil may vary from car to car and from driver to driver. To obtain results of general interest, we must compare the oils in several different cars and with several different drivers. If we choose 4 car models and 4 drivers, we will have 16 car–driver combinations. The type of car and driving habits are so influential that we consider each of these 16 combinations a block. If we then complete test drives with 4 oils (in random order) in each block, we need 4×16, or 64, test drives.

However, combinatorics gives us a way to test the effects of cars, drivers, and oils that requires only 16 test drives. Call the oils A, B, C, and D and assign *one* oil to each of the 16 car–driver combinations in the arrangement shown in Figure 5.9. Study this arrangement: each oil appears 4 times, exactly once in each row (for each car) and also exactly once in each column

(for each driver). This setup can, in fact, show how each oil performs with each car and with each driver.   ▲

An arrangement like that in the motor oil example is called a **Latin square.** Figure 5.9 is a 4×4 Latin square because it arranges 4 kinds of objects (the oils) in a 4×4 square array so that each object appears exactly once in each row and in each column. There are several ways of assigning the labels A, B, C, D so that each appears once in each row and once in each column; that is, there are several different 4×4 Latin squares. Exercise 43 asks you to find one that is not the same as Figure 5.9. Mathematicians studied Latin squares and catalogued many of them long before the statistical design of experiments was invented. The use of Latin squares to design experiments illustrates how the study of pure mathematics, originally pursued simply out of curiosity, often turns out to have practical importance.

The Latin square is a clever arrangement of comparisons. It is almost the opposite of a random assignment. However, the randomization principle of statistical design still governs the application of Latin squares. The arrangement of the labels in rows and columns that appears in Figure 5.9 is the Latin square itself. In applying this Latin square to an experiment, it is essential to assign the four drivers at random to the columns, the four cars at random to the rows, and the four motor oils at random to the labels A, B, C, D. Then the laws of probability will tell us whether the gas mileage figures we obtain using various oils differ by a greater amount than can be accounted for by chance. If so, the differences are statistically significant.   ◀

▶ STATISTICS IN PRACTICE

There is more to the wise use of statistics than a knowledge of such statistical techniques as Latin-square designs. A statistician must know when the

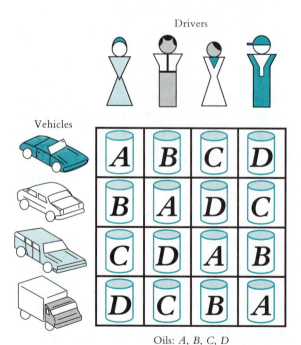

Drivers

Vehicles

Oils: *A, B, C, D*

Figure 5.9 The layout of a Latin-square design to compare four motor oils in four cars with four drivers.

techniques are applicable and appropriate. In the motor oil experiment, we knew that cars and drivers would influence gas mileage. A Latin square allowed us to compare different motor oils without getting confused by different kinds of cars and different drivers; we had three known variables. However, in a study of heart disease or cancer, we don't even know all the important influences. In such cases the Latin square does not apply.

Moreover, good data production requires more than a well-designed sample or experiment. We must also *measure* the variables of interest. That is easy in the case of gas consumption, but measuring "intelligence" or "attitude toward abortion" is much harder. We must decide exactly what we want to measure and design a procedure for measuring it.

Even with a good design and careful procedures for measurement, other sources of error can creep in. Some of the subjects in a random sample of people may not be at home. Others may misunderstand the questions, refuse to cooperate, or not tell the truth. In experiments with human subjects, it is often hard to apply realistic treatments. A psychologist who studies stress by exposing student volunteers to an artificial situation must ask if a few hours of laboratory stress can simulate months or years of hard living and elicit the same responses.

When we are planning a statistical study, we must also face some *ethical questions*. Does the knowledge gained from an experiment or study justify the possible risk to the subjects? In the Physicians' Health Study, doctors gave their informed consent to take either aspirin, beta carotene, or placebos, in any combination, as prescribed by the study designers. When it became clear that men taking aspirin had fewer heart attacks, the experiment was stopped so that all the subjects could take advantage of this new knowledge. In Spotlight 5.5 the directors of the Physicians' Health Study explain why randomized comparative experiments are the mainstay of medical,

agricultural, and many other kinds of research, and when such clinical trials are justified. Practical and ethical problems are never far from the surface when statistics is applied to real problems.

▶ REVIEW VOCABULARY

Bias   A systematic error that tends to cause the observations to deviate in the same direction from the truth about the population whenever a sample or experiment is repeated.

Block   A group of experimental subjects that is homogeneous on one or more variables, for example, a same-sex group. Division into blocks controls for the effects of that variable. Complex experimental designs often assign treatments to subjects at random separately within each of several blocks.

Confounding   Two variables are confounded when their effects on the outcome of a study cannot be distinguished from one another.

Control group   A group of experimental subjects who are given a standard treatment or no treatment (such as a placebo).

Double-blind experiment   An experiment in which neither the experimental subjects nor the persons who interact with them know which treatment each subject received.

Experiment   A study in which treatments are applied to people, animals, or things in order to observe the effect of the treatment.

Histogram   A graph that displays how often various outcomes occur by means of bars. The height of each bar is the frequency of occurrence of an outcome or group of outcomes.

Latin square   An arrangement of k kinds of objects in a $k \times k$ square array, so that each kind appears exactly once in each row and exactly once in each column.

Placebo effect   The effect of a dummy treatment (such as an inert pill in a medical experiment) on the response of subjects.

Population   The entire group of people or things that we want information about.

Randomized comparative experiment   An experiment to compare two or more treatments in which people, animals, or things are assigned to treatments by chance.

Sample   A part of the population that is actually observed and used to draw conclusions, or inferences, about the entire population.

Simple random sample   A sample chosen by chance, so that every possible sample of the same size has an equal chance of being the one selected.

Statistical significance   An observed effect is statistically significant if it is so large that it is unlikely to occur as a chance outcome of producing the data in the absence of a real effect in the population from which the data were drawn.

Table of random digits   A table whose entries are the digits 0, 1, 2, 3, 4, 5, 6, 7, 8, 9 in a completely random order. That is, each entry is equally likely to be any of the 10 digits, and no entry gives information about any other entry.

Voluntary response survey   A sample survey in which the sample chooses itself by responding to a general invitation to write or call with their opinions. Such a survey is usually strongly biased.

▶ SUGGESTED READINGS

BOX, GEORGE E. P., WILLIAM G. HUNTER, AND J. STUART HUNTER: *Statistics for Experimenters*, Wiley, New York, 1978, chapters 4, 7, and 8. This more advanced text places greater emphasis on concepts and on experimental design than most books at a similar level. It is a good source for more information about such topics as experiments with several factors and Latin squares.

FREEDMAN, DAVID, ROBERT PISANI, ROGER PURVES, AND ANI ADHIKARI: *Statistics*, 2nd ed., Norton, New York, 1991, chapters 1, 2, 19, and 20. Excellent, but rather lengthy, conceptual discussion with good examples. Slightly higher in level than *For All Practical Purposes*.

MOORE, DAVID S.: *Statistics: Concepts and Controversies*, 3rd ed., Freeman, New York, 1991, chapters 1 and 2. Written for liberal arts students, this book provides more extensive discussion at about the same level as *For All Practical Purposes*.

TANUR, JUDITH M.: "Samples and surveys," in David C. Hoaglin and David S. Moore (eds.), *Perspectives on Contemporary Statistics*, Mathematical Association of America, Washington, D.C., 1992, pp. 55–70. This essay describes the practice of sample surveys at a relatively nontechnical level.

▶ EXERCISES

In each of Exercises 1 to 3, identify the *population* about which information is desired and the *sample* that is actually observed. If the exact population is not specified by the information given, complete the description of the population in a reasonable way.

   1. Ms. Caucus is her party's candidate in the Second Congressional District of Indiana. The party wants to know what percent of registered voters would vote for Ms. Caucus if the election were held tomorrow. A polling firm contacts 800 voters, of whom 456 say they would vote for Ms. Caucus.

   2. Home canners sometimes can vegetables in used mayonnaise jars to avoid buying special canning jars. *Organic Gardening* magazine wondered what percent

of mayonnaise jars would break when used for canning. They obtained 100 mayonnaise jars and canned tomatoes in them. Only 3 of the jars broke.

3. A maker of electronic instruments buys 4-megabyte RAM memory chips from a supplier. The company wants to know the percent of substandard chips made by the supplier, so it tests all chips received and keeps records. Last year 32,000 out of 400,000 chips received failed to meet standards.

■ 4. A magazine for health foods and organic healing wants to establish that large doses of vitamins will improve health. They ask readers who have regularly taken vitamins in large doses to write in, describing their experiences. Of the 2754 readers who reply, 93% report some benefit from taking vitamins.

Is the sample proportion of 93% probably higher than, lower than, or about the same as the percent of all adults who would perceive some benefit from large vitamin intake? Why? (In answering these questions, you have identified a source of bias in the sampling method.)

■ 5. A television station cancels a program aimed at black audiences because a rating service shows that only 12% of the total viewers in the program's time slot watch it. The ratings are based on a telephone survey of households in the station's market.

Is the 12% rating probably higher than, lower than, or about the same as the true percent of all viewers who watch the program? Why? (In answering these questions, you have identified a source of bias in the rating service's survey.)

■ 6. A newspaper advertisement for *USA Today: The Television Show* said

Should handgun control be tougher? You call
the shots in a special call-in poll tonight.
If yes, call 1-900-720-6181
If no, call 1-900-720-6182

Why is this opinion poll almost certainly biased?

■ 7. Sampling from a list that contains only part of the population is a common cause of bias in sampling. In each of the following examples, explain why this source of bias may be present.

   a. To assess public opinion on a proposal to reduce welfare and unemployment payments, a polling firm selects a sample by random-digit dialing (RDD). RDD uses a machine that dials residential telephone numbers at random.
   b. To assess the reaction of her constituents to the same proposal, a member of Congress uses her free-mailing privilege to send a questionnaire to every registered voter in her district.

■ Discussion exercise.

8. You are worried about the problem of false credentials being offered by candidates for employment at your firm. You decide to investigate some of the credentials at random from now on. Use line 123 of Table 5.1 to choose a simple random sample of four candidates from the following group for investigation. Be sure to say how you labeled the candidates.

Adams	Edwards	Martinez	Russell
Alvarez	Frank	Michel	Sanguillen
Bartkowsky	Hoffer	Miller	Toon
Bishop	Hohenstein	Ogden	Tran
Borchardt	Jack	Pierce	Ungarn
Chan	Kodaira	Pollack	Vlasov
Cleveland	LeMay	Riersol	Wang
Drasin	Marsden	Rubin	Weinstein

9. Your class in Ancient Ugaritic Religion is poorly taught and has decided to complain to the dean. The class decides to choose three of its members at random to carry the complaint. The class list appears below. Choose a simple random sample of three using Table 5.1 starting at line 105.

Anderson	Gutierrez	Patnaik
Aspin	Green	Pirelli
Bennett	Harter	Rao
Bock	Henderson	Rider
Breiman	Hughes	Robertson
Chen	Johnson	Rodriguez
Dixon	Kempthorne	Siegel
Edwards	Landis	Tompkins
Fuller	Liang	Vandegraff
Grant	Olds	Williams

10. You must allocate five tickets to a rock concert among 25 clamoring members of your club. Choose five at random to receive the tickets, using line 135 of Table 5.1 (ignore the asterisks).

Agassiz	Darwin	Herrnstein	Myrdal	Vogt*
Binet*	Epstein	Jerison*	Perez*	Went
Blumenbach	Ferri	Lombrosco	Spencer*	Wilson
Chase*	Goddard*	Moll*	Thomson	Yerkes
Cuvier*	Hall	McKim*	Toulmin	Zimmer

■ 11. Let us illustrate sampling variability in a small sample from a small population. Ten of the 25 club members listed in Exercise 10 are female. Their names are marked with asterisks in the list.

  a. Draw five at random 20 times, using a different part of Table 5.1 each time. Record the number of females in each of your samples. Make a histogram to display your results. What is the average number of females in your 20 samples?

b. Do you think the club members should suspect discrimination if none of the five tickets go to women?

▲ 12. Random digits can be used to *simulate* the results of random sampling. Suppose that you are drawing simple random samples of size 25 from a large bowl of beads and that 20% of the beads in the bowl are dark. To simulate this experiment, let 25 consecutive entries in Table 5.1 stand for the 25 beads in your sample. The digits 0 and 1 stand for dark beads, while other digits stand for light beads. This is an accurate imitation of the sampling experiment because 0 and 1 make up 20% of the 10 equally likely digits. Simulate the results of 50 samples by counting the number of 0s and 1s in the first 25 entries in each of the 50 rows of Table 5.1. Note that the percent of dark beads in each sample is just four times the count; for example, 7 dark beads out of 25 is 28%. Make a histogram like Figure 5.4 to display the results of your 50 samples. Is the truth about the population (20% dark) near the center of your histogram? What are the smallest and largest percents you obtained in your 50 samples?

13. A student wishes to study the opinions of faculty at her college on the advisability of setting up a state board of higher education to oversee all colleges in the state. The college has 380 faculty members.

a. What is the population in this situation?

b. Explain carefully how you would choose a simple random sample of 50 faculty members.

c. Use Table 5.1, starting at line 135, to choose *only the first five* members of this sample.

14. The number of students majoring in political science at Ivy University has increased substantially without a corresponding increase in the number of faculty. The campus newspaper plans to interview 25 of the 450 political science majors to learn student views on class size and other issues. You suggest a simple random sample. Explain carefully how you would choose this sample. Then use Table 5.1, starting at line 120, to select *only the first five* members of your sample.

■ 15. The advice columnist Ann Landers regularly invites her readers to respond to questions asked in her newspaper column. On one occasion, she asked, "If you had it to do over again, would you have children?" Almost 10,000 people wrote in, of whom 70% said "No." Shortly afterward, a national poll asked a random sample of 1400 people the same question; 90% of this sample said "Yes." Which of these polls is more trustworthy, and why?

16. An opinion poll asks a sample of 1450 adults whether they jog regularly; 224 say "Yes." What percent of the sample jog? The margin of error for this

---

▲   Advanced exercise.

poll is ± 3%. What interval are you confident covers the percent of all adults who jog? (This margin of error will hold in 95% of all samples, that is, with probability 0.95.)

17. A sample survey asks a sample of 1324 adults whether they believe that life exists on other planets; 609 say "Yes." What percent of the sample believes in extraterrestrial life? The polling organization announces a margin of error (holding with probability 0.95) of ± 3%. What conclusion can you draw about the percent of all adults who believe that life exists on other planets?

18. National opinion polls such as the Gallup poll usually take weekly samples of about 1500 people. This sample size gives a margin of error of about ± 3 percentage points (that is, 95% of all such samples are accurate within ± 3%). Just before a presidential election, however, the polls often increase the size of their samples to about 4000 people. Is the margin of error now more than ± 3%, less than ± 3%, or still equal to ± 3%? Why?

The studies in Exercises 19 to 21 may produce invalid data because of confounding of outside influences with the treatment of interest. Explain in each case how confounding could influence the outcome.

■ 19. A college student believes that rose hip tea has remarkable curative powers. To demonstrate this, she and several friends visit a local nursing home several times a week, talking with the residents and serving them rose hip tea. After a month, the head nurse reports that the residents visited are indeed more cheerful and alert.

■ 20. A language teacher believes that study of a foreign language improves command of English. He examines the records at his high school and finds that students who elect a foreign language do indeed score higher on English achievement tests.

■ 21. A job-training program is being reviewed. Critics claim that because the unemployment rate in the manufacturing region affected by the program was 8% when the program began and 12% four years later, the program was ineffective.

■ 22. It has been suggested that there is a "gender gap" in political party preference in the United States, with women more likely than men to prefer Democratic candidates. A political scientist asks each of a group of men and a group of women whether they voted for the Democratic or Republican candidate in the last Congressional election. Explain carefully why this study is *not* an experiment.

■ 23. A study of the effect of living in public housing on family stability and other variables in poverty-level households was carried out as follows. The researchers obtained a list of all applicants for public housing during the previous year. Some applicants had been accepted, while others had been turned down by the housing authority. Both groups were interviewed and compared. Was this study an experiment? Why or why not?

■ 24. Before a new variety of frozen muffins is put on the market, it is subjected to extensive taste testing. People are asked to taste the new muffin and a competing brand, and to say which they prefer. (Both muffins are unidentified in the test.) Is this an experiment? Why or why not?

25. Will reducing blood-cholesterol levels prevent heart attacks? You have available a drug that will lower blood cholesterol. You also have 3000 men aged 50 to 65 who are willing to participate in a study. Outline the design of an experiment to settle the question. Your outline should follow the model of Figure 5.5. Be sure to give the sizes of the treatment groups and to indicate the outcomes you will examine.

■ 26. Does regular exercise reduce the risk of a heart attack? Several ways of studying this question suggest themselves.

　　a. A researcher takes a sample of 2000 men in their forties who recently suffered their first heart attack. He matches each with another man of the same age, occupation, and other demographic characteristics who has not had a heart attack. Both groups are questioned about their past exercise habits. Is this an experiment? Is it a prospective study? Explain your answers.

　　b. Another researcher finds 2000 men over 40 who exercise regularly and have not had heart attacks. She matches each with a similar man who does not exercise regularly, and she follows both groups for 10 years. Is this an experiment? Is it a prospective study? Explain your answers.

　　c. You have 4000 men over 40 who have not had a heart attack and who are willing to participate in a study. Outline the design of an experiment to investigate the effect of regular exercise on heart attacks.

■ 27. Should either or both the experiments in Exercises 25 and 26c be double-blind studies? Explain your answer.

■ 28. An experiment that claimed to show that meditation lowers anxiety proceeded as follows. The experimenter interviewed the subjects and rated their levels of anxiety. Then the subjects were randomly assigned to two groups. The experimenter taught one group how to meditate, and they meditated daily for a month. The other group was simply told to relax more. At the end of the month, the experimenter interviewed all the subjects again and rated their anxiety levels. The meditation group now had less anxiety. Psychologists said that the results were suspect because the ratings were not blind. Explain what this means and how lack of blindness could bias the reported results.

29. Some investment advisors believe that charts of past trends in the prices of securities can help predict future prices. Most economists disagree. In an experiment to examine the effects of using charts, business students trade (hypothetically) a foreign currency at computer screens. There are 20 student subjects available, named for convenience *A, B, C, . . . , T*. Their goal is to make as

much as possible, and the best performances are rewarded with small prizes. The student traders have the price history of the foreign currency in dollars in their computers; they may or may not also have software that highlights trends. Describe a design for this experiment and carry out the randomization required by your design.

30. Ignoring all practical difficulties and moral issues, outline the design of an experiment that would settle the question of whether cigarette smoking causes lung cancer.

31. In a test of the effects of persistent pesticides, researchers will feed a diet contaminated with DDT to rats for 60 days after weaning. They then will measure the rats' nerve responses to assess the effects of the DDT.
   a. Explain why the experimenter should also study a control group of rats that are fed the same diet uncontaminated with DDT.
   b. If 20 newly weaned male rats are available, outline the design of the experiment and use Table 5.1 starting at line 123 to carry out the randomization.

32. A college allows students to choose either classroom or self-paced instruction in a basic statistics course. To compare the effectiveness of self-paced and regular instruction, someone proposes administering the same final exam to all students in both versions of the course and comparing the average score of those who took the self-paced option with the average score of students in regular sections.
   a. Explain why confounding makes the results of that study worthless.
   b. Given 30 students who are willing to use either regular or self-paced instruction, outline an experimental design to compare the two methods of instruction. Then use Table 5.1 starting at line 108 to carry out the randomization.

33. Below are the names of 20 patients who have consented to participate in a trial of surgical treatments for angina. Outline an experiment to compare surgical treatment with a placebo (sham surgery) and use Table 5.1, beginning at line 101, to do the required randomization. (Ignore the asterisks.)

Ashley	Cravens*	Lippmann	Strong*
Bean*	Dorfman	Mark*	Tobias
Block	Epstein	Morton*	Voison*
Burt	Huang*	Popkin	Washington
Chavez*	Kidder	Spearman	Williams

34. Unknown to the researchers in Exercise 33, the eight subjects whose names are marked by asterisks will have a fatal heart attack during the study period. We can observe how sampling variability operates in a randomized experiment by keeping track of how many of these eight subjects are assigned to the group that will receive the new surgical treatment. Carry out the random assign-

ment of 10 subjects to the treatment group 20 times, keeping track of how many asterisks are on the names you choose each time. Then make a histogram of the count of heart attack victims assigned to the treatment. What is the average number in your 20 tries?

▲ 35. Explain clearly the advantage of using several thousand subjects, rather than just 20, in the experiment of Exercise 33.

■ 36. A randomized comparative experiment examined whether a calcium supplement in the diet reduces the blood pressure of healthy men. The subjects received either a calcium supplement or a placebo for 12 weeks. The researchers concluded that "the blood pressure of the calcium group was significantly lower than that of the placebo group." "Significant" in this conclusion means statistically significant. Explain what statistically significant means in the context of this experiment, as if you were speaking to a doctor who knows no statistics.

■ 37. The finanical aid office of a university asks a sample of students about their employment and earnings. The report says that "for academic year earnings, a statistically significant difference was found between the sexes, with men earning more on the average. No significant difference was found between the earnings of black and white students." Explain both of these conclusions, for the effects of sex and of race on average earnings, in language understandable to someone who knows no statistics.

▲ 38. Is the number of days a letter takes to reach another city affected by the time of day it is mailed and whether or not the ZIP code is used? Describe briefly the design of a two-factor experiment to investigate this question. Be sure to specify the treatments exactly and to tell how you will handle outside variables, such as the day of the week on which the letter is mailed.

▲ 39. The previous exercise illustrates the use of a statistically designed experiment to answer questions that arise in everyday life. Select a question of interest to you that an experiment might answer and carefully discuss the design of an appropriate experiment.

● 40. Explain carefully how you would randomly assign the 20 subjects named in Exercise 33 to the four treatments in the Physicians' Health Study. (Assign 5 of the 20 to each group.) Then enter Table 5.1 at line 120 to carry out the randomization.

● 41. New varieties of corn with altered amino acid patterns may have higher nutritive value than standard corn, which is low in the amino acid lysine. Researchers conduct an experiment to compare a new variety, called floury-2, with normal corn. They mix corn–soybean meal diets using each type of corn at each of three protein levels, 12% protein, 16% protein, and 20% protein. There are thus six diets in all. Ten one-day-old male chicks are assigned to each diet, and

● Optional exercise.

their weight gains after 21 days are recorded. The weight gain of the chicks is a measure of the nutritive value of their diet.

   a. This experiment has two factors. What are they?
   b. Outline the design of the experiment. Be sure to use randomization. (You need not actually carry out the randomization required by your design.)

● 42. An engineer wants to study the effect of the speed of a conveyor belt carrying electronic circuit boards on the performance of a wave-soldering machine that simultaneously solders all the connections on a board as the conveyor moves the board through a standing wave of molten solder. The speeds to be compared are 20, 25, and 30 feet per minute. The outcome variable is the number of improperly soldered connections among the 2000 connections on a circuit board.

   a. The engineer plans to process 10 boards at each conveyor speed. Why should she assign the speeds at random to the 30 boards, rather than simply process the first 10 at 20 feet per minute, the second 10 at 25 feet per minute, and so on?
   b. Outline the design of a randomized comparative experiment, beginning with boards numbered 1 through 30 in the order in which they will be soldered.
   c. Enter Table 5.1 at line 130 to carry out the randomization required. List the sequence of 30 conveyor speeds that the engineer will use when she carries out the experiment.

● 43. Find a 3 × 3 Latin-square arrangement of three treatments A, B, and C. Then find a 4 × 4 Latin-square arrangement of four treatments A, B, C, and D that is different from the one illustrated in Figure 5.9.

● 44. You wish to compare the durability of four types of wood treatment by burying the treated wood in soil for one year, then measuring decay. Because the variety of wood and the soil type also influence decay, you choose four wood species and four locations with different soil types. Lay out a 4 × 4 Latin square as the basis for an experimental design. Then randomly assign the wood species to the rows, the soil types to the columns, and the four types of wood treatment to the letters in the body of the Latin square.

▶ WRITING PROJECTS

1. The Current Population Survey (CPS) is the most important sample survey of the federal government. The CPS provides monthly information on employment and unemployment, and gathers information on many other economic and social issues on a less frequent basis by varying the questions asked each month. The sampling design of the CPS is described in the Bureau of Labor Statistics (BLS) *Handbook of Methods,* which is updated from time to time.

Locate the BLS *Handbook of Methods* in the library. Write a clear description of the multistage sampling design used for the CPS. Because the sampling design makes use of the new idea of *stratified sampling,* you may also wish to read the discussion of stratified sampling in *Statistics: Concepts and Controversies* (see the Suggested Readings) or another text.

2. Articles in the press often describe medical findings based on an experiment. The conclusion of the Physicians' Health Study that taking aspirin regularly helps prevent heart attacks is an example. Find an article in a newspaper or magazine that deals with a recent medical study. Describe the purpose and design of the study. Was it an experiment? Does the article mention a control group? Does it mention random assignment of the subjects? What were the conclusions of the study, and how well grounded do you think they are?

3. Choose an issue of current interest to students at your school. Prepare a short (no more than five questions) questionnaire to determine opinions on this issue. Choose a sample of about 25 students, administer your questionnaire, and write a brief description of your findings. Also write a short discussion of your experiences in designing and carrying out the survey.

(Although 25 students are too few to be statistically confident of your results, this project centers on the practical work of a survey. You must first identify a population; if it is not possible to reach a wider student population, use students enrolled in this course. Did the subjects find your questions clear? Did you write the questions so that it was easy to tabulate the responses? At the end, did you wish you had asked different questions?)

4. Although sample surveys raise fewer ethical issues than experiments with human subjects, here are some ethical questions you might address in a brief essay. How much advance information should respondents be given in order to decide whether to participate? (Perhaps you expect to spend 10 minutes, but the survey takes an hour. Perhaps questions about sex and drugs appear without warning.) Should respondents always be told who is sponsoring the poll? (If so, will knowing that the Republican National Committee is the sponsor affect their answers?) Should a poll always offer to send respondents a copy of the final report so they can see how their information is being used? (That's expensive.) All agree that an individual's responses are never released, only statistical summaries of all the responses — but should responses be anonymous, so that no one at the polling organization knows whose answers these are? (That prevents follow-up if a person doesn't respond on the first try, and rules out in-person interviews.) Write an essay expressing your opinions on one or more of these issues. You can find background information in *Statistics: Concepts and Controversies.*

Chapter

6 Describing Data

The proliferation of data is a prominent feature of modern society. Data, or numerical facts, are essential for making decisions in almost every area of our lives. Like the proverbial trees in the forest, however, these numbers threaten to overwhelm us unless we can take control of them through careful organization and interpretation. A corporate data base, for example, contains an immense volume of data — on sales, personnel, inventories, customer accounts, equipment, and other topics. Modern scientific activities such seismic exploration for oil can produce hundreds of thousands of numbers each day. All these data must somehow be interpreted, a task that requires us to present the numbers so that their message is immediately clear.

To use data for human purposes, we must compress, summarize, and describe them. A few numbers computed from the data — averages, percents, and the like — can be very helpful. Numbers computed from the data are also the raw material of *statistical inference,* the science of drawing conclusions from data with the aid of the mathematics of probability.

The method of formal statistical inference is to ask specific questions in advance and then col-lect data to answer those specific questions. However, this neat process is not always possible. We may have to analyze a mass of data collected for other purposes, such as government or corporate records, before we know whether the data can help answer our questions. In other cases, we may not even know what questions we should ask.

Exploratory data analysis is the art of letting the data speak, of seeing patterns in data that we may not have anticipated. Exploratory analysis is informal; unlike inference, it does not seek answers to specific questions. However, exploratory data analysis is not at all a diversion from the mainstream of statistics. Even in the most carefully planned experiment, exploring the data is an essential first step. It may reveal unsuspected errors or an important effect that was not anticipated.

Exploratory analysis of data combines numerical summaries with graphical display. We can grasp pictures more easily than columns of numbers, so the most powerful tools of data analysis are graphs. As long as our emphasis is on description rather than inference, pictorial display of data occupies first place.

Pictorial display of data is not a new idea. In 1861, Charles Minard, a French engineer, used an elaborate and original graph to show the impact of harsh winter conditions on Napoleon's troops during the unsuccessful invasion of Russia in 1812. Some 422,000 French soldiers entered Russia and 100,000 reached Moscow, but only 10,000 straggled back. The band across the map of Eastern Europe in Minard's graph (Figure 6.1) shows the route of Napoleon's Grand Army. The width of the band is proportional to the number of surviving troops. As the number of survivors dwindles, the river of soldiers narrows to a trickle. At the bottom of his graph, Minard showed the temperature during the winter retreat from Moscow. The falling temperatures and shrinking army march together in that famous defeat.

Visual representations of data abound in books and news media, although few are as imaginative or as striking as Minard's. The progress of computer graphics has given a new emphasis to pictorial display of data in business, medicine, and technical fields. The computing power of machines allied with the unique ability of the human eye and brain to recognize visual patterns provides powerful new tools for data analysis.

For example, in order to discover where oil reservoirs lie, geologists probe the structure of the earth with explosions. The shock waves of these explosions bounce off underlying rock strata and are reflected up to seismic sensors scattered around the exploration site. A single explosion can produce half a million separate data points, and a skilled seismic crew operating by helicopter can set off up to 20 such tests a day. This abundance of data has to be organized into meaningful and easily understood patterns.

The results of the tests are processed numerically and then presented using the tool of computer graphics (see Figure 6.2). An oil expert could look at this display and instantly interpret it as a promising site for further exploration. The computer does the calculating and graphics, then a trained eye "sees" the oil in the graphically dis-

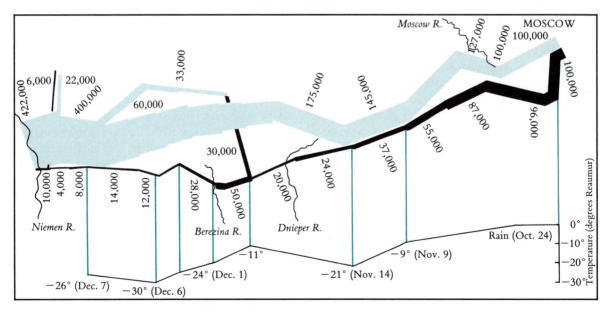

Figure 6.1 Redrawing of Charles Minard's 1861 graph of Napoleon's Russian campaign. (In the Reaumur temperature scale, water boils at 80°R and freezes at 0°R.)

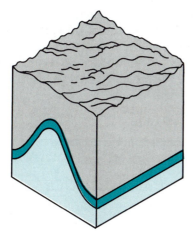

Figure 6.2 Underground structures reconstructed from seismic data.

played data. Such analysis is an effective collaboration between the computational power of machines and the ability of humans to see and understand.

In this chapter we use both numbers and pictures to explore data. We will organize our thinking by three principles:

1. Move from describing a single variable to describing relations among several variables.

2. Use graphical display in combination with numerical description.

3. Look first for an overall pattern in the data, then for important deviations from this pattern.

The pattern of values of a numerical variable is its **distribution.** Data analysis begins with graphical displays of a single variable's distribution.

▶ DISPLAYING DISTRIBUTIONS

Baseball fans have long memories for the statistics of the game (Spotlight 6.1). For example, how many home runs does it take to lead the league? Table 6.1 gives the American League leaders and their home run totals from 1972 to 1991. We can get a quick picture of the distribution of the league-leading home run total by making a **dotplot.** First, draw a horizontal axis marked off to span the range of the data. Then mark each observation with a dot above the axis. Here is the result:

TABLE 6.1   American League Home Run Leaders, 1972–1991

Year	Player	Home runs	Year	Player	Home runs
1972	Dick Allen	37	1982	Thomas and Jackson	39
1973	Reggie Jackson	32	1983	Jim Rice	39
1974	Dick Allen	32	1984	Tony Armas	43
1975	Scott and Jackson	36	1985	Darrell Evans	40
1976	Graig Nettles	32	1986	Jesse Barfield	40
1977	Jim Rice	39	1987	Mark McGwire	49
1978	Jim Rice	46	1988	Jose Canseco	42
1979	Gorman Thomas	45	1989	Fred McGriff	36
1980	Reggie Jackson	41	1990	Cecil Fielder	51
1981	Four players	22	1991	Canseco and Fielder	44

We see, for example, that 32 and 39 each led the league three times during these 20 seasons. The data cover the range from 32 to 51 without any strong pattern. A systematic pattern is often hard to see in small data sets such as this. But there is a clear **outlier,** an individual observation that falls outside the pattern of the remaining data. In 1981, it took only 22 home runs to lead the league. Outliers often point to errors in recording the data or to unusual circumstances. In fact, the 1981

SP⬤TLIGHT 6.1   Baseball: The Great Statistics Game

▶ ▶ ▶ ▶ ▶ ▶ ▶ ▶ ▶ ▶ ▶ ▶ ▶

Baseball fans have a love affair with statistics. There's absolutely no doubt about it: it's the greatest statistics game there is, probably because baseball goes back so far in this country: it's embedded in most of the population. They really understand a home run and an RBI, a batting average and an earned run average — all those basic statistics that have been with baseball throughout its history. The basics have never changed, so people know and love them.

Dick Bresciani
Red Sox Chief Statistician

George Brett of the Kansas City Royals hit .390 in 1980, the highest major league batting average since Ted Williams's average of .406 in 1941. [Kansas City Royals.]

There is a whole lore of baseball history involving statistics. The great thing is to compare the players of old with the players of today. Many times on talk shows or panels people will say, "Could Dave Winfield or Jim Rice or George Brett have played with a Ty Cobb or a Mickey Cochran or a Ted Williams or a Joe DiMaggio?" What you have to argue is statistics. You have to go back and examine DiMaggio's years in the big leagues. You look at what he did year by year: he was on average a .300-and-some hitter; he drove in so many runs; he did thus-and-so defensively in the outfield. The statistics are all that remain of the career of that player. So you lay them out against a Brett or a Winfield and try to compare them. That is the fun of the game.

Lou Gorman
Red Sox General Manager

baseball season was interrupted by a players' strike that reduced the number of games played from the usual 162 to 108.

Dotplots are quick to draw and work well for small numbers of observations. They become awkward when there are many observations or when the observations do not have well-separated values. (Whole-number values like counts of home runs are ideal for dotplots.) When a dotplot is not satisfactory, we display a distribution by a more formal type of graph, a **histogram.** To see how histograms work, let's ask another baseball question: How well do major league batters hit?

EXAMPLE: Drawing a Histogram

The simplest measure of how well a baseball player hits is his batting average, which is simply the proportion of times at bat that the player gets a hit. However, the batting average of a player who has been at bat only a few times may be due to luck—either good or bad. To elimi-

nate these cases, we consider only players who have been at bat 200 or more times in a season. Figure 6.3 is a histogram of the distribution of batting averages for all 167 American League players who batted 200 or more times in the 1980 season. We can learn from this example how to draw a histogram of any distribution.

Making a Histogram

1. Divide the range of the data into classes of equal width. In this case, each class covers a 10-point range of batting averages. The first two classes are:

$$.185 \leq \text{batting average} < .195$$
$$.195 \leq \text{batting average} < .205$$

Leave no space between classes and be careful about the endpoints of the classes; for example, .194 falls in the first class and .195 in the second.

2. Count the number of observations in each class. These counts are called **frequencies.** The

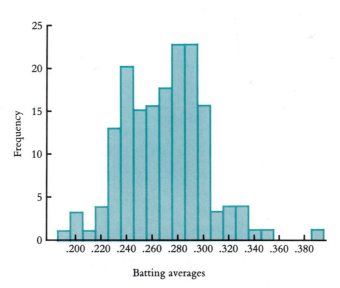

Figure 6.3 Histogram of 1980 American League batting averages.

TABLE 6.2   Frequencies for the Histogram in Figure 6.3

Class	Frequency	Class	Frequency
.185 to .194	1	.295 to .304	16
.195 to .204	3	.305 to .314	3
.205 to .214	1	.315 to .324	4
.215 to .224	4	.325 to .334	4
.225 to .234	13	.335 to .344	1
.235 to .244	20	.345 to .354	1
.245 to .254	15	.355 to .364	0
.255 to .264	16	.365 to .374	0
.265 to .274	18	.375 to .384	0
.275 to .284	23	.385 to .394	1
.285 to .294	23		

frequencies for the batting average classes are shown in Table 6.2.

3. Draw the histogram. The batting average scale is horizontal and the frequency scale vertical. Each bar represents a class. The base of the bar covers the class, and the bar height is the class frequency. The graph is drawn with no horizontal space between the bars (unless a class is empty, so that its bar has 0 height).   ▲

There is no single right number of classes in a histogram. You must use judgment. The goal is to give a clear picture of the distribution. Avoid using too few classes, which produces a skyscraper picture with all the observations bunched together; also avoid using too many classes, which gives a pancake histogram with few observations in each class. Figure 6.3 is a successful histogram because the overall pattern of the distribution is immediately apparent. Almost all major league regulars hit between about .225 and .305. A typical player hit about .270.

Another aspect of the histogram is also immediately apparent. The single observation at .390 is an outlier. Outliers are often the result of errors

and must be carefully investigated. It would be easy, for instance, to type .390 in place of .290 while recording the data. However, this outlier is not a mistake. George Brett of the Kansas City Royals hit .390 in 1980, the highest major league batting average since Ted Williams's average of .406 in 1941. Brett therefore deserves his isolated point on the histogram.

George Brett and his more-average colleagues illustrate an important principle in the exploration of data: always look for an *overall pattern*, and then look for *deviations* from that pattern. In shorthand language

$$Data = smooth + rough$$

The *smooth* in any data is an overall pattern. In this histogram, the smooth is the regular distribution of batting averages, with a few in the .220s, many between .240 and .290, then falling off to a few in the .330s and above. The *rough* is the outlier at .390. Both the smooth and the rough — the pattern and the exception — carry useful information.

The smooth in a histogram is the overall shape of the distribution. There are some shapes that are common enough to keep in mind when examining a histogram. A distribution is **symmetric** if the right and left halves are mirror images of each other. A **skewed** distribution has one tail that is much longer than the other. Distributions of real data will of course be only approximately symmetric. We consider Figure 6.3 (without George Brett) to be approximately symmetric.

EXAMPLE: How Many Operations Do Doctors Perform?

A study of the number of surgical procedures performed by physicians obtained the following data on the number of hysterectomies performed in a year by each of a sample of 15 male doctors in Switzerland.

27, 50, 33, 25, 86, 25, 85, 31,
37, 44, 20, 36, 59, 34, 28

Figure 6.4 is a histogram of this distribution; the classes are 20–29, 30–39, and so on. The distribution has a long right tail not matched by

corresponding observations to the left of the bulk of the data. That is, a few doctors perform many hysterectomies. We say that the distribution is *skewed to the right*. Notice that the direction of the skew is the direction of the long tail, not the direction in which most observations cluster.   ▲

The rough in a distribution often takes the form of *outliers* or *gaps*. The baseball strike of 1981 and George Brett's exceptional performance were responsible for outliers in the distributions of home runs and of 1980 batting averages. The next example features a gap.

EXAMPLE: Quality Control

Spotting patterns in quality control data can increase a manufacturer's productivity and profitability. Although Japanese products are known today for high reliability, this wasn't always the case. W. Edwards Deming, an American expert in quality control, was a leader in teaching Japa-

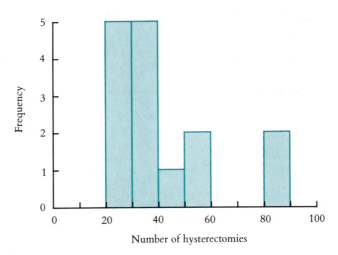

Figure 6.4 Histogram of hysterectomies performed by male Swiss doctors.

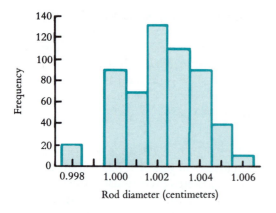

Figure 6.5 Deming's illustration of the effects of improper inspection: a histogram with a gap.

nese industrialists about statistical methods in the 1950s. His efforts were so successful that Japan's annual award for excellence in quality is named after him.

One of Deming's studies concerned the size of steel rods used in a manufacturing process. The histogram in Figure 6.5 summarizes the data. It displays the diameters of 500 steel rods, as measured by the manufacturer's inspectors. An overall pattern — the smooth — is apparent in the size of the rods: the distribution is approximately symmetric, centered at 1.002 centimeters and falling off rapidly both above and below. The departure from regularity — the rough — is the empty space at 0.999 centimeter.

One centimeter, precisely 1.000, is the lower specification limit for these rods. Rods that are any smaller than this will be loose in their bearings and should therefore be rejected. The empty 0.999 class in the histogram, together with the taller-than-expected bar in the 1.000 class, provides strong evidence that the inspectors were passing rods that fell just below this limit by recording them in the 1-centimeter class. The inspectors didn't realize that just $\frac{1}{1000}$ of a centimeter can be crucial. With better

training of the inspectors, the missing 0.999 class filled in and the distribution became quite regular. ▲

The strong visual impact of a graph helped pinpoint an important problem in the quality control example. Interpreting the smooth portion of a histogram can also be helpful. In this example, it might suggest that the entire process should be adjusted so that more of the distribution (that is, more of the steel rods produced) exceed 1 centimeter in diameter.

▶ NUMERICAL DESCRIPTION OF DISTRIBUTIONS

Dotplots and histograms display the overall shape of a distribution of values. We can describe specific aspects of this overall shape by a few carefully chosen numbers. Two important aspects of the overall shape of a distribution are its *center* (sometimes called *location*) and its *spread*. Center and spread are visible in a graph, but now we want to describe them using numbers.

One simple way to describe the center of a distribution is to find the number with half the values falling below it and the other half above. This is the **median** of the observations. The median can be regarded as the typical value. We will call the median M for short. Although the idea of the median as the midpoint of a distribution is simple, we need a precise rule for calculating the median.

Calculating the Median

1. Arrange all observations in order of size, from smallest to largest.

2. If the number n of observations is odd, the median M is the center observation in the ordered list. The location of the median is found by counting $(n + 1)/2$ observations up from the bottom of the list.

3. If the number n of observations is even, the median M is the average of the two center observations in the ordered list. The location of the median is again $(n + 1)/2$ from the bottom of the list.

Be sure to write down each individual observation in the data set, even if several observations repeat the same value. And be sure to arrange the observations by size before locating the median. The middle observation in the haphazard order in which the observations first come has no importance.

Another way to measure the center of a set of observations is to find their ordinary arithmetic average, called the **mean**. The mean is usually written as $\bar{x}$, which is read as "x-bar."

To find the *mean* of a set of observations, add the values and divide by the number of observations. If the observations are $x_1, x_2, \ldots, x_n$, their mean is

$$\bar{x} = \frac{x_1 + x_2 + \cdots + x_n}{n}$$

For an example of the calculation of the mean and the median, let's return to the Swiss doctors.

EXAMPLE: Calculating Mean and Median

Our sample data for male doctors were

27, 50, 33, 25, 86, 25, 85, 31,
37, 44, 20, 36, 59, 34, 28

The mean of this sample is

$$\bar{x} = \frac{27 + 50 + 33 + \cdots + 28}{15}$$

$$= \frac{620}{15} = 41.3$$

To find the median, first arrange the observations in order:

20, 25, 25, 27, 28, 31, 33, 34,
36, 37, 44, 50, 59, 85, 86

There are $n = 15$ observations, so the location of the median is

$$\frac{n + 1}{2} = \frac{16}{2} = 8$$

The median is the 8th observation in the ordered list. Therefore $M = 34$.

A sample of 10 female Swiss doctors was studied at the same time. The numbers of hysterectomies performed by these doctors (arranged in order) were

5, 7, 10, 14, 18, 19, 25, 29, 31, 33

Here $n = 10$ is even and the location of the median is

$$\frac{n + 1}{2} = \frac{11}{2} = 5.5$$

The location 5.5 means "halfway between the fifth and sixth observations in the ordered list." So the median is the average of these two observations,

$$M = \frac{18 + 19}{2} = 18.5$$

The typical female doctor performed many fewer hysterectomies than the typical male doctor. This was one of the important conclusions of the study.   ▲

This example illustrates an important difference between the mean and the median. The mean is strongly influenced by a few large observations, so that the mean of a right-skewed distribution is larger than the median. The median number of hysterectomies performed by the male doctors was 34, but the few large values (85 and 86) in the right tail of the distribution pull the mean up to 41.3. In practice, you must ask yourself whether the "typical value" (the median) or the "arithmetic average value" (the mean) is a better description of the center of the data.

The mean and median provide two different measures of the center of a distribution. But a measure of center alone can be misleading. The Bureau of Labor Statistics reports that in 1991 the median income of American families was $35,939. Half of all families earned less than $35,939, and half had higher incomes. But these figures do not tell the whole story.

EXAMPLE: Affluentia and Spartany

Imagine two countries with the same median family income, say $35,939. But these countries have vastly different economic systems. In Affluentia, all families earn nearly the same amount. In Spartany, on the other hand, there are extremes of wealth and poverty. Figure 6.6 displays the two distributions of pay. Despite

the fact that median income is identical in Affluentia and Spartany, income patterns in the two countries are dramatically different. Family incomes in Spartany have much greater spread or variation than in Affluentia. The median tells us nothing about the spread of a distribution. ▲

The simplest useful numerical description of a distribution consists of both a measure of center and a measure of spread. One way to measure spread is to calculate the *range,* the difference between the highest and lowest observations. For example, the highest and lowest 1980 American League batting averages were .390 and .188. The range is therefore .390 − .188, or .202. This example illustrates a weakness in the range as a measure of spread: it is determined by the two most extreme observations. George Brett alone adds 40 points to the range of 1980 batting averages. For a more useful indication of variability, one that is less influenced by extremes, we can limit the range to the middle 50% of observations.

We find this range by identifying points called **quartiles.** The *first quartile* is the value below which one-fourth of the observations fall. Three-fourths of the observations lie below the *third quartile.* The second quartile is the median, since two-fourths of the observations lie below the me-

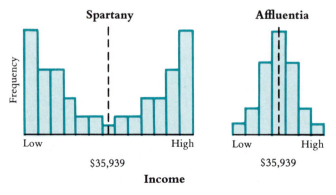

Figure 6.6 Two distributions with the same center but unequal spread.

dian. The exact rule for calculating the quartiles uses the rule for the median:

To calculate the quartiles, first locate the median in the ordered list of observations. The first quartile Q_1 is the median of the observations below the location of the overall median, and the third quartile Q_3 is the median of the observations above that location.

EXAMPLE: Calculating Quartiles

The numbers of hysterectomies performed by our sample of 15 male doctors were (arranged in order)

20 25 25 27 28 31 33 **34** 36 37 44 50 59 85 86

There are an odd number of observations, so the median is the middle one; it is the 8th in the list, $M = 34$. The median appears in bold type in the list. The first quartile is the median of the 7 observations to the left of the median. This is the 4th of these 7 observations, because in this case $(n + 1)/2 = (7 + 1)/2 = 4$. So $Q_1 = 27$. The third quartile is the median of the 7 observations to the right of the median, $Q_3 = 50$. The overall median is left out of the calculation of the quartiles when there are an odd number of observations.

Suppose that another sample gave the 10 observations

20 22 26 27 27 | 27 28 30 32 32

There are an even number of observations, so the median lies midway between the middle pair; its location is halfway between the 5th and 6th values, marked by | in the list above. Because both these values are 27, $M = 27$. The first quartile is the median of the first 5 observations, because these are the observations below the median's location. Be sure to note that it is the location of the median, not its numerical

value, that decides which observations are included in finding the quartiles. That location is halfway between the fifth and sixth observations, even though both these observations have the same value. Check that $Q_1 = 26$ and $Q_3 = 30$. ▲

A quick description of both the center and the spread of a set of data is provided by the **five-number summary.** This consists of the median, the two quartiles, and the two extremes (the smallest and largest individual observations). The five-number summary is always given in order from the smallest number to the largest. For example, we could calculate from the list of 167 American League batting averages that the five-number summary is: .188, .244, .271, .290, .390. This summary contains much useful information. For example, a typical major league regular player hit .271, and the middle half of all such players hit between .244 and .290. A graph helps to make the five-number summary more vivid.

A **boxplot** consists of a central box that spans the quartiles, with a line marking the median and whiskers extending out from the box to the extremes.

The box in a boxplot shows the range of the middle half of the data. Both the box and the span of the whiskers help display the spread of the distribution, while the median marks the center. You can draw the boxes in a boxplot either horizontally or vertically, as you prefer. In either case, be sure to accompany the plot by a scale marked off in units of the variable being described. The median, quartiles, and extremes are located on this scale in order to draw the boxplot.

Boxplots are particularly helpful for comparing several distributions. Figure 6.7, for example, displays the distributions of the number of hysterectomies performed in a year by our samples of male and female Swiss doctors. The scale in this example is simply the count of hysterectomies.

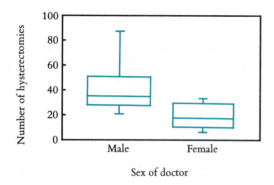

Figure 6.7 Side-by-side boxplots of the distributions of hysterectomies performed by male and female Swiss doctors.

The *variance* of n observations $x_1, x_2, \ldots, x_n$ is

$$s^2 = \frac{1}{n-1} \{(x_1 - \bar{x})^2 + (x_2 - \bar{x})^2 + \cdots + (x_n - \bar{x})^2\}$$

The *standard deviation s* is the square root of the variance s^2.

Conceptually, the variance s^2 is the average of the squared deviations $(x_i - \bar{x})^2$ of the individual observations from their mean. If the distribution is widely spread, at least some of these squared deviations will be large, so that s^2 and s are large. So the variance and the standard deviation do measure the spread of a distribution. You should rarely need to calculate s by hand, because most calculators allow you to obtain $\bar{x}$ and s from keyed-in data with the push of a button. Notice that the "average" in s^2 divides the sum by one fewer than the number of observations, $n - 1$ rather than n. The reason is that the deviations $x_i - \bar{x}$ always sum to exactly 0, so that knowing $n - 1$ of them determines the last one. Some calculators give you a choice between $n - 1$ and n in calculating s. The difference is small unless there are very few observations, but $n - 1$ is the usual choice in applied statistics.

Figure 6.7 quickly conveys a lot of information. You can tell at a glance that female doctors in general perform far fewer hysterectomies than men. In fact, the upper extreme for females falls below the male median. You can see, too, that the female distribution has less spread. In particular, it lacks the few very large observations that stretch out the upper whisker for the men. Figure 6.7 illustrates, once again, the effectiveness of visual displays. Before concluding that male doctors are more ready to perform hysterectomies, however, we need more information about the samples. Both males and females should be drawn from doctors with the same specialty and with practices of similar size. That was in fact the case for the Swiss study.

Though the five-number summary is the most generally useful numerical description of a distribution, it is not the most common. That distinction belongs to the combination of the mean with the **standard deviation.** The mean, like the median, is a measure of center; the standard deviation, like the quartiles and extremes in the five-number summary, measures spread. The standard deviation and its close relative, the *variance,* measure spread about the mean.

EXAMPLE: Calculating the Standard Deviation

The counts of hysterectomies performed by a sample of male Swiss doctors were

$$27, 50, 33, 25, 86, 25, 85, 31,$$
$$37, 44, 20, 36, 59, 34, 28$$

We found that the mean was $\bar{x} = 41.3$. So the variance is

$$s^2 = \frac{1}{n-1}\{(x_1 - \bar{x})^2 + (x_2 - \bar{x})^2 + \cdots$$
$$+ (x_n - \bar{x})^2\}$$
$$= \frac{1}{14}\{(27 - 41.3)^2 + (50 - 41.3)^2 + \cdots$$
$$+ (28 - 41.3)^2\}$$
$$= \frac{1}{14}\{204.49 + 75.69 + \cdots + 176.89\}$$
$$= \frac{5945.35}{14} = 424.67$$

The standard deviation is

$$s = \sqrt{424.67} = 20.61$$

A calculator gives the more exact results $\bar{x} = 41.3333$ and $s = 20.6074$.   ▲

More important than the details of hand calculation are the properties that determine the usefulness of the standard deviation:

▶ s measures spread about the mean, and should be used only when you choose the mean as your measure of center.
▶ s has the same units of measurement as the original observations. For example, if you measure lengths in centimeters, s is also in centimeters. This is one reason to prefer s to the variance s^2, which is in squared centimeters.
▶ $s = 0$ only when there is no spread, that is, when all observations have the same value. Otherwise $s > 0$.
▶ Like the mean $\bar{x}$, s is strongly influenced by a few extreme observations. For example, if we omit the two outliers 85 and 86 in the Swiss hysterectomy data, the standard deviation drops from 20.61 to 10.97.

We now have a choice between two numerical descriptions of the center and spread of a distribu-

tion: the five-number summary, or $\bar{x}$ and s. Because of the sensitivity of $\bar{x}$ and s to extreme observations, it is best to use the five-number summary to describe skewed distributions and reserve $\bar{x}$ and s for distributions that are roughly symmetric.

Although the standard deviation is widely used, it is not a natural or convenient measure of the spread of a distribution. Why, for example, should we make our measure of spread more sensitive to a few outliers by squaring the deviations, then take the square root at the end? Why not just average the distances of the observations from $\bar{x}$? The real reason for the popularity of the standard distribution is that it is the natural measure of spread for normal distributions, an important class of distributions presented in the next chapter.

▶ DISPLAYING RELATIONS BETWEEN TWO VARIABLES

In the examples we have looked at to this point, there was only one variable of interest, such as a hitter's batting average or the number of hysterectomies performed by a doctor. Now we examine data for two variables, emphasizing the nature and strength of the relationship between the variables. Because it is very hard to spot a relationship from columns of numbers, graphs are essential.

EXAMPLE: Natural Gas Consumption

Sue is concerned about the amount of energy she uses to heat her home. She keeps a record of the natural gas consumed over a period of nine months. Because the months are not all equally long, she divides each month's consumption by the number of days in the month to get cubic feet of gas used per day. Then from local weather records, Sue obtains the number of degree days for each month. She divides this total by the number of days in the month, giving the average number of degree days per day during the month. (Degree days are a measure of demand for heating; one degree day is accumulated for each degree that a day's average temperature falls below 65°F. For example, a day with an average temperature of 30°F gives 35 degree days.)

Table 6.3 shows Sue's data: nine measurements for each of the two variables. Looking at the numbers in the table, we can see that more degree days go with higher gas consumption. But the shape and strength of the relationship are not fully clear. Our problem is to display and interpret these data.   ▲

Data on two variables can be displayed in a **scatterplot**, a plot of all data points in two dimensions. Figure 6.8a is a scatterplot of Sue's data. Always plot the explanatory or causal variable on the horizontal scale and the outcome or response variable on the vertical scale. Degree days appear on the horizontal scale and gas consumption on the vertical scale because the weather affects gas consumption; gas consumption does not explain the weather.

Just as when we examined distributions of a single variable, we look for an overall pattern in a scatterplot and then for any striking deviations from that pattern. The relationship between degree days and gas consumption is very strong and takes the form of a straight line. We can represent the overall pattern of the relationship by drawing a straight line through the points of the scatterplot.

Figure 6.8b shows such a line. The smooth in our "data = smooth + rough" motto is represented by the line; the rough is the scatter of points about the line. In this case, the number of degree days explains most of the variation in gas consumption. The scattered points above and below the line reflect the effects of other factors, such as use of gas for cooking or turning down the thermostat when the family is away from home. These effects are relatively small. There are no outliers (points far outside the overall straight-line pattern) or other important deviations.

How can we draw a line through the points of a scatterplot? When the plot shows a strong straight-line relationship, it is easy to fit a line on the graph by using a transparent straightedge. This gives us a line on the graph, but not an equation for the line. There is also no guarantee that the line we fit by eye is the "best" line. There are statistical techniques for finding from the data the equation of the best line (with various meanings of "best"). The most common of these techniques, called *least squares regression*, is discussed in the next section. All statistical computer software packages and some calculators will calculate the least squares line for you, so that a line is often available with little work. You should therefore know how to use a fitted line even if you do not learn the details of how to get the equation from the data.

In writing the equation of a line, we use x for the explanatory variable because this is plotted on the horizontal or x axis, and y for the response variable. Any line has an equation of the form

TABLE 6.3   Sue's Household Consumption of Natural Gas Compared to the Need for Heat

	Oct	Nov	Dec	Jan	Feb	Mar	Apr	May	June
Degree days per day	15.6	26.8	37.8	36.4	35.5	18.6	15.3	7.9	0.0
Gas consumed per day (in cubic feet)	5.2	6.1	8.7	8.5	8.8	4.9	4.5	2.5	1.1

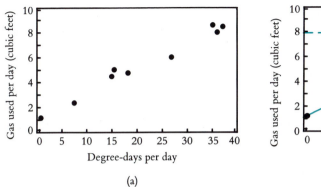 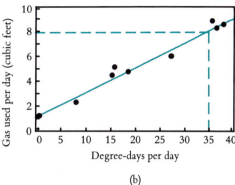

(a)   (b)

Figure 6.8 Natural gas consumption versus degree days. (a) A scatterplot. (b) A fitted line and its use for prediction.

$$y = b_0 + b_1 x$$

The number b_1 is the *slope* of the line, the amount by which y changes when x increases by one unit. The slope is usually important in a statistical setting, because it is the rate of change of the response y as x increases. The number b_0 is the *intercept*, the value of y when $x = 0$.

EXAMPLE: Interpreting Slope and Intercept

A computer program tells us that the least squares regression line computed from Sue's data is

$$y = 1.23 + 0.202x$$

The slope of this line is $b_1 = 0.202$. This means that gas consumption increases by 0.202 cubic foot per day when there is one more degree day per day. The intercept is $b_0 = 1.23$. When there are no degree days (that is, when the average temperature is 65 °F or above), gas consumption will be 1.23 cubic feet per day. The slope and

intercept are of course estimates based on fitting a line to the data in Table 6.3. We do not expect every month with no degree days to average exactly 1.23 cubic feet of gas per day. The line represents only the overall pattern of the data.   ▲

One of the most common goals in fitting a line to data is to predict the value of the response variable for a given value of the explanatory variable. A line drawn on a scatterplot can be used for making predictions with a straightedge and pencil. If the equation of the line is available, we can simply substitute the given value of the explanatory variable into the equation.

EXAMPLE: Predicting Gas Consumption

How much natural gas should Sue predict that she will consume in a month with 35 degree days per day? Figure 6.8b illustrates the use of the line drawn on the scatterplot. First locate 35 on the horizontal axis. Draw a vertical line up to the fitted line and then a horizontal line over to the gas consumption scale. As the figure shows, we predict that slightly more than 8 cubic feet per day will be consumed.

We can give a more exact prediction using the equation of the least squares regression line. This equation is

$$y = 1.23 + 0.202x$$

In this case, x is the number of degree days per day during a month and y is the number of cubic feet of gas consumed per day. Our predicted gas consumption for a month with $x = 35$ degree days per day is

$$y = 1.23 + (0.202)(35)$$
$$= 8.3 \text{ cubic feet per day}$$

This prediction will almost certainly not be exactly correct for the next month that has 35 degree days per day. But the past data points lie so close to the line that we can be confident that gas consumption in such a month will be quite close to 8.3 cubic feet per day. ▲

Sue had a practical reason for working through this exercise in statistics. She plans to add insulation to her house during the summer and at the end of next winter she will want to know how much she has saved in heating costs. She cannot simply compare before-and-after gas usage, because the winters before and after will not be equally severe. Nor can she do a comparative experiment to weigh the costs of insulated and uninsulated houses during the same winter, for she has only one house. Once Sue has next winter's degree-day data, however, she can use the fitted line to predict how much gas she would have used before insulating. Comparing this prediction with the actual amount used after insulation will show her savings.

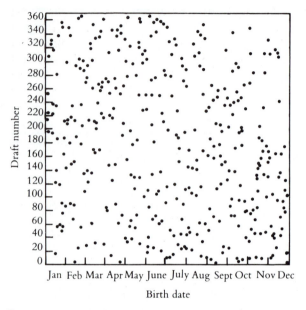

Figure 6.9 Scatterplot of draft selection number versus birth date for the 1970 draft lottery.

Scatterplots are a good tool for exploring the relationships between two or more variables, but often the overall pattern is not as simple as the straight-line pattern in Figure 6.8. Combining scatterplots with some of the numerical descriptions mentioned earlier can help us see the pattern in data.

Consider the case of the 1970 draft lottery. To eliminate distinctions among men eligible for the draft, Congress decided to allow chance to determine who would be selected for military service. The first draft lottery was held in 1970. Birth dates for all men born between 1943 and 1952 were to be drawn at random and assigned selection numbers in the order drawn. Men whose birth date was the first drawn (selection number 1) were drafted first. They were followed by men with numbers 2, 3, 4, and so on.

This procedure sounds fair, but the random drawing was mishandled. Birth dates were placed into identical capsules, and the capsules were placed in a drum for the drawing. But the capsules were not mixed thoroughly enough. December dates, which were added last, remained on top and had a greater chance of being drawn early. Thus, men born in December were drafted and sent to Vietnam in greater numbers than those born in January.

EXAMPLE: The 1970 Draft Lottery

Figure 6.9 is a scatterplot of all 366 birth dates and the selection numbers assigned to them by the 1970 lottery. Having a low selection number, nearer the bottom of this graph, means being drafted earlier. The alleged association between birth dates late in the year and low selection numbers isn't easily seen. The scatterplot alone is not very helpful because the graph looks much like a random scatter of points. A formal analysis by probability would show that a result this unfair to men born late in the year would happen in less than 1 in 1000 truly random lotteries. The inequity is really there, but to see it, we need to be more imaginative in looking at the data.

In Figure 6.10, we combine graphical methods with some basic numerical descriptions to get a more detailed picture of the draft lottery. Because medians and quartiles give a compact summary of both the center and spread of a distribution, we compute these for each month's selection numbers. We then replace each month's points on the scatterplot with a box whose ends are at the upper and lower quartiles. These boxplots allow us to instantly compare the distribution of selection numbers from month to month. A line within each box marks the median for the month. To make the picture even clearer, we can draw a line connecting the medians. The misfortune of men born late in the year is now evident. ▲

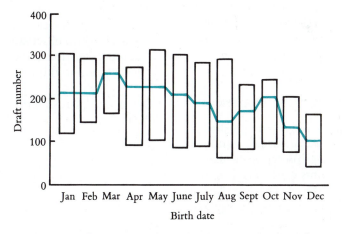

Figure 6.10 Monthly medians (line) and quartiles (boxes) for the 1970 draft lottery.

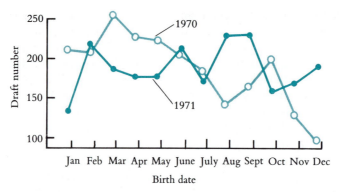

Figure 6.11   Median draft number by month for the 1970 and 1971 lotteries.

The media noticed this inequity, and in 1971 a new and genuinely random selection process was designed by statisticians at the National Bureau of Standards. Two drums instead of one were used in the 1971 lottery: one contained birth dates; the other, selection numbers. The capsules in both drums were thoroughly mixed, and then a randomly drawn birth date was paired with a randomly drawn selection number. When we add to our plot the medians of the reformed 1971 lottery, we can see the difference (Figure 6.11). There is a good bit of random variation in the 1971 medians, but no systematic trend as in 1970.

OPTIONAL ▶ **Least Squares Regression**

When a scatterplot shows a clear straight-line relation between an explanatory variable x and a response variable y, we want to describe the relation by a straight line. The points will rarely lie exactly on a line, so our problem is to find the line that best fits the points. To do this, we must first say what we mean by the "best fitting" line.

Suppose that we want to use our line to predict y for given values of x, as Sue used degree days to predict gas consumption. The error in our prediction is measured in the y, or vertical, direction. So we want to make the vertical distances of the points from the line as small as possible. A line that fits the data at all well will not pass entirely above or below the plotted points, so some of the errors will be positive and some negative. Their squares, however, will all be positive. The **least squares regression line** is the line that makes the sum of the squares of the vertical distances of the data points from the line as small as possible. The least squares idea says what we mean by the best fitting line. We must still learn how to find this line from the data. Given n observations on variables x and y, what is the equation of the least squares line? Here is the solution to this mathematical problem.

The least squares regression line of y on x calculated from n observations on these two variables is given by $y = b_0 + b_1 x$, where

$$b_1 = \frac{n\Sigma xy - (\Sigma x)(\Sigma y)}{n\Sigma x^2 - (\Sigma x)^2}$$
$$b_0 = \bar{y} - b_1 \bar{x}$$

Here Σ is just an abbreviation for "sum of," so that Σx means the sum of all the x values. The quantities $\bar{x}$ and $\bar{y}$ are the means of the x and y values. These algebraic formulas are shorthand for a series of operations that for clarity can be arranged in the form of a spreadsheetlike table, as the following example illustrates.

EXAMPLE: Gas Consumption Least Squares Line

The data for natural gas consumption y versus heating degree days x are given in Table 6.3. Table 6.4 contains the quantities needed to obtain the least squares regression line, beginning with the values of x and y in the first two columns. There are $n = 9$ observations. At the bottom of each column we write its sum. These sums are the building blocks for the least squares calculation. After writing down the x and y values, the steps in the calculation are:

STEP 1.   Sum the x and y columns and compute the two means. Here

$$\Sigma x = 193.9 \quad \text{and} \quad \bar{x} = \frac{193.9}{9} = 21.54$$

$$\Sigma y = 50.3 \quad \text{and} \quad \bar{y} = \frac{50.3}{9} = 5.59$$

STEP 2.   Calculate x^2 and xy for each (x, y) data point, enter in the proper column, and sum.

STEP 3.   Substitute $n = 9$ and the building block sums into the formulas for the slope b_1 and then for the intercept b_0.

$$b_1 = \frac{n\Sigma xy - (\Sigma x)(\Sigma y)}{n\Sigma x^2 - (\Sigma x)^2}$$

$$= \frac{(9)(1375.0) - (193.9)(50.3)}{(9)(5618.11) - (193.9)^2}$$

$$= \frac{2621.83}{12965.78} = 0.202$$

$$b_0 = \bar{y} - b\bar{x}$$

$$= \frac{50.3}{9} - 0.202\frac{193.9}{9} = 1.23$$

The equation of the least squares line is therefore

$$y = 1.23 + 0.202x. \quad \blacktriangle$$

It is more common in statistical practice to use a statistical calculator or computer software to obtain b_1 and b_0 with less work.   ◄

TABLE 6.4   **The Arithmetic of Least Squares**

x	y	x^2	xy
15.6	5.2	243.36	81.12
26.8	6.1	718.24	163.48
37.8	8.7	1428.84	328.86
36.4	8.5	1324.96	309.40
35.5	8.8	1260.25	312.40
18.6	4.9	345.96	91.14
15.3	4.5	234.09	68.85
7.9	2.5	62.41	19.75
0.0	1.1	0.0	0.0
$\Sigma x = 193.9$	$\Sigma y = 50.3$	$\Sigma x^2 = 5618.11$	$\Sigma xy = 1375.0$

▶ GRAPHICS IN MANY DIMENSIONS

Our graphical analysis of the draft lottery in Figures 6.9 to 6.11 illustrates the insight we can gain using only simple graphs and some basic statistical calculations. Looking at data in several ways can yield results that we wouldn't see using a single method such as a scatterplot.

But we have so far looked only at scatterplots for two variables. What if we want to display a third variable in the same plot? Because we have already used the horizontal and vertical directions of the graph, there is only one space dimension left, moving out of and into the page. Unfortunately, such three-dimensional scatterplots are very hard to see clearly unless color or motion (or both) are used to help us gain perspective. Computer graphics can add color and motion, allowing us to see a scatterplot in three dimensions.

Computer graphics makes it possible to see relations and detect outliers in high-dimensional data sets. Imagine a mass of points in space with a single outlier positioned far from the mass. From most viewing angles, an outlier would appear as part of the main group, blending invisibly with the mass of points. If, however, we change the viewing angle by rotating the plot, the outlier would eventually be seen apart, like the Death Star appearing from behind the moon in *Star Wars*. Consider, for example, a manufacturing process in which three key measurements are plotted in three dimensions. Figure 6.12 shows that from most points of view, the data appear as a cloud, like gnats swarming in space. But if we rotate the image, we find that there are two clouds, one of which was hidden behind the other from our first viewpoint. What is more, an outlier lies between the two clouds.

Remember that every point in this picture is positioned according to three measurements. If we tried to spot the outlier by looking at a long table with three columns of measurements for hundreds of points, we would quickly lose our bearings. This is a case where graphic display creates clarity out of the confusion of raw numbers.

Computer graphics allows us to move around the cloud of data points and look at it from any direction. Moreover, the computer can move around a ten-dimensional cloud of points almost as easily as a three-dimensional one. We cannot, of

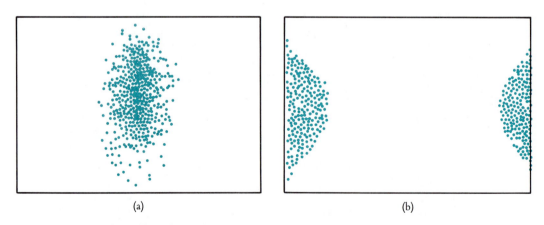

(a)   (b)

Figure 6.12 The effect of changing viewing angle in a three-dimensional scatterplot. (a) From this angle, the data form a single cloud. (b) From another angle, two clouds and an outlier are visible.

course, see ten dimensions. In fact, we do not directly see three dimensions on a piece of paper or video screen. In both cases, the computer presents a *projection* of the cloud of data onto the two-dimensional surface of the video screen. The computer can be instructed to compute and display changing projections, just as if we were moving around a many-dimensional cloud of data. Outliers and other important relations among the variables come clearly into view when they are scanned from the correct angle. To reduce the amount of time needed to discover the most meaningful viewing angles, the computer can even be programmed to search for interesting projections.

These multidimensional displays have practical applications. At Harvard University, for example, a graphics system designed by Professor Peter Huber aids geologists who are studying the pattern of earthquakes near the Fiji Islands (see Spotlight 6.2). For many years, the only view scientists had of earthquake epicenters was a standard map projection showing the two-dimensional location in latitude and longitude, as in Figure 6.13. Crucial data about the depths of the epicenters below the earth's surface were not reflected in the maps.

A computer graphics system allows earthquake locations to be seen in more detail. A fully three-dimensional view of the epicenters that adds depth to latitude and longitude can be presented and manipulated. This kind of picture permits geologists to examine the Fiji Islands earthquake pattern in light of plate tectonics, the geological theory of movements of vast plates that make up the earth's crust. These plate movements give rise to such geological events as the eruption of volcanos and the creation of mountain ranges.

Geologists explain that in the Fiji Islands, a plate moving in from the east has bent beneath one on the west so that the eastern plate dives straight down into the earth's mantle. The collision and resulting redirection of the plate account for most earthquakes in this region. The three-dimensional view that emerges as the computer continuously changes our viewing angle (see Figure 6.14) shows that earthquakes occur along the boundary between the plates deep inside the earth. The graphics system turns the image constantly, for a clearer view than a still picture provides. The detailed computer images enable geologists to spot wrinkles and other features in the surfaces of the plates.

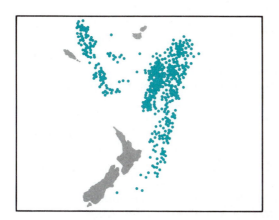

Figure 6.13 Earthquake epicenters in the Fiji Islands as seen from the earth's surface.

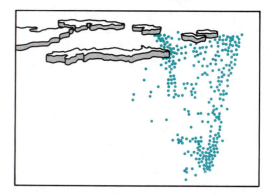

Figure 6.14 Earthquake epicenters after changing viewing angle to show depth beneath the surface.

SP TLIGHT 6.2   Visual Statistics: Eyeballing the Data

▶ ▶ ▶   ▶ ▶ ▶ ▶ ▶ ▶ ▶ ▶ ▶

Dr. Peter Huber, a statistician at the Massachusetts Institute of Technology, comments on the computer's ability to display complex, multidimensional arrangements of data, giving new meaning to the term *descriptive statistics:*

Dr. Peter Huber, MIT.

For most of the twentieth century, statistics focused on mathematical rigor and on small samples. In the past 10 years or so we have seen renewed interest in descriptive statistics, driven by the computer. It would be a mistake, however, to view this new emphasis as just a reaction of statistics to high tech. What we are witnessing now is a return to the descriptive statistics of the nineteenth century, completing unfinished business left over from that era. The computer makes it possible to do things that couldn't be done before.

In my view, descriptive statistics and mathematical statistics are complementary. In descriptive statistics, you see things, but you cannot test them in a formal way. In mathematical statistics, you may test for something preconceived, but you may overlook something you hadn't built into the test. So you simply have to do both so as to complement one with the other.

You also need subject specialists to complement the data analyst. In our experience, doing data analysis usually turns out to be a close conversational collaboration between the scientist (the one with the data) and the data analyst. They sit in front of the screen, discuss what they see, suggest the next action to be taken. Data analysis requires dynamic collaboration between the statistician and the subject-matter specialist.

The first step in any data analysis is data inspection, mainly to get familiar with the data and find extraordinary features. The next step is modification; enhance the picture by lines, maybe color groups, cluster and label selected points, maybe even fit some model to it. This leads to the most important step in data analysis: *comparison.* Without comparing things, you are not able to interact with your data. To help with interpretation, you almost always have to compare things — either several data sets or a data set and a model or different models for the same data set. Interpretation is the next step. And after interpretation, one usually has to begin another round of modeling. Very often the entire cycle starts again.

▶ REVIEW VOCABULARY

Boxplot   A graph of the five-number summary. A box spans the quartiles, with an interior line marking the median. Whiskers extend out from this box to the extreme high and low observations.

Distribution   The pattern of outcomes of a variable, listing the values that the variable takes and how often each value occurs.

Dotplot   A graphical display of the distribution of a variable for small data sets in which each observation is represented by a dot above its value on a scale of the variable being measured.

Exploratory data analysis   The practice of examining data for unanticipated patterns or effects, as opposed to seeking answers to specific questions.

Five-number summary   A summary of a distribution of values consisting of the median, the upper and lower quartiles, and the largest and smallest observations.

Frequency   The number of times an outcome or group of outcomes occurs in a set of data.

Histogram   A graph of the frequencies of all outcomes (often divided into classes) for a single variable. The height of each bar is the frequency of the class of outcomes covered by the base of the bar. All classes should have the same width.

Least squares regression line   A line drawn on a scatterplot that makes the sum of the squares of the vertical distances of the data points from the line as small as possible. The fitted line can be used for prediction.

Mean   The ordinary arithmetic average of a set of observations; to find the mean, divide the sum of all the observations by the number of observations summed.

Median   The midpoint of a set of data; half the observations fall below the median and half fall above.

Outlier   A data point that falls well outside the overall pattern of a set of data.

Quartiles   The first quartile of a distribution is the point with 25% of the observations falling below it; the third quartile is the point with 75% below it.

Scatterplot   A graph of the values of two variables as points in the plane; the value of the explanatory variable is plotted on the horizontal axis and the corresponding value of the response variable on the vertical axis.

Skewed distribution   A distribution in which observations on one side of the median extend notably farther from the median than do observations on the other side. In a right-skewed distribution, the larger observations extend farther to the right of the median than the smaller observations extend to the left.

Standard deviation   A measure of the spread of a distribution about its mean as center. It is the square root of the average squared deviation of the observations from their mean.

Symmetric distribution   A distribution described by a histogram or dotplot in which the part to the left of the median is roughly a mirror image of the part to the right of the median.

▶ SUGGESTED READINGS

CHAMBERS, JOHN M., WILLIAM S. CLEVELAND, BEAT KLEINER, AND PAUL A. TUKEY: *Graphical Methods for Data Analysis,* Wadsworth, Belmont, Calif., 1983. A detailed survey of modern graphical methods, many requiring a computer for effective use.

CLEVELAND, WILLIAM S.: *The Elements of Graphing Data,* Wadsworth, Monterey, Calif., 1985. A careful study of the most effective elementary ways to present data graphically, with much sound advice on improving simple graphs.

MOORE, DAVID S., AND GEORGE P. MCCABE: *Introduction to the Practice of Statistics,* 2nd ed., Freeman, New York, 1993. The first two chapters of this text provide a more extensive treatment of displaying and describing data for one and two variables. The material of this chapter is covered in more detail and much new material on both technique and interpretation is presented.

TUFTE, EDWARD R.: *The Visual Display of Quantitative Information*, Graphics Press, Cheshire, Conn., 1983. A beautifully printed book with classic examples such as Minard's work and suggestions for both statisticians and graphic artists.

VELLEMAN, PAUL F., AND DAVID C. HOAGLIN: "Data Analysis," in David C. Hoaglin and David S. Moore (eds.), *Perspectives on Contemporary Statistics*, Mathematical Association of America, Washington, D.C., 1992, pp. 19–39. A conceptual essay that presupposes knowledge of the basic techniques described in this chapter and in the text by Moore and McCabe just cited.

▶ EXERCISES

1. The data below are the number of days on which hail was observed at Evansville, Indiana, in each of 11 consecutive years:

$$4, 5, 3, 2, 4, 2, 0, 5, 0, 1, 1$$

Draw a dotplot of these data. Are there any outliers or other unusual features?

2. In an experiment on the effect of a drug on reaction time, a subject is asked to depress a button whenever a light flashes. Her reaction times for 11 trials are (in milliseconds)

$$96, 101, 102, 138, 93, 99, 107, 93, 95, 100, 100$$

Make a dotplot of these observations. Are there any outliers or other unusual features? What might have caused the outlier?

3. The following histogram displays data on the hour at which the first flash of lightning was observed each day during a study in Colorado. Describe this distribution: Is it roughly symmetric or distinctly skewed? Where is the center? Are there any outliers or gaps?

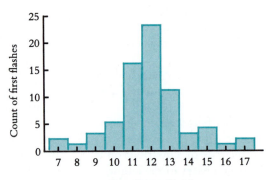

Hour from midnight

4. Members of a health maintenance organization (HMO) can make an unlimited number of visits to its member physicians for a fixed annual fee. The following histogram displays the distribution of the number of visits made by members of one HMO. (The vertical scale gives the proportion of the members in each class rather than the frequency.) Describe this distribution: Is it roughly symmetric or distinctly skewed? Are there any outliers or gaps?

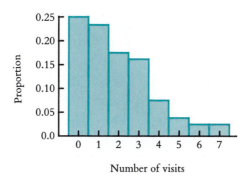

5. In 1798 the English scientist Henry Cavendish measured the density of the earth in a careful experiment with a torsion balance. Here are his 23 repeated measurements of the same quantity (the density of the earth relative to that of water) made with the same instrument:

5.36	5.62	5.27	5.46	5.53	5.57
5.29	5.29	5.39	5.30	5.10	5.79
5.58	5.44	5.42	5.75	5.34	5.63
5.65	5.34	5.47	5.68	5.85	

[Source: S. M. Stigler, "Do Robust Estimators Work with Real Data?" *Annals of Statistics,* 5:1055–1078 (1977).]

    a. Make a histogram of these data.
    b. Describe the distribution: Is it approximately symmetric or distinctly skewed? Are there gaps or outliers?

6. A fisheries researcher compiled the following data on lengths of 6-year-old white female crappies (in millimeters).

217	230	220	221	225	223
219	217	225	228	234	222
231	222	220	222	222	223
225	214	221	233	227	234
223	225	253	220	213	224
235	283	210	218	235	231

a. The data range from 210 to 283 mm. Group them into 5 classes of width 15 mm, starting with

$$210 \leq \text{length} < 225$$

as the leftmost class. Draw a well-labeled frequency histogram of the grouped data.

b. Describe the distribution: Is it roughly symmetric or clearly skewed? Are there gaps or outliers?

7. Return to the data on annual number of days with hail given in Exercise 1.
a. Find the mean and the median number of hail days.
b. Find the first and third quartiles of the number of hail days.

8. Find the five-number summary of the reaction times given in Exercise 2.

9. Here are the percentages of the popular vote won by the successful candidate in each of the presidential elections from 1948 to 1988.

Year	1948	1952	1956	1960	1964	1968	1972	1976	1980	1984	1988
%	49.6	55.1	57.4	49.7	61.1	43.4	60.7	50.1	50.7	58.8	53.9

a. Make a dotplot of these percents. Are there any outliers?
b. What is the median percent of the vote won by the successful candidate in presidential elections?
c. Call an election a landslide if the winner's percent falls at or above the third quartile. Find the third quartile. Which elections were landslides?

10. Find the mean and standard deviation of the reaction times in Exercise 2. Then find the mean and standard deviation when the outlier (138) is left out. What does your work show about the behavior of $\bar{x}$ and s? Explain why the mean is larger than the median.

11. A random sample of 12 Wisconsin farms showed the following soybean yields (bushels per acre).

41, 28, 34, 36, 26, 44, 39, 32, 40, 35, 36, 33

a. Make a dotplot of these data. Are there any outliers or other unusual features?
b. Find the mean and median yields.
c. Find the first and third quartiles of the yields.

12. Return to the data on lengths of fish given in Exercise 6.
a. Find the five-number summary of this distribution. Between what lengths does the middle 50% of the distribution lie?
b. From the shape of the distribution, do you expect the mean to be larger than the median, smaller than the median, or about the same as the median? Find the mean and verify your expectation.

c. Find the standard deviation. Based on the shape of the distribution, are $\bar{x}$ and s acceptable summary measures of center and spread?

13. Find the median and quartiles of Cavendish's measurements of the density of the Earth in Exercise 5. Then give a five-number summary. How is the symmetry of the distribution reflected in the five-number summary?

14. The mean of the 23 measurements in Exercise 5 is Cavendish's best estimate of the density of the Earth. Find this mean. Then find the standard deviation. (Because of the symmetry of the distribution, it can be summarized by $\bar{x}$ and s.)

15. Table 6.5 lists the percent of high school dropouts in each state. (More precisely, this is the percent of each state's residents of ages 16 through 19 who have not completed high school and are not currently enrolled in school.)
   a. Make a histogram of these data.
   b. Describe the overall shape of the distribution: Is it roughly symmetric, clearly skewed to the right, or clearly skewed to the left? There is an outlier. Which state produced the outlier? Can you suggest an explanation for the outlier?

TABLE 6.5   Percentage of 16- to 19-Year-Olds in Each State Who Are School Dropouts, 1991

State	Percent	State	Percent	State	Percent
Alabama	12.6	Kentucky	13.0	North Dakota	4.3
Alaska	9.6	Louisiana	11.9	Ohio	8.8
Arizona	14.3	Maine	8.4	Oklahoma	9.9
Arkansas	10.9	Maryland	11.0	Oregon	11.0
California	14.3	Massachusetts	9.5	Pennsylvania	9.4
Colorado	9.6	Michigan	9.9	Rhode Island	12.9
Connecticut	9.2	Minnesota	6.1	South Carolina	11.9
Delaware	11.2	Mississippi	11.7	South Dakota	7.1
D.C.	19.1	Missouri	11.2	Tennessee	13.6
Florida	14.2	Montana	7.1	Texas	12.5
Georgia	14.1	Nebraska	6.6	Utah	7.9
Hawaii	7.0	Nevada	14.9	Vermont	8.7
Idaho	9.6	New Hampshire	9.9	Virginia	10.4
Illinois	10.4	New Jersey	9.3	Washington	10.2
Indiana	11.4	New Mexico	10.8	West Virginia	10.6
Iowa	6.5	New York	10.1	Wisconsin	6.9
Kansas	8.4	North Carolina	13.2	Wyoming	6.3

16. The following table gives the survival times (in days) of 72 guinea pigs after they were infected by tubercle bacilli in a medical study. Make a histogram of these data. Is the survival-time distribution approximately symmetric or strongly skewed? Based on the shape of the distribution, would you prefer the five-number summary or $\bar{x}$ and s as a numerical description? Compute the numerical description you chose.

Guinea Pig Survival Times

43	45	53	56	56	57	58	66	67	73
74	79	80	80	81	81	81	82	83	83
84	88	89	91	91	92	92	97	99	99
100	100	101	102	102	102	103	104	107	108
109	113	114	118	121	123	126	128	137	138
139	144	145	147	156	162	174	178	179	184
191	198	211	214	243	249	329	380	403	511
522	598								

SOURCE: T. Bjerkedal, "Aquisition of Resistance in Guinea Pigs Infected with Different Doses of Virulent Tubercle Bacilli," *American Journal of Hygiene*, 72:130–148 (1960).

▲ 17. A common criterion for detecting suspected outliers in a set of data is as follows:
  1. Find the quartiles Q_1 and Q_3 and the *interquartile range IQR = Q_3 − Q_1*. The interquartile range is the spread of the central half of the data.
  2. Call an observation an outlier if it falls more than $1.5 \times IQR$ above the third quartile or below the first quartile.

Find the quartiles for the school dropout data in Table 6.5. Find the interquartile range *IQR*. Are there any outliers according to the $1.5 \times IQR$ criterion?

18. Find the mean and the median survival times for the guinea pigs in Exercise 16. Explain from the overall shape of the distribution the relationship between the two measures of location.

■ 19. Choose a set of interesting data from the *Statistical Abstract of the United States* or an almanac (for example, populations of the states or per capita incomes

▲   Advanced exercise.
■   Discussion exercise.

of nations). Make a histogram of the data and describe the pattern and any outliers. Then give a numerical description of the data.

■ 20. A study of the size of jury awards in civil cases (such as injury, product liability, and medical malpractice) showed that the median award in Cook County, Illinois, was about $8000. But the mean award was about $69,000. Explain how this great difference between two measures of location can occur.

■ 21. In 1991, two measures of the center of the distribution of prices for houses sold in Tippecanoe County, Indiana, were $67,750 and $81,884. Which of these numbers is the mean and which is the median? Explain your answer.

■ 22. Colleges announce an "average" Scholastic Aptitude Test score for their entering freshmen. Usually the college would like this "average" to be as high as possible. A *New York Times* article noted that "Private colleges that buy lots of top students with merit scholarships prefer the mean, while open-enrollment public institutions like medians." Use what you know about the behavior of means and medians to explain these preferences.

▲ 23. Give an example of a small set of data whose mean is larger than the upper quartile.

▲ 24. This is a standard deviation contest. You must choose four numbers from the whole numbers 0 to 10, with repeats allowed.
  a. Choose four numbers that have the smallest possible standard deviation.
  b. Choose four numbers that have the largest possible standard deviation.
  c. Is more than one choice possible in either part a or b? Explain.

25. Scores on the Stanford-Binet IQ test are approximately normally distributed with mean 100 and standard deviation 15. What is the variance of scores on this test?

26. Find the five-number summary for the home run data in Table 6.1 and make a boxplot of this distribution. Between what values do the middle half of league-leading home run totals lie?

27. Make a boxplot of the school dropout data in Table 6.5.

28. Joe DiMaggio played center field for the Yankees for 13 years. He was succeeded by Mickey Mantle, who played for 18 years. Here are the number of home runs hit each year by DiMaggio:

29, 46, 32, 30, 31, 30, 21, 25, 20, 39, 14, 32, 12

and by Mantle:

13, 23, 21, 27, 37, 52, 34, 42, 31, 40, 54, 30, 15, 35, 19, 23, 22, 18

Compute the five-number summary for each player, and make side-by-side boxplots of the home run distributions. What does your comparison show about the relative effectiveness of DiMaggio and Mantle as home run hitters?

■ 29. Do the southern states have higher school dropout rates than the rest of the country? Use Table 6.5 to answer this question. Decide which states you consider "southern." (You may want to look at a map.) Make side-by-side box-plots of school dropout percents for southern and nonsouthern states, and comment on the comparison of the two distributions.

30. The following table gives data on the lean body mass (kilograms) and resting metabolic rate for 12 women and 7 men who are subjects in a study of obesity. The researchers suspect that lean body mass (that is, the subject's weight leaving out all fat) is an important influence on metabolic rate.

    a. Make a scatterplot of the data for the female subjects. Which is the explanatory variable?

    b. Does metabolic rate increase or decrease as lean body mass increases? What is the overall shape of the relationship? Are there any outliers?

Obesity-Study Data

Subject	Sex	Mass	Rate	Subject	Sex	Mass	Rate
1	M	62.0	1792	11	F	40.3	1189
2	M	62.9	1666	12	F	33.1	913
3	F	36.1	995	13	M	51.9	1460
4	F	54.6	1425	14	F	42.4	1124
5	F	48.5	1396	15	F	34.5	1052
6	F	42.0	1418	16	F	51.1	1347
7	M	47.4	1362	17	F	41.2	1204
8	F	50.6	1502	18	M	51.9	1867
9	F	42.0	1256	19	M	46.9	1439
10	M	48.7	1614				

31. Compare the distribution of lean body mass among the male subjects in Exercise 30 with the distribution for female subjects by making side-by-side boxplots. Describe what the plots show.

▲ 32. When observations on two variables fall into several categories, you can display more information in a scatterplot by plotting each category with a different symbol or a different color. Add the data for male subjects to your scatterplot in Exercise 30, using a different color or symbol than you used for females. Does the type of relationship you found for females hold for men also? How do the male subjects as a group differ from the female subjects as a group?

33. The table on the opposite page lists the retail prices of several consumer items in Washington, D.C., and in Japan. Make a scatterplot with the U.S. price as the explanatory variable. Describe the overall pattern of the relationship between U.S. and Japanese consumer prices. Are there any items with prices that

fall clearly outside the overall pattern? If so, describe how these items differ from the overall pattern.

Retail Prices in Washington, D.C., and Japan

Item	Washington	Japan
First-run movie	$5.00	$13.00
Pound of rice	0.50	1.89
Big Mac hamburger	1.78	3.08
Electricity (kWh)	0.06	0.33
Gallon of gasoline	0.95	3.50
Cantaloupe	0.75	13.00
Dozen eggs	1.09	1.70
Subway	0.80	1.00
Quart of milk	0.55	1.48

SOURCE: *USA Today*, January 9, 1989.

34. We saw that the least squares regression line for the home heating data of Table 6.3 is $y = 1.23 + 0.202x$. Use this line to predict the daily gas consumption for this home in a month averaging 20 degree days per day and in a month averaging 40 degree days per day.

35. Concrete road pavement gains strength over time as it cures. Highway engineers use regression lines to predict the strength after 28 days (when curing is complete) from measurements made after 7 days. Let x be strength (in pounds per square inch) after 7 days and y the strength after 28 days. One set of data gave the least squares regression line to be

$$y = 1389 + 0.96x$$

A test of some new pavement after 7 days shows that its strength is 3300 pounds per square inch. Predict the strength of this pavement after 28 days.

36. The table on the following page shows the true number of calories in 10 common foods and the average number of calories estimated for these same foods in a sample of 3368 people.
   a. Make a scatterplot of these data, with true calories on the horizontal axis. Is there a general straight-line pattern? Which foods are outliers from the pattern?
   b. Fit a line by eye to the data (ignoring the outliers). If a food product contains 200 calories, what would you guess the general public's estimated calorie level for that food to be?

c. The least squares regression line, computed without dropping the outliers, is

$$y = 58.6 + 1.30x$$

Here y = estimated calories and x = true calories. Draw this line on
your scatterplot. (Hint: Use the equation to find y for two values of x,
then plot the two (x, y) points and draw the line through them.) Use
the least squares line to predict y when $x = 200$. The difference between your results in parts b and c reflects the influence of the outliers. It is often difficult to decide whether to include outliers in making a prediction.

True and Estimated Calories in 10 Common Foods

Product	True calories	Estimated calories
8 oz whole milk	159	196
5 oz spaghetti with tomato sauce	163	394
5 oz macaroni and cheese	269	350
1 slice wheat bread	61	117
1 slice white bread	76	136
2-oz candy bar	260	364
Saltine cracker	12	74
Medium-size apple	80	107
Medium-size potato	88	160
Cream-filled snack cake	160	419

SOURCE: *USA Today*, October 1983.

37. The scatterplot on the opposite page of the average SAT math score (y)
versus the average SAT verbal score (x) for high school seniors in each of the 50
states shows a strong straight-line pattern. The least squares regression line for
scores is

$$y = 27 + 1.03x$$

a. Do math SAT scores tend to be higher or lower than verbal SAT
scores? How can you tell?
b. The average verbal SAT score in New York was 422. Use the regression line to predict New York's average math score. New York's actual math score was 466. What is the error (observed score minus predicted score)?
c. The only outlier among the states is Hawaii, where $x = 393$ and $y = 471$. Is Hawaii's math score higher or lower than would be predicted
from its verbal score? Can you suggest why Hawaii might be an outlier?

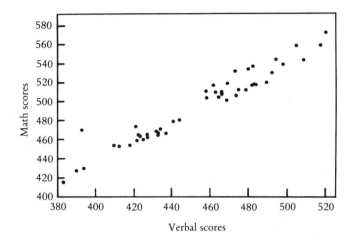

38. An old study in Iowa produced the following data on corn yield (bushels per acre) in 1910–1919 and value per acre in 1920 for farmland in 10 counties. (Both corn yields and land prices have changed since 1920!)

County	1	2	3	4	5	6	7	8	9	10
Yield	40	36	34	41	39	42	40	31	36	30
Value ($)	87	133	174	285	263	274	235	104	141	115

    a. Make a scatterplot of these data. There is an overall straight-line pattern with one outlier. Circle the outlier on your plot.
    b. Corn yield should help to predict the value of farmland. When the outlier is omitted, the least squares regression line of land value y on yield x is

$$y = -371.3 + 15.4x$$

Draw this line on your scatterplot. Use the equation to predict the value of land that yields 35 bushels per acre. Then use your graph to do this prediction as in Figure 6.8b and compare the two results.

39. A study of sewage treatment measures the oxygen demand of decomposing solid wastes. If y is the logarithm of the oxygen demand (milligrams per minute) and x is the total solids (milligrams per liter of waste), measurements on 20 occasions give the following data:

x	7.2	7.8	7.1	6.4	6.4	5.1	5.9	5.3	5.0	5.0
y	1.56	0.9	0.75	0.72	0.31	0.36	0.11	0.11	-0.22	-0.15

x	4.8	4.4	4.3	3.7	3.9	3.6	4.4	3.3	2.9	2.8
y	0.0	0.0	-0.09	-0.22	-0.4	-0.15	-0.22	-0.4	-0.52	-0.05

a. Make a scatterplot of these data. Is there an approximate straight-line relationship? Are there outliers?

b. Draw a fitted line on your scatterplot by eye. Use your line to predict the log of the oxygen demand y when $x = 4$ milligrams per liter of solids.

40. Manatees are large, gentle sea creatures that live along the Florida coast. Many manatees are killed or injured by power boats. The table below gives data on power boat registrations (in thousands) and the number of manatees killed by boats in Florida in the years 1977 to 1990.

Year	Boats (thousands)	Manatees killed	Year	Boats (thousands)	Manatees killed
1977	447	13	1984	559	34
1978	460	21	1985	585	33
1979	481	24	1986	614	33
1980	498	16	1987	645	39
1981	513	24	1988	675	43
1982	512	20	1989	711	50
1983	526	15	1990	719	47

a. Which is the explanatory variable? Make a scatterplot of these data. Describe the overall pattern and any outliers or other important deviations.

b. Fit a line to the scatterplot by eye and draw the line on your graph. If in a future year power boat registrations increase to 750,000, predict the number of manatees that will be killed by boats.

41. Suppose that in some far future year 2 million power boats are registered in Florida. Extend the fitted line you found in Exercise 40 and use it to predict manatees killed. Explain why this prediction is very unreliable. (Using a fitted line to predict the response to an x value outside the range of the data used to fit the line is called *extrapolation*. Extrapolation often produces unreliable predictions.)

■ 42. Table 6.6 gives the average Scholastic Aptitude Test (SAT) mathematics score for students in each state and also the percent of students in each state who take the examination, as of 1990. We expect that states in which most students take the test would have lower average scores than states in which only a few

TABLE 6.6   **Average SAT Scores by State, 1990**

State	Verbal score	Math score	% Taking SAT	State	Verbal score	Math score	% Taking SAT
Alabama	470	514	8	Montana	464	523	20
Alaska	438	476	42	Nebraska	484	546	10
Arizona	445	497	25	Nevada	434	487	24
Arkansas	470	511	6	New Hampshire	442	486	67
California	419	484	45	New Jersey	418	473	69
Colorado	456	513	28	New Mexico	480	527	12
Connecticut	430	471	74	New York	412	470	70
Delaware	433	470	58	North Carolina	401	440	55
District of Columbia	409	441	68	North Dakota	505	564	6
Florida	418	466	44	Ohio	450	499	22
Georgia	401	443	57	Oklahoma	478	523	9
Hawaii	404	481	52	Oregon	439	484	49
Idaho	466	502	17	Pennsylvania	420	463	64
Illinois	466	528	16	Rhode Island	422	461	62
Indiana	408	459	54	South Carolina	397	437	54
Iowa	511	577	5	South Dakota	506	555	5
Kansas	492	548	10	Tennessee	483	525	12
Kentucky	473	521	10	Texas	413	461	42
Louisiana	476	517	9	Utah	492	539	5
Maine	423	463	60	Vermont	431	466	62
Maryland	430	478	59	Virginia	425	470	58
Massachusetts	427	473	72	Washington	437	486	44
Michigan	454	514	12	West Virginia	443	490	15
Minnesota	477	542	14	Wisconsin	476	543	11
Mississippi	477	519	4	Wyoming	458	519	13
Missouri	473	522	12				

students take the test. This is true because in states where the SATs are not widely used, only students applying to selective colleges choose to take the SATs. Make an appropriate plot of these data and describe the relationship between these two variables. Is the expectation correct? Is there a straight-line relationship?

● 43. Calculate the least squares regression line of estimated calories y on true calories x from the data in Exercise 36. Verify that your result agrees with that given in Exercise 36c.

● 44. Calculate the least squares regression line of land value y on corn yield x for the data in Exercise 38, omitting county 1. Verify that your result agrees with that given in Exercise 38.

● 45. Find the equation of the least squares regression line of y on x for the sewage treatment data of Exercise 39. Use your line to do the prediction asked for in Exercise 39b.

● 46. Find the equation of the least squares regression line for the manatee data in Exercise 40. Use the equation to predict manatee deaths in a future year when 750,000 power boats are registered in Florida.

▶ WRITING PROJECTS

1. Part of analyzing data is to watch for implausible numbers. Here is part of a report on the problem of vacation cruise ships polluting the sea by dumping garbage overboard that appeared in *Condé Nast Traveler* magazine in June 1992:

> On a seven-day cruise, a medium-size ship (about 1,000 passengers) might accumulate 222,000 coffee cups, 72,000 soda cans, 40,000 beer cans and bottles, and 11,000 wine bottles.

Are these numbers plausible? Write a short essay arguing your position, with some arithmetic to back up your conclusion.

2. Table 6.5 records the school dropout rates for the states. Some states have higher rates than others. What factors do you think might account for this? For example, the high dropout rate for the District of Columbia suggests that a large urban population may contribute to a high state rate. Look in the *Statistical Abstract of the United States* for information you think is relevant. For example, the *Statistical Abstract* contains a table that gives the percent of each state's population that lives in metropolitan areas. (But "metropolitan areas" covers suburbs along with central cities and small cities as well as large.) Write an essay suggesting some explanations, and accompany it with data and graphs that explore at least one of your suggestions.

3. Graphs good and bad fill the news media. Some publications, such as *USA Today*, make particularly heavy use of graphs to present data. Collect several graphs (at least five) from newspapers and magazines (not from advertisements). Use them as examples in a brief essay about the clarity, accuracy, and attractiveness of graphs in the news. You can find information on what makes good graphs in the books by Tufte and Cleveland listed in the Suggested Readings.

● Optional exercise.

Chapter 7

Probability: The Mathematics of Chance

Have you ever wondered how gambling, which is a recreation or an addiction for individuals, can be a business for the casino? A business requires predictable revenue from the service it offers, even when the service is a game of chance. Individual gamblers may win or lose; they can never say whether a day at Lake Tahoe or Atlantic City will turn a profit or a loss. But the casino itself does not gamble. Casinos are consistently profitable, and lotteries are now a source of revenue for many state governments.

It is a remarkable fact that the aggregate result of many thousands of chance outcomes can be known with near certainty. The casino need not load the dice, mark the cards, or alter the roulette wheel. It knows that in the long run, each dollar bet will yield its five cents or so of revenue. It is therefore good business to concentrate on free floor shows or inexpensive bus fares to increase the flow of dollars bet. The flow of profit will follow.

Gambling houses are not alone in profiting from the fact that a chance outcome many times repeated is firmly predictable. For example, although a life insurance company does not know *which* of its policyholders will die next year, it can predict quite accurately *how many* will die. It sets

its premiums by this knowledge, just as the casino sets its jackpots.

A phenomenon is called **random** if the individual outcomes are uncertain but the long-term pattern of many individual outcomes is predictable.

To a statistician, "random" is not a synonym for "haphazard." Randomness is a kind of order that emerges only in the long run, in many repetitions. Many phenomena, both natural and of human design, are random. The lifespans of insurance buyers and the hair color of children are examples of natural randomness. Indeed, quantum mechanics asserts that at the subatomic level the natural world is inherently random. Probability theory, the mathematical description of randomness, is therefore essential to much of modern science.

Games of chance are examples of randomness deliberately produced by human effort. Casino dice are carefully machined, and their drilled holes — called pips — are filled with material equal in density to the plastic body. This guarantees that the side with six pips has the same weight as the opposite side, which has only one pip. Thus, each

Figure 7.1 The more dollars bet, the more money a casino is guaranteed to take in. (Las Vegas News Bureau.)

side is equally likely to land upward. All the odds and payoffs of dice games rest on this carefully planned randomness.

Statisticians and casino managers have the same vested interest in planned randomness, although statisticians use tables of random digits rather than dice and cards. The reasoning of statistical inference rests on planned randomness and on the mathematics of probability, the same mathematics that guarantees the profits of casinos and insurance companies. The mathematics of chance is the topic of this chapter.

▶ WHAT IS PROBABILITY?

The mathematics of chance, the mathematical description of randomness, is called the *theory of probability*. Probability describes the predictable long-run patterns of random outcomes.

Toss a coin in the air. Will it land heads or tails? Sometimes it lands heads and sometimes tails. We cannot say what the next outcome will

be. Perhaps you would argue that the coin looks balanced, so you think it has an equal chance of falling heads or tails on the next toss. Your personal probability of a head is then $\frac{1}{2}$. Probability as the expression of personal opinion is a reasonable idea. But we have in mind a different meaning, one based on *observation* of random phenomena.

Suppose that we toss a coin not once but 10,000 times. John Kerrich, an English mathematician, actually did this while interned by the Germans during World War II. Figure 7.2 shows Kerrich's results. His first 10 tosses gave 4 heads, a proportion of 0.4. The proportion of heads increased to 0.5 after 20 tosses, and to 0.57 after 30 tosses. Figure 7.2 shows how this proportion changes as more and more tosses are made. In a small number of tosses, the proportion of heads fluctuates — it is still essentially unpredictable. But many tosses produce a smoothing effect. A proportion of 0.507, or 50.7%, heads resulted after Kerrich threw the coin 5000 times. And in all 10,000 trials, he scored 5067 heads, again 50.7% of the total. After many trials, the proportion of heads settles down to a fixed number. This number is the probability of a head.

The **probability** of an outcome is the proportion of trials in which the outcome occurs in a very long run of trials.

The idea of probability involves using what would happen in many trials to describe the uncertain outcome of a single trial. This "definition" of probability would not satisfy either a mathematician or a philosopher, but it fixes in mind the idea we will deal with. Strictly speaking, Kerrich's 0.507 only estimates the probability that his coin will come up heads. The proportion after 100,000 tosses would be closer to the true probability. In fact, we can't even say that Kerrich's coin is weighted in favor of heads. There is still enough unpredictability in the results of 10,000 tosses that 70 excess heads could easily occur even if heads and tails were equally probable.

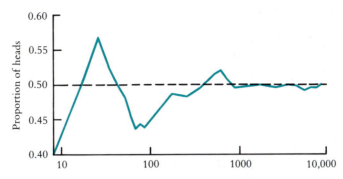

Figure 7.2 Percent of heads versus number of tosses in Kerrich's coin-tossing experiment. (David Freedman et al., *Statistics*, Norton, New York, 1978.)

▶ PROBABILITY MODELS

Gamblers have known for a long time that the fall of coins, cards, or dice stabilizes into definite patterns in the long run. France gave birth to the mathematics of probability when gamblers in the seventeenth century turned to mathematicians for advice (see Spotlight 7.1). The idea of probability rests on the observed fact that the average result of many thousands of chance outcomes can be known with near certainty. But a definition of probability as "long-run proportion" is vague. Who can say what "the long run" is? Instead, we give a mathematical description of *how probabilities behave*, based on our understanding of long-run proportions. This mathematical description of probability has the advantage that it applies equally well to probability thought of as personal assessment of chance. The same mathematics describes two different informal concepts of probability.

Our first task in assigning probabilities to outcomes is to list all the possible outcomes.

The set of all possible outcomes of a random phenomenon is called the **sample space,** *S*.

The sample space *S* may be very simple or very complex. When we toss a coin once, there are only two outcomes, heads and tails. So the sample space is $S = \{H, T\}$. If we draw a random sample of 1500 U.S. residents age 18 and over, as the Gallup poll does, the sample space contains all possible choices of 1500 of the over 190 million adults in the country. This *S* is extremely large — to count its members would require a number over 8000 digits long! Each member of *S* is a possible Gallup poll sample, which explains the term *sample space*.

EXAMPLE: Rolling Dice

Suppose that we roll a single die and observe the number of pips on the up face. The possible outcomes appear in Figure 7.3. We take the

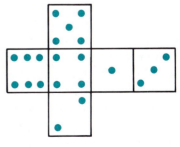

Figure 7.3 The possible outcomes for rolling one die.

SP TLIGHT 7.1   The Mathematical Bernoullis

Jakob Bernoulli

Johann Bernoulli

Few families have made more contributions to mathematics than the Bernoullis of Basel, Switzerland. No fewer than seven Bernoullis, over three generations spanning the years between 1680 and 1800, were distinguished mathematicians. Five of them helped build the new mathematics of probability.

Jakob (1654–1705) and Johann (1667–1748) were sons of a prosperous Swiss merchant, but they studied mathematics against the will of their practical father. Both were among the finest mathematicians of their times, but it was Jakob who concentrated on probability. Several seventeenth-century mathematicians had started the study of games of chance, concentrating on counting outcomes to find chances. Jakob Bernoulli was the first to see clearly the idea of a long-run proportion as a way of measuring chance. He proved that if in a very large population (say, size N), K mem-

bers have a property A, then the proportion of members of a sample of size n having property A must approach K/N (the probability of A) as the sample size n increases. This *law of large numbers* helped to connect probability to the study of sequences of chance outcomes observed in human affairs.

Johann's son Daniel (1700–1782) and Jakob and Johann's nephew Nicholas (1687–1759) also studied probability. Nicholas saw that the pattern of births of male and female children could be described by probability. Despite his own rebellion against his father's strictures, Johann tried to make his son Daniel a merchant or a doctor. Daniel, undeterred, became yet another Bernoulli mathematician. He studied mainly the mathematics of flowing fluids (later applied to designing ships and aircraft) and elastic bodies. In the field of probability, he worked to fairly price games of chance and gave evidence for the effectiveness of inoculation against smallpox.

The Bernoulli family in mathematics, like their contemporaries the Bachs in music, are an unusual example of talent in one field appearing in successive generations. Their work helped probability to grow from its birthplace in the gambling hall to a respectable tool with worldwide applications.

sample space, the set of these outcomes, to be

$$S = \{1, 2, 3, 4, 5, 6\}$$

What if we roll *two* dice, as many games of chance require? Figure 7.4 shows the possible combinations of up faces on the two dice in order. Think of the die on the left as being rolled first, then the die on the right. The sample space consists of these 36 outcomes.

In craps and other games, all that matters is the *sum* of the pips on the up faces. Let's change the random outcomes we are interested in: roll two dice and count the pips on the up faces. Now there are only 11 possible outcomes, from a sum of 2 for rolling a double one through 3, 4, 5, and on up to 12 for rolling a double six. The sample space is now

$$S = \{2, 3, 4, 5, 6, 7, 8, 9, 10, 11, 12\}$$

As this example shows, it is important when choosing a sample space to be clear which random outcomes you will look for. Rolling two dice and recording the up faces in order is

not the same as rolling two dice and recording only the sum of the two up faces. ▲

The next step in the mathematical description of chance is to assign probabilities to the outcomes that make up the sample space. There are many ways to assign probabilities, so it is convenient to start with some general rules that any assignment of probabilities to outcomes must obey. Let $P(s)$ stand for the probability of any outcome s in the sample space S. Here are the fundamental laws for assigning probabilities to outcomes:

LAW 1. Every probability $P(s)$ is a number between 0 and 1.

LAW 2. The sum of the probabilities $P(s)$ over all outcomes s in S is exactly 1.

These laws are based on our understanding of probability as long-run proportion. The first law reflects the fact that any proportion is a number between 0 and 1. If an outcome *never* occurs, it has probability 0; if it *always* occurs, the probability is 1; if it *sometimes* occurs, the proportion of trials

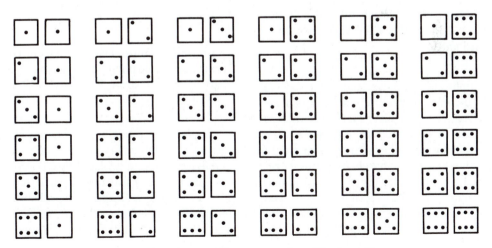

Figure 7.4 The possible outcomes for rolling two dice.

producing the outcome is a number between 0 and 1. The second law says that some outcome must always occur, so the sum of all their probabilities is 1. Laws 1 and 2 describe any legitimate assignment of probabilities to outcomes. An assignment that does not satisfy these laws does not make sense.

EXAMPLE: Rolling One Die

Let's return to tossing a single die to see what a probability model looks like. We have already seen that the sample space (see Figure 7.3) is

$$S = \{1, 2, 3, 4, 5, 6\}$$

Any assignment of probabilities $P(1)$, $P(2)$, . . . , $P(6)$ to these six outcomes must choose numbers between 0 and 1 that add to exactly 1. If the die is a carefully made casino model, each face should have the same probability of landing upward. Then the six probabilities are all the same and have sum 1, so each must be 1/6. The probabilities are

$$P(1) = P(2) = P(3) = P(4) = P(5) = P(6)$$
$$= 1/6, \text{ or about } 0.167$$

Dice and other chance devices that are carefully made so as to have equally likely outcomes are often called *fair*. ▲

It is important to understand what the two laws for outcome probabilities accomplish and what they leave unsaid. They do state which assignments of probabilities to outcomes make sense. But they do not tell us which assignment is correct, in the sense of accurately describing a real die; this can only be determined by actual trial. Professional dice are well described by the equal-probabilities model of the example. But cheap dice with hollowed-out pips fall unequally. The 6 face is lightest and is located opposite the 1 face, which

is heaviest. Thus, a cheap die might be described by a set of probabilities in which lighter faces are more likely to land upward, such as

$P(1) = 0.159$	$P(4) = 0.166$
$P(2) = 0.163$	$P(5) = 0.171$
$P(3) = 0.166$	$P(6) = 0.175$

These probabilities also satisfy Laws 1 and 2. They are legitimate, even if the die is not.

Assigning probabilities to individual outcomes is not enough. We also want to assign probabilities to **events,** which are collections of outcomes. For example, what is the probability of rolling an odd number in one toss of a fair die? The odd outcomes are 1, 3, and 5. The proportion of tosses on which one of these numbers comes up must be the sum of the proportions on which each alone comes up. So, thinking of probabilities as long-run proportions leads us to find the probability of any event by summing the probabilities of the outcomes that make up the event. In this case,

$$P(\text{outcome is odd}) = P(1) + P(3) + P(5)$$
$$= 3/6 = 0.5$$

We now have a complete probability model for rolling a fair die.

A **probability model** is a mathematical description of a random phenomenon consisting of two parts: a sample space S and a way of assigning probabilities to events.

Probability models for random phenomena with only finitely many possible outcomes have a simple form: assign probabilities to outcomes in a way that satisfies Laws 1 and 2, then find probabilities of events by adding up the probabilities of the outcomes that make up the event. You can state such a model by giving a table of the outcomes and their probabilities.

EXAMPLE: Household Size

A household is a group of people living together, regardless of their relationship to each other. Sample surveys such as the Current Population Survey select a random sample of households. Here is the probability model for the number of people living in a randomly chosen American household:

Household size	1	2	3	4	5	6	7
Probability	.236	.320	.181	.156	.069	.024	.014

These probabilities are the proportions for all households in the country, and so give the probabilities that a single household chosen at random will have each size. (The very few households with more than 7 members are placed in the 7 group.) Check that Laws 1 and 2 are satisfied. The probability that a randomly chosen household has more than two members is

$$
\begin{aligned}
P(\text{size} > 2) &= P(3) + P(4) \\
&\quad + P(5) + P(6) + P(7) \\
&= .181 + .156 + .069 + .024 \\
&\quad + .014 = .444 \quad \blacktriangle
\end{aligned}
$$

A simple random sample gives all possible samples an equal chance to be chosen. Dealing from a well-shuffled deck gives all possible card hands an equal chance to be the hand you are dealt. When randomness is the product of human design, it is often the case that the outcomes in the sample space are all equally likely. Laws 1 and 2 force the assignment of probabilities in this case:

If a random phenomenon has k possible outcomes, all equally likely to occur, then each individual outcome has probability $1/k$.

The probability of any event in the equally likely case is found as usual by adding the individual probabilities of the outcomes making up the event. Because each of these probabilities is the same $1/k$, we have this new rule:

When all outcomes have equal probabilities, the probability of any event A is

$$
P(A) = \frac{\text{number of outcomes in } A}{\text{number of outcomes in } S}
$$

EXAMPLE: Rolling Two Dice

Roll two fair dice and record the pips on each of the two up faces. The sample space consists of the 36 outcomes pictured in Figure 7.4. Because of the balance of the dice, these outcomes are all equally likely. So each has probability 1/36.

What is the probability of rolling a 5? The event "roll a 5" contains the four outcomes

and the probability is therefore 4/36, or about 0.111. What about the probability of rolling a 7? Look at Figure 7.4 and count six outcomes for which the sum of the pips is 7. The probability is therefore 6/36, or about 0.167.

Be certain that you understand that the method of finding probabilities by counting outcomes applies *only* when all outcomes are equally likely. The S shown in Figure 7.4 does have equally likely outcomes. But if we choose to use the sample space for rolling two dice *and counting the pips*, we get

$$
S = \{2, 3, 4, 5, 6, 7, 8, 9, 10, 11, 12\}
$$

These outcomes do *not* have equal probabilities. We just saw, for example, that the probability of 5 is 4/36 and that the probability of a 7 is 6/36.   ▲

When outcomes are equally likely, finding probabilities leads to the study of counting methods called **combinatorics.** Combinatorics is an important area of mathematics in its own right. Although we will not study combinatorics in detail, the following example uses the multiplication method that we called the *fundamental principle of counting* in Chapter 2.

EXAMPLE: Code Words

A computer system assigns log-in identification codes to users by choosing three letters at random. All three-letter codes are therefore equally probable. What is the probability that the code assigned to you has no "x" in it?

We must count the number of code words. There are 26 letters that can occur in each position in the word. Any of the 26 letters in the first position can be combined with any of the 26 letters in the second position to give 26 × 26 choices. (This is true because the order of the letters matters, so "ab" and "ba" are different choices.) Any of the 26 letters can then follow in the third position. The number of different codes is

$$26 \times 26 \times 26 = 17,576$$

Codes without an "x" are made up of the other 25 letters. There are

$$25 \times 25 \times 25 = 15,625$$

such codes. The probability that your code has no "x" is therefore

$$P(\text{no ``x''}) = \frac{\text{number of codes with no ``x''}}{\text{number of codes}}$$

$$= \frac{15,625}{17,576} = .889$$

Suppose that the computer is programmed to avoid repeated letters in the identification codes. Any of the 26 letters can still appear in the first position. But only the 25 remaining letters are allowed in the second position, so there are 26 × 25 choices for the first two letters in the code. Any of these choices leaves 24 letters for the third position. The number of different codes without repeated letters is

$$26 \times 25 \times 24 = 15,600$$

Codes with no "x" are allowed one fewer choice in each position; there are

$$25 \times 24 \times 23 = 13,800$$

such codes. The probability that your code has no "x" is then

$$P(\text{no ``x''}) = \frac{\text{number of codes with no ``x''}}{\text{number of codes}}$$

$$= \frac{13,800}{15,600} = .885$$

Eliminating repeats slightly decreases your chance of avoiding an "x." ▲

The example illustrates two facts about counting that are useful in finding probabilities:

A. Suppose we have a collection of n distinct items. We want to arrange k of these items in order, with repeats allowed. The number of possible arrangements is

$$n \times n \times \cdots \times n = n^k$$

B. Suppose we have a collection of n distinct items. We want to arrange k of these items in order, with *no* repeats allowed. The number of possible arrangements is

$$n \times (n-1) \times \cdots \times (n-k+1)$$

In the example, n (the number of letters available) is first 26 then 25, and k (the number of letters to be arranged to make a code) is 3. It is usually easier to think your way through the counting than to memorize the recipes.

EXAMPLE: Factorials

A jury of 7 students is seated in a row of 7 chairs to judge a speaking competition. In how many orders can the students sit?

Because each student can sit in only one chair at a time, no repeats are allowed. All students are seated, so n and k are both 7. The number of arrangements is therefore

$$7 \times 6 \times 5 \times 4 \times 3 \times 2 \times 1 = 5040$$

This number is often called 7!, read "seven factorial." In general, the factorial $n!$ is the product of the whole numbers from n down to 1, and is the number of different ordered arrangements of n distinct items. ▲

▶THE MEAN OF A RANDOM PHENOMENON

Suppose you are offered this choice of bets, each costing the same: bet A pays $10 if you win and you have probability 1/2 of winning, while bet B pays $10,000 and offers probability 1/10 of winning. You would very likely choose B even though A offers a better chance to win, because B pays much more if you win. It would be foolish to decide which bet to make just on the basis of the probability of winning. How much you can win is also important. When a random phenomenon has numerical outcomes, we are concerned with their amounts as well as with their probabilities.

What will be the average payoff of our two bets in many plays? Recall that the probabilities are the long-run proportions of plays on which each outcome occurs. Bet A produces $10 half the time in the long run and nothing half the time. So the average payoff should be

$$(\$10 \times \tfrac{1}{2}) + (\$0 \times \tfrac{1}{2}) = \$5$$

Bet B, on the other hand, pays out $10,000 on 1/10 of all bets in the long run. Bet B's average payoff is

$$(\$10,000 \times \tfrac{1}{10}) + (\$0 \times \tfrac{9}{10}) = \$1000$$

If you can place many bets, you should certainly choose B. Here is a general definition of the kind of "average outcome" we used to compare the two bets:

Suppose that the possible outcomes s_1, s_2, ..., s_m in a sample space S are numbers, and that p_j is the probability of outcome s_j. The **mean** μ of the random outcome is

$$\mu = s_1 p_1 + s_2 p_2 + \cdots + s_m p_m$$

Earlier, we met the mean $\bar{x}$, the average of n observations that we actually have in hand. The mean μ, on the other hand, describes the probability model rather than any one collection of observations. You can think of μ as a theoretical mean that says what average outcome we expect in the long run.

EXAMPLE: Mean Household Size

The probability model for the number of people living in a randomly chosen household is:

Household size	1	2	3	4	5	6	7
Probability	.236	.320	.181	.156	.069	.024	.014

The mean is therefore

$$\mu = (1)(.236) + (2)(.320) + (3)(.181)$$
$$+ (4)(.156) + (5)(.069)$$
$$+ (6)(.024) + (7)(.014)$$
$$= 2.63$$

In this case, the mean μ is the average size of all American households. If we took a random sample of (say) 100 households and recorded their sizes, we would call the average size for this sample $\bar{x}$. A second random sample would no doubt give a somewhat different value of $\bar{x}$. So $\bar{x}$ varies from sample to sample, but μ, which describes the distribution of probabilities, is a fixed number. ▲

The mean μ is an average outcome in two senses. The definition says that it is the average of the possible outcomes, not weighted equally but weighted by their probabilities. More likely outcomes get more weight in the average. An important fact of probability, the **law of large numbers,** says that μ is the average outcome in another sense as well. The law of large numbers states that if the random phenomenon is repeated a large number of times, the mean $\bar{x}$ of the actually observed outcomes will get closer and closer to μ. The proportion of trials on which each outcome occurs will similarly get closer and closer to the probability of that outcome. These facts can actually be proved mathematically for any assignment of probabilities that satisfies Laws 1 and 2. The law of large numbers brings the study of basic probability back to a natural completion. We first observed that some phenomena are random in the sense of showing long-run regularity. Then we used the idea of long-run proportions to motivate the basic laws of probability. Those laws are mathematical idealizations that can be used without interpreting probability as proportion in many trials. Now the law of large numbers tells us that in many trials the proportion of trials on which an outcome occurs will always approach its probability.

▶ SAMPLING DISTRIBUTIONS

Sampling is, in a way, a lot like gambling. Both rely on the deliberate use of chance. To see how probability applies to sampling, let's look again at the population of light and dark beads described in Chapter 5.

We have a large container filled with beads of the same size and shape. How can we estimate the percentage of dark beads in the box? We plunged a scoop with 50 bead-sized hollows into the container and drew out 50 beads. This is a simple random sample of size 50. Of the 50 beads in the sample, 12 — or 24% of the sample — were dark. From this we estimated that 24% of the population is made up of dark beads. Our sampling strategy illustrates a basic form of statistical inference: take a simple random sample and use the sample result to estimate the unknown truth about a population. The Gallup poll and the Current Population Survey carry out more elaborate versions of this method.

When we drop the scoop into the box once more, drawing a second sample, perhaps only 8 of the 50 beads, or 16%, will be dark. A third sample might yield 7 dark beads, or 14%. This chance variation is called **sampling variability.** We saw in Chapter 5 how sampling variability affects the results of sample surveys. Now we can use the laws of probability to describe it more precisely.

The sample space in the bead-sampling experiment is made up of all the possible samples of 50 beads drawn from the beads in the container. This sample space is very large. Random sampling guarantees that each possible sample is equally likely to be drawn. So we could find the probability of getting 16% dark beads by counting the possible samples that have 16% dark beads and dividing by the number of outcomes in the entire sample space. That's hopelessly tedious; we need a shortcut method that will find the probability without counting.

Because the proportion of many trials on which an outcome occurs gets close to the probability of that outcome, we can estimate the probability of getting 16% dark beads by actually observing the outcomes of many samples. Figure 7.5a records the results of 200 samples. For example, 21 of the 200 samples had exactly 16% (8 of 50) dark beads. So we estimate the probability of getting 16% dark beads to be about 21/200, or 0.105. The figure is a new kind of histogram in which the heights of the bars are the proportions of outcomes in each class rather than the counts of outcomes.

Look at the overall pattern of the histogram of estimated probabilities in Figure 7.5a. The percentages of dark beads range from 6% to 38%. The

histogram is fairly symmetric. The center is close to the 20% bar (representing 10 dark beads out of 50). This is also the outcome that occurred most often. There are no extreme outliers.

Statisticians call a number that is computed from a sample a **statistic.** The percentage of dark beads in our sample is a statistic. The histogram displays the sampling variability of this statistic. In fact, the histogram assigns probabilities to each value of the statistic. These probabilities make up the **sampling distribution** of the statistic. Of course, the probabilities are only approximate because they are based on only 200 trials. But the

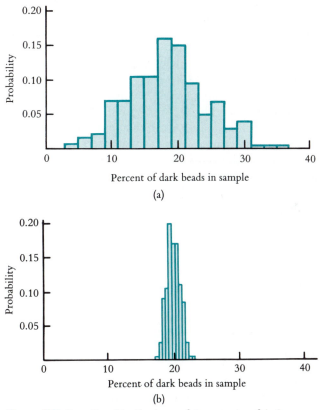

Figure 7.5 Sampling distributions of the percent of dark beads in samples of two sizes from the same population. (a) Sample size 50. (b) Sample size 1500.

overall pattern of the sampling distribution is already clear.

Let's try a second experiment in the hope of confirming the pattern. Remember that public opinion polls often rely on a sample of about 1500 people. We will take 200 simple random samples of 1500 beads each. To do this, we must imagine a very large container of beads. In fact, we instruct a computer to simulate the experiment. Figure 7.5b displays the outcomes, using the same scale as Figure 7.5a.

The histogram in Figure 7.5b for samples of size 1500 is taller and narrower than that in Figure 7.5a for samples of size 50. It is narrower because the sample variability is much smaller in the larger sample. All 200 outcomes fall between 17% and 23% dark beads. And if we add up the estimated probabilities for outcomes near the middle, we find there is probability 0.93 that a sample will fall between 18.25% and 21.75%. We can rely much more confidently on a single sample of size 1500 than on a single sample of size 50. In fact, all these samples were drawn from a population with 20% dark beads. Samples of size 1500 almost always give estimates quite close to this true value.

Our sampling experiment has both produced an approximate assignment of probabilities (without counting) and taught us a bit about how the sampling distribution behaves when we increase the size of the sample. The two following sections explore these two topics in more detail.

▶ Normal Distributions

Although they differ in variability, the histograms in Figure 7.5 have similar shapes in other respects. Both are symmetric, with centers close to 20%. The tails fall off smoothly on either side, with no outliers. Suppose that we represent the shape of each histogram by drawing a smooth curve through the tops of the bars. If we do this carefully — using the actual probabilities of the outcomes rather than estimates from only 200 samples — the two curves we obtain will be quite close to two members of the family of *normal curves*. The two normal curves appear in Figure 7.6.

Normal curves introduce a new way of describing probabilities. The assignment of probabilities to the values of a statistic can be described by a histogram like those of Figure 7.5. A histogram makes probability visible. The height of any bar is the probability of the outcomes spanned by the base of that bar. Because all bars have the same width, their area (height times width) is proportional to the probability. Normal curves can be thought of as approximations to a histogram of probabilities in which area is exactly equal to probability. Normal curves are easier to work with than histograms because many bars are replaced by a single smooth curve. Normal curves have the property that the total area under the curve is exactly 1, corresponding to the fact that all outcomes together have probability 1.

When probability is described by a normal curve, the probability of any interval of outcomes is the area under the normal curve above that interval.

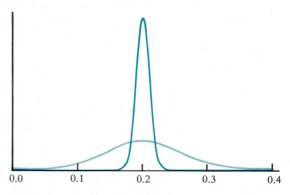

Figure 7.6 The normal curves that approximate the sampling distributions of the percent of dark beads for samples of sizes 50 (light curve) and 1500 (dark curve).

Our original method of assigning probabilities to events was to first give a probability to each individual outcome. Probability as area under a curve is the second important method of assigning probability, and is easier when there are many individual outcomes falling close together. Curves of different shapes describe different assignments of probability. We will emphasize the normal curves, because they describe probability in a number of important situations. An assignment of probabilities to outcomes by a normal curve is a **normal probability distribution.** Figures 7.5 and 7.6 demonstrate that the sampling distribution of a sample proportion from a simple random sample is close to a normal distribution. This is not just a matter of artistic judgment. It is a mathematical fact, first proved by Abraham DeMoivre in 1718. Other common statistics, such as the mean $\bar{x}$ of a large sample, also have sampling distributions that are approximately normal. A normal curve will not exactly describe a specific set of outcomes, such as 200 sample proportions. It is rather an idealized distribution that is convenient to use and gives a good approximation to the actual distribution of outcomes.

There is a close connection between describing an assignment of probability to numerical outcomes and describing a set of data. Histograms can be used for both tasks. Similarly, smooth curves such as the normal curves can replace histograms for describing large sets of data. Many sets of data are approximately described by normal distributions. The normal distributions therefore deserve more detailed study.

Normal curves can be specified exactly by an equation, but we will be content with pictures like Figure 7.6. All normal curves are symmetric and bell-shaped, with tails that fall off rapidly. The center of the symmetric normal curve is the center of the distribution in several senses. It is the mean μ for the assignment of probabilities. It is also the median in the sense that half the probability (half the area under the curve) lies on each side of the center. When probabilities are assigned as areas

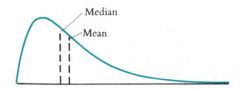

Figure 7.7 The mean of a skewed distribution is located farther toward the long tail than the median.

under a symmetric curve, the mean μ is also the median of the distribution. The mean and median of a skewed distribution are not equal. Figure 7.7, for example, shows a right-skewed distribution. The right tail of the curve is much longer than the left. The prices of new houses are an example of a skewed distribution — there are many moderately priced houses and a few extravagantly priced mansions out in the right tail. Those mansions pull the mean — or average — price up, so that it is greater than the median. For example, the mean price of a new house in 1989 was \$159,100, but the median price was only \$129,900.

As we saw in Chapter 6, even the most cursory description of data on a single variable should include a measure of spread in addition to a measure of center or location. What about the spread of a normal curve? *Normal curves have the special property that their spread is completely measured by a single number, the standard deviation.* We learned in the last chapter how to calculate the standard deviation from a set of observations. For normal distributions, the standard deviation (like the mean) can be found directly from the curve.

To find the **standard deviation** of a normal distribution, run a pencil along the normal curve from the center (the mean) outward. At first the curve falls ever more steeply as you go out; farther from the mean it falls ever less steeply. The two points where the curvature changes are located one standard deviation on either side of the mean.

With a little practice you can locate the change-of-curvature points quite accurately. For example, Figure 7.8 shows the distribution of heights of American women ages 18 to 24. The shape of the curve is normal, with mean (and median) height $\mu = 64$ inches. The two change-of-curvature points are at 61.5 inches and 66.5 inches. The standard deviation of the distribution is the distance of either of these points from the mean, or 2.5 inches.

The usual notation for the standard deviation of a distribution is σ, the Greek letter sigma. Just as for the mean μ, it is possible to find σ for any distribution directly from the assignment of probabilities. We will not do this, but will content ourselves with being able to find σ for normal distributions by looking at the curves. Just as for the mean, we distinguish between s, the standard deviation of a given set of observations, and σ, the standard deviation of a probability distribution.

In Chapter 6, we often used the quartiles to indicate the spread of a distribution. Because the standard deviation completely describes the spread of any normal distribution, it fixes both quartiles as follows:

The first quartile of any normal distribution is located 0.67σ below the mean; the third quartile is 0.67σ above the mean.

EXAMPLE: Heights of Young Women

The distribution of heights of young women, shown in Figure 7.8, is approximately normal with mean $\mu = 64$ inches and standard deviation $\sigma = 2.5$ inches. The quartiles lie 0.67σ, or

$$(0.67)(2.5) = 1.7 \text{ inches}$$

on either side of the mean. The first quartile is $64 - 1.7$, or 62.3 inches. The third quartile is $64 + 1.7$, or 65.7 inches. Figure 7.9 marks the quartiles on the normal curve. They contain between them the middle 50% of women's heights.   ▲

The mean and standard deviation of normal curves have a special property: the shape of a normal distribution is completely specified once μ and σ are given. A measure of center and a measure of spread are not sufficient to determine the exact shape of most distributions of data, but the mean and standard deviation are enough when the distribution is normal. Changing the mean of a normal curve does not change its shape; it only moves the curve to a new location. Changing the standard deviation does change the shape. A normal curve

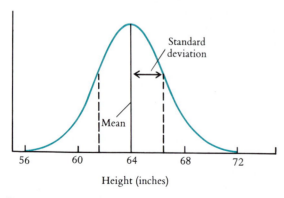

Figure 7.8 Locating the mean and standard deviation on a normal curve.

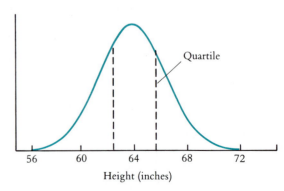

Figure 7.9 The quartiles of a normal distribution are located 0.67 standard deviation on either side of the mean.

with a smaller standard deviation is taller and narrower (has less spread) than one with a larger standard deviation. You can see this by comparing the two normal curves for our bead-sampling experiments in Figure 7.6. Both normal curves have the same mean, but the curve for samples of size 1500 has the smaller standard deviation.

Another consequence of the fact that the mean and standard deviation completely specify a normal distribution is that all normal distributions are the same when we record observations in terms of how many standard deviations they lie from the mean. In particular, the probability that an observation falls within one, two, or three standard deviations of the mean is the same for all normal distributions. The probability of an outcome falling within one standard deviation on either side of the mean is .68. If we go out two standard deviations from the mean, we find a probability of .95. Finally, the probability of falling within three standard deviations of the mean is almost 1, or .997 to be exact. These facts can be derived mathematically from the equation of a normal curve. They are not true for distributions with other shapes.

Figure 7.10 illustrates these facts expressed in terms of percents. Together, we call them the **68–95–99.7 rule for normal distributions.**

▶ 68% of the observations in any normal distribution fall within one standard deviation of the mean.

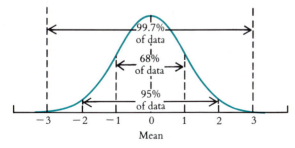

Figure 7.10 The 68–95–99.7 rule for normal distributions.

▶ 95% of the observations fall within two standard deviations of the mean.
▶ 99.7% of the observations fall within three standard deviations of the mean.

Using the three numbers in the 68–95–99.7 rule, we can quickly derive helpful information about any normal distribution. More detailed information can be gleaned from tables of areas under the normal curves, but the 68–95–99.7 rule is adequate for our purposes.

EXAMPLE: Heights of Young Women

The heights of American women between the ages of 18 and 24 are roughly normally distributed, with mean $\mu = 64$ inches and standard deviation $\sigma = 2.5$ inches. One standard deviation below the mean is $64 - 2.5$, or 61.5 inches. Similarly, one standard deviation above the mean is $64 + 2.5$, or 66.5 inches. The "68" part of the 68–95–99.7 rule says that about 68% of women are between 61.5 and 66.5 inches tall. Because two standard deviations are 5 inches, we know that 95% of young women are between $64 - 5$ and $64 + 5$, that is, between 59 and 69 inches tall. Almost all women have heights within three standard deviations of the mean, or between 56.5 and 71.5 inches. Very few women are 6 feet (72 inches) tall or over. ▲

EXAMPLE: SAT Scores

The distribution of scores on tests such as the Scholastic Aptitude Test (SAT) is close to normal. SAT scores are adjusted so that for a reference population of students, the mean score is $\mu = 500$ and the standard deviation is $\sigma = 100$. (The mean and standard deviation of the scores for students taking the test in any one year will differ from those of the reference population.) This information allows us to answer many questions about SAT scores.

▶ *How high must a student score to fall in the top 25%?*
The third quartile is $(0.67)(100) = 67$ points above the mean. So scores above 567 are in the top 25%.

▶ *What percent of scores fall between 200 and 800?*
Note that 200 and 800 are three standard deviations on either side of the mean. The 99.7 part of the $68-95-99.7$ rule says that 99.7% of all scores lie in this range. (In fact, 200 and 800 are the lowest and highest scores that are reported on the SAT. The few scores higher than 800 are reported as 800.)

▶ *What percent of scores are above 700?*
First note that 700 is two standard deviations above the mean. By the 95 part of the $68-95-99.7$ rule, 95% of all scores fall between 300 and 700 and 5% fall below 300 or above 700. Because normal curves are symmetric, half of this 5% are above 700. So a score above 700 places a student in the top 2.5% of the reference population. ▲

▶ THE CENTRAL LIMIT THEOREM

The significance of normal distributions is explained by a key fact in probability theory known as the **central limit theorem.** This theorem says that the distribution of any random phenomenon tends to be normal if we average it over a large number of independent repetitions. The central limit theorem allows us to analyze and predict the results of chance phenomena if we average over many observations.

We have already seen the central limit theorem at work in our bead-sampling experi-

ment. A single bead drawn at random is either dark or light. Only two outcomes are possible, and there is no normal curve in sight. However, the percent of dark beads when 50 beads are drawn at random roughly follows a normal distribution. You can think of the percent of dark beads as an average of "dark" or "light" over the 50 beads drawn. When 1500 beads are drawn, the percent of dark beads represents an average over a larger number of beads, and it is even closer to a normal curve.

Our sampling experiment showed that samples of 1500 beads have much less spread than samples of 50. Spread can be described by the standard deviation of the normal distribution of outcomes. The central limit theorem makes this explicit. Here is a more exact statement.

Central limit theorem. A sample mean or sample proportion from n trials on the same random phenomenon has a distribution that is approximately normal when n is large. The mean of this normal distribution is the same as the mean for a single trial. The standard deviation of the normal distribution is the standard deviation for a single trial divided by $\sqrt{n}$.

Pay attention to the important fact that the standard deviation of a mean or proportion decreases with the square root of the number of observations, $\sqrt{n}$. This is true for all values of n, not just when n is large enough that the central limit theorem says that the distribution is close to normal. For example, the height of a single young woman chosen at random has standard deviation $\sigma = 2.5$ inches. The mean height $\bar{x}$ of a sample of 5 young women therefore has standard deviation

$$\sigma_{\bar{x}} = \frac{\sigma}{\sqrt{n}}$$

$$= \frac{2.5}{\sqrt{5}} = \frac{2.5}{2.236}$$

$$= 1.118 \text{ inches}$$

The notation $\sigma_{\bar{x}}$ reminds us that this is the standard deviation of the distribution of $\bar{x}$, not the standard deviation σ of a single observation.

To cut the standard deviation of a sample mean or proportion in half, the sample size must be multiplied by four, not just by two. For example, an average over 100 observations ($\sqrt{100} = 10$) has a standard deviation 1/10 as large as the standard deviation of a single observation. An average over 25 observations ($\sqrt{25} = 5$) from the same population has standard deviation 1/5 as large as the standard deviation for an individual. Because 1/10 is half of 1/5, increasing the sample size from 25 to 100 cuts the standard deviation of the normal distribution of the sample mean in half. The same $\sqrt{n}$ effect applies to sample proportions. For example, a sample of 1500 beads is 30 times larger than a sample of 50 beads. So the standard deviation of the percentage of dark beads is $\sqrt{30}$, or 5.5, times smaller for samples of 1500 than for samples of 50. The two normal curves in Figure 7.6 display exactly this difference in standard deviations.

The central limit theorem can help to answer our opening question: How can gambling be a business for a casino? Let's look at just one of the many bets that a casino offers.

EXAMPLE: Red or Black in Roulette

An American roulette wheel has 38 slots, of which 18 are black, 18 are red, and 2 are green (see Figure 7.11). When the wheel is spun, the ball is equally likely to come to rest in any of the slots. Gamblers can place a number of different bets in roulette. One of the simplest wagers chooses red or black. A bet of one dollar on red will pay off an additional dollar if the ball lands in a red slot. Otherwise, the player loses. (When gamblers bet on red or black, the two green slots belong to the house.)

If we decide to bet on red, there are only two possible outcomes: win or lose. We win if the ball stops in one of the 18 red slots. We

Figure 7.11 A gambler may win or lose at roulette, but in the long run the casino always wins.

lose if it lands in one of the 20 slots that are black or green. Because casino roulette wheels are carefully balanced so that all slots are equally likely, the probabilities are

$$P(\text{win } \$1) = 18/38$$
$$P(\text{lose } \$1) = 20/38$$

The mean outcome of a single bet on red is found in the usual way:

$$\mu = (1)(\tfrac{18}{38}) + (-1)(\tfrac{20}{38})$$
$$= -\tfrac{2}{38} = -0.053$$

The law of large numbers says that in the long run gamblers will lose (and the casino will win) an average of 5.3 cents per bet. ▲

Just as when only one bead is selected from the container of beads, there is no normal curve in sight when only one bet is made on red in roulette. But the central limit theorem ensures that the average outcome of many bets follows a distribution that is close to normal. Suppose that we place 50 bets in an evening's play. The mean outcome is the average winnings $\bar{x}$, the overall gain (or loss) divided by 50. If we win 30 and lose 20 times, the

SP TLIGHT 7.2   State Lotteries

▶ ▶ ▶   ▶ ▶ ▶ ▶ ▶ ▶ ▶ ▶ ▶

Both public and private lotteries were common in the early years of the United States. After disappearing for a century or so, government-run gambling reappeared in New Hampshire and New York in the mid-1960s and has spread to almost all states outside the South. State lotteries sold over $15 billion worth of tickets in 1989 and continue to expand rapidly. Their growth is fed by constant advertising and by new games designed to hold the interest of the public. The most popular game in most states is lotto, in which players choose (for example) 6 numbers out of 54 in the hope of matching the randomly chosen winning numbers.

State lotteries differ from casino gambling in several respects. For one thing, the lotteries offer much poorer returns to their customers. Las Vegas and Atlantic City pay out between about 85% and 95% of the dollars bet, depending on the game. State lotteries on the average pay out 54% of the dollars bet; about 6% goes for advertising and expenses, and the remaining 40% flows into the state's treasury.

The states also pay lotto winners over time, usually 20 years. Because money earns interest over time, a jackpot advertised as $10 million actually costs the state only about $4.8 million.

Finally, gamblers can't rely on the central limit theorem to predict their long-run average lotto winnings as it predicts their long-run average winnings in roulette. Lotto offers one very large but very improbable jackpot, a few small prizes, and nothing to the remaining millions of tickets. The expected value of a $1 ticket is about 50 cents, but the variation in outcomes is so large that a gambler's average winnings over even thousands of tickets remain highly unpredictable. The central limit theorem remains true for any game of chance, of course. But when the variation in a single play is extremely large, no humanly possible number of plays, even aided by that $\sqrt{n}$, can reduce the variation in the average outcome enough for a useful prediction. The only compensation almost all lotto players will receive is the pleasure of imagining themselves rich.

overall gain is $10, an average winnings of $\bar{x} =$ $0.20 per bet. If we continue to gamble night after night, placing 50 bets each night, our average winnings per bet will vary from night to night. A histogram of these values will follow a normal distribution. Figure 7.12 shows the results of many trials of 50 bets each. The normal curve superimposed on the histogram is the distribution given by the central limit theorem in this case.

We know that the mean of the normal distribution in Figure 7.12 is the same as the mean of a single bet, -0.053. What is the standard deviation? The full spread of outcomes observed was -0.47 to 0.37, or 0.84 in all. By the 99.7 part of the 68–95–99.7 rule, the outcomes should span about three standard deviations on each side of the mean. The standard deviation is therefore about one-sixth of 0.84, or 0.14 (14 cents). Check this by locating the change-of-curvature points of the normal curve in the figure. From this combination of calculation and experiment we conclude that the average winnings in 50 bets follow approximately the normal distribution with mean −0.053 and standard deviation 0.14.

What will be the experience of an habitual gambler who places 50 bets per night? Almost all average nightly winnings will fall within three standard deviations of the mean, that is, between

$$-0.053 + (3)(0.14) = 0.367$$

and

$$-0.053 - (3)(0.14) = -0.473$$

The total winnings after 50 bets will therefore fall between

$$(0.367)(50) = 18.35$$

and

$$(-0.473)(50) = -23.65$$

The gambler may win as much as $18.35 or lose as much as $23.65. Gambling is exciting because the outcome, even after an evening of bets, is uncertain. It is possible to walk away a winner. It's all a matter of luck.

The casino, however, is in a different position. It doesn't want excitement, just a steady income. The house bets with all its customers — perhaps 100,000 individual bets on black or red in a week. The distribution of average customer winnings on 100,000 bets is very close to normal, and the mean is still the mean outcome for one bet, −0.053, a loss of 5.3 cents per dollar bet.

The central limit theorem says in addition that the standard deviation of the distribution of average winnings decreases with the square root of the number of bets over which we are averaging. Now, 100,000 is 2000 times as much as 50. So the standard deviation of the casino's distribution (average winnings over 100,000 bets) is

$$\sqrt{2000} = 44.72$$

times as small as the standard deviation of the gambler's distribution (average over 50 bets). The gambler has standard deviation 0.14; the casino therefore has standard deviation

$$\frac{0.14}{44.72} = 0.003$$

There you have it. The individual gambler will experience wide variation in winnings; he or she gets excitement. The casino experiences very little variation; it has a business. Here is what the spread in the casino's average result looks like after 100,000 bets:

$$\begin{aligned} \text{Spread} &= \text{mean} \pm 3 \text{ standard deviations} \\ &= -0.053 \pm (3)(0.003) \\ &= -0.053 \pm 0.009 \\ &= -0.044 \text{ to } -0.062 \end{aligned}$$

Because the casino covers so many bets, the standard deviation of the average winnings per bet becomes very narrow. And because the mean is negative, almost all outcomes will be negative. Thus, the gamblers' losses and the casino's winnings are almost certain to average between 4.4 and 6.2 cents for every dollar bet.

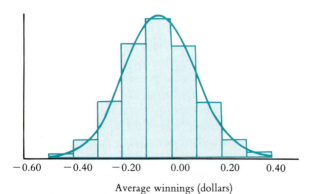

Figure 7.12 The distribution of winnings in repeated bets on red or black in roulette.

The gamblers who collectively placed those 100,000 bets will lose their money. We are now in a position to estimate the probable range of the losses:

$$(-0.044)(100,000) = -4400$$
$$(-0.062)(100,000) = -6200$$

The gamblers are almost certain to lose — and the casino is almost certain to take in — between $4400 and $6200 on those 100,000 bets. What's more, we have seen from the central limit theorem that the more bets that are made, the narrower the range of possible outcomes. That is how a casino can make a business out of gambling. The more money that is bet, the more accurately the casino can predict its profits.

▶ REVIEW VOCABULARY

Central limit theorem   The average of many independent random outcomes is approximately normally distributed. When we average n independent repetitions of the same random phenomenon, the resulting distribution of outcomes has mean equal to the mean outcome of a single trial and standard deviation proportional to $1/\sqrt{n}$.

Combinatorics   The branch of mathematics that counts arrangements of objects.

Event   Any collection of possible outcomes of a random phenomenon. An event is a subset of the sample space.

Law of large numbers   As a random phenomenon is repeated many times, the mean $\bar{x}$ of the observed outcomes approaches the mean μ of the probability model.

Mean of a probability model   The average outcome of a random phenomenon with numerical values, found by multiplying each possible outcome by its probability and then summing all the products.

Normal distributions   A family of probability models that assign probabilities to events as areas under a curve. The normal curves are symmetric and bell-shaped. A particular curve is completely described by giving its mean μ and its standard deviation σ.

Probability   A number between 0 and 1 that gives the long-run proportion of repetitions of a random phenomenon on which an event will occur.

Probability model   A sample space S together with an assignment of probabilities to events. If probabilities $P(s)$ are assigned to all outcomes s in S, they must satisfy two laws:

1. For every outcome s, $0 \le P(s) \le 1$.

2. The sum of the probabilities $P(s)$ over all outcomes s is exactly 1.

A probability model can also assign probabilities to events as areas under a curve; in this case, the total area under the curve must be exactly 1.

Random phenomenon   A phenomenon is random if it is uncertain what the next outcome will be, but each outcome nonetheless tends to occur in a fixed proportion of a very long sequence of repetitions. These long-run proportions are the probabilities of the outcomes.

Sample space   A list of all possible outcomes of a random phenomenon.

Sampling distribution   An assignment of probabilities to the possible values of a statistic. This distribution describes the sampling variability of the statistic.

Sampling variability   The random variability in the value of a statistic (such as a sample mean or proportion) when random samples are drawn repeatedly from the same population.

68–95–99.7 rule   In any normal distribution, 68% of the observations lie within one standard deviation on either side of the mean; 95% lie within two standard deviations of the mean; and 99.7% within three standard deviations of the mean.

Standard deviation of a probability model   A measure of spread that is particularly appropriate for normal distributions. The standard devia-

tion σ of a normal curve is the distance from the mean to the change-of-curvature points on either side.
Statistic   A number computed from a sample. In random sampling, the value of a statistic will vary in repeated sampling.

▶ SUGGESTED READINGS

MOSTELLER, FREDERICK, ROBERT E. K. ROURKE, AND GEORGE B. THOMAS: *Probability with Statistical Applications*, Addison-Wesley, Reading, Mass., 1970. A rich treatment of basic probability that requires only high school algebra, but is somewhat sophisticated.

OLKIN, INGRAM, LEON J. GLESER, AND CYRUS DERMAN: *Probability Models and Applications*, Macmillan, New York, 1980. This book is distinguished by an emphasis on the use of probability to describe real phenomena and by outstanding examples of modeling. In level it falls between Mosteller et al. and Snell.

SNELL, J. LAURIE: *Introduction to Probability*, Random House, New York, 1988. A calculus-based text aimed at undergraduate mathematics majors that is recommended here because of its excellent examples and historical remarks, and in particular because Snell makes good use of BASIC programs that are included in the text.

▶ EXERCISES

You can estimate an unknown probability by actually observing many repetitions of the random phenomenon in question. Exercises 1 to 5 produce rough estimates based on a small number of repetitions. You can see the random behavior in more detail by making a graph like Figure 7.2 rather than just reporting the final proportion of outcomes.

1. Toss a thumbtack on a hard surface 100 times. How many times did it land with the point up? What is the approximate probability of its landing point up?

2. Hold a penny upright on its edge under your forefinger on a hard surface, then snap it with your other forefinger so that it spins for some time before falling. Based on 50 spins, what is the probability of heads?

3. Open your local telephone directory to any page and note whether the last digit of each of the first 100 telephone numbers on the page is odd or even. How many of the digits are odd? What is the approximate probability that the last digit of a telephone number is odd?

4. The table of random digits (Table 5.1 on p. 150) was produced by a random mechanism that gives each digit a probability 0.1 of being a 0. What proportion of the first 200 digits in the table are 0s? This proportion is an estimate of the true probability, which in this case is known to be 0.1.

5. Pick up a book and open to any page. Count the words in the first complete paragraph on that page and note how many of them begin with a vowel. (If the paragraph contains fewer than 100 words, include the next paragraph as well.) What do you estimate to be the probability that a word chosen at random from this book begins with a vowel?

In each of Exercises 6 to 10, describe a reasonable sample space for the random phenomena mentioned. In some cases, more than one choice is possible.

6. Toss a coin 10 times.
   a. Count the number of heads observed.
   b. Calculate the percentage of heads among the outcomes.
   c. Record whether or not at least five heads occurred.

7. A female lab rat is about to give birth. You count the number of offspring in the litter. (We don't know how large rat litters can be, but you can set a reasonable upper limit if you want.)

8. A couple plans to have three children.
   a. Record the sex (M or F) of each child in order of birth.
   b. Record the number of girls.

9. Choose a student at random and record the number of dollars in bills (ignore change) that he or she is carrying. (We don't know the largest amount that a student could reasonably carry, so you will have to make a choice in stating the sample space.)

10. Subjects in a clinical trial are assigned at random to either the new treatment group or the control group. For the next subject, you record treatment or control, male or female, and smoker or nonsmoker.

11. Which of the following are legitimate probability models for tossing three (possibly unfair) coins? Explain your answer in each case.

Outcome	Model A	Model B	Model C	Model D
H, H, H	.125		.125	.250
H, H, T	.125	.375	.250	.125
H, T, H	.125		−.125	.250
H, T, T	.125	.125	.125	.125
T, H, H	.125		.125	.250
T, H, T	.125	.375	.125	.125
T, T, H	.125		.250	.250
T, T, T	.125	.125	−.125	.125

12. M&M candies come in several colors mixed together in a bag. Which of the following are legitimate probability models for drawing a single M&M and recording its color?

Color	Brown	Red	Yellow	Green	Orange	Tan
Model A	.3	.2	.2	.2	.1	.1
Model B	.2	.2	.2	.1	.1	.1
Model C	.3	.2	.2	.2	.1	0

13. Here is the distribution of the blood type of a randomly chosen black American. If this is to be a legitimate assignment of probabilities, what must be the probability of type AB blood?

Blood type	O	A	B	AB
Probability	.49	.27	.20	

14. Here is the distribution of marital status for American women aged 25 to 29 years. If this is to be a legitimate probability model, what must be the probability that a woman in this age group is married?

Outcome	Single	Married	Widowed	Divorced
Probability	.288		.003	.076

■ 15. A bridge deck contains 52 cards, four of each of the 13 face values ace, king, queen, jack, ten, nine, . . . , two. You deal a single card from such a deck and record the face value of the card dealt. Give an assignment of probabilities to these outcomes that should be correct if the deck is thoroughly shuffled. Give a second assignment of probabilities that is legitimate (that is, obeys Laws 1 and 2) but differs from your first choice. Then give a third assignment of probabilities that is *not* legitimate, and explain what is wrong with this choice.

16. Exactly one of Brown, Chavez, and Williams will be promoted to partner in the law firm that employs them all. Brown thinks that she has probability .25 of winning the promotion and that Williams has probability .2. What probability does Brown assign to the outcome that Chavez is the one promoted?

▲ 17. Suppose that A and B are events that have no outcomes in common, and thus cannot occur simultaneously. For example, in tossing three coins we could have $A = \{$First coin gives H$\}$ and $B = \{$First coin gives T$\}$. Starting from the fact that the probability of any event is the sum of the probabilities of the outcomes making up the event, explain why

$$P(A \text{ or } B \text{ occurs}) = P(A) + P(B)$$

must always be true for two such events.

18. If a fair die is rolled once, it is reasonable to assign probability 1/6 to each of the six faces. If we accept this probability model, what is the probability of rolling a number less than 3?

■ Discussion exercise.   ▲ Advanced exercise.

19. In the example on p. 219 we gave a probability model for rolling two fair dice and recording the two up faces that assigned equal probability to each of the 36 possible outcomes in Figure 7.4. Starting from this model, give a probability model for rolling two fair dice and recording the sum of the faces showing. Then use this model to answer the following questions:
    a. What is the probability of rolling a 7 or an 11?
    b. What is the probability of rolling a number 7 or greater?

20. Toss three coins and record heads or tails for each. Exercise 11 shows the eight members of the sample space. If the coins are fair, these outcomes are equally likely. Starting from this probability model, find the probability model for tossing three fair coins and counting the number of heads. What is the probability of at least two heads?

▲ 21. A computer assigns three-letter log-in identification codes at random as in the example on p. 222. If we take the vowels to be a, e, i, o, u, and y, what is the probability that your code contains no vowels if repeated letters are allowed? If no repeats are allowed?

▲ 22. The computer of the previous exercise is programmed to assign log-in codes of the form consonant–vowel–consonant. The consonants and vowels are both chosen at random and the consonants can repeat. What is the probability that your code does not contain an "x"?

▲ 23. Suppose that the computer of Exercise 21 assigns three-character log-in codes that may contain the digits 0 to 9 as well as letters, with repeats allowed. What is now the probability that your code contains no "x"? What is the probability that your code contains no digits?

▲ 24. Automobile license plate numbers in Indiana consist of seven characters. The first three describe the county in which the car is licensed, while the last four are digits assigned at random. You are hoping for a plate on which these four digits are identical (like 7777). What is your probability of receiving such a plate?

▲ 25. The personal identification numbers (PINs) for automatic teller machines, telephone calling cards, and the like, usually consist of four digits. You notice that most of your PINs have at least one 0, and you wonder if the issuers use lots of 0s to make the numbers easy to remember. What is the probability that a PIN chosen at random has at least one 0?

▲ 26. Automobile license plates in some states consist of three letters followed by three digits. How many different license plates are possible in such a state? Suppose the state did not require the letters to come first, so that any six letters or digits in any order could appear. How many different license plates would then be possible?

27. A monkey at a keyboard presses three keys and hits the letters a, g, and s in random order. How many possible three-letter "words" can the monkey type

▲ Advanced exercise.

**Classic fractals: von Koch Snowflake,
Sierpinski Gasket, and Julia Set**

Geometry is the mathematical language for shape, but the triangles and circles of Euclidean geometry offer a poor vocabulary for nature's complex forms. The mathematician Benoit Mandelbrot was the first both to recognize the obvious problem, "clouds are not spheres, mountains are not cones . . ." and to provide a solution in 1975 in terms of a new mathematical language: *fractal geometry*. In the early 1900s mathematicians conceived a number of shapes whose properties were so strange that they were dubbed *mathematical monsters* by their creators. These three classic shapes can be obtained by simple iterative procedures (adding or subtracting successively smaller triangles). They were originally thought to have no application in natural science, but it is now realized that they are basic fractals illustrating their unifying concept, *self-similarity*. [© R. F. Voss/IBM Research]

Fractal Surface Zoom
Euclidean shapes, such as a
circle, become smooth or
flat upon magnification.
Fractals, on the other
hand, retain interesting
detail at successively higher
magnification, and this
detail is often the same as
that at larger scales.
Blowups of portions yield
miniature copies of the
whole. *Self-similarity* is also
present in a statistical sense
in many natural shapes
from the twisted strings of
complex polymers to
microscopically rough surfaces; mountains and coast-
lines on the Earth's surface;
clouds; and the distribution
of galaxies in space.
Computer-generated forg-
eries of these objects are
constructed by adding ever
smaller variations.
Magnification of a single
small portion of the fractal
coastline cannot be distinguished from different portions at the original scale.
[© R. F. Voss/IBM
Research]

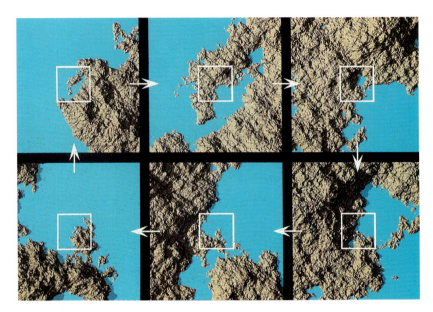

Fractal Planetrise
A readily recognized classic of fractal geometry composed of three separate fractal forgeries of natural shapes: the mountainous landscape (the same one in **Making Clouds out of Mountains**), the fractal planet constructed from a simple model of global earthquakes, and the craters on the landscape. [© R. F. Voss/IBM Research]

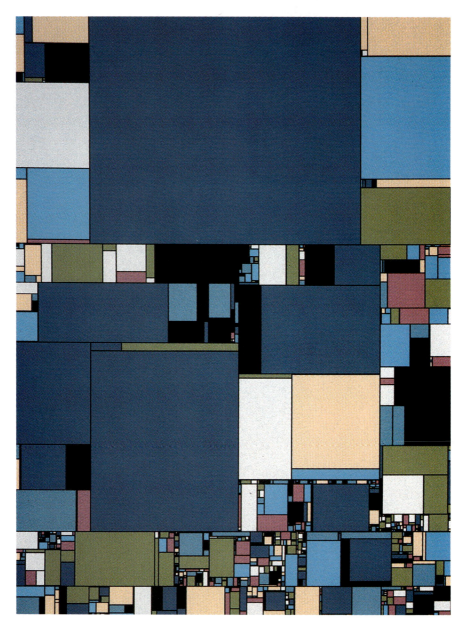

Rectangular Craters on the Moon
Self-similarity applies to collections of many objects as well as individual
shapes: the distribution of triangle sizes in the **von Koch Snowflake** and
the **Sierpinski Gasket**, island and lake sizes on the Earth, the branching
of trees, crater sizes on the Moon. As the size of an object decreases, the
number increases. Here, a Euclidean rectangle is randomly subdivided into
successively smaller rectangles with a natural fractal distribution that is the
same as the craters in **Fractal Planetrise**. [© R. F. Voss/IBM Research]

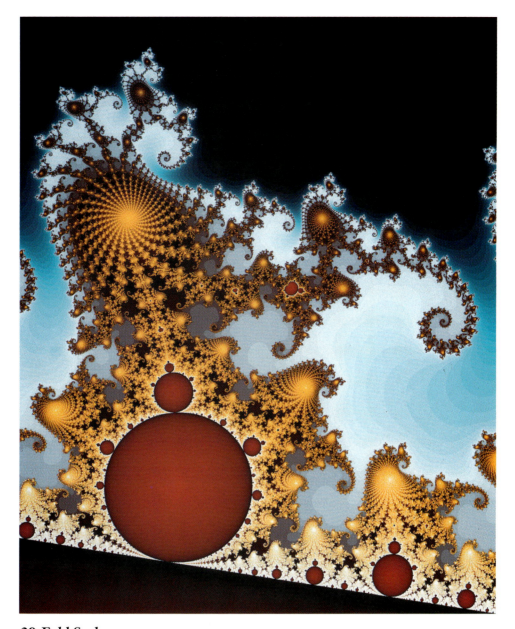

29-Fold Seahorse

The Mandelbrot set, or M-set, is one of the most beautiful, complex, and computed shapes known. It has become a modern icon for the mathematical fractal monsters. Upon magnification, its boundary yields increasingly wondrous detail, familiar reappearances of the whole, and infinite miniature replicas of Julia sets. The Mandelbrot set is a prime example of how a simple nonlinear repeated rule yields a complex fractal shape. Its study has produced important new scientific insights into chaos and the behavior of dynamic systems. [© R. F. Voss/IBM Research]

Percolation Animals
Simple mathematical models that produce fractal shapes have been of great importance to scientists in understanding nature. *Percolation* models fluid flow through irregular materials (such as water through coffee grounds or oil through sandstone) and electrical conduction in composite materials. In these models the irregular connected parts, shown here as the different colored animals, determine the behavior. [© R. F. Voss/IBM Research]

Making Clouds out of Mountains
Thanks to evolution in a natural world dominated by fractal shapes, the human perceptual system is particularly responsive to fractals. The same fractal shape can assume many familiar guises. Here, the planet of **Fractal Planetrise** becomes an arid moon. The boundary of the Mandelbrot set becomes galactic streamers, and an inverted mountain valley becomes its own invading cloud mass. [© R. F. Voss/IBM Research]

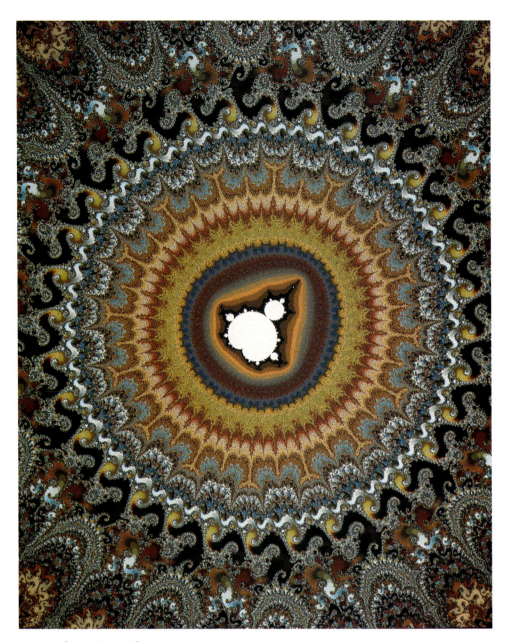

Avogadro's Magnification
Fractal geometry demonstrates that shapes may be classified according to how they behave under magnification. Euclidean shapes such as circles become smooth. Self-similar fractals such as the **von Koch Snowflake**, the **Sierpinski Gasket**, and the **Coastline Zoom** remain approximately unchanged, like much of nature. The Mandelbrot set, however, is even more complicated. Although magnification of its boundary yields infinite close copies of the entire set, the neighborhood around these replicas becomes increasingly intricate as the magnification is increased. Relative to this magnification of around Avogadro's number, 10^{23}, the original Mandelbrot set would be about 1,000,000 light years across. [© R. F. Voss/IBM Research]

using only these letters? Which of these are meaningful English words? What is the probability that the word the monkey typed is meaningful?

28. You are about to visit a new neighbor. You know that the family has four children, but you do not know their age or sex. Write down all possible arrangements of girls and boys in order from youngest to oldest, such as BBGG (the two youngest are boys, the two oldest girls). The laws of genetics say that all of these arrangements are equally likely.
    a. What is the probability that the oldest child is a girl?
    b. What is the probability that the family has at least three boys?
    c. What is the probability that the family has at least three children of the same sex?

29. What is the mean number of pips observed in rolling a single fair die?

30. In Exercise 19, you found a probability model for rolling two fair dice and counting the pips on the two up faces. What is the mean number of pips obtained?

31. In Exercise 20 you found the probability model for tossing three fair coins and counting the heads observed. Compute the mean number of heads.

32. Teachers in the Lost Valley Central School District are allowed up to 7 days of paid sick leave each year. Here is the distribution of the number of days of sick leave taken by the teachers last year. What is the mean number of days of sick leave that a teacher will take in a year?

Days taken	0	1	2	3	4	5	6	7
Percent of teachers	15	15	10	10	10	12	8	20

33. A study selected a sample of fifth-grade pupils and recorded how many years of school they eventually completed. Based on this study we can give the following probability model for the years of school that will be completed by a randomly chosen fifth grader:

Years	4	5	6	7	8	9	10	11	12
Probability	.010	.007	.007	.013	.032	.068	.070	.041	.752

    a. Verify that this is a legitimate probability model.
    b. What outcomes make up the event "The student completed at least one year of high school"? (High school begins with the ninth grade.) What is the probability of this event?
    c. What is the mean number of years of school completed?

34. In an experiment on the behavior of young children, each subject is placed in an area with five toys. The response of interest is the number of toys that the child plays with. Past experiments with many subjects have shown that the probability model for the number of toys played with is as follows:

Number of toys	0	1	2	3	4	5
Probability	.03	.16	.30	.23	.17	.11

a. What is the probability that a child will play with more than one toy during the experiment?
b. What is the mean number of toys a child will play with?

35. An American roulette wheel has 38 slots numbered 0, 00, and 1 to 36. The ball is equally likely to come to rest in any of these slots when the wheel is spun. The slot numbers are laid out on a board on which gamblers place their bets. One column of numbers on the board contains a multiple of 3, that is, 3, 6, 9, . . . , 36. A gambler places a $1 column bet that pays out $3 if any of these numbers comes up.
a. What is the probability of winning?
b. What are the mean winnings for one play, taking into account the $1 cost of each play?

36. Keno is a common casino game. The house chooses 20 numbers between 1 and 80 at random and gamblers attempt to guess some of the numbers in advance. As in roulette, a bewildering variety of Keno bets are available. Here are some of the simpler Keno bets. Give the mean winnings for each.
a. A $1 bet on "Mark 1 number" pays $3 if the single number you mark is one of the 20 chosen; otherwise, you lose your dollar.
b. A $1 bet on "Mark 2 numbers" pays $12 if both your numbers are among the 20 chosen. The probability of this is about 0.06. Is Mark 2 a more or a less favorable bet than Mark 1?

■ 37. Return to Exercises 10 and 11 of Chapter 5. Working as a team with other students, draw 100 simple random samples of size 5 from this population. Compute the sample proportion $\hat{p}$ of females in each sample. What probability model, based on your experiment, describes the sampling distribution of $\hat{p}$? Make a histogram of this distribution. Describe the shape of this distribution. In particular, does it appear roughly normal? Then find the mean number of females in a sample.

■ 38. The table on the opposite page contains the results of 100 repetitions of the drawing of a simple random sample of size 200 from a large lot of bearings,

■ Discussion exercise.

10% of which do not conform to the specifications. The numbers in the table are the percents of nonconforming bearings in each sample of 200.

8.5	11.5	9	13.5	7.5	8.5	9	6.5	8	9
10	7.5	9	8	10.5	8.5	9	9.5	8	11.5
10	9	9	8.5	9.5	6.5	13.5	11	11.5	13
8.5	6.5	8	7	12	11	8	10.5	12	10.5
15	12	8.5	7	8	8	8.5	12	10.5	8
8.5	11.5	9	11.5	11	12	11.5	11.5	10	9.5
10	9	10	12.5	8	12	12	12	7.5	11
11	8	14	7.5	11	4.5	9.5	8	9.5	9.5
12.5	12	10	7.5	10.5	12.5	12	9.5	9.5	10
14	9	8.5	8.5	12.5	8.5	8.5	9	9.5	9

Give an estimated sampling distribution for the sample proportion in this situation by recording each outcome and the proportion of trials on which it occurred. Make a histogram of the distribution and describe its shape. Is the center close to 10%? Is the distribution roughly symmetric? Does it appear approximately normal? Find the mean outcome from your distribution. Is it close to 10%?

39. The distribution of heights of adult American men is approximately normal, with mean 69 inches and standard deviation 2.5 inches. Draw a normal curve on which this mean and standard deviation are correctly located. (*Hint:* Draw the curve first, then mark the horizontal axis.)

40. Using the normal distribution described in Exercise 39 and the 68–95–99.7 rule, answer the following questions about the heights of adult American men.
    a. What percent of men are taller than 74 inches?
    b. Between what heights do the middle 95% of American men fall?
    c. What percent of men are shorter than 66.5 inches?

41. What are the quartiles of the distribution of heights of American men in Exercise 39?

▲ 42. The figure below is a probability distribution that is not symmetric. The mean and median do not coincide. Which of the points marked is the mean of the distribution, and which is the median?

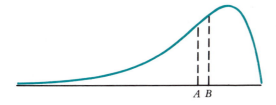

                A B

▲ Advanced exercise.

43. The concentration of the active ingredient in capsules of a prescription painkiller varies according to a normal distribution with $\mu = 10\%$ and $\sigma = 0.2\%$.
   a. What is the median concentration? Explain your answer.
   b. What range of concentrations covers the middle 95% of all capsules?
   c. What range covers the middle half of all capsules?

44. Answer the following questions for the painkiller in the previous exercise.
   a. What percent of all capsules have a concentration of active ingredient higher than 10.4%?
   b. What percent have a concentration higher than 10.6%?

45. Scores on the Wechsler Adult Intelligence Scale (a standard "IQ test") for the 20 to 34 age group are approximately normally distributed with $\mu = 110$ and $\sigma = 25$.
   a. About what percent of people in this age group have scores above 110?
   b. About what percent have scores above 160?

46. The army reports that the distribution of head circumference among soldiers is approximately normal with mean 22.8 inches and standard deviation 1.1 inches.
   a. What percent of soldiers have head circumference greater than 23.9 inches?
   b. What percent of soldiers have head circumference between 21.7 inches and 23.9 inches?

47. The length of human pregnancies from conception to birth varies according to a distribution that is approximately normal with mean 266 days and standard deviation 16 days.
   a. Between what values do the lengths of the middle 95% of all pregnancies fall?
   b. How short are the shortest 2.5% of all pregnancies?

48. The *deciles* of a distribution are the points having 10% (lower decile) and 90% (upper decile) of the observations falling below them. The lower and upper deciles contain between them the central 80% of the data. The lower and upper deciles of any normal distribution are located 1.28 standard deviations on either side of the mean. What score is needed to place you in the top 10% of the distribution of SAT scores in the reference population (normal with mean 500 and standard deviation 100)?

49. Based on the information in the two previous exercises, how short are the shortest 10% of human pregnancies?

50. A student makes a measurement in a chemistry laboratory and records the result in her lab report. When many students do this, the standard deviation of their individual measurements is $\sigma = 10$ milligrams. Suppose the student repeats the measurement 3 times and records the mean $\bar{x}$ of her 3 measurements. What is the standard deviation $\sigma_{\bar{x}}$ of the mean result?

51. A student organization is planning to ask a sample of 50 students if they have noticed AIDS education brochures on campus. The sample percentage who say "Yes" will be reported. Their statistical advisor says that the standard deviation of this percentage will be about 7%. What would the standard deviation be if the sample contained 100 students rather than 50?

■ 52. How many times must the student in Exercise 50 repeat the measurement to reduce the standard deviation of $\bar{x}$ to 5? Explain to someone who knows no statistics the advantage of reporting the average of several measurements rather than the result of a single measurement.

■ 53. How large a sample is required in the setting of Exercise 51 to reduce the standard deviation of the percentage who say "Yes" from 7% to 3.5%? Explain to someone who knows no statistics the advantage of taking a larger sample in a survey of opinion.

▶ WRITING PROJECTS

1. "France gave birth to the mathematics of probability when gamblers in the seventeenth century turned to mathematicians for advice." Two of the mathematicians in question were Pierre de Fermat and Blaise Pascal. Do some reading to learn more about the origins of probability and write a brief essay describing the roles of Fermat and Pascal. (One good source is Carl B. Boyer, *A History of Mathematics*, Wiley, New York, 1991.)

2. State-run lotteries are common in the United States and in other countries, as Spotlight 7.2 suggests. Write a brief essay describing current state lotteries in the United States. How much money do they take in? How is the money that is kept used? What are the trends in the games offered? What other forms of gambling are revenue-hungry states considering? (You can often find recent information about lotteries in the press. Consult, for example, the indexes to the *New York Times* in your library.)

3. Most people "overreact" to risks that have very low probability of occurring. The probability of dying from an airplane crash, a terrorist attack, or a tornado, for example, is extremely small. Yet public opinion and personal decisions often act as if these risks were as probable as death from an automobile accident or a heart attack. Write a brief essay describing how people assess risks, and what factors besides probability influence their actions. One reference is Richard J. Zeckhauser and W. Kip Vicusi, "Risk within reason," *Science*, Vol. 248, May 4, 1990, pp. 559–564.

■ Discussion exercise.

Chapter
8
Statistical Inference

Inference is the process of reaching conclusions from evidence. Evidence can come in many forms. In a murder trial, evidence might be presented by the testimony of a witness, by a record of telephone conversations, or by a weapon. Evidence can also be more subtle. For example, if we walk into an office filled with papers, journals, library books, class notes, and a computer terminal, we might infer that the office belongs to a college professor. In the case of statistical inference, the evidence is provided by data. Informal statistical inference is often based on graphical presentation of data. Formal inference, the topic of this chapter, uses the language of probability to say how confident we are that our conclusion is correct.

Informal evidence is sometimes compelling. The gap in the histogram of Figure 6.5 on p. 186, for example, demands an investigation of the inspection process. But in many cases it is difficult to reach a firm conclusion from informal evidence. We saw from Figure 6.10 on p. 195 that the 1970 draft lottery appeared to favor men born early in the year. However, that inequity was rather small, so small that it is not clearly visible in the scatterplot of Figure 6.9. We might well ask whether the 1970 outcome was simply due to chance rather

than to systematic bias in the lottery. After all, any lottery will show some deviation from perfect uniformity due to the play of chance.

The purpose of formal statistical inference is to verify appearance by calculation. Statistical inference can be compared to an engineer's calculation of the load on a beam — we are more confident after the mathematics is done than we are if the engineer merely says that the beam looks large enough. In the case of the 1970 draft lottery, calculation shows that in a truly random lottery a trend as strong as the one actually observed has probability less than 1 in 1000. This calculation of probability shows that the observed trend is strong evidence that the 1970 lottery was not random.

One of the most intriguing aspects of statistical inference is the fact that *chance* — which we usually associate with uncertainty — is the ally rather than the enemy of confident conclusions. At first glance, the opposite seems to be true. Suppose, for example, that the Gallup poll decided to take its weekly public opinion survey twice, separately and simultaneously selecting two random samples, sending out interviewers, and asking the same questions. Two random samples, each selecting 1500 of 190 million U.S. residents aged 18 and over, will contain different people. And these different people will hold somewhat different opinions. Thus Gallup's recent finding that 45% of Americans are afraid to go out at night for fear of crime really refers only to the 1500 people in one particular sample. If Gallup had taken a simultaneous second survey, no doubt the results would have been different. Random sampling may eliminate *bias*, but it can't eliminate *variability*.

How can we trust the results of a random sample, knowing that a second sample may yield a different result? For that matter, how can we trust the results of a randomized experiment? As we saw in Chapter 5, the Physicians' Health Study tested the effects of aspirin and beta carotene on reducing heart disease and cancer. We know, however, that different people react differently to drugs. What is more, the people in the four treatment groups were assigned by chance. A second trial with other participants would distribute drug reactions and persons with a high risk of cancer and heart disease differently among the treatments. Could the beneficial effect of aspirin in the Physicians' Health Study be just good luck in the random assignment of subjects? We must ask whether the conclusions drawn from this particular experiment are convincing or merely the result of chance.

This chapter addresses the issue of confidence in statistical conclusions. Formal statistical inference enables us to quantify our confidence in the results of random samples and randomized experiments, and thus to verify our impressions by calculation.

▶ CONFIDENCE INTERVALS

We use a simplified version of the Gallup crime survey to introduce an important type of statistical inference. Like most national sample surveys, the Gallup poll uses a complex multistage sampling design. Suppose that we instead drew a *simple random sample* of 1500 adults and discovered that 45% were afraid to go out at night because of crime. We

will call a **sample proportion** that refers to the 1500 people in this particular sample $\hat{p}$ (read as "p hat"). In this case, $\hat{p} = 45\%$. What we really want to know is the *population proportion*, the percent (call it p) of all adult Americans who stay home at night for fear of crime. To discuss statistical inference intelligently, it is essential to keep straight which numbers describe the sample and which describe the population.

A number such as p that describes a population is called a **parameter**; a number such as $\hat{p}$ that is calculated from a sample is called a **statistic.**

It is easy to remember that **p**arameters belong to **p**opulations and **s**tatistics belong to **s**amples because the first letters agree. In an inference problem, parameters are usually unknown. We do not know, for example, the true proportion p of all adults who stay home at night for fear of crime. We use the statistic $\hat{p}$, which we know because we actually interviewed the sample, to estimate the unknown p. *Our goal is not simply to estimate* p, *but to say how accurate our estimate is.* How close to the unknown p will the estimate $\hat{p}$ usually fall?

To answer this question, we turn to the *sampling distribution* of $\hat{p}$. This is the distribution of values taken by the sample proportion as it varies from sample to sample in a large number of repeated samples. If the sample is relatively large, such as our poll's sample of 1500 people, the sampling distribution will be very close to a normal curve like that in Figure 8.1. This figure illustrates several important facts about the sampling distribution. First, the mean of the curve is the true proportion p of people afraid to go out at night. This fact says that $\hat{p}$ has no bias or systematic error as an estimator of the unknown p. In repeated sampling our result will sometimes be high and sometimes low, but the long-run average result, the mean of the sampling distribution, will be correct. Of course, in practice we don't know the numerical value of the parameter p. But we now know that, whatever value p has, the observed

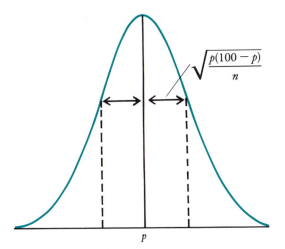

Figure 8.1 The sampling distribution of the sample proportion $\hat{p}$.

values of the statistic $\hat{p}$ cluster around it as shown in Figure 8.1.

Being correct on the average is not enough; a good estimator must also be highly repeatable in the sense of giving nearly the same answer in repeated samples. Repeatability is described by the spread of the sampling distribution, as measured by its standard deviation. Here are the facts for the sample proportion $\hat{p}$:

The standard deviation of the sampling distribution of $\hat{p}$ depends on the population proportion p and the sample size n. It is

$$\sigma_{\hat{p}} = \text{standard deviation of } \hat{p} = \sqrt{\frac{p(100-p)}{n}}$$

(Throughout this chapter, p and $\hat{p}$ are measured in percent.)

If we repeated the sampling many times, sending out waves of interviewers across the nation, each time we would get a value of the sample pro-

portion somewhere along the curve in Figure 8.1. To remind us that the standard deviation of this curve belongs to the sampling distribution of $\hat{p}$, we write it as $\sigma_{\hat{p}}$.

EXAMPLE: Sampling Distribution for the Crime Survey

Suppose that in fact 40% of all adults fear to go out at night because of crime. That is, suppose that $p = 40\%$. Take a simple random sample of size $n = 1500$ people. In repeated samples, the sample percent $\hat{p}$ will vary according to a normal distribution with

$$\text{Mean} = p = 40\%$$
$$\text{Standard deviation } \sigma_{\hat{p}} = \sqrt{\frac{p(100-p)}{n}}$$
$$= \sqrt{\frac{(40)(60)}{1500}}$$
$$= \sqrt{1.6} = 1.26\%$$

If, instead, the truth about the population is $p = 50\%$, the mean of the sampling distribution moves to 50% as well. The standard deviation changes to

$$\sigma_{\hat{p}} = \sqrt{\frac{p(100-p)}{n}}$$
$$= \sqrt{\frac{(50)(50)}{1500}} = \sqrt{1.67} = 1.29\%$$

Notice that this standard deviation does not change very much when p changes. That is, when we take a sample of the same size from different populations, the center of the sampling distribution of $\hat{p}$ moves to the true p for each population, but the spread stays about the same. The size of the sample is the major influence on the spread. Suppose that we took a sample of only $n = 400$ instead of 1500 people from a population for which $p = 40\%$. The mean of the distribution of $\hat{p}$ is 40% — this fact is not affected by the sample size — but the standard deviation increases to

$$\sigma_{\hat{p}} = \sqrt{\frac{p(100 - p)}{n}}$$
$$= \sqrt{\frac{(40)(60)}{400}} = \sqrt{6} = 2.45\%$$

Remember that in practice the value of the parameter p is unknown. We have now learned that if we can guess the value of p even roughly, we can obtain a quite accurate value for the standard deviation of the sample proportion $\hat{p}$. ▲

Our poll of 1500 people, in fact, found that $\hat{p} = 45\%$. This is our best guess for the population percent p. How close to the true p is our guess likely to be? Well, $\hat{p}$ varies normally. The 95 part of the 68–95–99.7 rule says that $\hat{p}$ falls within two standard deviations of the true p (the mean of the sampling distribution) in 95% of all samples. So our guess based on this one sample is likely to be within two standard deviations, that is, within

$$2\sigma_{\hat{p}} = 2\sqrt{\frac{p(100 - p)}{1500}}$$

of the true p.

The catch is that this standard deviation depends on the unknown p. Fortunately, as the example demonstrates, $\sigma_{\hat{p}}$ changes only slowly as p changes, as long as p is not very close to either 0% or 100%. Because $\hat{p}$ is close to p, we simply substitute $\hat{p} = 45\%$ for the unknown p in the formula for the standard deviation. To indicate that the standard deviation is estimated rather than known exactly, we call it $\hat{\sigma}_{\hat{p}}$.

EXAMPLE: Standard Error for the Crime Survey

We need to estimate the standard deviation of our observed sample proportion. The sample size was $n = 1500$, and for p we use the estimate $\hat{p} = 45\%$, based on our survey. The estimated standard deviation is

$$\hat{\sigma}_{\hat{p}} = \sqrt{\frac{(45)(55)}{1500}}$$
$$= \sqrt{1.65} = 1.28\%$$ ▲

Here at last is our conclusion: In 95% of all samples, the sample proportion $\hat{p}$ will fall within 2×1.28, or about 2.6%, of the unknown population proportion p. We took one sample and got $\hat{p} = 45\%$. So we conclude that the p lies in the interval

$$45\% \pm 2.6\%$$

or between 42.4% and 47.6%. We say that we are *95% confident* in this conclusion because we got the interval by calculating how close to p the sample proportion will lie in 95% of all samples. Our interval is a 95% *confidence interval* for estimating the unknown population proportion.

In mathematical terms, the probability is .95 that the sample proportion will fall within $\pm 2.6\%$ of the unknown true fraction of people in the total population afraid to go out at night because of crime. Figure 8.2 makes the idea clearer. The normal curve at the top of the figure is the sampling distribution of $\hat{p}$. When we take many samples, the actual values of $\hat{p}$ vary according to this distribution. The values of $\hat{p}$ observed in 25 samples appear as dots below the curve, together with the confidence intervals that extend out 2.6% on either side of the observed $\hat{p}$. The true population proportion p is marked by the vertical line. Although the intervals vary from sample to sample,

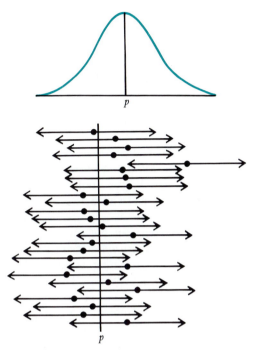

or miss the unknown true parameter. We don't know whether our sample is one of the 95% that hit or one of the 5% that miss. To say that our interval 45% ± 2.6% is a 95% confidence interval means "We got this interval by a method that catches the true parameter 95% of the time."

We have now accomplished two things: we have seen what "95% confidence" means, and we have actually found a 95% confidence interval for estimating a population proportion. Here, as a summary, is the recipe for this interval.

If a simple random sample of size n is drawn, then a 95% confidence interval for the population proportion p is

$$\hat{p} \pm 2\hat{\sigma}_{\hat{p}} = \hat{p} \pm 2\sqrt{\frac{\hat{p}(100 - \hat{p})}{n}}$$

Remember that both p and $\hat{p}$ are measured in percent. This recipe is only approximately correct, but is quite accurate when the sample size n is large.

Figure 8.2 The behavior of confidence intervals in repeated sampling.

all but one of these samples gave a confidence interval that covers the true p. To say that these are 95% confidence intervals is to say that the interval covers the true p in 95% of all samples, and misses in only 5%. Be sure you understand that this 95% and 5% refer to what would happen if we continued to take samples forever. In a small number of samples, the number of confidence intervals that fail to cover the true p may be a bit more or less than 5% of the samples. In Figure 8.2, one out of 25, or 4%, of confidence intervals fails to contain p.

A 95% **confidence interval** is an interval obtained from the sample data by a method that in 95% of all samples will produce an interval containing the true population parameter.

You can see in Figure 8.2 that a confidence interval from one particular sample can either hit

EXAMPLE: Germination of Seeds

A simple random sample of 100 seeds from a new lot is tested for germination; 87 of the 100 germinate. The sample proportion that germinates is

$$\hat{p} = \frac{87}{100} = 0.87 = 87\%$$

The 95% confidence interval for estimating the proportion p of all seeds in the lot that will germinate is

$$\hat{p} \pm 2\sqrt{\frac{\hat{p}(100 - \hat{p})}{n}} = 87 \pm 2\sqrt{\frac{(87)(13)}{100}}$$
$$= 87 \pm 2\sqrt{11.31}$$
$$= 87\% \pm 6.7\%$$

SP TLIGHT 8.1   How the Poll Was Taken

▶ ▶ ▶ ▶ ▶ ▶ ▶ ▶ ▶ ▶ ▶

In June of 1989, the *New York Times* conducted a national opinion poll on women's issues. In response to one of the questions, 41% of the women interviewed agreed that "all things considered, there are more advantages in being a man in America today." Among men, 30% agreed. The poll estimated that 37% of all adults share this opinion. We know that these are sample results that are subject to a margin of error when used to draw conclusions about the population as a whole. Here is the *Times*'s statement about the conduct of the poll (from the August 21, 1989, edition).

How the Poll Was Taken

The New York Times Poll on women's issues is based on telephone interviews conducted June 20 through 25 with 1,497 adults around the United States, excluding Alaska and Hawaii.

The sample of telephone exchanges called was selected by a computer from a complete list of exchanges in the country. The exchanges were chosen so as to assure that each region of the country was represented in proportion to its population. For each exchange, the telephone numbers were formed by random digits, thus permitting access to both listed and unlisted numbers. The numbers were then screened to limit calls to residences.

Women were sampled at a higher rate than men so that there would be enough women interviewed to provide statistically reliable comparisons among various subgroups of women. The results of the interviews with 1,025 women and 472 men were then weighted to their correct proportions in the population.

Results were also weighted to take account of household size and number of residential telephone lines and to adjust for varia-

We are 95% confident that between 80.3% and 93.7% of the entire lot will germinate. ▲

You should notice that the confidence interval depends on the size *n* of the sample, but *not* on the size of the population. This is true as long as the population is much larger than the sample. The confidence interval in the example works for a sample of 100 from a lot of 10,000 seeds as well as for a sample of 100 from a lot of 1,000,000 seeds. Put another way, what matters is how many seeds you examine, not what percent of the population you examine.

Any confidence interval has two essential pieces: the interval itself and the confidence level. The interval usually has the form

$$\text{statistic} \pm \text{margin of error}$$

The statistic (such as $\hat{p}$) estimates the unknown parameter, and the margin of error indicates how accurate this estimate is. In the germination example, the margin of error is $\pm 6.7\%$.

tions in the sample relating to region, race, age, and education.

A group of 978 of these respondents were interviewed a second time from July 25 through 30, after the Supreme Court's decision allowing states more freedom to restrict abortion. Respondents in the second survey amounted to 79 percent of the 1,236 randomly selected people who were to be asked to participate, but 258 declined or were not reached despite several attempts.

In theory, in 19 cases out of 20 the results based on either of such samples will differ by no more than three percentage points in either direction from what would have been obtained by seeking out all American adults.

The percentages reported are the particular results most likely to match what would be obtained by seeking out all adult Americans. Other possible percentages are progressively less likely the more they differ from the reported results.

The potential sampling error for smaller subgroups is larger. For example, for men it is plus or minus five percentage points in both the first and second surveys. For women it is plus or minus three percentage points in the first survey and plus or minus four percentage points in the second survey. For women aged 18 to 29 in the first survey, it is plus or minus six percentage points.

In addition to sampling error, the practical difficulties of conducting any survey of public opinion may introduce other sources of error into the poll.

The methods and margins of error described by the *Times* are typical of national opinion polls. Do note the mention of "practical difficulties" and "other sources of error" at the end of the description. The poll was no doubt unable to contact some members of the sample, even in repeated calls. Some of the respondents may have given a socially acceptable answer rather than express their true opinion. Households without telephones could not be included in the survey. These practical difficulties do lead to additional errors, and these errors are *not* included in the announced margin of error.

The confidence level states how confident we are that our interval contains the true parameter. Although 95% confidence is common, you can hold out for higher confidence, such as 99%, or be satisfied with lower confidence, such as 90%. Our 95% confidence interval was based on the middle 95% of a normal distribution. A 99% confidence interval requires the middle 99% of the distribution, and so is wider (has a larger margin of error). Similarly, a 90% confidence interval is shorter than a 95% interval obtained from the same data.

So there is a trade-off between how closely we can pin down the parameter (the margin of error) and how confident we can be in the result.

Understanding confidence intervals helps us read newspapers and listen to TV news broadcasts (see Spotlight 8.1). Sample survey results are common in the news, often with a margin of error attached. The margin of error, together with the basic result of the survey, in fact forms a confidence interval. A news report of our crime survey would say "The survey found that 45% of all

Americans are afraid to go out at night because of crime. The margin of error in the survey is plus or minus 2.6 percentage points." Although reputable sample surveys such as Gallup, Lou Harris, or the Current Population Survey always announce a margin of error to help us interpret their results, editors often cut this information from their stories. It is even more common for news reports to give the margin of error without the confidence level; we need to know both figures because higher confidence requires a larger margin of error. Almost all public opinion polls announce 95% confidence intervals. So if a story about an opinion poll gives a margin of error without a confidence level, you can usually assume 95%.

The Bureau of Labor Statistics, on the other hand, chooses to announce its unemployment-survey results at the 90% level of confidence. Basing its conclusions on the monthly Current Population Survey of 60,000 households, the bureau states that the unemployment rate they announce is within ±0.2% (two-tenths of 1 percent) of the figure they would get if they counted everyone. So when the headlines announce a 7.9% unemployment rate, the Bureau is saying—with 90% confidence—that between 7.7% and 8.1% of the labor force is out of work.

Opinion polls often have margins of error of about ±3%. The much smaller margin of error for the announced unemployment rate is due to the much larger sample interviewed by the Current Population Survey. Larger samples give smaller margins of error at the same confidence level. However, the square root of n that appears in the calculations shows that in order to reduce our margin of error by half, we need a sample size four times bigger. To obtain a very small margin of error, the Current Population Survey goes to the trouble to interview a sample of 60,000 people, compared with the Gallup poll's usual 1500. The Gallup poll can afford to be 3% off. The unemployment rate must be more exact because so many economic and political decisions depend upon it.

ESTIMATING A POPULATION MEAN

The statistician's tool kit contains many different confidence intervals, matching the many different population parameters that we may wish to estimate. We have met the confidence interval for estimating a population proportion p. Now we want to estimate a population mean. We have regularly used the **sample mean** $\bar{x}$ of a sample of observations to describe the center of a set of data. Now we will use the sample mean $\bar{x}$ to estimate the unknown mean μ of the entire population from which the sample is drawn. We use μ, the symbol for the mean of a probability distribution, for the population mean because it is the mean of the distribution of the results of drawing a single observation at random from the population. The sample mean $\bar{x}$ is a statistic that will vary in repeated samples, while the population mean μ is a parameter, a fixed number. Fortunately, the new confidence interval for estimating μ is quite similar to the familiar confidence interval for estimating p, because both intervals are based on a normal sampling distribution.

EXAMPLE: Scholastic Aptitude Test Scores

How well would high school seniors do on the mathematics part of the Scholastic Aptitude Test (SAT)? Although about a million students take the SAT each year, these students are planning to attend college and are not representative of the entire population of high school seniors. We therefore select a simple random sample of 500 seniors and administer the mathematics SAT to them. Their average score is $\bar{x} = 451$. What can we say about the mean score μ for the entire population, if we want to be 95% confident in our conclusion?   ▲

We once again use a statistic — the mean $\bar{x}$ of our sample — to estimate an unknown parameter — the mean score μ for the entire population. To give a confidence interval, we need to know the sampling distribution of $\bar{x}$. The central limit theorem tells us that this distribution is close to normal. The mean of the sampling distribution is the same as the mean μ of the population from which we drew our sample. That is, the sample mean has no bias or systematic error as an estimator of the unknown μ. To find the standard deviation, we need to know something about the population. SAT scores for any large population follow a distribution that is close to normal. Moreover, the tests are arranged so that the standard deviation for the population used to develop the tests is $\sigma = 100$. For SAT scores, the standard deviation σ may vary a bit among different populations of students. We will simplify our work by assuming that we know that $\sigma = 100$ for the population we are interested in.

Recall from Chapter 7 that the sampling distribution of the sample mean $\bar{x}$ has a standard deviation that decreases with the square root $\sqrt{n}$ of the sample size. Because individual SAT scores have a distribution that is close to normal, we don't even have to call on the central limit theorem. It's a fact that if the distribution of the population is normal, then the sampling distribution of $\bar{x}$ is also normal, no matter how small the sample is.

Suppose that a population is described by a normal distribution with mean μ and standard deviation σ. Draw a simple random sample of size n from this population. Then the sampling distribution of the sample mean $\bar{x}$ is normal with mean μ and standard deviation $\sigma_{\bar{x}} = \sigma/\sqrt{n}$.

Figure 8.3 shows the relation between the distribution of a single observation drawn from the population and the distribution of the mean of several (in this case 10) observations. The mean of several observations is less variable than individual

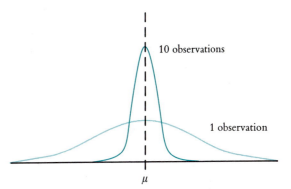

Figure 8.3 The sampling distribution of the sample mean $\bar{x}$ compared with the distribution of a single observation.

observations. Now we have in hand the facts we need to give a confidence interval for the mean SAT mathematics score for all high school seniors based on our sample of 500 students from this population.

EXAMPLE: Estimating the Mean SAT Score

The normal sampling distribution of $\bar{x}$ has mean equal to the unknown population mean μ. The standard deviation of the sampling distribution is

$$\sigma_{\bar{x}} = \frac{\sigma}{\sqrt{n}}$$

$$= \frac{100}{\sqrt{500}} = 4.47$$

By the 95 part of the 68 – 95 – 99.7 rule, $\bar{x}$ will fall within two standard deviations of μ in 95% of all samples. Two standard deviations is 2×4.47, or about 9 points. We observed $\bar{x} = 451$ in our sample. So we are 95% confident that the population mean μ lies in the interval

$$451 \pm 9$$

or between 442 and 460.

As in the case of the confidence interval for a population proportion p, we can give a recipe that summarizes our development. The confidence interval again has the form

$$\text{statistic} \pm \text{margin of error}$$

where the statistic is now the sample mean $\bar{x}$.

Suppose that a population is described by a normal distribution with unknown mean μ and known standard deviation σ. Draw a simple random sample of size n from this population and calculate the sample mean $\bar{x}$. A 95% confidence interval for the population mean μ is

$$\bar{x} \pm 2\sigma_{\bar{x}} = \bar{x} \pm 2\frac{\sigma}{\sqrt{n}}$$

Because of the central limit theorem, this recipe is also approximately correct when the population is not described by a normal distribution, if a large sample is drawn. Often in practice the standard deviation σ of the population is not known in advance. Then we must estimate σ by the standard deviation s of the sample. We will not concern ourselves with this situation. Here is another example of estimating a population mean.

EXAMPLE: Estimating Dust in Coal Mines

Because the mean of several observations is less variable than a single observation, it is good practice to take the average of several observations when accuracy is important. The amount of dust in the atmosphere of coal mines is measured by exposing a filter in the mine and then weighing the dust collected by the filter. The weighing is not perfectly precise; in fact, repeated weighings of the same filter will vary according to a normal distribution. The values that would be obtained in many weighings form

the population we are interested in. The mean μ of this population is the true weight (that is, there is no bias in the weighing). The population standard deviation describes the precision of the weighing; it is known to be $\sigma = 0.08$ milligram (mg). Each filter is weighed three times, and the mean weight is reported.

For one filter the three weights are

$$123.1 \text{ mg} \qquad 122.5 \text{ mg} \qquad 123.7 \text{ mg}$$

What is the 95% confidence interval for the true weight μ? First compute the sample mean

$$\bar{x} = \frac{123.1 + 122.5 + 123.7}{3}$$

$$= \frac{369.3}{3} = 123.1 \text{ mg}$$

Then the 95% confidence interval is

$$\bar{x} \pm 2\frac{\sigma}{\sqrt{n}} = 123.1 \pm 2\frac{0.08}{\sqrt{3}}$$

$$= 123.1 \pm (2)(0.046) = 123.1 \pm .09$$

or between 123.01 mg and 123.19 mg. ▲

▶ STATISTICAL PROCESS CONTROL

Statistical methods are widely used to gather social and economic information and in research on a wide variety of subjects. Most of our examples to this point, such as the Current Population Survey and the Physicians' Health Study, have illustrated these two types of application of statistics. Statistics is also heavily involved in the drive to improve the quality of manufactured products. Along with new technology, such as robots, and new management emphases, such as cooperating with workers and suppliers, statistical ideas are an important part of any manufacturer's efforts to compete in the worldwide marketplace. In this section we look at one simple but important statistical tool for monitoring and improving quality, the control chart.

At its Oklahoma City plant, AT&T Technologies manufactures the computerized electronic switches that interconnect our telephones. These switches are largely composed of complex electronic elements called *circuit packs*. AT&T needs efficient methods to check newly manufactured circuit packs for defects. The best strategy is to prevent defects by monitoring the manufacturing process to catch problems early rather than wait to inspect the product and fix defective circuit packs later (see Spotlight 8.2).

One important step in manufacturing a circuit pack is the soldering of the 2000 electrical connections that attach components to the printed wiring board. All 2000 connections are soldered at once as a conveyor carries the circuit pack through a wave of hot liquid solder. This wave-soldering operation is delicate. If the speed of the conveyor, the temperature of the solder, or other variables are not quite right, bad connections will appear both in the circuit pack and in our telephone conversations. AT&T therefore monitors the performance of the wave-soldering machine constantly and takes immediate action if something goes wrong.

To accomplish this goal, workers take a sample of five newly soldered circuit boards every hour and inspect them carefully for defects. Once a board is checked, the worker calculates a number that expresses the quality of soldering for that pack. A score of 100 represents the standard of quality that AT&T believes the process should attain. Lower scores represent poorer quality, while higher scores mean that the quality is higher than the target. The sample mean of the five quality scores is plotted on a **control chart**. This point is a sample estimate of the quality of that hour's production.

There will always be some chance variation in the mean quality scores over time. Any industrial process will produce some variability. Constant fiddling with the process in response to small vari-

ations is unnecessary and wasteful. The purpose of the control chart is to help us distinguish the usual natural variation in the process from the added variation that indicates that the process has been disturbed. When unusual variation is spotted, we look for a specific cause. The disturbance may be caused by a new operator who hasn't been properly trained or by a machine malfunction.

Just plotting the mean quality scores against time can be helpful. We can see if there is a trend up or down, for example. Or we can see whether the last point plotted falls outside the pattern of the earlier points, and so suggests that something has gone wrong. Figure 8.4 is a plot against time for the mean quality scores for 20 hourly samples. The graph appears to show that the level of quality dropped shortly after sample number 10. The horizontal center line at the target value 100 helps us see the trend. Once again, we would like to confirm this appearance by calculation. Adding the result of a simple calculation turns the plot against time into a control chart.

When the wave-soldering machine is performing its task properly, the quality index for individual circuit packs will vary according to a normal distribution. Let's imagine that we know from long experience that the mean of this distribution should be 100 and the standard deviation 4.

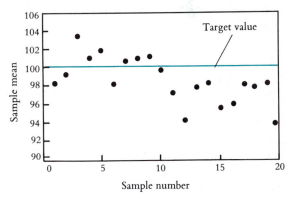

Figure 8.4 Plot of sample mean quality scores against time.

SP TLIGHT 8.2

Check the Process before the Product

▶ ▶ ▶ ▶ ▶ ▶ ▶ ▶ ▶ ▶ ▶ ▶ ▶

Process control engineer Connie Moore of AT&T.

At its Oklahoma City plant, AT&T uses statistical process control to ensure smooth production of circuit packs.

Connie Moore, a process control engineer at AT&T, gives her views on ensuring product quality.

If we looked at 100% of the product it would take more time, more people, and would not give us any better information about the process. There was a time when industry thought that a quality control department's function was to inspect quality at the end of the line. Now we know that the only reasonable philosophy is to build it right the first time.

I've been told that unless you measure how you're doing as you go along, you'll never know if you're done or if you succeeded. That's why I think that statistics and people are such an important combination. Statistics is the tool that tells us how we're doing as we go along and people are the force that drives us until we've succeeded.

We are plotting the mean $\bar{x}$ of five observations. What range of variation do we expect to see in the values of $\bar{x}$? We know that the distribution of $\bar{x}$ in many samples is itself normal, with mean 100 and standard deviation

$$\sigma_{\bar{x}} = \frac{\sigma}{\sqrt{n}}$$

$$= \frac{4}{\sqrt{5}} = 1.79$$

By the 95 part of the 68–95–99.7 rule, 95% of all values will fall within two standard deviations of the mean, that is, between

$$100 - (2)(1.79) = 96.42$$

and

$$100 + (2)(1.79) = 103.58$$

When we draw many samples, only 5% of the values of $\bar{x}$ will fall outside this range if the process is operating undisturbed. In particular, only 2.5% (half of 5%) of all samples will give an $\bar{x}$ less than 96.42. A mean quality score this low is good evidence that something has gone wrong with the wave-soldering process.

Figure 8.5 is the control chart for the observations from Figure 8.4. The dashed line is the control limit 96.42 that indicates when action should be taken. The control chart shows convincingly that the quality of the process has deteriorated. Not only are the means for samples 12, 15, 16, and 20 below the control limit, but the last 11 means are all below the center line. In the long run, only half these means should be below the center line if the process mean is really 100. It appears that the process quality shifted downward at about sample 9 or 10. In practice, the first out-of-control point at sample 12 would trigger an investigation to find

and correct the cause of this trend. Here is a summary of the steps in constructing a control chart such as the one in Figure 8.5.

Suppose that a process follows a normal distribution with mean μ and standard deviation σ when operating undisturbed. To monitor the process, take samples of size n at regular intervals. An $\bar{x}$ *control chart* for the process is a plot of the sample means against time with a solid *center line* at μ and dashed *control limits* at $\mu - 2\sigma/\sqrt{n}$ and $\mu + 2\sigma/\sqrt{n}$.

There are many variations on the control chart idea. Although most control charts have both upper and lower control limits lying at equal distances above and below the center line, only the lower limit is drawn in Figure 8.5. This is because only decreases in the quality index concern us in this example. Our control limits are placed two standard deviations out from the center line. It is more common in industry to place the control limits three standard deviations out in order to minimize the number of false alarms. Such limits contain 99.7% of all values of $\bar{x}$ if the process has not been disturbed. It is also common to keep control charts for statistics other than the sample mean $\bar{x}$.

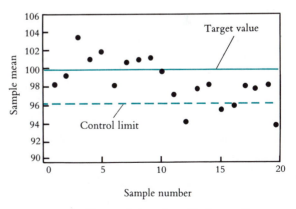

Figure 8.5 An $\bar{x}$ control chart for soldering quality.

More important than these details are the statistical ideas that are the basis for control charts. First, our goal is to distinguish expected from unexpected variation. Second, we use the normal sampling distribution of $\bar{x}$ and the $68-95-99.7$ rule to specify the range of expected variation. Third, we combine this formal inference with a graph of the data that can be used in the factory by people with little statistical training.

Why sample only five circuit boards each hour? The purpose of statistical process control is not to check the function of the circuit packs; they will be rigorously tested when completed. The goal is rather to monitor the soldering process and correct any malfunctions quickly. It is not practical to check every circuit pack at every stage of manufacture. Instead, statistical sampling techniques give a quick and economical way to keep the process running smoothly. Control charts based on samples not only help to keep the quality of the final product at a high level but also keep down costs by catching malfunctions quickly, allowing a faulty process to be corrected immediately. This eliminates the need to repair or scrap parts at the end of the assembly line.

EXAMPLE: Control Chart for a Machining Operation

An important operation in producing cast aluminum aircraft frame parts is the machining of the raw casting. When the machining process is operating in control, a critical dimension of the parts varies according to a normal distribution with mean $\mu = 1.50$ centimeters (cm) and standard deviation $\sigma = 0.20$ cm. A sample of four parts is measured each hour in order to keep an $\bar{x}$ control chart. Here are the data for the past 20 hours.

Hour	1	2	3	4	5
$\bar{x}$	1.60	1.54	1.31	1.45	1.40

Hour	6	7	8	9	10
$\bar{x}$	1.61	1.47	1.45	1.45	1.52

Hour	11	12	13	14	15
$\bar{x}$	1.49	1.66	1.60	1.57	1.73

Hour	16	17	18	19	20
$\bar{x}$	1.68	1.58	1.60	1.57	1.73

The control chart appears in Figure 8.6. The center line is at $\mu = 1.50$. The control limits are

$$\mu \pm 2\frac{\sigma}{\sqrt{n}} = 1.50 \pm 2\frac{0.20}{\sqrt{4}}$$
$$= 1.50 \pm 0.20$$
$$= 1.3 \quad \text{and} \quad 1.7$$

The points for hours 15 and 20 are outside the control limits. Moreover, the last 9 points all lie above the center line. This is very unlikely to occur if the mean remains at 1.5, so it is additional evidence that some outside cause has disturbed the process. A common criterion is to look for a disturbing cause when 8 straight points fall on the same side of the center line. This criterion would lead to action at hour 19. In this case, the point out of control at hour 15 would already have called for action. ▲

▶ THE PERILS OF DATA ANALYSIS

Statistical designs for collecting data may, like the Physicians' Health Study, involve experimentation. Or they may use sampling procedures such as those used in the Current Population Survey and also for process control. In both cases, we rely on randomization and the mathematics of probability

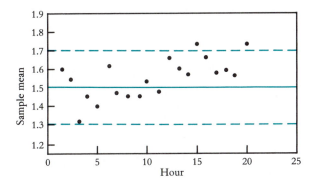

Figure 8.6 An $\bar{x}$ control chart for a machining operation.

to compute sampling distributions. From a sampling distribution we can obtain results that have known levels of confidence.

However, formal statistical inference, as reflected in levels of confidence, is secondary to well-designed data collection and to insight into the behavior of data. Inference is often unable to correct basic flaws in the data, such as the use of voluntary response samples. Moreover, the effects of *hidden variables* can make even an apparently clear inference misleading. We saw in Chapter 5 that a well-designed experiment can control for the effects of hidden variables. If an experiment is not possible, however, we may need to do some statistical detective work. Let's look at an example. Although the example is imaginary, it is based on a study of admissions to graduate programs at the University of California, Berkeley.

EXAMPLE: Admissions Discrimination?

Metro University has several limited-enrollment courses that admit only some of the students who apply. There are complaints about sex dis-

crimination in the admissions process. These complaints seem to be based on clear numerical evidence. Eighty men applied to limited-enrollment courses at Metro, and 35 were admitted. The percent of male applicants admitted was

$$\frac{35}{80} = 0.44 = 44\%$$

On the other hand, only 20 of the 60 women who applied were accepted. The success rate among women was

$$\frac{20}{60} = 0.33 = 33\%$$

Almost half the men, but only one-third of the women, were admitted. A probability calculation shows that this difference is much larger than could reasonably be expected to occur simply by chance. That is, formal inference backs up the appearance that a systematically higher percent of men are being admitted. Are men being favored over women? ▲

The data in the example are displayed in a **two-way table** in Table 8.1. When variables simply place subjects into categories, such as male – female or admit – deny, we cannot draw a scatter-

TABLE 8.1   Metro Admissions

	Male		Female	
	Count	Percent	Count	Percent
Admit	35	44	20	33
Deny	45	56	40	67
Total	80	100	60	100

plot to display the relationship between them. Instead we display the counts and percents in a two-way table and see the relationship by comparing percents. Comparing the percents of successful male and female applicants (44% and 33%) shows that men are more likely to be admitted. Metro is concerned. A closer look at admissions data is called for.

Investigation discloses that men and women tend to apply for different courses. The organic chemistry course is very rigorous. Forty women applied to it, but only 10 got in, or one-fourth. Twenty men also applied, and 5 were admitted—again, one-fourth. These figures do not seem to point to discrimination.

The only other limited course offered at Metro is the history and sociology of the TV sitcom. Compared with organic chemistry, this course is something of a soft option. Sixty men applied, with 30 admitted, and 10 out of 20 female applicants were accepted. Once again, we find no apparent discrimination because 50% of both groups were admitted. We can display the data in a pair of two-way tables, Tables 8.2 and 8.3.

When added together, these two tables give exactly the single two-way table for overall admissions, Table 8.1. The numbers are remarkable. Each program appears to make no distinction between men and women. Yet when we consider the overall totals, we see that 44% of the men who applied were admitted, compared with 33% of the women. This happened because two-thirds of the women applied to the course that is harder to get into, whereas three-quarters of the men signed up for the easier sitcom class. The hidden variable explains the apparent inequity.

This example provides a warning about statistical evidence, especially when an experiment was not done. Without walking through our hypothetical example, we would not suspect that data showing equality in each of several cases can appear as evidence of inequality when the cases are lumped together. First appearances can be deceiving, even in statistics. Statistical evidence not based on experiments can often show that an effect

TABLE 8.2 Organic Chemistry

	Male		Female	
	Count	Percent	Count	Percent
Admit	5	25	10	25
Deny	15	75	30	75

TABLE 8.3 TV Sitcom

	Male		Female	
	Count	Percent	Count	Percent
Admit	30	50	10	50
Deny	30	50	10	50

is present—more men than women get into limited courses at Metro—but does not show *why* the effect is present. At Metro, a hidden variable was responsible. In another case, an investigation might reveal discrimination. At a time when statistical evidence of all kinds is increasingly being used to formulate social policy and resolve legal disputes, it is crucial that we select, analyze, and interpret our data with great care.

Even if carefully collected data are properly analyzed, correct conclusions are not absolutely guaranteed. There is always some chance, however small, that random selection will lead to a false conclusion. The strength of statistical inference is that the chance of a false conclusion is known and can be controlled by setting the confidence level as high as we think necessary.

Statistics does not produce proof. But in a world where proof is always wanting and most evidence is uncertain, statistical evidence is often the best evidence available.

▶ REVIEW VOCABULARY

Confidence interval   An interval computed from a sample by a method that has a known probability of producing an interval containing the unknown parameter. This probability is called the *confidence level.* Confidence intervals usually have the form

statistic ± margin of error

Control chart   A graph showing the value of a statistic for successive samples (for example, one sample each hour or one sample each shift). The graph also contains a *center line* at the target value for the process parameter and *control limits* that the statistic will rarely fall outside of unless the process drifts away from the target. The purpose of a control chart is to monitor a process over time and signal when some unusual source of variation interferes with the process.
Parameter   A number that describes the population. In statistical inference, the goal is often to estimate an unknown parameter or make a decision about its value.
Sample mean   The mean (arithmetic average) $\bar{x}$ of the observations in a sample. The sample mean from a simple random sample is used to estimate the unknown mean μ of the population from which the sample was drawn.
Sample proportion   The proportion $\hat{p}$ of the members of a sample having some characteristic (such as agreeing with an opinion poll question). The sample proportion from a simple random sample is used to estimate the corresponding

proportion p in the population from which the sample was drawn.
Statistic   A number that describes a sample. A statistic can be calculated from the sample data alone, and does not involve any unknown parameters of the population.
Two-way table   A table showing frequencies (counts) or percentages of outcomes that are classified according to two variables (such as applicants classified by sex and admission decision).

▶ SUGGESTED READINGS

FREEDMAN, DAVID, ROBERT PISANI, ROGER PURVES, AND ANI ADHIKARI: *Statistics,* 2nd ed., Norton, New York, 1991. Chapters 21 and 23 discuss sampling distributions and confidence intervals for proportions and means.

MOORE, DAVID S., AND GEORGE P. MCCABE: *Introduction to the Practice of Statistics,* 2nd ed., Freeman, New York, 1993. Chapter 5 of this text discusses sampling distributions and control charts. Chapter 6 is devoted to the reasoning of formal inference, including confidence intervals. Later chapters present confidence intervals for use in many specific situations.

MOSES, LINCOLN E.: "The reasoning of statistical inference," in David C. Hoaglin and David S. Moore (eds.), *Perspectives on Contemporary Statistics,* Mathematical Association of America, Washington, D.C., 1992, pp. 107–122. This is a broad essay on the nature of inference that requires some knowledge of probability and should be read after one of the introductions just cited.

▶ EXERCISES

Identify each of the boldface numbers in Exercises 1 to 3 as either a *parameter* or a *statistic.*

   1. A random sample of male college students has a mean height of **64.5** inches, which is greater than the **63**-inch mean height of all adult American women.

2. A sample of students of high academic ability under 13 years of age was given the SAT mathematics examination, which is usually taken by high school seniors. The mean score for the females in the sample was **386**, whereas the mean score for the males was **416**.

3. About **6%** of all U.S. households are without telephones, but another **28%** have unlisted numbers.

4. Exercises 1 to 3 in Chapter 5 describe three samples. Find the sample proportion $\hat{p}$ in each case, first as a decimal fraction and then as a percent.

Each of the statistics described in Exercises 5 to 8 has a normal sampling distribution (at least approximately). Give the mean and standard deviation of the distribution in each case.

5. About 35% of residential telephones in the San Francisco area have unlisted numbers. A telephone sales organization uses random-digit dialing to dial a random sample of 200 residential telephone numbers. The percent of these that are unlisted is the statistic of interest.

6. A shipment of machined parts has a critical dimension that is normally distributed with mean 12 centimeters and standard deviation 0.01 centimeter. The acceptance sampling team measures a random sample of 25 of these parts; the sample mean $\bar{x}$ of the critical dimension for these parts is the statistic of interest.

7. The Acculturation Rating Scale for Mexican Americans (ARSMA) is a psychological test that evaluates the degree to which Mexican Americans have adapted to Anglo/English culture. The scores in a large population are normally distributed with mean 3.0 and standard deviation 0.8. A researcher gives the test to a random sample of 12 Mexican Americans. Their average score is the statistic of interest.

8. In a midwestern state, 84% of the households have Christmas trees at holiday time. A sample survey asks a random sample of 400 households "Did you have a Christmas tree this year?" The percent who say "yes" is the statistic of interest.

9. The standard deviation $\sigma_{\hat{p}}$ of a sample proportion $\hat{p}$ varies with the true value of the population proportion p. Fortunately, it does not vary greatly unless p is near 0% or 100%. Suppose that the size of the sample is $n = 1500$. Evaluate $\sigma_{\hat{p}}$ for $p = 30\%$, 40%, 50%, 60%, and 70%.

■ 10. The report of a sample survey of 1500 adults says, "With 95% confidence, between 27% and 33% of all American adults believe that drugs are the most serious problem facing our nation's public schools." Explain to someone who knows no statistics what the phrase "95% confidence" means in this report.

■ Discussion exercise.

■ 11. A Gallup poll of a random sample of 1540 adults asked, "Do you happen to jog?" Fifteen percent answered "yes." The news item stated that this poll has a 3% margin of error. Explain carefully to someone who knows no statistics what is meant by a "3% margin of error."

12. Suppose that the poll in Exercise 11 had used a simple random sample of size 1540, of whom 15% answered "yes." Give a 95% confidence interval for the percent of all adults who would have answered "yes" if asked.

13. The Forest Service is considering additional restrictions on the number of vehicles allowed to enter Yellowstone National Park. To assess public reaction, the Service asks a simple random sample of 150 visitors if they favor the proposal. Of these, 89 say "yes." Give a 95% confidence interval for the proportion of all visitors to Yellowstone who favor the restrictions. Are you 95% confident that more than half are in favor? Explain your answer.

14. A simple random sample of students at Upper Wabash Tech is asked whether they favor limiting enrollment in crowded majors as a way of keeping the quality of instruction high. The student government suspects that the plan will be unpopular among freshmen, who have not yet been admitted to a major. Here are the responses for freshmen and seniors.

	Favor	Oppose
Freshmen	40	160
Seniors	80	20

    a. Give a 95% confidence interval for the percent of all freshmen who support the plan.

    b. Give a 95% confidence interval for the percent of all seniors who support the plan.

15. Ronald Reagan was president from 1981 to 1988. He was generally quite popular. A Gallup poll of 1514 adults taken between July 30 and August 2, 1983 asked, "Do you approve of the way Ronald Reagan is handling his job as President?" Of these, 41% said "yes."

    a. If the poll had used a simple random sample, what would have been the margin of error in a 95% confidence interval?

    b. The actual margin of error for a Gallup poll of this same size is ±3%. Why does this not agree with your result in part **a**?

16. In August 1983, *Organic Gardening* magazine reported the results of a test to see whether mayonnaise jars would break when used for home canning. Here is their conclusion:

The mayonnaise jars didn't do badly—only 3 out of 100 broke. Statistically this means you'd expect between 0% and 6.4% to break.

Verify this statistical statement by giving a 95% confidence interval.

Exercises 17 to 20 are based on the following situation. A news report says that a national opinion poll of 1500 randomly selected adults found that 43% thought they would be worse off during the next year. The news report went on to say that the margin of error in the poll result is ±3 percentage points with 95% confidence.

■ 17. Which of the following sources of error are included in the poll's margin of error?

    a. The poll dialed telephone numbers at random and so missed all people without phones.

    b. The poll could not contact some people whose numbers were chosen.

    c. There is chance variation in the random selection of telephone numbers.

▲ 18. Would a 90% confidence interval based on the poll results have a margin of error less than ±3 percentage points, equal to ±3 percentage points, or greater than ±3 percentage points? Explain your answer.

▲ 19. If the poll had interviewed 1000 persons rather than 1500 (and still found 43% believing they would be worse off), would the margin of error for 95% confidence be less than ±3 percentage points, equal to ±3 percentage points, or greater than ±3 percentage points? Explain your answer.

▲ 20. Suppose that the poll had obtained the outcome 43% by a similar random sampling method from all adults in New York State (population 18 million) instead of from all adults in the United States (population 249 million). Would the margin of error for 95% confidence be less than ±3 percentage points, equal to ±3 percentage points, or greater than ±3 percentage points? Explain your answer.

▲ 21. In the text we used the sampling distribution of $\hat{p}$ and the 68–95–99.7 rule to give a 95% confidence interval for a population proportion p.

    a. Explain carefully why

$$\hat{p} \pm \sqrt{\frac{\hat{p}(100 - \hat{p})}{n}}$$

    is a 68% confidence interval for p.

    b. Give the recipe for a 99.7% confidence interval for p.

22. Use the result of Exercise 21b and the data in Exercise 11 to give a 99.7% confidence interval for the percent of all adults who jog. Compare the width of the 99.7% interval with that of the 95% confidence interval from Exercise 12. What is the reason for the difference in widths?

■ Discussion exercise.   ▲ Advanced exercise.

23. Use the result of Exercise 21**a** and the data in Exercise 13 to give a 68% confidence interval for the percent of visitors to Yellowstone who support restricting the number of vehicles allowed into the park. Compare the width of the 68% interval with that of the 95% interval from Exercise 13 and explain the difference in plain language.

24. Electrical pin connectors for use in computers are gold-plated for better conductivity. The specified plating thickness is 0.001 inch. Due to variations in the plating process, the actual plating thickness on different pins has a normal distribution with mean 0.001 inch and standard deviation 0.0001 inch.

    a. What range of plating thickness contains 95% of all pins?

    b. Quality control samples of four pins are taken regularly during production. The plating thickness is measured and the sample mean of the four measurements is recorded on a control chart. What range of plating thickness contains 95% of the recorded sample means?

25. Scores on the American College Testing (ACT) college admissions examination for the reference population used to develop the test vary normally with mean $\mu = 18$ and standard deviation $\sigma = 6$. The range of reported scores is 1 to 36.

    a. What range of scores contains the middle 95% of all students in the reference population?

    b. If the ACT scores of 25 randomly selected students are averaged, what range contains the middle 95% of the averages $\bar{x}$?

26. Errors in careful measurements often have a distribution that is close to normal. Experience shows that the error in a certain surveying method varies when a measurement is repeated according to a normal distribution with mean 0 (that is, the procedure does not systematically overestimate or underestimate the true distance) and standard deviation 0.03 meter. Each measurement is repeated 3 times and the mean of the three measurements is used as the final value.

    a. What is the distribution of the mean error $\bar{x}$ when this surveying method is used to measure many distances?

    b. Between what values do 95% of the errors fall?

27. A laboratory scale is known to have a standard deviation of $\sigma = 0.001$ gram in repeated weighings. Suppose that scale readings in repeated weighings are normally distributed, with mean equal to the true weight of the specimen. Three weighings of a specimen give (in grams)

3.412   3.414   3.415

Give a 95% confidence interval for the true weight of the specimen.

28. An instrument in a chemistry laboratory measures the concentration of trace substances in specimens. When the instrument makes repeated measurements on the same specimen, the readings are known to vary normally with standard deviation $\sigma = 0.03$. It is customary to make 3 readings and use the sample mean as the final result. For a particular specimen, the readings are

$$53.12 \qquad 53.08 \qquad 53.17$$

Give a 95% confidence interval for the mean of the distribution of readings for this specimen. (This mean is the true concentration if the instrument has no bias.)

29. The Family Adaptability and Cohesion Evaluation Scales (FACES) is a psychological test that measures two different aspects of family behavior. One of these is "cohesion," which is the degree to which family members are emotionally connected to each other. Suppose it is known that the cohesion scores for adults vary normally with standard deviation $\sigma = 5$. A researcher administers FACES to a sample of 33 adults in families with a runaway teenager. The mean cohesion score in this sample is $\bar{x} = 36.9$. Give a 95% confidence interval for the mean cohesion (as rated by an adult) of families with runaway teenagers.

30. A milk processor monitors the number of bacteria per milliliter in raw milk received for processing. A random sample of 10 one-milliliter specimens from milk supplied by one producer gives the following data:

5370	4890	5100	4500	5260
5150	4900	4760	4700	4870

Suppose it is known that the bacteria count varies normally and that the standard deviation is $\sigma = 265$ per milliliter. Give a 95% confidence interval for the mean bacteria count per milliliter in this producer's milk.

31. An automatic lathe machines shafts to specified diameters as part of a manufacturing operation. Due to small variations in the operation of the lathe, the actual diameters produced follow a normal distribution with standard deviation 0.0005 inch. The shafts now being produced are supposed to have a diameter of 0.75 inch. You measure 10 such shafts and find that they have a sample diameter of $\bar{x} = 0.7505$. Give a 95% confidence interval for the true mean diameter μ of the shafts being produced.

▲ 32. The upper and lower deciles of any normal distribution are located 1.28 standard deviations above and below the mean. (The lower decile is the point with probability 10% below it; the upper decile has probability 90% below it.)
   a. Use this information to give a recipe for an 80% confidence interval for a population proportion p based on the sample proportion $\hat{p}$ that is accurate for large sample sizes n.
   b. Give an 80% confidence interval for the proportion of visitors to Yellowstone favoring vehicle restrictions, using the data in Exercise 13.

▲ Advanced exercise.

▲ 33. The upper and lower deciles of any normal distribution are located 1.28 standard deviations above and below the mean.

      a. Use this information to give a recipe for an 80% confidence interval for the mean μ of a normal population based on the sample mean $\bar{x}$ of a simple random sample of size n.

      b. Give an 80% confidence interval for the mean FACES cohesion score in Exercise 29.

34. Give the center line and control limits for a control chart for means $\bar{x}$ of samples of size 4 in the gold-plating process described in Exercise 24. Use 2σ limits, as in the text example.

35. It is common for laboratories to keep a control chart for a measurement process based on regular measurements of a standard specimen. Suppose that you are maintaining a control chart for the scale in Exercise 27 by weighing a 5-gram standard weight three times at regular intervals. What should be the center line of your chart? What are the control limits if you decide to use 3σ limits?

36. The laboratory instrument of Exercise 28 is monitored by measuring a specimen with known concentration 50 each morning. The specimen is measured three times and an $\bar{x}$ control chart is kept. What should be the center line and 2σ control limits for this chart?

Use "3σ" limits in the control charts of Exercises 37 through 39. That is, use control limits that are three standard deviations on either side of the mean μ. Also look for runs of 8 or more consecutive observations on the same side of the center line. These are common control chart signals used in industry. Notice that the center line and control limits are the same for all three charts.

37. In the data set below are $\bar{x}$'s from samples of size 4 with $\mu = 101.5$ and $\sigma = 0.2$. Only random variation is present. Make an $\bar{x}$ chart of these data using the given μ and σ. Are any points out of control?

Sample	$\bar{x}$	Sample	$\bar{x}$
1	101.627	11	101.383
2	101.613	12	101.715
3	101.493	13	101.485
4	101.602	14	101.509
5	101.360	15	101.429
6	101.374	16	101.477
7	101.592	17	101.570
8	101.458	18	101.623
9	101.552	19	101.472
10	101.463	20	101.531

38. The following set of $\bar{x}$'s for samples of size 4 shows the effect of a shift in the standard deviation. The process has $\mu = 101.5$ and $\sigma = 0.2$ when it is undisturbed. The first 10 samples are taken in this condition. Then the process is disturbed so that for the last 10 samples $\mu = 101.5$ and $\sigma = 0.3$. Make a control chart for these data using the undisturbed μ and σ. Are any points out of control? Is the increase in σ visible in any way on the chart?

Sample	$\bar{x}$	Sample	$\bar{x}$
1	101.602	11	101.664
2	101.547	12	101.823
3	101.312	13	101.629
4	101.449	14	101.602
5	101.401	15	101.756
6	101.608	16	101.707
7	101.471	17	101.612
8	101.453	18	101.628
9	101.446	19	101.603
10	101.522	20	101.816

39. The following set of $\bar{x}$'s for samples of size 4 illustrates the effect of a steady drift in the mean of the population. The first 10 samples have $\mu = 101.5$ and $\sigma = 0.2$. Then the process is disturbed so that the last 10 samples have $\sigma = 0.2$ and μ increasing by 0.04 in each successive sample, reaching 101.7 at sample 15 and 101.9 at sample 20. Make an $\bar{x}$ chart for these data. Are any points out of control? Is the upward drift in μ visible in any way on the chart?

Sample	$\bar{x}$	Sample	$\bar{x}$
1	101.458	11	101.453
2	101.618	12	101.258
3	101.507	13	101.557
4	101.494	14	101.484
5	101.533	15	101.896
6	101.334	16	101.634
7	101.547	17	101.632
8	101.695	18	101.824
9	101.351	19	101.968
10	101.555	20	101.783

■ 40. The U.S. government publication *Science Indicators 1980* shows that the average salary of women in all science and engineering fields was only 77% of the average salary for all male engineers and scientists. But the same source shows that in every individual field of science and engineering, the average female salary was at least 92% of the average male salary. Explain how this apparent discrepancy can come about.

41. In a study of the effect of parents' smoking habits on the smoking habits of high school students, researchers interviewed students in eight high schools in Arizona. The results appear in the following two-way table. [From S. V. Zagona, (ed.), *Studies and Issues in Smoking Behavior*, University of Arizona Press, Tucson, 1967, pages 157–180.]

	Student smokes	Student does not smoke
Both parents smoke	400	1380
One parent smokes	416	1823
Neither parent smokes	188	1168

Describe the association between the smoking habits of parents and their high school children by computing and comparing several percents. Then summarize the results in plain language.

42. Here is a two-way table based on information in the 1991 *Statistical Abstract of the United States.*

Degrees Earned in 1990, by Level and Sex (Thousands)

	Bachelor's	Master's	Professional	Doctorate
Male	481	156	48	23
Female	536	163	27	13

    a. What percent of all bachelor's degrees were earned by women?
    b. How many master's degrees were awarded in 1990?
    c. What percent of all degrees earned by women were doctorates?
    d. Summarize in words the relation between the level of degrees and the sex of the recipient. Back your summary by computing and comparing appropriate percents.

■ Discussion exercise.

43. The following two-way tables of counts compare the batting records of two baseball players, Bill and Will. How well a batter hits may depend on whether the pitcher is right-handed or left-handed, so the tables record the type of pitcher (right or left) and the result (hit or out) for each time at bat.

Bill

	Right	Left
Hits	40	80
Outs	60	320

Will

	Right	Left
Hits	120	10
Outs	280	90

a. Show how the information in these tables can be combined to make a two-way table of batter (Bill or Will) by outcome (hit or out) for all times at bat. Which player gets a hit a higher proportion of the time? (The proportion of at bats in which a player gets a hit is his batting average.)

b. Who has the higher batting average against right-handed pitching? Who has the higher batting average against left-handed pitching?

c. Explain carefully, as if talking to a skeptical baseball manager, how it is possible for one player to do better against both right-handers and left-handers and yet have a lower overall batting average. Which hitter would you prefer to have on your team?

44. A community has two hospitals. Hospital A is a large medical center, while Hospital B is a fashionable spa for prosperous patients. An article in the local newspaper claims that a higher percent of surgery patients die at Hospital A than at Hospital B. The paper says that people who need surgery should go to Hospital B. The following two-way tables count patients who died or survived and add a hidden variable, the condition good or poor of the patients before surgery.

Hospital A

	Good	Poor
Died	6	57
Survived	594	1443

Hospital B

	Good	Poor
Died	8	8
Survived	592	192

a. Create percent columns for the tables. Use these percents to show that both a higher percent of patients in good condition and a higher percent of patients in poor condition survive at Hospital A than at Hospital B.

b. Show how the information in the tables on the opposite page can be combined to make a table of patient outcome (died or survived) by hospital (A or B). Show that, as the newspaper reported, a higher percent of patients survive at Hospital B.

c. Explain carefully, as if talking to a skeptical reporter, how Hospital A can have a poorer overall survival rate even though it does better than B for both classes of patients.

▶Writing Projects

1. The margin of error announced for a sample survey takes into account the chance variation due to random sampling. In practice, survey results can be in error for other reasons. Some subjects can't be contacted, others lie or don't remember information, and the wording of the questions will influence the responses. These are the "practical difficulties" mentioned in Spotlight 8.1.

Write a brief discussion of the most important practical difficulties encountered in opinion polls and other surveys of human populations. You will want to read more on the subject. Some sources are: Section 1.5 of *Statistics: Concepts and Controversies*; Section 19.6 of *Statistics* (see the Suggested Readings in Chapter 5 for both books); and the article by P. E. Converse and M. W. Traugott, "Assessing the accuracy of polls and surveys," *Science*, Vol. 234, 1986, pp. 1094–1098.

2. There are many confidence intervals for use in settings other than the two described in this chapter. Another common setting is to estimate the difference between *two* population proportions, p_1 and p_2. For example, p_1 could be the proportion of women and p_2 the proportion of men who stay home at night due to fear of crime. Read an account of the confidence interval for $p_1 - p_2$ in a statistical methods text (for example, in Section 8.2 of *Introduction to the Practice of Statistics*). Then explain carefully how this new confidence interval arises from the same reasoning we have used: find an estimate, discover that the sampling distribution of the estimate is at least approximately normal, learn the standard deviation of this normal distribution, and go out two standard deviations from the estimate to get 95% confidence.

Part III

Coding Information

Y ou've seen bar-coded numbers on everything from milk cartons to books. And you've seen long numbers on everything from credit cards to airline tickets. Of course these numbers function as identification numbers just as social security numbers do and serial numbers on currency do. However, because of inexpensive, fast, reliable computers, the identification numbers used by businesses today are more sophisticated than those of earlier years. Modern identification numbers have a built-in "check" to partially ensure that the numbers have been correctly entered into a computer or have been correctly scanned by an optical device. Identification numbers for people often contain a variety of personal information such as sex, date of birth, and portions of names.

Identification numbers are examples of codes. A **code** is a group of symbols that represents information. Codes existed thousands of years ago: hieroglyphics, the Greek alphabet, and Roman numerals. Many codes have been developed for a particular application: musical scores, the Morse code (see Spotlight 10.4), and the "genetic code" used to describe the make up of DNA. About 40 years ago an MIT graduate student named David Huffman invented a code that is now used in computers, high-definition television, modems, and even VCR Plus+ devices that automatically programs a VCR (see Spotlight 10.5). In recent decades coding schemes have been invented that do

Entomologist Stephen Buchmann developed a reliable, inexpensive way to track bees using the same technology that supermarkets use to speed up the checkout lines and keep track of inventory. He glued bar code labels onto the backs of 100 bees and placed a laser scanner above the hive. In the past, researchers marked bees with paint or tags, but the monitoring of activity required the presence of a human observer. (© Scott Camazine)

more than simply code data. For example, data from space probes, signals from compact discs, and modem, fax machine, and high-definition television transmissions are coded so that errors in the data that occur during transmission can be corrected (see Spotlight 10.1). Premium television services such as Home Box Office (HBO) and The Disney Channel code their signals so that only those with a decoding device receive the service.

In this portion of the text we examine some of the ways that information is coded.

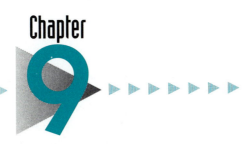

Identification Numbers and Bar Codes

Modern identification numbers have at least two functions. Obviously, an identification number should unambiguously identify the person or thing with which it is associated. Not obvious is a "self-checking" aspect of the number.

Look at the ISBN number printed on the back of this book. The number 0-7167-2378-6 distinguishes this book from all others. The last digit "6" is there solely to detect errors that may occur when the ISBN number is entered into a computer. Look at the bottom of the airline ticket shown in Figure 9.3. Notice the letters "ck" (for "check") above the last digit of the stock control number and above the last digit of the document number. They also are there for the purpose of error detection. Grocery items, credit cards, express mail, magazines, personal checks, travelers checks, soft drink cans, and many other items you encounter daily have identification numbers that code data and include check digits for error detection. In this chapter we examine some of the methods that are used to assign identification numbers and check digits.

Let us begin by considering the U.S. Postal Service money order shown in Figure 9.1 with the serial number 39539887571 (the 1 at the end is there for "checking" purposes).

The first ten digits of the eleven-digit number 39539887571 simply identify the money order. The last digit, 1, serves as an **error-detecting code** or mechanism. Let us see how this mechanism works. The eleventh (last) digit of a Postal Service money order number is the remainder obtained when the number represented by the first 10 digits is divided by 9. In our example the last digit is 1 because $3953988757 = 9 \times 439332084 + 1$. Now suppose instead of the correct number, the number 39559887571 (an error in the fourth position) were entered into a computer programmed for error detection of money orders. The machine would divide the entered number (excluding the eleventh digit) by 9 and find a remainder of 3. Since the last digit of the entered number is 1 rather than 3, the entered number cannot be correct. This crude method of error detection will not detect the mistake of replacing a "0" with a "9" or vice versa. Nor will it detect the transposition of digits such as 39359887571 instead of 39539887571 (the digits 3 and 5 have been transposed). (Spotlight 9.1 shows a transposition error.)

Figure 9.1 Money order customer receipt with identification number 3953988757 and appended check digit 1. The check digit is the remainder upon division by 9.

American Express (see Figure 9.2) and VISA travelers checks also utilize a check digit determined by division by 9. In these cases, the check digit is chosen so that the sum of the digits, including the check digit, is evenly divisible by 9.

EXAMPLE: The American Express Travelers Check

The American Express travelers check with identification number 210687315 has check digit 3 because $2 + 1 + 0 + 6 + 8 + 7 + 3 + 1 + 5 = 33$ and $33 + 3$ is evenly divisible by 9. ▲

The scheme used on airline tickets, Federal Express mail, UPS packages, Avis, and National rental cars assigns the remainder upon division by 7 as the check digit (see Figure 9.3). For example, the check digit for the number 540047 is 4 since $540047 = 7 \times 77149 + 4$. This method will not detect the substitutions of 0 for a 7, 1 for an 8, 2 for a 9, or vice versa. However, unlike the Postal Service method, it will detect transpositions of adjacent digits with the exceptions of the pairs 0, 7; 1,

8; or 2, 9. For example, if 5400474 were entered into a computer as 4500474 (the first two digits are transposed), the machine would determine that the check digit should be 3 since $450047 = 7 \times 64292 + 3$. Because the last digit of the entered number is not 3, the error has been detected.

The scheme used on grocery products, the so-called **Universal Product Code (UPC)**, is more sophisticated. Consider a number such as 0 38000 00127 7 found on the bottom of a box of corn flakes. The first six digits identify the manufacturer, the next five the product, the last is a check. Suppose this number were entered into a computer as 0 58000 00127 7 (a mistake in the second position), how would the computer recognize the mistake?

The computer is programmed to carry out the following computation: Add the digits in positions 1, 3, 5, 7, 9, 11 and triple the result; then add this tally to the sum of the remaining digits. If the result doesn't end with a 0, the computer knows the entered number is incorrect.

For the incorrect corn flakes number, we have $((0 + 8 + 0 + 0 + 1 + 7) \times 3) + (5 + 0 + 0 + 0 + 2 + 7) = (16 \times 3) + 14 = 62$. Since 62 doesn't end with 0, the error is detected. Notice that had we used the correct digit, 3, in the second

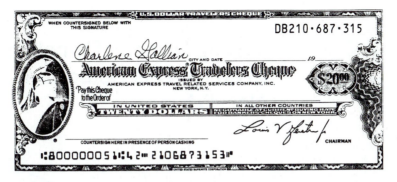

Figure 9.2 Travelers check with identification number 210687315 and check digit 3. The check digit is chosen so that the sum of all the digits including the check digit is evenly divisible by 9. Notice that the check digit is included as part of the identification number along the bottom, but not included in the upper right corner. The numbers along the bottom are read by a computer.

position instead of 5, the sum would have ended in a 0 as it should. This simple scheme detects *all* single-position errors and about 89% of all other kinds of errors.

The U.S. banking system uses a variation of the UPC scheme to append check digits to the numbers assigned to banks. Each bank has an eight-digit identification number $a_1 a_2 \cdots a_8$ together with a check digit a_9 so that a_9 is the last digit of $7a_1 + 3a_2 + 9a_3 + 7a_4 + 3a_5 + 9a_6 + 7a_7 + 3a_8$. The numbers 7, 3, and 9 used in this formula are called **weights**. The weights were carefully chosen so that all single-digit errors and most transposition errors are detected.

Figure 9.3 Airline ticket with identification number 121399538132 and check digit 5; stock control number 2551016950 and check digit 6.

▶ ▶ ▶ ▶ ▶ ▶ ▶ ▶ ▶ ▶ ▶

SP TLIGHT 9.1

Wrong Number: Demolition Crew Wrecks House at 415, Not 451

▶ ▶ ▶ ▶ ▶ ▶ ▶ ▶ ▶ ▶ ▶ ▶

It was a case of mistaken identity. A transposed address that resulted in a bulldozer blunder.

City orders had called for demolition on Tuesday of the boarded-up house at 451 Fuller Ave. SE. . . .

But when the dust settled, 451 Fuller stood untouched. Down the street at 415 Fuller Ave. SE, only a basement remained.

SOURCE: Doug Guthrie, *The Grand Rapids Press*, December 5, 1990.

EXAMPLE: Bank Identification Number

The North Shore Bank of Commerce of Duluth, Minnesota, has the number 091900106 on the bottom of all its checks (see Spotlight 9.2). The check digit 6 is the last digit of $7 \cdot 0 + 3 \cdot 9 + 9 \cdot 1 + 7 \cdot 9 + 3 \cdot 0 + 9 \cdot 0 + 7 \cdot 1 + 3 \cdot 0 = 106$. ▲

One of the most efficient error-detection methods is one used by all major credit card companies as well as many libraries, blood banks, photofinishing companies, German banks, and the South Dakota driver's license department. It is called **Codabar.** Say a bank intends to issue a credit card with the identification number 3125600196431. It must then add an extra digit

for error detection. This is done as follows: The digits in positions 1, 3, 5, 7, 9, 11, and 13 are added and the result is doubled: $(3 + 2 + 6 + 0 + 9 + 4 + 1) \times 2 = 50$. Next, count the number of digits in positions 1, 3, 5, 7, 9, 11, and 13 that exceed 4 and add this to the total. For our example, only 6 and 9 exceed 4, so the count is 2 and our running total is 52. Now add in the remaining digits: $52 + (1 + 5 + 0 + 1 + 6 + 3) = 68$.

The check digit is whatever is needed to bring the final tally to a number that ends with 0. Since $68 + 2 = 70$, the check digit for our example is 2. This digit is appended to the end of the number the bank issues for identification purposes. Errors in input data are detected by applying the same algorithm to the input, including the check digit. If the correct number is entered into a computer, the result will end in a zero. If the result doesn't end with a zero, a mistake has been made. This method allows computers to detect 100% of single-position errors and about 89% of other common errors.

Besides detecting errors, the check digit offers partial protection against fraudulent numbers. A person who wanted to create a phony credit card, bank account, or driver's license number would have to know the check digit scheme I've described here in order to fool the computer.

Thus far we have not discussed any schemes that detect 100% of single errors and 100% of transposition errors. The **International Standard Book Number (ISBN)** method used throughout the world is one that detects all such errors. (See the back of this book.)

A correctly coded ten-digit ISBN number $a_1a_2 \cdots a_{10}$ has the property that $10a_1 + 9a_2 + 8a_3 + 7a_4 + 6a_5 + 5a_6 + 4a_7 + 3a_8 + 2a_9 + a_{10}$ is evenly divisible by 11. Consider the ISBN number of the book you are now reading: 0-7167-2378-6. The initial digit "0" indicates that the book is published in an English-speaking country (not that the book is written in English). The next block of digits — 7167 — identifies the publisher, W. H. Freeman and Company. (The number of digits in the block can vary from one publisher to another.) The third block — 2378 — is assigned by the publisher and identifies the particular book. The last digit 6 is the check digit.

Let us verify that this number is a legitimate possibility. We must compute $10 \cdot 0 + 9 \cdot 7 + 8 \cdot 1 + 7 \cdot 6 + 6 \cdot 7 + 5 \cdot 2 + 4 \cdot 3 + 3 \cdot 7 + 2 \cdot 8 + 6 = 220$. Since $220 = 11 \cdot 20$ it is evenly divisible by 11, and no error has been detected.

How can we be sure that this method detects 100% of the single-position errors? Well, let us say a correct number is $a_1a_2a_3a_4a_5a_6a_7a_8a_9a_{10}$ and a mistake is made in the second position. (The same argument applies equally well in every position.) We can write this incorrect number as $a_1a_2'a_3a_4a_5a_6a_7a_8a_9a_{10}$, where $a_2' \neq a_2$. Now in order for this error to go undetected it must be the case that $10a_1 + 9a_2' + 8a_3 + 7a_4 + 6a_5 + 5a_6 + 4a_7 + 3a_8 + 2a_9 + a_{10}$ is evenly divisible by 11. Then, since both $10a_1 + 9a_2 + 8a_3 + 7a_4 +$ $6a_5 + 5a_6 + 4a_7 + 3a_8 + 2a_9 + a_{10}$ and $10a_1 + 9a_2' + 8a_3 + 7a_4 + 6a_5 + 5a_6 + 4a_7 + 3a_8 + 2a_9 + a_{10}$ are divisible by 11, so is their difference:

$$(10a_1 + 9a_2 + 8a_3 + \cdots + 1a_{10})$$
$$-(10a_1 + 9a_2' + 8a_3 + \cdots + 1a_{10})$$
$$= 9(a_2 - a_2')$$

Because a_2 and a_2' are distinct digits between 0 and 9 their difference must be one of $\pm 1, \ldots, \pm 9$. Thus the only possibilities for the number $9(a_2 - a_2')$ are $\pm 9, \pm 18, \pm 27, \pm 36, \pm 45, \pm 54, \pm 63, \pm 72, \pm 81, \pm 90$, and none of these is divisible by 11. So, the error is detected.

Since this method, in contrast to the others we have described, detects all single-position errors and all transposition errors, why is it not used more? Well, it does have a drawback. Say the next title published by Freeman is to have 1910 for the third block (all Freeman books begin with 0-7167-). What check digit should be assigned? Call it a. Then $10 \cdot 0 + 9 \cdot 7 + 8 \cdot 1 + 7 \cdot 6 + 6 \cdot 7 + 5 \cdot 1 + 4 \cdot 9 + 3 \cdot 1 + 2 \cdot 0 + a = 199 + a$. Since the next integer after 199 that is divisible by 11 is 209, we see that $a = 10$. But appending 10 to the existing nine-digit number would result in an eleven-digit number instead of a ten-digit one. This is the only flaw in the ISBN scheme. To avoid this flaw publishers use an X to represent the check digit 10. As a result not all ISBN numbers consist solely of digits (some end with X). Publishers could avoid this inconsistency by simply refraining from using numbers that require an X.

Many identification numbers utilize both alphabetic and numeric characters. One of the most prevalent of these was developed in 1975 and is called Code 3-out-of-9 (the name derives from the method of bar coding of the characters), or simply **Code 39**. Code 39 permits the use of the 26 uppercase letters A through Z and the digits 0 through 9. Because Code 39 has been chosen by the Department of Defense, the automotive com-

SP TLIGHT 9.2   Bank Checks

What do the string of numbers at the bottom of a check represent? Here is the answer.

0919   the bank's Federal Reserve District, office, and state or special collection arrangement

0010   the bank's identification number

6   the check digit

1 05 7199   the checking account number

0419   the check number

panies, and the health industry for use by their suppliers, it has become the workhorse of nonretail business. A typical example of a Code 39 number is 210SA0162322ZAY. The last character is the "check." The check character is determined by assigning the letters A through Z the numerical values 10 through 35, respectively. The original number, composed of the digits 0 through 9 and letters A through Z, is now converted to a string a_1, $a_2, \ldots, a_{14}, a_{15}$, where the a_i are integers between 0 and 35. The check character a_{15} is chosen so that $15a_1 + 14a_2 + 13a_3 + \cdots + 2a_{14} + a_{15}$ is divisible by 36. Finally, a_{15} is converted to its alphabetical counterpart if necessary.

EXAMPLE: Code 39 Number 210SA0162322ZA

Let us examine the Code 39 method for the number 210SA0162322ZA. Here is how we determine the check character. First we convert

the alphabetic characters to their numeric counterparts: $210SA0162322ZA \rightarrow 2, 1, 0, 28, 10, 0, 1, 6, 2, 3, 2, 2, 35, 10$. Then we compute

$$15 \cdot 2 + 14 \cdot 1 + 13 \cdot 0 + 12 \cdot 28$$
$$+ 11 \cdot 10 + 10 \cdot 0 + 9 \cdot 1 + 8 \cdot 6$$
$$+ 7 \cdot 2 + 6 \cdot 3 + 5 \cdot 2 + 4 \cdot 2$$
$$+ 3 \cdot 35 + 2 \cdot 10 = 30 + 14 + 0$$
$$+ 336 + 110 + 0 + 9 + 48 + 14 + 18$$
$$+ 10 + 8 + 105 + 20 = 722.$$

Now we select a_{15} so that $722 + a_{15}$ is divisible by 36. Since 722 divided by 36 has a remainder of 2 ($722 = 36 \cdot 20 + 2$), we choose a_{15} as 34. Finally, we convert 34 to Y. Thus the number becomes 210SA0162322ZAY.

In many applications of Code 39 the seven special characters -, ., space, $, /, +, and % are permitted. These characters are assigned the numerical values 36 through 42, respectively. In these applications the check character is determined by the remainder upon division by 43 instead of 36. ▲

ZIP CODE NATIONAL AREAS

The first digit of the ZIP Code divides the country into 10 large groups of states numbered from 0 in the Northeast to 9 in the far west.

Figure 9.4 ZIP code scheme. (Adapted from U.S. Postal Service 1988 National Five-Digit ZIP Code and Post Office Directory)

EXAMPLE

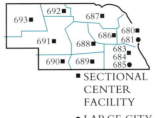

■ SECTIONAL
  CENTER
  FACILITY
● LARGE CITY
○ POST OFFICE

Within these areas, each state is divided into an average of 10 smaller geographic areas, identified by the second and third digits of the ZIP Code.

WHAT YOUR ZIP CODE MEANS

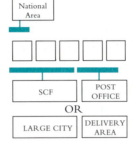

The fourth and fifth digits identify a local delivery area.

▶ THE ZIP CODE

Identification numbers occasionally **encode** geographical data. The **ZIP code,** social security numbers, and telephone numbers are prime examples. In 1963 the U.S. Postal Service numbered every American post office with a five-digit ZIP code. The numbers begin with zeros at the point farthest east—00601 for Adjuntas, Puerto Rico—and work up to nines at the point farthest west—99950 for Ketchikan, Alaska (see Figure 9.4). (In 1983 the U.S. Postal Service added four digits to the ZIP code.) Here's what the first five digits mean. Let's use one of the ZIP codes for Duluth, Minnesota, as an example:

55812

5   The first digit represents one of 10 geographical areas, usually a group of states.

The numbers begin at points farthest east (0) and end at points farthest west (9).

58   The second two digits, in combination with the first, identify a central mail-distribution point known as a sectional center. The location of a sectional center is based on geography, transportation facilities, and population density; although just four centers serve the entire state of Utah, there are six of them to take care of New York City.

12   The last two digits indicate the town, or local post office. The order is often alphabetical for towns within a delivery area—for example, towns with names beginning with "A" usually have low numbers. (There are many exceptions to this, such as towns that came into existence after the ZIP code scheme was created.) In many cases the largest city in a region will be

Andersonville	23911	Hampden-Sydney	23943
Boydton	23917	Kenbridge	23944
Buckingham	23921	Keysville	23947
Burkeville	23922	Lunenburg	23952
Charlotte Court House	23923	Meherrin	23954
Chase City	23924	Nottoway	23955
Clarksville	23927	Pamplin	23958
Crewe	23930	Phenix	23959
Cullen	23934	Prospect	23960
Darlington Heights	23935	Red Oak	23964
Dillwyn	23936	Rice	23966
Drakes Branch	23937	Skipwith	23968
Dundas	23938	Victoria	23974
Farmville	23901	Wylliesburg	23976
Green Bay	23942		

Figure 9.5 Postal ZIP codes in Farmville, Virginia, area. (Courtesy of Robb Koether.)

given the digits 01 and surrounding towns assigned succeeding digits alphabetically (see Figure 9.5).

When four digits are added after a dash, for example, 55812-1234, the number is called the **ZIP + 4 code.** Mail with ZIP + 4 coding benefits from cheaper bulk rates, being easier to sort with automated equipment. It's also helpful for businesses that wish to sort the recipients of their mailings by geographical location. The first two numbers of the 4-digit suffix represent a delivery sector, which may be several blocks, a group of streets, several office buildings, or a small geographical area. The last two numbers narrow the area further: They might denote one floor of a large office building, a department in a large firm, or a group of post office boxes. For businesses that receive an enormous volume of mail the ZIP + 4 code permits automation of in-house mailroom sorting. For example, the first seven digits of all mail sent to the University of Minnesota, Duluth, are 55812-24. The school has designated nine pairs of digits for the last two positions to direct

the mail to the appropriate dormitory or apartment complex. Even President Clinton has a ZIP + 4 code for mail from his friends (see Spotlight 9.3).

▶ **BAR CODES**

In modern applications bar codes and identification numbers go hand in hand. Bar coding is a method for automated data collection. It is a way to rapidly, accurately, and efficiently transmit information to a computer. A **bar code** is a series of dark bars and light spaces that represent characters. To **decode** the information in a bar code, a beam of light is passed over the bars and spaces via a scanning device, such as a hand-held wand or a fixed beam device. The dark bars reflect very little light back to the scanner, whereas the light spaces reflect much light. The differences in reflection intensities are detected by the scanner and converted to strings of 0s and 1s that represent specific numbers and letters. Such strings are called a **binary coding** of the numbers and letters. Any system for representing data with only two symbols is a **binary code.**

ZIP Code Bar Code

The simplest bar code is the **Postnet code** used by the U.S. Postal Service, which is commonly found on business reply forms (see Figure 9.6). For a ZIP + 4 code there are 52 vertical bars of two possible lengths (long and short). The bars at the beginning and end are always long and together provide a frame for the remaining 50 bars. In blocks of five, the 50 bars within the frame represent the ZIP + 4 code and a tenth digit for error correction. Each block of five is composed of exactly two long bars and three short bars, according to the pattern shown in Figure 9.7. The conversion from bars to 0s and 1s is done by assigning each short bar a 0 and each long bar a 1. In this way, the bar code is transformed into a binary code.

This scheme is called a **2-out-of-5 code.** It was first used in the 1940s by Bell Telephone Laboratories in computers that used a tape input

Figure 9.6 ZIP + 4 bar code and binary representation.

Nine-Digit ZIP Code for President's Friends

Every day an enormous volume of mail arrives at the White House addressed to the president. Shortly after President Clinton took office he sent a letter to his friends asking them to use a "secret" nine-digit ZIP code "to ensure that correspondence from you reaches them [the president's family] immediately." The letter adds, "Hearing from their friends on a regular basis is important to the Clintons."

SOURCE: *The Wall Street Journal,* March 19, 1993.

---

with two holes per row in five possible positions.

The tenth digit of a Postnet code number is a check digit chosen so that the sum of the nine digits of the ZIP + 4 code and the tenth one is evenly divisible by 10. That is, the check digit C for the ZIP + 4 code $a_1 a_2 \cdots a_9$ is the digit with the property that the sum $(a_1 + a_2 + \cdots + a_9 + C)$ is evenly divisible by 10. For example, the ZIP + 4 code 80321-0421 has the check digit 9, since $8 + 0 + 3 + 2 + 1 + 0 + 4 + 2 + 1 = 21$ and $21 + 9 = 30$ is evenly divisible by 10.

Because each digit is represented by exactly two long bars and three short ones, any error in reading or printing a single bar would result in a block of five with only one long bar or three long bars. In either case the error is detected. (This is the reason behind the choice of five bars to code each digit rather than four bars. With five bars per digit, there are exactly 10 arrangements composed of two long bars and three short bars. Any misreading of a single bar in such a block is therefore recognizable, since it does not match any other of the blocks for the 10 digits. In contrast, with four bars per digit, some of the 10 digits would have to be coded with one long bar and three short ones and other digits would be coded with two long bars and two short bars. In this case, misreading one of the three short bars might result in a block that represents another digit.) And since the block location of the error is known, the check digit permits the correction of the error. Let's look at an example of an incorrectly printed bar code and see how the error is correctable.

The scanner ignores the framing bars and reads the remaining bars in blocks of five as shown

Decimal Digit	Bar Code	Binary Code
1	ıı‖	00011
2	ıı‖ı	00101
3	ıı‖ı	00110
4	ıİıİı	01001
5	ıİıİı	01010
6	ıİ‖ıı	01100
7	‖ııİ	10001
8	‖ıİı	10010
9	‖İ‖ı	10100
0	‖‖ıı	11000

Figure 9.7 The Postnet bar code and Postnet binary code.

below. (We have underscored the incorrect block for readability.)

$$\text{||ul|l||uu||u||u||u|u||uuu|h||ul||uuu||l|l||ull}$$

3 0 7 2 2 **?** 9 0 1 7

Since the sixth block has only one long bar, it is an incorrect one. To correct the error, the computer linked with the bar code scanner sums the remaining 9 digits to obtain 31. Since the sum of all ten digits must be divisible by 10, the correct value for the sixth digit must be 9.

Effective March 1993, large organizations and businesses that want to receive reduced rates for ZIP + 4 bar-coded mail are required to use a 12-digit bar code called the *delivery point bar code*. This code permits machines to sort a letter into the order it will be delivered by the carrier. (Mail for the first location on a mail route occurs first, mail for the second location on a route occurs second, and so on.)

The new 12-digit bar code uses the Postnet bar scheme to code the 12-digit string composed of the 9-digit ZIP + 4 number followed by the last two digits of the street address or box number and a check digit chosen so that the sum of all 12 digits is evenly divisible by 10. For example, a letter addressed to 1738 Maple Street with ZIP + 4 code 55811-2742 would have the Postnet bar code for the digits 558112742384 (38 is from the street address and 4 is the check digit).

The UPC Bar Code

The bar code that people encounter most often is the Universal Product Code (UPC). It first appeared on grocery items in 1973 and has since spread to most retail products. The UPC bar code translates a twelve-digit number into bars that can be quickly and accurately read by a laser scanner. The number has four components—two five-digit numbers sandwiched between two single digits—as shown in Figure 9.8.

Here is what the four components represent:

5   The first digit (called the *number system character*) identifies the kind of product. A 0, 6, or 7 is assigned to all nationally branded products except the following: a 2 signals random weight items, such as cheese or meat; a 3 means drug and certain health-related products; a 4 means products marked for price reduction by the retailer; a 5

Figure 9.8 UPC identification number 5 38000 51150 4. The initial 5 indicates the number is a manufacturer's coupon. The block 38000 identifies the manufacturer as Kellogg's. The block 51150 identifies the product. The last digit, 4, is a check digit.

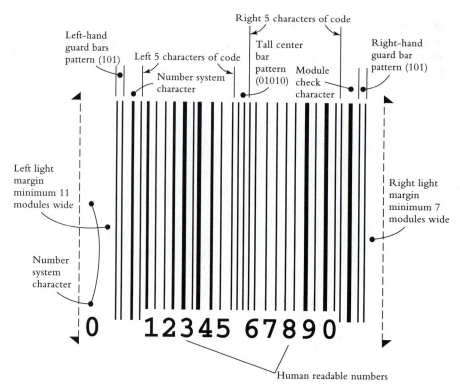

Left-hand guard bars pattern (101)

Left 5 characters of code

Number system character

Right 5 characters of code

Tall center bar pattern (01010)

Module check character

Right-hand guard bar pattern (101)

Left light margin minimum 11 modules wide

Right light margin minimum 7 modules wide

Number system character

0 12345 67890 0

Human readable numbers

Figure 9.9 UPC bar code format.

signals cents-off coupons (see Figure 9.8); a 9 indicates the item is a book. (Actually, a UPC number beginning with a 9 is an extended version of the UPC code called the EAN (European Article Numbering) code. It has a lead digit followed by two six-digit blocks. The check digit is included as part of the product number. The EAN code follows the same rules as the UPC code.)

38000   The next five digits represent the manufacturer. This number is assigned by the Uniform Code Council in Dayton, Ohio.

51150   The next five digits are assigned by the manufacturer to represent the product, and can include size, color, or other important information (but not price).

4   The final digit is the check digit. It signals the computer if one of the other digits is incorrect. This digit is often not printed, but it is always included in the bar code.

Each digit of the code is represented by a space divided into seven modules of equal width. How these seven modules are filled depends on the digit being represented and whether it is part of the manufacturer's number or the product number. In every case there are two light "spaces" and two dark bars of various thicknesses that alternate. A UPC code has on each end two long bars of one module thickness separated by a light space of one module thickness. These three modules are called the *guard bar patterns* (see Figure 9.9). The guard bar patterns define the thickness of a single module of each type; they are not part of the identification number. The manufacturer's number and the product number are separated by a center bar pattern consisting of the following five modules: a light space, a (long) dark bar, a light space, a (long) dark bar, and a light space (see Figure 9.9). The center bar pattern is not part of the identification number, but merely serves to separate the manufacturer's number and product number. Figure 9.10 shows how the digits 6 and 0 in a manufacturer's number are coded.

Observe the following pattern in Figure 9.10: a light space of one module thickness, a dark bar of one module thickness, a light space of one module thickness, a dark bar of four modules thickness. Symbolically such a pattern of light spaces and dark bars is represented as 0101111. Here each 0

Figure 9.11 Translation of bar and space modules into binary code (see top of bars). The guard pattern defines a single module thickness for a bar and a space.

means a one module thickness light space and each 1 means a one module thickness dark bar. Figure 9.11 illustrates how the thicknesses of the spaces and bars are translated into a binary code.

The code for the digits in the product number (the block of five digits on the right side) can be obtained from the code for the digits in the manufacturer's number (the block of digits on the left side) and vice versa by replacing each 0 by a 1 and each 1 by a 0. Thus the code 0111011 for 7 in a manufacturer's number becomes 1000100 in the product number. Table 9.1 shows the codes for all digits. Notice that each manufacturer's number has an odd number of 1s, whereas each product number has an even number of 1s. This permits a computer linked with an optical scanner to determine whether the bar code was scanned left to right or right to left. (If the first block of digits has an even number of 1s for each digit, the scanning is being done right to left.) Thus scanning can be done in either direction without ambiguity.

There is an abbreviated version of the UPC code called **Version E** that is often used in special circumstances, such as for round containers (for

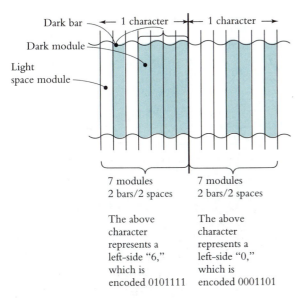

Figure 9.10 UPC bar coding for a left-side 6 and left-side 0.

TABLE 9.1 **Binary UPC Coding**

Digit	Manufacturer's Number	Product Number
0	0001101	1110010
1	0011001	1100110
2	0010011	1101100
3	0111101	1000010
4	0100011	1011100
5	0110001	1001110
6	0101111	1010000
7	0111011	1000100
8	0110111	1001000
9	0001011	1110100

Figure 9.12 Version E of the UPC code.

example, soft drink cans), small items, and magazines (see Figure 9.12 for an example). Details about this code are too complicated to present here. Suffice it to say that there are four algorithms for reducing the 11-digit UPC number (not counting the check digit) to the 7-digit Version E number, depending on the manufacturer's number, and ten schemes for converting the digits to bars, depending on the check digit. Exercises 23 and 24 explain how two of the four check digit schemes work.

Code 39 Bar Code

The Code 39 bar code is a variation on the "2-out-of-5" scheme used for the ZIP + 4 bar code. In this case each alphanumeric character corresponds to a string of three 1s and six 0s (hence the name "3-out-of-9" or just "39"). The number 1, for instance, is coded as by the string 100100001. Whereas the ZIP + 4 bar code utilizes bars of two lengths (long and short), the Code 39 bar code uses bars and spaces of two thicknesses — narrow and wide (a wide bar is three times the thickness of a narrow bar). Each character is represented by a total of nine alternating bars and spaces, with the thickness of each determined by the presence of a

0 or a 1. A 0 in the string is represented by a narrow bar or space, while a 1 in the string is represented by a wide bar or space. The digits in positions 1, 3, 5, 7, and 9 determine the thickness of the bars, whereas the digits in positions 2, 4, 6, and 8 determine the thickness of the spaces.

EXAMPLE: Code 39 Bar Coding of 1 or A

Since the Code 39 binary code for 1 is 100100001, its bar coding is wide bar, narrow space, narrow bar, wide space, narrow bar, narrow space, narrow bar, narrow space, wide bar (see Figure 9.13).

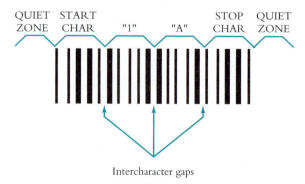

Figure 9.13 Code 39 bar code for message 1A.

Since the Code 39 binary code for A is 100001001, its bar coding is wide bar, narrow space, narrow bar, narrow space, narrow bar, wide space, narrow bar, narrow space, wide bar (see Figure 9.13). ▲

In a Code 39 bar symbol each character is separated by an intercharacter gap that can be from one to five times the width of a narrow space. The Code 39 scheme utilizes a special start/stop character denoted by * to indicate the beginning and end of the bar code. The binary code for * is 010010100. Figure 9.13 shows the complete Code

39 bar code encoding for the message 1A including the start and stop symbol and the intercharacter gaps. Table 9.2 gives the bar code and the Code 39 binary code for all characters.

With properly designed scanning equipment and high-quality symbol printing, the Code 39 bar code has an error rate of about one out of 70 million characters scanned. Even with better quality dot matrix printing for bar codes, the error rate is less than one in three million characters scanned. When the check digit is included, undetected errors occur about once in every 45 trillion characters scanned.

TABLE 9.2   Bar and Binary Representation for Code 39

Char.	Pattern	Bars	Spaces	Char.	Pattern	Bars	Spaces
1		10001	0100	M		11000	0001
2		01001	0100	N		00101	0001
3		11000	0100	O		10100	0001
4		00101	0100	P		01100	0001
5		10100	0100	Q		00011	0001
6		01100	0100	R		10010	0001
7		00011	0100	S		01010	0001
8		10010	0100	T		00110	0001
9		01010	0100	U		10001	1000
0		00110	0100	V		01001	1000
A		10001	0010	W		11000	1000
B		01001	0010	X		00101	1000
C		11000	0010	Y		10100	1000
D		00101	0010	Z		01100	1000
E		10100	0010	-		00011	1000
F		01100	0010	.		10010	1000
G		00011	0010	SPACE		01010	1000
H		10010	0010	*		00110	1000
I		01010	0010	$		00000	1110
J		00110	0010	/		00000	1101
K		10001	0001	+		00000	1011
L		01001	0001	%		00000	0111

The * symbol denotes a unique start/stop character that must be the first and last character of every bar code symbol.

The Scanner

Most scanners use a red laser to translate the bars and spaces into time intervals. The scanner distinguishes the spaces from the bars by the amount of light reflected (spaces reflect more light). Since it takes twice as long to read a double-thickness bar or space than a single-thickness bar or space, the bars and spaces are converted to time intervals that are then translated into digits. Although the thickness of each module may vary from one product to the next, the guard bars for each bar code define the module thickness for that particular bar code. For instance, a single-thickness dark bar on the UPC code for Tic-Tacs may be one-quarter the width of a single-thickness dark bar on a milk carton. But the dark guard bar for Tic-Tacs is also one-quarter the width of a dark guard bar for a milk carton. Thus it is the relative thickness of the bars that is the determining feature.

Other Bar Codes

Although the ZIP + 4 code bar code, the UPC bar code, and Code 39 are the three code types you most frequently encounter, there are more than 50 other bar codes in use. Each has its own particular advantages and disadvantages. Figure 9.14 shows the most common ones.

OPTIONAL ▶ ENCODING PERSONAL DATA

Here is a social security number: 189-31-9431. What information about the holder can be deduced from this number? Only that the holder obtained it in western Pennsylvania (see Spotlight 9.4). Here is a New York driver's license number: S095074571828089. What information about the holder can be deduced from this number? This time we can determine the birthday, sex, and much about the person's name.

These two examples illustrate the extremes in coding personal data. The social security number has no personal data encoded in the number. It is entirely determined by the place and time it is issued, not the individual to whom it is assigned. In contrast, in some states the driver's license numbers are entirely determined by personal information about the holders (see Spotlight 9.5). It is no coincidence that the social security numbering scheme predates computers. Agencies that have large data bases that include personal information

Figure 9.14 The most popular bar code symbologies.

SP TLIGHT 9.4   Social Security Numbers

▸ ▸ ▸ ▸ ▸ ▸ ▸ ▸ ▸ ▸ ▸ ▸ ▸ ▸

The first three digits of social security numbers show where the number was applied for. Changes in population have forced some numbers to be moved or assigned out of sequence over the years.

001 – 003	New Hampshire	387 – 399	Wisconsin	526 – 527	Arizona
004 – 007	Maine	400 – 407	Kentucky	& 600 – 601	
008 – 009	Vermont	408 – 415	Tennessee	528 – 529	Utah
010 – 034	Massachusetts	416 – 424	Alabama	530	Nevada
035 – 039	Rhode Island	425 – 428	Mississippi	531 – 539	Washington
040 – 049	Connecticut	& 587 – 588		540 – 544	Oregon
050 – 134	New York	429 – 432	Arkansas	545 – 573	California
135 – 158	New Jersey	433 – 439	Louisiana	& 602 – 626	
159 – 211	Pennsylvania	440 – 448	Oklahoma	574	Alaska
212 – 220	Maryland	449 – 467	Texas	575 – 576	Hawaii
221 – 222	Delaware	468 – 477	Minnesota	577 – 579	District of Columbia
223 – 231	Virginia	478 – 485	Iowa	580	Virgin Islands
232 – 236	W. Virginia	486 – 500	Missouri	580 – 584	Puerto Rico
232, 237 – 246	N. Carolina	502	N. Dakota	& 596 – 599	
247 – 251	S. Carolina	503 – 504	S. Dakota	586	Guam
252 – 260	Georgia	505 – 508	Nebraska	586	American Samoa
261 – 267	Florida	509 – 515	Kansas	586	Philippines
& 589 – 595		516 – 517	Montana	700 – 728	*through July 1, 1963,*
268 – 302	Ohio	518 – 519	Idaho		*reserved for railroad*
303 – 317	Indiana	520	Wyoming		*employees*
318 – 361	Illinois	521 – 524	Colorado		
362 – 386	Michigan	525 & 585	New Mexico		

SOURCE: *Social Security Administration.*

such as names, sex, and dates of birth find it convenient to encode these data into identification numbers. Examples of such agencies are the National Archives (where census records are kept), genealogical research centers, the Library of Congress, and the department in charge of motor vehicle management in each state.

There are many methods in use to encode personal data, such as name, sex, and date of birth. Perhaps the most conspicuous applications of

SP⬤TLIGHT 9.5   Encoding New York Driver's License Numbers

▶ ▶ ▶ ▶ ▶ ▶ ▶ ▶ ▶ ▶ ▶

Prior to September of 1992, the state of New York utilized a complex algorithm based on an individual's name, and day, month, and year of birth to assign driver's license numbers. The first character of the number is the first character of the last name (J for Jones, S for Smith). The next twelve digits are determined by converting the second, third, fourth, and fifth letters of the last name, the first three of the first name, and the middle initial to the corresponding position on the alphabet (A converts to 1, B converts to 2, ..., Z to 26, a blank to 0). Let us denote these alphabetic positions by $n_1, n_2, \ldots, n_8$. That is, n_1 is the alphabet position of the second letter of the last name, n_2 is the alphabet position of the third letter of the last name, and so on.

If the last name has 4 or more characters (so that $n_3 > 0$), the formula for the first 12 digits of the license number is

$$10{,}017{,}758{,}323 \cdot n_1 + 371{,}538{,}441 \cdot n_2 \\ + 13{,}779{,}585 \cdot n_3 + 510{,}355 \cdot n_4 \\ + 19{,}657 \cdot n_5 + 729 \cdot n_6 + 27 \cdot n_7 \\ + n_8 - 385{,}829{,}132.$$

(When the resulting number has fewer than 12 digits, we fill in with 0s on the left.)

Applying this formula to Alvy J. Singer we have

i	n	g	e	a	l	v	j
n_1	n_2	n_3	n_4	n_5	n_6	n_7	n_8
9	14	7	5	1	12	22	10

Figure 9.15 Wisconsin driver's license.

these methods are those used to assign driver's license numbers in some states. Coding license numbers solely from personal data enables automobile insurers, government entities, and law enforcement agencies to determine the number from the personal data.

Illinois, Florida, and Wisconsin, for instance, encode the surname, first name, middle initial, date of birth, and sex by a quite sophisticated scheme (see Figure 9.15). The first four characters of the license number are obtained by applying the **Soundex Coding System** to the surname as follows:

So, the formula gives

$$10{,}017{,}758{,}323 \cdot 9 + 371{,}538{,}441 \cdot 14$$
$$+ 13{,}779{,}585 \cdot 7 + 510{,}355 \cdot 5$$
$$+ 19{,}657 \cdot 1 + 729 \cdot 12 + 27 \cdot 22$$
$$+ 10 - 385{,}829{,}132$$
$$= 0950745571828.$$

If the last name has 3 characters (for example, Lee), the term $-385{,}829{,}132$ must be replaced by $-385{,}318{,}778$. But if it has exactly 2 characters (for example, Ho), then $-385{,}829{,}132$ must be replaced by $-371{,}539{,}194$. Here is the formula for Julia Ho (no middle name):

o	—	—	—	j	u	l	—
n_1	n_2	n_3	n_4	n_5	n_6	n_7	n_8
15	0	0	0	10	21	12	0

$$10{,}017{,}758{,}323 \cdot 15 + 371{,}538{,}441 \cdot 0$$
$$+ 13{,}779{,}585 \cdot 0 + 510{,}355 \cdot 0$$
$$+ 19{,}657 \cdot 10 + 729 \cdot 21$$
$$+ 27 \cdot 12 + 0 - 371{,}539{,}194$$
$$= 1498950047854.$$

In the cases where the first name consists of exactly one letter (for example, D. Denise Aiken, J. Thomas Bucker), we must add 26 to the number produced by the formula.

The next three digits of the driver's license number are determined by the day and month of birth, and the sex. For a male born on day d and month m we append $63m + 2d$ to the 12 digits computed from the name (July 13 gives $63 \cdot 7 + 2 \cdot 13$). For a female we append $63m + 2d + 1$. The last two digits of the number represent the year of birth (42 for 1942, 63 for 1963). So, for Alvy J. Singer, born on January 13, 1935, we have S095074571828808935. For Julia Ho, born on October 2, 1950, we have H1498950478563550. In cases where this algorithm yields identical numbers for more than one person, a "tie breaking" digit is inserted before the two digits representing the year of birth.

1. Delete all occurrences of h and w. (For example, Schworer becomes Scorer and Hughgill becomes uggill.)

2. Assign numbers to the remaining letters as follows:

a, e, i, o, u, y → 0
b, f, p, v → 1   1 → 4
c, g, j, k, q, s, x, z → 2   m, n → 5
d, t → 3   r → 6

3. If two or more letters with the same numeric value are adjacent, omit all but the first. (For example, Scorer becomes Sorer and uggill becomes ugil.)

4. Delete the first character of the original name if still present. (Sorer becomes orer.)

5. Delete all occurrences of a, e, i, o, u, and y.

6. Retain only the first three digits corresponding to the remaining letters; append trailing zeros if less than three letters remain; pre-

SPOTLIGHT 9.6   Census Records at the National Archives

▶ ▶ ▶   ▶ ▶ ▶ ▶ ▶ ▶ ▶ ▶ ▶ ▶

One of the best places to look for information pertaining to family history is the old censuses that are kept by the National Archives in Washington, D.C. By law, census records are open to the public 72 years after the census was taken. The 1920 census material was available to researchers beginning in 1992. The data from the 1880, 1900, 1910, and 1920 censuses (records from 1890 were destroyed by fire) were put on cards during the 1930s as a Works Progress Administration project. This information was coded using the Soundex system so that names that sound alike regardless of how they are spelled are grouped together. On old documents family names were so often misspelled — especially those that were not of British origin — that genealogists say several variations of a name may apply to one set of ancestors.

To look for a surname on the index, the researchers must work out the Soundex code. This code together with the state identifies a

page number on microfilm where the data are located. A typical census Soundex card is shown. Note the Soundex code in the upper left-hand corner (B350).

cede the three digits by the first letter of the surname.

Figure 9.16 shows three examples.

What is the advantage of this method? It is an error-correcting scheme. Indeed, it is designed so that likely misspellings of a name nevertheless result in the correct coding of the name. For example, frequent misspellings of the name Erickson

are: Ericksen, Eriksen, Ericson, and Ericsen. Observe that all of these yield the same coding as Erickson. If a law enforcement official, a genealogical researcher, a librarian, or an airline reservation agent wanted to pull up the file from a data bank for someone whose name was pronounced "Erickson," the correct spelling isn't essential since the computer searches for records that are coded as E-625 for all variations of spellings. (The

```
              Step 1        Step 2
    Schworer   →   Scorer    →    Scorer
                                  220606
  Step 3       Step 4       Step 5     Step 6
   →  Sorer   →   orer   →   rr   →   S-660
      20606       0606       66
```

```
              Step 1        Step 2
    Hughgill   →   uggill    →    uggill
                                  022044
  Step 3       Step 4       Step 5     Step 6
   →  ugil   →   ugil   →   gl   →   H-240
      0204       0204       24
```

```
            Step 1              Step 2
  Schmidlapper  →  Scmidlapper  →   Scmidlapper
                                    22503401106
  Step 3         Step 4         Step 5       Step 6
   →  Smidlaper  →  midlaper  →  mdlpr  →  S-534
      250340106     50340106     53416
```

Figure 9.16   The Soundex Coding System.

Soundex system was designed for the Census Office when much census data were obtained orally, see Spotlight 9.6. Airlines use a somewhat different system called the Davidson Consonant scheme.)

The next three digits of an Illinois, Florida, or Wisconsin license number are determined by summing numbers that correspond to the first name and middle initial (see Tables 9.3 and 9.4). As shown in Table 9.4, the scheme for doing this begins with the block 000 for the letter A and makes jumps of 20 for especially common names and each subsequent letter of the alphabet. The values assigned to the middle initial are given in Table 9.3.

Thus Aaron G. Schlecker would be coded as S-426-007 (S-426 from Schlecker; 000 for Aaron + 7 for "G"). Anne P. Schlecker would be coded as S-426-055 (040 for Anne + 15 for "P").

The last five digits of Illinois and Florida numbers capture the year and date of birth as well as the sex. In Illinois, each day of the year is assigned a three-digit number in sequence beginning with 001 for January 1. However, each month is assumed to have 31 days. Thus, March 1 is given 063, since both January and February are assumed to have 31 days. These numbers are then used to identify the month and day of birth of male drivers. For females, the scheme is identical except January 1 begins with 601. The last two digits of the year of birth, separated by a dash (probably to obscure the fact that they represent the year of birth), are listed in the fifth and fourth positions from the end of the driver's license number. Thus, a male born on October 9, 1940, would have the last five digits 4-0288, while a female born on the same day would have 4-0888. When necessary, Illinois adds an extra character to avoid duplications. Among the 9,382,662 licenses on file in Illinois on January 1, 1987, this occurred in 14,856 instances. Of these, 55 numbers corresponded to three individuals (excluding the extra digit). No number corresponded to more than three people. The scheme to identify birth date and sex in Florida is the same as in Illinois except each month is assumed to have 40 days and 500 is added for women. For example, the five digits 4-9583 belong to a woman born on March 3, 1949.   ◄

TABLE 9.3   Illinois, Florida, Wisconsin Middle Initial Code

0	None	10	J
1	A	11	K
2	B	12	L
3	C	13	M
4	D	14	N, O
5	E	15	P, Q
6	F	16	R
7	G	17	S
8	H	18	T, U, V
9	I	19	W, X, Y, Z

TABLE 9.4 Illinois, Florida, Wisconsin Given Name or First Initial Code

000	A	500	K
020	Albert, Alice	520	L
040	Ann, Anna, Anne, Annie, Arthur	540	M
060	B	560	Margaret, Martin
080	Bernard, Bette, Bettie, Betty	580	Marvin, Marv
100	C	600	Melvin, Mildred
120	Carl, Catherine	620	N
140	Charles, Clara	640	O
160	D	660	P
180	Donald, Dorothy	680	Patricia, Paul
200	E	700	Q
220	Edward, Elizabeth	720	R
240	F	740	Richard, Ruby
260	Florence, Frank	760	Robert, Ruth
280	G	780	S
300	George, Grace	800	T
320	H	820	Thelma, Thomas
340	Harold, Harriet	840	U
360	Harry, Hazel	860	V
380	Helen, Henry	880	W
400	I	900	Walton, Wanda
420	J	920	William, Wilma
440	James, Jane, Jayne	940	X
460	Jean, John	960	Y
480	Joan, Joseph	980	Z

▶ REVIEW VOCABULARY

Bar code　A code that employs bars to represent information.

Binary code　A coding scheme that uses two symbols, usually 0 and 1.

Binary coding　A string of 0s and 1s that represents a number, letter, or symbol.

Codabar　An error-detection method used by all major credit card companies, many libraries, blood banks, and others.

Code　A group of symbols that represents information together with a set of rules for interpreting the symbols.

Code 39　An alphanumeric bar code that is widely used on nonretail items.

Decoding　Translating code into data.

Encoding　Translating data into code.

Error-detecting code　A code in which certain types of errors can be detected.

International Standard Book Number (ISBN)　A ten-digit identification number used on books throughout the world that contains a check digit for error detection.

Postnet code   The bar code used by the U.S. Postal Service for the ZIP code.

Scanner   An optical device used to read bar codes.

Soundex Coding System   An encoding scheme for surnames based on sound.

2-out-of-5 code   A bar code that uses blocks of five bars consisting of two long bars and three short bars.

Universal Product Code (UPC)   A bar code and identification number that is used on most retail items. The UPC code detects 100% of all single-digit errors and most other types of errors.

UPC Version E   An abbreviated version of the UPC code. The UPC Version E employs a bar coding scheme that is different than the UPC bar code.

Weights   Numbers used in the calculation of check digits.

ZIP code   A five-digit code used by the U.S. Postal Service to divide the country into geographical units to speed sorting of the mail.

ZIP + 4 code   The nine-digit code used by the U.S. Postal Service to refine the ZIP code into smaller units.

▶ SUGGESTED READINGS

COLLINS, D. J., AND N. WHIPPLE: *Using Bar Code*, Data Capture Institute, Duxbury, Mass., 1990. This book contains extensive information on bar codes.

DAVIDSON, L.: "Retrieval of Misspelled Names in an Airline Passenger Record System," *Communications of the Association for Computing Machinery*, 5:169–171 (1962). This article describes the method used by airlines (the Davidson Consonant Code) to store and retrieve passenger names.

GALLIAN, J.: "Assigning Driver's License Numbers," *Mathematics Magazine*, 64:13–22 (1992).

This article describes various methods used by the states to assign driver's license numbers. Several of these methods include check digits for error detection.

GALLIAN, J.: "The Mathematics of Identification Numbers," *The College Mathematics Journal*, 22:194–202 (1991). This article is a comprehensive survey of check digit schemes that are associated with identification numbers.

GALLIAN, J., AND S. WINTERS: "Modular Arithmetic in the Marketplace," *The American Mathematical Monthly*, 95:548–551 (1988). This article provides a more detailed analysis of the check digit schemes presented in this chapter. In particular, the error-detection rates for the various schemes are given.

GREENFIELD, R.: "An Experiment to Measure the Performance of Phonetic Key Compression Retrieval Schemes," *Methods in Information in Medicine*, 16:230–233 (1977). This article compares the performance of the Soundex scheme and the Davidson Consonant Scheme.

HARMAN, C. K., AND R. ADAMS: *Reading Between the Lines*, Helmers Publishing, Peterborough, N.H., 1989. This book contains extensive information on bar codes.

PHILIPS, LAWRENCE: "Hanging on the Metaphone," *Computer Language*, 7(12):39–43 (December 1990). This article describes a sound-based retrieval algorithm that in some respects is superior to Soundex.

ROUGHTON, KAREN, AND DAVID A. TYCKOSEN: "Browsing with Sound: Sound-based Codes and Automated Authority Control," *Information Technology and Libraries*, 4(2):130–136 (June 1985). The article explains the Soundex code and the Davidson Consonant Code and their uses. The Davidson Consonant Code is a sound-based retrieval algorithm used by systems to retrieve the names of airline passengers.

►EXERCISES

1. Determine the ZIP + 4 code and check digit for each of the following Postnet bar codes:

a.

|.|.|....||.|.|.|.|....||...|.|||....||.|..|.|.||

b.

|.|.||....||.|..||..|||..||...||..|.|.|.|...|||

2. In each Postnet bar code below, exactly one mistake occurs (i.e., a long bar appears instead of a short one or vice versa). Determine the correct ZIP code.

a.

|..|.|||..|..||..|..|.||....|.|..|.||..||..|.||

b.

|.|..|.|..|..|..||..|.|.|.|..|.|..|.||..|.|||..|

c.

|.|.|..|.||..|..||..|.|..|..|.|..|||....||..|||

3. Explain why any two errors in a particular block of five bars in a Postnet is always detectable. Explain why not all such errors can be corrected.

4. Determine the check digit for a money order with identification number 7234541780.

5. Suppose a money order with the identification number and check digit 21720421168 is erroneously copied as 27750421168. Will the check digit detect the error? Explain your reasoning.

6. Determine the check digit for the United Parcel Service (UPS) identification number 873345672.

7. Determine the check digit for the Avis rental car with identification number 540047.

8. In Florida, the last three digits of the driver's license number of a female with birth month m and birth date b are $40(m - 1) + b + 500$, whereas for a male the formula is $40 (m - 1) + b$. For both males and females, the fourth and fifth digits from the end give the year of birth. Determine the dates of birth of people with the numbers whose last five digits are 42218 and 53953.

9. For New York driver's license numbers prior to September 1992 the last two digits were the year of birth. The three digits preceding the year encoded the sex and month and day of birth. For a woman with birth month m and birth date b the three digits were $63m + 2b + 1$. For a man with birth month m and birth date b the three digits were $63m + 2b$. Determine birth months, birth dates, and sexes of drivers with the digits 248 and 601 preceding the year.

10. Determine the check digit for the UPC number 05074311502.

11. Determine the check digit for the ISBN number 0-669-19493.

12. Use the bank scheme to determine the check digit for the number 09190204.

13. Use the Codabar scheme to determine the check digit for the number 3125600196431.

14. Determine the check character for the Code 39 number 210SA0162305ZA. (Assume the code uses only 36 characters.)

15. Determine the check character for the number 3050-0000 HEAD using the 43-character Code 39 scheme. (Be sure to include hyphen and space as characters.)

16. Use the "2-out-of-5" code to code 173.

17. Is there any mathematical reason for a check digit to be at the end of an identification number? Explain your reasoning.

● 18. Determine the Soundex code for Smith, Schmid, Smyth, and Schmidt.

● 19. Determine the Soundex code for Skow, Sachs, Lennon, Lloyd, Ehrheart, and Ollenburger.

● 20. Determine the Illinois driver's license number for James P. McCartney, born on July 18, 1942.

● 21. Determine the Florida driver's license number for James P. McCartney, born on July 18, 1942.

22. Suppose the first block of a UPC bar code following the guard bar pattern a scanner reads is 1000100. Is the scanner reading left-to-right or right-to-left?

23. Use the fact that the check digit a_8 for a UPC Version E identification number $a_1a_2a_3a_4a_5a_6a_7$, where a_7 is 0, 1, or 2, is chosen so that $a_1 + a_2 + 3a_3 + 3a_4 + a_5 + 3a_6 + a_7 + a_8$ is divisible by 10 to determine the check digit for the following E Version numbers:

     a. 0121690   b. 0274551   c. 0760022

24. Use the fact that the check digit a_8 for a UPC Version E identification number $a_1a_2a_3a_4a_5a_6a_7$, where a_7 is 4, is chosen so that $a_1 + a_2 + 3a_3 + a_4 + 3a_5 + 3a_6 + a_8$ is divisible by 10 to determine the check digit for the following numbers:

     a. 0754704   b. 0774714

25. Form all possible strings consisting of exactly three a's and two b's and arrange the strings in alphabetical order (for example, the first two possibilities are *aaabb* and *aabab*). Do you see any relationship between your list and the binary "2-out-of-5" code used for the Postnet bar code?

26. Explain why the bank scheme will detect the error $751 \cdots \rightarrow 157 \cdots$, but the UPC scheme will not.

● Optional exercise.

27. Explain how a computer can determine whether a Code 39 bar code is being read left-to-right or right-to-left.

28. What can you say about a Code 39 character whose space binary representation is 0100?

29. Determine the identification number represented by the following Code 39 bar code.

30. Suppose a Code 39 bar code represents an identification number composed entirely of digits (in particular, alphabetic characters are not used). Explain why you may ignore the spaces in determining the digits represented by the bars?

▲ 31. The ISBN 0-669-03925-4 is the result of a transposition of two adjacent digits not involving the first or last digit. Determine the correct ISBN.

▲ 32. The state of Utah appends a ninth digit a_9 to their eight-digit driver's license number $a_1 a_2 \cdots a_8$ so that $9a_1 + 8a_2 + 7a_3 + 6a_4 + 5a_5 + 4a_6 + 3a_7 + 2a_8 + a_9$ is divisible by 10.

   a. If the first eight digits of a Utah license number are 14910573, what is the ninth digit?

   b. Suppose a legitimate Utah license number 149105767 is miscopied as 149105267. How would you know a mistake was made? Is there any way you could determine the correct number? Suppose you know the error was in the seventh position, could you correct the mistake?

   c. If a legitimate Utah number 149105767 were miscopied as 199105767, would you be able to tell a mistake was made? Explain.

   d. Explain why any transposition error involving adjacent digits of a Utah number would be detected.

▲ 33. The Canadian province of Quebec assigns a check digit to an 11-digit driver's license number $a_1 a_2 \cdots a_{11}$ so that $12a_1 + 11a_2 + 10a_3 + 9a_4 + 8a_5 + 7a_6 + 6a_7 + 5a_8 + 4a_9 + 3a_{10} + 2a_{11} + a_{12}$ is divisible by 10. Criticize this method. Describe all single-digit errors that are undetected by this scheme. How does the transposition of two adjacent digits of a number affect the check digit of a number?

▲ 34. Suppose the check digit a_{10} of ISBN numbers were chosen so that $a_1 + 2a_2 + 3a_3 + 4a_4 + 5a_5 + 6a_6 + 7a_7 + 8a_8 + 9a_9 + 10a_{10}$ is divisible by 11 instead of the way described in the chapter. How would this compare with the actual check digit?

▲ Advanced exercise.

▲ 35. Suppose the check digit a_9 for bank checks were chosen to be the last digit of $3a_1 + 7a_2 + a_3 + 3a_4 + 7a_5 + a_6 + 3a_7 + 7a_8$ instead of the way described in the chapter. How would this compare with the actual check digit?

36. Below is an actual identification number and bar code from a roll of wallpaper. What appears to be wrong with them? Speculate on the reason for the apparent violation of the UPC format.

5 011419 194056

Building Regulations: 1985 Class 0
FINE ART WALLCOVERINGS LTD.
HOLMES CHAPEL, CHESHIRE

MADE IN ENGLAND
FABRIQUE EN ANGLETERRE

37. Determine the New York driver's license number for Leonard Zelig (no middle name) born on February 5, 1940 using the formula in Spotlight 9.5.

38. Along the upper edge of postage-paid business reply envelopes and cards, two inches from the right side, are six vertical bars arranged in one of two patterns. By examining samples, deduce what these two patterns signify.

39. Along the upper edge of many addressed envelopes (not postage-paid) provided by businesses as a convenience for their customers, two inches from the right side, are two pairs of narrowly spaced vertical bars with a single vertical bar in between. By examining samples, deduce what these bars signify.

40. Examine the bar code on the back of six recently published books that begin with the digit 9. How does this bar code format differ from the UPC code used on grocery items? How is the UPC number related to the ISBN number? How is the last digit of the UPC number (the check digit) determined?

41. Conduct the following experiment four times: Take any ZIP + 4 number and rearrange the 9 digits in some way. In each case compare the check digit for the rearranged number with the check digit for the original. Make a conjecture about the check digit of a rearranged number and of the original number.

42. The state of Washington encodes the last two digits of the year of birth into driver's license numbers (in positions 8 and 9) by subtracting the two-digit number from 100. For example, a person born in 1942 has 58 in positions 8 and 9, whereas a person born in 1971 has 29 in positions 8 and 9. Speculate on the reason for subtracting the birth year from 100.

43. Driver's license number assignment schemes that utilize personal data occasionally produce the same number for different people. Speculate about circumstances under which this is more likely to occur.

▲ 44. Consider a UPC number in which the digits 7 and 2 appear consecutively (that is, the number has the form $\cdots 72 \cdots$). Will the error caused by transposing these digits (that is, the number is taken as $\cdots 27 \cdots$) be detected?

What if the digits 6 and 2 are transposed instead? For the UPC scheme, state the general criterion for the detection of an error of the form

$$\cdots ab \cdots \rightarrow \cdots ba \cdots$$

● 45. Apply the Soundex code to common ways of misspelling your name. Do they give the same code as your name does?

■ 46. The Canadian postal system has assigned each geographical region a 6-character code composed of alternating letters and digits, such as P7B5E1 and K7L3N6. Discuss the advantages this scheme has over the U.S. 5-digit ZIP code.

●■ 47. Think of as many instances as you can where a data bank coded with Soundex could be an advantage.

● 48. Determine the total number of codes possible using the Soundex scheme.

► WRITING PROJECTS

1. Prepare a report on coded information in your location. Possibilities for investigation include: driver's license numbers in your state; student ID numbers and bar codes at your school; bar codes used by your school and city libraries. Identify the coding schemes and, when possible, determine whether a check digit is employed. Include samples. The suggested readings for this chapter contain information that will assist you.

2. Prepare a report on the driver's license coding schemes used by Minnesota, Michigan, Maryland, and Washington (the first three states use the same method). The reference entitled "Assigning Driver's License Numbers" has the information you will need.

3. Imagine that you are employed by a small company that does not now use identification numbers and bar codes for its employees or products. Prepare a report that discusses the various methods and makes a recommendation.

4. The Davidson Consonant code and Metaphone are two text-retrieval algorithms based on sound that are alternatives to Soundex. Prepare a report on either of these encoding schemes. The suggested reading by Roughton and Tyckosen and the one by Philips contain the information you will need.

5. Prepare a report on the Code 128 bar code. Include in your report a description of at least three instances where you found this code in use. Include samples if possible. The suggested reading by Harman and Adams and the one by Collins and Whipple contain the information you will need.

■ Discussion exercise.   ▲ Advanced exercise.   ● Optional exercise.

Chapter
10
Transmitting Information

Data stored in computers may be modified by naturally occurring radiation; information transmitted from communication satellites and space probes is subject to a variety of electromagnetic interference; compact discs are corrupted by dust, dirt, scratches, and fingerprints; magnetic tapes deteriorate; and entry of data into computers by humans is subject to frequent error. In this chapter we present mathematical methods for detecting or correcting these errors. We also illustrate a way data can be coded to reduce transmission time and storage space, and ways to securely transmit secret messages.

▶ BINARY CODES

Because of the way in which computers are built, in high technological applications such as compact disc players, fax machines, high-definition television, modems, and signals sent back from space probes, data are represented as strings of 0s and 1s rather than the usual digits 0 through 9 and letters A through Z. Recall from Chapter 9 that a system

for coding data with 0s and 1s is called a *binary code*. There are many binary codes in use. The one preferred for microcomputers is the **American Standard Code for Information Interchange (ASCII)**. In the ASCII format each character is coded in decimal form (that is, using digits 0,1, . . . , 9) and in binary form as a string consisting of eight 0s and 1s. A sample of ASCII characters and their decimal and binary codings is shown in Table 10.1

The idea behind error-correction schemes is simple and one you often use. To illustrate, suppose you are reading the employment section of a newspaper and you see the phrase "must have a minimum of bive years experience." Instantly you detect an error since "bive" is not a word in the English language. Moreover, you are fairly confident that the intended word is "five." Why so? Because the English language inherently has a built-in redundancy that often allows you to infer the intended meaning when mistakes are made.

Over the past 40 years mathematicians and engineers have devised highly sophisticated

TABLE 10.1   Partial List of Printable Characters and Their ASCII Codes

Character	ASCII code (decimal)	ASCII code (binary)	Character	ASCII code (decimal)	ASCII code (binary)
0	48	0011 0000	:	58	0011 1010
1	49	0011 0001	;	59	0011 1011
2	50	0011 0010	<	60	0011 1100
3	51	0011 0011	=	61	0011 1100
4	52	0011 0100	>	62	0011 1110
5	53	0011 0101	?	63	0011 1111
6	54	0011 0110	@	64	0100 0000
7	55	0011 0111	A	65	0100 0001
8	56	0011 1000	B	66	0100 0010
9	57	0011 1001	C	67	0100 0011

schemes to build in redundancy in messages composed of 0s and 1s (see Spotlight 10.1). As a simple example let us say that our set of messages consists of the 16 possible 4-tuples of 0s and 1s (as shown in the left column of Table 10.2).

We build redundancy in these messages with the aid of the Venn diagram in Figure 10.1. Begin by placing the four message digits in the four overlapping regions I, II, III, IV, with the digit in position 1 (starting at the left of the sequence) in region I, the digit in position 2 in region II, and so

on. For regions V, VI, and VII assign 0 or 1 so that the total number of 1s in each circle is even. See Figure 10.2.

Now suppose the message 1001, which we have encoded using the Venn diagram as 1001101, is received as 0001101 (an error in the first position). How would we know an error was made? We place each digit from the received message in its appropriate region as in Figure 10.3.

Noting that in both circles A and B there is an odd number of 1s, we instantly realize that some-

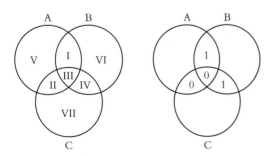

Figure 10.1 Venn diagram for message 1001.

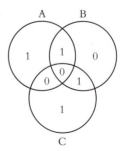

Figure 10.2 Venn diagram for encoded message 1001101.

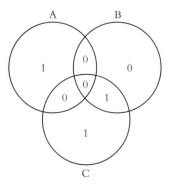

Figure 10.3 Venn diagram for received message 0001101.

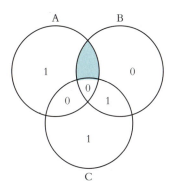

Figure 10.4 Circles A and B but not C have wrong parity.

thing is wrong, since the intended message had an even number of 1s in each circle. How do we correct the error? Since circles A and B have the wrong parity (parity refers to the oddness or evenness of a number; even integers have **even parity**; odd integers have **odd parity**) and C does not, the error is located in the portion of the diagram in

TABLE 10.2

Message		Code Word
0000	→	0000000
0001	→	0001011
0010	→	0010111
0100	→	0100101
1000	→	1000110
1100	→	1100011
1010	→	1010001
1001	→	1001101
0110	→	0110010
0101	→	0101110
0011	→	0011100
1110	→	1110100
1101	→	1101000
1011	→	1011010
0111	→	0111001
1111	→	1111111

circles A and B, but not in circle C. See Figure 10.4; that is, region I. Here we also see the advantage of using only 0s and 1s to encode data. If you have only two possibilities and one of them is incorrect, then the other one must be correct. Since the 1 in region I is incorrect, we know 0 is correct. The encoding for all 16 messages using the Venn diagram is shown in the right column of Table 10.2.

This method, called the **Venn diagram method,** will correctly decode any received message that has at most one error. If a received message has two or more errors, the method will not always yield the correct message.

▶ ENCODING WITH PARITY-CHECK SUMS

In practice, Venn diagrams are not used to encode strings of 0s and 1s. Rather, the messages are encoded by appending extra digits determined by the parity of various sums of certain portions of the messages (see Spotlight 10.2). We illustrate this method for the 16 messages shown in the left column of Table 10.2.

Our goal is to take any string $a_1a_2a_3a_4$, and append three check digits $c_1c_2c_3$ so that any single

SP⬤TLIGHT 10.1   The Ubiquitous Reed–Solomon Codes

▶ ▶ ▶ ▶ ▶ ▶ ▶ ▶ ▶ ▶ ▶ ▶ ▶

In this Age of Information, no one need be reminded of the importance not only of speed but also of accuracy in the storage, retrieval, and transmission of data. Machines *do* make errors, and their non-man-made mistakes can turn otherwise flawless programming into worthless, even dangerous trash. Just as architects design buildings that will remain standing even through an earthquake, their computer counterparts have come up with sophisticated techniques capable of counteracting digital disasters. Many of these techniques are based on mathematical ideas. One of the mathematical ideas underlying the current error-correcting techniques for everything from computer hard disk drives to CD players was first introduced in 1960 by Irving Reed and Gustave Solomon, then staff members at MIT's Lincoln Laboratory.

Irving Reed and Gustave Solomon at the Jet Propulsion Laboratory in 1989 monitor the encounter of Voyager 2 with Neptune. (Photo by Rex Ridenoure, NASA/JPL)

Reed – Solomon codes (plus a lot of engineering wizardry, of course) made possible the stunning pictures of the outer planets sent back by the space probe *Voyager 1 and 2*. They make it possible to scratch a compact disc and still enjoy the music. And in the not-too-distant future, they will enable the profit mongers of cable television to squeeze more than 500 channels into their systems, making a vast wasteland vaster yet.

"When you talk about CD players and digital audio tape and now digital television, and various

other digital imaging systems that are coming — all of those need Reed – Solomon [codes] as an integral part of the system," says Robert McEliece, a coding theorist in the electrical engineering department at Caltech.

Why? Because digital information, virtually by definition, consists of "bits" — 0s and 1s — and a physical device, no matter how capably manufactured, may occasionally confuse the two. *Voyager 2*, for example, was transmitting data at incredibly low power — barely a whisper — over billions of miles. Disk drives pack data so densely that a read/write head can (almost) be excused

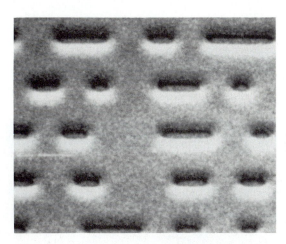

Microscope photograph (magnification 2500×) showing the molded pits on the bottom of a compact disc. When the laser beam that focuses on the revolving disc strikes a pit, its light scatters and the level of light reflected back changes. This change (along with that resulting from a transition from pit to land) is interpreted as binary 1. The varying path lengths between transitions in the level of reflected light are interpreted as corresponding numbers of 0s. Errors in this pattern of binary digits are corrected by Reed–Solomon codes. (Photograph courtesy of 3M Prerecorded Optical Media)

excused if it can't tell where one bit stops and the next one (or zero) begins. Careful engineering can reduce the error rate to what may sound like a negligible level — the industry standard for hard disk drives is 1 in 10 billion — but given the volume of information processing done these days, that "negligible" level is an invitation to daily disaster. Error-correcting codes are a kind of safety net — mathematical insurance against the vagaries of an imperfect material world.

In 1960, the theory of error-correcting codes was only about a decade old. The basic theory of reliable digital communication had been set forth by Claude Shannon in the late 1940s. At the same time, Richard Hamming introduced an elegant approach to single-error correction and double-error detection. Through the 1950s, a number of researchers began experimenting with a variety of error-correcting codes. But the Reed–Solomon paper, McEliece says, "hit the jackpot."

The payoff was a coding system based on groups of bits rather than individual 0s and 1s. That feature makes Reed–Solomon codes particularly good at dealing with "bursts" of errors. Current implementations of Reed–Solomon codes in CD technology are able to cope with error bursts as long as 4000 consecutive bits.

Reed, now a professor of electrical engineering at the University of Southern California, is still working on problems in coding theory. Solomon, recently retired from Hughes Aircraft Company, consults for the Jet Propulsion Laboratory. Reed was among the first to recognize the significance of "finite fields," which are important objects in modern abstract algebra, as the basis for error-correcting codes.

"In hindsight it seems obvious," he recently said. However, Reed added, "coding theory was not a subject when we published the paper." The two authors knew they had a nice result; they didn't know what impact the paper would have.

Three decades later, the impact is clear. The vast array of applications, both current and pending, has settled the question of the practicality and significance of Reed–Solomon codes. Billions of dollars in modern technology depend on ideas that stem from Reed and Solomon's original work.

SOURCE: Adapted with permission from *SIAM News,* January 1993, p. 1. © by SIAM. All rights reserved.

SP TLIGHT 10.2   Neil Sloane

▶ ▶   ▶ ▶ ▶ ▶ ▶ ▶ ▶ ▶ ▶ ▶ ▶ ▶

In the middle of Neil Sloane's office, which is in the center of AT&T Bell Laboratories, which in turn is at the heart of the Information Age, there sits a tidy little pyramid of shiny steel balls, each about an inch across, stacked up like oranges at a neighborhood grocery.

For Sloane the arrangement of balls in the pyramid is anything but trivial. He has been pondering different ways to pile up balls of one kind or another for most of his 51 years. Along the way he has become one of the world's leading researchers in the field of sphere packing, a once obscure mathematical recreation that has become indispensable to modern communications. Without it we might not have modems or compact discs or satellite photos of Neptune.

"Computers would still exist," says Sloane. "But they wouldn't be able to talk to one another." And the technology of the digital age depends on machine conversation.

To exchange information rapidly and correctly, machines must code it — that is, translate sounds, images, or any other information into strings of 0s and 1s. The more efficient the coding scheme, the more data you can store on a disk, or the faster you can send information through a communications channel. As it turns out, designing a code is a lot like packing spheres: both involve cramming things together into the tightest possible arrangement. Sloane, fittingly, is also one of the world's leading coding theorists, not least because he has studied the shiny steel balls on his desk so intently.

Here's how a code might work. Imagine, for example, that you want to transmit a child's drawing, one in which the ambitious young artist

Neil Sloane at work, wearing his famous "Code-mart" t-shirt (952 points in a sphere).

used every one of the 64 colors found in a jumbo box of Crayola crayons. For transmission, you could code each of those colors as a number — say, the integers from 1 to 64. Then you could divide the image into many small units, or pixels, and assign a code to each one based on the color it contains. The transmission would then be a steady stream of those numbers, one for each pixel.

In digital systems, however, all those numbers would have to be represented as strings of 0s and 1s. Because there are 64 possible combinations of 0s and 1s in a six-digit string, you could handle the entire Crayola palette with 64 different six-digit "code words." For example, 000000 could represent the first color, 000001 the next color, 000010 the next, and so on.

Limiting the signal to 0s and 1s has a distinct advantage in electronics: it's easy to tell the two digits apart. Physically, the two numbers are represented by variations in electric current. Each 1 in a code word might arrive at the receiver as a spike of increased current, and each 0 as a drop in current.

But if some noise pops up in transmission (as it often does, whether the signal is a digitized crayon drawing or a photo of Jupiter), your Crayola code might run into problems. In a noisy signal two different code words might look practically the same. A bit of noise, for example, might shift a spike of current to the wrong place, so that 001000 looks like 000100. The receiver might then color someone's eye green, even though the artist clearly intended it to be purple.

An elegant and efficient way to keep the colors straight in spite of noise is to add more information to the six-digit code words. Expanding them to ten digits would allow 1024 possible combinations of 0s and 1s. The receiver, programmed to know the 64 permissible combinations, could now spot any other combination as an error introduced by noise. It would automatically correct the error to the "nearest" permissible color. Such a scheme is called an *error-correcting code.*

In fact, says Sloane, "an error-correcting code for the sixty-four crayons would need ten digits. That would allow me to correct one error per word. If any of those ten digits were wrong — a one for a zero or a zero for one — you could still figure out what the right crayon was."

Why ten digits and not eight or nine? The answer, Sloane explains, can be found through sphere packing. Clearly, you want your code words to be as short as possible, with a minimum number of digits; each pulse of current requires energy and takes extra time to transmit. At the same time, you need code words long enough to put numerical "distance" between the permissible ones, so that the receiver can correct errors. By defining the minimum necessary distance between code words as the diameter of a hypothetical sphere, you can make this coding problem identical to the sphere-packing problem. With 64 possible crayons, Sloane says it's not until the code words are ten digits long that enough numerical "space" opens up between them to insert a complete error-correcting code.

Sloane solves the problem by envisioning the six-digit code words as six-dimensional addresses; then he raises the number of dimensions until he has the space he needs to insert his error-correcting code. Thinking of code numbers this way lets him use his spatial intuition. "What the code is doing is picking those points that are not too close together," he explains. The rigid diameter of his hypothetical spheres enforces a minimum separation between addresses. That satisfies the error-correcting requirement of transmission. To conserve power, he then tries to pack the imaginary spheres as tightly as possible around the center of some grid. When he's found the tightest packing, the address of each sphere's center gives him the most efficient list of code words for a given type of transmission.

SOURCE: Adapted from an article by David Berreby, *Discover,* October 1990.

error in any of the seven positions can be corrected. This is done as follows: choose

$$c_1 = 0 \text{ if } a_1 + a_2 + a_3 \text{ is even;}$$
$$c_1 = 1 \text{ if } a_1 + a_2 + a_3 \text{ is odd;}$$
$$c_2 = 0 \text{ if } a_1 + a_3 + a_4 \text{ is even;}$$
$$c_2 = 1 \text{ if } a_1 + a_3 + a_4 \text{ is odd;}$$
$$c_3 = 0 \text{ if } a_2 + a_3 + a_4 \text{ is even;}$$
$$c_3 = 1 \text{ if } a_2 + a_3 + a_4 \text{ is odd}$$

The sums $a_1 + a_2 + a_3$, $a_1 + a_3 + a_4$, and $a_2 + a_3 + a_4$ are called **parity-check sums**. They are so named because their function is to guarantee that the sum of various components of the encoded message is even. Indeed, c_1 is defined so that $a_1 + a_2 + a_3 + c_1$ is even. (Recall that this is precisely how the value in region V was defined.) Similarly, c_2 is defined so that $a_1 + a_2 + a_4 + c_2$ is even, and c_3 is defined so that $a_2 + a_3 + a_4 + c_3$ is even.

Let us revisit the message 1001 we considered in Figure 10.1. Then $a_1a_2a_3a_4 = 1001$ and

$$c_1 = 1 \text{ since } 1 + 0 + 0 \text{ is odd;}$$
$$c_2 = 0 \text{ since } 1 + 0 + 1 \text{ is even;}$$

and

$$c_3 = 1 \text{ since } 0 + 0 + 1 \text{ is odd}$$

So, because $c_1c_2c_3 = 101$, we have $1001 \rightarrow$ 1001101.

Now how is the intended message determined from a received encoded message? This process is called **decoding**. Say, for instance, that our message 1000, which has been encoded using parity-check sums as $u = 1000110$, is received as $v = 1010110$ (an error in the third position). We simply compare v with each of the sixteen code words (that is, the possible correct messages) in Table 10.2 and decode it as the one that differs from v in the fewest positions. (Put another way, we decode v as the code word that agrees with v in the most positions.) In the situation that there is more than one code word that differs from v in the fewest positions, we do not decode. To carry out this comparison it is convenient to define the **distance** between two strings of equal length as the number of positions in which the strings differ. For example, the distance between $v = 1010110$ and $u = 1000110$ is 1, since they differ in only one position (the third). In contrast, the distance between 1000110 and 0111001 is 7, since they differ in all 7 positions. Thus our decoding procedure is simply to decode any received message v as the code word v' that is "nearest" to v in the sense that among all distances between v and code words, the distance between v and v' is a minimum. (If there is more than one possibility for v', we do not decode.) Table 10.3 shows the distance between $v = 1010110$ and all sixteen code words. From this table we see that v will be decoded as u since it differs from u in only one position, while it differs

TABLE 10.3

v	1010110	1010110	1010110	1010110	1010110	1010110	1010110	1010110
code word	0000000	0001011	0010111	0100101	1000110	1100011	1010001	1001101
distance	4	5	2	5	1	4	3	4

v	1010110	1010110	1010110	1010110	1010110	1010110	1010110	1010110
code word	0110010	0101110	0011100	1110100	1101000	1011010	0111001	1111111
distance	3	4	3	2	5	2	6	3

from all others in the table in at least two positions. This method is called **nearest-neighbor** decoding. Assuming that errors occur independently, the nearest-neighbor method decodes each received message as the one it most likely represents.

The scheme we have just described was first proposed in 1948 by Richard Hamming, a mathematician at Bell Laboratories (see Spotlight 10.3). It is one of an infinite number of codes that are called the **Hamming codes.**

Strings of 0s and 1s obtained from all possible k-tuples of 0s and 1s by appending $n - k$ extra 0s and 1s using parity-check sums as illustrated earlier are called (n, k) **binary linear codes.** The strings are called **code words.**

You should think of an (n, k) binary linear code as a set of n-tuples where each n-tuple is composed of two parts: the message part, consisting of k digits, and the redundancy part, consisting of the remaining $n - k$ digits. Where there is no possibility of confusion, it is customary to denote an n-tuple $(a_1, a_2, \ldots, a_n)$ more concisely as $a_1 a_2 \cdots a_n$, as we did in Table 10.2.

Given a binary linear code, how can we tell if it will correct errors and how many errors it will detect? It is remarkably easy. We examine all the code words to find one that has the fewest number of 1s excluding the code word consisting entirely of 0s. Call this minimum number of such 1s in any nonzero code word the **weight** of the code and denote it by t. If t is odd, the code will correct up to $(t - 1)/2$ errors; if t is even, the code will correct up to $(t - 2)/2$ errors. If we prefer simply to detect errors rather than to correct them (as is often the case in applications), the code will detect up to $t - 1$ errors.

Applying this test to the code in Table 10.2, we see that the weight is 3, so it will correct any $(3 - 1)/2 = 1$ error or it will detect up to $3 - 1 = 2$ errors. Be careful here. We must decide in advance whether we want our code to *correct* single errors or *detect* double errors. It can do whichever we choose, but not both. To understand why

we can't do both, consider the received word 0001010 and the code in Table 10.2. The intended message could have been 0000000, in which case two errors were made (likewise for the intended messages 1011010 and 0101110), or the intended message could have been 0001011, in which case one error was made. But there is no way for us to know which of these possibilities occurred. If our choice were error correction, we would assume, perhaps mistakenly, that 0001011 was the intended message. If our choice were error detection, we would not decode.

Here is another example. Let the set of messages be {000, 001, 010, 100, 110, 101, 011, 111} and append three check digits c_1, c_2, and c_3 using

$$c_1 = 0 \text{ if } a_1 + a_2 + a_3 \text{ is even;}$$
$$c_1 = 1 \text{ if } a_1 + a_2 + a_3 \text{ is odd;}$$
$$c_2 = 0 \text{ if } a_1 + a_3 \text{ is even;}$$
$$c_2 = 1 \text{ if } a_1 + a_3 \text{ is odd;}$$
$$c_3 = 0 \text{ if } a_2 + a_3 \text{ is even;}$$
$$c_3 = 1 \text{ if } a_2 + a_3 \text{ is odd}$$

For example, if we take $a_1 a_2 a_3$ as 101 we have

$$c_1 = 0 \text{ since } 1 + 0 + 1 \text{ is even;}$$
$$c_2 = 0 \text{ since } 1 + 1 \text{ is even;}$$
$$c_3 = 1 \text{ since } 0 + 1 \text{ is odd}$$

Then, encode 101 by appending 001, i.e., 101→ 101001. The entire code is shown in Table 10.4.

Since the minimum number of 1s of any nonzero code word is 3, this code will either correct any single error or detect any double error, whichever we choose.

Data Compression

Codes such as an (n, k) binary code and the ASCII code are fixed-length codes. In a fixed-length code each code word is represented by the same number

SP⬤TLIGHT 10.3   Richard W. Hamming

▷ ▷ ▷ ▷ ▷ ▷ ▷ ▷ ▷ ▷ ▷ ▷ ▷ ▷

Richard W. Hamming was born in Chicago, Illinois, on February 11, 1915. He graduated from the University of Chicago with a B.S. degree in mathematics. In 1939, he received an M.A. degree in mathematics from the University of Nebraska, and in 1942, a Ph.D. in mathematics from the University of Illinois.

During the latter part of World War II, Hamming was at Los Alamos, where he was involved in computing atomic-bomb designs. In 1946, he joined Bell Telephone Laboratories, where he worked in mathematics, computing, engineering, and science.

When Hamming arrived at Bell Laboratories the Model V computer there had over 9000 relays and over fifty pieces of teletype apparatus, occupied about 1000 square feet of floor space, and weighed some 10 tons. (In computing power it equals some of today's hand-held calculators.) The input was entered in the machine via a punched paper tape, seven-eighths of an inch wide, which had up to six holes per row. Each row was read as a unit. If the sensing relays expected a two-out-of-six code (i.e., two holes per row), they would prevent further computation if more or less than two holes appeared in a given row. Similar checks were used in nearly every step of a computation. When such a check failed, two results were possible, depending upon whether the machine was set for "daytime" use with operating personnel present or for "nighttime" or "weekend" operation. In the first mode, a check failure stopped the machine and sounded an alarm. In the second, a check failure switched the machine immediately to other work. In the first case, the check failure was located by an elaborate check light panel.

Richard W. Hamming.

However, in the second case, the problem simply had to be rerun. This inefficiency, especially because of the latter case, led Hamming to investigate the possibility of automatic error correction.

After two weekends in a row in which his stuff had been dumped Hamming said, "Damn it, if the machine can detect an error, why can't it locate the position of the error and correct?"

In 1950, Hamming published his famous paper on error-detecting and error-correcting codes, resulting in the use of the Hamming codes in many modern computers and a new branch of information theory.

SOURCE: Adapted from T. Thompson, *From Error-correcting Codes through Sphere Packing to Simple Groups,* The Mathematical Association of America, 1983, and from J. Gallian, *Contemporary Abstract Algebra,* 3rd ed., D. C. Heath, 1994.

TABLE 10.4

Message		Code Word
000	→	000000
001	→	001111
010	→	010101
100	→	100110
110	→	110011
101	→	101001
011	→	011010
111	→	111100

A	. —	N	— .
B	— . . .	O	— — —
C	— . — .	P	. — — .
D	— . .	Q	— — . —
E	.	R	. — .
F	. . — .	S	. . .
G	— — .	T	—
H	. . . .	U	. . —
I	. .	V	. . . —
J	. — — —	W	. — —
K	— . —	X	— . . —
L	. — . .	Y	— . — —
M	— —	Z	— — . .

Figure 10.5 Morse code.

of digits (or symbols). In contrast, the Morse code, designed for the telegraph (see Spotlight 10.4), is a variable-length code; that is, a code in which the number of symbols for each code word can vary.

Notice that in the Morse code (Figure 10.5) the letters that occur most frequently have the shortest coding, whereas the letters that occur the least frequently have the longest coding. By assigning the code in this manner, telegrams could convey more information per line than would be the case for fixed-length codes or an arbitrarily assigned variable-length coding of the letters. The process of encoding data so that the most frequently occurring data are represented by the fewest symbols is called **data compression.** Figure 10.6 shows a typical frequency distribution for letters in English-language text material.

Let us illustrate the principles of data compression with a simple example. Biologists are able to describe genes by specifying sequences composed of the four letters A, T, G, and C, which represent the four nucleotides adenine, thymine, guanine, and cytosine, respectively. One way to encode a sequence such as AAACAGTAAC in fixed-length binary form would be to encode the letters as

$$A \rightarrow 00 \quad C \rightarrow 01 \quad T \rightarrow 10 \quad G \rightarrow 11$$

The corresponding binary code for the sequence AAACAGTAAC is then

$$0000000100111000001$$

	A	B	C	D	E	F	G	H	I	J	K	L	M
Percentage:	8	1.5	3	4	13	2	1.5	6	6.5	0.5	0.5	3.5	3

	N	O	P	Q	R	S	T	U	V	W	X	Y	Z
Percentage:	7	8	2	0.25	6.5	6	9	3	1	1.5	0.5	2	.25

Figure 10.6 A widely used frequency table for letters in normal English usage.

SPOTLIGHT 10.4   The Morse Code

▶ ▶ ▶ ▶ ▶ ▶ ▶ ▶ ▶ ▶ ▶ ▶ ▶

People have been transmitting messages by signs and sounds for centuries. More than 2000 years ago the Greeks used torch signals to represent the letters of the alphabet and regular torch stations were set up throughout Greece. Up to the invention of the telegraph, mirrors, flags, and mechanical devices with moving arms were used to transmit messages. There was once a 1200-mile system of stone towers from Saint Petersburg to the frontier of Russia for such communication purposes. The precursors of modern-day transmission are the telegraph and the Morse code.

In 1844 Samuel Morse sent the first public telegram: "What has God wrought!" Morse invented both the telegraph and a code consisting of dots, dashes, and spaces that were sent as electric pulses over a wire. The "Morse code" was also used in flash-lamp communication in marine navigation.

Samuel Morse (The Granger Collection, New York)

On the other hand, if we knew from experience that the hierarchy of occurrence of letters is A, C, T, and G (that is, A occurs most frequently, C second most frequently, and so on) and that A occurs much more frequently than T and G together, the most efficient binary encoding would be

$$A \rightarrow 0 \quad C \rightarrow 10 \quad T \rightarrow 110 \quad G \rightarrow 111$$

For this encoding scheme the sequence AAACAGTAAC is encoded as

$$0001001111100010$$

Notice that this binary sequence has 20% fewer digits than our previous sequence, in which each letter was assigned a fixed length of 2 (16 digits versus 20 digits). However, to realize this savings, we have made decoding more difficult. For the binary sequence using the fixed length of two symbols per character we decode the sequence by taking the digits two at a time in succession and converting them to the corresponding letters. For the compressed coding, we can decode by examining the digits in groups of three.

EXAMPLE: Decode 0001001111100010

Consider the compressed binary sequence 0001001111100010. Look at the first three digits: 000. Since our code words have 1, 2, or 3 digits and neither 00 nor 000 is a code word, the sequence 000 can only represent the *three* code words 0, 0, and 0. Now look at the next three digits: 100. Again, since neither 1 nor 100 is a code word, the sequence 100 represents the *two* code words 10 and 0. The next three digits, 111, can only represent the code word 111 since the other three code words all contain at least one 0. Next consider the sequence 110. Since neither 1 nor 11 is a code word, the sequence 110 can only represent 110 itself. Continuing in this fashion we can decode the entire sequence to obtain AAACAGTAAC.

The following observation can simplify the decoding process for compressed sequences. Note that 0 only occurs at the end of a code word. Thus each time you see a 0, it is the end of the code word. Also, because the code words 0, 10, and 110 end in a 0, the only circumstances under which there are three consecutive 1s is when the code word is 111. To quickly decode a compressed binary sequence using our coding scheme, insert a comma after every 0 and after every three consecutive 1s. The digits between the commas are the code words. ▲

EXAMPLE: Code AGAACTAATTGACA and Decode the Result

Recall: A → 0, C → 10, T → 110, and G → 111. So,

AGAACTAATTGACA →
0111001011000110110111010

To decode the encoded sequence we insert commas after every 0 and after every occurrence of 111 and convert to letters:

$$0,111,0,0,10,110,0,0,110,110,111,0,10,0$$
$$\text{A,G,A,A,C,T,A,A,T,T,G,A,C,A} \quad ▲$$

Modern data compression schemes were first invented in the 1950s (see Spotlight 10.5). They are now routinely used by modems and fax machines for data transmissions and by computers for data storage. In many cases data compression results in a savings of up to 50% on telephone charges or storage space.

Application to Computers

The 0s and 1s stored in computer memory are represented by the presence or absence of negative electric charges in the silicon chips making up the memory. These chips are subject to constant bombardment from alpha particles from nearly all materials. When an energetic alpha particle penetrates a site where a 1 is stored the 1 can change to a 0. In large computers such an occurrence can be expected every few days. To protect the integrity of the data in the memory, computers employ error-correcting codes. To illustrate the power of these codes let us consider a modest-size memory bank with the capacity of a million 8-digit binary words. Without an error-correcting scheme an error due to an alpha particle would occur on average about every 43 days. But with an error-correcting code, such as the one we have described in this chapter, an 8-digit binary word will be incorrect only when two or more alpha particles strike the same word. One can expect this to occur about once every 63 years!

▶ CRYPTOGRAPHY

Thus far we have discussed ways in which data can be encoded to detect errors or correct errors in transmission. In many situations there is also a

SP TLIGHT 10.5   Profile — David A. Huffman

▶ ▶ ▶ ▶ ▶ ▶ ▶ ▶ ▶ ▶ ▶ ▶ ▶ ▶

Large networks of IBM computers use it. So do high-definition television, modems, and a popular electronic device that takes the brain work out of programming a videocassette recorder. All these digital wonders rely on the results of a 40-year-old term paper by a modest Massachusetts Institute of Technology graduate student — a data compression scheme known as Huffman encoding.

In 1951 David A. Huffman and his classmates in an electrical engineering graduate course on information theory were given the choice of a term paper or a final exam. For the term paper, Huffman's professor, Robert M. Fano, had assigned what at first appeared to be a simple problem. Students were asked to find the most efficient method of representing numbers, letters, or other symbols using binary code. Besides being a nimble intellectual exercise, finding such a code would enable information to be transmitted over a computer network or stored in a computer's memory.

Huffman worked on the problem for months, developing a number of approaches, but none that he could prove to be the most efficient. Finally, he despaired of ever reaching a solution and decided to start studying for the final. Just as he was throwing his notes in the garbage, the solution came to him. "It was the most singular moment of my life," Huffman says. "There was the absolute lightning of sudden realization."

David A. Huffman (Matthew Mulbry)

That epiphany added Huffman to the legion of largely anonymous engineers whose innovative thinking forms the technical underpinnings for the accoutrements of modern living — in his case, from facsimile machines to modems and a myriad of other devices. "The Huffman code is one of the fundamental ideas that people in computer science and data communications are using all the time," says Donald E. Knuth of Stanford University, who is the author of the multivolume series *The Art of Computer Programming.*

Huffman says he might never have tried his hand at the problem — much less solved it at the age of 25 — if he had known that Fano, his professor, and Claude E. Shannon, the creator of information theory, had struggled with it. "It was my luck to be there at the right time and also not have my professor discourage me by telling me that other good people had struggled with the problem," he says.

Like many codes, including the one named after Samuel Morse, Huffman's creation tried to find a way to assign the shortest codes to those characters used most, the longest codes being reserved for those used rarely if at all.

By systematically employing codes of varying length, Huffman's idea may reduce by half or even more the number of code symbols that would be needed if the codes were of a fixed length.

When presented with his student's discovery, Huffman recalls, Fano exclaimed in his thick Italian accent: "Is that all there is to it!"

Products that use Huffman code might fill a consumer electronics store. A recent entry on the shop shelf is VCR Plus+, a device that automatically programs a VCR and is making its inventors wealthy.

Although others have used Huffman's code to help make millions of dollars, Huffman's main compensation was dispensation from the final exam. He never tried to patent an invention from his work and experiences only a twinge of regret at not having used his creation to make himself rich. "If I had the best of both worlds, I would have had recognition as a scientist, and I would have gotten monetary rewards," he says. "I guess I got one and not the other."

If Huffman were just starting his career, patent attorneys would surely be knocking on his door. Patenting of algorithms is still subject to endless judicial debate. But a lawyer today would tell Huffman to "cloathe" his code in silicon, that is, produce a patentable microchip that contains his code programmed into memory. "I bet I could write an application that would be considered patentable," says Richard H. Stern, a patent attorney who was chief of the intellectual property section of the U.S. Department of Justice from 1970 to 1978.

But Huffman has received other compensation. Textbooks on data communications and other digital arts include sections on Huffman code. Huffman has received several awards from the Institute of Electrical and Electronics Engineers. A few years ago an acquaintance told him that he had noticed that a reference to the code was spelled with a lowercase "H." Remarked his friend to Huffman, "David, I guess your name has finally entered the language."

SOURCE: Adapted from an article by Gary Stix, *Scientific American,* September 1991.

desire for security against unauthorized interpretation of coded data (that is, a desire for secrecy). The process of disguising data is called **encryption. Cryptology** is the study of methods to make and break secret codes. Access to computers is controlled by **passwords,** which are stored in encrypted form in the computers. Banking transactions, military transmissions, and intelligence information are encrypted. (One of the earliest methods for encrypting information is attributed to Julius Caesar, who used it to communicate with his soldiers. See Exercise 18.) Premium television services such as HBO, Showtime, and The Disney Channel also have a need to prevent their television signals to local cable operators and satellite dish subscribers from being received free by satellite dish owners. The method used by these services involves a monthly password, a subscriber sequence called a **key,** and the addition of binary sequences. We add two binary sequences $a_1 a_2 \cdots a_n$ and $b_1 b_2 \cdots b_n$ as follows:

$$
\begin{array}{c}
a_1 a_2 \cdots a_n \\
+\, b_1 b_2 \cdots b_n \\
\hline
c_1 c_2 \cdots c_n
\end{array}
$$

where $c_i = 0$ if $a_i = b_i$ and $c_i = 1$ if $a_i \neq b_i$. Equivalently, $c_1 = 0$ if $a_i + b_i$ is 0 or 2 and $c_i = 1$ if $a_i + b_i$ is 1. (Add a_i and b_i in the ordinary way, but replace 2 by 0.)

EXAMPLE: Sum of Binary Sequences

$$
\begin{array}{c}
11000111 \\
+\, 01110110 \\
\hline
10110001
\end{array}
\qquad
\begin{array}{c}
00111011 \\
+\, 01100101 \\
\hline
01011110
\end{array}
\qquad
\begin{array}{c}
10011100 \\
+\, 10011100 \\
\hline
00000000
\end{array}
$$

The data security method we describe hinges on the fact that the sum of two binary

sequences $a_1 a_2 \cdots a_n + b_1 b_2 \cdots b_n = 00 \cdots 0$ if and only if the sequences are identical.

Beginning in 1984 HBO scrambled its signal. To unscramble the signal, a cable system operator or dish owner who pays a monthly fee has to have a password that is changed monthly. The password is transmitted along with the scrambled signal. Although HBO uses binary sequences of length 56, we will illustrate the method with sequences of length 8. Let us say that the password for this month is p. Each subscriber of the service is assigned a sequence uniquely associated with him or her called a **key.** Let us say that the list of keys issued by HBO to its customers is $k_1, k_2, \ldots$. HBO transmits the password p and the sequences $k_1 + p, k_2 + p, \ldots$ (that is, one sequence for each authorized user). A microprocessor in each subscriber's decoding box adds its key, say k_i, to each of the sequences. That is, it calculates $k_i + (k_1 + p), k_i + (k_2 + p), \ldots$. As it does so, the microprocessor compares each of these calculated sequences with the correct password p. When one of the sequences matches p, the microprocessor will unscramble the signal. Notice that the correct password p will be produced precisely when k_i is added to $k_i + p$, since $k_i + (k_i + p) = (k_i + k_i) + p = 00 \cdots 0 + p = p$ and $k_i + (k_j + p) \neq p$ when $k_j \neq k_i$. (That is, key k_i "unlocks" the sequence $k_i + p$ and no other.) If a subscriber with key k_i fails to pay the monthly bill, HBO can terminate the service by not transmitting the sequence $k_i + p$ the next month. ▲

EXAMPLE: Encryption and Decoding

Let us say that the password for this month is $p = 10101100$ and your key is $k = 00111101$.

One of the sequences transmitted by HBO is $k + p$:

$$
\begin{array}{r}
00111101 \\
+\ 10101100 \\
\hline
10010001
\end{array}
$$

Your decoder box adds your key $k = 00111101$ to each of the sequences received. Eventually, it finds the sequence obtained by adding the password to your key (namely, $p + k = 10010001$) and calculates

$$
\begin{array}{r}
00111101 \\
+\ 10010001 \\
\hline
10101100
\end{array}
$$

to match the password p. Once the password has been matched, the decoder descrambles the signal.

One might suspect that a computer hacker could find the password by simply trying a large number of possible keys until one "unlocks" the password. But with sequences of length 56 there are 2^{56} possible keys, of which only a few million are used by HBO for its subscribers. The number 2^{56} is so large (it exceeds 72 quadrillion), however, that even if one tries a billion possible keys, the chance of finding one that works is essentially 0. ▲

OPTIONAL ▶ **PUBLIC KEY CRYPTOGRAPHY**

In the mid-1970s Ron Rivest, Adi Shamir, and Len Adelman devised an ingenious method that permits each person who is to receive a secret message to publicly tell how to scramble messages sent to him or her. And even though the method used to scramble the message is known publicly, only the person for whom it is intended will be able to unscramble the message.

Before describing the method, it is convenient to introduce a new notation. For any positive integers a and n we write $a \bmod n$ (read: "a modulo n" or just "$a \bmod n$") to be the remainder when a is divided by n. Thus,

$$
\begin{aligned}
3 \bmod 2 &= 1 \text{ since } 3 = 1 \cdot 2 + 1; \\
6 \bmod 2 &= 0 \text{ since } 6 = 2 \cdot 3 + 0; \\
4 \bmod 3 &= 1 \text{ since } 4 = 1 \cdot 3 + 1; \\
15 \bmod 3 &= 0 \text{ since } 15 = 5 \cdot 3 + 0; \\
12 \bmod 10 &= 2 \text{ since } 12 = 1 \cdot 10 + 2; \\
37 \bmod 10 &= 7 \text{ since } 37 = 3 \cdot 10 + 7; \\
98 \bmod 85 &= 13 \text{ since } 98 = 1 \cdot 85 + 13; \\
342 \bmod 85 &= 2 \text{ since } 342 = 4 \cdot 85 + 2; \\
62 \bmod 85 &= 62 \text{ since } 62 = 0 \cdot 85 + 62
\end{aligned}
$$

Calculations involving mod n are called **modular arithmetic.** One rule of modular arithmetic we need is

$$(ab) \bmod n = ((a \bmod n)(b \bmod n)) \bmod n$$

EXAMPLE: Modular Arithmetic

$$
\begin{aligned}
(17 \cdot 23) \bmod 10 &= ((17 \bmod 10) \cdot \\
&\quad (23 \bmod 10)) \bmod 10 \\
&= (7 \cdot 3) \bmod 10 \\
&= 21 \bmod 10 = 1 \\
(22 \cdot 19) \bmod 8 &= ((22 \bmod 8) \cdot \\
&\quad (19 \bmod 8)) \bmod 8 \\
&= (6 \cdot 3) \bmod 8 = 2
\end{aligned}
$$

We now describe the Rivest, Shamir, and Adelman method. Recall that a **prime number** is an integer greater than 1 whose only divisors are 1 and itself. The idea is based on the fact that there exist efficient methods for finding very large prime numbers (say, about 100 digits long) and for multiplying large numbers, but no one knows an efficient algorithm for factoring large integers (say, about 200 digits long). So

SP^OTLIGHT 10.6   Claude E. Shannon

Probably no single work in this century has more profoundly altered man's understanding of communication than C. E. Shannon's "A mathematical theory of communication" first published in 1948. There resulted theorems of great power, elegance, generality and beauty. They have shed much understanding on the elusive true nature of the communication process and have delineated its interest limitations.

David Slepian, *Key Papers in the Development of Information Theory*, IEEE Press, New York, 1974

Claude E. Shannon was born on April 30, 1916, in Petoskey, Michigan, and grew up in nearby Gaylord. He received a B.S. degree in electrical engineering from the University of

Claude E. Shannon (The MIT Museum)

the person who is to receive the message finds a pair of large primes p and q and chooses an integer r so that r and the least common multiple of $p - 1$ and $q - 1$ have no common divisors other than 1. This person calculates $n = pq$ and publicly announces that a message M is to be sent to him or her as M^r mod n. Although r, n, and M^r are available to everyone, only the person who knows how to factor n as pq will be able to decipher the message.

To present a simple example that nevertheless illustrates the essential features of the method, say we wish to send the message

"IBM." We convert the message to digits by replacing A by 1, B by 2, . . . , and Z by 26. So the message IBM becomes 9213. The person to whom the message is to be sent has picked two primes p and q, say, $p = 5$ and $q = 17$ (in actual practice p and q would have a hundred or so digits), and a number r that has no divisors in common with the least common multiple m of $p - 1 = 4$ and $q - 1 = 16$ other than 1, say, $r = 3$, and published $n = pq = 85$ and r in a public directory. The receiver also must find a number s so that $r \cdot s = 1$ mod m (this is where the knowledge of p and q is necessary). That is,

Michigan in 1936. In 1940, he received an M.S. degree in electrical engineering and a Ph.D. degree in mathematics, simultaneously, from the Massachusetts Institute of Technology. In his master's thesis he discovered a mathematical way for analyzing electrical switching circuits. After spending 1941 at the Institute for Advanced Study at Princeton, Shannon went to Bell Laboratories as a research mathematician. There he made major contributions to coding theory, cryptography, computing circuit design, and information theory, a field he founded.

Shannon's theory defined information. It explained crucial relationships among the elements of a communications system — signal power, bandwidth (the frequency range of an information channel), and noise (static).

In 1957, Shannon became a faculty member at MIT with the dual title of Professor of Communication Science, in the Department of Electrical Engineering, and Professor of Mathematics. In 1958, he was named to the Donner Chair of Science. Shannon served as a consultant to Bell Laboratories until 1972.

In 1966, Shannon received the National Medal of Science from President Lyndon B. Johnson. The American Institute of Electrical Engineers awarded him its Alfred Nobel Prize. He has received the Medal of Honor from the Institute of Electrical and Electronics Engineers, the Research Corporation Award, the Morris Liebmann Memorial Prize from the Institute of Radio Engineers, the Stuart Ballantine Medal from the Franklin Institute, and the Medal of Honor from Rice University.

Shannon has received honorary degrees from Yale University, University of Michigan, Princeton University, University of Pittsburgh, Edinburgh University in Scotland, and Northwestern University. He is a member of the National Academy of Sciences, the National Academy of Engineering, and has written more than fifty technical articles and coauthored two books. His hobbies include chess, playing clarinet, listening to music, and building electronic gadgets.

SOURCE: From J. Gallian, *Contemporary Abstract Algebra,* 3rd ed., D. C. Heath, 1994.

$3 \cdot s = 1 \bmod 16$. This number is 11. (There is a simple algorithm for finding the number s.) We consult this directory to find n and r, then send the "scrambled" numbers $9^3 \bmod 85$, $2^3 \bmod 85$, and $13^3 \bmod 85$ rather than 9, 2, and 13, and the receiver will unscramble them. Thus we send

$$9^3 \bmod 85 = 49$$
$$2^3 \bmod 85 = 8$$

and

$$13^3 \bmod 85 = 72$$

Now the receiver must take the numbers he or she receives, 49, 8, and 72, and convert them back to 9, 2, and 13 by calculating $49^{11} \bmod 85$, $8^{11} \bmod 85$, and $72^{11} \bmod 85$.

The calculation of $49^{11} \bmod 85$ can be simplified as follows:[1]

$$49 \bmod 85 = 49;$$
$$49^2 \bmod 85 = 2401 \bmod 85 = 21;$$

---

[1] To determine $49^2 \bmod 85$ with a calculator, enter 49×49 to obtain 2401, then divide 2401 by 85 to obtain 28.247058. Finally, enter $2401 - 28 \times 85$ to obtain 21.

$49^4 \bmod 85 = 49^2 \cdot 49^2 \bmod 85$

$\qquad = 21 \cdot 21 \bmod 85 = 441 \bmod 85$

$\qquad = 16 \bmod 85;$

$49^8 \bmod 85 = 49^4 \cdot 49^4 \bmod 85$

$\qquad = 16 \cdot 16 \bmod 85 = 1$

So $49^{11} \bmod 85 = (49^8 \bmod 85) \cdot$
$(49^2 \bmod 85)(49 \bmod 85)$

$$= (1 \cdot 21 \cdot 49) \bmod 85$$
$$= 1029 \bmod 85$$
$$= 9 \bmod 85$$

Thus, the receiver has correctly determined the code for "I." The calculations for $8^{11} \bmod 85$ and $72^{11} \bmod 85$ are left as exercises for the reader. Notice that without knowing how pq factors, one cannot find the least common multiple of $p-1$ and $q-1$ (in our case 16), and therefore the s that is needed to determine the intended message.

The procedure just described is called the **RSA public key encryption scheme** in honor of Rivest, Shamir, and Adelman, who discovered the method. The algorithm is summarized below. (In practice, the messages are not sent one letter at a time. Rather, the entire message is converted to decimal form with A represented by 01, B by 02, . . . , and a space by 00. The message is then broken up into blocks of uniform size and the blocks are sent. See step 2 under Sender below.)

Receiver

1. Pick very large primes p and q and compute $n = pq$.

2. Compute the least common multiple of $p-1$ and $q-1$; let us call it m.

3. Pick r so that it has no divisors in common with m other than 1 (any such r will do).

4. Find s so that $rs = 1$ modulo m (there is always exactly one such s between 1 and m).

5. Disregard p and q.

6. Publicly announce n and r, but keep s secret.

Sender

1. Convert the message to a string of digits. (In practice, the ASCII decimal code is used.)

2. Break up the message into uniform-size blocks of digits appending 0s in the last block if necessary; call them $M_1, M_2, \ldots, M_k$. For example, for a string such as 2105092315 we could use $M_1 = 2105$, $M_2 = 0923$, and $M_3 = 1500$.

3. Check to see that the greatest common divisor of each M_i and n is 1. If not, n can be factored and the code is broken. (In practice the primes p and q are so large that they exceed all M_i so this step may be omitted.)

4. Calculate and send $R_i = M_i^r \bmod n$.

Receiver

1. For each received message R_i, calculate $R_i^s \bmod n$.

2. Convert the string of digits back to a string of characters.

Why does this method work? Because of a basic property of modular arithmetic and the choice of r. It so happens that the number m has the property that for each x having no common divisors with n except 1, we have $x^m = 1 \bmod n$. So, because each message M_i has no common divisors with n except 1, and r was chosen so that $rs = 1 + mt$ for some t, we have, modulo n,

$$R_i^s = (M_i^r)^s = M_i^{rs} = M_i^{1+mt}$$
$$= M_i(M_i^m)^t = M_i 1^t = M_i \quad \blacktriangle$$

◄

▶Review Vocabulary

American Standard Code for Information Interchange (ASCII)   A decimal and binary coding scheme for representing character data.

Binary linear code   A code consisting of words composed of 0s and 1s obtained by using parity-check sums to append check digits.

Code words   Words from a binary linear code.

Cryptology   The study of how to make and break secret codes.

Data compression   The process of encoding data so that the most frequently occurring data are represented by the fewest symbols.

Decoding   The process of translating encoded data into the original data.

Distance between two messages   The number of positions in which two messages differ.

Encryption   The process of encoding data to protect against unauthorized interpretation.

Even parity   Even integers are said to have even parity.

Hamming codes   An infinite family of error-correcting codes that detect all single errors.

Key   A string used to decode data.

Modular arithmetic   Addition and multiplication involving modulo n.

Nearest-neighbor decoding   A method that decodes a message as the code word that agrees with the message in the most positions.

Odd parity   Odd integers are said to have odd parity.

Parity-check sums   Sums of digits whose parities determine the check digits.

Password   A word used to encode data.

Prime number   An integer greater than 1 whose only divisors are 1 and itself.

RSA public key encryption   A method of encoding that permits each person to announce publicly the means by which secret messages are to be sent to him or her.

Venn diagram method   A method of encoding and decoding binary strings using diagrams.

Weight   The minimum number of 1s that occur among all nonzero code words of a code.

▶Suggested Readings

DENEEN, L.: "Secret Encryption with Public Keys," *The UMAP Journal*, 8:9–29 (1987). This article describes several ways in which modular arithmetic can be used to code secret messages. Methods range from a simple scheme used by Caesar to the RSA public key encryption method.

HALDANE, R. A.: *The Hidden World*, St. Martin's, New York, 1971. Introduction to cryptography.

KAHN, DAVID: *Codebreakers: The Story of Secret Writing*, Macmillan, New York, 1967. A monumental, illustrated history.

KORHEIM, ALAN G.: *Cryptography: A Primer*, Wiley, New York, 1981. An introduction to cryptography.

MCELIECE, R.: "The Reliability of Computer Memories," *Scientific American*, 252(1):88–95 (1985). This article discusses error-correcting codes in computer memories.

PETERSON, W.: "Error-Correcting Codes," *Scientific American*, 206(2):96–108 (1962). This article gives a lucid discussion of the Hamming (15, 11) binary code.

RICHARDS, I.: "The Invisible Prime Factor," *American Scientist*, 70:176–179 (1982). This article explains how elementary number theory and modular arithmetic can be used to test whether an integer is prime, and how prime numbers can be used to create secret codes.

THOMPSON, T.: *From Error-Correcting Codes through Sphere Packing to Simple Groups*, Mathematical Association of America, Washington, D.C., 1983. Chapter 1 of this award-winning book gives a fascinating historical account of the origins of error-correcting codes.

▶EXERCISES

1. Use the Venn diagram method to verify the code word in Table 10.2 for the message 0101.

2. Use the Venn diagram method to decode the received messages 0111011 and 0100110.

3. Find the distance between each of the following pairs of words:

    a. 11011011 and 10100110
    b. 01110100 and 11101100

4. Referring to Table 10.2, use the nearest-neighbor method to decode the received words 0000110 and 1110100.

5. If the code word 0110010 is received as 1001101, how is it decoded using the Venn diagram method?

6. Suppose a received word has the Venn diagram arrangement shown below.

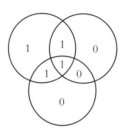

What can we conclude about the received word?

7. Determine the (6, 3) binary linear code whose parity-check sums are $a_2 + a_3$, $a_1 + a_3$, and $a_1 + a_2$. (That is, $c_1 = 0$ if $a_2 + a_3$ is even; $c_1 = 1$ if $a_2 + a_3$ is odd; and similarly for c_2 and c_3.)

8. Let C be the code

{0000000,1110100,0111010,0011101,1001110,0100111,1010011,1101001}

What is the error-correcting capability of C? What is the error-detecting capability of C?

9. Find all code words of the (7, 4) binary linear code whose parity-check sums are $a_2 + a_3 + a_4$, $a_2 + a_4$, and $a_1 + a_2 + a_3$. Will this code correct any single error?

10. Consider the binary linear code

$$C = \{00000,10011,01010,11001,00101,10110,01111,11100\}$$

Use nearest-neighbor decoding to decode 11101 and 01100. If the received word 11101 has exactly one error, can you determine the intended code word? Explain your reasoning.

11. Construct a (6, 3) binary linear code with parity-check sums $a_1 + a_2$, $a_2 + a_3$, and $a_1 + a_3$. Decode each of the received words

$$001001,011000,000110,100001$$

by the nearest-neighbor method.

12. Suppose the weight of a binary linear code is 6. How many errors can the code correct? How many errors can the code detect?

13. How many code words are there in an (8, 5) binary linear code?

14. Add the following pairs of binary sequences:

   a. 10111011 and 01111011
   b. 11101000 and 01110001

15. All binary linear codes have the property that the sum of two code words is another code word. Use this fact to determine which of the following sets cannot be a binary linear code.

   a. {0000,0011,0111,0110,1001,1010,1100,1111}
   b. {0000,0010,0111,0001,1000,1010,1101,1111}
   c. {0000,0110,1011,1101}

16. Let $v = a_1 a_2 \cdots a_n$ and $u = b_1 b_2 \cdots b_n$ be binary sequences. Explain why the number of 1s in $v + u$ is the same as the distance between u and v.

17. Extend the code words listed in Table 10.2 to eight digits by appending a 0 to words of even weight and a 1 to words of odd weight. What is the error-detecting and error-correcting capability of the new code?

18. The *Caesar cipher* encrypts messages by replacing each letter of the alphabet with the letter shown beneath it

$$\text{ABCDEFGHI J KLMNOPQRS TUVWXYZ}$$
$$\text{DEFGHI J KLMNO P QRSTUVWXY Z ABC}$$

Use the Caesar cipher to encrypt the message RETREAT. Determine the intended message corresponding to the encrypted message DWWDFN.

● 19. Use the RSA scheme with $p = 5, q = 17$, and $r = 3$ to determine the numbers sent for the message VIP.

● Optional exercise.

● 20. Use the RSA scheme with $p = 5$, $q = 17$, and $r = 3$ to decode the received numbers 52 and 72.

● 21. In the RSA scheme with $p = 5$, $q = 17$, and $r = 5$, determine the value of s.

● 22. Try using the RSA scheme with $p = 5$, $q = 17$, and $r = 4$. Explain why the scheme does not work in this situation.

● 23. Assume the letters of the alphabet have been converted to integers according to the correspondence $A \rightarrow 0$, $B \rightarrow 1$, $C \rightarrow 2$, . . . , $Z \rightarrow 25$. Write a formula that describes the result of applying the Caesar cipher (see Exercise 18) to the integer x. (Hint: Use modular arithmetic.)

▲ 24. Explain why no (6, 3) binary linear code can correct all possible double errors.

▲ 25. A (4, 2) *ternary* code is formed by starting with all possible pairs of 0s, 1s, and 2s and appending two extra digits that are also 0s, 1s, or 2s. Form a ternary code by appending to each message $a_1 a_2$ the check digits $c_1 c_2$ using:

$$c_1 = 0 \text{ if } a_1 + a_2 \text{ is 0 or 3; } c_1 = 1 \text{ if } a_1 + a_2 \text{ is 1 or 4;}$$
$$c_1 = 2 \text{ if } a_1 + a_2 \text{ is 2;}$$
$$c_2 = 0 \text{ if } 2a_1 + a_2 \text{ is 0, 3, or 6; } c_2 = 1 \text{ if } 2a_1 + a_2 \text{ is 1 or 4;}$$
$$c_2 = 2 \text{ if } 2a_1 + a_2 \text{ is 2 or 5.}$$

▲ 26. Use the ternary code in Exercise 25 and the nearest-neighbor method to decode the received word 1211.

▲ 27. A (6,4) ternary code is formed by starting with all possible 4-tuples of 0s, 1s, and 2s and appending two extra digits that are also 0s, 1s, and 2s. How many code words are there in a (6, 4) ternary code? How many possible received words are there in a (6, 4) ternary code?

■ 28. Discuss the relative merits of a (7, 3) binary linear code with weight 3 and a (4, 2) ternary linear code with weight 3.

▲ 29. Suppose we code a five-symbol set $\{A, B, C, D, E\}$ into binary form as follows:

$$A \rightarrow 0, \quad B \rightarrow 10, \quad C \rightarrow 110, \quad D \rightarrow 1110, \quad E \rightarrow 1111.$$

Convert the sequence *AEAADBAABCB* into binary code. Determine the sequence of symbols represented by the binary code 01000110100011111110.

▲ 30. Devise a variable-length binary coding scheme for a six-symbol set $\{A, B, C, D, E, F\}$. Assume that A is the most frequently occurring symbol, B is the second most frequently occurring symbol, and so on.

● Optional exercise.   ▲ Advanced exercise.   ■ Discussion exercise.

31. Judging from the Morse code, what are the three most frequently occurring consonants in English text material? What is the most frequently occurring vowel?

32. In English, the letter H occurs more often than D, G, K, and W, but in Morse code H has a longer code than D, G, K, and W. Speculate on the reason for this apparent violation of data compression principles.

33. Explain why the Morse code must include a space after each letter but fixed-length codes do not.

● 34. For each part below explain how modular arithmetic can be used to answer the question.
   a. If today is Wednesday, what day of the week will it be in 16 days?
   b. If a clock (with hands) indicates that it is now 4 o'clock, what will it indicate in 37 hours?
   c. If a military person says it is now 0400, what time would it be in 37 hours? (Instead of A.M. and P.M., military people use 1300 for 1 P.M., 1400 for 2 P.M., and so on.)
   d. If it is now July 20, what day will it be in 65 days?
   e. If the odometer of an automobile reads 97,000 now, what will it read in 12,000 miles?

▶Writing Projects

1. Prepare a report on cryptography. Discuss at least three methods of encryption. Discuss the interface between computers and cryptography.

2. Prepare a report on applications of modular arithmetic. Explain the calculation of the check digits described in Exercises 7, 9, and 25 with modular arithmetic. Use modular arithmetic to describe the error-detection schemes used in Chapter 9.

3. Prepare a report on the early history of error-correcting codes. The suggested reading by Thompson has the information you will need.

Part IV

Social Choice and Decision Making

A revolution currently taking place in the field of mathematics is the successful use of mathematics as a fundamental tool to study human beings — their behavior, values, interactions, conflicts, organizations, fair allocations, and decision making, as well as their interface with modern technology and complex organizations. This latter revolution could eventually prove to be as far reaching as the turning of mathematics to study physical objects and their motion some three centuries ago. As mathematics and computers play an increasingly important role in understanding our social institutions, a new profession is emerging devoted to thinking mathematically about human affairs.

In particular, human decision making is being influenced profoundly by modern mathematics, and several particular mathematical subjects have been created primarily to assist in arriving at good decisions. Many aspects involved in arriving at a decision are, of course, nonquantitative in nature. These may relate to history, past experience, instinct, judgment, morality, and so forth. As a consequence one often refers to decision making as an art rather than as a science. On the other hand, many ingredients in contemporary decision making are mathematical in nature, and one can also view this activity as a scientific subject.

A decision maker will begin by listing the options over which he or she has some control and the likely outcomes

People registering to vote at political rally. (Tony Stone Worldwide, Ltd/Robert E. Daemmrich.)

resulting from these choices. The person may attempt to identify all relevant variables and the relationships among them, and may associate quantitative measures when possible. Moreover, the decision maker must clarify his or her own values, identify desired goals, and spell out explicitly any limiting resources or social constraints. One then seeks the best possible result obtainable.

The situation is typically confounded by a variety of different uncertainties involving data and forecasting that typically cannot be completely resolved in advance. The effect that other decision makers may have on the outcome, and the best contingent responses to their moves, should be predetermined. Various ethical concerns such as fairness may well need consideration, and ways to ascertain group opinions may be necessary. Finally, decision makers must study the social and political context in which the decision will be implemented.

Several different mathematical subjects have been introduced since World War II, many for the purpose of assisting individuals or groups in arriving at good or equitable decisions. As an illustration, a dozen major aspects of making decisions with the corresponding mathematical specialties are listed in the table at the bottom of this page. All

of these fields, except for continuous optimization, statistics, and probability theory, have developed mostly in recent decades and in the context of mathematics applied to human actions and organization. Topics 8, 9, and 10 in the table are the subject matter of Part IV, whereas topics 11 and 5 were considered in Parts I and II, respectively.

In this part of the text we will illustrate a few of the mathematical techniques available to assist decision makers. In Chapter 11 we discuss the important problem of social choice. How does a group of individuals, each with his or her own set of values, select one outcome from a list of possibilities? This problem arises frequently in any democratic society, and even in more authoritarian institutions where decisions are made by more than one person. We will learn that all voting systems have inherent flaws and that agendas designed for ascertaining the collective group will are often subject to manipulation. Group decision making is inherently a strategic encounter, and every citizen should be aware of the difficulties and pitfalls that can arise in this arena.

In Chapter 12 we consider decision-making bodies in which the individual voters or parties do not have equal power. In particular we will look at weighted voting systems such as stockholders in a

A decision maker's concerns	Related mathematical subjects
1. Identify and measure strategic variables	Theory of measurement
2. Understand a complex system	System analysis, graph theory
3. Quantify one's preferences	Utility theory
4. Formulate objectives and constraints	Mathematical programming
5. Acquire data and forecast results	Statistics
6. Determine the most efficient outcomes	Optimization theories
7. Deal with uncertainty	Probability theory
8. Resolve conflicts	Game theory
9. Group decision mechanisms	Social choice theory
10. Equity considerations	Fair division theory
11. Make decisions using a multidisciplinary approach	Operations research, management science
12. Make decisions in an institutional setting	Policy science

corporation or political parties in a national assembly in which the voters cast different numbers of votes. We first observe that power in such systems is not necessarily proportional to the voters' weights. The notion of power is of fundamental importance in political science, although it is typically difficult to quantify. We will nevertheless describe two popular indices for measuring power for weighted voting systems. It allows one to assign weighted votes to the legislators from constituencies of varying sizes so that the individuals in the districts are represented in an equitable manner.

A general theme throughout this part concerns the idea of fairness in decision making. In Chapter 13 this becomes most explicit. Here we describe some fair division schemes in which a group of individuals with different values can be assured of each receiving what he or she views as a fair share when dividing up objects like cakes or goods such as estates.

In Chapter 14 we discuss the apportionment problem, which occurs when we are required to approximate fractional quantities with integers. Apportionment problems often arise when re-sources must be allocated in an equitable manner, for instance, when college administrators must decide the number of faculty positions to be allocated to each college and department or management must decide how to distribute salary increases. Perhaps the most controversial apportionment problem, however, involves political representation, most notably the equitable allocation of seats in the U.S. House of Representatives to the states.

Chapter 15 introduces the mathematical field called game theory, which describes situations involving two or more decision makers seeking different goals. Game theory provides a collection of models to assist in the analysis of conflict and co-operation. It prescribes optimal strategies for games of total conflict in which one's gain is another's loss. It also provides insights into purely cooperative situations as well as encounters of partial conflict that involve aspects of both competition and cooperation. Some particular games such as those known as "prisoners' dilemma" and "chicken" provide us with insights into certain social paradoxes that we routinely meet in our daily lives.

Social Choice: The Impossible Dream

The basic question of *social choice*, of how groups can best arrive at decisions, has occupied social philosophers and political scientists for centuries. Social-choice theory arose to help explain voting and other decision-making processes. Voting is a subject that lies at the very heart of representative government and participatory democracy. In theory and practice, voting poses difficult problems.

The fundamental problem is to turn individual preferences for different outcomes into a single choice by the group as a whole. This situation arises whenever government representatives pass a bill, stockholders decide on a course of action, a political party nominates a presidential candidate, or a community elects members to serve on the school board. Few people realize that the voting method they use can significantly affect the outcome of an election, and that the voting mechanism they use is typically subject to manipulation.

The first type of voting that often comes to mind is that of majority rule. In the case of **majority rule** each voter votes for one candidate, and the candidate receiving over half of the total votes is declared the winner. However, this method is truly effective solely in elections in which only

two candidates are competing for a single office. The process is then quite simple: the candidate with the larger share of votes wins. But where there are more than two candidates, it is possible that the person with the highest tally will not actually hold a majority of all votes cast. In this chapter we will be concerned entirely with decision making in cases where there are three or more possible outcomes.

Most democracies have developed a variety of voting procedures to single out particular options from a longer list of feasible alternatives. How do we decide which of these voting schemes should be used?

All voting methods have some inherent faults, and any method of voting we use will occasionally give rise to rather paradoxical results. So no single voting method is either universally applicable or the best overall: for each situation we have to select one of the many available voting schemes, knowing that the method we choose may greatly influence the result of the election.

The large number of different voting procedures from among which we can choose include majority rule, plurality wins, elimination and runoffs, sequential pairwise comparisons, various weighted or scoring schemes, approval voting, and a host of various other partitioning schemes that choose successively between subsets of potential outcomes. All these methods can at times exhibit shortcomings (see Spotlight 11.1, pp. 336–337).

The outcome of an election can be affected by the type of voting method employed or by the order or formatting of the questions within a given method. The agenda itself can be a significant factor in determining which motion survives, and agendas are usually subject to manipulation. By an **agenda** we mean an ordering or list of alternatives for consideration. For example, insincere amendments can be introduced for diversionary purposes or to kill off popular motions at an early stage. Voters may sometimes benefit by falsifying their preferences and misleading others. The final outcome of an election can be altered by many such strategic-voting maneuvers.

As we focus on decision-making procedures in this chapter, we will see that it is no easy matter to ascertain the "true will of the people." The problem of unifying individual preferences into one particular choice for the whole group is essentially unsolvable. We now examine a few actual cases to illustrate these difficulties in voting.

▶ THE ROMAN SENATE

Almost 2000 years ago, the Roman historian Pliny the Younger was grappling with a voting dilemma:

A debate arose in the Senate concerning the freedmen of the consul Afranius Dexter; it being uncertain whether he killed himself, or whether he died by the hands of his freedmen; and again, whether they killed him from a spirit of malice, or of obedience.

It appears that there were questionable circumstances surrounding the death of Consul Dexter. If his freedmen (former slaves and servants) did in fact execute him, it is not clear whether it was an act of "mercy killing," according to Dexter's own wishes, or murder. His death may also have been the result of suicide. The three possible circumstances are shown in Figure 11.1, with appropriate verdicts for the freedmen indicated in parentheses. Pliny continues in his letter:

One of the senators (it is of little purpose to tell you I was the person) declared that he thought these freedmen ought to be put to the question, and afterwards released. The sentiments of another were, that the freedmen should be banished, and of another, that they should suffer death. It was impossible to reconcile such a diversity of opinions.

Pliny describes three groups in the Roman Senate:

▶ Group *A* believed the freedmen were innocent and thus favored their *acquittal*.
▶ Group *B* considered them guilty to some extent and thought the appropriate punishment was *banishment*.
▶ Group *C* believed the freedmen guilty of the crime of murder and felt that they should be *condemned* to death.

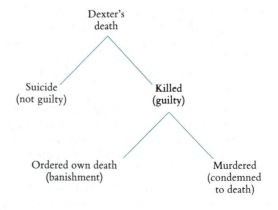

Figure 11.1 How was Dexter killed?

We can use A, B, and C to represent these three groups, and a, b, and c to denote the three actions: acquittal, banishment, and condemnation. To be more specific, let's suppose that these three groups consist of 40%, 35%, and 25% of the senators, respectively. We can arrange this information in a table:

	Group		
	A	B	C
Preferred action	a	b	c
Percent	40	35	25

The outcome of this trial could depend as much on the voting procedure the senators use as on the kind of information they have, as well as the strategies or arguments they employ.

METHOD 1.   In the **plurality** method of voting the candidate with the most votes is declared the winner. The winner need not have a majority of the votes cast.

If the prisoners' fate were determined by a plurality, then at first glance it looks as if they would go free: in a plurality vote, the position with the most votes is declared the winner. Our table shows that outcome a is favored by the largest number of senators, 40%, including Pliny himself.

However, the senators may choose to vote differently. What if they know the distribution of potential votes in advance? Suppose they know that the numbers 40%, 35%, and 25% would be in favor of a, b, and c, respectively. Perhaps those in group C would then compromise on their hard-line position (the death penalty) and vote instead for banishment. The result would be 40% for a and $35\% + 25\% = 60\%$ for b. Banishment would then carry the day.

It is not unreasonable for group C to vote in this way, that is, in favor of outcome b rather than their first choice, c. Such **strategic,** or **insincere, voting** is not at all uncommon.

How does strategic voting occur in this case? Suppose that we have additional information about the groups A, B, and C. Assume, for example, that we know their second and third choices as well. It seems reasonable that group A would prefer a over b and b over c and that group C would prefer the reverse order: c over b and b over a. It may well be that those in group B, who favor banishment b, would be divided in their second choice. Some may prefer acquittal a to the harsh penalty c, whereas others who are against the death penalty may nevertheless feel that the guilty must be punished in some manner, even if it means execution. For the sake of simplicity, however, let's assume that *all* members of group B favor b over a and a over c. This **preference schedule** is summarized as follows:

	Group		
	A	B	C
First choice	a	b	c
Second choice	b	a	b
Third choice	c	c	a
Percent	40	35	25

It is fairly clear that the senators in group A have no real choice other than to vote for acquittal. So we'll assume that A votes for a. Group B is free to choose either a, b, or c, as is group C. We can represent these strategies and the resulting outcomes as follows:

		C votes		
A votes	a	a	b	c
B votes	a b c	a a a	a b a	a a c

SP TLIGHT 11.1   Unattainable Ideals

▶ ▶ ▶ ▶ ▶ ▶ ▶ ▶ ▶ ▶ ▶ ▶ ▶ ▶

A popular view of mathematics holds it to be a collection of factual statements (theorems, propositions) that are deduced as true from more basic premises (axioms, hypotheses), plus a variety of more mechanical techniques (algorithms) used to compute actual solutions to a particular problem. In contrast to this notion, however, mathematicians also devote significant efforts to discovering illustrations (counterexamples) that show that some conjectured facts are not true. Moreover, some of the most important mathematical discoveries demonstrate that a certain presumed situation does not exist or that some highly desirable outcome cannot always be attained.

The Pythagoreans of ancient Greece knew that the number $\sqrt{2}$ was *irrational,* in the sense that it could not be written as a ratio p/q for any two integers p and q. Such insights often have the effect of terminating the search for some ideal state or for certainty within a subject. They may place theoretical limits on what can be achieved by means of the scientific method. Four outstanding cases from the twentieth century follow.

In 1927, the German physicist Werner K. Heisenberg (1901 – 1976) announced the *uncertainty principle,* which states that it is impossible to determine precisely both the position and momentum of a particle at a given time. (The product of the uncertainties in these two variables always exceeds a particular constant.) This result played havoc with the popular deterministic philosophy.

In the table on the preceding page, the three rows correspond to the three choices for group B, the three columns headed by a, b, and c indicate the three options for group C, and the nine entries in the table itself are the outcomes when the corresponding strategies are chosen by these groups. For example, if A voted for a, B voted for b, and C voted for c, the result would be the boldfaced **a**, found in the second row and third column in this table. This result, in which each senator has voted for his most preferred outcome, is called **sincere voting.**

A closer examination of the table shows that both groups B and C will benefit if C votes for b instead of c. If group C switches its vote, C will achieve its second choice b instead of its third choice a, and group B will achieve its first choice b.

			C votes	
A votes	**a**	a	b	c
		a	a	a
B votes		**b**	a	b	a
		c	a	a	c

This analysis suggests that group C should vote *insincerely* and select b. Groups B and C will in effect have formed a *coalition against A.* However, neither collusion nor communication between these two groups need take place to bring this about. Not only do both C and B benefit from voting for b, but once they have done so, there is no possible further switch that can do better for either. Thus this action is self-reinforcing, since no

In 1931, the Austrian-American mathematician Kurt Gödel (1906 – 1978) published his paper on "formally undecidable propositions." (It proved that given any set of axioms, there would always be statements within the system ruled by these axioms that could be neither proved nor disproved on the basis of these axioms.) This showed that the paradoxes that had been disturbing mathematical logicians for the previous half-century were unavoidable. Gödel showed, against the hopes of some, that the totality of mathematics could not be deduced from any single system of axioms.

In 1951, Kenneth J. Arrow (1921 –) listed five highly desirable properties that one would expect any reasonable voting system to possess. He then went on to prove that no possible voting method could satisfy these properties in all situations. (See Spotlight 11.2.) Every voting scheme will, at times, exhibit shortcomings. This *impossibility theorem* destroyed the dream of social philosophers who had sought fair and nonmanipulative social-choice mechanisms for more than a century. It also forced social scientists to use more rigorous methodologies in their analyses.

In 1980, Michel L. Balinski (1933 –) and H. Peyton Young (1945 –) showed that there is no general method for rounding a set of fractions to integers with a given sum that will always satisfy three very natural conditions. Therefore, allocating seats to states in the U.S. House of Representatives or seats to parties in parliament has no completely satisfactory solution. The two-century search by the U.S. Congress and other representative bodies was doomed from the start (see Chapter 14). More generally, attempts to apportion discrete objects in an equitable manner can result in undesirable allocations.

---

voting group can now deviate unilaterally and expect to gain from its action. (Technically, we refer to these choices as being in *equilibrium*, a concept that arises in Chapter 15.)

Sequential Voting

One major objection to plurality voting is that when there are more than two outcomes, the final outcome may be favored by less than half of those voting — only 40% in the case of *A*. In order to make sure that the ultimate decision receives a majority vote, it may be necessary to resort to a *runoff* election or to some other type of **sequential voting** — a procedure that requires a majority vote at each step. We now consider two such additional agendas that the Roman Senate could have used.

In the first of these agendas, the Senate will vote first between innocent and guilty; and only when a guilty verdict results will they decide the punishment, *b* or *c*. This agenda is depicted in Figure 11.2.

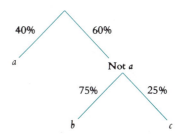

Figure 11.2 A runoff election.

For this scheme, assuming group A always votes for acquittal a, a loses to "not a" 40% to 60% (35% + 25%) on the first round, and b beats c by 75% (40% + 35%) to 25% on the second ballot. You can see that banishment will win if the voters vote sincerely at each decision point. There is no maneuvering or collusion that A or C can undertake in this case that will produce a better outcome for them. Whereas insincere voting was the optimal strategy in the case of plurality voting, the best strategy in this case is to vote in a sincere manner.

In the second of the sequential agendas, we assume that the Senate moves to decide *first* upon the appropriate punishment, b or c, before it addresses the question of guilt. This agenda is pictured in Figure 11.3. Sincere voting would result in b winning over c by 75% to 25% on the first round, and then b winning over a by 60% to 40%. It appears as though b should win.

However, it is not clear that A will vote sincerely: group A may well vote for outcome c on the first ballot, which could result in the middle position b being eliminated on the first vote by 65% (40% + 25%) to 35%. After eliminating b, group A would change from c to a on the second ballot; thus, preference a would ultimately prevail over c by 75% (40% + 35%) to 25% (see Figure 11.4). When forced to choose between a and c, group B should go with a, their second choice, rather than c, their least-preferred outcome.

On the other hand, the groups may not vote this way after all. C is well aware that group A, in an effort to eliminate b at an early stage, may vote

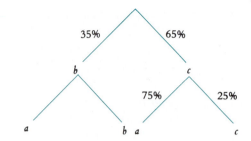

Figure 11.4 An outcome under strategic voting.

insincerely on the first round. In effect, A and C, representing the extreme positions a and c, have temporarily united to eliminate the middle position b. However, C need not go along with this ploy by A. To avoid eliminating b, C may actually vote for b instead of c at the first tally. In short, in the first round A may vote for c rather than a or b and C may actually vote for b rather than c.

Sequential voting can take many different forms, especially when selecting one from among several different candidates. The resulting outcome may well depend upon ordering of the issues, as well as upon strategic voting.

Pairwise Comparisons

Let's make a fourth attempt to resolve Pliny's problem. Consider what would happen if we held an election between each pair of outcomes: a versus b, b versus c, and c versus a. Under sincere voting,

- ▶ b beats a 60% to 40%
- ▶ b beats c 75% to 25%
- ▶ a beats c 75% to 25%

It seems as though b should be the winner because b beats either of the other positions when they meet head to head. Such a winner, if it exists, is called a Condorcet winner.

METHOD 2. A candidate who wins over every other candidate in a head-to-head ballot is called a

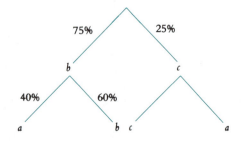

Figure 11.3 The outcome under sincere voting.

Condorcet winner. (Such a winner may not exist.)

Once again, however, those in group A could frustrate the victory for outcome b by *pretending* to alter their "preference" schedule from

$$
\begin{array}{ccc}
\underline{A} & & \underline{A} \\
a & \text{to} & a \\
b & & c \\
c & & b
\end{array}
$$

That is, they will vote for c over b whenever this pair comes up for a vote. As a result, we wind up with

▶ b beats a 60% to 40%
▶ a beats c 75% to 25%
▶ c beats b 65% to 35%

So we end up in a tie, indicated by the cycle

▶ b beats a beats c beats b

This last result is said to violate the **law of transitivity,** which says that if x is preferred to y, and if y is preferred to z, then x is preferred to z. Our example shows the **paradox of voting,** or the **Condorcet paradox:** even if individuals hold to the law of transitivity, the voters as a group may not satisfy it. It is, however, only one of a variety of different paradoxes that occur in voting situations (see Spotlight 11.1, pp. 336–337, and Spotlight 11.2, pp. 340–341).

Having discussed only four of many possible agendas that could have been used by the Roman Senate in the trial of Dexter's freedmen, we can appreciate Pliny's dilemma: the fate of the freedmen depends to a large extent on game playing!

Rank and Score

Many elections ask each voter to submit a complete ranking (preference schedule) of all the candidates. The goal is to arrive at a final group rank ordering of all the contestants that best expresses the desires of the electorate. The purpose is not only to determine the winner, say, the class valedictorian, but also to arrive at who finished second, third, and so on, as in the case of one's rank in his or her senior class. In other applications, such as an election to a hall of fame, the first few finishers each receive the award, while the remaining nominees are "also-rans." One common mechanism for achieving this objective is to assign points to each voter's rankings and then to sum these for all voters to obtain the total points for each candidate. If there are ten candidates, for example, then we could assign 10 points to each first-place vote for a given candidate, 9 points for each second-place vote, 8 for each third, and so forth. The candidate with the highest total number of points is the winner. Subsequent positions are assigned to those with the next highest tallies.

METHOD 3. A voting method that assigns points in a descending manner to each voter's subsequent ranking and then sums these points to arrive at a group's final ranking is called a **Borda count.**

Let us return to Pliny's problem in the Roman Senate example and apply a Borda count. According to this agenda they could score 3 points for each senator's first choice, 2 points for each second choice, and 1 point for each third choice. The tally for the three outcomes is as follows:

a: $(3)(40\%) + (2)(35\%) + (1)(25\%) = 2.15$
b: $(3)(35\%) + (2)(40\%) + (2)(25\%) = 2.35$
c: $(3)(25\%) + (1)(40\%) + (1)(35\%) = 1.50$

We conclude that b (banishment) is the winner, a is second, and c is the losing position. In this situation, however, only outcome b will be implemented, and the rankings of a and c are irrelevant. We will see in what follows that Borda counts are vulnerable to insincere voting as well as to the particular number of points assigned to each posi-

SP●TLIGHT 11.2   Kenneth J. Arrow

▶ ▶ ▶ ▶ ▶ ▶ ▶ ▶ ▶ ▶ ▶ ▶ ▶ ▶ ▶

For centuries, mathematicians have searched for a perfect voting system. Finally, in 1951, economist Kenneth Arrow proved that finding an absolutely fair and decisive voting system is impossible. Arrow is the Joan Kenney Professor of Economics, as well as a professor of operations research, at Stanford University. In 1972, he received the Nobel Memorial Prize in Economic Science for his outstanding work in the theory of general economic equilibrium. His numerous other honors include the 1986 von Neumann Theory Prize for his fundamental contributions to the decision sciences. He has served as president of the American Economic Association, the Institute of Management Sciences, and other organizations. Dr. Arrow talks about the process by which he developed his famous impossibility theorem and his ideas on the laws that govern voting systems:

Kenneth Arrow.

My first interest was in the theory of corporations. In a firm with many owners, how do the owners agree when they have different opinions, for example, about the prospects of the company? I was thinking of stockholders. In the course of this, I realized that there was a paradox involved — that majority voting can lead to cycles. I then dropped that discussion because I was frustrated by it.

I happened to be working with The RAND Corporation one summer about a year or two later. They were very interested in applying concepts of rationality, particularly of game theory, to military and diplomatic affairs. That summer, I

felt not like an economist but instead like a general social scientist or a mathematically oriented social scientist. There was tremendous interest in game theory, which was then new.

Someone there asked me, "What does it mean in terms of national interest?" I said, "Oh, that's a very simple matter," and he said, "Well, why don't you write us a little memorandum on the subject." Trying to write that memorandum led to a sharper formulation of the social-choice question, and I realized that I had been thinking of it earlier in that other context.

I think that society must choose among a number of alternative policies. These policies may be thought of as quite comprehensive, covering a number of aspects: foreign policy, budgetary policy, or whatever. Now, each individual member of the society has a preference, or a set of preferences, over these alternatives. I guess that you can say one alternative is better than another. And these individual preferences have a property I call *rationality* or *consistency,* or more specifically, what is technically known as *transitivity:* if I prefer *a* to *b*, and *b* to *c*, then I prefer *a* to *c*.

Imagine that society has to make these choices among a set. Each individual has a preference ordering, a ranking of these alternatives. But we really want society, in some sense, to give a ranking of these alternatives. Well, you can always produce a ranking, but you would like it to have some properties. One is that, of course, it be responsive in some sense to the individual rankings. Another is that when you finish, you end up with a real ranking, that is, something that satisfies these consistency, or transitivity, properties. And a third condition is that when choosing between a number of alternatives, all I should take into account are the preferences of the individuals among those alternatives. If certain things are possible and some are impossible, I shouldn't ask individuals whether they care about the impossible alternatives, only the possible ones.

It turns out that if you impose the conditions I just stated, there is no method of putting together the individual preferences that satisfies all of them.

The whole idea of the axiomatic method was very much in the air among anybody who studied mathematics, particularly among those who studied the foundations of mathematics. The idea is that if you want to find out something, to find the properties, you say, "What would I like it to be?" [You do this] instead of trying to investigate special cases. And I was really accustomed to this approach. Of course, the actual process did involve trial and error.

But I went in with the idea that there was some method of handling this problem. I started out with some examples. I had already discovered that these led to some problems. The next thing that was reasonable was to write down a condition that I could outlaw. Then I constructed another example, another method that seemed to meet that problem, and something else didn't seem very right about it. Then I had to postulate that we have some other property. I found I was having difficulty satisfying all of these properties that I thought were desirable, and it occurred to me that they couldn't be satisfied.

After having formulated three or four conditions of this kind, I kept on experimenting. And lo and behold, no matter what I did, there was nothing that would satisfy these axioms. So after a few days of this, I began to get the idea that maybe there was another kind of theorem here, namely, that there was no voting method that would satisfy all the conditions that I regarded as rational and reasonable. It was at this point that I set out to prove it. And it actually turned out to be a matter of only a few days' work.

It should be made clear that my impossibility theorem is really a theorem [showing that] the contradictions are possible, not that they are necessary. What I claim is that given any voting procedure, there will be some possible set of preference orders for individuals that will lead to a contradiction of one of these axioms.

But you say, "Well, okay, since we can't get perfection, let's at least try to find a method that works well most of the time." Then when you do have a problem, you don't notice it as much. So my theorem is not a completely destructive or negative feature any more than the second law of thermodynamics means that people don't work on improving the efficiency of engines. We're told you'll never get 100% efficient engines. That's a fact—and a law. It doesn't mean you wouldn't like to go from 40% to 50%.

tion in a ranking. So we still have not arrived at a definitive answer to Pliny's dilemma. Now let us turn to some contemporary scoring systems that will illustrate some difficulties that may arise when using Borda counts.

EXAMPLE: The Football Poll

A poll by 25 sports announcers is used to rank the football teams from among the four following universities: Miami (of Florida), Notre Dame, Penn State, and Southern California. They elect to assign 3 points to each announcer's first choice, 2 points to a second, 1 to a third, and 0 for a fourth. There are $4! = 4 \times 3 \times 2 \times 1 = 24$ possible rankings, but assume that only the five in the following table appear:

	Number of announcers					
Choice	8	6	5	4	2	Points
First	Mi	ND	PS	SC	ND	3
Second	ND	Mi	Mi	PS	PS	2
Third	PS	SC	ND	Mi	SC	1
Fourth	SC	PS	SC	ND	Mi	0

We calculate the total points for each team as follows:

Mi: $(3)(8) + (2)(6) + (2)(5)$
$$+ (1)(4) + (0)(2) = 50$$
ND: $(2)(8) + (3)(6) + (1)(5)$
$$+ (0)(4) + (3)(2) = 45$$
PS: $(1)(8) + (0)(6) + (3)(5)$
$$+ (2)(4) + (2)(2) = 35$$
SC: $(0)(8) + (1)(6) + (0)(5)$
$$+ (3)(4) + (1)(2) = 20$$

The resulting ranking is (1) Miami, 50 points; (2) Notre Dame, 45; (3) Penn State, 35; and (4) Southern California, 20.

However, this poll is vulnerable to insincere voting. For example, what if three of the six voters who ranked ND over Mi over SC over PS suspected prior to the poll that Mi would edge out ND, and they decided to vote insincerely in an attempt to have ND come out on top? They could move Mi from second to fourth place in their rankings. This would take 6 points away from Mi (and add a total of 6 for SC and PS) and thus result in ND becoming the winner. However, such strategic voting may not end there. Thirteen of the seventeen voters who placed Mi over ND could be of like mind and place ND lower in their preference schedules. Many other possibilities for strategic voting appear in a poll of this type. ▲

Another serious difficulty with Borda counts is that the outcome may well depend upon the scale of numbers selected, as is indicated in the following illustration.

EXAMPLE: The Horse Show

The four horses A, B, C, and D are finalists in a show in which they are rated evenly on four attributes W, X, Y, and Z. They place first, second, third, and fourth in each category according to the following table:

Horse	Attributes			
	W	X	Y	Z
A	First	Third	Second	Third
B	Second	First	Fourth	Second
C	Fourth	Fourth	First	First
D	Third	Second	Third	Fourth

If one uses a Borda count that awards 3, 2, 1, and 0 points for a first, second, third, and fourth, respectively, then we see that A and B are tied for the championship with 7 points apiece:

$$A: (3)(1) + (2)(1) + (1)(2) = 7$$
$$B: (3)(1) + (2)(2) + (1)(0) = 7$$
$$C: (3)(2) + (2)(0) + (1)(0) = 6$$
$$D: (3)(0) + (2)(1) + (1)(2) = 4$$

On the other hand, if we score the points as 5, 3, 1, and 0, respectively, then horse B is the winner with 11 points:

$$A: (5)(1) + (3)(1) + (1)(2) = 10$$
$$B: (5)(1) + (3)(2) + (1)(0) = 11$$
$$C: (5)(2) + (3)(0) + (1)(0) = 10$$
$$D: (5)(0) + (3)(1) + (1)(2) = 5$$

If points are instead assigned as 5, 2, 1, and 0, then horse C wins:

$$A: (5)(1) + (2)(1) + (1)(2) = 9$$
$$B: (5)(1) + (2)(2) + (1)(0) = 9$$
$$C: (5)(2) + (2)(0) + (1)(0) = 10$$
$$D: (5)(0) + (2)(1) + (1)(2) = 4$$

If the points are awarded as 5, 3, 2, and 1, as is often the case in such animal competitions, then the result is a three-way tie for first:

$$A: (5)(1) + (3)(1) + (2)(2) + (1)(0) = 12$$
$$B: (5)(1) + (3)(2) + (2)(0) + (1)(1) = 12$$
$$C: (5)(2) + (3)(0) + (2)(0) + (1)(2) = 12$$
$$D: (5)(0) + (3)(1) + (2)(2) + (1)(1) = 8$$

Which of the horse(s) should be declared the winner? ▲

Borda counts are also used to score competitions in which several teammates can enter an event, and so more than one competitor can contribute points to the same team from a single event.

EXAMPLE: The Track Meet

The results of a track and field meet among the three schools A, B, and C are summarized in the following table:

	Events				
Place	Dash	Run	Hurdles	Jump	Throw
First	A	B	A	A	B
Second	A	C	B	C	C
Third	B	C	A	C	C
Fourth	C	C	C	A	C

If the finishing positions are scored 4, 3, 2, and 1, respectively, then school C (which has strong depth) is the victor with 19 points, to 18 for A and 13 for B. On the other hand, if the top places are scored 5, 3, 2, and 1, then A wins with 21 (because of more firsts) to C's 19 points, and B again is last with 15 points. There may be more than four finishers in the events and each is assigned a position in the ranking, but only the first four places count in the scoring.

Ties (or indifferences in preference) can also be incorporated into a Borda count by dividing the accumulated points evenly among the tied contestants. For example, if a remeasurement indicated that A and C were really tied for first (and second) place in the Jump, then they would each receive $(4 + 3)/2$ or 3.5 points in the 4, 3, 2, 1 scoring system. Then team C would have 19.5 points and A would receive 17.5. In the 5, 3, 2, 1 scoring scheme, the final result would then be 20 for A, 20 for C, and 15 for B. ▲

▶Bogus Amendments

We can use still another simple election situation to show how diversionary amendments, when strategically introduced, can mislead some voters into acting against their own interest.

Assume that three representatives A, B, and C each have the choice of voting in favor of or against a new bill N. Voting against this new law means that the old law O will prevail. Assume that two of the three voters do prefer the proposed bill N over the existing law O, as indicated in this table of preferences:

	Voter		
	A	B	C
First choice	N	N	O
Second choice	O	O	N

In a direct comparison between the outcomes N and O, N will win by a vote of 2 to 1. Nevertheless, voter C may attempt to defeat N by the following maneuver. He proposes to modify the new bill N with an amended version called M. C selects the amendment so that A prefers M most of all, whereas B prefers M least of all. This may be done, perhaps, by merely shifting some of the proposed reward in bill N from B to A. Meanwhile C pretends to prefer O over M and M over N. The schedule of preferences now becomes:

	Voter		
	A	B	C
First choice	M	N	O
Second choice	N	O	M
Third choice	O	M	N

The new agenda and voting appear in Figure 11.5. When voting between N and M is sincere at

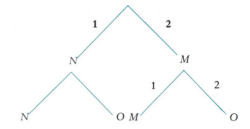

Figure 11.5 Voting on a bogus amendment.

the first decision point, M wins over N by 2 to 1. At the second step, O beats M by 2 to 1. Voter C has tricked the others into defeating N and maintaining the status quo O. Voter A should have noticed this tactic and resisted the temptation to initially vote for the fleeting amendment M.

In the course of a long, complex agenda and heated debate, we must be continuously on guard to avoid being manipulated into voting against our own long-range interests. Those designing the agenda can often rig it in their own favor. For example, one contingent might stack up a larger number of popular outcomes and pit them against a *single* highly desired one in an attempt to eliminate this single outcome at an early stage of the agenda. As a general rule of thumb, it's best to enter the more preferred outcomes at a later stage of the agenda. The chances of survival may increase when there are fewer competing alternatives and fewer remaining votes to be taken.

▶Everyone Wins

We have seen how the outcome of an election may very well depend on the voting procedure or agenda as well as on strategic choices by the voters. To show an extreme case of how the method chosen might affect the results in a realistic situation, we will consider an example of a political party convention at which five different voting schemes are adopted. Assume that there are 55 delegates to this national convention, at which five of the party members, denoted by A, B, C, D, and E, have been nominated as the party's presidential candidate. Each delegate must rank all five candidates according to his or her choice. Although there are $5! = 5 \times 4 \times 3 \times 2 \times 1 = 120$ possible rankings, many fewer will appear in practice because electors typically split into blocs with similar rankings. Let's assume that our 55 delegates submit only six different preference schedules, as indicated in the following table:

	Number of delegates					
	18	12	10	9	4	2
First choice	A	B	C	D	E	E
Second choice	D	E	B	C	B	C
Third choice	E	D	E	E	D	D
Fourth choice	C	C	D	B	C	B
Fifth choice	B	A	A	A	A	A

We see from the preceding table that the 18 delegates who most favor nominee A rank D second, E third, C fourth, and B fifth. Although A has the most first-place votes, he is actually ranked last by the other 37 delegates. Note that the 6 electors who most favor nominee E split into two subgroups of 4 and 2 because they differ between B and C on their second and fourth rankings. We will assume that our delegates must stick to these preference schedules throughout the following five voting agendas. That is, we will not allow any delegate to switch preference ordering in order to vote in a more strategic manner.

1. *Plurality.* If the party were to elect its candidate by a simple plurality, nominee A would win with 18 first-place votes, in spite of the fact that A was favored by less than one-third of the electorate and was ranked dead last by the other 37 delegates.

2. *Sequential: Winners' runoff.* On the other hand, if the party decided that a runoff election

should be held between the top two contenders (A and B), who together received a majority of the first-place votes in the initial plurality ballot, then candidate B outranks A on 37 of the 55 preference schedules and is declared the winner in the runoff.

3. *Sequential: Losers eliminated.* Another approach that could be used is holding a sequence of ballots and eliminating at each stage the nominee with the fewest first-place votes. The last to survive this process becomes the winning candidate. We see in our example that E, with only 6 first-place votes, is eliminated in the first round. E can then be deleted from our table of preferences, and *all* 55 delegates will vote again on successive votes. On the second ballot, the number of first-place votes for the 4 remaining nominees is

$$\begin{array}{cccc} A & B & C & D \\ 18 & 16 & 12 & 9 \end{array}$$

Thus, D is eliminated. Note that the 6 delegates who most favored E earlier now vote for their second choices, that is, 4 for B and 2 for C. On the third ballot the 9 first-place votes for D are reassigned to C, their second choice, giving

$$\begin{array}{ccc} A & B & C \\ 18 & 16 & 21 \end{array}$$

Thus, B is eliminated. On the final round, 37 of the 55 delegates favor C over A, and therefore C wins by this method.

4. *Borda count.* Given that they now have the complete preference schedule for each delegate, the party might instead choose to use a straight Borda count to pick the winner. This could be done, for example, by assigning 5 points to each first-place vote, 4 points for each second, 3

points for a third, 2 points for a fourth, and 1 point for a fifth. The highest total score of

$$191 = (5)(9) + (4)(18)$$
$$+ (3)(12 + 4 + 2)$$
$$+ (2)(10) + (1)(0)$$

is achieved by D, who then wins. Note that A has the lowest score (127) and B the second worst (156).

5. *Condorcet*. In the Condorcet method, each nominee is matched head-to-head with every other. There are 10 such competitions, and each candidate appears in 4 of them. Assuming sincere voting, we can easily see that E wins out over

▶ A by a vote of 37 to 18
▶ B by a vote of 33 to 22
▶ C by a vote of 36 to 19
▶ D by a vote of 28 to 27

In this case, the Condorcet method does produce a winner, namely, E.

In summary, our political party has employed five different common voting procedures and has come up with five different winning candidates. We see from this illustration that those with the power to select the voting method may well determine the outcome. Moreover, we have not considered the many possibilities for strategic voting or for the formation of political coalitions for this example. These latter possibilities could alter outcomes and greatly complicate the analysis. (See Exercises 6d to 8c.)

Approval Voting

One voting method that shows great promise for electoral reform is called **approval voting**. It is particularly suitable for elections in which several candidates typically compete, such as the party primary elections for the President of the United States. Many existing multicandidate-election procedures should be reviewed with a mind toward adopting this simple and practical method.

In approval voting, each voter is allowed to give *one* vote to each of the candidates on the multicandidate slate. No limit is set on the number of candidates an individual can vote for: voters can approve of as many choices as they like and show disapproval by withholding a vote on that candidate. This system replaces the traditional "one person, one vote" by "one candidate, one vote."

The winner in approval voting is the candidate who receives the largest number of approval votes. This approach is also appropriate in situations where more than one candidate or outcome may win, for example, in electing new members to an exclusive society such as the National Academy of Sciences or the Baseball Hall of Fame.

In past years, several important officials in New York State have been elected with much less than a majority of the vote. Clearly, some of these contests would have been reversed if approval voting had been used. The 1970 U.S. Senate race in New York State gave James Buckley 39%, Richard Ottinger 37%, and Charles Goodell 24% of the vote. Buckley may well have been last in approval voting or been the Condorcet loser in a head-to-head battle with either of the other two candidates. Similarly the result in 1980 was Alfonse D'Amato 45%, Elizabeth Holtzman 44%, and incumbent Jacob Javits 11%. Polls indicated that more of Javits's supporters preferred Holtzman to D'Amato and that she probably would have won in a runoff or under approval voting. John Lindsay was reelected mayor of New York City in 1969 with only 42% of the vote. In 1977, Edward Koch beat Mario Cuomo for mayor in a runoff after they initially won only 19.8% and 18.6% of the vote, respectively. (Four others also got over 10% of the vote in that initial election.) Sequential voting or approval voting might have given quite a different view of the "will of the people" and even have selected different winners in these cases.

Approval voting may well prove to be particularly effective in presidential primaries when several contestants are entered. The results of the

1980 New Hampshire elections for the Republican Party were

Ronald Reagan	50%
George Bush	23%
Howard Baker	13%
Others	14%

An ABC News exit poll indicated that approval voting might have given this tally:

Ronald Reagan	58%
George Bush	39%
Howard Baker	41%

It is very possible that such results would have delayed the withdrawal of Senator Baker from the race. Perhaps he would have become Vice President in 1981 and President in 1989.

Approval voting, like any other voting method, is not entirely free of faults. It, too, is subject to strategic manipulation and can give rise to counterintuitive outcomes. Analyses to date indicate that it is no more vulnerable to insincere voting than other known methods. On the other hand, it is practical, simple, and easy to implement. It gives the voters greater freedom in expressing themselves without requiring more complicated ranking schemes, which have their own inherent problems. Approval voting has a lot in its favor and seems ripe for widespread implementation.

We know from **Arrow's impossibility theorem** (Spotlight 11.2, pp. 340–341) that there can never be a perfect voting system. To select a voting system is to compromise between the different shortcomings inherent in each. Nonetheless, social-choice theorists continually strive to create (or perhaps rediscover) better voting schemes in an attempt to minimize such flaws.

▶REVIEW VOCABULARY

Agenda An ordering or list of alternatives for consideration. Often used in sequential voting.
Approval voting Each voter indicates approval or disapproval for each candidate or issue on a ballot, as opposed to voting for only one candidate or issue.

Arrow's impossibility theorem The discovery by Kenneth J. Arrow that any voting system can give undesirable outcomes.
Borda count Assigning points to voters' preferences and summing the points for each candidate to determine the winner.
Condorcet (or voting) paradox Candidate A beats B, B beats C, and C beats A.
Condorcet winner A candidate who beats every other candidate in a one-on-one ballot.
Insincere voting Voting contrary to one's true preferences in an attempt to obtain a better outcome in the long run.
Law of transitivity If A wins over B and B wins over C, then A must win over C.
Majority More than half of the votes cast.
Plurality The case where a candidate with the most votes in a multicandidate race is declared the winner. The number of votes for the winner could be less than half.
Preference schedule A list of possible outcomes in the order a voter most prefers them.
Sequential voting A voting procedure in which successive ballots are taken for the purpose of eliminating some candidates or issues before the final vote.
Sincere voting Voting in a manner consistent with one's preference schedule. One always votes for the most-preferred candidates or outcomes on each ballot.
Strategic voting Voting insincerely on a ballot in an attempt to achieve a more preferable outcome than could have resulted by voting sincerely.

▶SUGGESTED READINGS

BLACK, DUNCAN: *The Theory of Committees and Elections*, Kluwer, Dordrecht, 1986. The historical highlights and developments of voting methods in the nineteenth and twentieth centuries are traced in this economist's volume.

BRAMS, STEVEN J., AND PETER C. FISHBURN: *Approval Voting*, Birkhäuser, Boston, 1982. This volume is a research-level work on developments in the recently popular (but rediscovered) method now called approval voting. However, the first chapter is a highly readable and superb introduction to this voting method and its uses.

DAVIS, MORTON D.: *Mathematically Speaking*, Harcourt Brace Jovanovich, New York, 1980. In Chapter 6 there is an excellent introduction to the problem of voting, which includes an elementary discussion of the properties desired of any voting method (Arrow's axioms), which no voting system can achieve in general.

FARQUHARSON, ROBIN: *Theory of Voting*, Yale University Press, New Haven, 1969. Some of the examples presented in this chapter, as well as other illustrations, are discussed in more detail in this elementary but historically important monograph.

LUCE, R. DUNCAN, AND HOWARD RAIFFA: *Games and Decisions*, Wiley, New York, 1957. (Also available as a Dover paperback.) In Chapter 14 there is a more technical introduction to Arrow's axioms and theorem, as well as a proof of his famous impossibility theorem, which is suitable for upper-division undergraduates.

MALKEVITCH, JOSEPH, AND WALTER MEYER: *Graphs, Models and Finite Mathematics*, Prentice-Hall, Englewood Cliffs, N.J., 1974. In Chapter 10 there is an excellent introduction to the problem of voting, including a discussion of the properties desired of any voting method (Arrow's axioms).

PELEG, BEZALEL: *Game Theoretical Analysis of Voting in Committees*, Cambridge University Press, New York, 1984. This technical volume accepts the strategic nature of group decision making as a given and uses existing subjects, such as game theory, to illustrate how to better compete when such instances arise.

ROBERTS, FRED S.: *Discrete Mathematical Models*, Prentice-Hall, Englewood Cliffs, N.J., 1976. In Chapter 10 there is a more technical introduction to Arrow's axioms and theorem, as well as a proof of his famous impossibility theorem, which is suitable for upper-division undergraduates.

SEN, A. K.: *Collective Choice and Social Welfare*, Elsevier, New York, 1970. Describing the social-choice perspective via normative models and the axiomatic approach, this volume is concerned with the conditions under which particular voting schemes guarantee desirable outcomes.

▶ ## EXERCISES

1. How many different ways can a voter:
   a. rank 3 choices (when ties are not allowed)?
   b. rank 4 alternatives (without ties)?
   c. rank n potential outcomes (without ties)?

2. How many different ways can a voter:
   a. rank 3 choices when ties are not allowed, but *incomplete* rankings can be submitted (e.g., a first choice without giving a second or third choice)?
   b. rank 3 choices when ties are allowed and *complete* rankings are required?

3. Consider the trial in the Roman Senate example when the preference schedule of the Roman senators is given by the following table. The group favoring banishment now splits into two factions B and B' with different second and third choices.

	Group of senators			
	A	*B*	*B'*	*C*
First choice	*a*	*b*	*b*	*c*
Second choice	*b*	*a*	*c*	*b*
Third choice	*c*	*c*	*a*	*a*
Percent	40	20	15	25

    a. What verdict would result if they used the sequential agenda in Figure 11.2 in the text, and the senators voted sincerely at each step?

    b. What verdict would result if they used the sequential agenda in Figure 11.3 in the text, and the senators voted sincerely at each ballot?

    c. What verdict would result if they used a Borda count which assigned 3, 2, and 1 points for a first, second, and third choice, respectively, and the senators voted sincerely?

    d. What verdict would result if they used the Condorcet method, and the senators voted sincerely in each pairwise comparison?

▲   e. Discuss the possibilities for strategic voting in cases a, b, c, and d.

4. What would be the outcome in the trial in the Roman Senate example if some senator moved that they first vote on whether they wanted the verdict of banishment or not? (If *b* did not win on this first ballot, then they would vote between *a* and *c*.)

    a. Assume the senators vote sincerely at each stage.

▲   b. Assume that the voters vote strategically (insincerely) whenever it is to their advantage to do so.

5. The 10 members of a party's platform committee must pick one issue to receive the highest priority in the upcoming campaign. The three contenders are defense *D*, education *E*, and health *H*, and their preference schedules are as follows:

	Number of members		
	4	3	3
First choice	*D*	*E*	*H*
Second choice	*E*	*H*	*D*
Third choice	*H*	*D*	*E*

    a. Which issue wins if they first vote between *E* and *H*, and then vote between this initial winner and *D*?

▲ Advanced exercise.

    b. Which issue wins if they first vote between D and E, and then vote between this initial winner and H?

    c. Could those who most prefer E vote insincerely in some way to change the outcomes in case a or b in a way that benefits them?

    d. Which issue wins if they use a Borda count that scores 3 points, 2 points, and 1 point for each first choice, second choice, and third choice, respectively?

    e. Could those who most prefer H vote insincerely in some way so as to change the outcome in case d to their advantage?

6. To be elected to the Baseball Hall of Fame a player must be retired for five years and receive a vote from 75% of some 420 actual voters. (A few eligible voters often do not cast a ballot.) In the election for January 1993 there were 423 voters and the top five finishers (and their number of votes) were:

Reggie Jackson (396)
Phil Nickro (278)
Orlando Cepeda (252)
Tony Perez (233)
Steve Garvey (176)

    a. Who was elected in 1993?

    b. How many more votes would Perez have needed to have been elected?

    c. What percent of the voters, who did not vote for Garvey, would have had to change and voted for him in order for Garvey to have been elected?

▲   d. Is there any way in which a voter can vote in an insincere manner to help or hurt the chances of some player?

7. A player remains on the ballot for the Baseball Hall of Fame for 15 years provided he receives 5% of the votes cast each year. Some other players (and their votes) in the 1993 election were:

Mickey Lolich (43)
Thurman Munson (40)
Rusty Staub (32)
Bill Maddock (19)
Roy Cey (8)

Which of these five players meets the 5% cutoff criterion (for the 423 votes cast) for remaining on the ballot for the 1994 election?

8. The Academic Standards Committee of a college must vote on the future of Assistant Professor Jones who is up for reappointment the next academic year. Their choices are:

    a. Promote Jones to associate professor with tenure,

    b. Renew Jones's appointment as an assistant professor for a two-year term,

    c. Terminate Jones after the subsequent academic year.

▲ Advanced exercise.

The nine-member committee breaks into three groups A, B, and C, consisting of 4, 3, and 2 people, respectively, and has the following preference schedules.

	Groups		
	A	B	C
Number of people	4	3	2
First choice	a	b	c
Second choice	b	a	b
Third choice	c	c	a

    a. Describe how this is essentially a (modern-day) problem that is the same as the Roman Senate example in the text.

    b. Consider all the voting methods used in the text for the Roman Senate example and determine Jones's fate in each case.

▲   c. Consider Jones's fate in the cases where some voting group(s) "benefit" by using insincere voting.

9. One hundred voters who are to elect one of the three candidates A, B, or C have the following preference schedules:

	Number of voters			
	38	30	25	7
First choice	A	C	B	B
Second choice	B	A	C	A
Third choice	C	B	A	C

    a. Which candidate wins an election using the plurality method?

    b. Who wins if there is a runoff election between the top two finishers in the initial plurality ballot in case a?

    c. Who would win in cases a and b, respectively, if the 7 voters who prefer B over A and A over C were to switch their preference ranking to A over B over C?

    d. If the 45 voters who now prefer A over B over C (after the switch made in case c) knew everyone's preference schedule, could they vote more strategically to ensure a victory for A when the voting method in case b is used?

▲ Advanced exercise.

10. Thirteen students decide to vote on whether to play baseball B, soccer S, or volleyball V at their picnic. Their schedule of preference is as follows:

	Number of students			
	5	2	4	2
First choice	B	S	V	V
Second choice	S	V	B	S
Third choice	V	B	S	B

a. Which sport wins if they use the plurality method?
b. Which one wins if they use a Borda count that assigns 3, 2, and 1 points to each first, second, and third choice, respectively?
c. Which one wins if they first eliminate the one with the fewest first-place votes and hold a runoff between the other two?
d. Which one wins if they first eliminate the one with the most last-place votes and have a runoff between the other two? Is the method decisive in this case?
e. Which one wins in case d if the last two students misrepresent their preference ranking and pretend it is V over B over S rather than V over S over B as listed in the table?
f. Would there be a Condorcet winner if the students did vote sincerely?

11. One hundred sports writers with the following preference schedules are to pick the best college football team among Alabama A, Michigan M, and Washington W:

	Number of writers		
	52	38	10
First choice	W	M	A
Second choice	M	W	M
Third choice	A	A	W

a. Which team wins if the election is by a Borda count that assigns 3, 2, and 1 points to each first, second, and third choice, respectively?
b. If those who most favor Michigan suspected that Washington would win, and thus voted insincerely for Alabama as their second choice, what would the outcome be?
c. If the supporters of Washington believed that the insincere voting in case b might take place, could they still vote so as to guarantee that Washington wins?

12. Eleven students must decide whether to dine together at a Chinese, Italian, or Mexican restaurant. Their preference schedules are as follows:

	Number of students		
	5	2	4
First choice	Chinese	Mexican	Italian
Second choice	Mexican	Italian	Mexican
Third choice	Italian	Chinese	Chinese

    a. What choice will the group make if they vote sincerely according to the following methods:
       (1) the plurality method
       (2) eliminating the restaurant with the fewest first-place votes and having a runoff between the other two
       (3) eliminating the restaurant with the most last-place votes and having a runoff between the other two
    b. Is any restaurant a Condorcet winner?
    c. What choice will be made if they use a Borda count that assigns x points to each first choice, y points to each second choice, and z points to each third choice when
       (1) $x = 3, y = 2,$ and $z = 1$?
       (2) $x = 4, y = 2,$ and $z = 1$?
       (3) $x = 5, y = 2,$ and $z = 1$?
▲   d. Is there any way to pick the points x, y, and z in case c with $x > y > z$ so that the Italian restaurant wins the Borda count?

13. The result of a swim meet between the four schools A, B, C, and D is given by the following table:

	Event				
Place	Sprint	Distance	Relay	Medley	Dive
First	B	A	B	A	B
Second	D	D	D	C	D
Third	D	D	C	B	C
Fourth	C	B	A	D	D
Fifth	C	B	B	C	D

    a. How do the teams rank in this competition if the finishing positions are scored 5, 4, 3, 2, and 1, respectively?

▲ Advanced exercise.

b. How do the teams rank if the positions are scored 5, 3, 2, 1, and 0, respectively?

c. If the first three finishers in the Dive event were actually tied for first place, what would the teams' scores be
(1) in case a?
(2) in case b?

d. If the first two finishers in the Dive were disqualified and the new order of finish in this event is C, D, D, A, and A, then what are the teams' scores
(1) in case a?
(2) in case b?

14. The result of a gymnastics meet between three teams A, B, and C is given by the following table:

| | Event | | | |
Place	Beam	Floor	Vault	Bars
First	A	C	C	B
Second	A	B	B	C
Third	C	A	A	A
Fourth	B	C	A	B

a. How do the teams rank in this competition if the finishing positions are scored 4, 3, 2, and 1, respectively?

b. How do the teams rank if the positions are scored 3, 2, 1, and 0, respectively?

c. If the first two finishers in the Bars event were actually tied for first place, what would the teams' scores be
(1) in case a?
(2) in case b?

d. If the first finisher in the Floor was disqualified and the new order of finish in this event is B, A, C, and A, then what are the teams' scores
(1) in case a?
(2) in case b?

15. Ten board members vote by approval voting on eight candidates for new positions on their board as indicated in the following table. An X indicates an approval vote. For example, voter 1, in the first column, approves of candidates A, D, E, F, and G, and disapproves of B, C, and H.

	Voters									
Candidates	1	2	3	4	5	6	7	8	9	10
A	X	X	X			X	X	X		X
B		X	X	X	X	X	X	X	X	
C			X					X		
D	X	X	X	X	X		X	X	X	X
E	X		X		X		X		X	
F	X		X	X	X	X	X	X		X
G	X	X	X	X	X			X		
H		X		X		X		X		X

a. Which candidate is chosen for the board if just one of them is to be elected?
b. Which candidates are chosen if the top four are selected?
c. Which candidates are elected if 80% approval is necessary and at most four are elected?
d. Which candidates are elected if 60% approval is necessary and at most four are elected?

16. The 45 members of a school's football team vote on three nominees A, B, and C by approval voting for the award of "most improved player" as indicated in the following table. An X indicates an approval vote.

	Number of voters							
Nominee	7	8	9	9	6	3	1	2
A	X			X	X		X	
B		X		X		X	X	
C			X		X	X	X	

a. Which nominee is selected for the award?
b. Which nominee gets announced as runner-up for the award?
c. Note that two of the players "abstained," that is, approved of none of the nominees. Note also that one person approved of all three of the nominees. What would be the difference in the outcome if one were to "abstain" or "approve of everyone"?

17. Given that three members of a four-person committee prefer a newly proposed bill N over the old existing law O, can you suggest the type of amendment M to N that the advocate of O should propose in an attempt to defeat N (and M). Note that M or N must receive three or four of the four votes cast in order to pass, whereas the existing law wins on a tie vote of 2 to 2.

18. Consider the example in the text section "Everyone Wins," in which 55 delegates at a national convention vote on five candidates for the party's presidential nominee (pp. 344–345). Determine the winning candidate if they used a voting method that eliminates the loser at each step: at each ballot, eliminate the candidate with the most last-place votes, and then continue with successive ballots with all 55 delegates voting each time.

19. Assume that the members A, B, and C of a three-person committee have the following preference schedules over the three possible outcomes a, b, and c:

	Member		
	A	B	C
First choice	a	b	c
Second choice	b	c	a
Third choice	c	a	b

Each member can vote secretly for one outcome, and the majority rules. Furthermore, A is the chairman and has the power to break tie votes.

    a. What would the result be if each member voted sincerely for his or her most-preferred outcome?

    b. What do you expect to actually happen in this situation?

    c. Can you explain why this example is often referred to as *the chairman's paradox?*

20. Consider the following class project: pick some upcoming election involving more than two alternatives. For example, select a few of the leading candidates for a major party's presidential nominee. Compare the class results for the following different voting methods:

    a. Vote for only one candidate and select the winner by the plurality method.

    b. If the winner in case a does not have a majority, then hold a runoff ballot.

    c. Have each voter provide his or her preference schedule (that is, each ranks the candidates) and then select the winner by a Borda count.

    d. Use the method of approval voting where the winner is the one with the largest number of approval votes.

    e. Is there a Condorcet winner?

▶WRITING PROJECTS

1. In the 1992 presidential election, the final results were:

Candidates	Number of votes	Percentage of votes
Clinton	43,727,625	43
Bush	38,165,180	38
Perot	19,236,411	19

Making reasonable assumptions about voters' preference schedules discuss how the election might have turned out under the different voting methods discussed in this chapter.

2. Frequently in presidential campaigns, the winner of the first few primaries is given front-runner status that can lead to the nomination of his or her party. Frequently there are several candidates running in early primaries such as New Hampshire. Consider a recent election (for example, the 1976 or 1992 Democratic primaries), and discuss how the nominating process might have proceeded through the campaign if approval voting had been used to decide primary winners.

Chapter
12

▶▶▶▶▶▶▶▶▶▶▶▶▶▶▶▶

Weighted Voting Systems

In some voting situations, the "one-person, one-vote" principle does not apply. For example, when the shareholders of a public corporation elect a board of directors, each shareholder is entitled to one vote per share owned. Shareholders who own relatively large numbers of shares usually have greater influence in such an election than the small shareholders do.

A **weighted voting system** is a decision-making procedure in which the participants have varying numbers of votes. Examples of such systems include shareholder elections and the election of the President of the United States by the Electoral College (see Spotlight 12.1, page 362). Some legislative bodies have such strong party discipline that each legislator always votes as dictated by his or her party. These legislatures are weighted voting systems in which the participants are the political party organizations, each of which is entitled to a number of voters equal to the size of its delegation in the legislature.

The *power* of a participant in a weighted voting system can be roughly defined as the ability of the participant to influence a decision. There are several ways to measure mathematically the power of

a participant, or of a bloc of participants, in a weighted voting system. We study two such measures: the *Banzhaf power index* and the *Shapley–Shubik power index*. Either of these indices provides a much more accurate measure of a participant's power than the number of votes that the participant is entitled to cast.

▶ HOW WEIGHTED VOTING WORKS

In 1958, the Board of Supervisors of Nassau County, New York, consisted of six supervisors from five municipalities. Two of the supervisors were elected at large from the city of Hempstead, which had more than half of the county's population. To compensate for the unequal populations of the municipalities, the supervisors were given weighted votes, as described in Table 12.1.

The total number of votes assigned to the supervisors was 30, and a simple majority (16 votes) was required to pass a measure. Since the two Hempstead supervisors controlled 18 votes be-

TABLE 12.1 Weighted Voting, Nassau County Board of Supervisors, 1958

Municipality	Number of votes
Hempstead }	{ 9
Hempstead	9
North Hempstead	7
Oyster Bay	3
Glen Cove	1
Long Beach	1
Total	30

tween them, they would have had the power to pass any measure without consulting their colleagues from the smaller municipalities. However, the Nassau County Charter contained a provision requiring that any measure must have the support of supervisors from two different municipalities in order to pass. This provision complicates the analysis of power on the Board of Supervisors, so we will ignore it for the moment, and return to it later.

To pass a measure, the two Hempstead supervisors can vote together, or one of the Hempstead supervisors can vote with the North Hempstead supervisor. If one of the Hempstead supervisors should sponsor a bill, he or she will quickly find out that it is not worthwhile to lobby the supervisors of Oyster Bay, Glen Cove, and Long Beach to obtain their support. Between them, these supervisors have only five votes, so even if they added their votes to the sponsor's nine votes, the total would be only 14—not enough to pass the measure. If the three supervisors from the smaller communities joined the North Hempstead supervisor, their votes would total only 12. No bill can pass without the support of at least two of the Hempstead and North Hempstead supervisors, and if it has the support of two of these supervisors, it will pass without the help of anyone else.

In this situation, the three supervisors from Oyster Bay, Glen Cove, and Long Beach have no voting power. They might influence the decision-making process by serving on committees, by introducing bills, and by participating in the debate, but they are essentially disenfranchised by the Board's voting system. A voter whose vote will never be needed to pass any measure, or to defeat any measure, is called a **dummy.**

The voting system used by the Nassau County Board of Supervisors is similar to systems that have been used by many other legislative bodies in the state of New York. It has a relatively small number of participants, which makes it easier to analyze than most other systems. As populations have shifted, the number of votes assigned to the supervisors has changed several times. For example, in the 1960s, Oyster Bay and the two Hempstead representatives shared equal power, while North Hempstead joined Glen Cove and Long Beach in dummy status (see Table 12.2). In 1965, the voting system was analyzed in a law review article entitled "Weighted Voting Doesn't Work," by John F. Banzhaf III (see Suggested Readings, p. 388), who offered his opinion that it was unfair and unconstitutional.

In his article, Banzhaf described a new mathematical model for weighted voting. His was not the first; another model had been developed ten years earlier by Lloyd S. Shapley and Martin Shubik (see Spotlight 12.2, p. 364). However,

TABLE 12.2 Weighted Voting, Nassau County Board of Supervisors, 1964

Municipality	Number of votes
Hempstead }	{ 31
Hempstead	31
North Hempstead	21
Oyster Bay	28
Glen Cove	2
Long Beach	2
Total	115

Banzhaf's model has attracted the interest of the courts. As recently as 1992, it was a central issue in litigation involving the Nassau County Board of Supervisors (see Spotlight 12.3, pp. 372–373).

Notation for Weighted Voting

To describe a weighted voting system, we need to first specify the set of voters. If the voters are A, B, C, . . . , then the **weights** $w(A)$, $w(B)$, $w(C)$, . . . , which are the numbers of votes that these voters cast, must be specified. Finally, the total number q of votes necessary to pass a measure, called the **quota,** must be specified. The shorthand notation

$$[q : w(V_1), w(V_2), \ldots , w(V_n)]$$

describes a weighted voting system with n voters V_1, V_2, . . . , V_n, with weights $w(V_1)$, . . . , $w(V_n)$ and with quota q.

Thus, the voting system used by the Nassau County Board of Supervisors in 1958 is expressed as

$$[q : w(H_1), w(H_2), w(N), w(B), w(G), w(L)]$$
$$= [16 : 9, 9, 7, 3, 1, 1]$$

where H_1 and H_2 are the two Hempstead supervisors, N is the North Hempstead supervisor, and B, G, and L are the Oyster Bay, Glen Cove, and Long Beach supervisors, respectively.

A set of voters that has joined together to vote in favor of an issue, or to oppose an issue, is called a **coalition.** The coalition may consist of all the voters, or any subset of the voters. It may consist of just one voter, or it may even be *empty*. For example, if the voting body is unanimously in favor of a motion, then the coalition opposing the motion is empty.

A coalition of voters in favor of a measure is a **winning coalition** if the sum of its weights equals or exceeds the quota q. If the voting system is to reach an unambiguous decision, it is important not to permit two winning coalitions in opposition to each other. For this reason, we will require $q > \frac{1}{2} w$, where w is the sum of the weights of all participants in the voting system. Thus, if one coalition is winning, with a vote total $t \geq q$, that coalition has more than half of the votes. The voters who did not join the coalition will then have less than half of the votes, and cannot possibly win.

A **blocking coalition** is a subset of voters opposing a motion, with enough votes to defeat it. In a voting system with total weight w and quota q, any coalition with weight more than $w - q$ is a blocking coalition. Thus, in the Nassau County Board of Supervisors of 1958, where $w = 30$ and $q = 16$, any coalition with more than $30 - 16 = 14$ votes is a blocking coalition.

In a voting system with four voters, each with one vote, and a quota of three votes to pass a measure, any coalition of two voters opposing a measure will be a blocking coalition, although these voters could not pass any measure they favored without being joined by a third voter.

Suppose a jury in a criminal trial has 12 members. To pass a measure to convict or to acquit, a coalition must include every juror. If it is impossible to form a winning coalition of 12 jurors, a mistrial is declared, and the prosecution has the right to demand a new trial. Every coalition with at least one juror is a blocking coalition. Since each juror can block a measure on his or her own, we say that he or she has **veto power.**

EXAMPLES: Illustrations of Weighted Voting Systems

1. Consider a small corporation owned by two people, A and B, who possess 60% and 40% of the stock, respectively. If measures are allowed to pass by a simple majority, we express this voting system as

$$[q : w(A), w(B)] = [51 : 60, 40]$$

The interesting feature of this example is that shareholder A has all of the power. When one

SP TLIGHT 12.1   The Electoral College

▶ ▶ ▶ ▶ ▶ ▶ ▶ ▶ ▶ ▶ ▶ ▶ ▶

The Electoral College is the institution that officially elects the President of the United States. When voters go to the polls for a presidential election, they are not actually voting for one of the presidential candidates; they are voting for representatives in the Electoral College, who are called electors.

According to the United States Constitution, each state is to elect a number of electors equal to the size of its delegation in the United States Congress. That is the number of representatives plus the number of senators, which ranges from 3 (for states with small populations) to 54 (for the most populous state, California). The Constitution made no provision for voters in Washington, D.C., until the Twenty-third Amendment was adopted in 1961. The capital city is currently entitled to three electors.

If the Electoral College operated as intended by the framers of the Constitution, the less populous states would have an advantage over the larger states. On the average, each state is given one seat in the House of Representatives for each 540,000 citizens. Thus, state A with a population of 540,000 would have one representative and two senators — therefore it would have three electors. State B, with 25 times the population of state A, would have 25 seats in the House, but still just two senators, so it would be entitled to 27 electors in all. Thus, state B would have only *nine* times as many electors as state A.

The larger states have adopted a perfectly legal way to shift most of the power to elect the President to themselves: the unit rule. Several states passed laws requiring that all of their electors should vote as a unit. Thus, for example, all of California's 54 electors must vote for the same candidate. To ensure that they do this, all of California's electors are selected by the presidential candidate who received the plurality in the state's general election. We will see why this rule shifts the

---

individual's weight equals or exceeds the quota, that individual can pass any measure without consulting anyone else. We call such an individual a **dictator.** A coalition favoring a measure is a winning coalition if and only if A, the dictator, belongs to it. Also, a coalition opposing a measure is a blocking coalition if and only if it includes the dictator.

2. Let us examine a second company, with three shareholders, A, B, and C, who hold 49%, 48%, and 3% of the stock, respectively. Thus

$$[q:w(A), w(B), w(C)] = [51:49, 48, 3]$$

There is no dictator; indeed, this company is more "democratic" than one might expect. Any coalition of two or more shareholders has a simple majority, so the power is equally divided among the three shareholders. Although C owns only 3% of the stock, he has equal influence.

balance of power to the large states in Spotlight 12.4.

With the unit rule, the Electoral College becomes a weighted voting system. There are 51 participants with weights ranging from 3 to 54, and a total weight of 538. The quota required to elect is a simple majority of 270. Since there may well be more than two candidates, it might be assumed that occasionally no one would achieve a majority. Indeed, there have been many elections where no candidate received a majority of the popular vote, the most recent being in 1992. The Constitution provides that if no candidate receives the quota of votes for election in the Electoral College, then the House of Representatives shall choose the President in a voting system where each state's delegation is given one vote. The House has not elected a President since 1824, when it selected John Quincy Adams in a contest that also involved Andrew Jackson, William Crawford, and Henry Clay. Jackson accused Clay of a "corrupt bargain" by which he threw his support to Adams in exchange for being appointed Secretary of State.

The unit rule has been responsible in large part for preventing repetitions of the nasty 1824 election. Although a third party candidate, such as H. Ross Perot in 1992, can acquire a significant percentage of the popular vote, he or she is unlikely to achieve a plurality in any state, and will thus receive no electoral votes. With only two candidates before the Electoral College, the only way that neither would meet the quota would be for both to receive 269 votes. The unit rule thus forces a decision even when the voters are indecisive. However, it can cause the election of a candidate over another candidate who received more popular votes. This has happened twice: in 1876 when Samuel J. Tilden received a *majority* of the popular vote, but lost by one vote in the Electoral College to Rutherford B. Hayes, and in 1888 when Benjamin Harrison defeated Grover Cleveland, who received a plurality of the popular vote.

If we disapprove of outcomes where the winner according to the popular vote loses in the Electoral College, perhaps we should abolish the College and elect the candidate who receives a plurality of the popular vote. A constitutional amendment with this purpose was approved by the House of Representatives in 1969 by a wide margin. It fell to a filibuster in the Senate — opposed by senators from small states, who had the most to gain by passing it.

His weighted vote, when added to the weighted vote of either A or B, will form a majority.

3. A third company has shareholders A, B, C, and D. Shareholders A, B, and C each own 26% of the stock, while D holds the remaining 22%. The voting system for this corporation is

$$[q:w(A), w(B), w(C), w(D)]$$
$$= [51:26, 26, 26, 22]$$

Although D's share of the company is not much less than the shares of any of the other three shareholders, D is a dummy. It is impossible for D to turn a losing coalition into a winning coalition by joining it. The power in this company is equally divided among A, B, and C. ▲

These examples illustrate that power need not be even approximately proportional to one's share of the vote. We now examine the relationship that

SP TLIGHT 12.2   Power Indices

Lloyd S. Shapley

John F. Banzhaf III

Martin Shubik

The first widely accepted numerical index for assessing power in voting systems was proposed in 1954 by a mathematician, Lloyd S. Shapley, currently at the University of California, Los Angeles, and an economist, Martin Shubik of Yale University. Their index was derived from the Shapley value, a fundamental concept in game theory that is used in mathematical economics. A particular voter's power as measured by this index is proportional to the number of different *permutations* (or orderings) of the voters in which he or she has the potential to cast the pivotal vote — the vote that first turns from losing to winning.

The Banzhaf power index was introduced in 1965 by a law professor at George Washington University and well-known consumer advocate, John F. Banzhaf III. This index is the one most often cited in court rulings, perhaps because Banzhaf brought several cases to court and continues to file *amicus curiae* briefs when courts evaluate weighted voting systems. A voter's Banzhaf index is the number of different possible voting *combinations* in which he or she casts a swing vote — a vote in favor of a motion that is necessary for the motion to pass, or a vote against a motion that is essential for its defeat.

exists between the voter's weights and the quota, on the one hand, and the power of each voter, on the other.

▶ THE BANZHAF POWER INDEX

We have seen from our examples that participants in a weighted voting system cannot take their frac-

tion of the total vote as a meaningful indication of their share of voting power. Power is the ability to win. An individual can appear frequently on the winning side, however, without being powerful. For example, if a professional athletic team usually wins, that does not mean that all members of the team can demand high salaries. The high-salaried players are those who are crucial to winning. Similarly, the real significance of a vote is whether it is essential to victory.

One reasonable measure of voting power is the frequency with which a participant's vote can swing coalitions from the losing column to the winning column. This measure is a count of the number of different ways that the participant alone can turn defeat into victory, or vice versa. In other words, our measure of power is the number of different ways that the participant can join a losing coalition and thereby make it a winning coalition. This is identical to the number of *distinct* winning coalitions to which this participant belongs, and that would lose if he or she defected. In any winning coalition, a member who can cause the coalition to lose if he or she abandons the coalition and votes with the opposition is called a **swinger,** because he or she casts a swing vote. A swinger in a blocking coalition is a member whose vote is crucial to that coalition's purpose; if he or she defects, the coalition will no longer be able to prevent a motion from passing.

A participant's **Banzhaf power index** is the number of distinct winning coalitions in which the participant is a swinger, plus the number of distinct blocking coalitions in which he or she is a swinger. It often happens that the same set of voters can be a winning coalition or a blocking coalition, if the voters in the set all support a measure or oppose a measure. In such a set, some voters may be counted as swingers twice: once for being swingers in a winning coalition and again for being swingers in a blocking coalition. In some of the citations at the end of this chapter, terms such as *pivotal voter* and *critical voter* are used in place of swinger.

EXAMPLE: A Three-Person Committee

A committee has a chairperson A, with two votes, and two other members, B and C, each of whom has one vote. The quota for passing a measure is three votes. We can express this

weighted voting system as

$$[q : w(A), w(B), w(C)] = [3 : 2, 1, 1]$$

This voting system is equivalent to one in which each member has an equal vote, but the chairperson has veto power. The winning coalitions are all those whose weights sum to 3 or 4: $\{A, B\}$, $\{A, C\}$, and $\{A, B, C\}$. The coalitions $\{A, B\}$ and $\{A, C\}$ are **minimal winning coalitions;** that is, each member of these coalitions is a swinger. Every winning coalition must contain one of the minimal winning coalitions as a subset. The coalitions that are not winning are **losing:** $\{B, C\}$, $\{A\}$, $\{B\}$, $\{C\}$, and $\varnothing$ (the empty coalition).

The chairperson A has veto power, and so is a swinger in each of the three winning coalitions. This means that if she defects from any of these coalitions, it becomes a losing coalition. She is also a swinger in three blocking coalitions: $\{A\}$, $\{A, B\}$, and $\{A, C\}$. She is not a swinger in the blocking coalition $\{A, B, C\}$, since $\{B, C\}$, with a total weight of 2, would still block if she defected. The members B and C have equal power. Neither is a swinger in the winning coalition $\{A, B, C\}$, since both would have to defect to turn that coalition into a losing one. Member B is a swinger in $\{A, B\}$ as a winning coalition, since A cannot pass a motion by herself, but not a swinger in $\{A, B\}$ as a blocking coalition, since A can veto a motion by herself. Similarly, C is a swinger in $\{A, C\}$, as a winning coalition, but not as a blocking coalition. Finally, both B and C are swingers in the blocking coalition $\{B, C\}$. Thus, A has a Banzhaf power index of 6, while B and C each have Banzhaf power indices of 2. According to the Banzhaf model, A has *three* times as much power as B (or C), even though her vote has only twice the weight.

To summarize, the Banzhaf index for this voting system is (6, 2, 2). Notice that in the preceding example, A was a swinger in three winning coalitions and three blocking coalitions, while B and C were each members of one winning coalition and one blocking coalition. These were not coincidences. If a voter who is a swinger defects from a winning coalition to join the opposing coalition, the opposing coalition becomes a blocking coalition, and the same voter is now a swinger in the blocking coalition. If a swinger defects from a blocking coalition, that coalition would no longer have enough weight to block, and the opposing coalition would win. Thus, every voter is a swinger in exactly the same number of blocking coalitions as he or she is a swinger in winning coalitions. Knowing this, we could determine a participant's Banzhaf index by simply counting the winning coalitions in which he or she is a swinger, and double the result. ▲

▶ COMPUTING THE BANZHAF POWER INDEX

In a weighted voting system with no more than four voters, it is not difficult to calculate the Banzhaf power index by a brute force method. We merely list all of the theoretically possible ways that the participants can vote; that is, all the different **combinations** of yes and no votes. If there are n voters, there will be 2^n such combinations (see "How to Count Combinations," on page 369). Thus, with three voters, there are eight combina-

tions, and with four voters, sixteen combinations. We then examine each combination to see if it is winning or blocking, and determine the swing voters in each winning or blocking combination. This method is impractical with large numbers of voters. For example, with the U.S. Electoral College, where the participants are the fifty states and the District of Columbia, there are $2^{51} = 2{,}251{,}799{,}813{,}685{,}248$ combinations to examine. If we could examine one million combinations per second, we would finish the job in 71.4 years. Since the weights change every ten years with reapportionment, calculations would be obsolete before they were finished. This is another instance of the *combinatorial explosion*, which frequently thwarts brute force computations.

We use the brute force method to determine the Banzhaf index of the voting system

$$[q:w(A), w(B), w(C)] = [3:2, 1, 1]$$

for the three-person committee that was presented earlier. Table 12.3 lists the eight combinations of voters, according to whether they vote yes (Y) or no (N). Whether the issue will pass (P) or fail (F) is indicated in the outcome columns to the right of the combinations. If the issue passes, the coalition that voted Y is a winning coalition; if it fails, the voters who voted N form a blocking coalition.

TABLE 12.3 Combinations of Votes in the Three-Person Committee

Member:	A	B	C			
Weight:	2	1	1			
	Combinations			Votes	Pass	Fail
	Y	Y	Y	4	P	
	Y	Y	N	3	P	
	Y	N	Y	3	P	
	Y	N	N	2		F
	N	Y	Y	2		F
	N	Y	N	1		F
	N	N	Y	1		F
	N	N	N	0		F

Each row of the table is examined to see which of the voters are swingers. This means checking each yes and no vote in each combination to determine whether a switch of one vote will change the result.

For example, the first combination

A	B	C	
Y	Y	Y	P

results in passing the measure by a unanimous vote. If voter A changes her vote from yes to no

A	B	C	
Y	Y	Y	P
↓			
N	Y	Y	F

then the outcome changes to "defeat." We will indicate that A is a swinger in this combination by circling the Y that indicates her yes vote:

A	B	C	
Ⓨ	Y	Y	P

On the other hand, if only voter B switches his vote from yes to no in the first combination, the result remains the same: the issue passes, 3 to 1:

A	B	C	
Y	Y	Y	P
	↓		
Y	N	Y	P

For the same reason, voter C is not a swing voter in this combination. Now let us consider the second combination:

A	B	C	
Y	Y	N	P

If voter A changes her vote from yes to no the measure will no longer pass:

A	B	C	
Ⓨ	Y	N	P
↓			
N	Y	N	F

Furthermore, if voter B changes his vote, the measure will be defeated:

A	B	C	
Y	Ⓨ	N	P
	↓		
Y	N	N	F

Therefore A and B are swingers in the second combination. Voter C is not a swinger, since if she changes her vote, the outcome will not change:

A	B	C	
Y	Y	N	P
		↓	
Y	Y	Y	P

We proceed in the same way with each row of Table 12.3, determining each swinger and circling the corresponding Y or N. Thus, in the third combination, A and C are swingers:

A	B	C	
Y	N	Ⓨ	P
		↓	
Y	N	N	F

A	B	C	
Y	N	Ⓨ↓	P
Y	N	N	F

In the fourth row, voters B and C are swingers in a blocking coalition:

A	B	C	
Y	Ⓝ↓	N	F
Y	Y	Y	P

A	B	C	
Y	N	Ⓝ↓	F
Y	N	Y	P

In row 5, only A is a swinger:

A	B	C	
Ⓝ↓	Y	Y	F
Y	Y	Y	P

In rows 6 and 7, voter A is the only swinger in blocking coalitions:

A	B	C	
Ⓝ↓	Y	N	F
Y	Y	N	P

A	B	C	
Ⓝ↓	N	Y	F
Y	N	Y	P

There are no swingers in row 8.

Table 12.4 summarizes these calculations. If we count the number of circles in each voter's column, we arrive at a power index of (6, 2, 2).

EXAMPLE: A Corporation with Four Shareholders

Consider a weighted voting system

$$[q:w(A), w(B), w(C)] = [51:40, 30, 20, 10]$$

This could be four shareholders, A, B, C, and D in a corporation, owning 40%, 30%, 20%, and 10% of the stock, respectively. A simple majority (taken here as 51%) is necessary to pass a measure.

We find the Banzhaf power index by listing the $2^4 = 16$ distinct combinations of yes (Y) and no (N) for the shareholders. Each combination is shown in one of the rows of Table 12.5. The total percent of yes votes for each combination is indicated at its right. The issue passes (P) or fails (F), depending on whether or not the percent of yes votes meets the quota of 51%. Each vote in each combination must be examined to determine whether or not it is a swing vote. Will the change of this vote alter the result? If it will, the vote is a swing vote and it is circled. By counting the number of swing votes in each shareholder's column, we find that the Banzhaf power index is (10, 6, 6, 2). Notice that while B and C own different amounts of stock, they have the same voting power.

Again, we have counted each swinger in winning and in blocking coalitions. We could have made the process more efficient by counting only the swing votes in winning coalitions or, alternatively, we could have counted only the swing votes in blocking coalitions.

To restrict our attention to the winning coalitions, consider only the rows that show the measure passing. A voter is a swinger in a vot-

TABLE 12.4 The Swingers in Each Combination of Votes in the Three-Person Committee

Member:	A	B	C			
Weight:	2	1	1			
	Combinations			Votes	Pass	Fail
	(Y)	Y	Y	4	P	
	(Y)	(Y)	N	3	P	
	(Y)	N	(Y)	3	P	
	Y	(N)	(N)	2		F
	(N)	Y	Y	2		F
	(N)	Y	N	1		F
	(N)	N	Y	1		F
	N	N	N	0		F
Number of swingers	6	2	2			

ing combination given by one of these rows if (1) the voter voted Y, and (2) the measure would fail if the voter switched his or her vote to N. In Table 12.5, counting either the swing votes in winning coalitions, or the swing votes in blocking coalitions, will yield half the Banzhaf index: (5, 3, 3, 1). ▲

▶ How to Count Combinations

How do we know that if there are n voters, there is a total of 2^n voting combinations? To answer this question, we use the multiplication principle. Each voter has two options: to vote either yes or no, and the voters are independent of each other. Therefore, the number of ways n voters can cast their votes is

$$\underbrace{2 \times 2 \times 2 \times \cdots \times 2}_{n \text{ factors}} = 2^n$$

How many of these combinations consist of, say, 6 yes votes and $n - 6$ no votes? (Of course, the answer is "none" if $n < 6$, so we will assume $n \geq 6$.)

If we decide to keep track of the order in which the six yes voters cast their votes, then there are n voters that could cast the first vote. The first voter cannot vote again, so there are $n - 1$ voters that could cast the second yes vote. Similarly there are $n - 2$ voters that could cast the third yes vote, $n - 3$ that could cast the fourth, $n - 4$ that could cast the fifth, and $n - 5$ that could cast the sixth. Again, by the multiplication principle, there are

$$n \times (n - 1) \times (n - 2) \times (n - 3) \times (n - 4) \times (n - 5)$$

ways that exactly six voters could vote yes, *if we keep track of the order in which they vote*. The same six voters can vote in $6 \times 5 \times 4 \times 3 \times 2 \times 1 = 720$ different orders, so if we want to find the number of ways that there could be six yes votes without reference to order, we must divide by $6 \times 5 \times 4 \times 3 \times 2 \times 1$ to get

$$C_6^n = \frac{n \times (n - 1) \times (n - 2) \times (n - 3) \times (n - 4) \times (n - 5)}{6 \times 5 \times 4 \times 3 \times 2 \times 1}$$

TABLE 12.5   The Swingers in Each Combination of Votes in the Four-Stockholder Corporation

Stockholder:	A	B	C	D			
Percent ownership:	40	30	20	10			
	Combinations				Votes	Pass	Fail
	Y	Y	Y	Y	100	P	
	(Y)	Y	Y	N	90	P	
	(Y)	(Y)	N	Y	80	P	
	(Y)	(Y)	N	N	70	P	
	(Y)	N	(Y)	Y	70	P	
	(Y)	N	(Y)	N	60	P	
	Y	(N)	(N)	Y	50		F
	Y	(N)	(N)	N	40		F
	N	(Y)	(Y)	(Y)	60	P	
	(N)	Y	Y	(N)	50		F
	(N)	Y	(N)	Y	40		F
	(N)	Y	N	N	30		F
	(N)	(N)	Y	Y	30		F
	(N)	N	Y	N	20		F
	N	N	N	Y	10		F
	N	N	N	N	0		F
Number of swingers	10	6	6	2			

(Circled votes indicate swingers.)

combinations with six yes votes. There is nothing special about the number 6 here; to find the number of combinations with k yes votes, one forms the quotient

$$C_k^n = \frac{n \times (n-1) \times (n-2) \times \cdots \times (n-k+1)}{k \times (k-1) \times (k-2) \times \cdots \times 1}$$

This quotient is called the number of combinations of n items taken k at a time. When speaking, people often refer to C_k^n as "n choose k."

Another common notation for C_k^n is $\binom{n}{k}$.

EXAMPLE: The European Community, 1958

In 1958, the European Community consisted of six countries: Belgium (B), Denmark (D), France (F), Germany (G), Italy (I), and Luxembourg (L). The Council of Ministers employed a weighted voting system that gave F, G, and I four votes, B and D two votes, and L one vote. The quota for passing a measure was a two-thirds majority, or $11\frac{1}{3}$; thus twelve votes were needed to pass a measure, so this system is

$$[q : w(F), w(G), w(I), w(B), w(D), w(L)]$$
$$= [12 : 4, 4, 4, 2, 2, 1]$$

When analyzing any voting system, the first step is to search for dummies. Notice that if L belongs to any coalition, then the total weight of that coalition must be an odd number, since the weights of all the other countries are even. Hence, any winning coalition including L will have a weight of at least 13. If L defects, the coalition will still win. It follows that L is never a swinger; in other words, L is a dummy.

When computing the Banzhaf index of the other countries, we will ignore L. Any winning coalition without L will still be a winning coalition if L joins, and the same countries will be swingers in both. France F is a swinger in the winning coalitions $\{F, B, N, G\}$, $\{F, B, N, I\}$, and $\{F, G, I\}$, which have total weight 12, as well as $\{F, G, I, B\}$ and $\{F, G, I, N\}$, which have weight 14. Thus F is a swinger in five winning coalitions that do not include L. Also, G and I have the same Banzhaf index as F. Because they have only two votes each, B and N can be swingers only in coalitions that have a total weight of 12. These are $\{F, G, B, N\}$, $\{F, I, B, N\}$, and (G, I, B, N); hence B and N are each swingers in three winning coalitions not including L.

By doubling the number of winning coalitions not including L, in which a country is a swinger, we obtain the total number of winning coalitions in which the country is a swinger. Doubling again takes the blocking coalitions into account, and we have the Banzhaf index (20, 20, 20, 12, 12, 0) for this voting system. ▲

► EQUIVALENT VOTING SYSTEMS

Weighted voting systems are but one type of decision-making process that can be classified as a voting system. Consider, for example, a country whose government consists of a king, a council

with three members, and a parliament with five members. All measures must first be considered by parliament. Any measure that is approved by a vote of at least three members of parliament is sent to the council, which can either accept the measure by a majority vote of at least two, or reject the measure, in which case it will be a dead issue. Any measure that is passed by the parliament and the council is sent to the king, who may sign the measure into law, or veto it. If the king should veto the measure, it is dead, for a king's veto cannot be overridden. While this voting system is not a weighted voting system, it has one structure in common with weighted voting systems: winning coalitions. A winning coalition consists of the king, at least two councilors, and at least three members of parliament. We will see later how to count the swingers in winning coalitions and determine the Banzhaf power index for this voting system (with a total of nine voters, there are $2^9 = 512$ coalitions in all—an unpleasant number to deal with by brute force).

In a weighted voting system, the purpose of the weights is to specify which coalitions are winning and which are losing. However, any voting system can be described by simply listing all of the winning coalitions. If there are just two voters, A and B, how many really different voting systems are there? We can agree that the empty coalition ($\varnothing$) is always a losing coalition, and the unanimous coalition ($\{A, B\}$) must be a winning coalition. Therefore there are only three *distinct* voting systems involving A and B: in the first, unanimous consent is required for each measure, so the only winning coalition is $\{A, B\}$. In the second, A is the dictator, and the winning coalitions are $\{A\}$ and $\{A, B\}$. The third voting system has winning coalitions $\{B\}$ and $\{A, B\}$, and B is the dictator. Of course, there is an infinite number of ways that we can assign weights to the voters and a quota for passing measures in this two-voter system. However, there are only three ways to distribute the voting power: A as dictator, B as dictator, or consensus rule.

SPOTLIGHT 12.3   Litigation and the Banzhaf Index — A Mathematical Quagmire

▶▶ ▶ ▶ ▶ ▶ ▶ ▶ ▶ ▶ ▶ ▶

In the 1965 law review article that introduced the Banzhaf index, John F. Banzhaf III focused his attention on the weighted voting system used by the Board of Supervisors of Nassau County, New York. The article inspired legal action against several elected bodies that employ weighted voting systems.

The reason that weighted voting systems exist is best understood by looking at the example of Nassau County. Each significant community within the county should be represented on the Board of Supervisors. At the same time, democratic principles and the Fourteenth Amendment of the U.S. Constitution — as interpreted by the Supreme Court in the "one-person, one-vote" decisions (*Baker v. Carr* and *Reynolds v. Sims*) — require that each voter have approximately the same representation on the Board. The obvious solution, to divide the county into districts approximately the size of the smallest community, is impractical. According to the 1950 census, Glen Cove and Long Beach had populations of less than 20,000, while the county's population was approximately 1,000,000. If supervisorial districts had an average population of 20,000, the Board would have 50 members instead of the six that were envisioned. Weighted voting seemed an ideal compromise between the contradictory needs to have a small board that represented

the smaller communities and that at the same time conformed to the "one-person, one-vote" rule.

The first legal challenge to weighted voting occurred in Washington County, New York. In a 1967 case decided by the New York courts, *Iannucci v. Board of Supervisors,* a weighted voting system similar to the Nassau County system, was ruled unconstitutional. However, the New York State Court of Appeals drew a corollary from Banzhaf's work that provided a way to "fix" weighted voting systems. In its opinion the court said:

Ideally, in any weighted voting plan, it should be mathematically possible for every member of the legislative body to cast the decisive vote on legislation in the same ratio which the population of his constituency bears to the total population.

In other words, the court required that a legislator's *Banzhaf power index,* rather than his or her voting weight, should be proportional to the population of the community that he or she represents. The court went on in the *Iannucci* decision to note that expert opinion and detailed computer analyses would be needed to justify any weighted voting system, and predicted that the courts would eventually be dragged into a "mathematical quagmire."

Nassau County's turn in court came in a series of five cases starting in 1968. Initially, the courts ordered the county to assign its supervisors voting weights in accordance with the *Iannucci* ruling. The county attempted to comply and even passed a law codifying the ruling. The 1972 weights were as follows:

$$[q : w(H_1), w(H_2), w(N), w(B), w(L), w(G)]$$
$$= [71 : 35, 35, 23, 32, 3, 2]$$

except that for issues that would normally require a two-thirds majority, the quota was set at 92. Residents of Hempstead sued: a majority of the county's population lived in that city, and yet their supervisors could not pass a motion. The plan was overturned in the Nassau County court, but the state Court of Appeals allowed the voting system to stand because it met the *Iannucci* standard. The U.S. Supreme Court refused to hear a further appeal. Two further lawsuits protesting against the Nassau County voting system failed to dislodge it in the early 1980s.

The Supreme Court did hear a related case in 1989, *Morris v. Board of Estimate.* The Board of Estimate of New York City was composed of three elected city officials (the Mayor,

Comptroller, and President of the City Council) and the presidents of the five boroughs of the city. The city officials were each given two votes, each borough president had one vote, and the quota was six. The complaint was that the boroughs varied widely in population. Residents of Brooklyn, which is much larger than Staten Island, were underrepresented on the Board. The increased weight given to each city official's vote was intended to partially compensate for the discrepancy. The Supreme Court ruled that the composition of the Board of Estimate violated the Fourteenth Amendment. At the same time, it rejected an aspect of Banzhaf's analysis that is discussed in Spotlight 12.4: his square-root rule.

The Supreme Court's rejection of Banzhaf's square-root rule (which was not closely related to the case against the weighted voting system in use by Nassau County) prompted a fifth lawsuit to overturn the Nassau County voting system, *Jackson v. Nassau County Board of Supervisors.* This case was the mathematical quagmire that the New York State Court of Appeals feared in its *Iannucci* decision. There was conflicting expert testimony on whether the weights were fair by the *Iannucci* standard. There was the slightly irrelevant Supreme Court pronouncement against Banzhaf's method in *Morris.* Finally, there was a brief written by Professor Banzhaf, claiming that the plaintiff (Jackson) was right in saying the voting system was unfair, but that his reasoning was incorrect. In April 1993, the U.S. District Court for Eastern New York rejected the Board's voting system as inherently unfair, thus overturning *Iannucci.* Professor Banzhaf got it right when he entitled his 1965 law review article, "Weighted Voting Doesn't Work."

We will say that two voting systems involving the same numbers of voters are **equivalent** if there is a way for the voters in one system to exchange places with the voters in the other system without changing the winning coalitions. Every coalition that was winning before the switch should be winning after the switch. For example, the weighted voting systems $[q:w(A), w(B)] = [50:49, 1]$ and $[q:w(C), w(D)] = [4:3, 3]$ are equivalent because in each system, unanimous support is required to pass a measure. The exchange

$$A \longleftrightarrow C$$
$$B \longleftrightarrow D$$

takes the single winning coalition $\{A, B\}$ of the first system to the single winning coalition $\{C, D\}$ of the second.

For a second example of equivalent voting systems, consider $[q:w(A), w(B)] = [2:2, 1]$ and $[q:w(A), w(B)] = [5:3, 6]$. In the first, A is a dictator, while in the second, B dictates. The winning coalitions for the first system are $\{A\}$ and $\{A, B\}$ and correspond to the winning coalitions $\{B\}$, $\{A, B\}$ of the second under the switch $A \longleftrightarrow B$, and so the systems are equivalent. This shows that "equivalent" does not mean "the same." There is a distinction between the system where A dictates to B, and the system where B dictates to A. They are equivalent because in each case there is a dictator.

Every voting system with only two voters is equivalent to either a system that has a dictator or one that requires consensus. The number of distinct systems increases as the number of voters increases. Table 12.6 lists all five of the distinct three-voter systems. Each of these systems can be presented as a weighted voting system, and suitable weights are given in the table. Each of these systems can be considered as a four-voter system by simply making the fourth voter a dummy, and there are nine additional four-voter systems in which each voter has power, for a total of 14 distinct systems with four voters.

▶ FINDING THE WEIGHTS

If a voting system is not explicitly described as a weighted voting system, it may or may not be equivalent to a weighted system. We will look at two voting systems; for one we will find an equivalent weighted voting system, and we will show that there is no weighted voting system that is equivalent to the other.

EXAMPLE: Australia

Australia has six states: New South Wales (N), Northern Territory (T), Queensland (Q), South Australia (S), Victoria (V), and Western Australia (W). Some national decisions are made by the following system: the states are each given one vote, and the Federal Government (G) has two votes. The total number of votes is eight, an even number, which makes ties possible. By law, all ties are settled in favor of the Federal

TABLE 12.6 Voting Systems with Three Participants

System	Minimal winning coalitions	$[q:w(A), w(B), w(C)]$	Banzhaf index
Dictator	$\{A\}$	$[3:3, 1, 1]$	$(8, 0, 0)$
One dummy	$\{A, B\}$	$[4:2, 2, 1]$	$(4, 4, 0)$
Majority rule	$\{A, B\}, \{A, C\}, \{B, C\}$	$[2:1, 1, 1]$	$(4, 4, 4)$
Majority with chair veto	$\{A, B\}, \{A, C\}$	$[3:2, 1, 1]$	$(6, 2, 2)$
Unanimous consent	$\{A, B, C\}$	$[3:1, 1, 1]$	$(2, 2, 2)$

Government. Minimal winning coalitions thus consist of the Federal Government with two states, or any coalition of five of the six states.

Although the Federal Government is officially given two votes, it has, in effect, three. If five states are aligned against the government, the extra vote will bring the government's coalition to a total of four, and the five states will still win, 5 – 4. On the other hand, if the government has two states on its side, the additional vote will give the government's coalition the victory, 5 – 4, without needing to break a tie. It follows that the Australian voting system is equivalent to the weighted system

$$[q:w(G), w(N), w(T), w(Q), w(S), w(V), w(W)]$$
$$= [5:3, 1, 1, 1, 1, 1, 1] \quad \blacktriangle$$

One way to compute the Banzhaf power index for this system is by brute force: a list of the $2^7 = 128$ combinations of votes is made, and each vote is examined to see if it is a swing vote. A computer might be helpful in this task if one is a skillful programmer. We will compute the power index by another method below.

EXAMPLE: The King, His Council, and the Parliament

We can express this voting system as a "product" of three simpler systems. The first system is the king alone, who either approves or disapproves. The second is the council, which is a group of three acting by majority rule. The third is the parliament, which has five members, and also acts by majority rule. A measure passes if and only if it is approved by all three of these systems. Notice that the total number of participants in this decision-making process is nine.

We will show that there is no weighted voting system with nine voters that is equivalent to this product system. This is done by assum-

ing that there *is* a weighted voting system that has the same winning and losing coalitions, and showing that a contradiction results. Such a weighted voting system would take the form

$$[q:w_K, w_C, w_C, w_C, w_P, w_P, w_P, w_P, w_P]$$

where w_K is the weight of the king's vote, w_C is the weight of each councilor's vote, and w_P is the weight of each parliament member's vote. Any coalition that includes the king, two councilors, and three members of parliament is winning, so

$$w_K + 2w_C + 3w_P \geq q \qquad (12.1)$$

On the other hand, a coalition consisting of the king, only one councilor, and four members of parliament will lose in the council; thus

$$q > w_K + w_C + 4w_P \qquad (12.2)$$

We can combine the inequalities (12.1) and (12.2) to conclude that

$$w_K + 2w_C + 3w_P > w_K + w_C + 4w_P$$

or

$$w_C > w_P \qquad (12.3)$$

To reach our contradiction, we need to consider one more losing coalition: the king, his entire council, and two members of parliament. The entire weight of this coalition is less than the quota, so

$$q > w_K + 3w_C + 2w_P \qquad (12.4)$$

By combining inequalities (12.1) and (12.4), we obtain

$$w_K + 2w_C + 3w_P > w_K + 3w_C + 2w_P$$

which can be simplified to

$$w_P > w_C \qquad (12.5)$$

Since the inequalities (12.3) and (12.5) contradict each other, we conclude that no weighted voting system is equivalent to the voting system consisting of the king, his council of three, and the parliament of five. Nevertheless, it is possible to compute the Banzhaf power index for this voting system, and this is done below. ▲

EXAMPLE: The Banzhaf Power Index for the Nassau County Board of Supervisors in 1958

Table 12.1 on page 360 gives the weighted vote allotted to each supervisor in the 1958 Nassau County Board. To prevent Hempstead's two supervisors from assuming a dictatorship, the county's charter contained a provision saying no measure shall be adopted without the support of supervisors from at least two municipalities. With this provision, the county's voting system is no longer a weighted voting system, since, as a consequence, the two Hempstead supervisors alone form a losing coalition with 18 votes, while one Hempstead supervisor can team up with the North Hempstead supervisor to win with 16 votes.

We will denote the two Hempstead supervisors as H_1 and H_2. The other supervisors are North Hempstead (N), Oyster Bay (B), Glen Cove (G), and Long Beach (L). To have enough votes to win, a coalition must contain H_1 and H_2, or one Hempstead supervisor and N. Furthermore, $\{H_1, N\}$ and $\{H_2, N\}$ are winning coalitions, but $\{H_1, H_2\}$ is not. Therefore, the minimal winning coalitions are as follows:

$$\{H_1, N\} \quad \{H_2, N\} \quad \{H_1, H_2, B\}$$
$$\{H_1, H_2, G\} \quad \text{and} \quad \{H_1, H_2, L\}$$

The supervisor H_1 will be a swinger in any winning coalition that

(i) Contains $\{H_1, N\}$ and does not contain H_2,

(ii) Contains H_2 and at least one of B, G, or L, but not N.

The set $\{B, G, L\}$ has $2^3 = 8$ subsets, including $\varnothing$. The union of any one of these subsets with $\{H_1, N\}$ will give a coalition of type (i); thus there are 8 coalitions of this type. We can form coalitions of type (ii) by taking the union of $\{H_1, H_2\}$ with any *nonempty* subset of $\{B, G, L\}$. Therefore, there are 7 coalitions of type (ii). The Banzhaf index of H_1 is thus $2 \times (8 + 7) = 30$. Of course, H_2 has the same amount of power.

The supervisor N will be a swinger in any winning coalition that contains one, but not both of the Hempstead supervisors. These coalitions are formed by taking the union of $\{N, H\}$ with any subset of $\{B, G, L\}$ ($\varnothing$ is allowed), where H denotes one of the Hempstead supervisors. In addition, N is a swinger in $\{H_1, H_2, N\}$. It follows that the Banzhaf index of N is $2 \times (2^3 + 2^3 + 1) = 34$.

The supervisors B, G, and L each have a Banzhaf index of 2, since each is a swinger only when he or she forms a winning coalition with the two Hempstead supervisors or a blocking coalition by joining forces with the rest of the county against the two Hempstead supervisors. The Banzhaf index for this voting system is therefore (30, 30, 34, 2, 2, 2). The provision that prevents $\{H_1, H_2\}$ from being a winning coalition has two side effects. One was probably intentional: to give B, G, and L some voting power. The second side effect, to make N more powerful than H_1 or H_2, seems unintentional.

Is this voting system *equivalent* to a weighted voting system? The Banzhaf index gives a clue to the answer. Since N is more powerful than H_1 or H_2, N's vote must have more weight. Let us increase N's weight to 10, and leave the weights of the other supervisors

TABLE 12.7   Minimal Winning Coalitions, Nassau County Supervisors, 1958

Coalition	Corrected weights						Total
	H_1	H_2	N	B	G	L	
$\{H_1, N\}$	9		10				19
$\{H_2, N\}$		9	10				19
$\{H_1, H_2, B\}$	9	9		3			21
$\{H_1, H_2, G\}$	9	9			1		19
$\{H_1, H_2, L\}$	9	9				1	19

unchanged. The minimal winning coalitions, and their new weights, are shown in Table 12.7. The new quota must not exceed the total weight of any of these coalitions. Therefore $q \leq 19$. On the other hand, $\{H_1, H_2\}$ is not a winning coalition, and since its weight is 18, the quota cannot be less than 19. Thus, the weighted voting system

$$[q: w(H_1), w(H_2), w(N), w(B), w(G), w(L)]$$
$$= [19: 9, 9, 10, 3, 1, 1]$$

has the same winning coalitions as the voting system actually used by the Nassau County Board of Supervisors in 1958.   ▲

EXAMPLE: The Banzhaf Power Index for Australia

Since each state has the same weighted vote, each has the same amount of power. Let us consider the State of New South Wales (N), which will be a swinger in the following winning coalitions:

(i) Any coalition containing N and exactly four other states, but not the government,

(ii) Any coalition containing the government, N, and exactly one other state.

We will only count winning coalitions in which N is a swinger (N's Banzhaf index will be twice this number when we include blocking coalitions). A coalition that does not contain one of the two coalitions (i) or (ii) will be a losing coalition, and we will ignore it. Any coalition containing one of coalitions (i) or (ii), as well as at least one other voter, will have more than the quota of 5 votes. It will remain in the winning column if N defects, so N will not be a swinger. Thus, to determine the Banzhaf power index for N, we have to count the coalitions of types (i) and (ii). To form a coalition of type (i), N must be joined by four of the other five states. There are $C_4^5 = 5$ combinations of the other five states with four members; therefore the number of type (i) coalitions is 5. To form a coalition of type (ii), the government and N must be joined by one of the other 5 states. There are $C_1^5 = 5$ ways to choose the other state, so the number of type (ii) coalitions is also 5. We conclude that the Banzhaf index for N (and each of the other states as well) is $2 \times (5 + 5) = 20$.

The government will be a swinger in any winning coalition in which it is joined by two, three, or four other states. If the government defects from any of these winning coalitions, it will cause that coalition to lose. If there are more than four states in the coalition, the government's vote would not be needed to pass, so the government would not be a swinger. The number of coalitions of each size is equal to the number of combinations of the six states with k yes votes, where $k = 2, 3$, and 4. Thus, the government's Banzhaf power index is

$$2 \times (C_2^6 + C_3^6 + C_4^6)$$
$$= 2 \times \left(\frac{6 \times 5}{2 \times 1} + \frac{6 \times 5 \times 4}{3 \times 2 \times 1} + \frac{6 \times 5 \times 4 \times 3}{4 \times 3 \times 2 \times 1} \right)$$
$$= 2 \times (15 + 20 + 15) = 100$$

Therefore the Banzhaf power index for this voting system is (100, 20, 20, 20, 20, 20, 20). With this system, the government has five times as much voting power as any state. ▲

EXAMPLE: A King's Power

We will determine the Banzhaf power index for the voting system in which the king must agree with majorities of both his three member council and the five member parliament to pass a measure. The king is a swinger in every winning coalition, since he has veto power. There are $C_2^3 + C_3^3 = 4$ ways to form a majority coalition (which would have two or three members) in the council, and

$$C_3^5 + C_4^5 + C_5^5 = 10 + 5 + 1 = 16$$

ways to form a three, four, or five member majority coalition in the parliament. By the multiplication principle, there are $4 \times 16 = 64$ winning coalitions, and thus the Banzhaf power index of the king is $2 \times 64 = 128$.

Now let us determine the power index for a councilor. He or she will be a swinger in any coalition that includes the king, one other councilor, and three, four, or five members of parliament. There are $C_1^2 = 2$ ways to choose the other councilor, and, as before, 16 ways to choose a majority coalition in parliament. A councilor's Banzhaf index is therefore $2 \times 2 \times 16 = 64$.

A member of parliament will be a swinger in any coalition that includes the king, two or three councilors, and exactly two other members of parliament. There are 4 ways to get a majority coalition in the council (this was determined when we found the Banzhaf power index of the king), and $C_2^4 = 6$ ways to choose the other two members of parliament. By the multiplication principle, the Banzhaf power index for a member of parliament is $2 \times 4 \times 6 = 48$. The power index for the voting system as a whole is therefore (128, 64, 64, 64, 48, 48, 48, 48, 48). The king is twice as powerful as a councilor, and $2\frac{2}{3}$ times as powerful as a member of parliament. ▲

▶ THE SHAPLEY–SHUBIK POWER INDEX

In some political situations, coalitions are built one voter at a time. The most important voter in the sequence is the one who turns the coalition from a losing coalition into a winning coalition. In 1954, a power index based on this idea was introduced by Lloyd Shapley and Martin Shubik. To calculate the index, one considers **permutations** of voters. A permutation is an ordered list of all of the voters, and every possible ordering must be considered to determine the Shapley–Shubik index. The number of permutations is found by applying the multiplication principle. There are n voters who could be first on a list. When the first voter is known, there are $n - 1$ left who could be the second. As we form the list, the number of voters available for the next position decreases until finally only one voter is left to be last on the list. According to the multiplication principle, the number of permutations is then the **factorial** of n:

$$n! = n \times (n - 1) \times (n - 2) \times \cdots \times 2 \times 1$$

Each permutation can be considered to represent a spectrum of opinion on an issue. The first voter of the permutation is the most committed, the second is perhaps slightly less so, and so on until we reach the last voter, for whom the measure is anathema. A coalition is built, starting with the first voter. If the coalition is to become a winning coalition, its total weight will eventually have to reach the quota, and the voter in the permutation whose vote would make the coalition a winning coalition (if he or she could be induced to join) is called the **pivotal voter** in the permutation. Each permutation has exactly one pivotal voter. For example, consider the three-person committee, with the weighted voting system

$$[q : w(A),\, w(B),\, w(C)] = [3 : 2,\, 1,\, 1]$$

Table 12.8 shows the 3! = 6 permutations of the members, A, B, C. Next to each permutation, the total weight of the first voter, of the first two voters, and of all three voters is shown. The first number in this sequence to exceed the quota is underlined, and the corresponding pivotal voter's symbol is circled. We see that A is pivotal in four permutations, while B and C are each pivotal in one. The **Shapley–Shubik power index** of a voter is the fraction of the permutations in which that voter is pivotal. Thus in the three-person committee, the Shapley–Shubik index for A is $\frac{4}{6}$, and B and C each have a Shapley–Shubik index of $\frac{1}{6}$. According to the Shapley–Shubik model, the chairperson

of this committee, A, has four times as much voting power as an ordinary member. Recall that the Banzhaf power index for the same committee was $(3, 1, 1)$, so according to the Banzhaf model, A is three times as powerful as B or C.

EXAMPLE: The Corporation with Four Shareholders

Let us calculate the Shapley–Shubik power index for the corporation whose shareholders A, B, C, and D own 40%, 30%, 20%, and 10% of the stock, respectively. This weighted voting system is presented as

$$[q : w(A),\, w(B),\, w(C),\, w(D)]$$
$$= [51 : 40,\, 30,\, 20,\, 10]$$

There are 4! = 24 permutations to consider; they are shown in Table 12.9. In ten of the permutations, A is the pivot, B and C are each pivots in six, and D is the pivot in two permutations. Therefore, the Shapley–Shubik index for this voting system is

$$\left(\frac{10}{24}, \frac{6}{24}, \frac{6}{24}, \frac{2}{24}\right) = \left(\frac{5}{12}, \frac{1}{4}, \frac{1}{4}, \frac{1}{12}\right)$$

The Banzhaf power index gives the same ratios of power in this case. In most cases, however, the two indices do not agree, and they may not even be approximately the same. ▲

How to Compute the Shapley–Shubik Index

If there is a large number of voters with different weights, it is not a simple task to calculate the Shapley–Shubik index. If there are five voters, then there will be 5! = 120 permutations, enough to make the task of counting the number of pivots for each voter unpleasant. It is not even remotely practical to list the 51! permutations of the 50

TABLE 12.8 Permutations and Pivotal Voters for the Three-Person Committee

Permutations			Weights		
A	$\circledB$	C	2	$\underline{3}$	4
A	$\circledC$	B	2	$\underline{3}$	4
B	$\circledA$	C	1	$\underline{3}$	4
B	C	$\circledA$	1	2	$\underline{4}$
C	$\circledA$	B	1	$\underline{3}$	4
C	B	$\circledA$	1	2	$\underline{4}$

TABLE 12.9   Permutations and Pivotal Voters for the Four-Person Corporation

Permutations				Weights				Pivot			
A	(B)	C	D	40	70	90	100		B		
A	(B)	D	C	40	70	80	100		B		
A	(C)	B	D	40	60	90	100			C	
A	(C)	D	B	40	60	70	100			C	
A	D	(B)	C	40	50	80	100		B		
A	D	(C)	B	40	50	70	100			C	
B	(A)	C	D	30	70	90	100	A			
B	(A)	D	C	30	70	80	100	A			
B	C	(A)	D	30	50	90	100	A			
B	C	(D)	A	30	50	60	100				D
B	D	(A)	C	30	40	80	100	A			
B	D	(C)	A	30	40	60	100			C	
C	(A)	B	D	20	60	90	100	A			
C	(A)	D	B	20	60	70	100	A			
C	B	(A)	D	20	50	90	100	A			
C	B	(D)	A	20	50	60	100				D
C	D	(A)	B	20	30	70	100	A			
C	D	(B)	A	20	50	60	100		B		
D	A	(B)	C	10	50	80	100		B		
D	A	(C)	B	10	50	70	100			C	
D	B	(A)	C	10	40	80	100	A			
D	B	(C)	A	10	40	60	100			C	
D	C	(A)	B	10	30	70	100	A			
D	C	(B)	A	10	30	60	100		B		

states plus the District of Columbia that make up the U.S. Electoral College — 51! is a 67-digit number. By using an ingenious counting method, the Shapley–Shubik index of the Electoral College has been calculated (an article by John P. Lambert is cited in Suggested Readings, p. 389).

In an n-voter weighted voting system in which all voters have the same weight, all voters will be pivots in the same number of permutations. Thus they will all have the same Shapley–Shubik power index, $1/n$. The next simplest case is one where all of the voters but one have the same weight (as in the Australia example).

EXAMPLE: The Shapley–Shubik Index of the Australian Voting System

The brute force method that we have employed in calculating the Shapley–Shubik index so far will be too time-consuming for this seven-voter system, because we would have to examine 7! = 5040 permutations. Notice that all of the states will have the same power index. If we can find the government's power index, we can calculate each state's power index algebraically.

Suppose that x is the government's index, and y is the power index of New South Wales (and of the other five states). Since the sum of the Shapley–Shubik indices of all the voters is 1,

$$x + 6y = 1$$

When x has been determined, we can calculate y by solving this equation.

The federal government will be a pivotal voter in the following types of permutations:

$$N \quad T \quad \textcircled{G} \quad Q \quad S \quad V \quad W$$
$$N \quad T \quad Q \quad \textcircled{G} \quad S \quad V \quad W$$

and

$$N \quad T \quad Q \quad S \quad \textcircled{G} \quad V \quad W$$

In order to be pivotal, the government must be in either third, fourth, or fifth position in the permutation. In the permutations shown in the preceding illustration, the states were listed in an arbitrary order. Since there are six states, there are actually 6! ways to reorder them, each giving three permutations in which the government is a pivot. The government's Shapley–Shubik index is therefore $\frac{3 \times 6!}{7!}$. We resist the temptation to "simplify" this expression by multiplying out the 6! to get 720, and the 7! to get 5040. This would be a poor strategy, because quotients involving factorials can usually be simplified by canceling. We will write the quotient as

$$\frac{3 \times \cancel{6} \times \cancel{5} \times \cancel{4} \times \cancel{3} \times \cancel{2} \times \cancel{1}}{7 \times \cancel{6} \times \cancel{5} \times \cancel{4} \times \cancel{3} \times \cancel{2} \times \cancel{1}}$$

where the cancellation has been indicated. It follows that the government's Shapley–Shubik power index is $\frac{3}{7}$. Letting y denote the index for one of the six states, we have

$$\frac{3}{7} + 6y = 1$$

so $y = \frac{1}{6}(1 - \frac{3}{7}) = \frac{2}{21}$. The Shapley–Shubik index for this weighted voting system is therefore

$$\left(\frac{3}{7}, \frac{2}{21}, \frac{2}{21}, \frac{2}{21}, \frac{2}{21}, \frac{2}{21}, \frac{2}{21} \right)$$

Since $\frac{3}{7} \div \frac{2}{21} = \frac{9}{2} = 4\frac{1}{2}$, the Shapley–Shubik model indicates that the government has four and one-half times as much power as an individual state. This is in close agreement with the Banzhaf model, which held that the government was five times as powerful as an individual state (p. 377).   ▲

The calculation in this example was simplified by the following **factorial cancellation rule:** If $k < n$, then

$$\frac{k!}{n!} = \frac{1}{n \times (n-1) \times \cdots \times (k+1)}$$

This rule works because

$$n! = n \times (n-1) \times \cdots \times (k+1) \times k$$
$$\times (k-1) \times \cdots \times 1$$
$$= [n \times (n-1) \times \cdots \times (k+1)] \times k!$$

Using this rule, we calculated $\frac{3 \times 6!}{7!} = \frac{3}{7}$. Of course, we could have done that calculation without knowing the factorial cancellation rule. However, factorial cancellation plays an essential part in the following example.

EXAMPLE: A Corporation with 9001 Shareholders

A corporation has one shareholder who owns 10% of the stock. The remaining 90% of the stock is evenly divided among 9000 other shareholders. How much voting power does the 10% shareholder have? Our calculation will as-

SPOTLIGHT 12.4 Large or Small: Which States Are Favored in the Electoral College System?

Citizens of small states are proportionally much better represented in the Electoral College than citizens of large states are, but does that make them more powerful? Karl Mundt, a former Senator from South Dakota, wrote in 1968:

> The Electoral College, operating under the general ticket or unit rule ("winner take all") method, is, in my estimation, the most unfair, inaccurate, uncertain, and undemocratic institution of all.

Obviously, Senator Mundt felt that the overrepresentation enjoyed by South Dakota in the Electoral College did not compensate for the disadvantage imposed by the unit rule.

The unit rule has two effects: first, it makes the Electoral College into a weighted voting system with 51 participants (the states). As we know, a state's power may be poorly related to its voting weight (the number of electors that it has). The second effect is upon the role of the individual voter. If a resident of South Dakota casts a vote, he or she has more of a chance of affecting the outcome in South Da-

kota, and thus delivering three electoral votes to the candidate of his or her choice, than a resident of California would have, because California is much more populous. However, California has 54 electoral votes, not 3. If the resident of California casts a swing vote, he or she will make a bigger difference than the South Dakota voter will.

The Electoral College is about as close as we will get to a weighted voting system that fulfills the assumptions of the Banzhaf model for measuring voting power. The states have no influence on each other's votes, and if the election is close, each state will have a 50% chance of voting for either candidate. Unfortunately, it is not easy to compute the Banzhaf index of such a large system. However, the Shapley – Shubik index, which is believed to be more or less proportional to the Banzhaf index for the Electoral College, has been computed based on the data from the 1960 and 1970 censuses. In 1972 California was the most populous state, with 45 electoral votes, or 8.364% of the Electoral College. However, California's Shapley – Shubik index in the 1972 Electoral College was considerably more, 8.831%. Smaller states do not fare as well. The Shapley–Shubik index and the state's share of the electoral vote were closest to equality for the seventh most populous state, Michigan. Michigan had 21 votes, or 3.903% of the 1972 Electoral College, and Michigan's Shapley–Shubik index was 3.917%. Six states and the District of Columbia had the minimum number of electoral votes, 3, or 0.558% of the Electoral College. Each had a Shapley–Shubik index of 0.541%. These differences are not dramatic; indeed, they are not enough to offset the small states' advantage of having a larger number of electoral votes in proportion to their population.

The unit rule's main benefit to the larger states was first described by John F. Banzhaf III in a law review article entitled "One Man, 3.312 . . . Votes." Banzhaf determined the probability that an individual voter would be a swinger in a general election in his or her state. Assume that the number of voters participating in the election is odd. We will put this number equal to $2m + 1$. To be a swinger, the voter would have to be one of exactly $m + 1$ people voting for the Republican candidate, or one of the same number of voters voting for the Democrat. He or she would be in a coalition with m of the $2m$ other voters. The number of such coalitions is C_m^{2m}; this must be multiplied by 2 since either party might be favored. Even in the case of a very small state, m would be more than 100,000, and we know that for large values of m, C_m^{2m} is approximately equal to $\frac{2^{2m}}{\sqrt{\pi m}}$.

Therefore, the probability that our voter will be a swinger in some coalition favoring one of the two candidates (Banzhaf did not consider multicandidate elections) is

$$\frac{2 \times C_m^{2m}}{2^{2m+1}} \approx \frac{1}{\sqrt{\pi m}}$$

In other words, a voter's ability to influence the outcome of an election is inversely proportional to the square root of the voting population of his or her state. This is Banzhaf's square root rule.

In a large state, such as California, there might be 15,000,000 voters. We will say 15,000,001 voters, because it is simpler to work with odd numbers (although the results turn out to be the same with even numbers), so that $m = 7,500,000$. A voter's probability of casting a swing vote in the large state would be $\frac{1}{\sqrt{7,500,000 \times \pi}} = 0.0206\%$. Now consider a small state with 300,001 voters, and $m = 150,000$. Here, the probability of being a swinger is $\frac{1}{\sqrt{m\pi}} = 0.1457\%$. The resident of the small state is more likely to be a swinger, but can only swing three or four electoral votes. The voter in the larger state has one-seventh the chance of being a swinger, but swings 15 times as many electoral votes. It is this that makes the large state voter more powerful. By combining the two effects of the unit rule, Professor Banzhaf calculated that in the 1964 election, a voter in New York had approximately 3.312 times as much power as a voter in the District of Columbia.

sume that the small shareholders are fully participating in the process (an unusual occurrence), and we will determine the Shapley–Shubik index. We will encounter one of the largest numbers to be seen in a mathematics text in the process — 9001! — which would occupy more than seven pages of this book when written out.

Each of the small shareholders owns $\frac{90}{9000} =$ 0.01% of the stock. Since the 10% shareholder can only be pivotal if he casts his vote after more than 40%, but not more than 50%, of the shares have voted, his vote will be pivotal in permutations where he is in position 4002 through position 5001 — a total of 1000 positions. Each of these voting positions for the 10% shareholder occurs in 9000! permutations, so the Shapley–Shubik index for this shareholder is

$$1000 \times \frac{9000!}{9001!} = \frac{1000}{9001}$$

This is one instance where the Banzhaf and Shapley–Shubik indices widely disagree. We will see later that the Banzhaf index gives almost 100% of the power to the 10% shareholder, in contrast to the approximately 11% granted to that shareholder by the Shapley–Shubik index. ▲

▶ COMPARISON OF THE BANZHAF AND SHAPLEY–SHUBIK INDICES

Deciding which power index best describes the distribution of power in a particular voting system is a subjective judgment. The heart of the issue is the distinction between permutations and combinations. A permutation represents a possible range of opinion concerning an issue, so the Shapley–Shubik index is more appropriate if there is a spectrum of opinion about most issues before the vot-

ing body. The Banzhaf index should be preferred if the typical issue has no middle ground.

In many legislatures, the political dynamic is too complex to be accurately modeled by either index. When there are many points of view to consider, the opinions that are represented cannot be strictly ordered between two extremes. In these cases, the Banzhaf and Shapley–Shubik indices provide measurements of voting power from two points of view.

To understand the difference between the two indices more deeply, let us look again at Table 12.8, where the permutations of members of the three-person committee are displayed. If we remove all members to the right of the pivotal voter in each permutation, the result will be a winning coalition, in which the pivotal voter is a swinger. Thus, the table shows the coalition $\{A, B\}$ twice (in the first and third permutations). In the first permutation, we have A Ⓑ, while the second shows B Ⓐ. Thus, both A and B, who are swingers in the coalition, receive credit for being pivots. Similarly, the coalition $\{A, C\}$ appears twice, so that each member is counted as a pivot. Now consider the coalition $\{A, B, C\}$, in which only A is a swinger. This coalition is also listed twice: as B C Ⓐ, and as C B Ⓐ. The voter A gets credit for being a pivot in this coalition twice: once if the coalition starts forming with B and again if the coalition starts forming with C. It is this "extra credit" that explains the greater power ascribed to A by the Shapley–Shubik model.

In voting systems that have more voters, the multiple counting of swingers in winning coalitions becomes even more pronounced. Let us consider a voting system with n members. Suppose a voter, B, is a swinger in some winning coalition that has k members. Let $A_1, A_2, \ldots, A_{k-1}$ be the other members of this coalition, and let C_1, $C_2, \ldots, C_{n-k}$ be the voters who did not join the coalition. Then B is the pivot in the permutation

$$A_1 \quad A_2 \cdots A_{k-1} \quad Ⓑ \quad C_1 \quad C_2 \cdots C_{n-k}$$

This shows that every swinger is a pivot in some permutation. As we have seen, however, a swinger can be a pivot in more than one permutation. Now we will see precisely how many ways B gets credit for being a pivot as a result of being a swinger in this particular coalition. There are $(k-1)!$ ways to reorder the voters $A_1, \ldots, A_{k-1}$, and each such reordering produces another permutation in which B is a swinger for the same coalition. This is not all, however: there are also $(n-k)!$ ways to reorder the voters $C_1, \ldots, C_{n-k}$ who didn't join the coalition. If we reorder these voters, it has no effect on the winning coalition, or B's position as pivot. By the multiplication principle, B gets credit for being a pivot as a result of being a swinger in the same coalition $(k-1)! \times (n-k)!$ times. This effect is greatest if k is relatively small or if $n-k$ is relatively small. For example, if there are $n = 10$ voters, and B is a swinger in a winning coalition $\{A, B\}$, then B will be counted as a pivot $1! \times (10-2)! = 40,320$ times. Thus, B would be counted the same number of times as a pivot if B is a swinger in a winning coalition with nine members. On the other hand, if B is a swinger in a five-member winning coalition, then B is only counted as a pivot $(5-1)! \times (10-5)! = 2880$ times.

OPTIONAL ▶ **INTERPRETING THE INDICES AS PROBABILITIES**

It is natural to interpret the Shapley–Shubik power index as a probability. In this model, the measure of the power of a voter A is

$$\frac{\text{Number of permutations in which } A \text{ is the pivot}}{\text{Total number of permutations}}$$

If it is assumed that each permutation of the voters is equally likely, then a voter's Shapley–Shubik index is the probability that he or she will be the pivotal voter. To evaluate the validity of the Shapley–Shubik model for a particular voting system, it is necessary to evaluate the assumption on which it is based: are all permutations of voters equally likely? The permutation that applies in a given voting situation represents the spectrum of opinion on the issue among the voters. The voter that is most committed to the issue is first, and the voter most opposed to the issue will be last. Unless the first voter is a dictator, or the last voter has veto power, neither will be pivots. Instead, the pivot is usually a moderate on the issue. If the same voters in a legislature usually take the extreme positions, then the assumption that all permutations are equally likely does not hold, and the Shapley–Shubik index will ascribe to the extremists more power than they actually have. However, political alignments will change over the years. Without knowing who will be occupying the seats in a legislature, we can more easily accept the Shapley–Shubik model over the long term.

The Banzhaf model can also be interpreted in terms of probability. If we assume that all voting combinations are equally likely, then the probability that A will be a swinger in a winning or blocking coalition is equal to

$$\frac{\text{Number of voting combinations in which } A \text{ is a swinger}}{\text{Total number of voting combinations}}$$

With n voters, there are 2^n voting combinations in all; therefore, the probability that A will be a swinger is obtained by dividing A's Banzhaf index by 2^n.

The example that most clearly delineates the difference between the Shapley–Shubik index and the Banzhaf index is the 9001 shareholder corporation (p. 381). We have seen that the 10% shareholder has about 11% of the voting power, as measured by the Shapley–Shubik index. According to the Banzhaf index, that shareholder has almost all of the power, and the probabilistic interpretation of the Banzhaf index makes it clear why this is so. The underlying assumption, that all coalitions are equally likely, would be valid if each shareholder, large or small, let a coin toss decide his or her vote. The 10% shareholder will be a swinger unless his coalition is joined by fewer than 4000 small shareholders (which would make it a losing, nonblocking coalition), or if his coalition is joined by at least 5000 small shareholders (which would make it a blocking or winning coalition with no swingers).

Suppose that 9000 people each toss a coin, and we count the number of heads. The expected number of heads is $\frac{1}{2} \times 9000 = 4500$, and the standard deviation (see Chapter 8) is $\sqrt{\frac{1}{2} \times \frac{1}{2} \times 9000}$, which is approximately 47. By the 68–95–99.7 rule, 68% of the time there will be 4500 ± 47 heads, or between 4453 and 4547 heads. Ninety-five percent of the time there will be between 4406 and 4594 heads, and 99.7% of the time, between 4359 and 4641 heads. The probability of getting between 4000 and 5000 heads is approximately the same as the probability of an outcome within *ten* times the standard deviation from the expected number of heads. This probability turns out to be 99.99999999999999999985%. This is the probability that the 10% shareholder will be a swinger, if everyone votes by tossing a coin. Under these circumstances, however, a small shareholder has practically no chance of being a swinger, since he or she would have to be in a coalition with either 5000 other small shareholders, or 4000 other shareholders and the 10% shareholder.

EXAMPLE: Majority Rule and Unanimous Consent

The most common voting systems that we encounter are majority rule, where each voter has the same weight, and unanimous consent, where all voters must approve a measure for it to pass. Trial juries, for example, require unanimous consent.

Let us consider what the Shapley–Shubik model has to say about majority rule. In any permutation of n voters, the pivot will be the voter in position $\frac{n}{2} + 1$ (if n is even), or position $\frac{n+1}{2}$ (if n is odd). If all permutations are equally likely, then each voter will have the same probability to be in the pivot position. Each voter will have a Shapley–Shubik index equal to $\frac{1}{n}$. Under unanimous rule, the pivot position is held by the last voter — the voter who is the least enthusiastic about the issue. Each voter has the same chance to occupy that position also, so each voter's Shapley–Shubik index is $\frac{1}{n}$ with unanimous rule.

Let us now determine the Banzhaf index for each voter under majority rule. To simplify the discussion, we will assume an odd number $n = 2m + 1$ of voters. A minimal winning coalition will then have $m + 1$ voters, and a minimal blocking coalition will also have $m + 1$ voters. Every voter in one of these minimal coalitions is a swinger, and larger coalitions have no swingers. A voter A can be joined by any group of exactly m voters out of the $2m$ other voters to form a minimal winning coalition. Therefore, A is a swinger in C_m^{2m} winning coalitions, and the same number of blocking coalitions (each $(m + 1)$-member coalition must be counted twice, once if it casts its votes in favor of a measure, and again if it votes against a measure). His or her probability of being a swinger is

$$2 \times \frac{C_m^{2m}}{2^{2m+1}}$$

For example, if $n = 11$, a minimal winning coalition will have 6 members. Since $C_5^{10} = 252$ and $2^{11} = 2048$, the probability that A will be a swinger is $2 \times \frac{252}{2048} = 0.246$, approximately. As n gets larger, a voter's chance of being a swinger decreases. For example, if there are 21 voters, the probability is

$$2 \times \frac{C_{10}^{20}}{2^{21}} = 2 \times \frac{184{,}756}{2{,}097{,}152} = 0.176$$

approximately. With a large number of voters, it is cumbersome to evaluate C_m^{2m}, but there is a formula that gives a good approximation:

$$C_m^{2m} \approx \frac{2^{2m}}{\sqrt{m\pi}}$$

where $\pi = 3.14159 \ldots$ is the ratio between a circle's circumference and its diameter. (The story of how π turns up in this context is a long one.) Thus, if A is a voter in a majority-rule system with $2m$ voters, A's chance of being a swinger is

$$\frac{2C_m^{2m}}{2^{2m+1}} = \frac{1}{2^{2m}} C_m^{2m}$$

$$\approx \frac{1}{2^{2m}} \frac{2^{2m}}{\sqrt{m\pi}}$$

$$= \frac{1}{\sqrt{m\pi}}$$

This approximation is rather good even when $m = 10$; it estimates the probability as 0.178, rather than the correct value of 0.176. As m increases, the approximation improves (see Table 12.10).

With unanimous rule, there is only one winning coalition, which consists of the entire set of voters. All coalitions except $\varnothing$ are blocking coalitions, but a voter will be a swinger in only one blocking coalition: the coalition consisting of that voter alone. Thus, each voter is a swinger in just two coalitions, and has a probability of $2/2^n = 1/2^{n-1}$ of being a swinger. This probability is *much* less than the probability of being a swinger under majority rule. ▲

This example illustrates another interesting distinction between the Shapley–Shubik and Banzhaf models. As far as the Shapley–Shubik index is concerned, the majority-rule system and the unanimous-rule system provide the same distribution of power: all voters are equal. All voters

TABLE 12.10   **Exact and Approximate Probabilities That a Voter in a Majority-Rule System Will Be a Swinger**

Number of voters ($2m + 1$)	Exact probability ($C_m^{2m}/2^{2m}$)	Approximate probability ($1/\sqrt{m\pi}$)
11	0.2461	0.2523
21	0.1762	0.1784
41	0.1254	0.1262
81	0.0889	0.0892
161	0.0630	0.0631
321	0.0446	0.0446

have equal power in these two voting systems with the Banzhaf model, but in the majority-rule system, each voter is more likely to make a difference than he or she would be in the unanimous-rule system. ◀

▶REVIEW VOCABULARY

Banzhaf power index   A numerical measure of power for participants in a voting system. A participant's Banzhaf index is the number of winning or blocking coalitions in which he or she is a swinger.

Blocking coalition   A set of participants in a voting system that can prevent a measure from passing by voting against it.

C_k^m A set with m elements has C_k^m subsets with k elements. This number, referred to as "m choose k," is given by the formula

$$C_k^m = \frac{m!}{k!(m-k)!}$$

which can be simplified by using the factorial cancellation rule:

$$C_k^m = \frac{m \times (m-1) \times \cdots \times (m-k+1)}{k \times (k-1) \times \cdots \times 1}$$

Coalition A set — consisting of some, all, or none of the participants in a voting system — that has united to vote either in favor of or against a measure.

Dictator A participant in a voting system who can pass any issue even if all other voters oppose it, and block any issue even if all other voters approve it.

Dummy A participant that has no power in a voting system. A dummy is never a swinger in any winning or blocking coalition, and is never the pivot in any permutation.

Equivalent voting systems Two voting systems are equivalent if there is a way for all of the voters of the first system to exchange places with the voters of the second system and preserve all winning coalitions.

Factorial If n is a positive integer, the factorial of n (denoted $n!$) is the product of all the positive integers less than or equal to n. It is usually a big number: 10! is a seven digit number, 9000! is a seven page number.

Factorial cancellation rule A quotient involving factorials can be simplified by using the formula

$$\frac{k!}{n!} = \frac{1}{n \times (n-1) \times \cdots \times (k+1)}$$

where $k < n$.

Losing coalition A coalition that does not have the voting power to pass a measure alone.

Minimal blocking coalition A blocking coalition that will not block if any member defects. Each member is a swinger.

Minimal winning coalition A winning coalition that will become losing if any member defects. Each member is a swinger.

Permutation A specific ordering from first to last of the elements of a set; for example, an ordering of the participants in a voting system.

Pivot The first voter in a permutation who, with his or her predecessors in the permutation, will form a winning coalition. Each permutation has one and only one pivot.

Quota The minimum number of votes necessary to pass a measure in a weighted voting system.

Shapley–Shubik power index A numerical measure of power for participants in a voting system. A participant's Shapley–Shubik index is the number of permutations of the voters in which he or she is the pivot, divided by the number of permutations ($n!$ if there are n participants).

Swinger A member of a winning coalition whose vote is essential for the coalition to win, or a member of a blocking coalition whose vote is essential for the coalition to block.

Veto power A voter has veto power if no issue can pass without his or her vote. A voter with veto power is a one-person blocking coalition.

Voting combination A list giving each participant's vote on an issue. When there are n voters, there is a total of 2^n voting combinations; of these, there will be C_k^n voting combinations with exactly k yes votes.

Weight The number of votes assigned to a voter in a weighted voting system, or the total number of votes of all voters in a coalition.

Weighted voting system A voting system in which the participants can have different numbers of votes. It can be represented as $[q:w(A_1), w(A_2), \ldots, w(A_n)]$, where $A_1, \ldots, A_n$ are the voters, $w(A_1), \ldots, w(A_n)$ represent the numbers of votes held by these voters, and where q is the quota necessary to win.

Winning coalition A set of participants in a voting system that can pass a measure by voting for it.

▶ SUGGESTED READINGS

BANZHAF, JOHN F., III: "Weighted Voting Doesn't Work," *Rutgers Law Review*, 19:317–343 (1965). The author defines the Banzhaf index, and uses it to show that the weighted voting system in use by the Nassau County Board of Supervisors was unfair.

BANZHAF, JOHN F., III: "One Man, 3.312 . . . Votes: A Mathematical Analysis of the Electoral College," *Villanova Law Review*, 13:304–332 (1968). The author shows that voters residing in the more populous states have more power to influence the outcome of a presidential election. The same issue has commentaries on Banzhaf's analysis (pages 333–346). Another commentary appears in the same *Review*, 14:86–96 (1968).

BARRETT, CAROL, AND HANNA NEWCOMBE: "Weighted Voting in International Organizations," *Peace Research Reviews*, vol. II (1968), Canadian Peace Research Institute, Oakville, Ontario. A list of international organizations that employ weighted voting systems is included in this article.

BRAMS, STEVEN J.: *Game Theory and Politics*, Free Press, New York, 1975. Chapter 5 treats the Shapley–Shubik and Banzhaf indices.

———: *Paradoxes in Politics: An Introduction to the Nonobvious in Political Science*, Free Press, New York, 1976.

———: *The Presidential Election Game*, Yale University Press, New Haven, 1978.

BRAMS, STEVEN J., W. F. LUCAS, AND P. D. STRAFFIN, JR. (EDS.): *Political and Related Models. Modules in Applied Mathematics*, vol. 2, Springer-Verlag, New York, 1983. Chapters 9–11 are devoted to measuring power in weighted and other types of voting systems. The Banzhaf and Shapley–Shubik indices are the focus of chapters 9 and 11; chapter 10 is about an index based on counting minimal winning coalitions.

DUBEY, PRADEEP, AND L. S. SHAPLEY: "Mathematical Properties of the Banzhaf Power Index," *Mathematics of Operations Research* 4:99–131 (1979). A technical article that presents an axiomatic treatment of the Banzhaf index. The introduction summarizes the history of power indices.

GOLDBERG, SAMUEL: *Probability in Social Science*, Birkhäuser, Boston, 1983. Chapter 1 is devoted to the Shapley–Shubik index.

Iannucci v. Board of Supervisors of Washington County 20 N.Y. 2d 244, 251, 229 N.E. 2d 195, 198, 282 N.Y.S. 2d 502, 507 (1967). This code will help a law librarian find this case for you. It opened a "mathematical quagmire."

LAMBERT, JOHN P.: "Voting Games, Power Indices, and Presidential Elections," *UMAP Journal*, 9(3):213–267 (1988).

LUCAS, WILLIAM F.: *Fair Voting: Weighted Votes for Unequal Constituencies*, COMAP: HistoMAP Module 19, Lexington, Mass., 1992. An introduction to the power indices with emphasis on the historical aspects.

Morris v. Board of Estimate 489 U.S. 688 (1989). In this case, the Supreme Court ruled that the Banzhaf square root rule was not a realistic measure of a citizen's voting power, and could not be used to justify weighted voting in the Board of Estimate.

OWEN, GUILLERMO: *Game Theory*, 2nd ed., Academic Press, Orlando, Fla., 1982.

SCHRODT, PHILIP A.: "Simulation of Weighted Voting: The Banzhaf Index," *Byte* 9(3):138–154 (1984). This article gives a "Monte Carlo" estimate for the Banzhaf index of the Electoral College. Since it is clearly impossible to count all voting combinations of the 51 participants in the Electoral College, the author counts the swingers in a random sample of voting combinations.

SHUBIK, MARTIN (ED.): *Game Theory and Related Approaches to Social Behavior*, Wiley, New York, 1964. This book reprints Shapley and Shubik's 1954 paper, in which their index was introduced, as chapter 9. Chapter 10, originally published in 1960 and 1962, is an analysis by I. Mann and L. S. Shapley of the Electoral College. Chapter 11 is a study of power in the U.S. Congress based on the Shapley–Shubik index, originally published in 1956 by R. D. Luce and A. A. Rogow. Chapter 12 is a more general discussion of the measurement of power, by J. Harsanyi. It is a reprint of a 1962 paper.

▶**EXERCISES**

1. a. List the 16 possible combinations of how four voters, A, B, C, and D, can vote either yes (Y) or no (N) on an issue.
   b. List the 16 subsets of the set $\{A, B, C, D\}$.
   c. How do the lists in parts a and b correspond to each other?
   d. In how many of the combinations in part a is the vote
      (1) 4 Y to 0 N?
      (2) 3 Y to 1 N?
      (3) 2 Y to 2 N?

■ 2. Will a blocking coalition turn into a winning coalition if every voter in the coalition votes Y? Consider the following examples:
   a. A committee with 9 members, each with one vote, where majority rules.
   b. A committee with 12 members, each with one vote, where majority rules.
   c. A jury with 9 members in a criminal trial, where a unanimous decision is necessary to convict or to acquit.

3. For each of the following weighted voting systems, list the following

(1) All minimal winning coalitions.

(2) All winning coalitions containing the voter A.

(3) All minimal blocking coalitions.

(4) All blocking coalitions containing the voter A.

(5) All losing coalitions containing the voter A.

(6) All dummy voters.
   a. $[q:w(A), w(B)] = [51:52, 48]$.
   b. $[q:w(A), w(B), w(C)] = [2:1, 1, 1]$.
   c. $[q:w(A), w(B), w(C)] = [3:2, 2, 1]$.
   d. $[q:w(A), w(B), w(C)] = [8:5, 4, 3]$.
   e. $[q:w(A), w(B), w(C), w(D)] = [51:45, 43, 8, 4]$.
   f. $[q:w(A), w(B), w(C), w(D)] = [51:28, 27, 26, 19]$.
   g. $[q:w(A), w(B), w(C), w(D)] = [16:10, 10, 10, 1]$.
   h. $[q:w(A), w(B), w(C), w(D), w(E)] = [21:10, 10, 10, 10, 1]$.

■ 4. Comparing the voting systems in parts g and h of Exercise 3 reveals a paradox. Explain it.

5. Describe the minimal winning coalitions for the weighted voting system

$$[q:w(A), w(B), w(C), w(D)] = [51:28, 24, 24, 24].$$

6. For the voting system in Exercise 5, list all winning coalitions in which
   a. A is a swinger.
   b. B is a swinger.

■ Discussion exercise.

7. For the voting system in Exercise 5, list all permutations of the voters in which

    a. A is the pivot.

    b. B is the pivot.

8. Calculate the Banzhaf index for the voting system in Exercise 5.

9. Calculate the Shapley–Shubik index for the system in Exercise 5.

10. Calculate the Banzhaf index for each of the weighted voting systems in Exercise 3.

11. Calculate the Banzhaf index for each of the following weighted voting systems:

    a. $[q:w(A), w(B), w(C), w(D)] = [51:28, 24, 24, 24]$.

    b. $[q:w(A), w(B), w(C), w(D)] = [6:4, 3, 2, 1]$.

12. Before being declared unconstitutional by a federal district court in 1993, the weighted voting system of the Nassau County Board of Supervisors was changed several times. The weights in use since 1958 were as follows:

Year	$[q:w(H_1), w(H_2), w(N), w(B), w(G), w(L)]$
1958	$[16:9, 9, 7, 3, 1, 1]$
1964	$[58:31, 31, 21, 28, 2, 2]$
1970	$[63:31, 31, 21, 28, 2, 2]$
1976	$[71:35, 35, 23, 32, 2, 3]$
1982	$[65:30, 28, 15, 22, 6, 7]$

where H_1 is the presiding supervisor, always from Hempstead. The second supervisor from Hempstead is H_2, and N, B, G, and L are the supervisors from North Hempstead, Oyster Bay, Glen Cove, and Long Beach, respectively.

■   a. From 1970 on, more than a simple majority was required to pass any measure. Give an argument in favor of this policy from the viewpoint of a supervisor who would benefit from it, and an argument against the policy from the viewpoint of a supervisor who would lose some power.

    b. In which years were some supervisors dummy voters?

    c. Suppose that the two Hempstead supervisors always vote together. In which years are some of the supervisors dummy voters?

    d. Assume that the two Hempstead supervisors always agree, so that the Board is in effect a five-voter system. Determine the Banzhaf index of this system in each year.

▲   e. Assume that the supervisors (including the two from Hempstead) vote independently. Calculate the Banzhaf index in each year.

▲ Advanced exercise.

■   f. The table below gives the 1980 census for each municipality, the number of votes assigned to each supervisor, and the Banzhaf index for each supervisor in 1982. Do you think the voting scheme is fair?

Nassau County Board of Supervisors, 1982

Supervisor from	Population	Number of votes	Banzhaf power index
Hempstead (Presiding)	738,517	30	30
Hempstead		28	26
North Hempstead	218,624	15	18
Oyster Bay	305,750	22	22
Glen Cove	24,618	6	2
Long Beach	43,073	7	6
Totals	1,321,582	108	104

Number of votes required to pass a measure: 65.

13. The Nassau County Board of Supervisors had a higher quota for votes that require a two-thirds majority. In 1982, this quota was 72.
    a. Assume that the two Hempstead supervisors vote together, and determine the Banzhaf index for each municipality.
▲   b. Assume that all supervisors vote independently, and determine the Banzhaf index.
■   c. Is there any justification for using the quota 72 to represent a two-thirds majority? Is the distribution of power proportional to population (see table in Exercise 12f)?

14. The New York City Board of Estimate consists of the Mayor, the Comptroller, the City Council President, and the presidents of each of the five boroughs. It employed a voting system in which the city officials each had two votes, and the borough presidents each had one; the quota to pass a measure was 6. This voting system was declared unconstitutional by the U.S. Supreme Court in 1989 (see *Morris v. Board of Estimate*, listed in the Suggested Readings section).
    a. Describe the minimal winning coalitions.
    b. Are these coalitions the same as the minimal blocking coalitions?
▲   c. Determine the Banzhaf power index.

15. An objection to the voting system in use by the New York Board of Estimate (see Exercise 14) is that the more populous boroughs, Brooklyn (K) and Manhattan (H), had exactly the same power as the smaller boroughs, Queens (Q), Bronx (X), and Staten Island (S). One proposal that addressed this objection

■ Discussion exercise.   ▲ Advanced exercise.

was to give the city officials each 3 votes, the borough presidents from Brooklyn and Manhattan 2 votes each, and leave the remaining borough presidents with 1 vote each. The quota would be increased to 9, so that this voting system had the form

$$[q:w(M), w(C), w(P), w(K), w(H), w(Q), w(X), w(S)]$$
$$= [9:3, 3, 3, 2, 2, 1, 1, 1]$$

    a. Describe the minimal winning coalitions.
    b. Are these the same as the minimal blocking coalitions?
▲   c. Determine the Banzhaf power index.

16. Here is another proposed weighted voting system for the Board of Estimate (see Exercises 14 and 15):

$$[q:w(M), w(C), w(P), w(K), w(H), w(Q), w(X), w(S)]$$
$$= [71:35, 35, 35, 11.3, 7.3, 9.6, 6.0, 1.8]$$

    a. Find a simpler system of weights that yields the same voting system.
▲   b. Calculate the Banzhaf power index.

17. A corporation has four shareholders and a total of 100 shares. The quota for passing a measure is the votes of shareholders owning 51 or more shares. The number of shares owned are as follows:

A	48 shares
B	23 shares
C	22 shares
D	7 shares

All transactions must be in whole numbers of shares; sales of fractional shares are not permitted.

    a. List the minimal winning coalitions.
    b. How many shares can A sell, without changing the set of minimal winning coalitions, to one of the other shareholders, or to E, who currently owns none of the stock? (Notice that A may be able to sell more to D than to B, etc.)
    c. How many shares can D sell, without changing the set of minimal winning coalitions, to B, C, D, or E? Again, it is conceivable that D would be able to sell more to one stockholder than to another.
    d. How many shares can D sell to $B, C, D,$ or E without becoming a dummy?
    e. How many shares can B sell to C without changing the set of minimal winning coalitions?

18. A five-member committee has the following voting system. The chairperson can pass or block any motion that she supports or opposes, provided that at least one other member is on her side.

■   a. Show that this voting system is equivalent to the weighted voting system

$$[q:w(C), w(M_1), w(M_2), w(M_3), w(M_4)] = [4:3, 1, 1, 1, 1]$$

     b. Calculate the Banzhaf index for this voting system.

     c. Calculate the Shapley–Shubik index for this voting system.

19. A nine-member committee has a chairperson and eight ordinary members. A motion can pass if and only if it has the support of the chairperson and at least two other members, or if it has the support of all eight ordinary members.

     a. Find an equivalent weighted voting system.

     b. Determine the Banzhaf index.

     c. Determine the Shapley–Shubik index.

20. Which of the following voting systems are equivalent to weighted voting systems? Find the weights and quota for those that are.

     a. A committee of three faculty and the Dean. To pass a measure, at least two faculty members and the Dean must vote yes.

     b. A committee of three faculty, the Dean, and the Provost. To pass a measure, two faculty, the Dean, and the Provost must vote yes.

     c. A three-member faculty committee and a three-member administration committee vote separately on each issue. The measure passes if it receives at least two votes in each of the committees.

21. Which of the following voting systems is equivalent to the voting system in use by the corporation in Exercise 17?

     a. $[q:w(A), w(B), w(C)] = [3:1, 1, 1, 1]$

     b. $[q:w(A), w(B), w(C)] = [3:2, 1, 1, 1]$

     c. $[q:w(A), w(B), w(C)] = [5:3, 1, 1, 1]$

     d. $[q:w(A), w(B), w(C)] = [5:3, 2, 1, 1]$

     e. $[q:w(A), w(B), w(C)] = [5:3, 2, 2, 2]$

22. Determine the Banzhaf and Shapley–Shubik power indices for the corporation in Exercise 17.

23. The United Nations Security Council has five permanent members: China, France, Russia, the United Kingdom, and the United States, and ten other members that serve two-year terms. To resolve a dispute not involving a member of the council, nine votes, including the votes of each of the permanent members, is required. (Thus, each permanent member has veto power.)

     a. Show that this voting system is equivalent to the weighted voting system in which each permanent member has 7 votes, each ordinary member has 1 vote, and the quota is 39.

     b. Compute the Banzhaf index for the Security Council.

     c. Compute the Shapley–Shubik index for the Security Council. (This is harder than computing the Banzhaf index.)

     d. Which index is more appropriate for measuring power in the Security Council?

24. A committee has a chairperson and six ordinary members. It uses majority rule, except that the chairperson is only allowed to vote when it is necessary to break a tie. Give an equivalent weighted voting system for the committee.

25. The committee in Exercise 24 is reduced in size: now there are just five ordinary members. The rules are the same. Give an equivalent weighted voting system for the committee.

26. Consider a four-person voting system with voters A, B, C, and D. The winning coalitions are

$$\{A, B, C, D\} \quad \{A, B, C\} \quad \{A, B, D\} \quad \{A, C, D\} \quad \{A, B\}$$

    a. List the minimal winning coalitions.
    b. List the minimal blocking coalitions.
    c. Determine the Banzhaf power index for this voting system.
    d. Determine the Shapley–Shubik power index for this voting system.
    e. Find an equivalent weighted voting system.

27. Consider the n-person voting system in which each participant has one vote and a simple majority wins. In the notation for weighted voting systems, this system can be expressed as

$$[q:w(V_1), \ldots, w(V_n)] = \left[\frac{n+1}{2} : 1, \ldots, 1 \right]$$

Assume that all voting combinations are equally likely. What is the probability that a voter will be a swinger, when $n = 1, 2, 3, 4, 5, 6$, or 7? When $n = 2m + 1$ is an odd number, we know that each voter is a swinger in $2C_m^{2m}$ combinations. How many swings does a voter have when $n = 2m$ is even?

28. A corporation has 100 shares of stock outstanding. There are eighty shareholders who own one share each, and one shareholder who owns 20 shares. To pass an issue, owners representing 51 shares must vote yes. Determine the Shapley–Shubik index of each shareholder.

■ 29. Explain why the large shareholder of the corporation in Exercise 28 has almost all of the power, as measured by the Banzhaf index.

■ 30. The Vice President of the United States is allowed to break ties in the U.S. Senate. How does his or her Banzhaf power index compare with that of an individual senator?

■ 31. How many *distinct* (nonequivalent) voting systems with four voters can you find? Systems that have dummies don't count. The challenge is to find all nine. Hint: Each voting system can be specified by listing all of the minimal winning coalitions. If there are no dummies, each voter belongs to at least one minimal winning coalition. Remember that two winning coalitions cannot be disjoint, if the system is to be decisive.

● 32. The ABC College Student Senate is a five-member body with simple majority rule; each member has one vote. Two members of the Senate, A and B, have a pact to vote together.
    a. In how many voting combinations do A and B cast the same vote?
    b. Say that $\{A, B\}$ is a swinger in a voting combination if the result of the vote would be reversed if both A and B defected. In how many of the voting combinations in part a is $\{A, B\}$ a swinger?

c. In how many voting combinations in part a would Senator C, who is not involved in a pact, be a swinger?

d. How does A's pact with B affect the probability that A will be a swinger? How does it affect C's chance of being a swinger?

● 33. Senators A and B have quarreled in the ABC College Student Senate (see Exercise 32). In the future, they will never be on the same side of *any* issue.

a. In how many voting combinations will A and B be on opposite sides?

b. In how many of the voting combinations in part a is A a swinger?

c. In how many of the voting combinations in part a is C a swinger?

d. How does the quarrel affect the distribution of power, as compared to independent voting?

● 34. This problem is like Exercise 32, in that two members of the Student Senate have formed an alliance. However, this senate has only four members, and Senators C and D, who are not involved in the pact, are demoralized.

a. Show that the voting system in effect is

$$[q:w(\{A, B\}), w(C), w(D)] = [3:2, 1, 1]$$

and compute the Banzhaf index.

b. Compute the Shapley–Shubik index.

c. Senators C and D quarrel, and resolve to oppose each other on every issue. Assuming that A and B maintain their alliance, determine the Banzhaf index and the probability that each voter ($\{A, B\}$ is considered to be a single voter here) will be a swinger.

d. Two voters are said to be on opposite sides of a permutation if one voter appears after the pivot, and the other does not. If there is a quarrel, the Shapley–Shubik index of a voter will be the number of permutations in which the voter is the pivot, and the quarrelers are on opposite sides, divided by the number of permutations in which the quarrelers are on opposite sides. Determine the Shapley–Shubik index for the Student Senate with the quarrel.

● 35. This is a version of a paradox discovered by Professor Steven Brams. A senate has seven members, whom we label A–G. Senators A and B have formed an alliance, and Senators C and D have also. Would it be advantageous for Senators E, F, and G to form an alliance? Compare the distribution of power in the weighted voting systems

$$[q:w(\{A, B\}), w(\{C, D\}), w(E), w(F), w(G)] = [4:2, 2, 1, 1, 1] \qquad (12.6)$$

with the system

$$[q:w(\{A, B\}), w(\{C, D\}), w(\{E, F, G\})] = [4:2, 2, 3] \qquad (12.7)$$

In which system do E, F, and G have the greatest share of power as measured by the

a. Banzhaf power index?

b. Shapley–Shubik power index?

● Optional exercise.

● 36. A legislative body has n members, and a two-thirds majority is required to amend the by-laws. What is the probability that your representative will be a swinger on a vote to amend the by-laws? If it is convenient, you may assume that $n = 3m + 1$ for some integer m (for example, n might be equal to 100; then $m = 33$).

● 37. Here is a useful approximation, which is reasonably good provided that k and $n - k$ are both bigger than 10:

$$C_k^n \approx \sqrt{\frac{n^{2n+1}}{2\pi k^{2k+1} (n - k)(2n - 2k + 1)}}$$

Use this approximation and your answer to Exercise 36 to estimate the probability that your legislator will be a swinger in a by-law amendment vote when $n = 40$, $n = 100$, and $n = 400$.

▶ WRITING PROJECTS

1. The Electoral College. The world's most important weighted voting system is the Electoral College (see Spotlights 12.1 and 12.4). Three alternate methods to elect the President of the United States have been proposed:

▶ *Direct election.* The Electoral College would be abolished, and the candidate receiving a plurality of the votes would be elected. Most versions of this system include a runoff election or a vote in the House of Representatives in cases where no candidate receives more than 40% of the vote.

▶ *District system.* Every presidential candidate would be entitled to choose one elector from each congressional district in which he or she received a plurality of the vote in the general election. In addition, a candidate would choose two electors from every state that he or she carried. In effect, the unit rule would be retained for the District of Columbia and for states with a single congressional district. Larger states would typically have electors representing both parties.

▶ *Proportional system.* Each state, and the District of Columbia, would have fractional electoral votes assigned to each candidate in proportion to the number of popular votes received. Thus, if Candidate A received 203,567 popular votes out of 458,922 cast in the District of Columbia, which has 3 electoral votes, then A would receive

$$\frac{203,567}{458,922} \times 3 = 1.3307294$$

electoral votes. Obviously, there would be no actual electors involved in the process.

Should the present Electoral College, operating under the unit rule, be replaced by one of these systems? A starting point to answer this question is the article by John Banzhaf III, "One Man, 3.312 . . . Votes" (1968). Another reference is the article, "The A Priori Voting Strength of the Electoral College," by I. Mann and L. S. Shapley. This can be found in the anthology edited by Martin Shubik (1964). Another reference is *The Presidential Election Game*, by Steven Brams (1978), which contains useful references to Senate hearings on Electoral College reform. The Banzhaf index for the Electoral College in the 1984 and 1988 elections was estimated by Philip A. Schrodt (1984), and the Shapley–Shubik index for these elections can be found in the article by John Lambert (1988).

2. Bandwagons. In a political convention, it is useful to join the winning coalition before it has enough votes to win. Since everyone knows this, when a coalition reaches a critical size, a *bandwagon* develops and the coalition grows rapidly. Give a detailed account of an occurrence of the bandwagon phenomenon in a recent political convention of legislative session. The book by Steven Brams has a discussion of bandwagons. Another recommended source would be chapter 11, by Philip D. Straffin, Jr., of *Module in Applied Mathematics*, edited by Steven J. Brams, W. F. Lucas, and P. D. Straffin, Jr. (1983).

3. Weighted Voting in _____. Choose an organization that uses weighted voting, and explain the organization's reasons for using weighted voting. If you can, compute the Banzhaf and Shapley–Shubik indices for the system. If they differ significantly in their allocation of power, which, if either, represents the true balance of power better?

13 Fair Division

Achieving fairness is an important aspect of decision making. How can a group of students divide up a pizza so that each perceives his or her share as fair? How should an estate be divided equitably among the heirs? Can the head of a department fairly allocate such tasks as teaching assignments and committee activities? The goal in such fair-allocation problems is to have all persons feel that they obtain a fair and unbiased share of the available benefits or losses, in light of the various limitations present.

In practice we often appeal to someone with authority or experience to answer such questions. Mother will surely divide the cake into fair portions. Managers, we assume, will assign work in a just manner. Labor and management will call in an expert arbitrator to detail the resolution of a dispute. Failing agreement between a husband and wife, a judge may dictate the division of the assets in divorce proceedings.

Another common approach to fairness is first to devise some appropriate measure of inequity and then attempt to find an allocation that minimizes inequity — either the largest individual inequity or the total group inequity. Some examples

of this approach to fairness appear in Chapter 12. Still other approaches take a more statistical view of the problem and seek to achieve even distributions in some average sense, for example, over repeated allocations.

Our approach in this chapter differs from all these. We wish to arrive at methods for fair division that involve only the participants themselves and that satisfy individuals with *different* value systems. We do not want to assume that each person assigns the same worth or utility to objects or tasks, nor do we want some third party to guess what another's values might or should be. The items to be divided need be neither uniform (homogeneous) nor symmetrical in character. We also want to avoid talking about equity in terms of statistical notions. For example, it is hardly fair to suggest that we flip a coin, giving you all the cake if it is heads and me the whole cake if it is tails, even though we each have an "expected value of half a cake."

In general, a **fair-division problem** consists of n individuals, called **players,** whom we indicate by the numbers $1, 2, \ldots, i, \ldots, n$. In some instances the players will be called by their names or indicated by letters A, B, C, ... We use the letter i when addressing an arbitrary or generic player. The n participants must partition some set S of goods (or losses) into n disjoint parts S_1, S_2, . . . , S_i, . . . , S_n. The objective is to find subsets S_i so that each person i considers the share S_i as fair in his or her own personal value system.

We will illustrate some **fair-division schemes** for two different cases. First, we will examine the **continuous case,** when the S to be divided is *finely divisible,* like a cake. Second, we will consider the more difficult **discrete case,** where S consists of various *indivisible goods.* For example, an estate's house, furniture, silverware and china, artworks, and automobiles are objects that cannot be further subdivided. In both cases, discrete and continuous, we will consider some fair-division procedures that ensure "reasonable" allocations of goods.

Figure 13.1 A fair-division scheme can help us decide how to best divide a cherry pie among any number of individuals. [Photo by John Paul Endress/The Stock Market.]

► THE CONTINUOUS CASE

Let us consider first the fair division of objects that are finely divisible into a great variety of different parts. Examples include cake, land, money, and a pile of sand. We will begin with the case of only two participants.

Two Players

A traditional technique for dividing a finely divisible object S in a fair manner between two players A and B is "one cuts, the other chooses":

1. Player A divides the set S into two pieces S_1 and S_2.

2. Player B picks either piece, S_1 or S_2.

3. Player A is given the piece not selected by player B.

Because the roles of players A and B, the *cutter* and the *chooser,* are asymmetrical, a player may prefer one role, say, the chooser, over the other. It is not uncommon to flip a coin at the start to decide whether player A or B will be the cutter.

This method (with or without the coin flip) assures that each of the players can realize a fair piece, provided that two axioms are met:

1. Each player is able to divide the set S into two parts so that *either* one of the pieces is acceptable by that player as fair.

2. Given *any* division of S into two parts, each player will find at least one of the pieces acceptable.

These axioms apply to the positions of cutter and chooser, respectively. The first guarantees that the cutter will be satisfied with whatever piece is remaining, whereas the second assures the chooser of finding at least one piece agreeable, no matter how the cutter divides object S.

If these axioms are not in fact realized, then a fair result may not be reached using this method. For example, if one player will accept only the full cake as fair and the other wants a (nonempty) piece, then no method can arrive at a fair split. Any fair-division scheme must presume some reasonable conditions similar to axioms 1 and 2.

It is important to note that neither our axioms nor our method assume that players can assign numerical measures to pieces of S or that they must delineate all acceptable pieces before the cut is made. The method requires only that the operational requirements in axioms 1 and 2 are met and each player is able to decide whether a given piece is fair. This is an example of applying mathematical concepts to practical problems without the explicit use of numbers.

We should also note that the object of a fair-division scheme is to assure that (under certain assumptions) each player can act so as to obtain a fair share. It is not intended to guarantee a fair piece to a player who decides to "gamble" and thus acquire *more* than a fair share. A gambler may try to make use of information (or merely guess) about the value systems of the other players in an attempt to gain more. Consider a brother and sister who use the two-person method to divide the cake left over from dinner. Assume that the brother, Joel, is the cutter and that he "knows" that his sister, Laura, craves the icing on the cake. Joel may well cut the cake into two pieces: S_1 is a large piece with little icing, and S_2 is a small piece with a good deal of icing. He hopes that Laura will select S_2 and leave him with S_1, which he views as more than half of the cake. On the other hand, Laura may well "teach" her brother a lesson and select S_1 instead; although she receives a less desired piece this time, she lets Joel know that he cannot expect to get away with such gambling in any future divisions.

Three Players

There are several extensions of the preceding two-person divide-and-choose scheme to deal with in the case of three (or more) participants. We denote the three players by 1, 2, and 3. Their goal is to divide a commodity S into three disjoint parts and to allocate the parts to the players so that each player views his or her part as fair. One way is to apply the **lone-divider method**.

EXAMPLE: The Lone Divider

1. Player 1 divides the set S into three pieces S_1, S_2, and S_3.

2. Player 2 denotes which of the pieces in step 1 are acceptable (as fair).

3. Player 3 similarly indicates which of the pieces in step 1 are acceptable.

Player 3 is typically unaware of the selection made by player 2 in step 2 (although this is not essential to arrive at a fair division). Steps 2 and 3 may well take place simultaneously, in the sense that neither player 2 nor 3 is aware of the other's choice. ▲

Two different situations may arise out of this three-step process. In *Case A* we should be able to make a feasible assignment of fair pieces from the three original parts S_1, S_2, and S_3 cut by player 1 in step 1. For example, let's assume that player 2 says that only piece S_1 is acceptable, and player 3 indicates that either piece S_1 or S_3 is acceptable. (We stress again that different players may hold different opinions regarding what is fair.) This information can be summarized in the following 3-by-3 table (matrix) with entries of 0 or 1.

		Pieces		
		S_1	S_2	S_3
	1	1	1	1
Players	2	1	0	0
	3	1	0	1

An entry of 1 in a column means that the above piece is acceptable to the player in the corresponding row, whereas a 0 means it is unacceptable (unfair) in that player's judgment. The three 1s in the first row of the table state that player 1 has divided

S into three pieces so that any one of the three is acceptable to him or her.

A feasible assignment of fair pieces to the three players in this illustration is as follows:

Player 2 must receive piece S_1

Player 3 can then receive piece S_3

Player 1 finally receives piece S_2

This assignment is indicated by circling three 1s, as in the following table.

		S_1	S_2	S_3
	1	1	①	1
Players	2	①	0	0
	3	1	0	①

Note that there is exactly one circle in each row and each column. Note also that this assignment is the only possible fair one for this particular array of 0s and 1s. A pictorial view of *Case A* is shown in Figure 13.2.

In *Case B* it may not be possible to assign the three original pieces S_1, S_2, and S_3 cut by player 1 in a fair manner. In other words, though both player 2 and player 3 find S_1 acceptable, both find pieces S_2 and S_3 unacceptable. In such instances, we must extend the lone-divider method to find an assignment strategy that is acceptable to all three players.

EXAMPLE: The Lone-Divider Plus

4. Assign *one* of the three pieces S_1, S_2, or S_3 to player 1, but assign a particular piece S_i ($i = 1$, 2, or 3) that both player 2 and player 3 found unacceptable.

5. Players 2 and 3 then enter into a two-person, fair-division procedure for dividing the set $T = S - S_i$ consisting of the remaining two pieces. ▲

1. Set S is to be divided between three players.

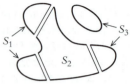

2. Player 1 divides S into three pieces S_1, S_2, and S_3.

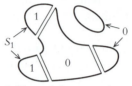

3. Player 2 denotes only piece S_1 as acceptable: (1,0,0).

4. Player 3 denotes pieces S_1 and S_3 as acceptable: (1,0,1).

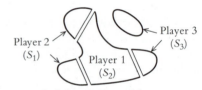

5. A fair division of the three pieces to Players 1, 2, and 3.

Figure 13.2 *Case A:* A three-person fair division using the lone-divider method.

Case B can be illustrated by the following table:

		S_1	S_2	S_3
	1	1	1	1
Players	2	1	0	0
	3	1	0	0

In this illustration we see that both pieces S_2 and S_3 have a 1 in the first row, and 0s in the second and third rows. Since players 2 and 3 both view the pieces S_2 and S_3 as unacceptable, they should be willing to have *one* of these two pieces assigned to player 1. We can accordingly assign piece S_2 (or S_3) to player 1. Then players 2 and 3 undertake a two-person, fair-division scheme for the remaining two parts, S_1 and S_3 (or S_1 and S_2, respectively). These remaining two sets are now viewed as a single commodity T that will be redivided fairly into two new pieces T_1 and T_2 for players 2 and 3.

This extension of the lone-divider method will allow each player to obtain a fair piece (in this person's own value system) whenever the following three axioms hold.

1. Each player is able to divide the set S into three parts so that *any* one of the pieces is acceptable as fair.

2. Given *any* division of the set S into three parts, each player will find at least one of the three pieces acceptable.

3. Any two players who view a piece S_i as unacceptable can obtain a fair share from a two-person, fair-division scheme over the remaining set $T = S - S_i$.

Axiom 1 applies to the player who cuts S into three pieces. It implies that one row in the table of 0s and 1s will consist of *all* 1s. (This was row 1 in our illustration just given.) Axiom 2 implies that our table will have *at least* one 1 in *each* row. Axiom 3 is an "additivity assumption." It says "the whole has the same value as the sum of its parts." It assures that the remaining two players in *Case B* see enough value in the diminished set $T = S - S_i$ to realize a fair piece from step 5.

We should emphasize that players 2 and 3 in the lone-divider method do not typically designate just their *one* most desired piece from the three cut by player 1. For example, assume that player 3 viewed the respective pieces S_1, S_2, and S_3 as 40%, 25%, and 35% of the value of the set S. Player 3 should declare both S_1 and S_3 as acceptable, and not just the best piece S_1. Otherwise, the piece S_3 might go to player 1, leaving players 2 and 3 to

divide the set $S - S_3$ consisting of the pieces S_1 and S_2. Player 3 may view $S - S_3$ as only 65% (= 40% + 25%) of S, and he may not obtain his fair share of 33 1/3% from this 65%.

As a final consideration, let us imagine a 3-by-3 table whose entries are 0s or 1s. Assume that the three entries in the first row are all 1s, and that the second and third rows each have at least one 1 as an entry. There are 7 possibilities for the second and third rows: (1,0,0), (0,1,0), (0,0,1), (1,1,0), (1,0,1), (0,1,1), and (1,1,1,). Since axiom 2 rules out a row of three zeros, (0,0,0), it follows that there can be $7^2 = 49$ different tables, given axioms 1 and 3. Only three out of these 49 possibilities lead to *Case B* in the lone-divider method. The other 46 alternatives result in the simpler *Case A*. In other words, steps 4 and 5 are unlikely to arise often when using this three-person, fair-division scheme.

An Alternative Method for Multiple Players

Other fair-division schemes have been proposed for three or more players, along with corresponding axioms to assure that acceptable distributions can always be reached. One such "divide and choose" approach for the case of *n* participants is called the **last-diminisher method.** Let's see how this method works for seven students who want to divide a chocolate cake:

1. Student 1 cuts any piece he or she views as fair from the cake.

2. Student 2 can "pass" on this piece, or diminish the piece cut if he or she views it as too large.

3. Students 3, 4, 5, 6, and 7 each have the right — but not the obligation — to further diminish the remaining piece as their turns come.

4. The piece is assigned to the *last* player who elected to diminish it, who then exits from the game. This is student 1 if all the other players passed, that is, if no one chose to challenge the piece as unfair.

SPOTLIGHT 13.1   Fifty Years of Cake Cutting

▶ ▶ ▶   ▶ ▶ ▶ ▶ ▶ ▶ ▶ ▶ ▶ ▶ ▶

Although the cut-and-choose method of dividing a cake undoubtedly dates back to antiquity, it is still in use today, as when a parent suggests to two squabbling children how to divide up a dessert or the Halloween candy. But it also has more serious uses. For example, the Law of the Sea Treaty specifies that whenever a developed country wants to mine a portion of the seabed, the country must propose a division of this portion into two tracts. An international mining company called the Enterprise, funded by the developed countries but representing the interests of developing countries, chooses the tract it prefers and the divider receives the other tract. In this manner, developed countries preserve parts of the seabed for commercial development by the developing countries, which they could not otherwise afford by themselves.

The modern era of cake cutting really began with the investigations of the Polish mathematician Hugo Steinhaus during World War II. His research, and that of dozens of others over the past half century, involved dealing with, among other things, two fundamental difficulties:

1. Allocation schemes that work in the context of two or three players often do not generalize easily to the context of four or more players.

2. Constructive allocation procedures that yield envy-free allocations (meaning that each player receives a piece of cake that he or she would not trade for the piece received by anyone

else) are considerably harder to obtain than procedures that yield fair allocations (meaning that each of the players thinks he or she receives at least a share of the cake equal to the total value of the cake divided by the number of players).

The mathematics inspired by these two difficulties over the past 50 years constitutes a rather elegant corner of the large and important area of fair division. Steinhaus's investigations in the 1940s led to his observation that there is a rather natural extension of cut-and-choose to the case of three players. This is essentially the lone-divider method discussed on pages 402–403, and further illustrated in Exercise 1. Steinhaus, however, did not make use of matrix notation to keep track of which players approved of which pieces and so was not led to the natural generalization of his solution to any number of players, found by Harold W. Kuhn of Princeton University in 1967.

Unable to extend his scheme from three players to four players, Steinhaus proposed the problem to some Polish colleagues. Two of them, Stefan Banach and Bronislaw Knaster, solved this problem in the mid-1940s by producing the last-diminisher method (see page 403 and Exercise 4).

In addition to the fair-allocation procedures of Banach, Knaster, and Kuhn, there are at least two other well known constructive procedures for obtaining a fair allocation among four or more players. One of these is due to

Arlington M. Fink of Iowa State University and the other (and most efficient in terms of the number of cuts needed) to the Israeli mathematicians Shimon Even and Azaria Paz at the Technion in Haifa, and, independently but later, to Jack M. Robertson and William A. Webb of Washington State University.

In 1986, Douglas R. Woodall of the University of Nottingham in England provided an algorithmic procedure to implement a 1946 observation of Knaster: if we have at hand a piece of cake that at least two of the players value differently, then there is a scheme in which *every* player will think he or she received *strictly more* than $1/n$ of the value of the cake. Additional contributions along these lines have been made by Theodore P. Hill of the Georgia Institute of Technology and Jerzy Legut of the Institute of Mathematics, Technical University of Wroclaw in Poland.

One other constructive procedure of note — although different in flavor from the others — is the 1961 recasting by Lester E. Dubins and Edwin H. Spanier of the University of California at Berkeley of the last-diminisher method in terms of a "moving-knife scheme" (illustrated in Exercise 14). The trade-off here involves giving up the "discrete" nature of the last-diminisher method in exchange for the conceptual simplicity of the moving knife.

Although the *existence* of an envy-free allocation (even for $n \geq 4$ players) was known to Steinhaus in the 1940s, the first *constructive* procedure for producing an envy-free allocation among three players was not found until around 1960. At that time, John L. Selfridge of Northern Illinois University, and, later but independently, John H. Conway of Princeton University, found an elegant scheme for producing an envy-free allocation among three players.

Although never published by either, the scheme was quickly and widely disseminated by Richard K. Guy of the University of Calgary and others; eventually it appeared in several treatments of the problem by different authors.

In 1980, a moving-knife procedure for producing an envy-free allocation among three players was found by Walter R. Stromquist of Daniel Wagner Associates. Shortly thereafter, another scheme, capable of being recast as a moving-knife solution for the three-player case, was found by a law professor at the University of Virginia, Saul X. Levmore, and a former student of his, Elizabeth Early Cook.

In 1992, Steven J. Brams, a political scientist at New York University, and Alan D. Taylor, a mathematician at Union College, succeeded in finding a constructive procedure for producing an envy-free allocation among four or more players.

Beyond the division of an estate among heirs (illustrated in the text and exercises), this envy-free trimming procedure is potentially applicable to other real-world fair-division problems. For example, if no single party has a majority of seats in a parliamentary democracy, several parties will usually try to form a coalition government that does have a majority of seats, in which case they must agree on which party gets what cabinet posts. (Presumably, larger parties are entitled to more seats, which can be built into the procedure.) Insofar as the parties differ on what posts they consider most valuable — a "green" party, for example, might be most interested in an environmental post — the parties face the same problem that confronts heirs in dividing up an estate. Because the trimming procedure results in an envy-free allocation of the posts, it should facilitate the coalition members' reaching a satisfactory agreement.

5. The preceding steps are repeated with the remaining six players, using the original cake less the piece that was assigned to the last diminisher.

6. This process is repeated for five players, then four players, then three, until only two players remain in the game.

7. The final two players can continue in the same manner or decide to use the "cut-and-choose" method.

The reader may want to list the assumptions (axioms) under which this method will guarantee that a fair division can always result. For example, it might be essential that the values of all the deleted parts always add up to the total value of the cake. (The whole is equal to the sum of its parts.) That is, the act of dismembering the whole into many pieces has not lessened any student's appetite for the part he or she eventually receives. In the case of a crumbly cake and actual slicing, this assumption may well fail. On the other hand, if the object being divided is land, the preliminary cuts may only be made by marking on a map, which would hardly devalue the land itself. As in all mathematical models, the conclusions may not apply if the assumptions are not satisfied.

▶ THE PROBLEM OF ENVY

"One cuts, the other chooses" (p. 400) has a property that none of the other procedures we have discussed so far possesses: it can assure each player of a piece of cake he or she considers the largest or tied for the largest. In the case of only two players, this means that each player can get what he or she perceives to be at least half the cake, no matter what the other player does.

To offer this guarantee, however, the cutter must play "conservatively" by dividing the cake exactly in half, according to his or her valuation of it. That way, whatever piece the chooser selects, the cutter is assured of getting one-half.

Of course, by selecting what he or she thinks is the larger of the two pieces, the chooser will get more than half, unless he or she values the cake exactly the same as the cutter. In the latter case, the chooser, too, will get exactly one-half.

Because both the cutter and chooser think they get at least half the cake, they will not envy each other. More generally, we define a fair-division procedure to be **envy-free** if each player has a strategy that can guarantee him or her a piece of cake that is at least as large as any other player's piece (as each player perceives the allocations), no matter what the other players do.

The lone-divider method that we illustrated for three players is not envy-free in *Case B*, when players 2 and 3 both find one piece unacceptable. Although these players will not envy each other's piece when one cuts and the other chooses, player 1 may think that this is not a 50-50 split. Indeed, if player 1 divided the cake initially into what he or she thought was three equal pieces (axiom 1 says this is possible), an unequal split of the remaining 2/3 of the cake by players 2 and 3 means that player 1 will prefer the larger of these two pieces to the 1/3 he or she got. Consequently, player 1 will envy the person who got this larger piece.

To try to sidestep this problem, assume that player 1 does not divide the cake initially into three equal pieces but instead has a most preferred piece, or two acceptable pieces. If both players 2 and 3 consider as acceptable two different pieces that include player 1's most preferred piece or the two acceptable pieces, then we have *Case A*. But now player 1 will envy one or both of these players, so here, as well, the lone-divider method is not envy-free.

Neither is the last-diminisher method envy-free. In our earlier example, assume student 7 is the last diminisher of student 1's piece. Then students 1 through 6 will not envy student 7. But when the procedure is applied again for students 1 through 6, student 7, who has exited the game, may think that the student who was the last diminisher and got the piece in the next round received a bigger piece than he or she did. But because student 7 is out of the game, he or she cannot prevent this student—or any other student in remaining rounds of the game—from walking off with a bigger piece. Hence, students who exit may envy students who remain.

One of the earliest results in the mathematical history of fair division, proved by Hugo Steinhaus in 1949, is that there *exists* an envy-free allocation of a cake among n people (for every n) if each person's preferences are given by what is technically known as a "countably additive measure." Steinhaus's proof, however, is only an *existence proof:* it provides "no clue as to how to accomplish such a wonderful partition," as Kenneth Rebman wrote in 1979, and, moreover, relies on some fairly deep theorems of analysis.

The first breakthroughs in solving the problem of obtaining envy-free allocations *constructively*—by specifying a step-by-step procedure, or algorithm, for actually dividing up a cake—were made in the 1960s and 1970s. For the specific case of $n = 3$, different solutions were found independently by several mathematicians. Some involved moving several knives across a cake simultaneously and were quite complicated (a procedure with a *single* moving knife, which does *not* guarantee envy-freeness, is given as Exercise 14 on p. 423), while another involved making at most five cuts and then combining certain pieces. (We describe the latter procedure in the next section.) No general procedure was found for $n > 3$ until 1992, when Steven J. Brams, a political scientist at New York University, and Alan D. Taylor, a mathematician at Union College, invented an envy-free procedure that works for $n \geq 3$ and can be implemented with a finite number of cuts.

The Solution to the *n*-Person Envy-free Problem

We start with a cake and n people. The point we wish to arrive at is an envy-free allocation of the entire cake among the n people in a finite number

Figure 13.3 In 1992 political scientist Steven J. Brams *(left)* and mathematician Alan D. Taylor *(right)* developed an algorithm that provides an envy-free division of a cake among n people, each of whom may have a different way of measuring the value of fractional parts of the cake.

of steps. This task may seem formidable; however, quite often in mathematics, an important part of solving a problem involves breaking the problem into identifiable parts. In this case, let us call our starting point A, and the final point we wish to reach D. Now let us identify some appropriate in-between points B and C that make going from A to D—via B and C—more manageable.

We take this approach in presenting the main ideas behind both the $n = 3$ solution to the envy-free problem and the recent $n > 3$ solution. Our first in-between point is the following:

Point B: Getting a constructive envy-free allocation of *part* of the cake.

We start with the case of $n = 3$. Can we constructively obtain three pieces of cake, whose union may not be the whole cake, which can be given to the 3 people so that each thinks he or she received a piece at least tied for largest? This turns out to be quite easy, with the solution due to John Selfridge and John Conway, who arrived at it independently around 1960. The following process and strategies do the trick:

1. Player 1 cuts the cake into three pieces he considers to be the same size. He hands the three pieces to player 2.

2. Player 2 trims at most one of the three pieces so as to create at least a two-way tie for largest. Setting the trimmings aside, player 2 hands the three pieces (one of which may have been trimmed) to player 3.

3. Player 3 now chooses, from among the three pieces, one that he considers to be at least tied for largest.

4. Player 2 next chooses — from the two remaining pieces — one that he considers to be at least tied for largest, with the proviso that if he trimmed a piece in step 2, and player 3 did not choose this piece, then player 2 must now choose it.

5. Player 1 receives the remaining piece.

Let us reconsider the five steps of this trimming procedure to assure ourselves that each player experiences no envy. Recall that player 1 cuts the cake into three pieces, and player 2 trims one of these three pieces. Now player 3 chooses, and as the first to choose, he certainly envies no one. Player 2 created a two-way tie for largest, and at least one of these two pieces is still available after player 3 selects his piece. Hence, player 2 can choose one of the tied pieces he created and will envy no one. Finally, player 1 created a three-way tie for largest and, because of the proviso in step 4, the trimmed piece is not the one left over. Thus, player 1 can choose an untrimmed piece and therefore will envy no one.

   To illustrate this procedure, suppose that all three players view the cake as having 18 units of "value," with each unit of value represented by a small square. Suppose, however, that the players value various parts of the cake differently (or that player 1 views the cake as being perfectly rectangular, whereas players 2 and 3 see it as being

skewed in opposite ways). We represent this pictorially as follows:

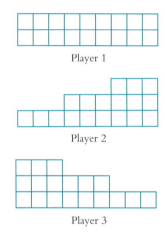

Player 1

Player 2

Player 3

   In step 1, player 1 cuts the cake into three pieces he considers to be the same size (or value). From player 1's point of view, this yields:

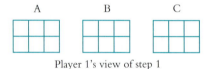

Player 1's view of step 1

He now hands the pieces to player 2, who also views the vertical cuts as being made just to the right of the third and sixth "columns" of squares. Player 2, however, sees the three pieces quite differently:

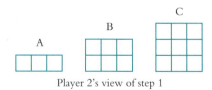

Player 2's view of step 1

   In step 2, player 2 trims at most one of the three pieces so as to create at least a two-way tie for largest. In our illustration, this would mean trimming a piece from C to yield a piece C' that is the same size (or value) as B. Although the most natural way to do this is to remove the top row of squares from C, we will henceforth make the cuts in our illustration correspond to *vertical* lines. Thus, step 2, from player 2's point of view, is as follows:

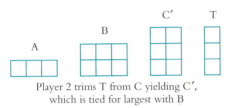

Player 2 trims T from C yielding C′, which is tied for largest with B

Player 2 now sets the trimmings T aside and hands the three pieces (A, B, and C′ in our illustration) to player 3. To see how player 3 views the result of the first two steps, we simply impose the vertical cuts players 1 and 2 made in steps 1 and 2 on player 3's view of the cake (from the beginning of this discussion). Thus, we have cuts after the third, sixth, and eighth columns of squares, which are the right-hand boundaries of A, B, and C′. (In all of our illustrations and exercises, we will have the same number of columns in the views of the different players.) Here, player 3 views what he is receiving as follows:

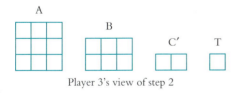

Player 3's view of step 2

Player 3 will now choose, from among the three pieces, one that he considers to be at least tied for largest. In our illustration, this will be A (which he thinks is 9 units of value).

Player 2 next chooses—from the two remaining pieces—one that he considers to be at least tied for largest, with the proviso that if he trimmed a piece in step 2, and player 3 did not choose this piece, then player 2 must now choose it. In our illustration, player 2 thinks B and C′ are tied for largest, and both are available, since player 3 took A. Thus, the proviso dictates that player 2 must choose C′.

Finally, player 1 receives the remaining piece, which is B in our illustration. (Notice that, without the proviso, player 1 could have been stuck with the trimmed piece C′, which he thinks is only 4 units of value.)

A naive attempt to generalize to $n = 4$ what we have done for $n = 3$ would proceed as follows: We would begin by having player 1 cut the cake into four pieces he considers to be the same size. Then we would have players 2 and 3 trim some pieces (but how many?) to create ties for the largest. Finally, we would have the players choose from among the pieces—some of which would have been trimmed—in the following order: player 4, player 3, player 2, player 1.

Alas, this approach fails because player 1 could be left in a position of envy. In order to understand how the approach could fail, consider how many pieces player 3 might have to trim in order to create a sufficient supply of pieces tied for largest so that he is guaranteed to have one available when it is his turn to choose. Player 3 might have to trim one piece to create a two-way tie for largest. Player 2 might need to trim two pieces to create a three-way tie for largest (since, if there were only a two-way tie for largest, player 3 might further trim one of these pieces and player 4 might choose the other). This leaves player 1 in a possible position of envy, because we could have a situation where player 2 trims two pieces and player 3 trims a third piece, and player 4 then chooses the only untrimmed piece. If this happens, player 1—by being forced to choose a trimmed piece—will definitely envy player 4.

All is not lost, however, since there are slight modifications of the Selfridge–Conway procedure that will work for arbitrary n. We describe one such modification for the case $n = 4$. The new idea we introduce is to have the first player cut the cake into more pieces than there are players. The procedure works as follows:

1. Player 1 cuts the cake into *five* pieces she considers to be the same size. She hands the five pieces to player 2.

2. Player 2 trims at most two of the five pieces so as to create at least a *three-way tie* for largest. Setting the trimmings aside, player 2 hands the five pieces — one or two of which may have been trimmed — to player 3.

3. Player 3 trims at most one of the five pieces she has been handed so as to create at least a *two-way tie* for largest. This, of course, may involve further trimming of a piece that player 2 already trimmed in step 2. She sets her trimmings aside with those of player 2, handing the further altered collection of five pieces to player 4.

4. Player 4 now chooses from among the five pieces — some of which may have been trimmed by player 2 and/or player 3 — a piece that she considers to be at least tied for largest. (The remaining steps now reverse the order of initial play.)

5. Player 3 chooses next — from among the four remaining pieces — a piece that she considers to be at least tied for largest, with the proviso that if she trimmed a piece in step 3, and player 4 did not choose this piece, then player 3 must choose it now.

6. Player 2 chooses next — from among the three remaining pieces — a piece that she considers to be at least tied for largest, with the proviso that if she trimmed a piece or pieces in step 2, and one of these is still available, then she must now choose such a piece.

7. Player 1 now chooses — from the remaining two pieces — one that was not trimmed.

This time check this procedure on your own to be certain that it achieves an envy-free allocation of *part* of the cake (see Exercise 9). Notice, however, that for $n = 4$ we not only have the trimmings from steps 2 and 3 left over but also one of the five pieces, perhaps now trimmed, with which we started.

We can illustrate the preceding procedure in a manner similar to what we did for the $n = 3$ case.

That is, suppose we have four players, all of whom view the cake as having 20 units of value (with each unit of value represented by one small square). Assume the four players view (or value) the cake as follows:

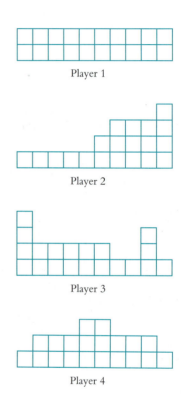

Player 1

Player 2

Player 3

Player 4

In step 1, player 1 cuts the cake into five pieces she considers to be the same size (or value). From player 1's point of view, this yields:

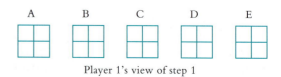

Player 1's view of step 1

She now hands the five pieces to player 2, who views them as follows:

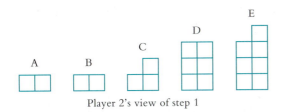

Player 2's view of step 1

In step 2, player 2 trims at most two of the pieces to create at least a three-way tie for largest. In our illustration, this means trimming D and E down to the size of C (that is, 3 units of value), which can be achieved by trimming off the right-hand columns of both D and E. Pictorially:

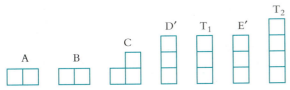

Player 2 trims T_1 from D yielding D′, and she trims T_2 from E yielding E′. Notice that she thinks C, D′, and E′ are all tied for largest (three units of value).

Player 2 now sets the trimmings aside, and hands the five pieces (A, B, C, D′, E′ in our illustration) to player 3. Pictorially, player 3 views what she is receiving as follows:

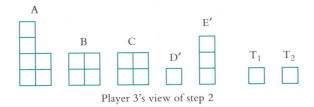

Player 3's view of step 2

In step 3, player 3 trims at most one of the pieces to create at least a two-way tie for largest. In our illustration, this means trimming A down to

the size of B (and C, which happens to be the same size as B). Pictorially:

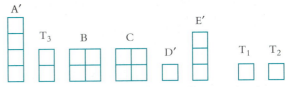

Player 3 trims T_3 from A yielding A′. Notice that she thinks A′ and B (and C, for that matter) are tied for largest.

Player 4 will now choose, from among the five pieces (A′, B, C, D′, E′) one that she considers to be at least tied for largest. This would be C, which she thinks is 6 units.

Player 3 next chooses one of the two (or more) pieces she has tied for largest, subject to the proviso that if one that she trimmed is available, she must choose it. In our illustration, she must choose A′.

Player 2 chooses next, subject to the same kind of proviso as constrained player 3's choice. Thus player 2 must choose D′ or E′ (say, D′, for definiteness).

Finally, player 1 must choose a piece that was neither trimmed nor chosen. In our illustration, this would be B, the only available untrimmed piece.

So far we have gone from point *A* to point *B*: starting with a cake and *n* players, we have constructively obtained (in finitely many steps) an envy-free allocation of part of the cake. Our next step is a fairly small one:

Point C: Getting a constructive envy-free allocation of *all* the cake, but using infinitely many steps.

For *n* = 3, we obtained an envy-free allocation of all the cake, except the part T that player 2 trimmed from one of the pieces. It turns out that T can now be allocated straightforwardly among the three players in such a way that the resulting allocation of the whole cake is envy-free. (This is the rest of the Selfridge–Conway procedure.) The process we are about to describe, however, is *not* infinite; the infinite scheme comes into play when we turn to $n = 4$.

The key observation for the $n = 3$ case is that player 1 will not envy the player who receives the trimmed piece, even if that player is given all of T. Recall that player 1 created a three-way tie and received an untrimmed piece. The union of the trimmed piece and the trimmings yields a piece that player 1 considers to be exactly the same size as the one he received. Thus, assume that it is player 3 who receives the trimmed piece (it could as well be player 2). Then player 1 will not envy player 3, however T is allocated.

The next step ensures that neither player 2 nor player 3 will envy another player when it comes time to allocate T. Let player 2 cut T into three pieces he considers to be the same size. Let the players choose which of the three pieces they want in the following order: player 3, player 1, player 2.

To see that this yields an envy-free allocation, notice that player 3 envies no one, because he is choosing first. Player 1 does not envy player 2, because he is choosing ahead of him; and player 1 does not envy player 3 because, as pointed out earlier, player 1 will not envy the player who receives the trimmed piece. Finally, player 2 envies no one, because he made all three pieces of T the same size.

Hence, for $n = 3$, we have an envy-free allocation of all the cake except T, followed by an allocation of T that gives an allocation of *all* the cake, at least if the preferences are what are called "weakly additive."

For $n > 3$, however, the desired *finite* scheme does not come so easily. What does come quite easily, though, is an infinite scheme that yields a more constructive proof of Steinhaus's original existence theorem (still more constructive is a finite scheme, which we discuss presently).

The intuition behind the infinite scheme is simple: one applies the scheme that yields an envy-free allocation of part of the cake over and over again, with each application yielding an envy-free allocation of part of what was left over from the previous application. Eventually the whole cake is allocated.

The next question for the $n = 4$ case is how to get from point C to point D, where the whole cake is allocated in a finite number of steps. That is, how do we make the infinite scheme finite? Our approach mirrors what we did in the $n = 3$ case: if two players have different preferences, then we arrange the partial-allocation scheme so that each player thinks he received a piece of cake *strictly* larger than the other player.

Eventually, when the size of the crumb is sufficiently small (in the eyes of these two players), then neither will care if the other player should get the whole crumb. This is just a glimpse of what is needed; full details can be found in the article by Brams and Taylor (see Suggested Readings).

Thus, there is a solution to the problem of envy-freeness that obviates the need for an endless process of finer and finer divisions. Practically speaking, however, it is only necessary to know that the trimmings become progressively smaller; then one can stop this procedure, which we call the **trimming procedure,** when what remains no longer matters much to the players. (For more on the shrinkage of the leftover portion of a cake, see Exercise 13.)

Dividing Up the Chores

We conclude by briefly discussing a problem that is the dual of allocating a "good," like cake, which we postulated everyone wants more of rather than less. Suppose one must divide up a "bad," like chores, such as mowing the lawn, grocery shopping, and washing dishes, for which players have different preferences about those they least want to do, but everyone wants to do fewer chores rather than more.

The trimming procedure is applicable to the chores problem, but it works in reverse: players "add on" rather than "trim" pieces. Like the trimming procedure, they do so by creating ties, though this time for the smallest rather than the largest piece.

A difficulty here is that players must first create a supply of cake (set of chores) to be used in the adding-on process, which players then draw upon to create the requisite number of ties for smallest. They then choose from among these (tied) smallest pieces in a manner analogous to that of the trimming procedure.

Now it is the *least* of the bad rather than the *most* of the good that is allocated. As before, the procedure is multistage, because larger pieces always remain to be allocated, but the total left over decreases after each round.

Space precludes our giving a more detailed explanation of the add-on procedure for chore division, but like the trimming procedure for goods, it yields an envy-free allocation. Equitably sharing bads, from household chores to civic duties like taxes (involuntary) and military service (voluntary), is probably as pervasive a problem as equitably sharing goods.

▶ THE DISCRETE CASE

In many fair-division problems, including some inheritances, some of the objects to be allocated cannot be further subdivided into smaller parts. One approach in such cases is to attempt to assign numerical values, such as dollar amounts, to the objects and then divide the total sum into fair ratios. The final allocation can then be achieved by assigning either the objects themselves or the dollar equivalents. This typically requires monetary side payments between players. A great number of auction and bidding schemes have been introduced to force participants to honestly reveal their individual monetary values for specific objects.

Two Players

In the case of two players dividing one object into equal shares, we can arrive at a fair division as follows:

1. Players A and B enter sealed bids of amounts a and b, representing their respective honest evaluations of the object.

2. The object is awarded to the higher bidder, say B, whose bid was $b > a$.

3. Then the higher bidder B pays the lower bidder A the amount of

$$\frac{a}{2} + \frac{b-a}{4} = \left(\frac{a}{2} + \frac{b}{2}\right) \div 2$$

As a consequence, player B realizes an amount that B values as

$$b - \left(\frac{a}{2} + \frac{b-a}{4}\right) = \frac{b}{2} + \frac{b-a}{4}$$

It follows that each player receives half of his or her own evaluation (which is his or her fair share) plus a surplus of $(b - a)/4$. Paradoxically, when the participants in a fair division assign different values, one can arrive at a split that gives each of them *more* than a fair share.

EXAMPLE: A Two-Person Inheritance

Consider the case where Alice and Barbara inherit equal shares in a house; they bid $100,000 and $150,000, respectively. They see Alice's fair share as $50,000 and Barbara's as $75,000. Our method assigns the house to Barbara and has her pay $62,500 to Alice. Barbara ends up with a house worth $150,000 to her, less $62,500 paid out, for a net gain of $87,500. Similarly, Alice receives her fair share ($50,000) plus $12,500, for a total of $62,500. So each receives her fair share plus $12,500 (see Table 13.1). ▲

TABLE 13.1   Dividing a House*

Players	Alice	Barbara
Fair division:		
1. Totals bid on the house:	100,000	150,000
2. Fair share ($\frac{1}{2}$ of line 1):	50,000	75,000
3a. House awarded:	—	House
3b. At the higher bid:	0	150,000
4. Remaining claim (lines 2 − 3b):	50,000	−75,000
5. Total surplus: 25,000		
6. Share of surplus ($\frac{1}{2}$ of line 5):	12,500	12,500
		House
7. Final settlement (lines 3a + 4 + 6):	+62,500	−62,500

* Numbers represent dollar amounts.

The two-person example can easily be extended to any number of players, as we see in the next example.

Multiple Players

We illustrate one of the possible ways the discrete fair-division method described previously can be extended to more than two persons.

EXAMPLE: A Four-Person Inheritance

Consider the case of four children, Ann, Bob, Carol, and Don, with equal shares who inherit their parents' estate consisting of three objects, a house, a summer cabin on the lake, and a boat. Assume that each heir enters a secret bid (in dollars) on each of the objects as indicated by the top three lines of Table 13.2.

The next row in this table, denoted as line 1, gives the sum of each heir's three bids. These amounts are taken as their respective evaluations for the whole estate. The numbers in line 2 are one-fourth of those in line 1. They represent the fair shares due the heirs in monetary terms in this case of equal division. Note that these numbers are not all the same! They do, nevertheless, represent one-fourth of the total estate's value for each heir according to his or her own evaluation, and we do not wish to impose some other evaluation upon any of these individuals. In line 3 we denote the items that are awarded to the highest bidders. (One could use a random device to pick the awardees if necessary to break any ties at this point.) Line 4 indicates the differences between line 2 and the declared values of the goods awarded in line 3. This represents the additional value due each person and is negative whenever the value of the goods on line 4 exceeds the fair share on line 2. Such negative numbers are payments to be made *into* the "kitty" and become available for further distribution to the heirs. The sum of the numbers on line 4 is never positive, and the absolute value of this sum is called the *surplus* and is recorded on line 5. We distribute this surplus in four equal shares of 19,125 to each of the participants as indicated on line 6. Line 7 sums up the goods awarded in line 3 plus the final amounts of money the heirs receive. A negative amount here indicates a side payment to compensate for goods obtained whose worth

TABLE 13.2   **Fair Division of an Estate by Sealed Bid**

Players	Ann	Bob	Carol	Don
Bids on:				
House	120,000	200,000	140,000	180,000
Cabin	60,000	40,000	90,000	50,000
Boat	30,000	24,000	20,000	20,000
Fair division:				
1. Sum of the bids:	210,000	264,000	250,000	250,000
2. Fair shares:	52,500	66,000	62,500	62,500
3. Objects awarded:	Boat	House	Cabin	—
At the high bids:	30,000	200,000	90,000	0
4. Remaining claims:	22,500	−134,000	−27,500	62,500
5. Total surplus: 76,500				
6. Share of surplus:	19,125	19,125	19,125	19,125
7. Final settlements:	Boat	House	Cabin	
	+41,625	−114,875	−8,375	+81,625

exceeds one's fair share plus his or her surplus. Observe that the sum of all the numbers on this final line is zero. Note that each heir ends up with $19,125 in value in addition to what he or she declared as a fair share. This fair-division scheme shows again that whenever some participants have different evaluations of some objects, there is an allocation in which everyone of them obtains *more* than a fair share. ▲

Exercises 16 to 20 show some variations and extensions that can be made in this particular fair-division scheme.

In many cases the heirs to an estate do not obtain equal shares. The following example shows how our multiperson auction method can be modified to handle such cases.

EXAMPLE: Unequal Shares

Consider our previous example where Ann, Bob, Carol, and Don inherited their parents' house, cabin, and boat. However, assume now

that the four heirs' respective shares of the estate are 40%, 30%, 20%, and 10%. If the four children enter the same bids as before, then our modified fair-division procedure is given in Table 13.3. ▲

▶ **APPLYING THE TRIMMING PROCEDURE TO INDIVISIBLE GOODS**

Although we have used the metaphor of cake cutting throughout our earlier discussion on the problem of envy, the *idea* of successive trimming is nonetheless one that is eminently applicable to problems of fair division other than parceling out the last crumbs of a cake. The main practical problem in applying the trimming procedure is, as we have seen, that many fair-division problems involve the discrete case: *indivisible goods*, which cannot be divided up at all, much less trimmed in fine amounts.

Although the trimming procedure, as such, is not applicable to allocation problems involving in-

TABLE 13.3   **Fair Division with Unequal Shares**

Players	Ann (40%)	Bob (30%)	Carol (20%)	Don (10%)
Bids on:				
House	120,000	200,000	140,000	180,000
Cabin	60,000	40,000	90,000	50,000
Boat	30,000	24,000	20,000	20,000
Fair division:				
1. Sum of bids:	210,000	264,000	250,000	250,000
2. Fair shares (% of 1):	84,000	79,200	50,000	25,000
3a. Objects awarded:	Boat	House	Cabin	—
3b. At the highest bids:	30,000	200,000	90,000	0
4. Remaining claims:	54,000	−120,800	−40,000	25,000
5. Total surplus: 81,800				
6. Share of surplus (% of 5):	32,720	24,540	16,360	8,180
7. Final settlement:	Boat	House	Cabin	
	+86,720	−96,260	−23,640	+33,180

divisible goods, it can be adapted to such problems under certain conditions. The key condition is that there be a sufficient quantity of more divisible goods, like small items or — even better — money, which can be trimmed in lieu of the indivisible good.

Take, for example, the problem of dividing up an estate, in which a house is the single big item. Assume there are four heirs, but only one thinks the house is worth more than one-fifth of the estate. If this person is the one to make the initial division, and if, in addition, he knows that the other heirs do not value the house so highly, he can begin by dividing up the estate into five pieces, with one piece being just the house. If none of the other three heirs thinks this indivisible piece has to be trimmed on the first round — even after other trimmings are made — then the house can, in effect, be "reserved" for the heir who thinks it is the most valuable piece. ▲

This example illustrates how one player's knowledge of the preferences of the others need not always be exploitative but can, instead, facilitate the search for a solution. It will not always be apparent, however, precisely what information players should reveal and what they should hide (as is true in most negotiations). But because the trimming procedure has certain safeguards built in — in particular, allocating in stages in addition to ensuring envy-freeness in each — players probably can afford to be more open about their preferences than if these safeguards were absent.

It is interesting to recall that when the allies agreed in 1944 to partition Germany into sectors after World War II (first stage), they at first did not reach agreement about what to do with Berlin. Subsequently, they decided to partition Berlin itself into sectors (second stage), even though this city fell 110 miles within the Soviet sector. Berlin was simply too valuable a "piece" for the Western allies (Great Britain, France, and the United States) to cede to the Soviets, which suggests how,

TABLE 13.4   Fair Division of an Estate by Trimming

	Heirs			
Item	1	2	3	4
1. House (H)	50	50	50	50
2. Boat (B)	20	10	10	10
3. Car (C)	10	20	10	10
4. Furniture (F)	10	10	10	10
5. Piano (P)	10	0	10	10
6. Art (A)	0	10	10	10
Total point valuation of estate	100	100	100	100

after a leftover piece is trimmed off, it can be subsequently divided under the trimming procedure.

Yet what if a large piece like Berlin is not divisible? In the settlement of an estate, this might be the house, as we suggested earlier, which may be worth half the estate to the claimants. In this situation, there may be no alternative but to sell this big item and use the proceeds to make the remaining estate more liquid or, in our terms, "trimmable."

EXAMPLE: Dividing Up an Estate

To illustrate the trimming procedure in the case of an estate, assume the estate is composed of six items. Four heirs have valuations for each item that are indicated by points that sum to 100 for each in Table 13.4. Notice the following features of this example: (1) all heirs consider H to be worth half the estate; (2) setting H aside, for heirs 1 and 2 the most and least valuable items differ, whereas for heirs 3 and 4 their valuations are the same yet different from those of heirs 1 and 2.

Assume no heir has sufficient resources to pay off the other three to get H. Accordingly, the heirs agree to sell H on the open market. Suppose they get exactly 50 for it, which is

what they agree it is worth. (If they get less— say, 40—this would change their totals to 90 but would not affect the trimming process in a fundamental way.) After the sale, H becomes 50 (divisible) points rather than a single indivisible item.

We start with heir 1 (it could be any of the heirs, who this time we will assume are all women). She begins by dividing the estate into five parts (Items 2 through 6 in Table 13.4), as prescribed by the trimming procedure discussed on pages 414–415. For her the following parts are all worth 20 when the 50 divisible points earned upon the sale of H are added on. We underscore to indicate a tie for largest:

Heir 1:   <u>B</u>   <u>C + 10</u>   <u>F + 10</u>   <u>P + 10</u>   <u>A + 20</u>

If heir 2 goes next, she must create at least a three-way tie for largest. Because she initially assigned 20 points to C and 10 points to A, C + 10 and A + 20 will be the largest parts for her (each is worth 30), and she will trim each by 10 to create a tie with her next-largest item (F + 10), which is worth 20:

Heir 2:   B   <u>C</u>   <u>F + 10</u>   P + 10   <u>A + 10</u>

If heir 3 goes next, she must create at least a two-way tie for largest. But because F + 10, P + 10, and A + 10 are all worth 20 to her, she need do no trimming, but her ties are different from heir 2's:

Heir 3:   B   C   F + 10   P + 10   A + 10

Now heir 4 must choose a part from the five that she considers to be at least tied for largest. This part will be F + 10, P + 10, or A + 10, because her preferences are the same as heir 3's. Assume she chooses A + 10, and heir 3 next chooses P + 10. Then heir 2 has two remaining pieces, F + 10 and C, that she considers tied for largest. But since C is the result of her trimming a piece (namely, C + 10), heir 2 must take it (C) because it is available. Finally, assume heir 1 chooses B. Then the first-stage allocation to heirs (1, 2, 3, 4) is {B, C, P + 10, A + 10}, leaving F + 10 and the 20 trimmed by heir 2 for the second stage of the procedure.

Now all the heirs value F equally (10), but none can divide F + 30 (F + 10 + 20) into five equal parts as heir 1 did at the beginning of the first stage of the procedure. Thus, we can see that there could be a problem of indivisibility at the second stage in some instances, which would necessitate selling other items to provide proceeds that can be divided in later stages.

In our example, however, there is no such necessity, because F + 30 can be exactly divided into *four* equal parts for the four heirs:

F   10   10   10

Adding these four equal parts, in this order, to the previous first-stage envy-free allocation, we obtain as a final envy-free allocation {B + F, C + 10, P + 20, A + 20} for heirs (1, 2, 3, 4), as shown below:

Heir	Parts	Point valuation
1	B + F	20 + 10
2	C + 10	20 + 10
3	P + 20	10 + 20
4	A + 20	10 + 20

Thus, our trimming procedure guarantees envy-freeness as long as there are enough divisible goods to ensure that no player ever has to trim an indivisible good. Generally speaking, this means that there cannot be a single item, like a house, highly valued by all the players at the start, or a comparatively large item at a later stage. If there is, such an item could be sold off to a nonplayer, as we assumed in our example. Alternatively, a player who desires it could pay off the other players in a negotiated settlement, which is simply another way of introducing more divisible goods into the system.

Still another way of lending divisibility to the trimming procedure is for all the players to make injections of cash. If, when combined with the indivisible goods, the cash gives enough "cushion" to each indivisible good so as to make pieces trimmable when necessary, then it may be possible to accomplish the trimming without selling off any of the indivisible goods.

In our example, this could be done if each heir contributed 150/4 = 37.5. Then heir 1 could divide the estate into five pieces, each worth 50:

Heir 1:   H   B + C + F + P + A   50   50   50

Since all the other players also value these pieces at 50 each, there would be no need for trimming in the first stage. One of the 50 pieces could be saved for the second stage and then immediately divided evenly among the four heirs.

Unlike the bidding procedure, indivisible goods under the trimming procedure need only be sold off when necessary to ensure sufficient liquidity, or each player need contribute only a relatively small amount to create sufficient divisibility. ▲

▶ REVIEW VOCABULARY

Continuous (divisible) case A fair-division problem in which the object to be divided has no indivisible components and can be finely divided into parts; examples include time, land, money, or sand.

Discrete (indivisible) case   A fair-division problem in which some parts of the objects to be divided, such as the house and cars of an estate, cannot be finely divided into arbitrarily small parts in any manner.

Envy-free   A fair-division procedure is said to be envy-free if each player has a strategy that can guarantee him or her a piece of cake that is at least as large as any other player's piece (as each player perceives the allocation), no matter what the other players do.

Fair-division problem   To divide up some gains or losses into n separate parts so that each of n people considers the part he or she receives as a fair allocation.

Fair-division scheme   A method for solving a fair-division problem. Each participant in the procedure must have a way to realize a piece that he or she views as fair in his or her own value system. Any such scheme must be based on certain reasonability assumptions in order to guarantee that a fair division will exist.

Last-diminisher method   A fair-division method for dividing a continuous object among any number of players. This method was introduced by Stefan Banach and Bronislaw Knaster in the 1940s.

Lone-divider method   A fair-division scheme that allows three players to divide a continuous commodity S among themselves in such a way that each can guarantee that the piece he or she receives is "acceptable." It was developed by

Hugo Steinhaus in the 1940s and generalized by Harold Kuhn in the 1960s.

Player   A participant in a fair-division scheme.

Trimming procedure   A fair-division scheme for producing an envy-free allocation of a continuous object among any finite number of players.

▶ SUGGESTED READINGS

BRAMS, S. J., AND A. D. TAYLOR: "An Envy-Free Cake Division Protocol," *American Mathematical Monthly*, in press. Brams and Taylor describe in detail the finite version of their envy-free procedure for $n = 4$; in addition, they review earlier work on "protocols" (step-by-step procedures) that led up to their constructive solution of the envy-freeness problem for $n \geq 3$.

DUBINS, L. E., AND E. H. SPANIER: "How to Cut a Cake Fairly," *American Mathematical Monthly*, 68:1–17 (1961). This article gives some extensions of the simple fair-division concepts introduced in this chapter.

KUHN, H. W.: "On Games of Fair Division," in Martin Shubik (ed.), *Essays in Mathematical Economics*, Princeton University Press, Princeton, N.J., 1968, pp. 29–37. This article provides extensions of, and additional references to, the fair-division concepts presented in this chapter.

STEINHAUS, H.: *Mathematical Snapshots*, Oxford University Press, Oxford, 1960. A brief but significant introduction to both fair division and apportionment is provided in this popular book on interesting mathematical topics.

▶ EXERCISES

1. Consider the following six 3-by-3 tables with 0 and 1 entries, where the columns correspond to three pieces $S1$, $S2$, and $S3$, and the rows correspond to three players 1, 2, 3. An entry of 1 in the tables indicates an acceptable piece for the corresponding player, and a 0 indicates an unacceptable piece. Using the lone-divider method, show a fair allocation for each table by circling three 1s so

that you obtain exactly one 1 in each row and exactly one 1 in each column.

(a) 1 1 1   (b) 1 1 1   (c) 1 1 1   (d) 1 1 1
    1 0 0   1 0 0   1 0 0   1 1 0
    0 1 0   0 1 1   1 1 1   1 1 0

                   (e) 1 1 1   (f) 1 1 1
                       1 1 0   1 1 1
                       1 0 1   1 1 0

How many different fair allocations are there for each of these six tables?

2. Suppose that players 1, 2, and 3 view a cake as follows:

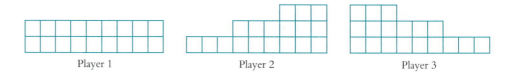

Notice that each player views the cake as having 18 square units of area (or value). Assume that each player regards a piece as acceptable if and only if it is at least $\frac{18}{3} = 6$ square units of area (his or her "fair share"). Assume also that all cuts made correspond to vertical lines.

    a. Provide a total of three drawings to show how each player views a division of the cake by player 1 into three pieces he or she considers to be the same size or value. Label the pieces A, B, and C.

    b. Identify two of these pieces that player 2 finds acceptable, and two that player 3 finds acceptable.

    c. Show that a feasible assignment of fair pieces can be achieved by letting the players choose in the order: player 3, player 2, player 1. Indicate how many square units of value each player thinks he or she received. Is there any other order in which players can choose pieces (in this example) that also results in a feasible assignment?

3. Suppose that players 1, 2, and 3 view a cake as follows:

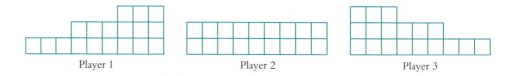

    a. Provide a total of three drawings to show how each player views a division of the cake by player 1 into three pieces he or she considers to be the same size or value. Label the pieces A, B, and C. (We are still assuming that all cuts correspond to vertical lines, so this will require a cut along a vertical center line of some of the squares.)

b. Show that neither player 2 nor player 3 finds more than one of the three pieces acceptable (with "acceptable" defined as in Exercise 2).

c. Identify a single piece that player 2 and player 3 agree is *not* acceptable. (There are actually two such pieces; for definiteness, find the one on the right.)

d. Assume that players 2 and 3 give the piece from part c to player 1. Suppose they reassemble the rest and players 2 and 3 divide it between themselves using cut-and-choose (with a single vertical cut). Determine what size piece each of the three players will think he or she received (1) if player 2 cuts and player 3 chooses, and (2) if player 3 cuts and player 2 chooses.

4. Suppose players 1, 2, and 3 view a cake as in Exercise 2. Illustrate the last-diminisher method (still restricting attention to vertical cuts and, furthermore, assuming that the piece potentially being diminished is a piece off the left side of the cake) by following steps a–h below:

a. Draw a picture showing the third of the cake (6 squares) that player 1 will slice off the cake.

b. Determine if player 2 will pass or further diminish this piece. If he or she would further diminish it, make a new drawing.

c. Determine if player 3 will pass or further diminish this piece. If he or she would further diminish it, make a new drawing.

d. Determine who receives the piece cut off the cake and what size or value he or she thinks it is. (Actually, we *knew* what size the person receiving this first piece would think it was, assuming he or she followed the prescribed strategy. How did we know this?)

e. Finish the last-diminisher method using cut-and-choose on what remains, with the lowest-numbered player who remains doing the cutting.

f. Redo step e with the other player doing the cutting.

g. Redo step e, but with the last two players using the last-diminisher method directly, instead of cut-and-choose (with the order as in step e).

h. Redo step g with the order reversed.

5. Redo Exercise 4, but with the players viewing the cake as in Exercise 3.

6. Suppose players 1, 2, and 3 view the cake as in Exercise 3. Illustrate the envy-free procedure for $n = 3$ (yielding an allocation of part of the cake) by following steps a–c below. Again, restrict attention to vertical cuts.

a. Provide a total of three drawings to show how each player views a division of the cake by player 1 into three pieces he or she considers to be the same size or value. Label the pieces A, B, and C. (This is the same as Exercise 3a.)

b. Redraw the picture from player 2's view, and illustrate the trimming of piece A that he or she would do. Label the trimmed piece A' and the actual trimmings T.

c. Indicate which piece each player would choose (and what he or she thinks its size is) if the players choose in the order: player 3, player 2, player 1 according to the envy-free procedure on page 408. Does the proviso in step 4 come into play here?

7. Apply the remainder of the Selfridge–Conway procedure from page 412 to what was obtained in Exercise 6 by completing a–c below:
   a. Draw a picture of T from each player's view.
   b. The procedure calls for the player (other than player 1) who did not receive the trimmed piece to divide T into three pieces he or she considers to be the same size. Here, that would be player 2. Illustrate this division, and label the pieces X, Y, Z.
   c. Indicate which parts of T (and the sizes or values) the players will choose when they go in the order: player 3, player 1, player 2.

8. Suppose players 1, 2, 3, and 4 view a cake as follows:

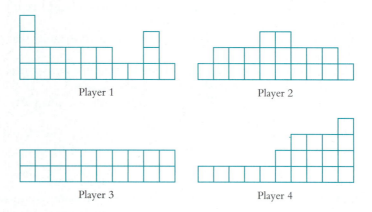

Player 1   Player 2

Player 3   Player 4

Mimic what was done on pages 410–411 to illustrate the procedure yielding an envy-free allocation of part of the cake among the four players.

■ 9. Consider the trimming procedure for four people described on pages 409–410. Explain why each player experiences no envy.

10. In generalizing the $n = 4$ envy-free procedure to arbitrary n, player 1 cut the cake into $2^{n-2} + 1$ pieces. For $n = 4$, we had $2^{n-2} + 1 = 5$. Suppose that $n = 5$. Then $2^{n-2} + 1 = 9$. Determine how many pieces each player must trim to make the procedure work for $n = 5$.

▲ 11. Suppose that $n = 5$, and we begin by having player 1 cut the cake into 16 pieces of the same size. Suppose player 2 creates an 8-way tie, then player 3 creates a 4-way tie, and, finally, player 4 creates a 2-way tie. Let the players now choose in the following order: player 5, player 4, player 3, player 2, player 1.

■   Discussion exercise.   ▲   Advanced exercise.

Show that no provisos about choosing trimmed pieces are needed to ensure envy-freeness.

12. Explain why, in the envy-free procedure for $n = 3$ on p. 412, player 1 will not envy the player who receives the trimmed piece, regardless of how the trimmings are allocated in the next step.

▲ 13. Consider the envy-free procedure for $n = 4$, wherein player 1 cuts the cake into five pieces he considers to be the same size. Since he (player 1) gets an untrimmed piece, he thinks the size of the leftover L_1 from the first stage is at most 4/5 of the original cake. (He may think it as little as 1/5 if the other players did no trimming, leaving only one of the five pieces he cut for the second stage.) If we now do the same thing with L_1, then he will think the size of the leftover L_2 from the second stage is at most $(4/5)(4/5) = 16/25 = .64$ of the cake. How large a portion of the cake will he think the maximum size of the eleventh leftover is? Is the size of this leftover less than 1/10 of the cake?

■ 14. This is one illustration of the moving-knife procedure introduced by L. E. Dubins and E. H. Spanier: A mother wishes to divide a large submarine sandwich among three children:

> STEP 1.   Mother passes the knife slowly from left to right over the top of the sandwich. The first child who says "Stop" is given the piece to the left of the knife and leaves the contest.
> STEP 2.   Mother continues to move the knife. The next child to say "Stop" receives the piece to the left and exits the game.
> STEP 3.   The piece to the right of the knife in step 2 is given to the remaining child.

> a. Is this a reasonable fair-division scheme?
> b. What assumptions need to be made in order for this to be a fair method?
> c. Can this approach be extended to more than three children?

■ 15. Discuss what assumptions (axioms) are necessary in order that the fair-division scheme presented in the text for continuous objects and multiple players (the last-diminisher method) will always result in a fair division.

16. If John bids $28,225 and Mary bids $32,100 on their aging parents' old classic car, which they no longer drive, how would you reach a fair division?

17. John and Mary inherit their parents' old house and classic car. John bids $28,225 on the car and $55,900 on the house. Mary bids $32,100 on the car and $59,100 on the house. How should they arrive at a fair division?

18. Can you modify your fair-division scheme in Exercise 17 so that both John and Mary receive one of the two objects while still considering the allocation as fair?

19. a. Describe a fair division for three heirs A, B, and C who inherit a house in the city, a small farm, and a valuable sculpture, and who submit

sealed bids (in dollars) on these objects as follows:

	A	B	C
House	145,000	149,999	165,000
Farm	135,000	130,001	128,000
Sculpture	110,000	80,000	127,000

    b. Describe a fair division for the three heirs A, B, and C if their shares are $\frac{5}{10}$, $\frac{3}{10}$, and $\frac{2}{10}$, respectively.

    c. Describe a fair division if their shares are $\frac{1}{2}$, $\frac{1}{4}$, and $\frac{1}{4}$, respectively.

20. a. Describe a fair division for three children E, F, and G who inherit equal shares of their parents' classic car collection and who submit sealed bids (in dollars) on these five cars as follows:

Cars	E	F	G
Duesenberg	18,000	15,000	15,000
Bentley	18,000	24,000	20,000
Ferrari	16,000	12,000	16,500
Pierce-Arrow	14,000	15,000	13,500
Cord	24,000	18,000	22,000

    b. Describe a fair division for E, F, and G if their shares are $\frac{1}{2}$, $\frac{1}{4}$, and $\frac{1}{4}$, respectively.

    c. Describe a fair division for E, F, and G if their shares are $\frac{3}{6}$, $\frac{2}{6}$, and $\frac{1}{6}$, respectively.

21. Suppose that four heirs have the following valuations for each of seven items in an estate, indicated by points that sum to 100 for each.

Item	Heirs			
	I	II	III	IV
Money (M)	40	40	40	40
Boat (B)	15	10	20	15
Car (C)	10	15	10	10
Furniture (F)	10	10	10	14
Piano (P)	17	10	10	5
Art (A)	5	15	10	10
Dog (D)	3	0	0	6

Use the trimming procedure to find an envy-free settlement of the estate among the four heirs by completing steps a–n below.

   a. Show how heir I can divide the estate into five parts that she considers to be of equal value. (Although there is more than one way to do this, choose the way that combines the art and dog, together with some money, to make one part, and then combine the car with some money to make another part.)

   b. Write down heir II's values of the five parts handed to her by heir I.

   c. Indicate how heir II can trim two of the parts to create a three-way tie for most valuable part. Make the trimming consist of money if possible.

   d. Write down heir III's values of the five parts handed to her (after the trimming) by heir II.

   e. Indicate how heir III can trim one of the parts to create a two-way tie for most valuable part.

   f. Indicate which part heir IV will now choose (and how many points of value she thinks it is worth).

   g. Indicate which part heir III will now choose (and how many points of value she thinks it is worth). Does the proviso about choosing a piece you trimmed if one is available come into play here?

   h. Indicate which part heir II will now choose (and how many points of value she thinks it is worth). Assume she really does not want the dog.

   i. Indicate which part heir I will now choose (and how many points of value she thinks it is worth).

   j. In dividing up the trimmings and the piece not chosen, assume the art is sold for 10 units of money. Show that this yields:

Item	Heirs I	II	III	IV
Money (M)	32	32	32	32
Dog (D)	3	0	0	6

   k. Show how heir I can divide the estate into five parts that she considers to be of equal value.

   l. Explain why neither heir II nor heir III will do any trimming.

   m. Indicate which part each will choose when they go in the order: heir IV, heir III, heir II, heir I.

   n. What is the obvious thing to do with what is left?

Exercises 22 to 26 apply to the following alternative fair-division method for dividing discrete objects (which differs from the one presented in the text): Each of the n participants makes a sealed bid on each object. Each object is then assigned to the highest bidder. Each participant will finally obtain either objects or cash equal in value to $\frac{1}{n}$th of the *sum* of the *highest* bids for the objects.

22. Do Exercise 16 using this alternative fair-division method.

23. Do Exercise 17 using this alternative fair-division method.

24. Do Exercise 18 using this alternative fair-division method.

25. Do Exercise 19 using this alternative fair-division method.

26. Do Exercise 20 using this alternative fair-division method.

27. Pablo Picasso left his enormous estate to six heirs with shares as follows:

Person	Relationship	Share
Jacqueline	Wife	$\frac{11}{32}$
Claude	Child	$\frac{3}{32}$
Paloma	Child	$\frac{3}{32}$
Maya	Child	$\frac{3}{32}$
Bernard	Grandson	$\frac{6}{32}$
Marina	Granddaughter	$\frac{6}{32}$

Can you suggest a reasonable fair-division scheme for the heirs that takes into account the huge numbers of objects involved?

▶ Writing Projects

1. One of the most important differences between the three-person and the *n*-person envy-free procedures is that the latter procedure may take more than two stages. And of course, the more stages there are, the more cuts and trimmings that may be necessary. Do you consider this a serious practical problem, or is it mainly a theoretical problem? Why?

2. Besides the division of an estate among heirs, or the allocation of cabinet positions to political parties in a coalition government, what other possible applications of the trimming procedure can you think of? Give examples of situations not in the text in which it is "bad" things like chores—rather than good things like cake (for most of us, anyway)—that you might want to allocate in an envy-free way? (*Optional:* In applying the trimming procedure to bad things, how might you create ties for worst with, say, next-worst by adding things to the worst piece?)

Chapter 14

Apportionment

The delegates who wrote the U.S. Constitution in 1787 intended the federal legislature to consist of two houses: the Senate, with two senators per state, and the House of Representatives, which "shall be apportioned among the several states within this union according to their respective Numbers . . ." (Article I, Section 2). Thus, each state's representation in the House is to be proportional to its population.

The apportionment problem arises because it is unlikely that any state's fair share of the House will be a whole number. For example, in the census of 1790, the population of the United States was found to be 3,615,920. The most populous state was Virginia, with 630,560 people, while the least populous was Delaware, with 55,540. With 105 seats in the House to apportion, Virginia's fair share would be

$$\frac{\text{Population of Virginia}}{\text{Population of the United States}} \times 105$$

or

$$\frac{630,560}{3,615,920} \times 105 = 18.310$$

Delaware's apportionment would be

$$\frac{55,540}{3,615,920} \times 105 = 1.613$$

Should the numbers be rounded to give Delaware two representatives and Virginia 18? Virginians would find this inequitable, since each of their 18 representatives would have to account for 35,031 constituents, while each of the two Delaware representatives would represent 27,770 people.

The **apportionment problem** arises when we are required to round fractions so that their sum is maintained at some constant value (see Spotlight 14.2, page 432, regarding fractions and their rounding). This constraint appears in many situations, for example, in cases where the fractions correspond to numbers of persons (who must be assigned in whole numbers), as we have seen in the apportionment of the House of Representatives. The same constraint occurs when statistical tables must be presented with percentages that must be expressed as whole numbers and the sum must be maintained at 100%.

Figure 14.1 The signing of the Constitution. (Commissioned by the Pa., Del., N.J. state societies, Daughters of the American Revolution. Independence National Historic Park Collection. Copyright Louis Glanzman.)

Figure 14.2 A university must resolve many apportionment problems, such as arise in class scheduling. (Photo by Robert Cohen. © 1987 Photographic Services, University of Delaware. All rights reserved.)

EXAMPLE: College Enrollments

Consider a university with 20,000 students and six colleges, as in Table 14.1. The exact percents appear in the third column. The fourth and fifth columns show the percents rounded to one decimal place and to whole numbers, respectively. We use the normal rounding procedure—rounding fractions up when they are greater than or equal to one-half—for these percentages. In each case, the total of the rounded percentages is not equal to 100%. Alternate *apportionments* that realize the sum of 100% for these percentages appear in the last two columns. ▲

Apportionment problems often arise when resources are to be allocated in an equitable manner. College administrators decide how many faculty positions to allocate to each college and de-

TABLE 14.1   University Enrollment by Colleges

College	Number of students	Exact percent	Rounded percent	Integer percent	Apportioned percent	
Arts and Sciences	6716	33.580	33.6	34	33.6	34
Engineering	4832	24.160	24.2	24	24.1	24
Agriculture	4093	20.465	20.5	20	20.5	20
Business	3211	16.055	16.1	16	16.0	16
Law	852	4.260	4.3	4	4.3	4
Architecture	296	1.480	1.5	1	1.5	2
Totals	20,000	100.000	100.2	99	100.0	100

partment. A school district must allocate teachers to schools in proportion to enrollments. However, the apportionment problem that provokes the greatest interest is in the political arena.

▶ POLITICAL APPORTIONMENT

In some parliaments, the number of seats assigned to a party is to be proportional to the number of votes that the party received. The party's share of parliament typically will not be a whole number; so it must be replaced by a nearby integer value.

In an apportionment problem, the **quota** is the exact share that would be allocated if a whole number were not required. The **apportionment** is the integer that is actually allocated. For example, if a party obtains 41.23% of the votes in a national election for a parliament with 120 seats, its quota is the number q, that is, 41.23% of 120. Therefore

$$q = (0.4123)(120) = 49.476$$

The number of individual representatives assigned to this party — its apportionment — must be a whole number, such as 49, 50, or some other integer value.

As we have seen, the apportionment problem arises in allocating seats in the U.S. House of Representatives to states in proportion to their populations. Since 1920, the House has had 435 voting members. The population of Montana, for example, was 803,655 according to the 1990 census. The total population of the 50 states was recorded in the census as 249,022,783. Thus, the population of Montana is

$$\frac{803,655}{249,022,783} \times 100\% = 0.3227\%$$

of the U.S. population. Therefore, Montana's quota q is 0.3227% of 435. This yields

$$q = (0.003227)(435) = 1.404 \text{ seats}$$

Figure 14.3 The U.S. House of Representatives in session in Washington, D.C.

SP TLIGHT 14.1   Number Systems

▶ ▶ ▶ ▶ ▶ ▶ ▶ ▶ ▶ ▶ ▶ ▶ ▶ ▶

The concept of number is one of the most fundamental notions in mathematics. The *natural numbers*, which are also called the *whole* numbers, the *counting* numbers, or the *positive integers*, are the numbers 1, 2, 3, 4, 5, 6, 7, ... The nineteenth-century German mathematician Leopold Kronecker (1823–1891) attested to the fundamental role of the natural numbers when he said, "God made the whole numbers: all else [in mathematics] is the work of man."

Arithmetic with the natural numbers is limited. Addition and multiplication are possible, but the respective inverse operations, subtraction and division, cannot always be executed. For example, if 9 is subtracted from 6, or if 1 is divided by 2, the result will not be a natural number. We confront this problem by *extending* the number system. By introducing 0 and the negative numbers −1,

−2, −3, . . . , we create the system of *integers,* and make any subtraction possible. Thus, $6 − 9 = −3$. However, the quotient of two integers is not necessarily an integer, so the integers are *not closed* under division.

Another important extension of the natural numbers is the *positive fractions*. These are the ratios p/q of natural numbers p and q. The number p is called the *numerator* and q the *denominator* of the fraction p/q. These ratios are also called *positive rational numbers*. Any such fraction p/q can be divided by another fraction r/s according to the rule

$$\frac{p}{q} \div \frac{r}{s} = \frac{p \times s}{q \times r}$$

and the quotient will again be a rational number. In other words, the positive fractions are *closed under division*, although they are not closed under subtraction.

The current method for apportioning seats in the House, called the **Hill–Huntington method,** assigns Montana one seat from 1993 to 2002 (see Spotlight 14.3).

The House of Representatives is the best known and most frequently studied case of political apportionment. Some half-dozen different apportionment methods have been seriously considered for use by the Congress. In addition to the Hill–Huntington method, three other methods have been implemented, those of **Thomas Jefferson, Alexander Hamilton,** and **Daniel Webster.** John Quincy Adams also proposed an apportionment method, but it was never implemented.

The method of apportionment used for the House of Representatives can affect the outcome of a presidential election. The electoral college, which formally elects the president of the United States, gives each state a number of votes equal to the size of its congressional delegation, including representatives and senators. Each state votes as a bloc, casting all of its electoral votes for the candidate who received the plurality of popular votes within the state. In 1873, the House was malapportioned; the Hamilton method was then in effect, but it was incorrectly applied. Rutherford B. Hayes won the 1876 election with 185 electoral votes. His opponent, Samuel J. Tilden, garnered a

Fractions can be expressed in *decimal notation,* as in these examples:

$$\frac{1}{4} = \frac{25}{100} = 0.25$$

$$\frac{5}{4} = 1\frac{1}{4} = 1\frac{25}{100} = 1.25$$

$$\frac{8}{3} = 2.666666\cdots$$

$$\frac{7}{12} = 0.583333\cdots$$

$$\frac{1}{7} = 0.142857142857142857\cdots$$

Notice that the decimal representations of some fractions, such as one-fourth (0.25), terminate, whereas others, such as eight-thirds (2.6666 · · ·) go on without end. However, all rational numbers expressed in their decimal form will either terminate or else become infinitely repeating. For example, one-seventh will continue to repeat the block of digits 142857 unendingly.

Good notation facilitates the use of a number system. The ancient Greek mathematicians, despite their extraordinary accomplishments, used notation that was poorly suited for calculation. Use of the familiar Hindu–Arabic decimal notation was a great leap forward in the history of mathematics. It greatly simplified performing the elementary operations of arithmetic.

However, as even the ancient Greeks knew, fractions are not sufficient to represent all "real" measurements. For example, some ratios that arise naturally cannot be represented as rational numbers. The ratio of the diagonal of a square to one of its sides is the irrational number $\sqrt{2} = 1.4142146\cdots$. The ratio of the circumference of a circle to its diameter, denoted by the Greek letter π, is the irrational number $\pi = 3.1415927\cdots$. Irrational numbers cannot be represented as terminating decimals or as repeating decimals.

---

majority of the popular votes, but only 184 electoral votes, and was defeated. If the House had been correctly apportioned according to the Hamilton method, the result of the 1876 election would have been reversed, and Tilden would have won.

The fascinating history of apportionment of the U.S. House of Representatives is told in a delightful and important book, *Fair Representation: Meeting the Ideal of One Man, One Vote,* by Michel L. Balinski and H. Peyton Young. These authors advocate the use of the Webster method in place of the Hill–Huntington method, which was adopted by Congress in 1941 and has been the law since then.

Figure 14.4 In the 1970s mathematicians Michel L. Balinski *(left)* and H. Peyton Young *(right)* analyzed apportionment methods and recommended use of the Webster–Willcox method.

SP TLIGHT 14.2   Rounding Fractions

▶ ▶ ▶ ▶ ▶ ▶ ▶ ▶ ▶ ▶ ▶ ▶ ▶ ▶ ▶

In many practical applications of arithmetic, we are forced to *round* fractions. To round a fraction is to approximate it with a terminating decimal. For many purposes, it is sufficient to approximate the fraction $\frac{1}{3}$ with 0.33333, or $\frac{1}{7}$ with 0.14286. In these two examples, the fraction was rounded *down*, or *truncated*, to 5 decimal places. To round a ratio down to k decimal places, the first k digits to the right of the decimal point are determined, and any remaining digits are omitted. It is also possible to round *up*, by increasing the last digit retained by 1. For example, $\frac{1}{3}$ rounded up to three decimal places, is equal to 0.334, and 0.99899, rounded up to four decimal places, is equal to 0.9990.

If your savings account at the local bank grows at the annual rate of $6\frac{2}{3}\% = 0.066666\cdots$, then the interest that accrues after 3 months on a deposit of $1000 is given by multiplying the time period ($\frac{1}{4}$ year) times the interest rate, expressed as a decimal ($0.066666\cdots$) times the deposit ($1000). Thus, the interest due is

$$\frac{1}{4} \times 0.066666\cdots \times \$1000$$

$$= \$16.66666\cdots$$

Typically, the bank will round this amount down to two decimal places and credit $16.66 to your account, keeping the remaining two-thirds of a cent for itself. If the bank were computing interest on a loan, it would round *up*, and charge you $16.67 in interest for the quarter. Customers rarely complain about these practices, although the accumulated "roundings," as they are called, from many accounts add up to a considerable sum for the bank.

In most practical applications, the decision to round up or down is based on which gives the better approximation. If a number is to be rounded to two decimal places, it is rounded up if the third digit to the right of the decimal point is 5, 6, 7, 8, or 9, and rounded down if that digit is 0, 1, 2, 3, or 4. When this is done, the impact of rounding on the result of a calculation is usually insignificant.

When a ratio is rounded to zero decimal places, we say that it has been rounded to an integer value. There is a special notation for the integers that result from rounding a positive number x. We let $\lfloor x \rfloor$ denote the integer obtained by rounding x down; it is the largest integer that is not greater than x. For example, $\lfloor \pi \rfloor = 3$. The integer obtained by rounding a ratio x up is denoted $\lceil x \rceil$; thus $\lceil \pi \rceil = 4$. Notice that if x is not an integer, $\lceil x \rceil = \lfloor x \rfloor + 1$; however, if x is an integer, then it does not change when it is rounded up or down. Hence for an integer x, $\lfloor x \rfloor = \lceil x \rceil$.

▶ APPORTIONMENT METHODS

Although the apportionment problem appears in many contexts other than the House of Representatives, the terminology that we will use in discussing it will refer to *states*, *populations*, and a *house size*. Let n be the number of states, and let the populations of these states be denoted

$$p_1, p_2, \ldots, p_i, \ldots p_n$$

We denote the house size by h. Thus, each state's fair share, or *quota* q_i, is given by the formula

$$\frac{q_i}{h} = \frac{p_i}{p} \quad \text{or} \quad q_i = h\frac{p_i}{p}$$

where $p = p_1 + p_2 + \cdots + p_n$ is the total population of all n states. The number of representatives assigned to a state should be close to its quota. An apportionment is given by n integers

$$a_1, a_2, \ldots, a_i, \ldots, a_n$$

with a_i representing the number of representatives apportioned to state i. Since it is impossible to have a negative number of representatives, $a_i \geq 0$, and since the total number of representatives is equal to the house size,

$$a_1 + a_2 + \cdots + a_i + \cdots a_n = h$$

In our example on college enrollments, assume that there is a student representative assembly with 100 seats. How many seats should be assigned to each college? Since the total number of students enrolled in the university is 20,000, the quota q_i for the ith college is given by

$$q_i = \frac{hp_i}{p} = \frac{100p_i}{20,000} = \frac{p_i}{200}$$

where p_i is the number of students in the ith college. These numbers q_i are the same as the exact percents listed in the third column of Table 14.1. Our objective is to select six integers a_i that sum to 100 and approximate corresponding quotas q_i.

Hamilton's Method

A natural way to do this apportionment is called the *method of largest fractions*, or the *method of Alexander Hamilton*. With this method, each state must receive either its **lower quota,** which is the integer part of its quota, or its **upper quota,** which is obtained by increasing the lower quota by 1, unless the quota is already a whole number, in which case the upper and lower quotas are the same. In the college enrollment example, the lower quotas are

$$33, 24, 20, 16, 4, \text{ and } 1$$

and their sum is 98. As a preliminary step, the Hamilton method assigns to each state its lower quota of representatives. Unless each quota is an integer, the total number of seats assigned at this point will be less than the house size h, and this leaves a number of additional seats to be apportioned. The Hamilton method assigns these additional seats, one each, to those states whose quotas

Figure 14.5 The apportionment method of largest fractions, also known as the Hamilton method, was named for Alexander Hamilton. (*Alexander Hamilton*, John Trumbull; National Gallery of Art, Washington, D.C.; Andrew W. Mellon Collection.)

SP TLIGHT 14.3   Legal Challenges to Apportionment

▶ ▶ ▶ ▶ ▶ ▶ ▶ ▶ ▶ ▶ ▶ ▶ ▶

During the first two centuries of apportionment of the House of Representatives, millions of controversies were referred to the courts. However, until 1991, no state had ever challenged the method used to apportion the House in the federal courts. In that year, the Census Bureau reported the new apportionment that will be in effect for the congressional elections in the years 1992–2000. Several states lost representatives: New York lost three, and Ohio and Pennsylvania lost two apiece. However, the greatest percentage loss was sustained by Montana, whose apportionment decreased from two to one. Montana sued the U.S. Government in Federal District Court to regain that seat. Represented by its attorney general, Montana made a convincing argument. Since there was no precedent for its suit on congressional apportionment, Montana referred to two famous cases concerning state apportionment: *Baker v. Carr* and *Wesberry v. Sanders. Baker* was the 1962 case in which the

U.S. Supreme Court made its "one person, one vote ruling." Tennessee had drawn its state legislative district boundaries in such a way that the populations of the districts differed significantly. The court found that this practice was unfair to the residents of the large districts, and required Tennessee to redraw its district boundaries so that the populations of the district would be as nearly equal as possible. Two years later, in *Wesberry,* the Supreme Court issued a similar ruling, that all congressional districts within a state must be as nearly equal in population as possible.

Montana's argument was that these precedents should be applied when apportioning House seats to the states. The correct apportionment, in Montana's view, would be the one that met the *Baker* and *Wesberry* criterion of having districts as nearly equal in population as possible. The apportionment method that minimizes the difference in district size is the Dean method. Therefore, Montana

have the largest fractional parts. For the college enrollment example, the quotas with the largest fractional parts are 33.580 (the College of Arts & Sciences), with fractional part 0.580; and 1.480 (the College of Architecture), with fractional part 0.480. Each of these colleges would receive its upper quota of representatives. The Hamilton method therefore gives the apportionment

34, 24, 20, 16, 4, and 2

appearing in the last column of Table 14.1. Notice that $q_6 = 1.480$ was rounded up to 2, even though its fractional part was less than 0.5.

The Hamilton method is straightforward, but it has defects. The first presidential veto in U.S. history occurred when George Washington rejected an apportionment bill that implemented the Hamilton method. President Washington identified one of the defects in his veto message to Congress rejecting the Hamilton apportionment bill in

asked the court to require the Census Bureau to recompute the apportionments using the Dean method.

Let us see what would have happened if the Census Bureau had used the Dean method. With a population of 803,655, and one seat, Montana would have an "average" district size of 803,655. The State of Washington had gained a representative in the new apportionment, and now had 9 seats for a population of 4,887,941. Washington's average district size was therefore $4,887,941 \div 9 = 543,105$. The Montana district was 260,550 larger. Now suppose we take the disputed seat from Washington and give it to Montana. The average district in Montana would have a population of $803,655 \div 2 = 401,828$, while the average district size for Washington would be

$4,887,941 \div 8 = 610,993$. The average Washington district would be larger than the average Montana district by 209,165 people. Thus, by transferring Montana's second seat to Washington, the Census Bureau had *increased* the difference in district sizes. Montana argued that it should have the disputed seat.

In a split decision, the Montana Federal District Court agreed, and the U.S. Government appealed to the U.S. Supreme Court. In *Montana v. Department of Commerce,* the Supreme Court unanimously reversed the lower court. The opinion of the Court, written by Justice Stevens, demonstrated a full understanding of the issues of apportionment. Justice Stevens pointed out that *intra*state districts, which were the subject of the *Baker* and *Wesberry* cases, could be equalized in population by drawing district boundaries correctly. Since the Constitution forbids congressional districts from crossing state lines, some inequity is inevitable in congressional apportionment. While the opinion conceded that the Hill – Huntington method was not the only equitable apportionment method, it pointed out that in the selection process Congress had made a sincere effort to choose a fair and unbiased method.

1792. He said, in part, "there is no one proportion or division, which, applied to the respective numbers of the states, will yield the number and allotment of Representatives proposed by the bill." Washington objected that the numbers of representatives allotted to the states by the Hamilton apportionment were not as close to being proportional to the populations as they could have been.

A second defect of the Hamilton method is illustrated in the following example.

EXAMPLE: Salary Increments

A small college mathematics department has two full-time faculty, whose salaries are currently $43,100 and $42,150, and a part-time instructor, who is paid $10,000 per year. The total salary budget for this department is thus $95,250. The dean proposes to increase the salary budget by 5%, but requires that from now on all salaries must be rounded to multiples of

TABLE 14.2 **Apportioning Salaries**

Professor	Current salary	Increased by 5%	Lower quota	Apportioned salary
A	43,100	45,255	45,000	45,000
B	42,150	44,257	44,000	44,000
C	10,000	10,500	10,000	11,000
Totals	95,250	100,012	99,000	100,000

$1000. The new salary budget is therefore not $100,012, which is exactly 5% more than the old budget, but an even $100,000.

Table 14.2 shows the computation of the new salaries. The third column gives the salaries that would result from a straight 5% increase, without rounding. These numbers are the *quotas*. Thus, each faculty member's quota is 105% of his or her old salary. In the fourth column, the quotas are rounded *downward* to a multiple of $1000, to give the lower quotas. The sum of the lower quotas is $99,000, so there is an additional $1000 to be apportioned. It goes to the part-time instructor, C, because the fractional part of his salary, $500 out of $1000, is greatest. The apportioned salaries are shown in the fifth column of Table 14.2.

Professors A and B are displeased, because their salaries have been rounded down, giving each an effective salary increase of about 4.4%. The part-time instructor, on the other hand, has received a 10% raise. When this is pointed out to the dean, she responds by allocating an additional $1000 to the department's salary budget. This now amounts to a 6% increase, and the salary apportionments are recomputed as shown in Table 14.3.

Observe that with the new apportionment, the part-time instructor, C, has received no raise at all, even though he did receive a raise when the department's budget was increased by a smaller amount. ▲

A method of apportionment is said to be **house monotone** if an increase in the house size cannot cause any state's apportionment to decrease. The example that we have just considered shows that the Hamilton method is not house monotone. With a house size of 100 (the "house" in the example is actually the salary budget, in thousands of dollars), the apportionments were 45 for A, 44 for B, and 11 for C. When the house size increased to 101, A and B saw their apportionments increase to 46 and 45, respectively, while the apportionment for C *decreased* to 10.

For historical reasons, an apportionment method that is not house monotone is said to suffer the **Alabama paradox.** This name arose from an observation made after the census of 1880, when the Hamilton method was being used to apportion the house. The Bureau of the Census supplied Congress with a table of congressional apportionments for a range of different house sizes from 275 to 350. The table showed Alabama was entitled to 8 representatives if the house size was 299, but only 7 representatives if the house size was 300. Congress addressed this problem by choosing a house size, 325 seats, for which all could agree that the Hamilton method gave a fair apportionment.

By 1901, some members of Congress had discovered that the Alabama paradox made it possible to apportion the House of Representatives in a way that would minimize the representation of their political opponents. Based on the census of 1900, tables of apportionment for all house sizes

TABLE 14.3   **Apportioning Larger Salaries**

Professor	Current salary	Increased by 6%	Lower quota	Apportioned salary
A	43,100	45,686	45,000	46,000
B	42,150	44,679	44,000	45,000
C	10,000	10,600	10,000	10,000
Totals	95,250	100,965	99,000	101,000

between 350 and 400 were produced. The state of Colorado received 2 seats with a house size of 357, and 3 seats for house sizes 350–356 *and* 358–400. When the House Apportionment Committee proposed to make the house size equal to 357, it was accused in debate of manipulating the apportionment to Colorado's disadvantage. Congress ultimately voted to reject the 1901 apportionment bill based on Hamilton's method, and adopted the method of Webster in its stead. The Webster method is one of several divisor methods, which we study next.

Divisor Methods

In his message to Congress vetoing the 1793 apportionment bill, George Washington insisted that each state's apportionment should be proportional to its population. He turned to Thomas Jefferson, who had assisted him in writing the message, for a more acceptable method of apportionment. Jefferson's method was to choose a divisor d approximately the size of the average congressional district, and divide each state's population by that divisor. The apportionments are then obtained by discarding the fractional parts of the resulting quotients. For example, Jefferson took $d = 33,000$. His state, Virginia, had a population of 630,560, according to the 1790 census. Since

$$\frac{630,560}{33,000} = 19.108$$

Jefferson apportioned to Virginia 19 seats, discarding the fractional part 0.108. Jefferson's bill was passed by Congress and signed into law by the president in time to apportion the House for the 1794 election.

The 1794 House of Representatives had 105 members. To achieve that number, Jefferson had to choose his divisor carefully. For example, a divisor of 30,000 would have resulted in larger apportionments for several states, and a house size of 112, while a divisor of 36,000 would have given

Figure 14.6 Thomas Jefferson favored a method of apportionment biased in favor of states with large populations. (The Bowdoin College Museum of Art.)

TABLE 14.4 Apportioning the House of Representatives by Jefferson's Method

State	Population p_i	$d = 30000$ p_i/d	$d = 33000$ p_i/d	$d = 36000$ p_i/d	Apportionments		
					$d = 30000$	$d = 33000$	$d = 36000$
Virginia	630,560	21.02	19.11	17.52	21	19	17
Massachusetts	475,327	15.84	14.40	13.20	15	14	13
Pennsylvania	432,879	14.43	13.12	12.02	14	13	12
North Carolina	353,523	11.78	10.71	9.82	11	10	9
New York	331,589	11.05	10.05	9.21	11	10	9
Maryland	278,514	9.28	8.44	7.74	9	8	7
Connecticut	236,841	7.89	7.18	6.58	7	7	6
South Carolina	206,236	6.87	6.25	5.73	6	6	5
New Jersey	179,570	5.99	5.44	4.99	5	5	4
New Hampshire	141,822	4.73	4.30	3.94	4	4	3
Vermont	85,533	2.85	2.59	2.38	2	2	2
Georgia	70,835	2.36	2.15	1.97	2	2	1
Kentucky	68,705	2.29	2.08	1.91	2	2	1
Rhode Island	68,446	2.28	2.07	1.90	2	2	1
Delaware	55,540	1.85	1.68	1.54	1	1	1
Totals	3,615,920	120.53	109.57	100.44	112	105	91

smaller apportionments, a smaller house size of 91. See Table 14.4.

With the Jefferson method, the house size is determined by the divisor, with larger divisors yielding smaller houses. If the house size is assigned in advance, then to use the Jefferson method it is necessary to adjust the value of the divisor to obtain the correct house size. If the divisor that is chosen first produces a house that is too large, a larger divisor should be tried.

EXAMPLE: Apportioning the 1790 House of Representatives

If Thomas Jefferson had been required to apportion the House of Representatives so as to obtain exactly 100 seats, Table 14.4 indicates that he should try a divisor between 33,000 and

36,000, since the divisor 33,000 yields a house of 105 seats and 36,000 yields a house of 91 seats. Using 34,000 and 35,000 as trial divisors gives house sizes of 100 and 96 seats, respectively, as shown in Table 14.5. Thus, to achieve a house size of 100 seats, Jefferson could have taken 34,000 as the divisor. ▲

The method of Webster is also classified as **divisor method.** As with the Jefferson method, the population of each state is divided by a common number d, the divisor. The methods differ in the way fractions are handled. While the Jefferson method gives its apportionment for the ith state as the integer part of p_i/d, the Webster apportionment is p_i/d, rounded to the nearest integer (see Spotlight 14.2). Thus, if the fractional part of p_i/d is .5 or larger, the apportionment a_i is obtained by rounding p_i/d up (in the notation introduced in Spotlight 14.2, $a_i = \lceil p_i/d \rceil$), and if the fractional

TABLE 14.5 Adjusting the Divisors to Achieve a House Size of 100 Seats

State	Population p_i	$d = 34000$ p_i/d	$d = 35000$ p_i/d	Apportionments $d = 34000$	$d = 35000$
Virginia	630,560	18.55	18.02	18	18
Massachusetts	475,327	13.98	13.58	13	13
Pennsylvania	432,879	12.73	12.37	12	12
North Carolina	353,523	10.40	10.10	10	10
New York	331,589	9.75	9.47	9	9
Maryland	278,514	8.19	7.96	8	7
Connecticut	236,841	6.97	6.77	6	6
South Carolina	206,236	6.07	5.89	6	5
New Jersey	179,570	5.28	5.13	5	5
New Hampshire	141,822	4.17	4.05	4	4
Vermont	85,533	2.52	2.44	2	2
Georgia	70,835	2.08	2.02	2	2
Kentucky	68,705	2.02	1.96	2	1
Rhode Island	68,446	2.01	1.96	2	1
Delaware	55,540	1.63	1.59	1	1
Totals	3,615,920	106.35	103.31	100	96

part is less than .5, a_i is obtained by rounding down ($a_i = \lfloor p_i/d \rfloor$).

Jefferson's method was used to apportion the House of Representatives based on the censuses of 1790 through 1820. However, it was apparent that this method favors larger states, so much so that the largest state can receive more than its upper quota. This happened after the 1820 census, which recorded that New York had a population of 1,368,775. The total population of the United States was found to be 8,969,878. Since the House had 213 members, New York's quota was

$$\frac{1,368,775}{8,969,878} \cdot 213 = 32.503$$

Hamilton's method, which always allocates either the lower quota or the upper quota, would have apportioned 33 seats to New York. However, Jef-

Figure 14.7 Statesman and orator Daniel Webster (1782–1852), who developed a divisor method for apportioning the U.S. House of Representatives. (Hood Museum of Art, Dartmouth College, Hanover, N.H.; purchased through the Julia L. Whittier Fund.)

ferson's method, using the divisor $d = 39,900$, gave New York

$$\left\lfloor \frac{1,368,775}{39,900} \right\rfloor = 34 \text{ seats}$$

The discrepancy occurred again when the House was reapportioned following the 1830 census. Jefferson's method awarded New York 40 seats, although its quota was only 38.593. While many were troubled by the 1830 apportionment, the Jefferson method was not abandoned until 1842.

An apportionment method is said to satisfy the **quota condition** if in every situation each state's apportionment is equal to either its lower quota or its upper quota. In 1820, New York's upper quota was $\lceil 32.503 \rceil = 33$, and in 1830, New York's upper quota was $\lceil 38.593 \rceil = 39$, but the Jefferson method apportioned 34 and 40 seats, respectively, to New York in those years. It only takes one example like this to show that an apportionment method does not satisfy the quota condition. If the House had continued to use the Jefferson method, it would have violated the quota condition in every apportionment from 1820 until the present, *except* the apportionment based on the 1840 census.

With the Hamilton method, each state starts with its lower quota. States whose quotas have the largest fractional parts are given their upper quotas, until enough seats have been assigned to fill the house. There is no way for a state to receive less than its lower quota, or more than its upper quota, so the Hamilton method does satisfy the quota condition. This was obvious to the Congress in 1850, so it based its apportionment on the Hamilton method.[1]

We have seen that the Hamilton has another problem, lack of house monotonicity. The Jeffer-

son method favors larger states and does not satisfy the quota condition, but it is easy to see that it is house monotone. If the size of the house is increased, the apportionments of some states must be increased to fill the additional seats. The Jefferson method (or any other divisor method) accomplishes this by using a smaller divisor. Since the apportionment for each state is obtained by rounding down the quantity p_i/d, and each such quantity is increased when the divisor decreases, no apportionment will decrease. A method of apportionment that is not house monotone is susceptible to political manipulation, so this property of divisor methods is a strong reason to use them. In reverting to the Hamilton method, Congress was only replacing a problem with a worse problem.

Congress has never used a method of apportionment that is both house monotone and satisfies the quota condition. It would seem to be desirable to have such a method, and in the 1970s, the mathematicians Michel L. Balinski and H. Peyton Young set out to find one. Although they succeeded in doing so, the **quota method** that they discovered turned out to have an undesirable property, called the **population paradox**. A state's population could increase, while all other states maintained their populations, and yet that state could see its apportionment decrease. It has been shown that all apportionment methods, except divisor methods, have this unfortunate property.

To avoid the population paradox and the Alabama paradox, it is necessary to choose a divisor method. However, *no divisor method always satisfies the quota condition!* Thus, Balinski and Young have proved an impossibility theorem like the impossibility theorem of Kenneth Arrow that is discussed in Chapter 11. Their theorem forces us to conclude that the choice of an apportionment method is a political decision, because no method is mathematically perfect. The population and Alabama paradoxes are considered by political scientists to be more harmful than occasional violations of the quota condition, because they can be

---

[1] The origins of the Hamilton method had been forgotten in 1850, and the method was named for Congressman Samuel Vinton, who had rediscovered the method.

manipulated to a party's advantage. Thus, the search for the best apportionment method should be limited to divisor methods.

Which Divisor Method Is the Best?

We have so far encountered two divisor methods: the methods of Thomas Jefferson and Daniel Webster. Research done in the 1920s by the mathematician Edward V. Huntington focused on these methods, and three others. All divisor methods begin as follows: choose a "divisor" d. In practice, one can start with a divisor equal to the total population divided by the desired house size. According to the 1990 census, the U.S. population is 249,022,783 (not including the District of Columbia). The House of Representatives has 435 members, so we would start with

$$d = 249,022,783 \div 435 = 572,466$$

Let the populations of the fifty states be p_1, $p_2, \ldots, p_{50}$ and form the quotients p_1/d, $p_2/d, \ldots, p_{50}/d$. Each divisor method obtains a tentative apportionment by rounding these quotients in some consistent way. Thus, the Jefferson method rounds all of the quotients downward, while the Webster method rounds up when the fractional part of the quotient is greater than or equal to .5, and rounds down otherwise. Another method, originally advocated by John Quincy Adams, rounded all quotients upward. A fourth method, which has never been considered by the U.S. Congress, rounds up when the fractional part of the quotient is greater than .4, and rounds down otherwise. This method was invented by Condorcet, whose ideas on determining the winner of a multicandidate election are discussed in Chapter 11. The Hill–Huntington method, and another method that was originally proposed in 1830 by a Dartmouth College professor named James Dean, use more complicated rounding rules.

It is unlikely that the tentative apportionments obtained in this way will sum to the desired house

Figure 14.8 John Quincy Adams, who was our sixth president and later served in the House of Representatives (1831–48), advocated an apportionment that favored small states. (National Portrait Gallery, Washington, D.C./Art Resource, New York.)

size. If the sum is too large, a greater divisor is chosen and the tentative apportionments are recomputed. If the sum is too small, a smaller divisor is chosen. Trial and error eventually leads to a divisor that gives the desired house size.

Huntington evaluated divisor methods by what he called *measures of pairwise inequity*. For example, suppose that an apportionment gives the ith state a_i seats, and the population of that state is p_i. The quotient a_i/p_i is called the **representative share;** it represents the share of representation that is due to each citizen of that state. In an ideal "one person, one vote" apportionment, the representative share for citizens of the ith state would be the same as the representative share for citizens of the jth state. We know that perfect apportionment is impossible, and slight differences of representative share between states are to be expected with any apportionment. The difference of representative shares,

$$\frac{a_i}{p_i} - \frac{a_j}{p_j}$$

was one of Huntington's measures of pairwise inequity between states i and j. The states i and j are always ordered so that the representative share for state i is greater than or equal to the representative share for state j.

A measure of inequity converts an apportionment problem to an optimization problem. The optimum apportionment is the apportionment with the *least* inequity. If an apportionment is found such that it is impossible to transfer seats in such a way as to reduce the inequity between any pair of states, then that apportionment is optimum. At first sight, it would seem to be difficult to find the optimum apportionment. A brute force approach would start with a tentative apportionment, and transfer seats until an optimal apportionment is found. However, the optimum apportionment for any reasonable measure of pairwise inequity is given by some divisor method. For example, the Webster method gives the apportionment with the least inequity as measured by representative share.

EXAMPLE: Inequities in the 78th Congress

The 78th Congress was apportioned by the Hill–Huntington method, based on the 1940 census. According to that census, Michigan had a population of 5,256,106 and was apportioned 17 seats. Each citizen of Michigan was given a representative share of

$$\frac{17}{5,256,106} = 0.000003234 \text{ seat.}$$

Arkansas, with a population of 1,949,387, received 7 seats, so that each citizen of that state had a representative share of

$$\frac{7}{1,949,387} = 0.000003591 \text{ seat.}$$

Comparing these two states, we find that each Arkansan has 0.000000357 more of a seat than each Michigander. If a seat had been taken from Arkansas and given to Michigan, then the representative share for a Michigander would have been $18/5,256,106 = 0.000003425$, while each Arkansan would have been left with a representative share of $6/1,949,106 = 0.000003078$ seat. Now, the Michiganders have the advantage, but the difference,

$$\begin{array}{r} 0.000003425 \\ -\,0.000003078 \\ \hline 0.000000347 \end{array}$$

would be less than it was before the transfer was made. Therefore it would have been more equitable, when inequity is measured by comparing representative shares, to have given Arkansas 6 seats, and Michigan 18 seats.   ▲

The reader may wonder why the apportionment to Michigan and Arkansas had this inequity. Was politics or mathematics the cause? The answer is that it was a mixture of the two; see Spotlight 14.4. The apportionment is correct under the Hill–Huntington method, which uses a different measure of inequity. In his study of the divisor methods of apportionment, Huntington identified sixty-four different measures of inequity. Each of the measures of inequity was associated with one of the five divisor methods. Table 14.6 gives, for each divisor method, a measure of inequity for which the method is optimal. In this table, the formulas represent the amount of inequity between state i and state j, given that apportionments are a_i and a_j. It is assumed that with these apportionments, state i is favored over state j, so that the inequity will be a positive number. The table indicates that the measure of inequity for the Hill–Huntington method is *relative difference in district size*. If Michigan had 18 seats and Arkansas had 6 in the 78th Congress, then the average congressional

TABLE 14.6   The Five Divisor Methods of Huntington and Associated Measures of Inequity

Method	Measure of inequity	Formula
Jefferson	Representative deficiency	$a_i \left(\dfrac{p_j}{p_i} \right) - a_j$
Webster	Absolute difference in representative share	$\dfrac{a_i}{p_i} - \dfrac{a_j}{p_j}$
Hill–Huntington	Relative difference in district size	$\left(\dfrac{p_j}{a_j} - \dfrac{p_i}{a_i} \right) \div \dfrac{p_i}{a_i}$
Dean	Absolute difference in district size	$\dfrac{p_j}{a_j} - \dfrac{p_i}{a_i}$
Adams	Representative surplus	$a_i - \left(\dfrac{p_i}{p_j} \right) a_j$

district in Michigan would have contained 292,006 people. This number is obtained by dividing Michigan's 1940 population by the 18-seat apportionment. Arkansas congressional districts would have been larger, with an average size of 324,898 people. The **relative difference** in size is given by expressing the difference in the district sizes as a percentage of the smaller district. Thus, the relative inequity in Michigan's favor, according to this measure, would have been

$$\frac{324,898 - 292,006}{292,006} \times 100\% = 11.26\%$$

The actual apportionment of 17 seats for Michigan and 7 for Arkansas, favored Arkansas. The average Arkansas district contained 278,483 people, while Michigan districts averaged 309,183 people in size. The relative difference was

$$\frac{309,183 - 278,483}{278,483} \times 100\% = 11.02\%$$

Since the relative inequity was less when Michigan had 17 seats and Arkansas had 7, this was the preferred apportionment with the Hill–Huntington method.

Since each of the five measures of inequity in Table 14.6 leads to a slightly different apportionment, one way to decide which apportionment method to use is to decide, in a political debate, which measure of inequity is the most important. This was the approach taken when the Hill–Huntington method was adopted in 1941 to apportion the House of Representatives. Challenges to apportionments have also followed this approach (see Spotlight 14.3). Since there are valid arguments to be made for using absolute difference in representative share, relative difference in district size, and absolute difference in district size as the best measure of inequity, this debate will never end, and we can expect occasional changes in the official method to apportion the House.

Another approach to choosing the best apportionment method, advocated by Balinski and Young, reaches a definite conclusion. Some of the divisor methods are obviously *biased* in favor of small states or in favor of large states. The Jeffer-

SPOTLIGHT 14.4   Mathematics and Politics: A Strange Mixture

▶ ▶ ▶ ▶ ▶ ▶ ▶ ▶ ▶ ▶ ▶ ▶ ▶

The years 1900–1941 were the time of the most intense debate about congressional apportionment. During the nineteenth century manipulation of apportionment schemes was not unknown, but serious controversy was avoided by increasing the house size each decade so that no state suffered a loss of seats when other states gained. In 1900, the Hamilton method was finally rejected when attempts to manipulate it to advance parochial causes were revealed (see page 436), and the Webster method was used instead.

The first American to consider apportionment from a theoretical point of view was Walter Willcox (1861–1964), who strongly advocated the Webster method and had computed the apportionment of 1900. His arguments convinced the Congress to use the Webster method again in 1910, when 433 seats were apportioned. Following the nineteenth-century tradition, the number 433 was chosen so that no state would lose a seat in the reapportionment. Arizona and New Mexico

were admitted to the Union in 1912, and were each given one seat in the House. Every congressional reapportionment after 1912 has been based on a house size of 435. In 1911, Joseph Hill, a statistician at the Bureau of the Census, proposed the Hill (or Hill–Huntington) method. Edward V. Huntington, a mathematics professor at Harvard, strongly endorsed Hill's suggestion.

In 1920, two methods were in competition, Webster's against Hill and Huntington's. There were significant differences in the apportionments determined by the two methods. There was also disagreement on whether or not to increase the house size again. The result was Washington gridlock: no apportionment bill passed during the decade, and the 1910 apportionments were retained throughout the 1920s. In preparation for the 1930 census results, the National Academy of Sciences formed a committee to determine whether either method was biased in favor of large or small states. In 1929, the committee reported that the Huntington

son method favors the large states (this probably caused Jefferson to advocate the method, as his state, Virginia, was the largest state in 1790), and the Adams method favors small states. To see this, consider the 1990 census. California's quota was 51.9967 seats, so a reasonable apportionment method should allot 52 seats to that state. The Dean, Hill–Huntington, and Webster methods all

give California 52 seats. However, the Jefferson method gives 54 seats (two more than the upper quota!), and the Adams method gives California 50 seats. On the other hand, Rhode Island's quota was 1.7529, and every divisor method except the Jefferson method apportions 2 seats to that state. The Jefferson method gives Rhode Island one seat. South Dakota's quota was 1.2198, and every

Walter F. Willcox (Department of Manuscripts and University Archives, Cornell University Libraries.)

Edward V. Huntington (Courtesy of the Harvard University Archives).

method was the more neutral (this conclusion has been disputed, with good evidence, by Balinski and Young).

The 1930 census was remarkable in that the apportionments calculated by the Webster method for a house of 435 members were the same as the Hill–Huntington apportionments. The House was therefore reapportioned, but the method used could be claimed to be either one of the competing methods. The coincidence was almost repeated in the 1940 census, but there was one small difference. The Hill–Huntington method gave Michigan 17 seats, and Arkansas 7. Webster's method gave Michigan 18 seats, and Arkansas 6 (see page 442). At the time, Michigan was a predominantly Republican state, and Arkansas was in the Democratic column. The Democrats in Congress voted to adopt the Hill–Huntington method, and the Republicans voted for the Webster method. Since the Democrats had the majority, the Hill–Huntington method became the law.

method except the Adams method gives South Dakota 1 seat, while the Adams method apportions 2.

In 1980, Balinski and Young proved that the Webster method of apportionment is the only method that is completely unbiased toward small or large states. The Hill–Huntington and Dean methods are both slightly biased in favor of small states. The Jefferson and Adams methods have a much more noticeable bias. Balinski and Young have also pointed out that while no divisor method will always satisfy the quota condition, the Webster method is the divisor method that is *least* likely to violate it. In fact, neither the Hill–Huntington, the Dean, nor the Webster methods would have violated the quota condition in any apportionment of Congress since 1790.

An Application to Scheduling

Let us see how apportionment methods work when applied to another kind of allocation program.

A small senior high school has only one mathematics teacher, who teaches five classes each day. One hundred students preregister to take one of the following three mathematics courses: 51 students sign up for tenth-grade geometry, 30 for eleventh-grade algebra, and 19 for twelfth-grade calculus. Given that a total of five sections are to be offered, how many sections of each course should be taught? This is an apportionment problem.

If there are 100 students and five sections, the average class size will be 20. However, we cannot realize this average in each class, because 20 does not divide evenly into the numbers 51, 30, or 19. Instead, we compute the ideal fair share of sections (the quota q) for each subject. For example, the quota for geometry will be $q = \frac{51}{100} \times 5 = 2.55$. Similarly, the quotas for algebra and calculus will be 1.50 and 0.95, respectively. Table 14.7 summarizes this information.

The apportionment problem is to round these three quotas to whole numbers while maintaining the sum at 5. The Hamilton method first assigns to each subject the integer part of its quota, which is its *lower quota*. For this example, the lower quotas are the numbers 2, 1, and 0, respectively, and they are shown in the fifth column of Table 14.7. These account for only three of the five sections. The remaining two sections are assigned to the subjects whose quotas have the largest fractional parts. The calculus course gets one of the extra sections, since its quota has the largest fractional part, 0.95. The quota with the next highest fractional part is geometry's, so geometry gets the other extra section. The Hamilton apportionment is therefore 3, 1, and 1, as shown in the eighth column of Table 14.7. The average class sizes for the three subjects are then 17, 30, and 19, respectively.

Suppose that when the term begins, the actual enrollments in the three math courses have changed to 52, 33, and 15. The new quotas, integer parts, fractional parts, and apportionments are shown in Table 14.8.

In comparing Tables 14.7 and 14.8, you will observe that even though the number of students taking geometry has increased from 51 to 52, the number of geometry sections has decreased from 3 to 2. One section, formerly allocated to geometry, has been switched to algebra, which had a *larger* increase in enrollment. An apportionment method is **quota monotone** if a state's apportionment (in this case, the number of sections allotted) cannot decrease when its quota increases. We have just seen that the Hamilton method is not quota monotone. This is not a defect of just the Hamilton method, though, because there is no reasonable method of apportionment that is quota monotone.

Let us now apportion classes, according to the revised enrollments, with the divisor methods. For simplicity, we will omit the Hill–Huntington and

---

TABLE 14.7 Apportioning Course Sections

Grade	Course	Number of students	Quota q	Integer part	Fractional part	Fraction rounded	Hamilton apportionment	Course average
10	Geometry	51	2.55	2	0.55	1	3	17
11	Algebra	30	1.50	1	0.50	0	1	30
12	Calculus	19	0.95	0	0.95	1	1	19
	Totals	100	5.00	3	2.00	2	5	20

TABLE 14.8 **Apportioning Course Sections with Revised Enrollments**

Grade	Course	Number of students	Quota q	Integer part	Fractional part	Fraction rounded	Hamilton apportionment	Course average
10	Geometry	52	2.60	2	0.60	0	2	26
11	Algebra	33	1.65	1	0.65	1	2	16.5
12	Calculus	15	0.75	0	0.75	1	1	15
	Totals	100	5.00	3	2.00	2	5	20

TABLE 14.9 **Apportioning Classes with Divisor 20**

Course	Number of students	Quotient	Apportionment		
			Adams	Webster	Jefferson
Geometry	52	2.60	3	3	2
Algebra	33	1.65	2	2	1
Calculus	15	0.75	1	1	0
Totals	100	5.00	6	6	3

Dean methods, and concentrate on the Adams, Webster, and Jefferson methods. First consider a divisor equal to the average class size, 20. The numbers of times that 20 divides the enrollment figures of 52, 33, and 15 are given in the *Quotient* column of Table 14.9. The last three columns of the table show the apportionments given by the three methods. For the Adams method, the quotients are all rounded up; for the Webster method, each quotient is rounded to the nearest integer; and for the Jefferson method, the quotients are all rounded down. Because none of the totals—6, 6, and 3—meets the requirement of precisely five sections, 20 is not a suitable divisor. The Adams and Webster methods will require a larger divisor, and the Jefferson method will require a smaller divisor.

A divisor of 26 yields an Adams apportionment of two sections each of geometry and alge-

bra, and one section of calculus (see Table 14.10). This divisor can be viewed as the maximum class size for any single section (provided the group of students taking geometry is evenly split into two sections of 26). In fact, any divisor between 26 and 32 would do as well. If the divisor is less than 26, there will be more than two sections of geometry, resulting in more than five sections in all, and if the divisor is 33 or more, there will be only one algebra section, and therefore a total of fewer than five sections.

Table 14.10 indicates that 26 is too large a divisor to use with the Webster method, since a total of only four sections is apportioned with that divisor. To find an appropriate divisor we could try apportioning with divisors between 20 (which we know is too small) and 26 (which is too large). However, we can take a shortcut, by figuring out which course will get the fifth section. That

TABLE 14.10   Apportioning Classes with Divisor 26

Course	Number of students	Quotient	Apportionment	
			Adams	Webster
Geometry	52	2.00	2	2
Algebra	33	1.27	2	1
Calculus	15	0.58	1	1
Totals	100	3.84	5	4

course will not be calculus, because algebra, with a larger enrollment, has a higher priority for getting a second section.

Let d denote a divisor that apportions three sections to geometry. The geometry enrollment, divided by d, must be at least 2.5, so that the quotient will be rounded up to get 3. We can write this condition as the inequality

$$2.5 \leq \frac{52}{d}$$

Multiply through by d to get

$$(2.5)d \leq 52$$

and divide through by 2.5 to get $d \leq (52/2.5) = 20.8$. On the other hand, if algebra gets a second section, then its enrollment, divided by d, must be at least 1.5. In other words,

$$1.5 \leq \frac{33}{d}$$

which can be solved as before to get $d \leq 22$. Thus, if $d \leq 22$, algebra will get two sections, while if $d \leq 20.8$, geometry will also get three sections, and a total of six sections will be apportioned. It follows that *algebra will get a second section before geometry gets its third section*. The Webster apportionment is therefore two sections each for geometry and algebra, and one section for calculus.

Let us now calculate the Jefferson apportionment. Table 14.9 indicates that when the divisor $d = 20$ is used, only three sections are apportioned: two geometry sections and one algebra section. To increase the geometry apportionment to three, the quotient of the geometry enrollment by the divisor would have to be at least three, since apportionment is obtained by rounding this quotient down. Therefore, the divisor d must satisfy

$$3 \leq \frac{52}{d}$$

This inequality can be solved to get $d \leq 52/3 = 17.3333$. Similar reasoning shows that to increase the algebra apportionment to two sections, we will need a divisor d such that $2 \leq 33/d$, or $d \leq 16.5$. Finally, to increase the calculus apportionment to one section, the divisor would need to satisfy $1 \leq 15/d$, or $d \leq 15$. Here is the solution. If $d \leq 17.3333$ and $d > 16.5$, there will be three geometry sections, one algebra section, and no calculus sections: a total of four sections. If $d \leq 16.5$ and $d > 15$, there will be five sections in all: three geometry, two algebra, and no calculus. A divisor less than 15 produces more than five sections.

Table 14.11 is a summary of the solutions of this apportionment problem. All methods, with the exception of the Jefferson method, apportion two sections each to the geometry and algebra courses, and one section to calculus. The Jefferson

TABLE 14.11   Apportioning Classes by the Methods of Hamilton, Adams, Webster, and Jefferson

Course	Number of students	Quota	Apportionment Hamilton	Adams	Webster	Jefferson
Geometry	52	2.60	2	2	2	3
Algebra	33	1.65	2	2	2	2
Calculus	15	0.75	1	1	1	0
Totals	100	5.00	5	5	5	5

method again shows its bias in favor of large states by taking away the calculus section, and adding a geometry section.

OPTIONAL ▶ **PRIORITY LISTS FOR APPORTIONMENT**

In 1941, Congress enacted a law adopting the Hill–Huntington method for apportioning the House of Representatives. The apportionments based on the censuses of 1940 through 1990 have been determined by the U.S. Census Bureau without any further action by the Congress.

Technically, the Hill–Huntington method is a divisor method. However, the Census Bureau determines the apportionments by making a *priority list*. The list is constructed as follows. Since the Constitution requires that each state shall have at least one representative, each of the 50 states is given one seat at the outset. Since there are 435 seats to be apportioned, $385 = 435 - 50$ are left.

The state to receive the next seat is always the state with the highest priority. Let n_i denote the number of seats already apportioned to state i (at first, $n_i = 1$, but n_i increases as the state acquires additional seats), and let p_i denote the population of state i. The priority of state i is determined by evaluating the formula

$$R_i = \frac{p_i}{\sqrt{n_i(n_i + 1)}}$$

If state i has the highest priority at some point in the apportionment process, then it will receive the next seat (provided that all 435 seats have not already been apportioned). This will cause n_i to increase, and thus R_i *decreases*. It usually happens that another state will now have the highest priority for the following seat.

Let us see how the fifty-first through fifty-fourth seats were apportioned after the 1990 census. At the outset, each state has one seat, so $n_i = 1$ for $i = 1, \ldots, 50$. Therefore $\sqrt{n_i(n_i + 1)} = \sqrt{2}$, so for each state, $R_i = p_i/\sqrt{2}$. The state with the highest rank is the most populous state, California (population 29,839,250). We will make California state number 1. California gets the fifty-first seat, so n_1 increases from 1 to 2. The priority of California now decreases:

$$R_1 = \frac{29,839,250}{\sqrt{(2)(3)}} = 12,181,823$$

The second most populous state is New York; its population is 18,044,505. We will number New York as state 2, so that $n_2 = 1$. The priority of New York is given by comparing the new value of R_1 with

$$R_2 = \frac{18,044,505}{\sqrt{2}} = 12,759,392$$

Since now $R_2 > R_1$, the fifty-second seat goes to New York. Now $n_2 = 2$, so

$$R_2 = \frac{18{,}044{,}505}{\sqrt{(2)(3)}} = 7{,}366{,}638$$

State 3, the third most populous state, is Texas, with population $p_3 = 17{,}059{,}805$. Since $n_3 = 1$,

$$R_3 = \frac{17{,}059{,}805}{\sqrt{2}} = 12{,}063{,}104$$

California has higher priority for the fifty-third seat than Texas, because R_1 is slightly larger than R_3. Therefore, the fifty-third seat goes to California, and now $n_1 = 3$. We recompute R_1 as

$$R_1 = \frac{p_1}{\sqrt{(3)(4)}} = \frac{29{,}839{,}250}{\sqrt{12}} = 8{,}613{,}850$$

Texas is now the highest ranking state, and receives the fifty-fourth seat. This reduces R_3 to the value 6,964,636. Observe that although California now has three seats, and New York and Texas have two seats each, California still has a higher priority than either of those states for receiving the fifty-fifth seat. However, Florida has a still higher priority for that seat, and California actually receives the fifty-sixth seat.

It is possible to make a priority list giving the apportionments for any divisor method. To construct a priority list for the Webster method, start with a divisor $d = 2 \times p_1$, where p_1 is the largest state population. The quotient of p_1 by this divisor is $\frac{1}{2}$, so state 1 receives the first representative. The quotients of the populations of the remaining states by this divisor are less than $\frac{1}{2}$; they are rounded down to 0, and therefore no other seats are apportioned. Now decrease the divisor. State 1 will receive a second seat when the $p_1/d = 1.5$, because 1.5 is the smallest number that would be rounded up to 2. The equation $p_1/d = 1.5$ can be solved for d to obtain $d = p_1/1.5$. Of course, other

states may receive their first seats before d declines to that value. State 2, with population p_2, will receive a seat when $d = 2p_2$.

With the Webster method, state i will receive its $n + 1$st seat when $p_i/d = n + \frac{1}{2}$, which is the smallest number that can be rounded up to get $n + 1$. Solving this equation for d, we see that state i receives its $n + 1$st seat when

$$d = \frac{p_i}{n + \frac{1}{2}}$$

The number $R_i = p_i/(n_i + \frac{1}{2})$ determines the priority of state i for receiving $n_i + 1$ seats. Thus, the difference between the Webster and Hill–Huntington methods lies in the formulas used to compute the priorities. These formulas are called **ranking functions**. Ranking functions for each of the divisor methods that we have mentioned are given in Table 14.12. With the Hill–Huntington, Dean, and Adams methods, each state starts with one seat before the process begins, but with the Webster and Jefferson methods, the initial apportionment to each state is 0.

TABLE 14.12 Ranking Functions for the Divisor Methods

Method	Ranking function
Jefferson	$R_i = \dfrac{p_i}{n_i + 1}$
Webster	$R_i = \dfrac{p_i}{n_i + \frac{1}{2}}$
Hill–Huntington	$R_i = \dfrac{p_i}{\sqrt{n_i(n_i + 1)}}$
Dean	$R_i = \dfrac{p_i(2n_i + 1)}{2n_i(n_i + 1)}$
Adams	$R_i = \dfrac{p_i}{n_i}$

The ranking functions are the links between the divisor methods and their corresponding measures of inequity. To connect the representative share measure with the Webster method, let us consider two states, state 1 and state 2, that are competing for the next seat to be apportioned. The respective populations of the states are denoted p_1 and p_2; state 1 has already received n_1 seats, and state 2 has n_2 seats at this point. If state 1 gets the next seat, each of its citizens will have a representative share of $(n_1 + 1)/p_1$, while the representative share for a person in state 2 will remain at n_2/p_2. The difference in representative share if state 1 gets the next seat will be

$$D_{1,2} = \frac{n_1 + 1}{p_1} - \frac{n_2}{p_2}$$

Similarly, if state 2 gets the next seat, the difference in representative share will be

$$D_{2,1} = \frac{n_2 + 1}{p_2} - \frac{n_1}{p_1}$$

Assume that $D_{1,2} < D_{2,1}$; that is,

$$\frac{n_1 + 1}{p_1} - \frac{n_2}{p_2} < \frac{n_2 + 1}{p_2} - \frac{n_1}{p_1} \qquad (14.1)$$

Inequity in representative share will then be minimized by giving the next seat to state 1.

Let us rearrange the inequality (14.1). Put the terms involving state 1 on the left side, and those involving state 2 on the right. This yields

$$\frac{n_1}{p_1} + \frac{n_1 + 1}{p_1} < \frac{n_2 + 1}{p_2} + \frac{n_2}{p_2}$$

or, after simplifying,

$$\frac{2n_1 + 1}{p_1} < \frac{2n_2 + 1}{p_2} \qquad (14.2)$$

Since the inequalities (14.1) and (14.2) are equivalent, if inequality (14.2) holds, inequity in representative share will be minimized by assigning the next seat to state 1.

Recall that $R_1 = p_1/(n_1 + \frac{1}{2})$. Therefore,

$$\frac{1}{R_1} = \frac{n_1 + \frac{1}{2}}{p_1} = \frac{1}{2} \times \frac{2n_1 + 1}{p_1}$$

Similarly,

$$\frac{1}{R_2} = \frac{1}{2} \times \frac{2n_2 + 1}{p_2}$$

Thus, by multiplying both sides of the inequality (14.2) by $\frac{1}{2}$, we can replace that inequality by

$$\frac{1}{R_1} < \frac{1}{R_2} \qquad (14.3)$$

Inequality (14.3) is equivalent to inequalities (14.1) and (14.2). Therefore, if inequality (14.3) holds, the next seat should be given to state 1. But if $1/R_1$ is *less* than $1/R_2$, then R_1 must be *greater* than R_2. In other words, inequality (14.3) holds precisely when state 1 has a higher priority, as measured by the Webster ranking function, than state 2. We have thus verified that inequity, as measured by difference in representative share, will be minimized by apportioning seats by the Webster method. ◀

▶ Review Vocabulary

Adams method   An apportionment method invented by John Quincy Adams. It is a divisor method that rounds all positive fractions upward to the next integer.
Alabama paradox   An apportionment method suffers the Alabama paradox if it is possible for some state to lose a representative solely because the size of the house is increased.

Apportionment A whole number that is used to approximate the quota.

Apportionment problem To round a list of fractions to integers in a way that preserves the sum of the original fractions.

Condorcet method An apportionment method invented by the Marquis de Condorcet. All fractions greater than or equal to $\frac{2}{3}$ are rounded to the next integer; fractions less than $\frac{2}{3}$ are rounded down.

Dean method An apportionment method invented in 1830 by a mathematician, James Dean. It is a divisor method designed to make the district sizes as nearly equal as possible.

District size If a state has population p_i and apportionment a_i, the quotient p_i/d_i gives the average population of a congressional district in that state. This quotient is the district size.

Divisor method One of many apportionment methods in which the apportionments are determined by dividing the populations of the states by a number d, called the divisor, and rounding the resulting quotients to adjacent integer values. Divisor methods differ in how the decision to round up or down is made. The divisor methods mentioned in this chapter are the methods of Adams, Condorcet, Webster, Hill–Huntington, Dean, and Jefferson.

Hamilton method An apportionment method first advocated by Alexander Hamilton. It is the only nondivisor method studied in this chapter. This method assigns to each state the integer part of its quota (its lower quota), and then assigns an extra seat (to get the upper quota) to those states whose quotas have the largest fractional parts. The total number of extra seats assigned is the number required to reach the desired house size.

Hill–Huntington method A divisor method named for the statistician Joseph Hill and the mathematician Edward Huntington. This method minimizes the relative difference in district size; it also minimizes the relative difference in representative share.

House monotonicity An apportionment method is house monotone if no state's apportionment can decrease when the house size increases. A method that is house monotone is immune from the Alabama paradox.

Jefferson method An apportionment method invented by Thomas Jefferson. It is a divisor method that rounds all positive fractions downward, and takes the integer part of the quotient as the apportionment.

Lower quota The integer part of a state's quota.

Monotone A function is monotone if it will not decrease in value as a result of an increase in the variable.

Population paradox An apportionment method suffers the population paradox if a state's apportionment can decrease when only its population increases. All apportionment methods except divisor methods suffer this paradox.

Quota A state's quota in an apportionment problem is the number of seats it would receive if fractional seats could be awarded. The quota for state i is $q_i = (hp_i)/p$, where h is the house size, p_i is the population of the state, and p is the total population of all the states.

Quota condition An apportionment method satisfies the quota condition if it is impossible for a state to be apportioned less than its lower quota or more than its upper quota. All divisor methods fail in some cases to satisfy the quota condition.

Quota method An apportionment method invented by Michel Balinski and Peyton Young. This method satisfies the quota condition and is house monotone; however, it is not population monotone. It is only of theoretical interest and is not studied in this chapter.

Quota monotonicity An apportionment method is quota monotone if no state's apportionment can decrease when its quota increases. None of the methods discussed in this chapter has this property.

Ranking function The ranking function for a divisor method is used to construct a priority list for assigning seats. The function takes two variables, the population p_i and the number of seats already assigned, n_i. The value R_i is the largest divisor that will permit state i to receive its next seat. The state whose current R_i is largest has the highest priority.

Relative difference   The relative difference between two positive numbers is obtained by subtracting the smaller number from the larger, and expressing the result as a percentage of the smaller number. Thus, the relative difference of 120 and 100 is 20%.

Representative share   A state's representative share is the state's apportionment divided by its population. It is intended to represent the amount of influence a citizen of that state would have on his or her representative.

Upper quota   The whole number that results from rounding a state's quota upward.

Webster method   An apportionment method invented by Congressman Daniel Webster. All fractions greater than $\frac{1}{2}$ are rounded up to the next integer, and all fractions less than $\frac{1}{2}$ down. The Webster method minimizes the differences of representative share between the states.

▶ SUGGESTED READINGS

BALINSKI, M. L., AND H. P. YOUNG: *Fair Representation: Meeting the Ideal of One Man, One Vote,* Yale University Press, New Haven, 1982. In the 1970s, Balinski and Young analyzed apportionment methods in depth. Their point of view was to postulate the desirable properties of an apportionment method as axioms, and to deduce from the axioms the characteristics of the best method. This highly readable book combines a complete account of the interesting history of apportionment of the U.S. House of Representatives with the basic theory of apportionment.

BALINSKI, M. L., AND H. P. YOUNG: "The Webster Method of Apportionment," *Proceedings of the National Academy of Sciences, U.S.A.,* 77:1–4 (1980). A technical outline of the proof that the population monotonicity axiom is not compatible with the quota condition, and an analysis of bias that concludes that the Webster method is the unbiased divisor method.

BALINSKI, M. L., AND H. P. YOUNG: "The Huntington Methods of Apportionment," *SIAM Journal of Applied Mathematics,* 33:607–618 (1977). A theoretical discussion devoted to the divisor methods. A property of apportionment methods called *consistency* is introduced. A method is consistent if the apportionments of a group of states would not change if the rest of the states were removed, and the house size were reduced to the sum of the original apportionments of those states. It is proved that divisor methods are consistent, and that any apportionment method that is both consistent and house monotone is a divisor method.

BALINSKI, M. L., AND H. P. YOUNG: "The Quota Method of Apportionment," *American Mathematical Monthly,* 82:701–730 (1975). A description of the quota method, which is the only house monotone apportionment method that satisfies the quota condition.

Commonwealth of Massachusetts v. Mosbacher, 785 Federal Supplement 230 (District of Massachusetts 1992). This opinion concerns a suit by Massachusetts to increase its representation. The Commonwealth argued that the Bureau of the Census did not fairly assign federal employees who are stationed abroad to their home states, and this part of the opinion is not of interest to us. However, Massachusetts also claimed that the Hill–Huntington method was an unfair method of apportionment, and asked the court to replace that method with the Webster method. The discussion of this portion of the claim is Section D of the opinion, and starts on page 253. The Federal Supplement is available in law libraries.

HUNTINGTON, E. V.: "The Apportionment of the Representatives in Congress," *Transactions of the American Mathematical Society*, 30:85–110 (1928). Introduction to divisor methods of apportionment, relating each method to several of 64 "measures of inequity" between the states. The author also derives the rank function for each method, simplifying the actual calculation.

LUCAS, W. F.: "The Apportionment Problem," in S. J. Brams, W. F. Lucas, and P. D. Straffin, Jr.

(eds.), *Political and Related Models*, Springer-Verlag, New York, 1983, pp. 358–396. An introduction to apportionment, written at a somewhat more advanced level than the presentation in this text.

U.S. Dept. of Commerce v. Montana, 112 Supreme Court 1415 (1992). This opinion, available in any law library, gives the grounds for rejecting the Montana suit to replace the Hill–Huntington method with the Dean method.

▶ EXERCISES

1. Round the numbers in the following sum to whole numbers so that the total is maintained at 100.

$$14.3159$$
$$33.2792$$
$$26.3430$$
$$13.7491$$
$$\underline{12.3128}$$
$$100.0000$$

2. Round the numbers in the following sum to whole numbers so that the total is maintained at 1500.

$$214.7985$$
$$498.7880$$
$$395.7850$$
$$205.9365$$
$$\underline{184.6920}$$
$$1500.0000$$

3. A country has a parliament with 577 seats. In an election, the Democratic Socialists receive 323,829 votes, the Social Democrats receive 880,702 votes, the Christian Democrats receive 5,572,614 votes, the Greens receive 1,222,498 votes, and the Communists receive 111,224 votes. The number of seats won by each party is to be proportional to the number of votes cast in its favor. Calculate the quota for each party and apportion the seats by the Hamilton method.

4. Consider a small college with three divisions: arts, science, and business. There are 1500 students in all, with 690 in the arts division, 435 in the science

■ Discussion exercise.

division, and 375 in the business division. The student senate has five members, apportioned to the three divisions according to student population.

   a. What is the quota for each division?
   b. How would you apportion the seats to the divisions?
   c. The student populations in the next year are 555, 465, and 480, respectively. Recompute the quotas.
   d. How would you apportion the seats for the next year?
   e. Has any division gained students but lost representation in the second year?

5. Here is an apportionment method that should please everyone! Just give each state its upper quota.

   a. Show that the house size will be more than the planned house size h.
   b. Do you think California would be enthusiastic about this method, or would that state prefer to give each state its *lower* quota?
   c. Have you seen this method before?

6. A very small country has three states, with populations of 59, 76, and 14. Use the Hamilton method to apportion the seats of the National Legislature to the states according to their populations, with a house size of 35 seats. Repeat the calculation for 36, 37, 38, 39, and 40 seats. Suppose the people in the small state are communists, and the larger states are inhabited by people believing in private enterprise. Which house size would the large states prefer?

7. A country has five states, with populations 5,576,330, 1,387,342, 3,334,241, 7,512,860, and 310,968. Its house of representatives is apportioned by the Hamilton method.

   a. Calculate the apportionments for house sizes of 82, 83, and 84. Does the Alabama paradox occur?
   b. Repeat the calculations for house sizes of 89, 90, and 91.

8. Reconsider the college example given in the text (page 428).

College	Number of students	Percent	Percent rounded
Arts & Sciences	6716	33.580	34
Engineering	4832	24.160	24
Agriculture	4093	20.465	20
Business	3211	16.055	16
Law	852	4.260	4
Architecture	296	1.480	2
Totals	20,000	100.00	100

    a. What would the apportioned percents be if the apportionments were computed by the methods of (1) Jefferson, (2) Webster, and (3) Adams?

    b. For each of the methods in part a, give a value of the divisor d that will yield the correct house size (100).

    9. Apportion the small college student senate (see Exercise 4) by the methods of (1) Hamilton, (2) Jefferson, (3) Webster, and (4) Adams, using the data for the first year.

    10. Reapportion the small college senate as in Exercise 9, using the data for the second year. Do any of the methods display failure of quota monotonicity? That is, does any division's quota *increase* while its apportionment decreases?

    11. A country has six states with populations 27,774, 25,178, 19,947, 14,614, 9225, and 3292. Its House of Representatives has 36 seats. Find the apportionment using the methods of (1) Hamilton, (2) Jefferson, (3) Webster, and (4) Adams.

    12. Determine the relative difference between the numbers 5 and 7.

    13. Jim is 72 inches tall, and Alice is 65 inches tall. What is the relative difference of their heights?

    14. Professor Roe received a $1000 raise last year, and Professor Doe received a $2000 raise. What is the relative difference in the raises?

    15. In the 1991 apportionment of Congress, the average congressional district in Pennsylvania had a population of 567,843. The corresponding figure for New Jersey was 596,049.

    a. Which state is the more favored in this apportionment?

    b. What is the relative difference in the district sizes?

    16. According to the 1992 census, the population of Ohio was 10,887,325; Ohio was apportioned 19 House seats. The population of Kansas was 2,485,600, and Kansas received 4 House seats.

    a. Determine the average congressional district sizes for these states.

    b. Determine the relative difference in these district sizes.

    c. Suppose a seat were transferred from Ohio to Kansas, giving Ohio 18 seats and Kansas 5. What would the relative difference in district sizes now be?

●▲   d. Suppose that the governor of Kansas believes that the population of his state was undercounted. What population would be large enough to entitle Kansas to take a seat from Ohio, if the apportionment is by the Hill–Huntington method?

    17. The table on the opposite page shows populations and Hill–Huntington apportionments for several states, based on the 1970 census.

● Optional exercise.   ▲ Advanced exercise.   ■ Discussion exercise.

State	Population	Apportionment
California	20,098,863	43
Connecticut	3,050,693	6
Montana	701,573	2
Oregon	2,110,810	4
South Dakota	673,247	2

    a. Which state has the largest average district size, and which has the smallest?

    b. If a seat were transferred from the state with the smallest district size to the one with the largest, would the inequity be less (as measured by absolute difference in district size)?

    c. If a seat were transferred from the state with the smallest district size to the one with the largest, would the inequity be less (as measured by relative difference in district size)?

    d. Citizens of which state have the largest representative share? The smallest?

    e. If a seat were transferred from the state with the smallest representative share to the one with the largest, would the inequity be less (as measured by absolute difference in representative share)?

    f. If a seat were transferred from the state with the smallest representative share to the one with the largest, would the inequity be less (as measured by relative difference in representative share)?

18. A country has a 100-seat parliament with one major party, the National Party, and 10 splinter parties. In a recent election, the National Party received 87.85% of the vote. The splinter parties received the following percentages: 1.26%, 1.25%, 1.24%, 1.23%, 1.22%, 1.21%, 1.20%, 1.19%, 1.18%, and 1.17%.

    a. What is the best apportionment for the parliament?

    b. Compute the apportionments according to the methods of (1) Hamilton, (2) Jefferson, (3) Webster, (4) Condorcet, and (5) Adams.

    c. Do any of the methods in part b violate the quota condition? That is, does some party receive either more than its upper quota or less than its lower quota?

    d. Consider the Webster apportionment in part b. Compute the district size, as a percentage of the total vote, for each party. What is the largest inequity? What is the largest *relative* inequity? Could either the inequity or the relative inequity be reduced by transferring seats?

19. A country with a 100-seat parliament has an election in which one party captures 92.15% of the vote. Five splinter parties receive the following percent-

ages of the vote: 1.59%, 1.58%, 1.57%, 1.56%, and 1.55%. Answer all of the questions in Exercise 18.

20. Use the Condorcet method to apportion the 1794 House of Representatives, with the historical house size of 105. The populations of the states are given in Table 14.4.

21. A country that is governed by a parliamentary democracy has two political parties, the Liberals and the Tories. The number of seats awarded to a party is supposed to be proportional to the number of votes it receives in the election. Suppose the Liberals receive 49% of the vote. If the total number of seats in Parliament is 99, how many seats do the Liberals get under (1) the Hamilton method? (2) the Webster method? (3) the Jefferson method?

■ 22. A country with a parliamentary government has two parties, which capture 100% of the vote between them. Each party is awarded seats in proportion to the number of votes received.

    a. Show that the Webster and Hamilton methods will always give the same apportionment in this two-party situation.

    b. Show that the Hamilton method is house monotone when the seats are apportioned between two parties or states.

    c. Show that the Jefferson, Webster, and Adams methods all satisfy the quota condition when the seats are apportioned between two parties or states.

    d. Will the Jefferson and Adams methods also yield the same apportionments as the Hamilton method?

Exercises 23 to 26 concern the classroom scheduling example discussed in this chapter. The problem is to apportion a total of five class sections among three subjects, given the enrollment figures listed in the exercise. Apportionments should be calculated according to all of the following methods: (1) Hamilton, (2) Jefferson, (3) Webster, (4) Adams.

23. Geometry 43, Algebra 42, Calculus 10.

24. Geometry 55, Algebra 25, Calculus 20.

25. Geometry 67, Algebra 23, Calculus 5.

26. Geometry 76, Algebra 19, Calculus 20.

■ 27. Which divisor method would be most appropriate for apportioning sections to classes according to enrollments, as in the senior high school example?

■ 28. Would a relative measure of inequity or an absolute measure of inequity be the fairest when evaluating the apportioning of salaries as in the small mathematics department example?

■ 29. Which apportionment method do you think would be the more favorable to large states: Webster or Condorcet?

▲ Advanced exercise.   ■ Discussion exercise.

30. In *Massachusetts v. Mosbacher*, Massachusetts contested its 1991 apportionment, claiming a systematic census undercount of Massachusetts residents living abroad. Another issue in the suit was the claim by Massachusetts that the Hill–Huntington method of apportionment is unconstitutional, because it does not reflect the "one person, one vote" principle as well as the Webster method. Massachusetts sought an additional House seat that had been awarded to the state of Washington. Would Massachusetts have gained a seat from Washington if the Webster method had been used to apportion the House of Representatives in 1991? Use the following populations and Hill–Huntington apportionments:

	Population	Apportionment
Massachusetts	6,029,051	10
Washington	4,887,941	9

▲ 31. Let $q_1, q_2, \ldots, q_n$ be the quotas for n states in an apportionment problem, and let the apportionments assigned by some apportionment method be denoted $a_1, a_2, \ldots, a_n$. The *absolute deviation* for the state i is defined to be $|q_i - a_i|$; it is a measure of the amount by which the state's apportionment differs from its quota. The *total absolute deviation T* of the apportionment is the sum of the absolute deviations for all of the states. In mathematical notation

$$T = \sum_{i=1}^{n} |a_i - q_i|$$

Can you show that the Hamilton method always gives the least possible total absolute deviation?

■ 32. In this exercise, you are asked to invent a "new" apportionment method. Actually, you will not be the first to invent it. That honor belongs to Congressman William Lowndes of South Carolina, who first proposed the method in 1822. Let q_i denote the quota for the ith state, and suppose that some apportionment method awards that state a_i seats. In Exercise 31, the absolute deviation for that state was defined to be $|a_i - q_i|$. The *relative deviation R_i* is the absolute deviation expressed as a percentage of the quota q_i. Thus,

$$R_i = \frac{|a_i - q_i|}{q_i} \times 100\%$$

The relative deviation for the apportionment as a whole is the largest of these numbers R_i; it is the relative deviation of the state with the largest such deviation. Notice that these deviations can work in a state's favor or against a state.

    a. Devise the apportionment method that gives the least possible relative deviation as a whole. The method that satisfies this criterion *is not one of the methods described in this chapter.*

b. Would large states tend to favor your method?

■ 33. Which method do you think would minimize the *worst* absolute deviation (see Exercise 31 for the definition) for any single state in an apportionment?

● 34. Suppose that the Jefferson method is used to apportion the House according to the 1990 census. Use the ranking function $R_i = p_i/(n + 1)$ to decide the contest between Massachusetts and Washington for the last seat available to them. Given the population figures in Exercise 30, and that Massachusetts already has 10 seats, while Washington has 8, does Washington end up with the higher priority, or does Massachusetts?

●▲ 35. How much would have to be added to the population of Massachusetts for it to take a seat from Washington, if apportionment is done by
a. The Hill–Huntington method?
b. The Webster method?

● 36. Apportion the classes in Exercises 23 to 26 in such a way that the average section sizes for the courses are as nearly equal as possible.

● 37. Apportion the classes in Exercises 23 to 26 in such a way that the relative difference in average section sizes between the courses are as small as possible.

● 38. Apportion the student senate seats in the small college mentioned in Exercise 4 by (a) the Hill–Huntington method, and (b) the Dean method. Use the data for the first year.

● 39. Repeat Exercise 38, using the second year data. Do either of these methods exhibit failure of quota monotonicity?

▶ WRITING PROJECTS

1. Look up the history of the Lowndes apportionment method (see Exercise 32). What sort of problem did it address? What were the arguments against it? *Fair Representation* by Balinski and Young is a good place to start reading.

2. The 1990 census has been widely criticized for undercounting certain segments of the U.S. population. There are several articles on this subject in 1991 issues of the *Congressional Weekly*. How did the undercount affect apportionment?

3. Does the Hill–Huntington method best reflect the intentions of the Founding Fathers, as they were set down in the Constitution, and in their debate during the 1787 Constitutional Convention? Good sources of information here include the following publications listed under Suggested Readings: *Fair Representation*, by Balinski and Young; Balinski and Young's paper "The Webster Method of Apportionment"; and the two court opinions, *Massachusetts v. Mosbacher* and *Montana*. This writing project requires that you state your answer to the question and make a case for it.

Chapter 15

Game Theory: The Mathematics of Competition

Conflict is a central theme in human history and literature. It arises naturally whenever two or more individuals try to control the outcome of events. People compete in such situations because they have both freedom of choice and different values.

Game theory is a serious mathematical subject created to study situations involving conflict and cooperation. It is a new approach in that it brings scientific methods and the powerful tools of mathematics to bear on the topic. Game theory began in earnest in 1944 with the publication of *Theory of Games and Economic Behavior* by John von Neumann and Oskar Morgenstern. (See Spotlight 15.1, p. 462.)

A game situation arises when two or more individuals, called **players,** are each able to act freely and to select from a list of available options. These options are referred to as **strategies.** These choices in turn lead to various outcomes, called **payoffs.** Each player has various *preferences* among the resulting rewards or penalties.

The theory of games, then, is concerned with notions such as selection of the best strategies, equilibrium outcomes, bargaining and negotiations, formation and stability of coalitions, fair division, and resolution of conflict. This subject deals with the rules of the game, individual and coalition values, side payments, and repeated play, as well as various kinds of uncertainty and chance events. Game theory differs from the traditional subjects of statistics and probability in that it treats two or more individuals with different goals or objectives.

Many confrontations are primarily *noncooperative,* for example, those between combatants in warfare or competitors in sports. In these encounters, the adversaries' ultimate objectives are typically at cross-purposes: a gain for one means a loss for the other. Other social activities, such as those in economics or politics, typically have a large cooperative component. However, most human interactions involve a delicate mix of cooperative and noncooperative behavior. In business, for example, people cooperate to maintain a healthy economy even as they compete for shares in the marketplace.

In the following sections we present several simple examples of noncooperative *two-person* games. The first few will be games of complete or total conflict, and the last two will be games of only partial conflict.

SPOTLIGHT 15.1   Historical Highlights

As early as the seventeenth century, such outstanding scientists as Christiaan Huygens (1629 – 1695) and Gottfried W. Leibniz (1646 – 1716) proposed the creation of a discipline that would make use of the scientific method to study human conflict and interactions. Throughout the nineteenth century, several leading economists created simple mathematical examples to analyze particular illustrations of competitive encounters. The first general mathematical theorem in this subject was proved by the distinguished logician Ernst Zermelo (1871 – 1956) in 1912. It stated that any finite game with *perfect information,* such as checkers or chess, has an optimal solution in *pure* strategies; that is, no randomization or secrecy is necessary. A game is said to have perfect information if at each stage of the play, every player is aware of all past moves by himself and others as well as all future choices that are allowed. This theorem is an example of an *existence theorem:* it demonstrates that there must exist a best way to play such a game, but it does not provide a detailed plan for playing a complex game to achieve victory.

The famous mathematician F. E. Émile Borel (1871 – 1956) introduced the notion of a *mixed, or randomized,* strategy when he investigated some elementary duels around 1920. The fact that every two-person, *zero-sum* game must have optimal mixed strategies and an expected value for the game was proved by John von Neumann (1903 – 1957) in 1928. Von Neumann's result was extended to the existence of equilibrium outcomes in mixed strategies for multiperson *general-sum* games by John F. Nash, Jr. (1931 –) in 1951.

Modern game theory dates from the publication in 1944 of *Theory of Games and Economic Behavior* by the Hungarian-American mathematician John von Neumann and the Austrian-American economist Oskar Morgenstern (1902 – 1977). They introduced the first general

▶ GAME WITH OPTIMAL PURE STRATEGIES

For some noncooperative games with two players it is rather straightforward to determine the best strategies for the players to play. We begin with such a case.

EXAMPLE: A Location Game

Two young entrepreneurs, Henry and Lisa, plan to locate a new restaurant at a main-route intersection in the nearby mountains. They can agree on all aspects except one. Lisa likes low elevations, whereas Henry wants greater heights

John von Neumann. (Photo courtesy of The Institute for Advanced Study.)

Oskar Morgenstern. (Princeton University Archives.)

model and solution concept for multiperson cooperative games, which are primarily concerned with coalition formation (economic cartels, voting blocs, and military alliances) and the resulting distribution of gains or losses. Several other suggestions for a "solution" to such games have since been proposed. These include the value concept of Lloyd S. Shapley (1923 –   ), which relates to fair allocation and economic prices and serves as well as an index of voting power (see Chapter 12).

The French artist Georges Mathieu designed a medal for the Paris Musée de la Monnaie in 1971 to honor game theory. It was the seventeenth medal to "commemorate 18 stages in the development of Western consciousness." The first medal was for the Edict of Milan in A.D. 313. Game theory also has a mascot, the tiger, arising from the Princeton University tiger and the Russian abbreviation of the term "game theory" (ТЕОРИЯ ИГР).

—the higher up, the better. In this one regard, their preferences are diametrically opposed. What is better for Henry is worse for Lisa, and likewise what is good for Lisa is bad for Henry.

The layout for their location problem is shown in Figure 15.1. You can see that three routes, Avenue A, Boulevard B, and County

Road C, run in the east-west direction and that three highways, numbered 1, 2, and 3, run in the north-south direction. The table on the following page shows the altitudes at the nine corresponding intersections; the same information is in thousands of feet illustrated in three dimensions in Figure 15.2.

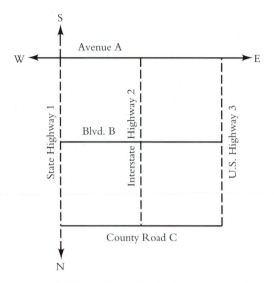

Figure 15.1 The road map for the location example.

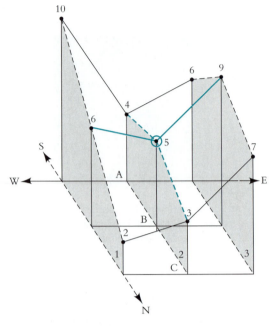

Figure 15.2 Three-dimensional road map showing Henry's and Lisa's selections (color). Underneath, the slanted map shows the map in Figure 15.1 in perspective.

	Highways		
Routes	1	2	3
A	10	4	6
B	6	5	9
C	2	3	7

Henry and Lisa agree to turn their decision into a competitive game, as follows: Henry will select one of the three routes, A, B, or C, and Lisa will simultaneously pick one of the three highways, 1, 2, or 3. The restaurant will then be located at the resulting intersection. The selections A, B, and C for Henry are his three strategies, and the choices 1, 2, and 3 are Lisa's

three strategies. (Such initial strategies are sometimes called **pure strategies** to distinguish them from the term "mixed strategies," which is introduced in the following section.)

Henry is pessimistic and considers the lowest altitude along each of the routes A, B, and C. He gets the numbers 4, 5, and 2, which are the respective *row minima*, indicated in the right-hand column of the table on the facing page. He notes that the highest of these values is 5. He can elect the corresponding route, B, and guarantee himself an altitude of at least 5000 feet. This number 5 in the right-hand column is referred to (for short) as the **maximin value,** because it is the maximum value of the minimum numbers from the various rows in our table. The corresponding strategy for Henry, Route B, is called his **maximin strategy** for this game.

	Routes	Lisa Highways 1	2	3	Row minima
Henry	A	10	4	6	4
	B	6	5	9	5
	C	2	3	7	2
	Column maxima	10	5	9	

Lisa likewise does a worst-case analysis and lists the highest — for her, the worst — elevations for each highway. These numbers, 10, 5, and 9, are the column maxima and are listed in the bottom line of this table. The best of these outcomes from her point of view is 5. If she picks Interstate Highway 2, then she is assured of an elevation of no more than 5000 feet. The number 5 appearing in the bottom row of our table is referred to as the **minimax value,** because it is the minimum value of the maximum numbers from the different columns in the table. The corresponding strategy for Lisa, Highway 2, is called her **minimax strategy.**

So Henry has a choice that will result in 5 or higher, and Lisa can choose so as to hold him down to 5 or less. The resulting height 5 at the intersection of Route B and Highway 2 is, simultaneously, the lowest value along Boulevard B and the highest on Interstate Highway 2. In other words, the maximin value and the minimax value are both equal to 5 for this location game. Such an outcome for a game is called a **saddle point,** or **mountain pass.** The reason for using these terms should be clear from the sad-

dle shape of the "payoff surface" in Figure 15.2. The "middle" point on a horse saddle is simultaneously the lowest point along the spine of a horse and the highest point between the rider's legs. As one drives across a continental divide (through a mountain pass) the car is typically at a high point on the road while at a low point along the divide itself.

The resolution of this contest is for Henry to pick B and Lisa to elect 2; then each settles for the resulting elevation of 5. This number 5 is called the **value** of the game. The selections B and 2 are the **optimal** (pure) **strategies,** and, along with the value 5, are referred to as the **solution** of the game.

There is no need for secrecy in the case of a game with a saddle point. Even if Henry were to reveal his choice of B in advance, Lisa would be unable to use this knowledge to exploit him. In fact both players can use the given information to compute the best strategies for their opponents as well as for themselves. In games with saddle points (in pure strategies) it turns out that players' worst-case analyses lead to the best possible solution. ▲

Another well-known game with a saddle point in pure strategies is tic-tac-toe (tick-tack-toe or tit-tat-toe). Two players alternately place an ✕ or an ○, respectively, in one of the nine unoccupied spaces in a 3-by-3 grid. The winner is the first player to have three ✕'s, or three ○'s, in either the same row, the same column, or a diagonal. An explicit list of all pure strategies for either the first or second moving player in this game is quite long. Young children initially find this game of some interest to play. Before long, however, they usually discover that either one of the players can always prevent the other player from winning. From then on the game "should" always end in a tie, and it becomes boring to continue to play.

EXAMPLE: The Restricted Location Game

Assume that in our location game that Henry and Lisa are informed by the county officials that it is against the law to locate a restaurant on either Boulevard B or Interstate Highway 2. These two choices are now forbidden. The resulting location game without these two strategies is described in the following table, where the payoffs are again expressed in thousands of feet.

	Highways	
Routes	1	3
A	10	6
C	2	7

Again, Henry and Lisa can each do a worst-case analysis. Henry is worried about the minimum number in each row, and Lisa is concerned with the maximum number in each column. These are listed in the right column and bottom row, respectively, in the following table. Henry sees from the row minima column that his maximin value is 6. He can guarantee a height of 6000 feet or more by choosing Route A. Lisa likewise observes that her minimax value is 7. She can keep the elevation of the restaurant down to 7000 feet or less by selecting Highway 3. There is a *gap* of 1 (= 7 − 6) between the maximin and minimax values. When the maximin value is less than the minimax value, as in this case, then a game does not have a saddle point in the original pure strategies.

		Lisa		
		Highways		Row
	Routes	1	3	minima
Henry	A	10	6	6
	C	2	7	2
	Column maxima	10	7	

If Henry does play his maximin strategy, Route A, and Lisa plays her minimax strategy, Highway 3, then the resulting payoff is 6. However, Henry may be motivated to "gamble" in this case. If he plays his alternate strategy, Route C, and Lisa remains with her conservative one, Highway 3, then the payoff is 7. Henry will have gained one unit (1000 feet) from 6 to 7. This is, however, a risk on his part. If Lisa suspected this move (Route C) by Henry, she might counter by selecting Highway 1. The payoff would then be 2, the best for Lisa and the worst for Henry. So Henry's gamble to gain one unit (6 to 7) has the risk that he might lose 4 units (6 to 2).

There is no incentive for Lisa to play her *non*minimax strategy (that is, to play Highway 1) in this restricted location game, if she does believe Henry will play his maximin strategy (Route A). This could only lead to a payoff of 10, which is worse than 6 from her viewpoint. ▲

Games, like our original 3-by-3 location game and tic-tac-toe, that are "antagonistic" games with saddle points in *pure* strategies are often referred to as *strictly determined* games. This is because each player can calculate a best strategy to play (as well as the optimal choice for the opposing player), and thus the resulting payoff (the value of the game) is known before the game is even played. Once the solution has been determined by mathematical analysis, or practical experience, there remains little interest in actually playing through the game itself.

On the other hand, most games, like the latter 2-by-2 restricted location game, do not have an optimal solution in pure strategies. These games do involve uncertainty and risk taking, and they are often played repeatedly and with great interest by many participants. We will see that there are "optimal" ways to play these latter games as well. However, their solution is based upon an additional statistical concept known as a *mixed strategy*, which is described below.

▶MIXED STRATEGIES

Most competitive games do not have a saddle point with optimal *pure* strategies, as discussed in our first location game example. There typically does not exist a pure strategy for a player that is dominant in the sense that it is the best one to use each time the game is played. On the contrary, players usually must maintain secrecy about their intentions. They must take care not to reveal what particular pure strategy they will select until the encounter actually takes place, when it is too late for the opposing player to alter his or her choice. In repeated play of most games a player will typically vary his or her selection of a pure strategy in order to surprise the opponent. This concealment and variability can be realized by employing the notion of a *mixed strategy*.

In parlor games like poker, players do not reveal their "down cards," and they often make use of the tactic known as "bluffing." In military engagements deception and stealth are often crucial to success. In many sporting events a team tries to mislead and surprise the opposition. A pitcher in baseball will not reveal the type of pitch he or she intends to throw in advance and will vary the type

of pitch throughout the game, to keep the batter "off balance." Let's consider an encounter between a pitcher and batter in baseball in more detail.

EXAMPLE: A Duel Game

Assume that a particular baseball pitcher can throw either a blazing fastball or a slow curve ball into the strike zone of a batter. The pitcher thus has two pure strategies: fast (denoted by F) and curve (C). The pitcher faces a given batter who attempts to guess before each pitch whether it will be a fastball or a curve. So the batter also has two pure strategies, which will likewise be denoted F and C. It is also known that this batter has the following batting averages:

▶ .300 if the batter guesses fast (F) and the pitcher throws fast (F)
▶ .200 if the batter guesses fast (F) and the pitcher throws curve (C)
▶ .100 if the batter guesses curve (C) and the pitcher throws fast (F)
▶ .500 if the batter guesses curve (C) and the pitcher throws curve (C)

A player's batting average is the number of times he hits safely divided by his recorded number of times at bat. An average of .300 means the batter has hit safely about 3 times out of 10 when in the given situation.

This game can be summarized in the following table.

| | | Pitcher | | Row minima |
		F	C	
Batter	F	.300	.200	**.200**
	C	.100	.500	.100
	Column maxima	**.300**	.500	

We see from the right-hand column in the table that the batter's maximin value is .200, which is realized when he selects his first strategy F. The batter can play it "safe" and always guess a fastball. This could result in his batting .200 (which is hardly enough for him to remain on the team). We also see from the bottom row of the table that the pitcher's minimax value is .300, which is obtained when he throws fast (F). Note that the batter's maximin value of .200 is less than the pitcher's minimax value of .300. Therefore, this game does not have a saddle point in pure strategies. There is a gap of .100(= .300 − .200) between these two values. Each player would like to play so as to "win" for himself as much of this .100 payoff in the gap as possible. That is, the batter would like to average more than .200 and the pitcher wants to hold the batter down to less than .300.   ▲

A Flawed Approach

If the batter and pitcher in our example limit their thinking to just *pure strategies,* they might well begin to reason along the following lines:

1. *Pitcher* (to himself): If I choose the strategy F, I hold the batter down to .300 (the minimax value) or less. However, the batter is likely to guess F because it guarantees him at least .200 (his maximin value) and it actually provides him with .300 against my F pitch. In this case the batter wins *all* of the .100 payoff in the gap.

2. *Batter* (to himself): I can figure out that the pitcher is reasoning as in step 1 and will try to surprise me with a C. So I should fool him and guess C. I would thus average .500. This will show him for trying to gamble and out-guess me!

3. *Pitcher:* But I know that the batter is thinking as in step 2, that is, of guessing C. So, on second thought, I should in turn really throw F.

This will lead to the average of only .200 for the batter, and teach him to not try to out-guess me.

4. *Batter:* On the other hand, I know that the pitcher knows what I am thinking in **2**, and so he is planning step 3. As a result, I should guess F after all, and obtain the batting average .300.

5. *Pitcher:* If I follow all of the earlier reasoning, I should surprise him and pitch C.

6. *Batter:* In light of the preceding five steps, I should fool the pitcher and guess C.

7. *Pitcher:* Since I know all of this, I should throw F.

8. *Batter:* So I should guess F.

9. *Pitcher:* So I should pitch C.

This type of cyclical reasoning can go on without end: I know that he knows that I know that he knows. . . . It provides no resolution of their decision problem. Clearly there is no pitch that is best in all instances. Both the pitcher and the batter can, in fact, do better than to always rely on their minimax and maximin strategies, respectively. The answer to their problem lies in the notion of a mixed strategy.

A Better Idea

The play of most games requires unpredictability and variability on the part of the participants. This element of surprise can be realized in practice by making use of a *mixed strategy.*

   A **mixed strategy** is a particular "randomization" over a player's list of pure strategies. Each one of the player's pure strategies is assigned some probability. This probability indicates the relative frequency with which the pure strategy will be played. The specific pure strategy that will be used in any given play of the game can be selected by means of some appropriate probabilistic mechanism. (Note that a pure strategy is a special case of a mixed strategy: The total probability of 1 is assigned to just this one pure strategy.) When a player resorts to such mixed strategies, however, the resulting outcome of the game is no longer predictable in advance, and it must be described in terms of the statistical notion of "average" or **expected value.**

If one of the n payoffs $s_1, s_2, \ldots , s_n$ will occur with the probabilities $p_1, p_2 \ldots , p_n$, respectively, then the *expected value E* is given by

$$E = p_1 s_1 + p_2 s_2 + \cdots + p_n s_n$$

We assume that the probabilities add to one, and that each probability p_i is never negative. That is, we assume that $p_1 + p_2 + \cdots + p_n = 1$, and that each $p_i \geq 0 (i = 1, 2, \ldots , n)$. To see how mixed strategies and expected values are used in the analysis of games, we turn to the "simplest" of all competitive games without a saddle point in pure strategies.

EXAMPLE: Matching Pennies

In the two-person version of matching pennies each of the two players simultaneously shows either a head H or a tail T. If the two coins match, with either two heads or two tails, then the first player (Player I) receives both coins. If the coins do not match, that is, if either one is an H while the other is a T, then the second player (Player II) lays claim to the two coins.

This game can be represented by the following table, called the **game matrix:**

		Player II	
		H	T
Player I	H	1	-1
	T	-1	1

The two rows correspond to Players I's two pure strategies: H and T. The two columns likewise give Player II's two pure strategies: H and T. The numbers in the table are the corresponding *winnings* for Player I and *losses* for Player II. If two H's or two T's are played, Player I wins 1¢ from Player II. When an H and a T are selected, Player I pays out 1¢ to Player II.

It is usually fruitless for one player to attempt to outguess the other in this game. They should instead resort to mixed strategies and use expected values to estimate their likely gains or losses.

The best thing for Player I to do is to randomly select H half of the time and T half of the time. This mixed strategy can be expressed as $(p_H, p_T) = (p_1, p_2) = (1 - p, p) = (\frac{1}{2}, \frac{1}{2})$ and it could be realized in practice by a flip of the coin. Player I's resulting expected value is

$$E = E(p, H) = \tfrac{1}{2}(1) + \tfrac{1}{2}(-1) = 0$$

whenever Player II plays H, and

$$E = E(p, T) = \tfrac{1}{2}(-1) + \tfrac{1}{2}(1) = 0$$

when Player II plays T. The symbol $E(p, H)$ indicates that Player I is using a mixed strategy determined by p and that Player II chooses the pure strategy H. Player I's average outcome of 0 is called the *value* of the game, but it must again be understood in a statistical sense. In a given play of the game, Player I will either win 1¢ or lose 1¢. However, his or her expectation over many plays of this **fair** game is 0. The best mixed strategy for Player II is likewise a half-half mix of H and T.

Player II gains nothing by knowing that Player I is using the *optimal* mixed strategy $(p_H, p_T) = (\frac{1}{2}, \frac{1}{2})$. However, Player I must not reveal to Player II whether an H or T will be displayed in any given play of the game, before Player II is committed to his or her own choice of an H or T. On the other hand, if Player II knew that Player I was using a particular *nonoptimal* mixed strategy $(1 - p, p)$ where $p \neq \frac{1}{2}$, then Player II could take advantage of this knowledge and increase his or her average winnings over time to above the value zero. (See Exercise 25.) ▲

EXAMPLE: Nonsymmetrical Matching

Players I and II can each show either heads H or tails T. When two H's appear, Player II pays $5 to Player I. When two T's appear, Player II pays $1 to Player I. When one H and one T are displayed, then Player II collects $3 from Player I.

This game is given by the following game matrix, which displays the payoff from Player II to Player I:

		Player II	
		H	T
Player I	H	5	-3
	T	-3	1

A worst-case analysis, like that which solved our initial location game, is of little help here. Player I may lose $3 whether he plays H or T. Player I's maximin value is -3. Player II can keep her losses down to $1 by always playing T (and thus avoiding the $5 penalty when two H's appear). Player II's minimax value is 1. However if Player II sticks to T and Player I knows this, then Player I will also play T in each game and always collect $1 from Player II. Can Player II do better than lose $1 in each play of the game?

Consider the situation where Player I uses a mixed strategy $(p_H, p_T) = (1 - p, p)$, which says play H with probability $1 - p$ and play T

with probability p, where $0 \leq p \leq 1$. Against Player II's pure strategy H, Player I's expected value is

$$E = E(p, H) = (5)(1 - p) + (-3p) = 5 - 8p$$

Against Player II's T, Player I's expectation is

$$E = E(p, T) = (-3)(1 - p) + 1p = -3 + 4p$$

These two linear equations in the two variables E and p are sketched in Figure 15.3. Note that the four "boundary points" where these two lines meet the vertical lines $p = 0$ and $p = 1$ are just the four payoffs appearing in the game matrix.

Player I's goal is to select p so as to achieve the highest value of E, against *both* strategies H and T by Player II, simultaneously. The best value of p, which will maximize Player I's return in both eventualities, occurs at the intersection of these two straight lines. We can thus solve these two equations in two unknowns to obtain the point $p = \frac{2}{3}$ and $E = -\frac{1}{3}$. The *optimal* mixed strategy $(1 - p, p)$ $(\frac{1}{3}, \frac{2}{3})$ for Player I is to pick H and T with probabilities $\frac{1}{3}$ and $\frac{2}{3}$, respectively. The expected value for Player I is $-\frac{1}{3}$.

A similar calculation for Player II results in the same optimal mixed strategy $(\frac{1}{3}, \frac{2}{3})$ and ex-

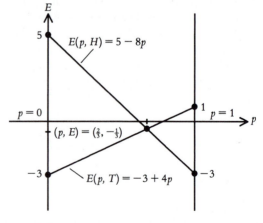

Figure 15.3 Solution for the nonsymmetrical matching game.

pected value $-\frac{1}{3}$. Recall that the payoffs for Player II are *losses*, so that this $-\frac{1}{3}$ means she *gains* $\frac{1}{3}$. This game is therefore *unfair*. It favors Player II, who will win $33\frac{1}{3}$ cents *on average* for each time the game is played. ▲

The last two examples are illustrations of what are called **zero-sum games.** The payoff to one player is the negative of the corresponding payoff to the other. The sum of their payoffs is zero. What one wins the other loses. These are also referred to as *strictly competitive* or *antagonistic* games. The two-person, zero-sum games are called **matrix games,** because they can be completely described by a square or rectangular table of numbers. These numbers represent the payoffs to Player I, while their negatives are the resulting payoffs to Player II.

The solution technique used in our last example works for any matrix game in which each player has only two viable strategies. We must use a slightly more involved method when only one of the two players has more than two strategies. However, one should always check first to see whether a game has a saddle point in pure strategies before employing any solution method for finding optimal mixed strategies.

EXAMPLE: A Duel Game Revisited

Let's return to the example of a duel between the pitcher and the batter. This encounter was described by the 2-by-2 matrix below, where F indicates fastball and C is for curve ball:

		Pitcher		
		F	C	
	F	.300	.200	$1 - q$
Batter	C	.100	.500	q
		$1 - p$	p	

The pitcher should use a mixed strategy $(p_1, p_2) = (p_F, p_C) = (1 - p, p)$. The probabilities $1 - p$ and p (where $0 \le p \le 1$) are indicated below the table and under the corresponding strategies, F and C, for the pitcher. If the pitcher plays a mixed strategy $(1 - p, p)$ against the two pure strategies F and C for the batter, he realizes the respective expected values:

$$E(F, p) = (0.3)(1 - p) + 0.2p = .3 - .1p$$
$$E(C, p) = (0.1)(1 - p) + 0.5p = .1 + .4p$$

One can solve these two equations in the two unknowns E and p to show that the intersection of these two lines occurs at the point $p = 0.4$ and $E = .260$. The pitcher should use the *optimal* mixed strategy that selects F with probability $1 - p = 0.6$ and C with probability $p = 0.4$. This choice will hold the batter down to an *expected* batting average $E(F, .4) = .3 - .1(0.4) = .260 = .1 - .4(0.4) = E(C, .4)$. The number .260 is the value of the game. It should be stressed that .260 is an average and it must be interpreted in a statistical manner.

Assume that the batter uses a mixed strategy $(p_1, p_2) = (p_F, p_C) = (1 - q, q)$, as indicated to the right of the game matrix. This mixed strategy when played against the pitcher's pure strategies, F and C, results in the respective expected values:

$$E(q, F) = (0.3)(1 - q) + 0.1q = .3 - .2q$$
$$E(q, C) = (0.2)(1 - q) + 0.5q = .2 + .3q$$

The intersection of these two lines is the point $q = 0.2$ and $E = .260$. The batter's optimal strategy is $(p_F, p_C) = (4/5, 1/5)$, and his expected batting average is again .260. ▲

We have seen that the *same* outcome of .260, the *value* of the game, occurs when the pitcher selects his optimal strategy (3/5, 2/5) as well as when the batter plays his optimal strategy (4/5, 1/5). This particular result holds true for any two-person, zero-sum game. There is a unique game value (the .260 in our example) and an optimal strategy for each player, so that either player alone realizes this value by playing his or her optimal strategy. This general result is the famous fundamental theorem for matrix games, which is also known as the **minimax theorem** of John von Neumann.

►LARGER SIZED GAMES

We have just shown how to compute the optimal mixed strategies and the value for 2-by-2 games, that is, for games in which each of the two players has two pure strategies. We now turn to a slightly larger game and show how our solution technique can be extended to games where only one of the two players has more than two pure strategies.

EXAMPLE: 3 by 3 Baseball

Consider the following duel that can take place between the pitcher and batter in the game of baseball. First, let's assume that the pitcher has enough control to throw the ball to whatever part of the strike zone he wishes. To simplify matters, let's suppose that he will select one of three pitches:

- ► HI: a fastball high and inside
- ► MD: a fastball down the middle
- ► LO: a fastball low and outside

These alternatives, HI, MD, and LO, are the pitcher's three strategies and are depicted in Figure 15.4.

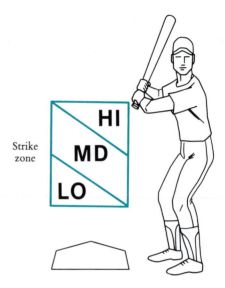

Figure 15.4 The three strike zones, which correspond to the pitcher's three strategies.

Next, let's assume that a particular batter is known to average

- ► .300 against high inside pitches
- ► .400 against pitches over the middle
- ► .200 against low outside pitches

However, the batter also has the option of outguessing the pitcher. He can choose to guess HI and step back from the plate as he swings. In this case his average becomes

- ► .400 against HI
- ► .200 against MD
- ► .000 against LO

Similarly, he can guess LO and average

- ► .000 against HI
- ► .300 against MD
- ► .400 against LO

The batter has three strategies (HI, MD, and LO), shown in the following game matrix.

		Pitcher		
		HI	MD	LO
	HI	.4	.2	.0
Batter	MD	.3	.4	.2
	LO	.0	.3	.4

We assume that all this information is known to both players. The numbers represent the corresponding probability that the batter will hit safely; they can serve as a measure of his likely reward. Whereas the batter's objective is to maximize the resulting payoff, the pitcher has the opposite goal: to minimize the likelihood that the batter will hit safely. What is good for the batter is bad for the pitcher, and vice versa. They are adversaries.

Clearly, this model is an oversimplification of real baseball. We have excluded from our analysis different types of pitches, pitches outside the strike zone, foul balls, bases on balls, and other possible strategies and outcomes. We could enlarge the game to make it more realistic, but that would only complicate our calculations without altering the nature of the analysis.

To solve this game we must answer the question, Which choices are best for the two players? The reader can observe that the batter's maximin value is .200, which he obtains when he uses his maximin strategy MD. The pitcher's minimax value is .400, which he realizes when he selects any minimax strategy. There is a gap of .200 (=.400 − .200) between the maximin and minimax values for this duel game. It follows that there is no optimal strategy using only pure strategies. Each players wants to "win" some of the gap, and each must employ mixed strategies (and the resulting expected values) in order to do so.

Let's look more closely at the pitcher's problem. He must determine the best probabili-

ties for each of his three pitches. These probabilities tell the pitcher the ratio of different types of pitches he should select. An old baseball adage says, "Pitch to the corners." That means that a pitcher should avoid throwing down the middle. (A mathematical analysis of the game matrix verifies this adage, showing that the pitcher should indeed assign probability zero to the MD. See Exercises 21 and 22.) So the pitcher should randomly choose between HI and LO only. That is, he should use a mixed strategy $(p_{HI}, p_{MD}, p_{LO}) = (1 - p, 0, p)$. Eliminating this middle column (MD) reduces the size of our game matrix to 3 by 2.

Next, consider the three outcomes that may occur, depending upon whether the batter guesses the pure strategy HI, MD, or LO:

1. If the batter guesses HI, then a random mix of HI and LO pitches will cause the batter to average somewhere between .400 and .000. This is indicated by the line labeled HI between the heights .400 and .000 in Figure 15.5. The

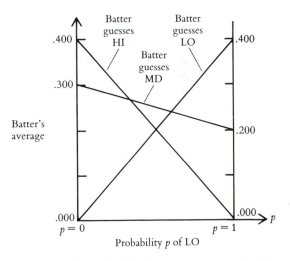

Figure 15.5 A graph of the batting averages, which depend on the batter's guesses (the three lines) and the pitcher's strategy given by p.

height of the sloping line indicates the expected batting average of the batter when he guesses HI, and the distance along the horizontal axis indicates the probability p with which the pitcher throws LO. The equation of the HI line is

$$E(\text{HI}, p) = (0.4)(1 - p) + 0p = 0.4 - 0.4p$$

2. If the batter guesses MD, his average will fall somewhere between .300 and .200, as shown in Figure 15.5. The equation of this MD line is

$$E(\text{MD}, p) = (0.3)(1 - p) + 0.2p = 0.3 - 0.1p$$

3. Similarly, if the batter guesses LO, his average will be between .000 and .400, as also shown in Figure 15.5. The equation of this LO line is

$$E(\text{LO}, p) = (0)(1 - p) + 0.4p = 0.4p$$

The pitcher wants to keep the batting average as low as possible against all three batting eventualities simultaneously. The pitcher is therefore concerned with the *highest* points in the graph, which are indicated by the blue curve in Figure 15.6. He should select his probability p so as to obtain the lowest point on this top curve. This can be done by finding the point at which the MD line and the LO line intersect.

The mathematical solution to this problem gives the result $p = 0.6$. The pitcher should throw LO with a probability $p = 0.6$ and HI with probability $1 - p = 0.4$. In other words, six-tenths of his pitches should be LO and four-tenths should be HI. This will result in holding the batter to an expected average of .240 or less.

A similar mathematical analysis would show that the batter would have an optimal mixed strategy if he guesses

▶ HI with probability 0
▶ MD with probability $q = 0.8$
▶ LO with probability $1 - q = 0.2$

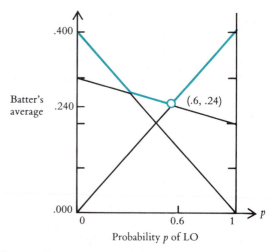

Figure 15.6 The pitcher's optimal mixed strategy is to select p so as to obtain the lowest point (circle) on the top curve (color).

This will guarantee him an average of .240 over the long run.

The number .240 is the value of this game. The pair of optimal mixed strategies, $p = 0.6$ for the pitcher and $q = 0.8$ for the batter, along with the resulting value .240, gives us the *solution* for this game. ▲

Again, it is essential for the players to act in an unpredictable manner—they must maintain secrecy. If the baseball pitcher were to "telegraph" his pitches and the batter were able to detect these signals in advance and react in time, then the batter could raise his batting average. On the other hand, if the batter were not mixing his guesses in an optimal way, then the pitcher could make use of this information to improve his own performance by throwing to the zone that the batter was least likely to guess. (See Exercises 21, 22, and 23.)

We have seen that the batter has a mixed strategy $(p_{HI}, p_{MD}, p_{LO}) = (0, 0.8, 0.2)$, which assures him of a batting average of .240, and that the pitcher has a mixed strategy $(0.4, 0, 0.6)$, which will hold the batter's average down to this *same* number .240, called the value of the game. This result is another illustration of the famous minimax theorem of John von Neumann. It states that for any matrix game there is a *unique* value that either player can realize by using an optimal mixed strategy. (Recall, however, that this value is a gain for Player I and a loss for Player II.)

Solving Matrix Games

Given any matrix game we should first go through the simple check to see whether it has a saddle point in *pure* strategies. List the minimum number from each row and list the maximum number from each column, as we did earlier in several examples. If the maximum of the row minima is equal to the minimum of the column maxima, then the resulting value and the corresponding pure strategies (which may not be unique) provide a solution for the game. The value will appear in the game matrix as the lowest number in its row and the largest in its column.

Next, we may be able to discard some strategies (i.e., eliminate some rows and columns in the matrix), because these strategies are bad ones and should never be played. If all the payoffs in one row of the game matrix are greater than or equal to the respective payoffs in another row, then the former strategy is said to **dominate** the latter strategy and the latter row can be eliminated. Similarly, if all the payoffs in one column are less than or equal to the corresponding payoffs in another column, then the former dominates the latter and the latter column can be deleted. For example, in our location game, Route B dominates Route C, and Highway 2 dominates Highway 3. Domination can also take more complicated forms. In our baseball game, for example, the pitcher did not consider using his middle pitch MD. A half-half mix of his pitches HI and LO results in average payoffs of .200, .250, and .200, which are less than or equal to, respectively, the .2, .4, and .3 realized by his MD pitch. Eliminating dominated rows and

columns reduces the size of the matrix game to be solved.

If we can use domination to reduce the size of a game matrix to n rows by two columns where $n > 2$, then we can solve for Player II's optimal mixed strategy $(1 - p, p)$, and the value of the game, by the same method used in our 3-by-2 baseball game. This geometric approach works when there are n lines as well as 3 lines (in Figures 15.5 and 15.6). We still find the value p that realizes the lowest point on the "upper curve" formed by the n lines involved. This requires solving a corresponding pair of two linear equations in two unknowns E and p, a topic that also arose in Chapter 4 on linear programming. Furthermore, there will then always be an optimal mixed strategy $(p_1, p_2, \ldots, p_n)$ for Player I that only involves (one or) *two* nonzero probabilities q_i and $q_j = 1 - q_i$, which are viable strategies to use against Player II's $(1 - p, p)$. (Analogous methods can be used to solve games with 2 rows and n columns.)

Several general algorithms have been developed since 1945 to solve any larger matrix games. Furthermore, the more recently developed subject of linear programming (see Chapter 4) is equivalent to the theory of matrix games. Any of the linear programming algorithms, such as the simplex method of G. B. Dantzig or the more recent method of N. K. Karmarkar, can also be used to solve matrix games. (See Suggested Readings for more details and references.)

Practical Applications

The element of surprise is essential in many encounters. Examples of the use of mixed strategies include various inspection procedures and auditing schemes. These should employ randomness to keep potential cheaters off guard as well as for statistical purposes.

Individual investigators or regulatory agencies often monitor certain accounts or actions to check for faults, errors, or illegal activities. The investigators include bank auditors, customs agents, insurance investigators, and quality control experts. The National Bureau of Standards is responsible for monitoring proper measuring instruments and for maintaining reliable standards. The Nuclear Regulatory Agency demands an accounting of dangerous (and expensive) nuclear materials as part of its safeguards program. The Internal Revenue Service wishes to identify those cheating on taxes. The military or intelligence service may wish to sneak in or intercept a weapon or secret agent that is in the midst of many decoys. Because it is prohibitively expensive to check out all cases, statistical methods must be used to check for violators. Many such encounters can be modeled as a competitive game, with the goal to obtain optimal mixed strategies for the inspector and the violator.

The notion of a "bluff," as in the game of poker, is also a viable strategy. For bluffing, game theory assigns an optimal probability with which one should bluff, given a particular situation. For example, in labor negotiations, a threat to strike is effective only if it is believed to some extent. (See Exercise 47.)

▶ NONZERO-SUM GAMES

The matrix games just presented were strictly competitive games: one player's gain was equal to the other player's loss. The sum of their joint payoffs was always a constant. Such games involve total conflict. We now consider two examples of games that are not entirely antagonistic. The players' goals are only in partial conflict. There is some mutual gain to be realized by both players if they could cooperate. Such cooperation typically involves elements of communication, trust, and the threat of enforcement. If these elements are lacking, however, we are in the realm of noncooperative games. The players' individual self-interest can then lead to less-than-optimal payoffs. We consider the following two games when played in this noncooperative mode.

The Prisoner's Dilemma

We first look at a game that has come to be known as the **prisoner's dilemma.** This elementary two-person game provides a simple explanation of the forces at work behind arms races, price wars, costly advertising campaigns, and many other similar escalations. The prisoner's dilemma illustrates a kind of social paradox that we frequently confront in the course of our everyday lives.

The term *prisoner's dilemma* was first assigned to this game by the Princeton mathematician Albert W. Tucker (1905–) in 1950. The game involves the scenario of two suspects in crime who are held *incommunicado*. Each is given one of two choices: to steadfastly maintain their individual or mutual innocence or to sign a confession accusing the partner of committing the crime. It is usually in the individual's self-interest to confess, that is, implicate one's partner and receive a reduced sentence. Yet when both confess, they each reach a bad outcome, that is, they are both found guilty. What is good for the prisoners as a pair — steadfast denial by both — is frustrated by their pursuit of individual rewards.

We can use this simple model for another crucial international problem: the arms race. Assume that there are two nations called Red and Blue. Each of these superpowers can independently select one of two policies:

(A) Noncooperation: arm heavily in preparation for any possible war contingency.

(D) Cooperation: disarm, or at least agree to a partial ban on armaments.

There are four possible outcomes:

(D, D) Both Red and Blue choose to disarm. Viewed as a whole, this is the most preferred social outcome for them, even in light of certain risks.

(A, A) Both nations elect to arm, which is taken as the worst possibility from the global perspective.

(A, D) Red decides to arm, whereas Blue elects to disarm. This amounts to unilateral disarmament by Blue, which is the most preferred of all outcomes to Red, but the least desired by Blue.

(D, A) Red disarms, whereas Blue arms. This is considered the worst result for Red and the best outcome for Blue.

This situation can be modeled by means of the following game matrix:

		Blue	
		A	*D*
Red	*A*	Arms race	Favors red
	D	Favors blue	Disarm

Here, Red's choice amounts to picking one of the two rows, whereas Blue's options correspond to the two columns. It may prove helpful to assign numerical payoffs to the four outcomes, as follows:

		Blue	
		A	*D*
Red	*A*	(2, 2)	(5, 0)
	D	(0, 5)	(4, 4)

For example, the pair of numbers (0, 5) in the second row and first column signifies a payoff of 0 to the row player Red and a payoff of 5 for the column player Blue. The least desired outcome is 0; the most preferred is 5. Only the relative magnitude, not the actual values, of these payoffs is essential for our analysis. The numbers are not intended to measure the absolute worth of unilateral disarmament or of a disarmed world compared with an armed one. They are used only to suggest the preferences of the players: we simply assume that a player prefers a larger numerical payoff to a smaller one.

Let's examine this arms race more closely. Should Red select strategy *A* or *D*? Red can see what will happen if Blue selects his first column *A*: Red will then receive a payoff of 2 for arming and 0 for disarming, so he will arm (first row). Similarly, Red notes the consequences if Blue were to select his second column *D*. In this case, Red will receive 5 for arming or 4 for disarming. Again Red decides in favor of his first row *A*. In either case, Red's first row gives him the more desired result. We say that the payoffs to Red in the first row *dominate* those in the second. There is always some advantage to Red to arm, whether Blue arms or disarms.

A similar argument leads Blue also to choose *A*, that is, to pursue a policy of arming. When each nation strives to maximize its own payoff independently, the pair is driven into the outcome *(A, A)*, with payoffs (2, 2). The better outcome *(D, D)*, with payoffs (4, 4), appears unobtainable when this game is played noncooperatively.

The outcome *(A, A)* is said to be in **equilibrium** because if either nation alone were to deviate from its choice of *A*, then it would be punished with the lower payoff 0, rather than 2. The forces involved prohibit only one nation from moving away from equilibrium *(A, A)*.

Even if both nations agree in advance to jointly pursue the globally optimal solution *(D, D)*, this outcome is unstable because if either nation alone reneges on the agreement and secretly arms, it will benefit. Each would thus be tempted to go back on its word and select *A*. After all, if you have no confidence in the trustworthiness of your opponent, you may be well advised to cover yourself against such a defection.

In real life, however, people often manage to avoid the noncooperative outcome in the prisoner's dilemma. The game is usually played within a larger context, where other incentives are at work. Moreover, the game is typically played on a

SP⬤TLIGHT 15.2   Repeated Play of Prisoner's Dilemma

▷ ▷ ▷ ▷ ▷ ▷ ▷ ▷ ▷ ▷ ▷ ▷ ▷

Robert Axelrod, professor of political science and public policy at the University of Michigan, is a well-known authority on the problem of the prisoner's dilemma. In the following interview he discusses the effects of playing the prisoner's dilemma game repeatedly.

If you play the game only once, there's no future to your interaction and so you might as well take the short-term gains. In one play there's no chance to reward or punish a defection by the other player: there's no hope that you'll get a mutual cooperation going.

A very important feature of the evolution of cooperation is that there is a long-term relationship. When the prisoner's dilemma is iterated, it gives the players an opportunity to base their current choices on the previous interactions they've had. So if the other player seems willing

Robert Axelrod,
University of Michigan.

to cooperate, you can cooperate yourself. If the other player has rarely been willing to cooperate, then it probably doesn't pay to cooperate. Re-

repeating basis—it is not a one-time affair. Elements such as reputation and trust also play a role. The players realize the mutual advantages in cooperation and may arrive at this point by slowly phasing down over time. They may also resort to other helpful measures, such as better communications channels, more reliable inspection procedures, truly binding agreements, or promptly enforced penalties for violators (see Spotlight 15.2, above).

The prisoner's dilemma nicely pinpoints the dynamics behind a frequently occurring social paradox. The resulting standoff, or noncoopera-

tive equilibrium outcome, is not as satisfactory a solution as were the optimal strategies derived in our previous constant-sum games. The cooperative outcome, or one that does not take into account immediate self-interest, is obviously the preferred solution in the long run.

The Game of Chicken

Let us look at one more two-person game of partial conflict, known as **chicken,** which leads to troublesome outcomes. Two drivers are approaching each other at high speeds. Each must decide at the

peating the game allows you to do things like base your own strategy on reciprocity. And therefore it allows you to try to mold the other player's behavior, to encourage him to cooperate with you.

I got the idea of a tournament because I was interested in determining a good way of playing prisoner's dilemma, because it captures some important features of the real world. No one seems to know exactly what's the best strategy. So I invited experts from a variety of fields — people who had written about prisoner's dilemma or game theory — and asked them what strategy they would use in this game. Then I played each with the other to see how well they would do. So I had something analogous to a computer chess tournament.

I was really surprised by the way it came out because the simplest of all the strategies submitted was the one that did best. That was "tit for tat" by Anatol Rapoport. This rule simply says to cooperate on the first move and then do what-

ever the other player did on the previous move. If the other player cooperated, you cooperate. If the other player defected, you defect. It works best for several reasons.

What the analysis shows is that an effective strategy is not to start defecting: never be the first to defect. But if the other side defects, it pays to be provokable. It also pays to be forgiving after you've been provoked, so as to keep the conflict as short as possible. It pays to respond promptly if someone does something you don't like.

I titled my book *The Evolution of Cooperation* to capture the analogy from biological evolution. It's an evolutionary study of cooperation, asking how it could get started in a world where there isn't any, how it can sustain itself, and how it can grow after it gets started. People are likely to continue to use strategies that are effective and to drop or change them if a strategy they use is not very effective. So in fact things tend to evolve toward more effective strategies.

---

last minute whether to swerve to the right or not swerve. There are several possible consequences:

1. Neither driver swerves, and the cars collide head-on. We assign this least preferred outcome a value of 0.

2. Both players swerve. Each loses some prestige by backing off at the brink, but they do remain alive. We give this outcome an intermediate value of 3.

3. One of the drivers swerves and badly loses face, whereas the other does not swerve and is viewed as the winner. Let us select the numeri-

cal payoffs of 1 for swerving and 5 for not swerving in this case.

These options can be summarized in the following game matrix:

		Driver 2	
		Swerve	Not swerve
Driver 1	Swerve	(3, 3)	(1, 5)
	Not swerve	(5, 1)	(0, 0)

If both players persist in their attempts to obtain the maximum payoff 5, then the resulting outcome is mutual disaster: the lowest payoff 0 for each. It is surely better for both drivers to simultaneously back down and obtain 3 each. But neither opponent wants to be in the position of being intimidated into swerving (for a payoff of 1) while the other does not give in (and appears as the winner with a payoff of 5).

Many superpower conflicts, prolonged labor disputes, and other power confrontations have elements in common with the game of chicken. We may find some comfort, however, in knowing that of the 78 essentially different two-by-two games of partial conflict, only chicken and the prisoner's dilemma give rise to such disturbing results, where players' following their immediate self-interests leads to such suboptimal results for the society as a group.

▶Review Vocabulary

Chicken   A common two-person symmetric game in which each player has two strategies: to swerve to avoid a collision or confrontation or else to continue straight ahead and cause a collision if the opponent has not chosen to swerve in the meantime. If both players refuse to swerve, disaster results; if only one backs down, the other wins.

Dominate   One of a player's strategies dominates another strategy when the first one results in higher (or equal) payoffs than the latter one, against any choice made by the opposing player.

Equilibrium   A set of strategies, one for each player, that is in equilibrium if no one player can unilaterally alter his or her strategy to obtain a better payoff.

Expected value   If one of the n payoffs s_1, $s_2, \ldots, s_n$ will occur with the respective probabilities $p_1, p_2, \ldots, p_n$, then the expected value E is

$$E = p_1 s_1 + p_2 s_2 + \cdots + p_n s_n$$

where $p_1 + p_2 + \cdots + p_n = 1$ and where each $p_i \geq 0$.

Fair   A zero-sum game is fair when the (expected) value of the game, obtained by using optimal strategies, is zero.

Game matrix   A rectangular array of numbers. The rows and columns correspond to the strategies for two players, respectively, and the numerical entries represent the resulting payoffs when particular strategies are selected.

Matrix game   A two-person, zero-sum game can be described by a matrix and is called a matrix game.

Maximin strategy   A pure strategy for Player I (the row player), which corresponds to the maximin value in a matrix game.

Maximin value   The largest number from the list (column) of the smallest numbers from each row in a matrix game.

Minimax strategy   A pure strategy for Player II (the column player), which corresponds to the minimax value in a matrix game.

Minimax theorem   The fundamental theorem for two-person, zero-sum games, stating that there always exist optimal mixed (randomized) strategies that enable both players to obtain the optimal expected value in the game.

Minimax value   The smallest number from the list (row) of the largest numbers from each column in a matrix game.

Mixed strategy   A strategy chosen in a probabilistic manner from a list of options (called pure strategies).

Optimal strategy   A particular strategy for a player (pure or mixed) that guarantees that the resulting payoff is the best one that this player can expect to achieve against all possible choices by the opposition.

Payoffs   The potential outcomes of a game, typically expressed as numbers.

Players   The participants who make strategic choices in a competitive encounter.

Prisoner's dilemma   A frequently occurring two-person symmetrical game in which each player has two strategies: cooperate or defect. The best outcome for the pair taken together occurs when they both cooperate. In the case where one cooperates and the other defects, the resulting payoffs are the worst and best possible, respectively.

Pure strategy   Each possible way for a player to play through a game is called a strategy, or pure strategy, to distinguish it from a mixed strategy.

Saddle point (mountain pass)   In a two-person, zero-sum game, a pair of strategies, one for each player, that are in equilibrium. Neither player alone can change strategy and achieve a higher payoff. The strategies and resulting value at a saddle point provide a solution for the game.

Solution   An optimal strategy for each player, along with the resulting value of the game.

Strategy   One of the possible ways to play a game; strategies are mixed or pure, depending on whether they are selected in a probabilistic manner (mixed) or not (pure).

Value   The payoff of a game, usually expressed numerically, that results when the players play optimally.

Zero-sum game   A game in which the payoff to one player is the opposite (or negative) of the payoff to the opposing player.

▶Suggested Readings

BRAMS, S. J., W. F. LUCAS, AND P. D. STRAFFIN, JR. (ED.): *Political and Related Models*, Springer-Verlag, New York, 1983. Chapter 4 by W. F. Lucas and L. J. Billera gives an elementary introduction, with many illustrations, to the multiperson cooperative games (the coalitional games) and related ideas on fair division.

HAMBURGER, HENRY: "*N*-person Prisoner's Dilemma," *Journal of Mathematical Sociology*, 3:27–48(1973). This article shows how the fascination of the two-person prisoner's dilemma increases as one goes to more than two players.

LUCE, R. DUNCAN, AND HOWARD RAIFFA: *Games and Decisions*, Wiley, New York, 1957; Dover, 1989. This venerable survey of most of early game theory presents the two-person zero-sum and nonzero-sum games in Chapters 4 and 5. Several different algorithms for solving the zero-sum case are mentioned in Appendix 6. The minimax theorem and its equivalence to the "duality theorem" in linear programming are given in Appendixes 2 and 5.

RAPOPORT, ANATOL, MELVIN GUYER, AND DAVID GORDON: *The 2 × 2 Game*, University of Michigan Press, Ann Arbor, 1976. This volume provides a detailed review of the 78 "different" two-by-two nonzero-sum games of which the prisoner's dilemma and chicken are the two most interesting and troublesome cases.

WILLIAMS, JOHN D.: *The Compleat Strategyst* (sic), McGraw-Hill, New York, 1954; revised edition, 1966; Dover, 1986. This gem, which contains many simple illustrations, is a humorous primer on the two-person, zero-sum games. The first and second editions provide two different solution procedures discovered by L. S. Shapley and R. Snow, and A. W. Tucker, respectively.

Game theory has found a great number of applications in a wide variety of fields. A mere glimpse into the uses of this subject in the respective areas of biblical studies, international relations, economics, biology, business, and political science can be found in the following (mostly popular) books. Brams's books in particular give many examples of two-person, nonzero-sum games and their applications.

BRAMS, STEVEN J.: *Biblical Games,* MIT Press, Cambridge, 1980.

————: *Superpower Games,* Yale University Press, New Haven, 1985.

CASE, JAMES H.: *Economics and the Competitive Process,* New York University Press, New York, 1979.

DAWKINS, RICHARD: *The Selfish Gene,* Oxford University Press, Oxford, 1976.

MCDONALD, JOHN: *The Game of Business,* Doubleday, Garden City, N.Y., 1975; Anchor, 1977.

ORDESHOOK, PETER J. (ED.): *Games Theory and Political Science,* New York University Press, New York, 1978.

▶ EXERCISES

Consider the following eight two-person, zero-sum games, where the payoffs represent gains to the row Player I and losses to the column Player II.

1. $\begin{bmatrix} 6 & 5 \\ 4 & 2 \end{bmatrix}$
2. $\begin{bmatrix} 0 & 3 \\ -5 & 1 \\ 1 & 6 \end{bmatrix}$
3. $\begin{bmatrix} 3 & 6 \\ 5 & 4 \end{bmatrix}$
4. $\begin{bmatrix} -2 & 3 \\ 1 & -2 \end{bmatrix}$
5. $\begin{bmatrix} -1 & 3 \\ 2 & 0 \end{bmatrix}$

6. $\begin{bmatrix} 13 & 11 \\ 12 & 14 \\ 10 & 11 \end{bmatrix}$
7. $\begin{bmatrix} -10 & -17 & -30 \\ -15 & -15 & -25 \\ -20 & -20 & -20 \end{bmatrix}$
8. $\begin{bmatrix} 6 & 5 & 6 & 5 \\ 1 & 4 & 2 & -1 \\ 8 & 5 & 7 & 5 \\ 0 & 2 & 6 & 2 \end{bmatrix}$

a. Which of these games have saddle points?
b. Find the optimal strategy for Player I and for Player II, and the value for those games given in step a.
c. List any bad strategies in these games, that is, ones the players should avoid because the resulting payoffs are dominated by the payoffs for some alternate strategy.

Solve the following three games of batter versus pitcher in baseball, where the pitcher can throw one of two pitches and the batter can guess either of these two pitches. The batter's batting averages are given in the game matrix.

9.

		Pitcher	
		Fastball	Curve
Batter	Fastball	.300	.200
	Curve	.100	.400

10.

		Pitcher	
		Fastball	Knuckleball
Batter	Fastball	.500	.200
	Knuckleball	.200	.300

11.

		Pitcher	
		Blooperball	Knuckleball
Batter	Blooperball	.400	.200
	Knuckleball	.250	.250

12. A businessman has the choice of either not cheating on his income tax or cheating and making $1000 if not audited. If caught cheating he will pay a fine of $2000 in addition to the $1000 he owes. He feels good if he does not cheat and is not audited (worth $100). If he does not cheat and is audited, he evaluates this at $-$100 (for the lost day). If he is willing to assume that this is a two-person, zero-sum game between himself and the tax agency, then what are the optimal strategies for each player and the expected value of the game?

13. When it is third down and short yardage to go for a first down in American football, the quarterback can decide to run the ball or pass it. Similarly, the other team can commit itself to defend more heavily against a run or a pass. This can be modeled as a 2-by-2 matrix game where the payoffs are the probabilities of obtaining a first down. Find the solution for this game.

		Defense	
		Run	Pass
Offense	Run	.5	.8
	Pass	.7	.2

14. Consider the game played between the opposing goalie and a soccer player who after a penalty is allowed a free kick. The kicker can elect to kick toward one of the two corners of the net or else aim for the center of the goal. The goalie can decide to commit in advance (after the kicker's decision) to either one of the sides or else remain in the center until he sees the direction of the

kick. This zero-sum game can be represented as follows, where the payoffs are the probability of scoring a goal:

		Goalie		
		Breaks left	Remains center	Breaks right
	Kicks left	.5	.9	.9
Kicker	Kicks center	1	0	1
	Kicks right	.9	.9	.5

If we assume that decisions between the left or right side are made symmetrically (i.e., with equal probabilities), then this game can be represented by a 2-by-2 matrix as follows, where $.7 = (\frac{1}{2})(.5) + (\frac{1}{2})(.9)$:

		Goalie	
		Remains center	Breaks side
Kicker	Kicks center	0	1
	Kicks side	.9	.7

Find the optimal strategies for the kicker and goalie and the value of this game.

15. You have the choice of either parking illegally on the street or else parking in the lot and paying $16. Parking illegally is free if the police officer is not patrolling, but you receive a $40 parking ticket if she is. However, you are peeved when you pay to park in the lot on days when the officer does not patrol, and you are willing to assess this outcome as costing $32 ($16 for parking plus $16 for your time, inconvenience, and grief). It seems reasonable to assume that the police officer ranks her preferences in the order (1) give you a ticket, (2) not patrolling with you parked in the lot, (3) patrolling with you in the lot, and (4) not patrolling with you parked illegally.

    a. Describe this as a matrix game, assuming that you are playing a zero-sum game with the officer.

    b. Solve this matrix game for its optimal strategies and its value.

    c. Discuss whether it is reasonable or not to assume that this game is zero-sum.

    d. Assuming that you play this parking game each working day of the year, how do you implement an optimal mixed strategy?

16. Describe how a pure strategy for a player in a matrix game can be considered as merely a special case of a mixed strategy.

▲ 17. a. Describe in detail *one* pure strategy for the player who moves first in the game of tic-tac-toe. (This strategy must tell how to respond to any manner in which the other player moves.) (Hint: You may wish to make use of the symmetry in the 3-by-3 grid in this game; that is, there is one "center" box, four "corner" boxes, and four "side" boxes.)
    b. Is your strategy in **a** optimal in the sense that it will guarantee this first player a tie (or win) in the game?

▲ 18. a. Describe in detail *one* pure strategy for the player who moves second in the game of tic-tac-toe.
    b. Is your strategy in step a optimal in the sense that it will guarantee the second player a tie (or win) in the game?

Exercises 19 to 24 refer to the 3-by-3 baseball example described in the text.

19. Show that the pitcher should avoid throwing to the middle zone. For example, show that his (nonoptimal) mixed strategy $(1 - p, 0, p) = (\frac{1}{2}, 0, \frac{1}{2})$ of throwing both HI and LO half of the time is better on average than using his pure strategy MD.

20. Show that if the pitcher uses his optimal mixed strategy $(1 - p, 0, p) = (0.4, 0, 0.6)$, then he does better on average than using his pure strategy MD, that is, $(0, 1, 0)$.

21. What strategy should the batter use if he knew that the pitcher was playing the (nonoptimal) mixed strategy $(0.5, 0, 0.5)$, that is, throwing a half-half mix of HI and LO?

22. What strategy should the batter use if he knew that the pitcher was playing the (nonoptimal) mixed strategy $(0.2, 0, 0.8)$?

23. Show that the batter should never select his pure strategy HI if he knew that the pitcher was using his optimal mixed strategy $(1 - p, 0, p) = (0.4, 0, 0.6)$.

24. What strategy should the pitcher use if he knew that the batter was playing the (nonoptimal) mixed strategy $(0, 1 - q, q) = (0, 0.5, 0.5)$?

25. In the matching pennies example consider the case where Player I favors heads H over tails T. For example, assume that Player I plays H three-fourths of the time and T only one-fourth of the time — a nonoptimal mixed strategy. What should Player II do if he or she knew this?

26. Assume that in the nonsymmetrical matching example that Player II is using the (nonoptimal) mixed strategy $(p, 1 - p) = (\frac{1}{2}, \frac{1}{2})$; that is, she is playing H and T with the same frequency. What should Player I do in this case if he knew this?

▲ Advanced exercise.

Find the optimal mixed strategies for both players and the value for the following eight n-by-2 matrix games:

27. $\begin{bmatrix} 0 & 6 \\ 2 & 4 \\ 3 & 0 \end{bmatrix}$
28. $\begin{bmatrix} -2 & 2 \\ 4 & -4 \\ 2 & 0 \end{bmatrix}$
29. $\begin{bmatrix} 6 & 8 \\ 4 & 10 \\ 4 & 5 \\ 5 & 2 \end{bmatrix}$
30. $\begin{bmatrix} -4 & 2 \\ -2 & 4 \\ 1 & 3 \\ 1 & 1 \end{bmatrix}$

31. $\begin{bmatrix} -2 & 1 \\ -1 & 0 \\ 1 & -1 \end{bmatrix}$
32. $\begin{bmatrix} 3 & 12 \\ 9 & 6 \\ 10 & 2 \end{bmatrix}$
33. $\begin{bmatrix} 4 & -1 \\ 3 & 1 \\ 1 & 1 \\ 0 & 2 \\ -1 & 3 \end{bmatrix}$
34. $\begin{bmatrix} 4 & -6 \\ -2 & -6 \\ -2 & 0 \\ -4 & 0 \\ -6 & 4 \end{bmatrix}$

The matrix games in the following four exercises have multiple optimal strategies for some players. Use the geometric approach for finding optimal mixed strategies and the value for a game, and describe the resulting range of optimal strategies in each case.

35. $\begin{bmatrix} 4 & 0 \\ 2 & 2 \end{bmatrix}$
36. $\begin{bmatrix} 4 & 0 \\ 2 & 2 \\ -3 & 1 \end{bmatrix}$
▲ 37. $\begin{bmatrix} 2 & -2 \\ 1 & -1 \\ -1 & 1 \end{bmatrix}$
▲ 38. $\begin{bmatrix} 0 & 6 \\ 8 & 2 \\ 2 & 5 \end{bmatrix}$

39. You plan to manufacture a new product for sale next year, and you can decide to make either a small quantity, in anticipation of a poor economy and few sales, or a large output, hoping for brisk sales. Your expected profits are indicated in the following table:

		Economy	
		Poor	Good
Quantity	Small	$500,000	$300,000
	Large	$100,000	$900,000

If you want to avoid risk and believe that the economy is playing an optimal mixed strategy against you in a two-person, zero-sum game, then what is your optimal mixed strategy and expected value? Discuss some alternative ways that you may go about making your decision.

40. On an overcast morning, deciding whether to carry your umbrella can be viewed as a game between yourself and nature as follows:

		Weather	
		Rain	No rain
You	Carry umbrella	Stay dry	Lug umbrella
	Leave it home	Get wet	Hands free

Let's assume that you are willing to assign the following numerical payoffs to these outcomes and that you are also willing to make decisions on the basis of expected values (that is, average payoffs):

$$\text{(Carry umbrella, rain)} = -2$$
$$\text{(Carry umbrella, no rain)} = -1$$
$$\text{(Leave it home, rain)} = -5$$
$$\text{(Leave it home, no rain)} = 3$$

a. If the weather forecast says there is a 50% chance of rain, should you carry your umbrella or not? What if you believe there is a 75% chance of rain?
b. If you are conservative and wish to protect against the worst case, what pure strategy should you pick?
c. If you are rather paranoid and believe that nature will pick an optimal strategy for this two-person, zero-sum game, then what strategy should you choose?
d. Another approach to this decision problem is to assign payoffs to represent what your *regret* will be after you know nature's decision. In this case, each such payoff is the best payoff you could have received under that state of nature, minus the corresponding payoff in the previous table.

		Weather	
		Rain	No rain
You	Carry umbrella	$0 = (-2) - (-2)$	$4 = 3 - (-1)$
	Leave it home	$3 = (-2) - (-5)$	$0 = 3 - 3$

What strategy should you select if you wish to minimize your maximum possible regret?

▲ Advanced exercise.

Consider the following six two-person, nonzero-sum games and discuss the players' possible behavior when these games are played in a noncooperative manner (i.e., with no prior communication or agreements). The first payoff is for the row player; the second, for the column player.

41.

	Player II	
Player I	(5, 5)	(1, 4)
	(3, 0)	(2, 2)

42.

	Player II	
Player I	(3, 5)	(4, 4)
	(1, 2)	(3, 1)

43. Battle of the sexes:

		She buys a ticket for:	
		Boxing	Ballet
He buys a ticket for:	Boxing	(4, 0)	(0, 0)
	Ballet	(1, 1)	(0, 4)

44.

	Player II	
Player I	(3, 5)	(2, 4)
	(1, 2)	(2, 1)

45.

	Player II	
Player I	(1, 3)	(4, 1)
	(0, 6)	(3, 5)

46.

	Player II	
Player I	(2, 4)	(1, 6)
	(4, 2)	(0, 1)

▲ 47. Consider the following miniature poker game with two players, I and II. Each antes $1. Each player is dealt either a high card H or a low card L, with probability one-half. Player I then folds or bets $1. If I bets, then Player II either folds, calls, or raises $1. Finally, if II raises, I either folds or calls.

Most choices by the players are rather obvious, at least to anyone who has played poker: if either player holds H, that player always bets or raises if he or she gets the choice. The question remains of how often one should bluff, that is, continue to play while holding a low card in the hope that one's opponent also holds a low card.

This poker game can be represented by the following matrix game where the payoffs are the *expected* winnings for Player I (depending upon the random deal) and the dominated strategies have been eliminated:

		Player II (when holding L)		
		Folds	Calls	Raises
Player I (when holding L)	Folds initially	−.25	0	.25
	Bets first and folds later	0	0	−.25
	Bets first and calls later	−.25	−.25	0

a. Are there any strategies in this matrix game that a player should avoid playing?
b. Solve this game.
c. Which player is in the more favored position?
d. Should one ever bluff?

▶ WRITING PROJECTS

1. In tennis one player often prefers to play from the baseline while her opponent prefers a serve and volley game (i.e., likes to come to the net). The baseline player attempts to hit passing shots. This player has a choice of hitting "down the line" or "crosscourt." The net player must often guess correctly which direction the ball will go in order to cover the shot. Formulate this situation as a duel game and discuss appropriate strategies for the players.

2. In Spotlight 15.2 we discuss repeated play of prisoner's dilemma. Try this yourself with several different opponents. Look at different strategies, including tit for tat. What do you observe about the psychology of the players? Which strategies seem to yield higher average scores? Do you agree with Professor Axelrod?

▲ Advanced exercise.

Part V

On Size and Shape

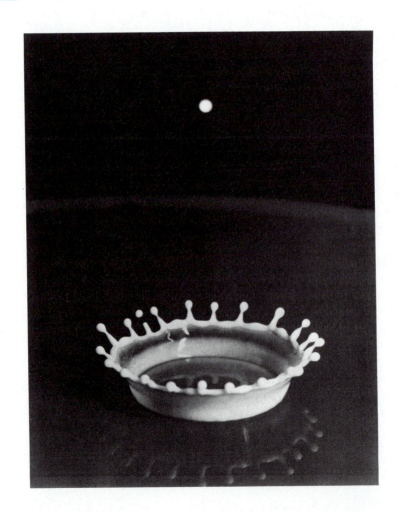

Mathematics is the study of patterns and relationships. It can be used to characterize the spiral growth of a sunflower's seeds, analyze designs on ancient pottery, design telescopes, measure the growth of populations, and calculate distances that are impossible to measure.

Mathematicians instinctively search for and classify numerical, geometrical, and even abstract patterns. In these chapters, we follow some of those searches, concentrating on geometrical patterns, but also looking at what geometry can express about some numerical patterns. Examining the underlying patterns will help explain why some of the objects in the world around us have the shapes that they do and will help you to recognize the same patterns occurring in contemporary problems.

Besides shape, size is a theme for these chapters. We investigate some BIG things: King Kong, mountains, large populations, astronomical distances, and even symmetries that extend infinitely in all directions. We look at how the shape of an animal changes as it grows, and how its size can greatly influence its form. We explore how savings accounts are similar to biological populations, and how the first can have an influence on the second — even leading to extinction of species. We calculate the distances through a mountain, around the earth, and to the sun, all on the basis of a very simple geometrical pattern.

"Milk Drop Coronet," by Harold Edgerton, 1957. The geometric symmetry of natural phenomena is strikingly revealed in this photograph of a drop of milk splashing from a smooth surface. The "coronet" has 22 prongs, arranged symmetrically like the vertices of a regular polygon with 22 sides. The white spot at the top is a tiny droplet of milk that rebounded directly upward.
(©The Harold Edgerton 1992 Trust.)

We find that specific geometrical shapes are essential to telescopes, suspension bridges, and the orbits of planets. And we enjoy the patterned beauty of African crafts and the prints of M. C. Escher.

Chapter
16
Growth and Form

Fantasy films have made us familiar with assorted giant creatures, including King Kong, Godzilla, and the 50-foot-high grasshoppers in *The Beginning of the End*. We also find supergiants in literature, such as the giant of "Jack and the Beanstalk," Giant Pope and Giant Pagan of *The Pilgrim's Progress*, and the Brobdingnagians of *Gulliver's Travels*.

Much as we appreciate those stories, even from an early age we don't really believe in monsters and giants. But could such beings ever exist? What problems would their enormous size cause them? How would they have to adapt in order to cope? (See Figure 16.1.)

Every species survives by adapting to its environment. In particular, it faces the **problem of scale:** how to adapt and survive at the different sizes from the beginning of life to the final size of a mature adult.

For example, consider the giant panda, which ranges from barely 1 lb at birth to 275 lbs in adulthood. A baby panda is at risk of being crushed by its mother; an adult panda needs to eat a great deal of food.

As a contrasting example, consider the horse. If a newborn foal weighed as little as a newborn panda, the foal would be too small to keep up with the moving herd and could not survive. An adult horse weighs much more than a panda and has to consume much more food; but the horse can move much more quickly and cover great distances, to take advantage of wide-ranging sources of sustenance.

There have been large land mammals (mammoths) and huge sea mammals (the blue whale) — not to mention the dinosaurs. But the tallest humans have been only 9 to 10 feet tall; the largest mammoth was 16 feet at the shoulder (about twice as tall as an elephant); and even the tallest dinosaur, *Supersaurus*, stood only 40 feet high.

But what about supergiants and utterly huge monsters? That they have never existed suggests that there are physical limits to size. In fact, with a few simple principles of geometry, we can show not only that lizards and apes of such size are impossible, but also that none of the living beings and objects in our world could exist, unchanged in shape, on a vastly different scale, larger or smaller.

Figure 16.1 Could King Kong actually exist? (The Museum of Modern Art/Film Stills Archive.)

▶Geometric Similarity

The powerful mathematical idea that we will use is *geometric similarity*. By geometric standards, two objects are **similar** if they have the same shape, regardless of the materials of which they are made. They may even be of different sizes. Corresponding angles must be equal, and corresponding dimensions must all have the same factor of proportionality.

For example, when a photo is enlarged, it is enlarged by the same factor in both the horizontal and vertical directions—in fact, in any direction whatever (such as a diagonal). We call this enlargement factor the **scaling factor.** In the photos in Figure 16.2, the scaling factor is 3: the enlarge-

ment is three times as wide and three times as high as the original. In fact, every pair of points goes to a new pair of points three times as far apart as the original ones.

We notice that the enlargement can be divided into $3 \times 3 = 9$ rectangles, each the size of the original. Hence, the enlargement has $3 \times 3 = 3^2 = 9$ times the area of the original. More generally, if the scaling factor is some general number M (not necessarily 3), the resulting enlargement will have an area $M \times M = M^2$ ("M squared") times the area of the original. Thus, the *area* of a scaled-up object goes up with the *square* of the scaling factor.

What about enlarging three-dimensional objects? If we take a cube and enlarge it by a scaling factor of 3, it becomes 3 times as long, 3 times as high, and 3 times as deep as the original (see Figure 16.3).

What about volume? The enlarged cube has 3 layers, each with $3 \times 3 = 9$ little cubes, each the same size as the original. Thus, the total volume is $3 \times 3 \times 3 = 3^3 = 27$ times as much as the original cube. In general, the *volume* of a scaled-up object goes up with the *cube* of the scaling factor. Thus, for an object enlarged by a scaling factor of M, the enlargement will have M^3 ("M cubed") $= M \times M \times M$ times the volume of the original. Like the relationship between surface area and M^2, this relationship holds even for irregularly shaped objects, such as science fiction monsters.

We observe, however, that the area of each face (side) of the enlarged cube is $3^2 = 9$ times as large as that of a face of our original cube, just as the area of the photo enlarged by a factor of 3 has 9 times the area of the original. Since this fact is true for all 6 faces, the total surface area of the enlarged cube is 9 times as much as the original.

More generally, for objects of any shape, the total *surface area* of a scaled-up object goes up with the *square* of the scaling factor. Thus, the surface area of an object scaled up by a factor of M is M^2 times the surface area of the original; this feature holds true even for irregular shapes.

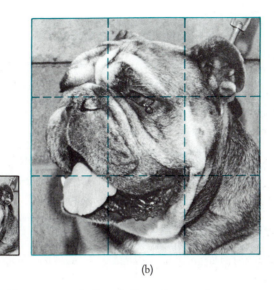

(a)   (b)

Figure 16.2 Two geometrically similar photographs. (Photo by Travis Amos.)

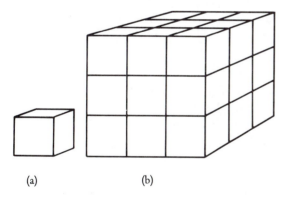

(a)   (b)

Figure 16.3 Cube (b) is made by enlarging cube (a) by a factor of 3.

Before we discuss scaling real three-dimensional objects, you should understand the pitfalls of the language for describing increases and decreases.

▶ THE LANGUAGE OF GROWTH, ENLARGEMENT, AND DECREASE

In 1976, the average price of a home in Madison [Wisconsin] was $38,323—about 108 per cent less than [in 1990]. [Madison Business (March 1991):38].

It certainly sounds as if home values in Madison have risen substantially, but could there be any reason to question a statement like this? Of course! Poor or imprecise language can confuse the meaning, and an author may want to exaggerate or minimize any increase or decrease involved.

How can you be completely accurate, without indulging in exaggeration? Consider, for instance, a photograph with area 10 sq in and enlarge it three times, so that it is three times as wide and three times as high. The enlargement has an area of 90 sq in. What are some ways that you might compare this area to the original area? Fill in the blank in the sentence: "The enlargement's area is ____ the area of the original." The phrases "nine times," "nine times as large as," and "80 sq in more than" are all simple, descriptive, correct, and unconfusing.

Comparative terms like "greater than," "more," and "larger" are also very familiar to us, but they can become confusing or misleading when used with numbers and percentages. Consider, for example, the phrases "nine times greater than," "nine times more than," and "nine times larger than." They all sound grander than the earlier phrases, because they use the comparative form or the word "more" in addition to the word "times." But "more," "larger," and "greater" refer to quantity *over and above* the original; so the phrases involving those words are properly interpreted as meaning "10 times as large" (the original plus nine times more).

The problem involved is easier to see if we ask what a phrase like "one times more than" could possibly mean. Consistent with the other misusages in the previous paragraph, the intended meaning is "the same as"—which isn't more at all!

Using both "times" and "more" in the same sentence is so confusing because "times" (or "as much as") refers to *multiplication* of the original amount, while "more" (or a comparative adjective, such as "greater") refers to *adding* on to the original amount. It's definitely confusing to use both at once! Compare two quantities by using just one or the other—*never use both "times" and "more" together.*

This problem is not solved by speaking of percent instead. In all the incorrect phrases we have considered, "nine times" could be replaced by "900%" and the same criticism would hold.

Percent has even further hazards for the careless user of language. An increase from one to nine is an increase of 800%, not 900%. In discussions of percent we also need to distinguish percent from percentage points: if support for the President has decreased from 60% to 30%, it has dropped 30 *percentage points* but decreased 50% (because the drop of 30 percentage points is 50% of the original 60 percentage points).

There are also perils associated with speaking of decreases. For a reduction from nine to one, the reduced quantity may be described incorrectly as "nine times less than" or "900% less than." Correct terminology is that the new amount is "one-ninth as much," "eight-ninths less than," "11% as much," and "89% less than" the original.

EXAMPLE: What about Those Homes?

The average price of a home in Madison in 1976 was $38,323, whereas it was $80,500 in 1990. The $80,500 figure represents 100% of the average price of a home. To get 108%, we multiply 1.08 times $80,500 to get $86,940. So "108 percent less" should be $86,940 less than the average price, or $80,500 − $86,940 = −$6,440. Well, that can't be right; the writer has misused the language. What could the writer have said instead that would have been clear and correct? The 1976 figure, $38,323, is about 0.48 times $80,500 (38,323 ÷ 80,500), or 48% of $80,500; so the writer could have said "about 48% of what it is in 1990" or "about 52% less than it is in 1990." What if the writer had wanted to use the 1976 figure as a base (i.e., the 100% for the calculation)? The 1990 price is about 2.10 times the 1976 figure, so the 1990 price is "210% of," or "110% more than," the 1976 price.

Caution: The dollar comparisons here, like many you see, may be misleading, since they do not take into account that a dollar was worth less in purchasing power in 1990 than in 1976, because of inflation. All such comparisons are best given in "constant" dollars. In fact, a 1976 dollar was worth about 2.2 of a 1990 dollar, or put the other way, a 1990 dollar was worth about 1/2.2 = 0.45 of a 1976 dollar. Thus, the 1976 price of an average home, $38,323, translates into 2.2 times $38,323, or about $84,000, in 1990 dollars. In other words, taking into account inflation, the average cost of a home in Madison in 1990 was *less* than in 1976.

Also, what does the author mean by average? "Average" may refer to either the mean or the median (see pp. 186–187); and there can be a big difference between the two for prices of houses (as noted on p. 227). The median is the measure generally used by government statisticians and economists to describe the center of a distribution of housing prices. Later in the article from which the quote about Madison housing is taken, the author notes that the median price of a Madison home in 1976 was $34,000, revealing that the earlier "average" must have been the mean. ▲

Scaling three-dimensional objects requires measuring physical quantities such as distance, weight, area, and volume. Before we consider the possibility of a King Kong or a building ten times as tall as the Sears Tower, we need to discuss the units in which those quantities are measured.

▶ MEASURING LENGTH, AREA, VOLUME, AND WEIGHT

We give here an introduction to the common units in which various physical quantities are measured, together with a handy table of conversion factors and examples of how to convert successfully from one system of units to another.

You are no doubt familiar with the common units of the *U.S. Customary System* of measurement and their abbreviations. But please pay close attention to the systematic way to convert from one unit to another, and to the expression of approximate numbers in scientific notation:

Distance:

$$1 \text{ mile (mi)} = 5280 \text{ feet (ft)}$$
$$1 \text{ foot (ft)} = 12 \text{ inches (in)}$$

Area:

1 square mile (sq mi)
$$= 1 \text{ mi} \times 1 \text{ mi}$$
$$= 5280 \text{ ft} \times 5280 \text{ ft}$$
$$= 27{,}878{,}400 \text{ ft} \times 1 \text{ ft}$$
$$= 28 \times 10^6 \text{ sq ft, approximately}$$
$$= 27{,}878{,}400 \times 1 \text{ ft} \times 1 \text{ ft}$$
$$= 27{,}878{,}400 \times 12 \text{ in} \times 12 \text{ in}$$
$$= 4{,}014{,}489{,}600 \times 1 \text{ in} \times 1 \text{ in}$$
$$= 4{,}014{,}489{,}600 \text{ sq in}$$
$$= 4 \times 10^9 \text{ sq in, approximately}$$

We also have

$$1 \text{ square mile} = 640 \text{ acres}$$

with

$$1 \text{ acre} = 43{,}560 \text{ sq ft}$$

Volume:

1 cubic mile (cu mi)
$$= 1 \text{ mi} \times 1 \text{ mi} \times 1 \text{ mi}$$
$$= 5280 \text{ ft} \times 5280 \text{ ft} \times 5280 \text{ ft}$$
$$= 147{,}197{,}952{,}000 \times 1 \text{ ft} \times 1 \text{ ft} \times 1 \text{ ft}$$
$$= 147 \times 10^9 \times 1 \text{ ft} \times 1 \text{ ft} \times 1 \text{ ft},$$
$$\text{approximately}$$
$$= 147 \times 10^9 \times 12 \text{ in} \times 12 \text{ in} \times 12 \text{ in}$$
$$= 147 \times 10^9 \times 12 \times 12 \times 12 \text{ cu in}$$
$$= 2.5 \times 10^{14} \text{ cu in, approximately}$$

For liquid measure, the customary unit in the United States is

$$1 \text{ U.S. gallon} = 231 \text{ cu in, exactly}$$

Weight:

$$1 \text{ ton (t)} = 2000 \text{ pounds (lb)}$$

There are other units (rods, light-years, bushels, ounces, etc.), but we will not consider them here.

The metric system was first proposed in France by Gabriel Mouton, Vicar of Lyons, in 1670 and was adopted in France in 1795. The fundamental unit of length, the *meter*, was originally defined to be one ten-millionth of the distance from the North Pole to the Equator, as measured on the meridian through Paris. Later, the meter was redefined as the distance between two lines marked on a platinum-iridium bar kept at the International Bureau of Weights and Measures, near Paris, when the bar is kept at a temperature of 0°C. Finally, in 1960 the meter was redefined in terms of a standard reproducible in any laboratory, namely, 1,650,763.73 times the wave length of the orange-red light emitted by atoms of the gas krypton-86 when an electrical charge is passed through them. All other units of length, area, and volume are *defined* in terms of the meter; for example, a centimeter is a hundredth of a meter. The metric unit of weight, the *kilogram*, is defined as the weight of a platinum-iridium standard.

In the metric system, we have

Distance:

$$1 \text{ kilometer (km)} = 1000 \text{ meters (m)}$$
$$= 100,000 \text{ centimeters (cm)}$$
$$= 1 \times 10^5 \text{ cm}$$
$$1 \text{ meter (m)} = 100 \text{ centimeters (cm)}$$

Area:

$$1 \text{ square meter (sq m, or m}^2\text{)}$$
$$= 1 \text{ m} \times 1 \text{ m}$$
$$= 100 \text{ cm} \times 100 \text{ cm}$$
$$= 10,000 \text{ sq cm (cm}^2\text{)}$$
$$= 1 \times 10^4 \text{ cm}^2$$

(Land is usually measured in a larger unit: 1 hectare = 10,000 m².)

Volume:

1 cubic meter (cu m, or m³)
$$= 1 \text{ m} \times 1 \text{ m} \times 1 \text{ m}$$
$$= 100 \text{ cm} \times 100 \text{ cm} \times 100 \text{ cm}$$
$$= 1,000,000 \text{ cu cm (cm}^3\text{)}$$
$$= 1 \times 10^6 \text{ cm}^3$$

(For liquid measure, the unit is 1 liter = 1000 cm³.)

Weight:

$$1 \text{ kilogram (kg)} = 1000 \text{ grams (g)}$$

The metric system also has other units (angstrom, metric tonne) that we will not consider here.

What are the conversions between the U.S. Customary System and the metric system? Since 1960, the fundamental units of the U.S. Customary System, the yard (for length) and the pound (for weight), have been *defined* in terms of metric units, so that we have

$$1 \text{ yd} = 0.9144 \text{ m, exactly}$$
$$1 \text{ lb} = 0.45359237 \text{ kg, exactly}$$

The conversions of other units are

Distance:

$$1 \text{ in} = 2.54 \text{ cm, exactly}$$
$$1 \text{ ft} = 12 \text{ in} = 12 \times 1 \text{ in} = 12 \times 2.54 \text{ cm}$$
$$= 30.48 \text{ cm, exactly}$$
$$1 \text{ mi} = 5280 \text{ ft} = 5280 \times 1 \text{ ft}$$
$$= 5280 \times 30.48 \text{ cm}$$
$$= 160,934.4 \text{ cm, exactly}$$
$$= 1.61 \times 10^5 \text{ cm, approximately}$$
$$= 1.61 \times 1 \text{ km}$$
$$= 1.61 \text{ km, approximately}$$

We can also go the other way:

$$1 \text{ m} = 100 \times 1 \text{ cm}$$

But how much is 1 cm in terms of inches? Since

$$1 \text{ in} = 2.54 \text{ cm}$$

we can divide both sides of this equation by 2.54, getting

$$\frac{1}{2.54} \text{ in} = 1 \text{ cm}$$

or, reading from right to left,

$$1 \text{ cm} = 0.393701 \text{ in, approximately}$$
$$= 0.4 \text{ in, approximately}$$

Hence

$$1 \text{ m} = 100 \times 1 \text{ cm} = 100 \times 0.393701 \text{ in}$$
$$= 39.3701 \text{ in, approximately}$$

For *weight* we have

$$1 \text{ lb} = 0.45359237 \text{ kg}$$

Hence

$$1 \text{ kg} = \frac{1}{0.45359237} \text{ lb}$$
$$= 2.205 \text{ lb, approximately}$$

The exercises offer practice in converting area, volume, and weight back and forth between the U.S. Customary and metric systems.

▶SCALING REAL OBJECTS

Why is the size of a three-dimensional object always limited to some degree? Real three-dimensional objects are made of matter, which has extension (volume) and substance (mass). **Mass** is the aspect of matter that is affected by forces, according to physical laws. For example, your mass reacts to the gravitational force of the earth by staying close to it (and your mass exerts an equal force on the earth that tends to keep the earth close to you). We perceive the mass of an object when we try to move it (as in throwing a ball). When we try to lift an object, we perceive its mass as weight, due to the gravitational force that the earth exerts on it. As we will see, gravity exerts an enormous effect on the size and shape that objects and beings can assume.

In Figure 16.4, we consider two cubes. The first is a cube of steel 1 foot on a side. The bottom face supports the weight of the entire cube. **Pressure** is force per unit area, so the pressure exerted on the bottom face by the weight of the cube is

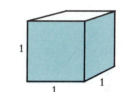

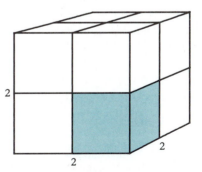

Figure 16.4 A cube of side 1 and a cube of side 2.

equal to the weight of the cube divided by the area of the bottom face, or

$$P = \frac{W}{A}$$

A cubic foot of steel weighs about 500 pounds (we say it has a **density** of 500 pounds per cubic foot), and the area of the bottom face is 1 square foot; so the pressure exerted on this face is therefore 500 pounds per square foot.

The second cube in Figure 16.4 is made of the same steel but is 2 feet on a side. The area of the bottom face has increased with the square of the scaling factor, so it is $2^2 \times 1 = 4$ square feet. As we learned on p. 494, volume goes up with the cube of the scaling factor. So this larger cube has a volume of $2^3 \times 1 = 8$ cubic feet. Because both cubes are made of the same steel, the larger cube has eight times as much steel as the smaller; hence it weighs eight times as much as the smaller cube, or $8 \times 500 = 4000$ pounds.

When we divide this weight by the area of the bottom face (4 square feet), we find that the pressure exerted on the bottom face is 1000 pounds per square foot, or twice the pressure on the bottom face of the original cube. This makes sense because over each 1 square foot area stands 2 cubic feet of steel.

If we scale up the original cube of steel up to a cube 3 feet on a side, we observe that the pressure on the bottom face triples. In general, if the scaling factor for the cube is M, the pressure on the bottom face will be M times as much.

EXAMPLE: What about a 10-Foot Cube?

If we scale the original cube of steel up to a cube 10 feet on a side, then the dimensions are

$$10 \text{ ft} \times 10 \text{ ft} \times 10 \text{ ft}$$

The total volume is

$$\begin{aligned} V &= \text{length} \times \text{width} \times \text{height} \\ &= 10 \text{ ft} \times 10 \text{ ft} \times 10 \text{ ft} \\ &= 1000 \text{ ft} \times 1 \text{ ft} \times 1 \text{ ft} = 1000 \text{ ft}^3 \\ &= 1000 \text{ cu ft} \end{aligned}$$

The weight of the cube is

$$\begin{aligned} W &= V \times \text{density} \\ &= 1000 \text{ cu ft} \times 500 \text{ lb/cu ft} \\ &= 500,000 \text{ lb} \end{aligned}$$

The area of the bottom face is

$$\begin{aligned} A &= \text{length} \times \text{width} \\ &= 10 \text{ ft} \times 10 \text{ ft} \\ &= 100 \text{ ft} \times 1 \text{ ft} \\ &= 100 \text{ ft}^2 = 100 \text{ sq ft} \end{aligned}$$

The pressure on the bottom face is

$$P = \frac{W}{A} = \frac{500,000 \text{ lb}}{100 \text{ sq ft}} = 5000 \text{ lb/sq ft}$$

This is 10 *times* — not "10 times *more* than" — the pressure on the bottom face of the original 1-foot cube.   ▲

At some scale factor, the pressure on the bottom face will exceed the steel's ability to withstand that pressure — and the steel will deform under its own weight. That point for steel is reached for a cube about 3 miles on a side — the pressure exerted by the cube's weight exceeds the resistance to crushing (ability to withstand pressure, or **crushing strength**) of steel, which is about 7.5 million lb/sq ft. Since a mile is 5280 feet, a 3-mile-long cube of steel would be more than 15,000 times as long as the original 1-foot cube; that is, the scaling factor is more than 15,000. The pressure on the bottom face of the cube would therefore be more than 15,000 times as much as for the 1-foot cube, or 15,000 × 500 lb/sq ft = 7.5 million lb/sq ft.

EXAMPLE: What about the Sears Tower?

The Sears Tower in Chicago is the world's tallest building, at 110 stories and 1454 ft (443 m), if we don't count radio and television broadcast towers. What is the pressure on its foundation?

The Sears Tower is made of reinforced concrete, which weighs about 160 lbs per cubic foot. Consider the walls of the tower. Over each square foot of foundation stands 1454 cubic feet of reinforced concrete wall, which weighs 1454 × 160 = 233,000 lbs. The pressure on the foundation, from the walls alone, is 233,000 lbs per square foot. That's not counting all the contents of the Tower, which also must be supported by the walls!

Could we have a Super Sears Tower that was 10 times as high? If we could, its foundation would have to be able to support 2.3 million pounds per square foot. ▲

▶ SORRY, NO KING KONGS

Unfortunately, the resistance of bone to crushing is not nearly as great as that of steel. This fact helps to explain why there couldn't be any King Kongs (unless they were made of steel!). A creature scaled up by a factor of 20 would weigh 20^3 = 8000 times as much. Though the weight increases with the cube of the scaling factor, the ability to support the weight—as measured by the cross-sectional area of the bones, like the area of the bottom face of the cube on p. 499—increases only with the square of the scaling factor.

These simple consequences of the geometry of scaling apply to other objects, natural and artificial, not only to supermonsters. Three hundred and fifty years ago, Galileo was able to give a good

estimate of how high the tallest trees could be (see Spotlight 16.1, p. 502). Also, we can estimate how high the tallest mountains could be.

EXAMPLE: How Tall Can a Tree Be?

Galileo suggested that no tree could grow taller than 300 feet. The world's tallest trees are giant sequoias, which grow only on the West Coast of the United States, and hence were unknown to Galileo. They grow to 360 feet (Figure 16.5).

What can limit the height of a tree? If the roots do not adequately anchor it, a tall tree can blow over. (This, in fact, happened in 1990 to the world's tallest tree, the Dyerville Giant, a giant sequoia in Humboldt Redwoods State Park in California.) The tree could buckle or snap under its own weight and the force of a strong wind. The wood at the bottom will

Figure 16.5 Even these giant sequoias may grow no taller than their form and materials allow. (Larry Ulrich.)

SP TLIGHT 16.1   Galileo and the Problem of Scale

▶ ▶ ▶ ▶ ▶ ▶ ▶ ▶ ▶ ▶ ▶ ▶ ▶

Galileo Galilei (1564–1642) was the first to describe the problem of scale, in 1638, in his *Dialogues Concerning Two New Sciences* (in which he also discussed the idea of the earth revolving around the sun):

You can plainly see the impossibility of increasing the size of structures to vast dimensions either in art or in nature; likewise, the impossibility of building ships, palaces, or temples of enormous size in such a way that their oars, yards, beams, iron-bolts, and, in short, all their other parts will hold together; nor can nature produce trees of extraordinary size because the branches would break down under their own weight, so also would it be impossible to build up the bony structures of men, horses, or other animals so as to hold together and perform their normal functions if these animals were to be increased enormously in height; for this increase in height can be accomplished only by employing a material which is harder and stronger than usual, or by enlarging the size of the bones, thus changing their shape until the form and appearance of the animals suggest a monstrosity.

To illustrate briefly, I have sketched a bone whose natural length has been increased three times and whose thickness has been multiplied until, for a correspondingly large animal, it would perform the same function which the small bone performs for its small animal. From the figures here shown you can see how out of proportion

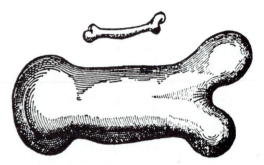

One bone, with another three times as long and thick enough to perform the same function in a scaled-up animal. (Illustration from Galileo's *Dialogues Concerning Two New Sciences.*)

the enlarged bone appears. Clearly then if one wishes to maintain in a great giant the same proportion of limb as that found in an ordinary man he must either find a harder and stronger material for making the bones, or he must admit a diminution of strength in comparison with men of medium stature; for if his height be increased inordinately he will fall and be crushed under his own weight. Whereas, if the size of a body be diminished, the strength of that body is not diminished in proportion; indeed the smaller the body the greater its relative strength. Thus a small dog could probably carry on his back two or three dogs of his own size; but I believe that a horse could not carry even one of his own size.

(Translated by Henry Crew and Alfonso De Salvo, and published by Macmillan, 1914, and Northwestern University, 1946)

begin to crush if there is too much weight above. Finally, there is a limit to how far the tree can lift water and minerals from the roots to the leaves.

Most of the trunk of a tree is dead wood; only a thin band in and under the bark is alive. Water and minerals flow from the roots to the leaves through a thin cylinder, at most an inch thick, immediately inside the bark. Much as the bottom of the steel cube must support the weight of the cube, the wood at the base of the trunk of a tree must support all of the wood above it. The tree must be able to overcome the pressure of this weight to "pump" water and minerals up to the leaves.

Consider a tree 360 ft high. What is the pressure on the wood at the base of the trunk? To make an easy but rough estimate, let's ignore the fact that the tree tapers toward the top. Since a cell of wood is largely water, we may think of a "stack" of cells as a column of water (this will give us an underestimate of the weight of the wood, as fresh wood contains minerals and organic compounds and weighs considerably more than water). So an equivalent question is: What is the pressure on the bottom of a 360-ft column of water? Over each square foot at the bottom, there will be 360 cu ft of water; we will find out how much that weighs. First, we translate 1 cu ft into metric measurement:

$$
\begin{aligned}
1 \text{ cu ft} &= 12 \text{ in} \times 12 \text{ in} \times 12 \text{ in} \\
&= 12 \times 2.54 \text{ cm} \times 12 \times 2.54 \text{ cm} \times \\
&\quad 12 \times 2.54 \text{ cm} \\
&= 28{,}316 \text{ cm}^3 \text{ (or cc)}
\end{aligned}
$$

The reason to convert to cubic centimeters is the convenient fact that water weighs almost exactly 1 gram per cubic centimeter. Now, 1 cu ft of water weighs about 28,300 g = 28.3 kg = 28.3 × 2.20 lb = 62 lb. Consequently, 360 cu ft of water weighs 360 × 62 lb = 22,000 lb, so the pressure on the bottom layer is 22,000 lb/ft².

A biological organism needs a safety factor of at least 2 to 4 times the minimum physical limits for its processes, so a tree 360 ft tall would need to have from 40,000 to 80,000 lb/ft² of upward pressure for water and minerals. Experiments show that tension in the string of water molecules from root to leaf ranges from 80,000 to 3.2 million lb/ft², for different kinds and heights of trees. Based on these considerations, even taller trees could exist.

What about the pressure on the bottom of the tree? At more than about 500 lb/in² (70,000 lb/ft²), the bottom of the tree would begin to crush under the weight above. On this point, our 360-ft sequoia has a safety margin of only about a factor of 3. (However, the tapering of the tree, combined with extra thickness near the base, make for a larger safety margin.)

Trees taller than 360 ft *might* be physically possible. The taller the tree, though, the greater the area from which it must draw water and minerals, for which nearby trees also compete. For that reason, evolution may select against extremely tall trees.   ▲

EXAMPLE: How High Can a Mountain Be?

Gravity and the physical characteristics of wood limit the height of trees. Gravity also limits the height of mountains. Mountains differ from one to another in composition and shape, and some assumptions about those features will be necessary in order to do any calculating. We want to make our assumptions as realistic as we can and still be able to calculate easily an estimate of how high a mountain can be. In effect, we build a simple mathematical model of a mountain.

Let's suppose that the mountain is made entirely of granite, a common material in many mountains, and assume that the granite has uniform density. Relevant facts about granite are that it weighs 165 lb/cu ft and it has a crushing strength of about 4 million lb/sq ft.

In the interests of both realism and simplicity, we assume that our model mountain is in the shape of a cone whose width at the base is the same as its height. Let's model Mount Everest: the tallest earth mountain, it is about 6 miles high. The base, then, is a circle with a distance across (or diameter) of 6 miles. The radius of the circle is half the diameter, so our model Everest has a radius of 3 miles measured at the base (Figure 16.6). Since we are taking such a round number for the height of Everest, we record as significant only the first two digits of the results of our calculations.

What does our model Everest weigh? The relevant formula is

$$\text{Weight} = \text{density} \times \text{volume}$$

We already know the density of granite (165 lb/cu ft), so to find the weight we are going to need to know how to calculate the volume of a cone. The formula is

$$\text{Volume} = \pi \times (\text{radius})^2 \times \frac{\text{height}}{3}$$

For our Everest, the radius is 3 miles and the height is 6 miles; π (pi) is about 3.14. Using those values in the formula, we find that our model Everest has a volume of about 57 cubic miles.

To find the weight of 57 cubic miles of granite, we need to do some conversion of

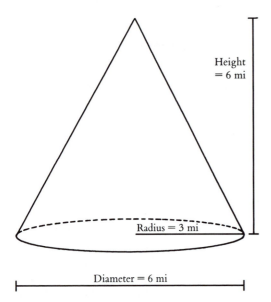

Figure 16.6 Model of Mt. Everest as a cone of granite.

Height = 6 mi

Radius = 3 mi

Diameter = 6 mi

units, since the density is given in pounds per cubic foot. Let's convert to units of feet:

$$
\begin{aligned}
1 \text{ cu mi} &= 1 \text{ mi} \times 1 \text{ mi} \times 1 \text{ mi} \\
&= 5280 \text{ ft} \times 5280 \text{ ft} \times 5280 \text{ ft} \\
&= 1.5 \times 10^{11} \text{ ft} \times 1 \text{ ft} \times 1 \text{ ft}, \\
&\qquad \text{approximately} \\
&= 1.5 \times 10^{11} \text{ cu ft, approximately}
\end{aligned}
$$

and

$$
\begin{aligned}
57 \text{ cu mi} &= 57 \times 1 \text{ cu mi} \\
&= 57 \times 1.5 \times 10^{11} \text{ cu ft}, \\
&\qquad \text{approximately} \\
&= 8.6 \times 10^{12} \text{ cu ft}, \\
&\qquad \text{approximately}
\end{aligned}
$$

So we have

Weight of mountain
$$
\begin{aligned}
&= 165 \text{ lb/cu ft} \times 8.6 \times 10^{12} \text{ cu ft} \\
&= 1.4 \times 10^{15} \text{ lb} \\
&= 1.4 \text{ quadrillion lb}
\end{aligned}
$$

Now that we know the weight of the mountain, we want to find out what the pressure is on the base of the cone and compare that with the crushing strength of granite. (Everest is standing, so if our model is any good, that pressure will be below the crushing strength.) Physics tells us that the weight of the mountain is spread evenly over the base of the cone (we are oversimplifying the geology underlying mountains). Since

$$\text{Pressure} = \frac{\text{weight}}{\text{area}}$$

we need to calculate the area of the base of the cone. The shape is a circle, and the familiar formula

$$\text{Area} = \pi \times (\text{radius})^2$$

gives an area of 28 square miles for a radius of 3 miles.

Once again, we need to convert to units of feet in order to express the pressure in pounds per square foot, the units in which we are given the crushing strength. We get

$$Area = 28 \text{ sq mi}$$
$$= 28 \times 1 \text{ mi} \times 1 \text{ mi}$$
$$= 28 \times 5280 \text{ ft} \times 5280 \text{ ft}$$
$$= 8 \times 10^8 \text{ sq ft, approximately}$$

Then

$$Pressure = \frac{weight}{area}$$
$$= \frac{1.4 \times 10^{15} \text{ lb}}{8 \times 10^8 \text{ sq ft}}$$
$$= 1.8 \times 10^6 \text{ lb/sq ft}$$
$$= 1.8 \text{ million lb/sq ft}$$

This number is safely below the crushing strength of granite, 4 million pounds per square foot.

For a mountain to come close to the limitation of the crushing strength of granite, it would have to be only about twice as high as Everest, or about 10 miles high. Other physical considerations suggest a maximum height of at most 15 miles. That no present mountains are that high is probably a consequence of the earth's high amount of volcanic activity and the structural deformation of the earth's crust.   ▲

What about mountains made of other materials — glass, ice, wood, old cars? They couldn't be nearly as high; the pressure would cause glass to flow, ice to melt, and old cars to compact. What about mountains on another planet? Their potential height depends on the gravity of the planet.

▶ SOLVING THE PROBLEM OF SCALE

A large change in scale forces a change in either materials or form. A major manifestation of the scaling problem is the tension between weight and the need to support it. For example, a real building or machine must differ from a scale model; the balsa wood or plastic of the model would never be strong enough to use for the real thing, and the materials in the scaled-up version must be aluminum, steel, or reinforced concrete. So one way to compensate for the problem of scale is to use stronger materials in the scaled-up object.

The other way to compensate is to redesign the object so that its weight is better distributed. Let's go back to our original cube. It supports all its weight on its bottom face. In the version scaled up by a factor of 3, each small cube of the bottom layer has a bottom face that is supporting that cube's weight plus the weight of the other two cubes piled on top of it.

Now, let's redesign the scaled-up cube, concentrating for simplicity only on the front face, with its nine small cubes. We take the three cubes on top and move them to the bottom, alongside the three already there. We take the three cubes on the second level, cut each in half, and put a half cube over each of the six ground-level cubes (see Figure 16.7). We have the same volume and weight we

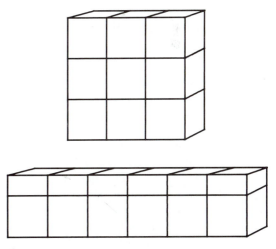

Figure 16.7 Nine small cubes rearranged to support greater weight.

started with, but now there is less pressure on the bottom face of each small cube. Of course, our new design is not geometrically similar to the object we started with—it's no longer a cube. By changing the proportions, we have given up the precise scaling of geometrical similarity, but we have managed to compensate for the scaling problem.

We observe in nature both strategies for adaptation to scaling: change of materials and change of form. Smaller animals generally do not have bony internal skeletons; larger animals generally do. Those animals made of similar materials but differing greatly in size, such as a mouse and an elephant, will most certainly differ in shape. If a mouse were scaled up to the size of an elephant, its legs could no longer support it. It would need the disproportionately thicker legs of the elephant, and the elephant's thick hide to contain its tissue.

Some dinosaurs, like *Supersaurus* (which weighed 30 tons), had special adaptations to lighten their weight, such as hollow bones, just as some birds have. (Hollow bones also turn out to be stronger, a paradox that Galileo analyzed. Of two bones of the same weight and length, the hollow one will be wider across at its midpoint, because of the air it contains; and the greater the width, the greater the resistance to fracture.)

▶FALLS, DIVES, JUMPS, AND FLIGHTS

The need to support weight can be thought of as a tension between volume and area. As we scale up an object, its volume and weight go up together, as long as we maintain a constant density (for example, no air bubbles introduced into our steel to make it into a Swiss cheese!). At the same time, the ability to support the weight goes up with the cross-sectional area, just as our steel cube had to be supported by its bottom face.

Area-volume tension has many other practical consequences, some of them related to our childhood fantasies. We can forget about humans "leaping tall buildings in a single bound," "soaring like an eagle," diving miles below the sea, and jumping from airplanes without parachutes.

EXAMPLE: Falls

Area-volume tension affects how creatures respond to falling, another of gravity's effects. A mouse may be unharmed by a 10-story fall, a cat by a two-story fall, but a human may well be injured just by falling down while running, walking, or even just standing.

What is the explanation? The energy acquired in falling is proportional to the weight of the falling object, hence to its volume. This energy must be absorbed either by the object or by what it hits or must be otherwise dissipated at impact—for example, as sound. The fall is absorbed over part of the surface area of the object, just as the weight of the cube was distributed over its base. With scaling up, volume—hence weight, hence falling energy—goes up much faster than area. As volume increases, the hazards of falling from the same height increase.   ▲

EXAMPLE: Dives

Whales can hold their breath and stay under water for as long as 20 minutes. Why can't we? Basically, because we aren't as large as whales. A mammal's breath-holding ability depends on two things: the volume of oxygen carried in the lungs, which volume is proportional to the volume of the mammal and hence to the cube of its length; and the rate at which oxygen is absorbed by the surface area of the lungs, which area is proportional to the square of the length of the mammal. We would therefore expect the limits of duration of dive to be proportional to the lung volume divided by the lung area, thus to the length of the animal; and we would be right. Although special adaptations may play a larger role for a particular species, and exceptional individuals can outperform the average, this case does illustrate a straightforward proportional relationship. Blue whales are about 16 times as long as human adults and can hold their breath about 16 times as long.   ▲

EXAMPLE: Jumps

A flea can jump about 2 feet vertically, many times its own height. Many people believe that if a flea were as large as a person, it could jump a thousand feet into the air. Imagining — against our earlier arguments — that there could be so large a flea, we know its limits: a scaled-up flea could jump about the same height as a small flea. The strength of a muscle is proportional to its cross-sectional area (see Spotlight 16.2, pp. 508–509). A jump involves suddenly contracting the muscle through its length, so it turns out that the ability to jump is proportional to the volume of muscle. But the volume of the flea and the volume of its leg muscles go up in proportion. Let's say that a real flea's leg muscles account for 1% of its body. If we scale the flea up to the size of a person (without any change in its form), the enlarged flea's leg muscles would still make up 1% of its body. For either flea, each bit of muscle has the same power: in a jump, it propels 100 times its own weight, and it can do so to the same height. Both the weight of the flea and the power of its legs go up proportionately.   ▲

EXAMPLE: Flight

Wouldn't it be nice to be able to fly? Well, you have to be able to stay up. The power necessary for sustained flight is proportional to the **wing loading,** which is the weight supported divided by the area of the wings. We know that in scal-

ing up, weight grows with the cube of the length of the bird or plane, and wing area with the square of the length. So the wing loading is proportional to the length of the flying object.

For example, if a bird or plane is scaled up proportionally by a factor of 4, it will weigh $4^3 = 64$ times as much but have only $4^2 = 16$ times as much wing area. So each square foot of wing must support 4 times as much weight.

Second, you have to keep moving. To stay level, an airborne object must fly fast enough to maintain the lift on the wings. The minimum necessary speed is proportional to the square root of the wing loading. Combining this fact with our first consideration, we conclude that the minimum speed goes up with the square root of the length. Our bird that was scaled up by a factor of 4 must fly $\sqrt{4} = 2$ times as fast.

Take, for instance, a sparrow, whose minimum speed is about 20 miles per hour. An ostrich is 25 times as long as a sparrow, so the minimum speed for an ostrich would be $\sqrt{25} \times 20 = 100$ miles per hour. Have you seen any flying ostriches lately? Heavy birds have to fly fast or not at all!

Of course, ostriches are not just scaled up sparrows, nor are eagles. The larger flying birds have disproportionately larger wings than a sparrow, to keep the wing loading down. The largest animal ever to have taken to the air was *Quetzalcoatlus northropi*, a flying reptile of 65 million years ago, with a wingspan of 36 feet and a weight of about 100 pounds.

You have to stay up, you have to keep moving — and you have to get up there. Here basic aerodynamics imposes further limits. Paleontologists originally had thought that *Quetzalcoatlus northropi* weighed 200 pounds and had a 50-foot wingspan. Even though that works out to just about the same wing loading as for 100 pounds and a wingspan of 36 feet, other considerations from aerodynamics show that at 200 pounds, the reptile wouldn't have been able to get off the ground.   ▲

SP TLIGHT 16.2 "Take That, King Richard!"

▶ ▶ ▶ ▶ ▶ ▶ ▶ ▶ ▶ ▶ ▶ ▶ ▶

Shakespeare, following the propaganda of the Tudor historians, painted Richard III as a humpbacked Machiavellian monster. Did Richard have an advantage in armored combat because he was short? That suggestion was made some years ago by one of the leading modern historians of the Tudor era, Garrett Mattingly.

Between a short man and a tall man, height increases by the linear dimension — from 5 feet 2 inches, say, to 6 feet — while the surface of the body increases as the square. Since it's . . . the surface of the body that the armorer must plate with steel, the armor of a short warrior, like Richard, would be lighter than a tall warrior's by a lot more than the few inches' difference in height would indicate. So Richard's notorious deadliness in battle would have been possible at least in part — or so Mattingly's speculation ran — because his armor, while protecting him as well as the big man's, left him less encumbered.

Did wearing armor give an advantage to the shorter warrior?

▶ KEEPING COOL (AND WARM)

Area-volume tension is also of crucial importance to a creature's maintenance of thermal equilibrium. Both warm-blooded and cold-blooded animals gain or lose heat from the environment in proportion to body surface area. A warm-blooded animal usually is losing heat; its equilibrium consumption, or food intake needed to maintain body heat, depends primarily on the amount of its surface area, the temperature of its environment, and the insulation provided by its coat or skin. Other factors being equal, a scaled-up mammal scales up its food consumption by *surface area* (proportional

After the lecture, someone said to Mattingly that he had grasped the right idea — but by the wrong end. Muscle power, the listener claimed, is a matter of bulk — and physical volume goes up by the cube, whereas the surface to be protected goes up by the square. So the large warrior should have more strength left over than the little guy after putting on his armor. And the large warrior, swinging a bigger club, can deliver a far more punishing blow — because the momentum of the club depends on its weight, which goes up with its volume, which means by the cube. Richard was at a terrible disadvantage.

But wait a minute, a second listener said. That's true about the club — but not about the muscles. The strength of a muscle is proportional not to its bulk but to the area of its cross section. And since the cross section of muscles obviously increases by the square, just as the surface of the body does, the big guy, plated out, has no more, or less, advantage over the little guy than if both were naked.

But hang on, a third person interjected — an engineer. That's right about the muscles, but it's not right about the armor. The weight of the armor increases not simply with the increase in the surface area that it must cover but slightly faster. The reason is that to obtain the same strength with a larger area of metal, there must be reinforcing ribs. Or else the metal must be significantly thicker overall. So maybe Richard had an advantage after all.

To maximize protection within the weight, the armorer adopted two strategies: variable thickness and deflection. The unexpected fact is that armor was made as thin as possible. Thickness, reinforcement, and structural stiffening of a large surface were concentrated where opponents' weapons were likely to hit. From these strong, shaped places, the metal tapered away, until the sheet steel was as thin as the lid of a coffee can at the sides of the rib cage beneath the arms, or across the fingers, or at the cheek of a helmet.

(Quoted from Horace F. Judson, *The Search for Solutions*, Johns Hopkins Univ. Press, 1987, pp. 54 - 56.)

to the square of the scaling factor), *not* by *volume* (proportional to its cube).

Mammals regulate their metabolism and maintain a constant internal body temperature. Cold-blooded animals, such as alligators or lizards, have a somewhat different problem. They absorb heat from the environment for energy, but they must also dissipate any excess heat to keep their temperature below unsafe levels. The amount of heat that must be gained or lost is proportional to total volume, because the entire animal must be warmed or cooled. But the heat is exchanged through the skin, so the rate is proportional to surface area.

Figure 16.8 *Dimetrodon* may have evolved a sail to absorb and dissipate heat efficiently. (Courtesy, Field Museum of Natural History.)

Dimetrodon was a large mammal-like reptile that roamed present-day Texas and Oklahoma 280 million years ago (see Figure 16.8). *Dimetrodon* had a great "sail" or fan on its back. As an individual grew, and as the species evolved, the sail grew. But it did not grow according to *geometric similarity*, the kind of growth we refer to as **proportional growth.** Instead, the area of the sail grew precisely in proportion to the volume of the animal, a fact that strongly suggests to paleontologists that the sail was a temperature-regulating organ that was able to absorb or radiate heat. So, an individual *Dimetrodon* twice as long would have eight (= 2³) times as much weight and volume and also a sail with eight times as much area. If it had grown according to geometric similarity, the sail would have been twice as high and twice as wide, and hence would have had only four times as much sail area. Larger specimens of *Dimetrodon* didn't look quite like scaled-up smaller ones; we would say that the sail grew disproportionately large compared to the rest of the animal.

Dimetrodon was a large animal, but the need for heat regulation is even more acute for smaller animals. Like human babies, small animals can lose heat quickly, because of their high ratio of surface

area to volume. Leading paleontologists now believe that birds (most of whom are quite small) evolved from dinosaurs and that feathers are modified reptilian scales. Though not a prevailing view, it has been hypothesized that the wings of birds and insects evolved originally not for flight but as temperature control devices.

▶ SIMILARITY AND GROWTH

Although a large change of scale forces adaptive changes in materials or form, within narrow limits —perhaps up to a factor of 20—creatures can grow according to a law of similarity. They grow in such a way that their shape is preserved. A striking example of such growth is that of the chambered nautilus (*Nautilus pompilius*). Each new chamber that is added onto the nautilus shell is larger but the same shape as the previous chamber, and the shape of the shell as a whole—an *equiangular,* or *logarithmic,* spiral—remains the same (see Figure 16.9).

Most living things grow over the course of their lives by a factor greater than 2. We've seen with *Dimetrodon* that a big specimen was not just a scaled-up small one. Nor is a human adult simply a scaled-up baby. Relative to the length of the body, a baby's head is much larger than an adult's. The arms of the baby are disproportionately shorter than an adult's. In the growth from baby to adult, the body does not scale up as a whole. But different parts of the body scale up, each with a different scale factor. That is, a baby's eyes grow at one rate to perhaps twice their original size, while the arms grow at another rate, to about four times their original size.

Although the laws for growth can be much more complicated than proportional growth (or even the allometric growth we discuss in the next section), more sophisticated mathematics—for example, differential geometry, the geometry of curves and surfaces—permits analysis of complex and interlocking scalings. For a model of the pro-

Figure 16.9 A chambered nautilus shell. (Photo by Nancy Rodger.)

cess in which a baby's head changes shape to grow into an adult head, we can use graph paper: we put a picture of the baby's skull on graph paper, then determine how to deform the grid until the pattern matches an adult skull (see Figure 16.10 and Spotlight 16.3, p. 512). The same idea lies at the heart of computerized "morphing," the process in which the face of one film character can be made to change smoothly into the face of another, with different scalings for different parts of the face.

OPTIONAL ▶ ALLOMETRY

If we measure the arm length or head size for humans of different ages and compare these measurements with body height, we observe that humans do not grow in a way that maintains geometric similarity. The arm, which at birth is one-third as long as the body, is by adulthood closer to two-fifths as long. The head of a newborn baby may be one-third of the baby's length, but an adult's head is usually close to one-seventh of the individual's height.

Ordinary graphing provides a way to test for differential growth. We can plot body height on the horizontal axis and arm length on the vertical axis. After doing so, we get a curve, which indicates that the height is not increasing the same amount each year. If the growth were propor-

tional, that is, according to geometric similarity, we would have gotten a straight line, indicating the same amount of growth every year.

Is there an orderly law by which we can relate arm length to height? Let's plot again, this time using a different scale. For this **logarithmic scale,** we mark off equal units, as usual. But instead of labeling the marked points with 0, 1, 2, 3, etc., we label them with the corresponding powers of 10: $10^0 = 1$, $10^1 = 10$, $10^2 = 100$, $10^3 = 1000$, etc.,

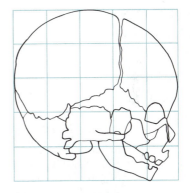

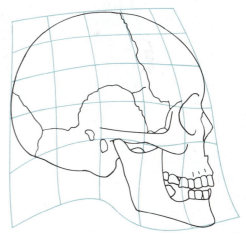

Figure 16.10 Modeling the changes in shape of a human head from infancy to adulthood. (From Richard C. Lewontin, "Adaptation." ©*Scientific American* 239(3): 220, (1978). All rights reserved.)

SP⬤TLIGHT 16.3   Helping to Find Missing Children

▶▶ ▶▶ ▶ ▶ ▶ ▶ ▶ ▶ ▶ ▶ ▶

It can be valuable to be able to predict what a developing organ will look like in the future. For example, what does a child look like now who was kidnapped 2 years ago, at age 3?

At the National Center for Missing and Exploited Children (NCMEC) in Arlington, Virginia, a computer and a more sophisticated version of our graph-paper technique are used to answer such questions. Computer age-progression specialists scan photographs of both the missing child at age 3 and an older sibling or a

biological parent at age 9 into a computer. Then the face of the 3-year-old is stretched, depending on age, to reflect craniofacial growth and merged with the image of the sibling or parent at 9 years old. The result is a rough idea of what the missing child may look like. As mathematicians and biologists refine their models of how faces change over time, this technique will improve. It may even become possible for a child to gain an idea of how he or she may look at age 40 or 65.

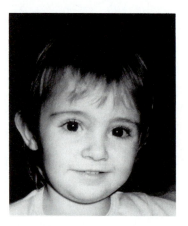

Child at age 3 when abducted.

Age-progression as a 9-year-old.

which are also called **orders of magnitude.** Plotting a point on such a scale is not easy, since the point midway between 1 and 10 is not 5.5, but instead is closer to 3. Special graph paper (available in most college bookstores) marks smaller divisions and makes it easier to plot; paper marked with log scales on both axes is called **log-log paper,** while *semilog* paper has a logarithmic scale

on just one axis. Also, many computer plotting packages can produce logarithmic scales.

To return to our problem: As you may imagine, we could use a logarithmic scale for either arm length or height, or both. If we use logarithmic scales for both, as in Figure 16.11, the data plot closely to a straight line. Actually, looking carefully, we can discern two different straight lines: a

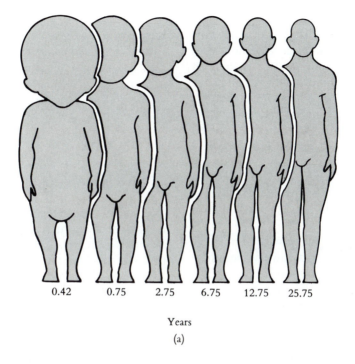

0.42   0.75   2.75   6.75   12.75   25.75

Years

(a)

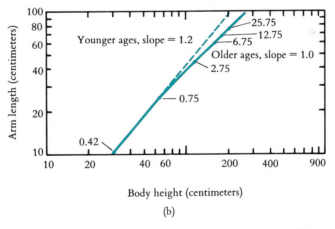

(b)

Figure 16.11 (a) The proportions of the human body change with age. (b) A graph of human body growth on log-log paper. The numbers shown beside the points indicate the age in years; they correspond to the stage of human development shown in part a. (From Thomas A. McMahon and John Tyler Bonner, *On Size and Life,* Scientific American Library, Freeman, New York, 1983.)

steeper one that fits early development, with slope 1.2, and a less steep one that fits development after 9 months of age, with slope 1.0. The change from one line to another at 9 months indicates a change in pattern of growth. The pattern after 9 months, characterized by the straight line with slope 1, is indeed proportional growth (sometimes called **isometric growth**). For the pattern before 9 months, we know from the slope (1.2) being greater than 1 that arm length is increasing relatively faster than height. That earlier growth also follows a definite pattern, called allometric growth.

Allometric growth is the growth of one feature at a rate proportional to a power of another. For the infant before 9 months, arm length grows allometrically with height. We have seen that in geometric scaling, area grows according to the square (second power) and volume according to the cube (third power) of length, so we can say that they grow allometrically with length.

If we denote arm length by y and height by x, a straight-line fit on log-log paper corresponds to the algebraic relation

$$\log_{10} y = B + a \log_{10} x$$

where a is the slope of the line and B is the point where the graph crosses the vertical axis. If we raise 10 to the power of each side, we get

$$y = bx^a$$

where $b = 10^B$. This equation describes a **power curve:** y as a constant multiple of x raised to a certain power.

For a slope $a = 1$, we get $y = bx$, which is a linear relationship describing proportional growth, that is, growth according to geometric similarity. On ordinary graph paper, proportional growth appears as a straight line, allometric growth as a curve. On log-log paper, both patterns appear as straight lines.

The technique of allometry has been used in the last few years by paleontologists to determine that all of the six known specimens of the earliest fossil bird *Archaeopteryx* are from the same species, and that the minute fossil and puzzling fish known as *Palaeospondylus* (found only in Scotland) is probably just the larval stage of some better-known fish.   ◄

► CONCLUSION

We have examined the problem of scale and noted that a large change in scale forces a change in either materials or form. A particular instance of the problem of scale is area-volume tension, and we have seen how an animal's size and geometric shape affect its abilities to move and to keep itself warm or cool.

In this chapter we have explored the limitations of life in three dimensions; in Chapter 20 we will see that dimensionality also imposes surprising limits on artistic creativity in devising patterns.

► REVIEW VOCABULARY

Allometric growth   A pattern of growth in which one feature grows at a rate proportional to a power of another feature.

Area-volume tension   The fact that in scaling up, volume increases faster than area.

Crushing strength   The maximum ability of a substance to withstand pressure without crushing or deforming.

Density   Weight per unit volume.

Isometric growth   Proportional growth.

Logarithmic scale   A scale on which equal divisions correspond to powers of 10.

Log-log paper   Graph paper on which both the vertical and the horizontal scales are logarithmic scales, that is, the scales are marked in orders of magnitude 1, 10, 100, 1000, . . . , instead of 1, 2, 3, 4, ...

Mass The aspect of matter that is affected by forces, according to physical laws.

Orders of magnitude Powers of 10.

Power curve A curve described by an equation $y = cx^n$, so that y is proportional to a power of x.

Pressure Weight divided by area.

Problem of scale As an object or being is scaled up, its area and its volume increase at different rates, forcing adaptations of materials or shape.

Proportional growth Growth according to geometric similarity.

Scaling factor The number by which each linear dimension of an object is multiplied.

Similar Two objects are geometrically similar if they have the same shape, regardless of the materials they are made of. They need not have the same size. Corresponding linear dimensions must have the same factor of proportionality.

Wing loading Weight supported divided by wing area.

▶SUGGESTED READINGS

CAMPBELL, R. B.: "Hercules' height," *The UMAP Journal*, 5(3): 265–269 (1984). Pythagoras calculated the height of Hercules (who must have been a legend already in Pythagoras's time!) by assuming proportional scaling.

DEWDNEY, A. K.: *200% of Nothing: An Eye-Opening Tour through the Twists and Turns of Math Abuse and Innumeracy*. Wiley, New York, 1993.

DRELA, MARK, AND JOHN S. LANGFORD: "Human-powered flight," *Scientific American* 253(5): cover, 144–151, 178 (November 1985).

DUDLEY, BRIAN A. C.: *Mathematical and Biological Interrelations*. Wiley, New York, 1977. Excellent and gentle extended introduction to graphing, scale factors, and logarithmic plots.

GOULD, STEPHEN JAY: "The origin and function of bizarre structures: Antler size and skull size in the 'Irish elk,' *Megaloceros giganteus*," *Evolution* 28(2): 191–220 (1974). Logarithmic plots solve a long-standing mystery.

———: "Size and shape," in *Ever Since Darwin*, Norton, New York, 1977, chapter 21.

———: "Not necessarily a wing," *Natural History*: 12–25 (October 1985). Also in *Bully for Brontosaurus: Reflections in Natural History*, Norton, New York, 1991, chapter 9, pp. 139–151.

HALDANE, J. B. S.: "On being the right size," in *Possible Worlds*. R. West. James R. Newman (ed.), reprinted in *The World of Mathematics*, Vol. 2, Simon & Schuster, New York, 1956, pp. 952–957. Also reprinted in John Maynard Smith (ed.), *On Being the Right Size and Other Essays by J. B. S. Haldane*, Oxford University Press, Oxford, 1985, pp. 1–8. Succinctly surveys area-volume tension, flying, the size of eyes, and even the best size for human institutions.

HILDEBRANDT, STEFAN, AND ANTHONY J. TROMBA: *Mathematics and Optimal Form*, Scientific American Library, New York, 1985.

HOUCK, MARILYN A., JACQUES A. GAUTHIER, AND RICHARD E. STRAUSS: "Allometric scaling in the earliest fossil bird, *Archaeopteryx lithographica*," *Science*, 247: 195–198 (January 12, 1990).

HUXLEY, JULIAN: "The size of living things," in *Man Stands Alone*, Harper Brothers, New York, 1942.

———: *Problems of Relative Growth*, Methuen, London, 1932. Reprinted by Dover, New York, 1972. Many semilog and log-log plots of biological relations.

MCMAHON, T. A., AND J. T. BONNER: *On Size and Life*, Scientific American Library, New York, 1983. Astonishingly beautiful and informative book on the effects of size and shape on living things.

PEARCE, PETER: *Structure in Nature is a Strategy for Design*, MIT Press, Cambridge, Mass., 1978.

SMITH, JOHN MAYNARD: *Mathematical Ideas in Biology*. Cambridge University Press, New York, 1968.

STEVENS, PETER S.: *Patterns in Nature*, Atlantic Monthly Press, Boston, 1974. Splendid treatment of the problem of scale and other physical phenomena in nature: flows, meanders, branching, trees, soap films, cracking, and packing.

THOMPSON, D'ARCY: *On Growth and Form*, Cambridge University Press, Cambridge, England, 1917, 1961. "A discourse on science as though it were a humanity" (J.T. Bonner), this was the first book to describe in quantitative terms the processes of growth and shaping of biological forms.

THOMSON, KEITH STEWART: "The puzzle of *Palaeospondylus*," *American Scientist*, 80: 216–219 (May/June 1992).

TREFIL, JAMES S.: "What would a giant look like?" in *The Unexpected Vista: A Physicist's View of Nature*, chapter 10, pp. 156–171, Scribners, New York, 1983. Explanation of the effects of scaling up. In Trefil's illustration on p. 162, however, the eyes of the giants are unrealistically large.

WENT, F. W.: "The size of man," *American Scientist*, 56(4): 400–413 (1968). Demonstrates a schism between the macroworld and the molecular world, by comparing human life with that of an ant.

WILLIAMS, CHRISTOPHER: *Origins of Form*, Architectural Book Publishing Company, New York, 1981.

WILLIAMS, ROBERT: *The Geometrical Foundation of Natural Structure: A Source Book of Design*, Dover, New York, 1979. Originally titled *Natural Structure*. Eudaemon Press, Moorpark, Calif., 1972.

► EXERCISES

Most of the exercises below require a calculator; one that offers square roots will suffice.

1. Suppose you are printing photographs from negatives of so-called 35-millimeter film, whose frames actually measure 24 by 36 millimeters, which is just under 1 inch by $1\frac{1}{2}$ inches.
   a. First you make some contact prints, which are exactly the same size as the negatives. What is the scaling factor of a contact print?
   b. One enlargement you want to make is to be three times as high and three times as wide as the negative. What is the scaling factor for this print? How does its area compare with the area of the negative?
   c. Considering the negative as measuring approximately 1 inch by $1\frac{1}{2}$ inches, what is the approximate scaling factor for a 4 by 6 print, that is, one that is 4 by 6 inches? What is the area of the print?
   d. The size of so-called 3 by 5 prints can vary, depending on whether the print has a border or not. For a common commercially made print, the size is about $3\frac{1}{16}$ by $4\frac{19}{32}$ inches. For such a print, what is the scaling factor of the enlargement from the negative?

    e. The cost of raw photographic paper is pretty close to exactly propor-
tional to the area of the paper. Suppose you are comparing the cost of
getting 3 by 5 enlargements versus 4 by 6 enlargements, and let's assume for the sake of simplicity that the prints are exactly 3 by 5 inches
and 4 by 6 inches. The smaller prints cost 17 cents each, and the
larger cost 50 cents each. From what you know about scaling factors
and their role in areas, what can you say about the relative cost of the
two kinds of prints?

    f. Based on the amount of paper used, what would you expect a 7 by 10
print to cost, considering the cost of the 3 by 5 prints in part e? Considering the cost of the 4 by 6 prints in part e?

  2. The area of a circle can be expressed in terms of the diameter (the distance across the center from one side to the other, or twice the radius) as

$$\text{Area} = \pi \times (\text{radius})^2 = \pi \times \left(\frac{\text{diameter}}{2} \right)^2$$

If we apply a scaling factor M to the diameter of a circle, then—as in the case of
the square we considered in the text—the area of the scaled circle changes with
M^2, the square of the scaling factor. A natural application of this idea, of course,
is to your local pizza parlor and the prices on its menu. The actual prices at the
pizza restaurant closest to Beloit College are $5, $6, $6.95, and $7.95, respectively, for small (10-inch), medium (12-inch), large (14-inch), and extra large
(16-inch) cheese pizzas.

    a. What is the scaling factor for an extra large pizza compared to a small one?

    b. How many times as large in area is the extra large pizza compared to
the small one?

    c. How much pizza does each size give per dollar? What "hidden" as-
sumptions are you making about how the pizzas are scaled up?

    d. The corresponding prices for a pizza with "the works" are $8.25,
$9.75, $11.95, and $13.95. Is there any size of these for which you get
more pizza per dollar than some size of the cheese pizzas?

  3. Toy trains, sometimes called model trains, come in various sizes or
gauges. Not all toy trains are exact scale models of real trains, but some are.

    a. HO-gauge toy trains are usually built to an exact scale of 1 to 87,
meaning that a part 1 foot long on the real train will be one
eighty-seventh of a foot long on the toy train. What is the scaling factor of an HO-gauge toy train?

    b. How does the volume of a real boxcar compare with the volume of an
HO-gauge scale model?

    c. O-gauge toy trains are built to a scale of approximately $\frac{1}{4}$ inch to a
foot, meaning that a part 1 foot long on the real train will be one-
fourth of an inch long on the toy train. (In fact, O-gauge trains tend to

be a little shorter than exact scale would demand, and their wheels are oversized compared to exact scale.) What is the scaling factor of an O-gauge toy train?

4. Doll houses and their furnishings are customarily built to a scale of exactly 1 inch to 1 foot, meaning that an item 1 foot long in a real house is 1 inch long in a doll house.

    a. What is the scaling factor for a doll house?

    b. If a doll house were made of the same materials as a real house, how would their weights compare?

5. Two geometric figures are *similar* if they have the same shape but not necessarily the same size. Indicate whether the geometric figures described below are always, sometimes, or never similar:

    a. Two squares

    b. Two isosceles triangles

    c. Two equilateral triangles

    d. Two pentagons

    e. Two regular pentagons

    f. Two rectangles

    g. A square and a rectangle

    h. Two circles

    i. A regular pentagon and a regular hexagon

    j. Two angles

6. Identify each of the following statements as either true or false:

    a. Every polygon is similar to itself.

    b. If polygon A is similar to polygon B and polygon B is similar to polygon C, then polygon A is similar to polygon C.

    c. Corresponding interior angles of similar polygons are congruent.

7. One of the famous problems of Greek antiquity was the *duplication of the cube*. Our knowledge of the history of the problem comes down to us from the third century B.C. from Eratosthenes of Cyrene, who is famous for his estimate of the circumference of the earth (see Chapter 18). According to him, the citizens of Delos were suffering from a plague. They consulted the oracle, who told them that to rid themselves of the plague, they must construct an altar to a particular god that would be geometrically similar to the existing one but double the volume.

    a. How would the volume of the new altar compare with the old if each of its linear dimensions were doubled?

    b. What should the scaling factor be for the new altar?

(The actual problem intended by the oracle was to construct with straightedge and compasses a line segment with this scaling factor as its length, a task that was shown in the nineteenth century to be impossible. Eratosthenes relates that the Delians interpreted the problem in this sense, were perplexed, and went to ask Plato about it; Plato told them that the god didn't really want an altar of double the volume but wished to shame them for their "neglect of mathematics and their contempt for geometry.")

8. The Susan B. Anthony dollar coin was a failure with the U.S. public, who found it too small and light. Suppose you have been put in charge of designing a new dollar coin that is to be made of the same material as the current U.S. 25-cent piece ("Liberty quarter") and weigh four times as much. A quarter can be described geometrically as a circular cylinder approximately $\frac{15}{16}$ inch in diameter and $\frac{1}{16}$ inch thick. Since your new dollar should weigh four times as much, it will need to have four times the volume of a quarter. (You may find it helpful that the formula for the volume of a cylinder is $\pi \times$ (diameter/2)$^2 \times$ height.)

   a. A member of your public advisory panel suggests that the requirements will be fulfilled if you just double the diameter and double the thickness. What do you tell this individual, in the most diplomatic terms?

   b. If you go along with the member's suggestion to double the diameter, how thick does the coin need to be?

   c. Another member of the board feels that the resulting coin would be too large in diameter to be convenient and proposes instead that you scale up the quarter proportionally (she took a course from the first edition of this book). What would the dimensions be for this new dollar?

9. Criticize the following statement, which appears on sacks of the product, and write a correct version:

   Erin's Own Irish sphagnum moss peat. It
   enriches your soil and makes your growing
   easier. Compressed to $2\frac{1}{2}$ times normal volume.

10. Criticize the following claims, which were cited in the *New York Times* of 9/25/87 and 10/21/87:

   a. A new dental rinse "reduces plaque on teeth by over 300%."

   b. An airline working to decrease lost baggage has "already improved 100% in the last six months."

   c. "If interest rates drop from 10% to 5%, that is a 100% reduction."

11. In Germany, the fuel efficiency of cars is measured in terms of liters of gasoline used per 100 kilometers traveled. On a recent trip there, driving a subcompact car, we averaged 6.5 liters per 100 km. What is the equivalent in miles per gallon?

12. A *light-year* is a measure of distance: the distance that light travels in a year.

   a. How long is a light-year in kilometers?

   b. In miles?

   c. In angstroms? (1 angstrom $= 10^{-10}$ m)

13. Consider a real locomotive that weighs 88 tons and an HO-gauge scale model of it. (See Exercise 3 above.)
    a. How much would an exact scale model weigh, in tons?
    b. What assumptions are involved in your answer to part a?
    c. How much would an exact scale model weigh, in pounds?
    d. In kilograms?
    e. In metric tonnes? (1 metric tonne = 1000 kg)

14. An ad for a software package for data analysis on the Apple II included a data set on tropical rain forests and deforestation. The data were given in hectares and were accompanied by the statement, "A hectare equals 10,000 square miles or 2471 acres." What conversion factors should have appeared instead?

15. Gasoline is sold in the United States by the U.S. gallon and in Canada by the liter. (1 U.S. gallon = 231 cu in; 1 liter = 1000 cm³) What is the equivalent cost, in U.S. dollars per U.S. gallon, for gasoline in Canada priced at 65 Canadian cents per liter, when one Canadian dollar exchanges for 78 cents U.S.?

16. In 1991, Edward N. Lorenz, a meteorologist who was an early researcher into chaos and dynamical systems, received the Kyoto Prize in Basic Sciences, consisting of a gold medal and ¥45 million. If U.S. \$1 = ¥125 at the time, what was the value of the cash award in U.S. dollars?

17. In connection with Eratosthenes' measurement of the circumference of the earth, Chapter 18 discusses the Greek measuring unit of a *stadium*. In *The American Heritage Dictionary*, Second College Edition (Houghton-Mifflin, Boston, 1982), we read for the second meaning of *stadium*: " An ancient Greek measure of distance . . . equal to about 185 kilometers, or 607 feet." The name of the unit came from the length of a racecourse that was a bit less than an eighth of a mile long. If the numbers in the definition are correct, what are the correct units that should have appeared?

18. The weight of a 1-foot cube of steel is 500 pounds. What is the pressure on the bottom face in
    a. Pounds per square inch?
    b. Atmospheres? (1 atmosphere = 14.7 pounds per square inch)

19. In an article on adding organic matter to soil, the magazine *Organic Gardening* (March 1983) said, "Since a 6-inch layer of mineral soil in a 100-square-foot plot weighs about 45,000 pounds, adding 230 pounds of compost will give you an instant 5% organic matter."
    a. What is the density of the mineral soil, according to the quotation?
    b. How does this density compare with that of steel?
    c. How do you think the quotation should be revised to be accurate?

20. A mature gorilla weighs 400 pounds and stands 5 feet tall.
    a. Give an estimate of its weight when it was half as tall.
    b. What assumptions are involved in your estimate?
    c. A mature gorilla's two feet together have a combined area of about 1

square foot. When the gorilla is standing on its feet, what is the pressure on its feet, in pounds per square inch?

21. Suppose King Kong is a gorilla scaled up with a scaling factor of 10.
    a. How much does the King weigh?
    b. What is the pressure on the King's feet, in pounds per square inch?

22. You may have wanted to have a waterbed, but found that waterbeds were not allowed in your building. Apart from the danger of flood if the bed should puncture or leak, there is the consideration of the weight.
    a. If a queen-size mattress is 80 inches long by 60 inches wide by 12 inches high, and water weighs 1 kg per liter, how much does the water in the mattress weigh in pounds?
    b. If the weight of the mattress and frame is carried by four legs, each 2 inches by 2 inches, what is the pressure, in pounds per square inch, on each leg?
    c. How does the pressure on the legs of the waterbed compare with the pressure that a person exerts on their feet — for example, a 130-lb person with a total foot area of about one-quarter of a square foot?
    d. If you aren't allowed to have a waterbed, how about a spa (hot tub)? Find the weight of the water in a spa that is in the shape of a cylinder 6 ft in diameter and 3.5 ft deep. (Hint: The volume of a cylinder is $\pi r^2 h$, where r is the radius and h is the height.)

23. What does the largest giant sequoia tree weigh? Model the tree as a (very elongated) cone, supposing that the tree is 360 ft high and has a circumference of 40 ft at the base, and that the density of the wood is 62 lb/cu ft.

24. (Adapted from "Animal Form or Keeping Your Cool," by George Knill and George Fawcett, *Mathematics Teacher*: 395–397 (May 1982). The movie *Them* features enormous ants (about 8 m long and about 3 m wide). We can investigate how feasible such a scaled-up insect is by considering its oxygen consumption. A common ant, which is about 1 cm long, needs about 24 milliliters of oxygen per second for each cubic centimeter of its volume. Since an ant does not have lungs, it must absorb the oxygen through its "skin," which it can do at a rate of about 6.2 milliliters per second per square centimeter. We may suppose that the tissues of a scaled-up ant would have the same need for oxygen for each cubic centimeter, and that its skin could absorb oxygen at the same rate, as a normal ant. Compared to a common ant, how many times as large is an enormous ant's
    a. length?
    b. surface area?
    c. volume?
    d. What proportion of such an ant's oxygen need could its skin supply? What can you conclude about the existence of such insects?

25. In the children's story *Peter Pan*, Peter and Wendy can fly. We can suppose that they are 4 feet tall, so they are about 12 times as tall as a sparrow is long. What should their minimum flying speed be?

26. Icarus of Greek legend escaped from Crete with his father, Daedalus, on wings made by Daedalus and attached with wax. Against his father's advice, Icarus flew too close to the sun; the wax melted, the wings fell off, and he fell into the sea and drowned. What must have been his minimum cruising speed? What assumptions does your answer involve?

▲ 27. Goliath (of David and Goliath, as related in the Bible in I Samuel 17:4) was "six cubits and a span." A span was originally the distance from the tip of the thumb to the tip of the little finger when the hand is fully extended, about 9 inches. What range of heights would this indicate for Goliath, in feet and inches? In centimeters?

▲ 28. According to classical Greek sources, Pythagoras (sixth century B.C.) used geometric scaling to model the height of Hercules, the most heroic figure in classical mythology, in the epic poems of Homer. Pythagoras compared the lengths of two racecourses, one (according to tradition) paced off by Hercules and the other by a man of average height. Both were 600 "paces" long, but the one established by Hercules was longer because of Hercules' longer stride (600 "Herculean" paces vs. 600 paces by a normal man). A normal man in the time of Pythagoras would have been about 5 ft tall.

    a. If the distance paced off by Hercules was 30% longer than the other racecourse, how tall was Hercules? What does your calculation assume?

    b. In fact, the ancient sources do not give the original data but only the two conflicting answers that Hercules was 4 cubits tall and 4 cubits and 1 foot tall. A cubit was supposed to be the distance between a person's elbow and the tip of the middle finger of the person's outstretched arm, much as a "foot" was originally the length of a person's foot. So the measurement depended on the person, though there was some attempt at standardization. Although the length of a Greek cubit is not known precisely, most estimates place it between 17 and 22 inches. What range does this give for the height of Hercules, in feet and inches? In centimeters?

■ 29. Recent years have seen the beginnings of human-powered controlled flight, in the *Gossamer Condor* and other superlightweight planes. The *Gossamer Condor* is far longer than an ostrich, but it flies at only 12 miles per hour. How can it?

■ 30. Jonathan Swift's Gulliver also traveled to Lilliput, where the Lilliputians were human-shaped, but only about 6 inches tall. In other words, they were geometrically similar in shape to ordinary human beings but only one-twelfth as tall. What would a Lilliputian weigh?

    Are Lilliputians ruled out by the size-shape and area-volume considerations in this chapter? If you think they are, what considerations do you find convincing? If not, why not?

▲ Advanced exercise.   ■ Discussion exercise.   ● Optional exercise.

■ 31. [Contributed by Charlotte Chell of Carthage College, Kenosha, Wisconsin.] A 6-ft indoor holiday tree needs four strings of lights to decorate it. How many strings of lights will be needed for an outdoor tree that is 30 ft high?

■ 32. What would you expect an individual *Quetzalcoatlus northropi* to weigh if it had half the wingspan of an adult? If an individual weighed half as much as an adult, what would you expect its wingspan to be?

● 33. Consider the data for weight and metabolic rate of various mammals:

	Weight (kilograms)	Calories per kilogram
Guinea pig	0.7	223
Rabbit	2	58
Human	70	33
Horse	600	22
Elephant	4,000	13
Whale	150,000	1.7

Graph these data on log-log paper. What can you conclude from your graph? [Note that you may have to "create" your own log scale on the weight axis, since the weights span six orders of magnitude. Commercial log paper is usually limited to 2, 3, or perhaps 5 orders of magnitude ("cycles").] If you don't have log-log paper available, use a calculator to take the logarithms of all the numbers and graph these values on ordinary graph paper; this applies to Exercise 34 too.

● 34. Listed in the table below are the winning times in the 1983 World Rowing Championships for rowing shells with one, two, four, and eight oars. The men's times are for 2000 meters, the women's for 1000 meters. Convert the times to speed. For men and women separately, plot speed versus number of oarsmen on ordinary graph paper, and then on log-log paper. Is the relationship proportional? Allometric?

Fit the best line you can through the log-log data and estimate its slope. Compare your results with the theoretical discussion and data in McMahon and Bonner's *On Size and Life*, Scientific American Library, 1983, pp. 42–47 and 67.

Event	Number of oars	Men (2000 meters)	Women (1000 meters)
Single sculls	1	6:49.75	3:36.51
Pairs without coxswain	2	6:35.85	—
Fours without coxswain	4	6:14.83	3:26.68
Eights (with coxswain)	8	5:34.39	2:56.22

▶WRITING PROJECTS

1. A human infant at birth usually weighs between 5 and 10 pounds and has a height (length) between 1 and 2 feet, with the shorter babies having the lesser weight. Considering the weight and height of an adult human, give an argument that human growth must not be just proportional growth.

2. Use algebra to demonstrate that for proportional growth, "area scales as volume to the two-thirds power."

3. The principle that area scales with the square of length, and volume with the cube, has important consequences for the depiction and interpretation of data in graphic form. Suppose we wish to indicate in an artistic way that the weekly income of a U.S. carpenter is twice that of a carpenter in (mythical) Rotundia. We draw one money bag for the Rotundian and another one "twice as Large" for the American. [Illustration from Darrell Huff, *How to Lie with Statistics,* Norton 1954, p. 69.]

What's the problem? Well, first, people tend to respond to graphics by comparing *areas*. Since the larger moneybag is twice as high and twice as wide as the smaller one, the image of it on the page has four times the area. Second, we are used to interpreting depth and perspective in drawings in terms of three-dimensional objects. Since the larger bag is also twice as thick as the smaller, it has eight times the volume. The graphic leaves the subconscious impression that the U.S. carpenter earns *eight* times as much, instead of twice as much.

With these ideas in mind, evaluate the depictions of data on the two following pages. [Illustrations reproduced or adapted from Edward R. Tufte, *The Visual Display of Quantitative Information,* Graphics Press, 1983, pp. 55, 57, and 70.]

a.

Comparative Annual Cost per Capita for care of Insane in Pittsburgh City Homes and Pennsylvania State Hospitals.

Pittsburgh Civic Commission, *Report on Expenditures of the Department of Charities* (Pittsburgh, 1911), p. 7.

b.

This line, representing 18 miles per gallon in 1978, is 0.6 inches long.

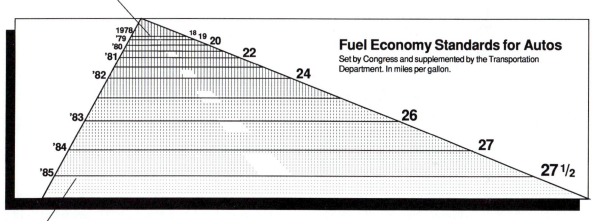

This line, representing 27.5 miles per gallon in 1985, is 5.3 inches long.

New York Times, August 9, 1978, p. D2

c.

1958—Eisenhower

1973—Nixon

1963—Kennedy

1978—Carter

1984—Reagan

1968—Johnson

1990—Bush

1993—Clinton

4. As in Writing Project 3, evaluate the depictions on the facing page and the page following. [Illustrations reproduced or adapted from Edward R. Tufte, *The Visual Display of Quantitative Information*, Graphics Press, 1983, pp. 62 and 69.]

a.

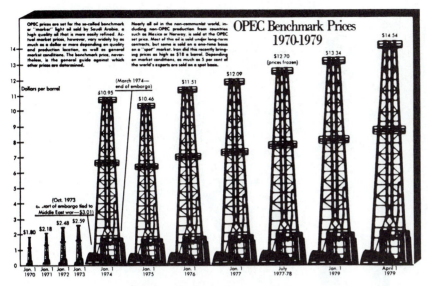

Washington Post, March 28, 1979, p. A–18.

b.

Time, April 9, 1979, p.57

c.

THE SHRINKING FAMILY DOCTOR
In California

Percentage of Doctors Devoted Solely to Family Practice

1964	1975	1990
27%	16.0%	12.0%

1: 4,232
6,212

1: 3,167
6,694

1: 2,247 RATIO TO POPULATION
8,023 Doctors

17

Geometric Growth

Many of the problems that we face relate to populations and their changes over time. We all have a stake in the problems associated with human population growth, such as hunger and disease. Our food supplies are affected by the growth and behavior of nonhuman populations such as bacteria, locusts, and rats. Even inanimate populations affect us. Growing "populations" of household refuse and nuclear waste pose disposal and storage issues, with accompanying environmental questions, while depletion of natural resources causes other concerns. Similarly, but more favorably, a growing population of dollars in a bank account can provide resources that enrich our lives.

Geometric growth is the key mathematical concept behind the growth of populations of all kinds. Understanding the mathematics of geometric growth is essential for realizing just how severe the economic and social issues caused by growth can become, and for measuring the effectiveness of policies to alter patterns of growth. To analyze the growth of a population, we will concentrate on the questions:

▶ How big is the population?
▶ How fast is it growing or shrinking?
▶ How is its structure or makeup changing?

We will use the term **population growth** to refer to both increases (positive growth) and decreases (negative growth) in population size. The term **population structure** refers to the divisions of a population into subgroups. For example, human populations are frequently described according to age structure, as in U.S. census data. For a specific problem it may be advantageous to break a population down according to economic, social, or educational criteria.

In this chapter we pay particular attention to the size of a population and to the way its size changes over time. We investigate models of population growth of two different kinds. In one kind, the population grows at a rate proportional to its current size, so that, for example, when it is twice as large, it is growing twice as fast. In the other kind of population, the increase in the population is the same in each time interval. For examples, we focus on two seemingly different kinds of populations — financial and biological — in order to illustrate the broad application of our models.

▶ GEOMETRIC GROWTH AND FINANCIAL MODELS

We begin our study with a population whose structure and behavior are relatively simple — the population of dollars in a bank account. We deposit money in a savings account; our primary concerns are the safety and the growth of such savings. Suppose that we deposit $1000 in an account that, we are told, "pays interest at a rate of 10%, compounded and paid annually." Assuming that we make no other deposits or withdrawals, how much is in the account after 1, 2, or 5 years?

The $1000 is usually called the **initial balance** or the **principal** of the account. At the end of one year, **interest** is added. The amount of interest is 10% of the principal, $100 in this case. So the balance at the beginning of the second year is $1100. During the second year the interest is also 10% — not of the *initial* balance of $1000 but of the *new* balance of $1100 — so at the end of the second year, 10% of $1100, or $110, is added to the account.

Notice that during the second year we earn interest on both the initial balance of $1000 and on the $100 interest earned during the first year. Interest that is paid on both the principal and on the accumulated interest is known as **compound interest.** The simple but remarkable consequence is that we receive more interest during the second year than during the first, that is, the account grows by a greater amount during the second year. At the beginning of the third year the account contains $1210, so at the end of the third year we

receive $121 in interest. Again this is larger than the amount we received at the end of the preceding year. Moreover, the increase during the third year

Third-year interest − second-year interest
$$= \$121 - \$110 = \$11$$

is larger than the increase during the second year

Second-year interest − first-year interest
$$= \$110 - \$100 = \$10$$

Thus, not only is the account balance increasing each year, but the amount added also increases each year. This type of growth is called **geometric growth** or **exponential growth.**

There is another way to pay interest, called **simple interest.** In this method, interest is paid only on the original balance, no matter how much interest has accumulated. With simple interest, for an account with $1000 and a 10% interest rate, we receive $100 interest at the end of the first year; so at the beginning of the second year, our account contains $1100, as before. But at the end of the second year, we again receive only $100; so at the beginning of the third year, our account contains

TABLE 17.1   The Growth of $1000: Compound Interest versus Simple Interest

Years	Amount in account from compounded interest	Amount from simple interest
1	1100.00	1100.00
2	1210.00	1200.00
3	1331.00	1300.00
4	1464.10	1400.00
5	1610.51	1500.00
10	2593.74	2000.00
20	6727.50	3000.00
50	117,390.85	6000.00
100	13,780,612.34	11,000.00

$1200. In fact, at the end of each year we receive just $100 in interest. Clearly this method yields less than if interest were compounded.

Although the simple-interest method is seldom used in today's competitive financial markets, we frequently observe this type of growth, called **simple growth** or **arithmetic growth,** in other contexts. If you withdraw the interest paid on an account at compound interest, then there is no difference — as far as the amount remaining in the account is concerned — between compound interest and simple interest.

The amounts in accounts paying interest at the rate of 10% per year with compound and simple interest are shown in Table 17.1 and in the graph in Figure 17.1. The impressive growth associated with compound interest is dramatically illustrated by these figures, as is the distinction between geometric and arithmetic growth. It is this distinction that led the demographer and economist Thomas Malthus (1766–1834) to his famous theory that human populations grow geometrically, whereas food supplies grow arithmetically (see Spotlight 17.1, p. 532).

Populations that grow by the same amount in each time interval follow the arithmetic growth model. The population of medical doctors in the United States grows arithmetically, since the fixed number of medical schools each graduate about the same total number of doctors each year. On the other hand, general human populations tend to grow geometrically because the number of children born increases as the population increases.

The situation of nuclear waste generated by a nuclear power plant is more complicated. The absolute volume of waste added each year depends on the fixed size and output of the power plant, not on the growing amount of waste in storage. Hence the volume of waste grows arithmetically. What about the total amount of *radioactive* material in the storage dump? The waste is a mixture of radioactive and nonradioactive substances; over time, the radioactive ingredients decay slowly into nonradioactive ones. While the radioactivity of waste already in storage is decreasing, new amounts of radioactive material are being added each year. The situation requires a hybrid model that incorporates negative geometric growth (radioactive decay) accompanied by positive arithmetic growth (adding to the dump). The situation is like turning on the faucet to the bathtub while leaving the drain hole open; what happens to the height of water in the tub depends on how fast water runs in versus how fast it can run out.

▶ THE MATHEMATICS OF GEOMETRIC GROWTH

Suppose we have the option of depositing our $1000 in a bank that pays interest at the rate of 10% per year but compounds the interest quarterly, that

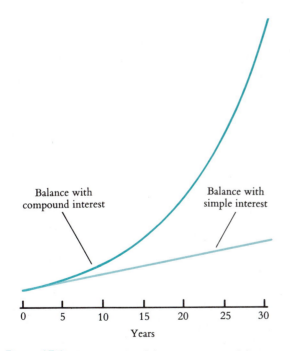

Figure 17.1 The growth of $1000: compound interest and simple interest.

SPOTLIGHT 17.1   Thomas Malthus

▶ ▶ ▶ ▶ ▶ ▶ ▶ ▶ ▶ ▶ ▶ ▶ ▶ ▶

Thomas Malthus (1766–1834), a nineteenth-century English demographer and economist, based a well-known prediction on his perception of the differences in the rates of growth of two populations, the human population and the "population" of food supplies.

His belief was that human populations increase geometrically, while food supplies increase arithmetically — so that the increase in food supplies will eventually be unable to match increases in population. He concluded, however, that over the long run there would be restrictions on the natural growth of human populations too, including war, disease, and starvation — hardly an optimistic forecast and, doubtless, responsible for the dreary image associated with his views.

Thomas Malthus. (The Bettman Archive/BBC Hulton.)

---

is, four times per year. With an interest rate of 10% per year, we are entitled to one-fourth of that rate, or 2.5%, each quarter. In this case, the quarter (3 months) is the **compounding period,** the time elapsing before interest is calculated.

Let's see what happens during the first year. At the end of the first quarter, we have the original balance plus $25 interest, so the balance at the beginning of the second quarter is $1025. During the second quarter we receive interest equal to 2.5% of $1025, or $25.63, so our new balance at the end of the second quarter is $1050.63. Continuing in this manner, we find that the balance at the end of the first year is $1103.81.

Note that even though the account was advertised as paying 10% interest, the interest we receive over the course of the year, $103.81, amounts to more than 10% of our principal — in fact, to 10.381%. The 10% rate is known as the **nominal (annual) rate** or **annual interest rate;** the 10.381% rate is the **effective (annual) rate** or **annual equivalent yield.**

If interest were compounded monthly (12 times per year) or daily (365 times per year), the resulting balance would be still larger. A comparison of yearly, quarterly, monthly, and daily compounding for an interest rate of 10% is shown in Table 17.2.

We will shortly summarize our results in a general formula. From now on, we express all interest rates as fractions. For example, 10% will be expressed as 0.10; to convert a percentage to a fraction, we divide the percentage by 100, which means moving the decimal point two places to the left. An interest rate of 0.10 is the same as 10%, or $\frac{10}{100}$; an interest rate r is the same as $100r\%$, or (the rate in %)/100.

We generalize from our observations for annual and monthly compounding. For annual compounding, we found that at the end of one year we have

Initial balance + interest
$$= \$1000 + \$1000(0.10)$$
$$= \$1000(1 + 0.10)$$

and for quarterly compounding we have at the end of the first quarter

Initial balance + interest
$$= \$1000 + \$1000(0.025)$$

and at the end of the second quarter

Initial balance + interest
$$= \$1000 + \$1000(0.025)$$
$$+ [\$1000 + \$1000(0.025)] \, (0.025)$$
$$= [\$1000 + \$1000(0.025)]$$
$$\times (1 + 0.025)$$
$$= \$1000(1 + 0.025)(1 + 0.025)$$
$$= \$1000(1 + 0.025)^2$$

The pattern continues in this way, so that we have $\$1000(1 + 0.025)^4$ at the end of the fourth quarter.

In the more general setting, with an initial balance of P and an interest rate $r \, (= 100r\%)$ per compounding period, we have at the end of the first compounding period

$$P + Pr = P(1 + r)$$

This amount can be viewed as a new starting balance. Hence, in the next compounding period the amount $P(1 + r)$ grows to

$$P(1 + r) + P(1 + r)r = P(1 + r)(1 + r)$$
$$= P(1 + r)^2$$

The pattern continues, and we reach the following conclusion:

If a principal P is deposited in an account that pays interest at the rate r per compounding period, then after n compounding periods the account contains $P(1 + r)^n$.

Given an annual interest rate, we can determine the amount in an account with interest compounded quarterly, monthly, or according to any other compounding period, by using our formula with an interest rate r per compounding period and with the total elapsed time expressed in terms of the number n of compounding periods.

EXAMPLE: Compounding Interest for 10 Years

Suppose we have a principal of $P = \$1000$ with interest at an annual rate of 10%. Using the formula $P(1 + r)^n$, we can determine the amount in the account after 10 years using several different compounding periods:

▶ *Annual compounding.* If the compounding is done once a year, then the interest rate of 10% per year gives $r = 0.10$, and we determine the amount in the account after 10 years to be

$$1000(1 + 0.10)^{10} = 1000(1.10)^{10}$$
$$= 1000(2.59374)$$
$$= 2593.74$$

TABLE 17.2   Comparing Compound Interest: The Value of $1000, at 10% Annual Interest, If Interest Is Compounded

Years	Compounded yearly	Compounded quarterly	Compounded monthly	Compounded daily	Compounded continuously
1	1100.00	1103.81	1104.71	1105.16	1105.17
5	1610.51	1638.62	1645.31	1648.61	1648.72
10	2593.74	2685.06	2707.04	2717.91	2718.28

▶ *Quarterly compounding.* If the interest is compounded every quarter, then $r = 0.10/4 = 0.025$, and after 10 years (or 40 quarters) the account contains

$$1000\left(1 + \frac{0.10}{4}\right)^{40} = 1000(1.025)$$
$$= 1000(2.68506)$$
$$= 2685.06$$

▶ *Monthly compounding.* Finally, if compounding is done monthly, then $r = 0.10/12 = 0.008333$. The amount in the account after 10 years, or 120 months, is

$$1000\left(1 + \frac{0.10}{12}\right)^{120} = 1000(1.008333)^{120}$$
$$= 1000(2.70704)$$
$$= 2707.04$$

These entries are found in the last row of Table 17.2. ▲

▶ A LIMIT TO COMPOUNDING

The rows in Table 17.2 illustrate a general trend: for a fixed interest rate, more frequent compounding results in larger ending balances. Thus, as you move from left to right in any row of the table, the amounts increase. The amounts in an account that result from more and more frequent compounding do not grow without bound, however; instead, they get closer and closer to an amount that can be determined in advance from the interest rate. This amount is shown in the far right column in each row. There is a limit to the growth. For a given rate of interest (corresponding to one row of Table 17.2), no matter how frequently the compounding is done, the account cannot grow beyond the amount shown in the far right column.

Why is this so? Basically, because the extra interest that comes from more frequent compounding is *interest on interest.* For example, in the first row of Table 17.2, the $3.81 extra interest from compounding quarterly instead of yearly is interest on the $100 interest. The $3.81 is less than 10% of the $100 because the $100 interest is not on deposit for the whole year, since only part of it is credited to the account (and hence begins earning interest) at the end of each quarter. As the frequency of compounding increases, smaller and smaller amounts of interest on interest are added, so there is a limit to the total that can accumulate.

Let's go beyond an intuitive understanding of this phenomenon and investigate it quantitatively. In particular, how can we determine the limiting amount in the far right column of Table 17.2? To make things simple, let's suppose that we have an initial balance of $1, and we keep track at all stages of even the smallest fractions of a dollar. (A bank would round down to the nearest cent at each compounding period.)

We first suppose the absurdly high interest rate of 100% per year compounded n times per year, and then examine interest rates closer to the ones that we usually see in stable economies. For an initial balance of $1, the amount that we have at the end of one year is $\$(1 + 1.00/n)^n$. As n increases, this amount, which is just $(1 + 1/n)^n$, gets closer and closer to a special number called e. This is illustrated in the following table, where the dots (ellipses) indicate that more decimal places follow.

n	$(1 + 1/n)^n$
1	2.0000000...
5	2.4883200...
10	2.5937424...
50	2.6915880...
100	2.7048138...
1,000	2.7169239...
10,000	2.7181459...
100,000	2.7182682...

The value of e is approximately 2.7182818. This is the maximum amount that an initial $1 can grow to in one year at 100% interest, the amount that is produced by continuous compounding.

We did the calculations earlier for an interest rate of 100%. For a general interest rate r, the amount at the end of one year that $1 grows to when compounded n times during the year is $(1 + r/n)^n$. The corresponding limiting amount, as n grows large, is e^r, corresponding to continuous compounding. The effective rate is $(e^r - 1)$.

EXAMPLE: Continuous Compounding of Interest

For $1000 at an annual interest rate of 10%, compounded n times in the course of a single year, the balance at the end of the year is $(1 + 0.10/n)^n$. This quantity gets closer and closer to $e^{0.1} = 1.10517$ . . . as the number of compoundings n is increased. No matter how frequently interest is compounded — daily, hourly, every second, infinitely often ("continuously") — the original $1000 at the end of one year cannot grow beyond $1105.17. The corresponding values for other years are shown in the last column of Table 17.2.

If a principal P is deposited in an account that pays interest at the (nominal) annual rate of $r = 100r\%$ compounded continuously, then after 1 year the account contains Pe^r; after m years, it contains Pe^{rm}. ▲

EXAMPLE: Daily Compounding— Not So Simple

Most banks compound savings interest daily, but they don't all use the same formula. There are 365 days in a year, so some banks use 365 compounding periods per year (366 in leap years); but other banks use 360 compounding periods, a holdover from the days when interest calculations were done by hand and it was computationally convenient to assume 12 months of 30 days each. Independently of the decision about the number of compounding periods, banks then decide to divide the nominal rate r by either 365 or 360 to get the daily interest rate for each compounding period.

Surprisingly, all four possible combinations of these two independent choices are allowed by law, leading to the four different daily compounding formulas in the top half of Table 17.3. For the same nominal rate r, the method most favorable to the investor is "365 over 360," the two methods "360 over 360" and "365 over 365" differ insignificantly, and the method "365 over 360" has the lowest effective rate. ▲

TABLE 17.3 Methods for Daily Compounding of Interest

Method	Compounding periods per year	Rate per period	Formula for one year	Effective rate	Effective rate for $r = 5\%$
360 over 360, daily	360	$r/360$	$P\left(1 + \dfrac{r}{360}\right)^{360}$	$\left(1 + \dfrac{r}{360}\right)^{360} - 1$	5.12674%
365 over 365, daily	365	$r/365$	$P\left(1 + \dfrac{r}{365}\right)^{365}$	$\left(1 + \dfrac{r}{365}\right)^{365} - 1$	5.12675%
365 over 360, daily	365	$r/360$	$P\left(1 + \dfrac{r}{360}\right)^{365}$	$\left(1 + \dfrac{r}{360}\right)^{365} - 1$	5.19977%
360 over 365, daily	360	$r/365$	$P\left(1 + \dfrac{r}{365}\right)^{360}$	$\left(1 + \dfrac{r}{365}\right)^{360} - 1$	5.05477%
360 over 360, continuously			$Pe^{360r/360} = Pe^r$	$e^r - 1$	5.12711%
365 over 365, continuously			$Pe^{365r/365} = Pe^r$	$e^r - 1$	5.12711%
365 over 360, continuously			$Pe^{365r/360}$	$e^{365r/360} - 1$	5.20014%
360 over 365, continuously			$Pe^{360r/365}$	$e^{360r/365} - 1$	5.05513%

EXAMPLE: Options for Continuous Compounding

Still other banks offer continuous compounding (and advertise the fact). What they are actually offering is continuous compounding within each compounding period. But not all continuous compounding is the same! Here, too, banks may elect any one of the four options of "360 over 360" and so forth to use in their formula (see the bottom half of Table 17.3). The options of "360 over 360" and "365 over 365" lead to exactly the same result as the continuous compounding that we investigated earlier. The other two methods produce slightly different results, because in effect they are working with different nominal rates. The differences in effective rate among the four options are much greater than the tiny differences resulting from

whether the compounding is done daily or continuously.

(Caution: You may not be able to reproduce exactly the results of Table 17.3 on your calculator, due to rounding at intermediate steps because of the calculator's limited number of digits of precision.) ▲

▶ GROWTH MODELS FOR BIOLOGICAL POPULATIONS

We can now use a *geometric* growth model to make rough estimates about sizes of human populations. Indeed, projections reported in the popular media are based on such a model; they use for the growth rate r the difference between the birth rate and the death rate, for which the technical term is the **rate of natural increase.** In the terminology that we have been using for financial models, this is the effective rate, so we may think of it as a growth rate that is compounded annually.

A shortcoming of such a model is that birth and death rates rarely remain constant for very long, so projections must be made with extreme care. In addition, we exclude the effect of net migration. In the short run, however, predictions based on the model may provide useful information. Let's apply this model to two questions about the U.S. population.

EXAMPLE: Predicting the U.S. Population

The population of the United States was about 255 million at mid-1990. It was increasing then at an average growth rate of 0.7% per year. What is the anticipated size of the U.S. population at the beginning of the year 2000? What will it be if the rate of natural increase turns out instead to be 0.4% per year, or 1.0% per year?

To answer these questions, we apply our geometric model with the initial population size equal to 255 million. In the first case $r = 0.007$ (because 0.7% equals 0.007). Using a year as the compounding period and the formula $P(1 + r)^n$, where $n = 10$, we find that the projected population size in the year 2000 is

Population in 2000
$= $(population in 1990) $(1 + $growth rate$)^{10}$
$= 255{,}000{,}000 \, (1 + 0.007)^{10}$
$= 255{,}000{,}000 \, (1.007)^{10}$
$= 255{,}000{,}000 \, (1.072247)$
$= 273{,}000{,}000$, approximately

(The result of our calculation can't be any more precise than the ingredients. Because our estimates of population and growth rate are rough approximations, we don't copy down all the digits that our calculator gives but instead round off.) ▲

In the same way, we find that with a growth rate of 0.4% per year, we predict a population in the year 2000 of 265 million, while a growth rate of 1.0% per year yields a predicted population of 282 million. It is clear that an uncertainty of three-tenths of one percentage point, or 0.003, in the growth rate of a population has major implications, even over fairly short time horizons. The presence or absence of 8 million people would have a significant impact on our social and economic systems. According to our model, the increase would be children (as opposed to adult immigrants), who would need schools built and teachers trained. Indeed, a great deal of concern over the long-range funding of the social security programs results from uncertainties over birth rates. Figure 17.2 gives a graph of the structure of the U.S. population expected in the year 2000, structured by age and sex.

Rates of natural increase in most Third World countries are well above those experienced by the United States and other industrialized nations. It is

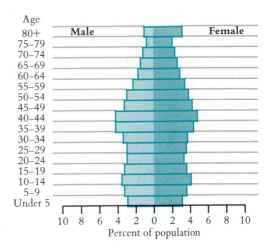

Figure 17.2 Graph of the projected population of the United States in the year 2000 grouped by age and sex and shown as a percentage of the total population. This projection is based on a geometric growth model.

SP TLIGHT 17.2   e — A Very Special Number

▶ ▶ ▶ ▶ ▶ ▶ ▶ ▶ ▶ ▶ ▶ ▶ ▶

The number e is similar to the number π in several respects. Both arise naturally, π in finding the area and circumference of circles, and e in compounding interest continuously (in other terms, e is the base for the system of "natural" logarithms). In addition, neither one is rational (expressible as the ratio of two integers, such as 7/2), nor is either algebraic (the solution of a polynomial equation with integer coefficients, such as $x^2 = 2$); we say that they are *transcendental* numbers. Finally, no pattern has ever been found in the digits of the decimal expansion of either number.

In addition to its fundamental importance in banking and population growth, the number e occurs naturally in several other common contexts as well.

A custom in some families is for each family member to buy a holiday gift for just one other family member (colloquially, "guy"). In advance of the holiday, all the members' names are put into a hat, and each member draws out a name at random. If anyone draws his or her own name, the drawing is annulled and is redone. What is the probability that the first drawing is successful in assigning each person someone else's name? (This problem is often called the *hat-check problem*, after a whimsical imaginary situation in which men who check their hats at a theater checkroom get hats back at random.) For a large family, the answer is approximately $1/e \approx 0.37$.

In other words, we can expect such a drawing to be successful only about 37% of the time. The chances for families with 2, 3, and 4 members are, respectively, 50%, 33%, and 38%; so the approximation $1/e$ is a good one even for small families.

The situation is usually complicated further by the additional restriction that the drawing is also annulled if any husband or wife draws the other's name. In this case, for a large family in which all members are paired off, the probability of a successful drawing turns out to be approximately $1/e^2$, or about 14%. The chances for a successful drawing with 2, 3, and 4 couples are, respectively, 17%, 11%, and 14%. The same results hold if, instead of the husband–wife restriction, no one can have the same "guy" as the previous year. If that rule is imposed *in addition* to the husband–wife restriction, then the chance of a successful drawing goes down to approximately $1/e^3$, about 5% — too small for the drawing ceremony to be any fun anymore!

not uncommon to find growth rates of 3% per year (and more) in developing nations. With its growth rate of 2.9%, Nigeria's population of 118 million in mid-1990 will become 157 million by the middle of the year 2000, a one-third increase. If such growth rates were to be maintained, the population would double in about 25 years. Projections of this sort are at the root of worldwide concern over our ability to provide sufficient food and other resources for all people.

▶ LIMITATIONS ON BIOLOGICAL GROWTH

Let's investigate more carefully the implications of using a geometric growth model to describe the growth of a population.

Examining Figure 17.3, we see that for a positive growth rate r, the size of the population increases as time increases, at least for all times shown in the graph. In fact, no matter how long a time span we consider, the model predicts that the population continues to grow, even to astronomical numbers.

Such predictions are clearly unreasonable for many situations. For instance, no biological population can continue to increase without limit. Its growth is eventually constrained by the availability of resources such as food, shelter, and psychological and social "space." As the population grows, it eventually reaches a level at which there are no resources for new members. A geometric growth model cannot describe forever the growth of such a population.

Let's now think about the way that actual biological populations behave. Their percentage growth rate is likely to depend on the size of the population and may actually decrease as popula-

tion size increases. (This happened fairly steadily to the U.S. population from 1865 to 1945.) For a sufficiently large population, the "growth" may even be negative. How can we account for a decline in the growth rate? One way this could occur is if, as the population size increases, the resources per individual decrease, and thus the energy available for growth and reproduction decreases. There may in fact be a maximum population size that can be supported by the available resources. Such a population size is called the **carrying capacity** of the environment. In one model for growth that takes into account the carrying capacity, we reduce the growth rate r by a factor that indicates how close the population size P is to the carrying capacity M:

$$\text{Growth rate} = r\left(1 - \frac{\text{population size}}{\text{carrying capacity}}\right)$$
$$= r\left(1 - \frac{P}{M}\right)$$

This method is known as the **logistic model** of population growth. We can check that this model has the properties we want. For small population sizes, that is, for values of the initial population P that are small relative to the carrying capacity M, the quantity P/M is small; therefore $1 - P/M$ is close to 1 and the growth rate is close to r.

EXAMPLE: Predicting the U.S. Population Using the Logistic Model

The U.S. population from 1790 to 1950 closely followed a logistic model with $r = 0.031$, $P =$ population in $1790 = 3,900,000$, and $M = 201$ million. For the early decades after 1790, the population was a small fraction of the "carrying capacity," and it grew at close to the rate r of 3.1% per year (a rate similar to many Third World countries today). By 1920 the U.S. population had reached 106 million, a little more

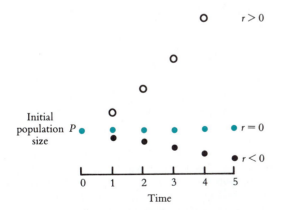

Figure 17.3 Projected population over five time units assuming a geometric growth model with growth rate r.

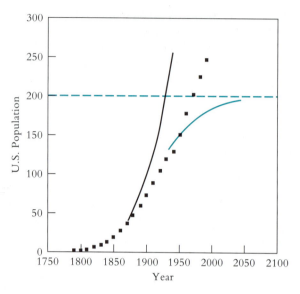

Figure 17.4 U.S. population by year. The points show actual growth; the black curve shows exponential (geometric growth) based on the early years of the United States; and the blue curve shows logistic growth based on figures through 1920.

than half of the "carrying capacity," and indeed the growth rate had slowed by about one-half, to 1.5% per year (see Figure 17.4).

Obviously, the mid-1990 U.S. population of 255 million exceeds the hypothesized "carrying capacity" of 201 million. Predicting the 1990 population on the basis of the 1950 population by using the logistic model would have resulted in great error. The postwar baby boom and increased immigration are two factors in the difference between prediction and fact. The predominant factor, however, is that the structure of the U.S. population changed, from a large proportion of people making their living on family farms to a highly urbanized society. Since the structure of the population changed, the model based on the prior structure is no longer valid under the new circumstances. ▲

For a logistic model, as the size of the population increases, the growth rate decreases—

because the term containing the population P has a negative sign. For a population size equal to the carrying capacity, that is, when $P = M$, the growth rate is zero.

If at any time the population size exceeds the carrying capacity, then the growth rate becomes negative (because $P > M$ in the formula for growth rate used in the logistic model) and the population size decreases. The carrying capacity refers to long-range capacity to support the population, so the population could exceed it for short periods of time. This could happen either because the population grows very rapidly and surges above the carrying capacity, or because of a sudden decrease in the food supply, thus lowering the carrying capacity, as happens to deer and other animals in winter.

The logistic model provides excellent predictions for the growth of some populations, particularly in laboratory environments. For example, the

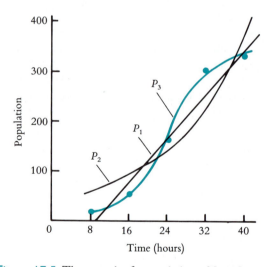

Figure 17.5 The growth of a population of fruit flies: P_1 is the best approximation by a simple growth model; P_2 is the best approximation by a geometric growth model; and P_3 is the best approximation by a logistic model. The blue points show the actual values. (Adapted from Daniel Maki and Maynard Thompson, *Mathematical Models and Applications*, Prentice-Hall, Englewood Cliffs, N.J., 1973, p. 431.)

graph in Figure 17.5 shows the growth of a population of fruit flies in a glass enclosure with a limited food supply. On the same coordinate system we show the predictions of an arithmetic population model (labeled P_1), a geometric population model (labeled P_2), and a logistic population model (labeled P_3). Predictions based on the logistic model come closest to the actual growth.

▶ NONRENEWABLE RESOURCES

One reason that we need to be concerned with the increasing population of people is that people use natural resources, some of which are renewable but many of which are not. In this section we model populations of nonrenewable resources; in the next section, we model populations of renewable resources.

Even without human population increases, we are faced with steadily dwindling populations of nonrenewable resources. A nonrenewable resource is one that does not tend to replenish itself; gasoline, coal, and natural gas are important examples. There is also no practical way to recover or reconstitute these resources after use. Some substances, such as aluminum or the sand used to make glass, are potentially recyclable; but to the extent that we do not recycle them, they too are nonrenewable.

For any nonrenewable resource, there is only a fixed amount S (in some convenient units) of it that is available to us. We are interested in the question:

How long will the supply ("population") of the resource last?

As long as the rate of use of the resource remains constant, the question is easy to answer. If we are now using U units per year and continue using U units per year, then the supply will last S/U years. This kind of calculation is the basis for statements such as "at the current rate of consumption, the U.S. coal reserves will last 300

years," or that the U.S. strategic reserve of gasoline (stored in salt domes in the South) would last 60 days. The figures involved will change with new additions to the reserves or with changes in the rate of use.

It is the change in the rate of use that we model here. In particular, the rate of use of resources tends to increase with increasing population and with increased "standard of living." For example, projections for use of electric power are often based on assumptions that the use will increase by some fixed percentage each year. This is the simplest situation (apart from constant usage), and one that we can analyze successfully to give interesting and important perspectives.

Suppose $U_1 = U$ is the rate of use of the resource in the first year (this year), and that usage increases $r = 0.05 = 5\%$ each year. Then the usage in the second year is $U_2 = U_1 + 0.05U_1 = 1.05U$, and usage in the third year is $U_3 = U_2 + 0.05U_2 = 1.05U_2 = 1.05(1.05U) = (1.05)^2U$. Generalizing, we see that usage in year i will be $(1.05)^{i-1}U$. Also, total usage over, for example, the next 5 years, will be

$$U + (1.05)^1U + (1.05)^2U \\ + (1.05)^3U + (1.05)^4U$$

If we move to general r and T years, then usage in the year i is $(1 + r)^{i-1}U$ and total usage is

$$U + (1 + r)U + (1 + r)^2U + (1 + r)^3U \\ + \cdots + (1 + r)^{T-1}U$$

which equals

$$U[1 + (1 + r) + (1 + r)^2 + (1 + r)^3 \\ + \cdots + (1 + r)^{T-1}]$$

The factor on the right involving r is known as a **geometric series,** because the terms of the series grow geometrically, by a constant proportion from term to succeeding term; here the proportion, or common ratio, is $(1 + r)$. There is a formula for the sum of such a series; we give it in an uncluttered form by considering a geometric series with common ratio x:

$$1 + x + x^2 + x^3 + \cdots + x^n = \frac{x^{n+1} - 1}{x - 1}$$

That this formula works for all x (except $x = 1$) can be seen by multiplying both sides by $(x - 1)$ and watching terms on the left cancel.

To fit our situation, we take $x = 1 + r$ and $n = T - 1$. Then we have, for the total usage over T years,

$$\frac{U[(1 + r)^{T-1} - 1]}{r}$$

To find out how long our supply S will last, we set the supply equal to the use over T years and then determine what T will be. We have

$$S = \frac{U(1 + r)^{T-1} - U}{r}$$

We perform some algebra to isolate the term involving T, getting

$$(1 + r)^T = \frac{S}{U}r + 1$$

At this point, to isolate T, we need to take the "natural logarithm" of both sides, denoted by "ln" here and by either "ln" or "LN" on your calculator (not "log," which stands for a different kind of logarithm). We get

$$\ln[(1 + r)^T] = T\ln(1 + r) = \ln\left[\frac{S}{U}r + 1\right]$$

which gives the complicated-looking

$$T = \frac{\ln[(S/U)\,r + 1]}{\ln(1 + r)}$$

Fortunately, a simple approximation will let us get rid of one of the logarithms. Most of the rates r that we will be interested in are between 0 and 10%; for r in this range, the value of $\ln(1 + r)$ is very close to r itself (try some values on your calculator! e.g., $\ln 1.05 = 0.0488$). We have (approximately)

$$T = \frac{1}{r}\ln\left[\frac{S}{U}r + 1\right]$$

The expression S/U represents how long the supply will last with constant use U. Michael Olinick, a mathematician at Middlebury College who has written about this model, has helped popularize the terms **static index** for S/U (corresponding to a static rate of use) and **exponential index** for the quantity T (reflecting the fact that it corresponds to an exponentially increasing rate of use).

EXAMPLE: U.S. Coal Reserves

We noted earlier that U.S. coal reserves are expected to last about 300 years at the current rate of use; so the static index for this resource is 300 years. How long will the supply last if the rate of use increases 5% per year? The corresponding exponential index is

$$T = \frac{1}{0.05}\ln[(300)(0.05) + 1]$$

$$= 20\ln 16 = 55 \text{ years}$$

That's quite a difference!   ▲

We must not take such projections exactly. Our estimates of supplies of a resource may underestimate how much there is, and in the future we may discover previously unknown sources or improve our technology to extract previously unavailable supplies. In addition, as supplies dwindle, the economic considerations of supply, demand, and price will come into play. We will never completely run out of oil; it will always be available "at a price."

However, we must not take such projections lightly, either, since we are discussing resources that, once used, are gone forever. Also, in any such projection, it is important to examine the assumptions, particularly since the exponential index can be substantially less than the static index.

▶RENEWABLE RESOURCES

A **renewable natural resource** is a resource that tends to replenish itself if we allow it to evolve without intervention. Obvious examples are fish, wildlife, and forests. We would like to predict how much of a resource we can harvest and still allow the resource to replenish itself.

We will concentrate on the subpopulation of individuals that have a commercial value. In the instance of a forest, the subpopulation might be trees of a commercially useful species and appropriate size. Because trees of different sizes have different commercial values, we will measure the size of the population in common units of equal value. For example, we measure a forest not by counting the trees but by estimating the number of board feet of usable timber. Similarly, we might measure the size of a fish population in terms of pounds rather than numbers of fish. When a population is measured in this way—in common units of equal value—we say that we are considering the **biomass** of the population.

Reproduction Curves

We use a figure called a **reproduction curve,** which predicts next year's population based on this year's population. A typical reproduction curve is shown in Figure 17.6. The reproduction

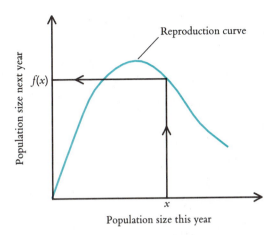

Figure 17.6   A typical reproduction curve.

curve represents the total change in the population's biomass from one year to the next, including the growth of continuing members, plus the addition of new members, minus losses due to death and other factors. Although the precise shape of the curve will vary from one population to another, reasonable biological conditions result in a curve of the general shape shown.

Let's take a closer look at the reproduction curve in Figure 17.6. The size of the population in the current year is measured on the horizontal axis; let x be a typical value. For a population of size x this year, its size *next* year is given by the height of the curve above the horizontal axis. This value is denoted by $f(x)$. (You can think of f as standing for "forthcoming.")

Figure 17.7 shows the same reproduction curve, only this time with the addition of a broken line inclined at a 45° angle with the horizontal axis. If you experiment with various choices for x, you will see that whenever the reproduction curve is above the broken line, then the next year's population is larger than this year's. In fact, the vertical distance from the broken line to the curve is the gain in population, or **natural increase,** which in algebraic terms is $f(x) - x$. Whenever the repro-

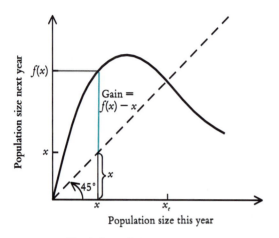

Figure 17.7 Depiction of the natural increase (gain) in population from one year to the next. The population size x_e is the equilibrium population size, for which the population one year later is the same, or $f(x_e) = x_e$.

duction curve is below the broken line, then the next year's population is smaller than this year's and $f(x) - x$ is negative. Note that for the special population size labeled x_e, the population at which the reproduction curve crosses the broken line, the population is exactly the same next year as this year. The population size x_e is known as the **equilibrium**, or **steady state, population size**. (Notice that if you project to both axis lines from where the reproduction curve and the broken line cross, the resulting figure is a square.)

The broken line provides a convenient way to trace the evolution of the population over a period of several years (see Figure 17.8). Begin with the first year's population on the horizontal axis, go up vertically to the curve; the height is the population in the second year. Proceed horizontally from the point on the curve over to the broken line; this point on the broken line is located directly above the point on the horizontal axis corresponding to the second year's population. Proceeding vertically from there to the curve then yields a height that is the population in the third year.

Figure 17.8 shows the results of several of these traces for the same reproduction curve, each starting from a different-size initial population, that is, from a different point on the horizontal axis. The resulting variation is quite surprising—it can even be "chaotic" in a very specific mathematical sense, and is an example of how what would otherwise appear as random behavior in fact follows a pattern corresponding to a very simple deterministic mathematical model.

Sustained-yield Harvesting

Many biological populations are harvested by predators (including humans). **Yield** is the amount harvested at each harvest. For the present discussion we will focus on a **sustained-yield harvesting policy,** that is, a harvesting policy that if continued indefinitely will maintain the same yield. Sustained-yield harvesting policies are obviously important to timber companies and other enterprises that plan to use a natural resource over a long period of time. Often a key question is to determine the best or optimal harvesting policy, and the answer is often a sustained-yield policy.

Under a sustained-yield harvesting policy, the population after each year's harvest will be the same. To achieve this stability, the amount h that is harvested must exactly equal the amount by which the population naturally increases each year. We recall that the population increases from x to $f(x)$ in one year, so the amount of natural growth is $f(x) - x$, which, to achieve sustained yields, must equal h. So for sustained yield, we must have $h = f(x) - x$, or equivalently $x = f(x) - h$ (see Figure 17.7).

Observe that depending on x, the value of h, which is the vertical distance from the broken line to the curve, can vary all the way from 0 (for $x = 0$; or for $x = x_e = $ the equilibrium population, in the absence of harvesting) up to some maximum value (for an x somewhere between 0 and x_e). A common problem facing a timber company or a fishery is to

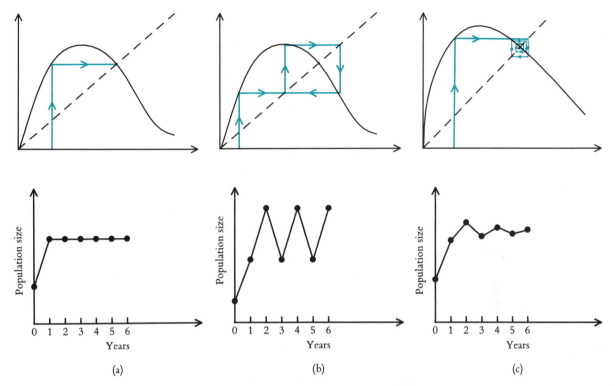

Figure 17.8 Examples of the dynamics, over time, for the same reproduction curve but different starting populations. (a) The population goes in one year to the equilibrium population and stays there year after year. (b) After initial adjustment, the population cycles between values over and under the equilibrium population. (c) The population spirals in toward the equilibrium population.

determine a harvesting policy that results in the largest possible sustainable yield, or **maximum sustainable yield.** In terms of our diagram, the goal is to select x so that the sustainable harvest is as large as possible. Without a description of the reproduction curve in numeric terms, the best we can do is estimate the value from our graph. The value of the population size that achieves maximum sustainable yield is shown as x_M in Figure 17.9. At this point, the vertical distance between the broken line and the reproduction curve — which represents the harvest — is as large as possible.

Considerations from Economics

Let's now consider how the costs of harvesting may complicate our analysis. We consider two models: one for a cattle ranch, and one for either a fishing boat or a tree farm.

In our models we assume that the price we receive for our harvest is the same for each harvested unit and does not depend on the size of our harvest. In effect, we are assuming that our operation is so small a part of the total market for the population in question that the size of our harvest will not substantially affect overall supply and

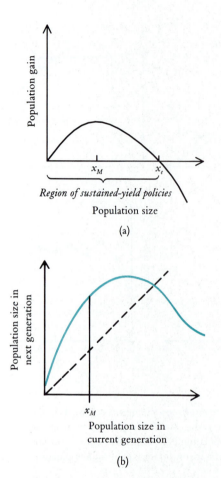

Figure 17.9 Determining the maximum sustainable yield, x_M, using a reproduction curve. (a) The natural increase, or sustainable yield, for each population size, with the maximum sustainable yield determined by the high point on the curve. The corresponding population size is x_M. (b) The reproduction curve, with the population size x_M corresponding to the maximum sustainable yield. The maximum sustainable yield is the greatest vertical distance from the 45° line to the reproduction curve.

hence price. We will let p denote the price that we obtain for each harvested unit, so that the total revenue from our operation is price per unit times number of units harvested, or $\$ph$.

Remember that we are concerned with determining the optimal harvesting level for a sustained-yield policy. We want to stay in business; therefore, we are not going to extinguish our resource for quick profits. In particular, for any given population size, we will harvest exactly the gain in population from one year to the next.

EXAMPLE: Cattle Ranching

For the cattle ranch, assume that the cost of harvesting a unit of the population is the same for each unit and does not depend on how many units we harvest. We include in the cost of harvesting all the costs of raising the cattle and delivering them to market. Let us say that it costs $\$c$ to harvest each unit. Since our cost does not depend on the population size, our cost curve is a horizontal line (Figure 17.10).

Now, if we harvest h units, the associated cost is $\$ch$. Then our net profit P from harvesting h units from a population consisting of x units is

Net profit P
= revenue − cost
= value of harvested units − harvest cost
= $ph - ch = (p - c)h$

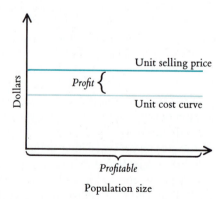

Figure 17.10 The unit cost, unit revenue, and unit profit of harvesting one unit, as a function of population size, for the cattle ranch.

As long as the *selling price* per unit is higher than the *harvest cost* per unit, we make a positive profit. The points of view of economics and biology agree, since the maximum profit occurs for the maximum sustainable yield. ▲

EXAMPLE: Fishing and Logging

Our key assumption for this model is that the cost of harvesting a unit of the population depends on how abundant the population is. This assumption incorporates the familiar principle of **economy of scale:** the cost of harvesting one unit decreases as the size of the population increases. For example, the same fishing effort yields more fish when fish are more abundant. Similarly, a logger's costs of harvesting one tree are less when that tree is in a stand of many harvestable trees than when it is in one with few; this is the logger's motivation for wanting to "clear-cut" large stands of trees. Since we assume that the cost goes down as the size of the population goes up, our cost curve looks something like Figure 17.11. The size of a population from which one unit is harvested is shown on the horizontal axis; the cost of harvesting a

single unit is measured on the vertical axis. The fact that the curve slopes downward and to the right is a reflection of the basic principle that the cost of harvesting a single individual is less in a large population than in a small population.

Our analysis proceeds as before, except now our unit cost is no longer constant but depends on x. So we write it as $c(x)$, and our net profit is

$$\text{Net profit } P = [p - c(x)] \times h$$

An optimal harvesting policy will depend on the relation between price and costs. There are two cases (Figure 17.12).

First, if the price we receive for a harvested unit is less than the cost of harvesting that unit for all population sizes, then it is impossible to make a positive net profit. The best we can do is to have a net profit of zero, by harvesting no units from our population.

The second case is of greater interest: Above a certain population size, the price we receive for one unit of the population is more than the cost of harvesting one unit. Now it is possible to generate a positive net profit (Figure 17.13). At the same time, the harvest is smaller for population sizes that are very large. There must be some population size, call it x_Q, that gives a maximum net profit. Using calculus, it can be shown that x_Q is actually larger than x_M, the population that gives the maximum sustainable yield, as shown in Figure 17.12. ▲

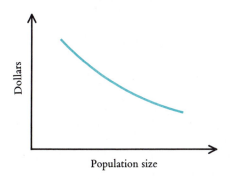

Figure 17.11   The unit cost, as a function of population size, for fishing or logging.

But our models fail to take into account a very critical feature of a modern economy that we have already concentrated on earlier in this chapter: the time value of money, as measured by the interest capital can earn. In the next section, we see one important explanation why biological populations are susceptible to overexploitation and even extinction.

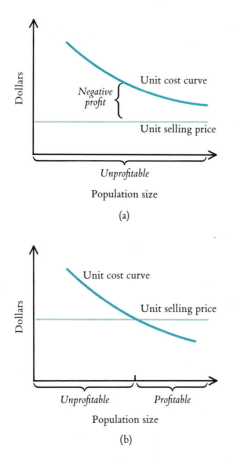

(a)

(b)

Figure 17.12 The unit cost, unit revenue, and unit profit of harvesting one unit, as a function of population size, for fishing or logging. (a) The market price is below harvesting cost for all population sizes. (b) The operation is profitable for populations above a certain minimum size.

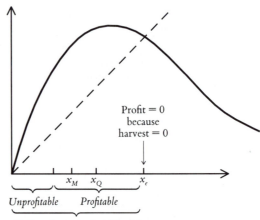

Figure 17.13 Region of profitability for sustained-yield policy, with the economically optimal population size x_Q marked.

Why Eliminate a Renewable Resource?

We might well ask why anyone would want to eliminate a renewable resource. In at least some instances, such as the case of the now-extinct passenger pigeon, populations have been completely harvested (see Spotlight 17.3, pp. 550–551). We can now apply our approach to understand why.

Sustained-yield policies operate under the assumption that some of the revenues will be received at a fairly distant future time. It is reasonable that the value of these revenues should be discounted to reflect the loss of income that would be earned if the funds were available for investment today. Rather than their current value, we should consider what economists call the **present value** of revenues to be received in the future. After all, a dollar today is worth more than a dollar you expect to receive a year from now.

To be specific, if funds are invested at an interest rate of $100r\%$ per year, with interest paid annually, the present value V of an amount A to be received n years in the future is related to A by the formula $A = V(1 + r)^n$. Hence $V = A/(1 + r)^n$. In this model, our goal is to maximize the sum of the present values of all future receipts of a sustained-yield harvesting policy. We refer to this sum as the present value of our return. The optimal harvesting policy will depend on the price p per unit harvested, the cost function $c(x)$, and the additional consideration of the interest rate r.

We don't delve into the details of the calculations here, but instead just give the results of the analysis.

Again there are several cases to consider. In the first case, suppose that the cost of harvesting $c(x)$ exceeds the price for all population sizes x. Then it is impossible to have a positive net profit, and the best we can do is have a return of zero. The optimal policy in this case is to harvest nothing.

For the second case, suppose that there is a population size x such that $c(x) = p$. Then there is a population size between x and x_e (the equilibrium population size) for which the present value of the total return is maximized. In particular, the population is maintained at that level (after harvesting) and is not completely harvested.

For the third case, suppose that p is larger than $c(x)$ for *all* population sizes x. Then the conclusion depends on the value of the interest rate r. If r is small, then the situation is the same as in the second case. On the other hand, if r is large, then it may be that the optimal harvesting policy is not to sustain the yield but to harvest the entire population immediately. This conclusion certainly corresponds to our intuition: if the price is high enough and if the proceeds can be invested at a sufficiently high rate of return (interest), then the most profitable course of action is to harvest everything — that is, extinguish the resource — and invest the proceeds.

Let's put this in the simplest and starkest terms. Suppose you own a population of a valuable resource, such as a forest, and the cost of harvesting is small relative to the value of the resource. If the rate at which the population is increasing is greater than the current interest rate on investments, it pays to let the forest keep on growing.

On the other hand, if the forest is growing more slowly than the interest rate, your economically optimal policy is to cut down all the trees now and invest the money. You could then start raising cattle on the land — and right there you have the scenario that is resulting in deforestation all over the world.

The sobering fact is that *very few economically significant renewable resources can sustain annual growth rates over 10%.* Many, like whales and most forests, have growth rates in the 4% to 5% range. These values — even a growth rate of 10% — are far below the return investors expect on their investment. For example, Wisconsin electric power utilities — far from being an exciting growth industry — are guaranteed by the state a profit of 14.25%; and venture capital firms expect to exceed 25% profit.

The concept of maximum sustainable yield is an attractive ideal if interest rates and expectations of investors are low enough. However, there are still difficult problems involved.

One problem is "the tragedy of the commons," discussed by ecologist Garrett Hardin. Several hundred years ago, English shepherds would graze their flocks together on common land. The grass of the commons could support only a fixed number of sheep. Each shepherd could reasonably think that adding just one or two more sheep to his flock would not overtax the commons; yet if each one were to do so, there could be a disaster, with all the sheep starving. Similarly, many natural products industries, such as fisheries, are a form of commons; a small amount of overexploitation by each harvester can produce disastrous results for all.

The major problem is how, in the presence of human greed, to anticipate and prevent overexploitation and possible extinction of a resource, common or not. Even to determine the maximum sustainable yield, experiments may be needed with harvesting at different levels; but it is likely to be politically impossible to force a harvesting industry to accept lesser profits from reducing harvesting now for the sake of experimentation to assure possible stability in the future.

A further complication is that in some industries, such as a fishery, growth of the population may be abundant one year but meager another, so that a steady yield cannot be sustained without damaging the resource. A few good years in a row

SPOTLIGHT 17.3   Extinction of the Passenger Pigeon

▶ ▶ ▶ ▶ ▶ ▶ ▶ ▶ ▶ ▶ ▶ ▶ ▶

Although once numbering in the billions, passenger pigeons are now extinct. (Courtesy, Field Museum of Natural History, Chicago.)

Historically, the utilization of a renewable resource has followed a characteristic pattern. First comes a stage of expanding harvests, perhaps based on a new use of the resource or on new harvesting technology. This is followed by concern for overutilization. Conservation measures are then adopted and the industry either stabilizes or collapses.

In some cases, the population has actually become extinct. For example, the passenger pigeon was once considered to be the world's most abundant land bird. Over a century ago, they numbered from 3 to 5 billion, traveling and nesting in huge flocks, mostly in eastern North America. But by 1914, the last remaining passenger pigeon had died at the Cincinnati Zoo.

The demise of the passenger pigeon can clearly be traced to expanding harvests, brought about by new technology — in this case, the development of the eastern railroad network and the telegraph. To understand how these developments were able to severely diminish such an abundant species, we can look at the ecological characteristics that were once key to the passenger pigeon's earlier success: colonization and nomadism.

Passenger pigeons were not solitary. They nested in colonies containing millions of pairs (the entire population consisted of perhaps fewer than a dozen flocks). The flocks were sometimes so immense that they were reported to have obscured the sun. (The largest flight ever recorded was estimated to contain 2.23 billion birds.) By traveling and nesting in such large groups, pigeons were able to literally "shield" themselves from predators. No matter where they nested, there were not enough local predators to significantly reduce their numbers. This concept is known as "predator satiation."

Passenger pigeons fed on large crops of nuts found in the deciduous forests of eastern (and occasionally midwestern) North America. Because the location of crops large enough to accommodate their numbers varied from year to year, passenger pigeons rarely nested in the same place two years in a row, with nesting sights ranging from New York and Pennsylvania to Michigan or Wisconsin. Thus, it was difficult to predict their location from one nesting season to the next.

By all accounts, the final decline of the passenger pigeon was rapid. The arrival of flocks of passenger pigeons had always meant food to the local people, but it was probably not until pigeons were actually harvested for market that the population began to markedly diminish. (Market harvesting began before 1800, but was not a major industry until 1840.)

While we can never know for certain, it is believed that the technological developments of the nineteenth century — namely, the railroad and the telegraph — increased the efficiency and scope of market harvesting to the point where it was ultimately responsible for the extinction of the passenger pigeon. By the time of the Civil War, the railroad network through America, east of the Mississippi, was complete. This network allowed the professional pigeoners, who numbered about 1000 in their heyday, rapid access to all major nesting colonies. It also provided a fast means of shipping barrels of pigeons to the big city markets in the east and midwest. The telegraph was able to keep professional pigeoners informed of the locations of nesting colonies. In fact, the entire operation was organized so efficiently that word of any pigeon nestings spread rapidly for hundreds of miles. Since the railroads benefited from the pigeon harvest, it is likely that they, too, helped to see that this information was transmitted.

The fact that passenger pigeons nested in gigantic colonies — which at one time had assured their safety from predators — now made them especially vulnerable to harvesting for market. People did not understand that such an abundant resource could ever be severely diminished. They also did not allow for undisturbed nesting sites so that the pigeons could replenish their numbers. Instead, harvests at the nesting sites were so efficient and complete that there were no successful nesting colonies for a period of over 10 years. The last known colonial nesting attempt occurred in 1887 in Wisconsin, but the site was rapidly abandoned by the birds, probably because of disturbances.

If the harvest had not occurred at the nesting colonies, it is unlikely that the adult population could ever have been exterminated. Or, if only the adults had been harvested, the species might have survived. But because the fat nestlings were especially prized, and because many birds were driven away from nesting sites by the violent hunting methods sometimes used (shooting, setting trees on fire), the adults could not replace themselves, and the fate of the passenger pigeon was sealed.

Passenger pigeons were once a renewable resource, but within a period of about 20 years — twice an individual pigeon's lifetime — they became extinct. Certain other species, such as the bison, have also been reduced to numbers below a level of economic significance.

tend to provoke increased investment in fishing capacity; then attempting to harvest at the same levels in succeeding normal or below-normal years results in overfishing. Exactly this scenario produced the collapses of the California sardine fishery in the 1930s and the Peruvian anchovy fishery in 1972 (though environmental factors may also have been involved).

▶ Review Vocabulary

Annual equivalent yield   Effective rate.

Annual rate   Nominal rate.

Arithmetic growth   Growth by a constant amount in each time period.

Biomass   A measure of a population in common units of equal value.

Carrying capacity   The maximum population size that can be supported by the available resources.

Compound interest   The method of paying interest on both the principal amount and the accumulated interest in an account.

Compounding period   The interval that elapses before interest is calculated on an account.

e The base for continuous compounding and geometric (exponential) growth; $e = 2.71828....$

Economy of scale   Costs per unit decrease with increasing volume.

Effective rate   A yearly rate of interest that is compounded only once per year.

Equilibrium population size   The population size for which the size next year will be exactly the same as the size this year.

Exponential growth   Geometric growth.

Exponential index   How long a fixed amount of a resource will last at a constant rate of use.

Geometric growth   Growth by a constant proportion in each time period.

Geometric series   A sum of terms, each of which is the same constant times the previous term, that is, the terms grow geometrically.

Initial balance   Initial deposit in a bank account.

Interest   Money earned on a bank account.

Logistic model   A particular population model that begins with near geometric growth but then tapers off toward a limiting population (the carrying capacity).

Maximum sustainable yield   The largest harvest that can be repeated indefinitely.

Natural increase   The growth of a population that is not harvested.

Nominal rate   The stated annual rate of interest, which is then subject to compounding.

Population growth   Change in population, whether increase (positive growth) or decrease (negative growth).

Population structure   The division of a population into subgroups.

Present value   The value today of money to be received in the future.

Principal   Initial balance.

Rate of natural increase   Birth rate minus death rate, the annual rate of population growth without taking into account net migration.

Renewable natural resource   A resource that tends to replenish itself if it is allowed to evolve without external intervention; examples are fish, forests, wildlife.

Reproduction curve   A curve that shows population size in the next year plotted against population size in the current year.

Simple growth   Arithmetic growth.

Simple interest   The method of paying interest on only the initial balance in an account and not on any accrued interest.

Static index   How long a fixed amount of a resource will last at a constantly increasing rate of use.

Steady state   The population size for which the size next year will be exactly the same as the size this year. The amount harvested is exactly equal to the natural increase.

Sustained-yield policy   A harvesting policy that can be continued indefinitely while maintaining the same yield.

Yield   The amount harvested at each harvest.

▶Suggested Readings

CHERFAS, JEREMY: "What price whales," *New Scientist*, 5:36–40 (June 1986). Offers history of the whaling industry and concrete facts about both the economics and the biology of whaling.

CLARK, COLIN: "Some socially relevant applications of calculus," *The Two-Year College Mathematics Journal*, 4(2):1–15 (Spring 1973). Gives a mathematical approach to animal resource economics.

——— : "The mathematics of overexploitation," *Science*, 181:630–634 (August 17, 1973).

COHEN, JOEL: "How many people can earth hold?" *Discover*, 13(11) 114–119 (November 1992).

FLASPOHLER, DAVID C., FRANK MASTRIANNA, AND RICHARD PULSKAMP: *The Consumer Price Index: What Does It Mean?* UMAP Modules in Undergraduate Mathematics and Its Applications: Module 639. COMAP, Inc., Lexington, Mass., 1983. Reprinted in *The UMAP Journal*, 4(3):293–320 (1983), and in *UMAP Modules: Tools for Teaching 1983*, COMAP, Lexington, Mass., 1984, 465–492.

HARDIN, GARRETT: "The tragedy of the commons," *Science*, 162:1243–1248 (1968).

KLEINBAUM, DAVID G., AND ANNA KLEINBAUM: *Adjusted Rates: The Direct Rate*. UMAP Modules in Undergraduate Mathematics and Its Applications: Module 330. COMAP, Inc., Arlington, Mass., 1980. Reprinted in *The UMAP Journal*, 1(1):49–80 (1980), and in *UMAP Modules: Tools for Teaching 1980*, Birkhäuser, Boston, 303–334. A beginning exploration into the structure of populations, which explains, for instance, the paradox of how a Third World country can have a lower mortality rate than the United States, yet have a higher mortality rate for every age group.

LINDSTROM, PETER A.: *Nominal vs. Effective Rates of Interest*. UMAP Modules in Undergraduate Mathematics and Its Applications: Module 474. COMAP, Inc., Arlington, Mass., 1988. Reprinted in *UMAP Modules: Tools for Teaching 1988*, edited by Paul J. Campbell, COMAP, Inc., Arlington, Mass., 21–53. A learning module, requiring no more background than this chapter, that teaches about the difference between nominal and effective rates of interest and how to calculate them. Gives examples of banks using particular options for calculating interest.

LUDWIG, DONALD, RAY HILBORN, AND CARL WALTERS: "Uncertainty, resource exploitation, and conservation: Lessons from history," *Science*, 260:17, 36 (April 2, 1993).

MAKI, D. P., AND M. THOMPSON: *Finite Mathematics*, 2nd ed., McGraw-Hill, New York, 1983, Chapter 11.   ·

OLINICK, MICHAEL: "Modelling depletion of nonrenewable resources," *Mathematical Computer Modelling*, 15(6):91–95 (1991).

OPHULS, WILLIAM, AND A. STEPHEN BOYAN, JR.: *Ecology and the Politics of Scarcity Revisited*, Freeman, New York, 1992.

PITT, DAVID E.: "Despite gaps, data leave little doubt that fish are in peril," *New York Times* (August 3, 1993), B7.

POPULATION REFERENCE BUREAU, INC.: Annual World Population Data Sheet, 777 14th St. NW, Suite 800, Washington, D.C. 20005.

SCHWARTZ, RICHARD H.: *Mathematics and Global Survival*, Ginn Press, Needham Heights, Mass., 1989.

▶EXERCISES

The exercises below require a scientific calculator [with buttons for powers and for exponential (EXP) and natural logarithm (LN) functions].

1. You deposit $1000 at 8% annual rate of interest. What will the balance be at the end of 1 year, and what is the effective annual yield, if the interest paid is
   a. simple interest?
   b. compounded annually?
   c. compounded quarterly?
   d. compounded continuously?

2. As in Exercise 1, but for $1000 at 3% annual rate of interest.

3. [Contributed by John Oprea of Cleveland State University.] Use your calculator to evaluate for $n = 1, 10, 100, 1000$, and $1,000,000$:

   a.
   $$\left(1 + \frac{1}{n}\right)^n$$

   b.
   $$\left(1 + \frac{2}{n}\right)^n$$

   c. As n gets large, what numbers are the expressions in parts a and b tending toward?

4. Use your calculator to evaluate for $n = 1, 10, 100, 1000$, and $1,000,000$:

   a.
   $$\left(1 - \frac{1}{n}\right)^n$$

   b.
   $$\left(1 - \frac{2}{n}\right)^n$$

   c. As n gets large, what numbers are the expressions in parts a and b tending toward?

5. A recent issue of *Computer Language*, a magazine for professional programmers, repeats an oft-heard claim that "the amount of information in the world doubles every three days." Presumably the claimant is referring to the amount of *data*, which can be quantified in terms of number of bits. (A *bit* is the smallest unit of storage in a computer.) Show that the claim is absolutely preposterous, by doing a little arithmetic and comparing your result with the estimated number of particles in the universe (10^{70}). In particular:
   a. Start with one bit of data and double the number of bits every third day. How long does it take to get past 10^{70}? (Hint: Don't just keep multiplying by 2 over and over. Convince yourself that since the amount of data increases by a factor of 2 every three days, then it in-

creases by a factor of $2^2 = 4$ every six days, a factor of $4^2 = 16$ every twelve days, a factor of $16^2 = 256$ every twenty-four days, and so forth.)

  b. Part **a** involves a lot of multiplying by 2, even if you do it efficiently. Another approach is to use the fact that $2^{10} = 1024$, which is approximately 1000. Thus, the amount of data increases by a factor of more than 1000 every $3 \times 10 = 30$ days, or every month (except February, but the 31-day months make up for it). By when will the total be sure to be past 10^{70}?

6. An old legend tells of a wizard who agreed to save a kingdom provided that the king would agree to a "modest" reward. The wizard asked to be given merely as much grain as would put one kernel on the first square of a chessboard, two kernels on the second, four on the third, eight on the fourth, and so forth, up through the sixty-fourth square. The king agreed, the wizard saved the kingdom, but the king was completely unable to honor the agreement. Why? (Hints: Notice that $1 = 2^1 - 1$, $1 + 2 = 2^2 - 1$, $1 + 2 + 4 = 2^3 - 1$, and generalize to arrive at a total for the number of kernels. A kernel of rice is about a quarter of an inch long and about a sixteenth of an inch wide and a sixteenth of an inch high. So about a thousand kernels will fit in a cubic inch (you should verify this calculation). Calculate the total volume of kernels.

7. You work for a firm that has a system of merit raises. If you put in a lot of overtime over the coming year, you estimate that you will probably qualify for a $1200 merit increase; but that doesn't seem like a lot of money compared to the time that you would have to put in. Suppose you also estimate that inflation is going to be a fairly constant 5% over the next year or two. Assume that you work hard and get the raise, and that the raise is paid as $100 per month.

  a. How much is the first $100 installment of the raise (which you will get a year from now) worth to you now, in today's dollars?

  b. Your friend Sally (who got a merit raise last year) reminds you that you get the additional $1200 not only this year but every year from now on. So, two years from now, you will still be benefiting from the raise at the end of this year. How much is the $100 in the monthly paycheck that you will get two years from today worth to you now, in today's dollars?

8. As in Exercise 7, but for an inflation rate of
  a. 7%;
  b. 3%

9. The total population of the less-developed countries (excluding China) was 3.0 billion in mid-1990, and the growth rate was 2.4% per year (this is an effective annual yield, so you may think of it as compounded annually). If this growth rate continues until mid-2000, what will be the size of the population then?

10. If the growth rate of the less-developed countries of Exercise 9 changes suddenly to 2% in 1995, what will be the size of the population in mid-2000?

11. If the growth rate of the less-developed countries of Exercise 9 decreases by one-twentieth of a percentage point per year from 1990 through 2000, what will be the size of the population in mid-2000?

12. Which population pyramid in the accompanying figure shows the highest birth rate?

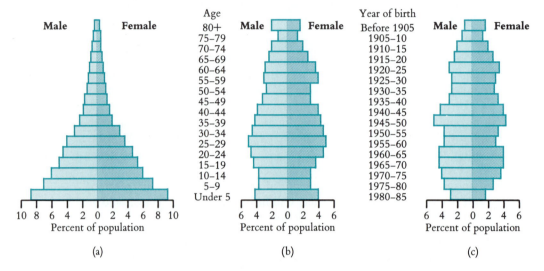

(a)   (b)   (c)

13. An advertisement for Paul Kennedy's book *Preparing for the Twenty-First Century* (Random House, 1993) asks: "By 2025, Africa's population will be: 50%, 150% or 300% greater than Europe's?" The population of Europe in mid-1993 was 500 million and was expected to stay constant through 2025. The population of Africa in mid-1993 was 720 million and was increasing at about 2.9% per year. What answer would you give to Kennedy's question?

14. Suppose a reproduction curve for a certain population is as in the accompanying figure, where the units are in thousands.

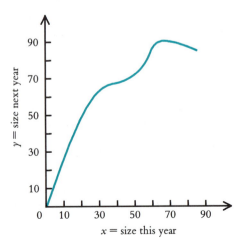

    a. Estimate the sustainable yield corresponding to a population of size 10 remaining after the harvest.

    b. Estimate the maximum sustainable yield.

15. A reproduction curve for a population is shown in the figure below. Estimate the equilibrium population size and the maximum sustainable yield.

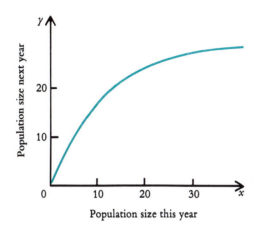

16. As in Exercise 15, but for the reproduction curve below.

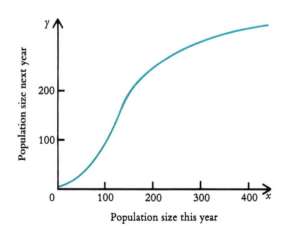

17. In 1990 the known global oil reserves totaled 917 billion barrels. Consumption, which had been 53.4 million barrels per day in 1983, rose an average of about 1.7% per year through 1990, when the consumption was about 60 million barrels per day.

    a. What is the static index for oil?

    b. If the rate of increase in consumption stays constant at 1.7%, what is the exponential index for oil?

    c. What considerations may affect these indexes over time?

18. Aluminum is the most abundant structural metal in the earth's crust. The world demand for new supplies of aluminum in 1983 was 16.5 million metric tons, while the known reserves were then 21,000 million metric tons.
   a. What was the static index for aluminum in 1983?
   b. In 1983, the demand for new aluminum was projected to increase at 4% per year at least through the year 2000. For that rate of increase, what was the exponential index for aluminum in 1983?
   c. What considerations may affect these indexes over time?

▲ 19. In some cases it is important to look at problems that are in a sense the reverse of those discussed in this chapter. In a financial setting such a question might be: How much do you need to deposit today in an account that pays interest at a known rate in order to have a specified amount at a particular time in the future? This question is crucial in certain financial planning considerations, for instance, in planning for a major purchase in the future.

Suppose that we ask for a specified amount, say A dollars, at a specified time in the future, say n years, and we know that the account pays interest at the rate $100r\%$ per year. The unknown quantity—namely, the amount that must be deposited today—will be denoted by P. Using our basic formula, we know that A, P, r, and n are related by $P = A/(1 + r)^n$. The quantity P is the present value of an amount A to be paid n years in the future.

Suppose that you will need $15,000 to pay for a year of college 8 years in the future, and you can buy a certificate of deposit whose interest rate of 10% compounded quarterly is guaranteed for that period. How much do you need to deposit?

▲ 20. As in Exercise 19, except that the interest rate is only 8%. How much do you need to deposit?

▲ 21. The situation described in Exercise 19 gives the rationale for the pricing of the so-called *zero-coupon bonds*. These securities pay no current interest but are sold at a substantial discount from redemption value. The difference between purchase price and redemption value provides income to the bondholder at the time of redemption or resale. If the interest rate in the economy is now 7%, what should be the price of a zero-coupon bond that will pay $10,000 8 years from now? (Use daily compounding.)

▲ 22. In times of economic inflation, prices behave like populations undergoing exponential growth.
   a. Suppose inflation proceeds at a level rate of 5% per year from 1990 through 1995. Find the cost in 1995 of a basket of goods that cost $1 in 1990.
   b. During price inflation, the value of the dollar behaves like a population undergoing exponential decay; all our formulas work, with a little adjustment. If we let i represent the effective annual rate of inflation, then what costs $1 now will cost $(1 + i)$ this time next year, and a

▲ Advanced exercise.

Color Plate 1
A Penrose nonperiodic tiling made with two rhombus shapes. [Tiling by
Roger Penrose.] (See pp. 709 – 716.)

Color Plate 2
A modification of a Penrose tiling by refashioning the kites and darts into
bird shapes. [Tiling by Roger Penrose.] (See pp. 709 – 716.)

Color Plate 3
A Penrose tiling of kites and darts in five colors.
A Penrose tiling can always be colored using
four colors, in such a way that two tiles that
share an edge have different colors. Whether a
Penrose tiling can be colored in such a way
using only three colors is an unsolved problem;
however, if one Penrose tiling can be colored
using three colors, then all Penrose tilings can.
[Tiling and coloring by Roger Penrose.] (See
pp. 709 – 716.)

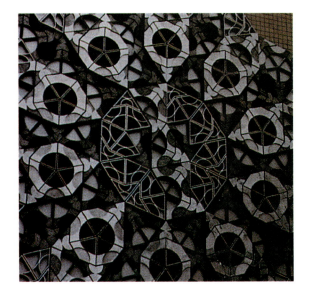

Color Plate 4
A Penrose tiling with specially marked tiles,
forming what is known as the cartwheel tiling.
[From Roger Penrose.] (See pp. 709 – 716.)

Color Plate 5

Six Fish Pattern. This pattern is based on M. C. Escher's similar pattern of fish, *Circle Limit III*. The difference is that six fish meet at left fin-tips in this pattern (instead of the four fish that meet at left fin-tips in *Circle Limit III*). In *Six Fish Pattern*, as in *Circle Limit III*, 12 fish form a fundamental region for the colored pattern; of course, one fish serves as a fundamental region for either pattern if color is disregarded. Also, as in *Circle Limit III*, each white backbone is a circular arc making acute angles with the bounding circle; thus in hyperbolic geometry, such an arc is not a hyperbolic line but is an equidistant curve from the circular arc with the same endpoints that is orthogonal (at right angles) to the bounding circle (and thus is a hyperbolic line). [Pattern designed by Douglas Dunham, University of Minnesota, Duluth.] (See pp. 633 – 636, 642 – 644.)

Color Plate 6

Seven Butterflies Pattern. This pattern is based on a repeating pattern of the Euclidean plane by M. C. Escher in which six butterflies meet at left front wingtips; increasing this number to seven results in a repeating pattern of the hyperbolic plane. A fundamental region for *Seven Butterflies Pattern*, as a colored pattern, consists of 168 butterflies; within any "ring" of butterflies of one color there are three different ways to arrange the other seven colors around the center. [Pattern designed by Douglas Dunham, University of Minnesota, Duluth.] (See pp. 633 – 636, 642 – 644.)

Color Plate 7

Escher No. 128 [*Bird*], from Escher's 1941–1942 notebook. [© 1967 M. C. Escher Foundation, Baarn, Holland. All rights reserved.] (See pp. 698–709.)

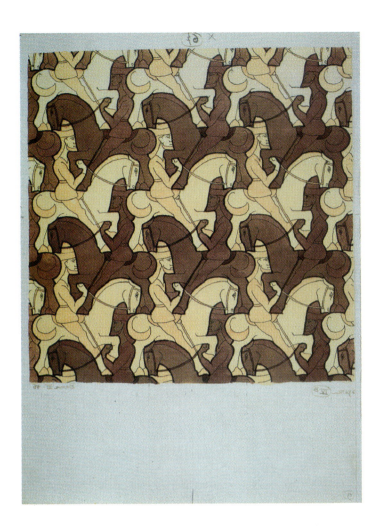

Color Plate 8a
Escher No. 67 [*Horseman*], from
Escher's 1941 – 1942 notebook.
(See pp. 698 – 709.)

Color Plate 8b
Sketch by Escher showing the
design of the tile for the *Horseman*
print.

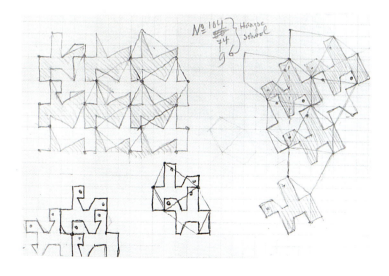

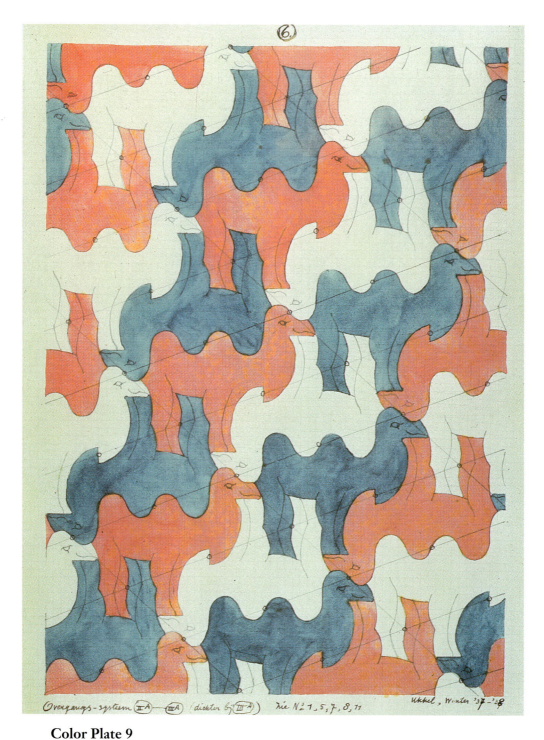

Color Plate 9
Escher No. 6 [*Camel*], from Escher's 1941 – 1942 notebook. [© 1937 – 1938
M. C. Escher Foundation, Baarn, Holland. All rights reserved.]
(See pp. 698 – 709.)

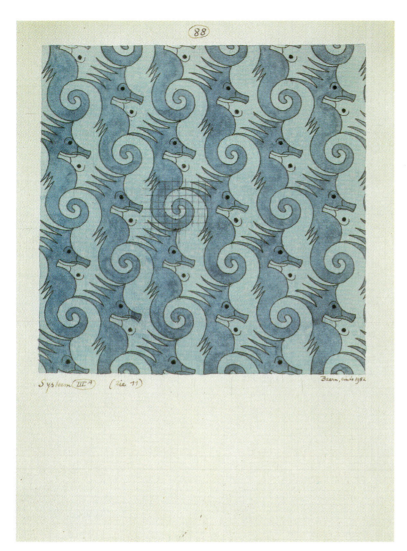

Systeem (III A)   (zie 11)   Baarn, einde 1952

Color Plate 10a
Escher No. 88 [*Sea Horse*],
from Escher's 1941 – 1942
notebook. [© 1952 M. C.
Escher Foundation, Baarn,
Holland. All rights reserved.]
(See pp. 698 – 709.)

Color Plate 10b
The geometric skeleton for No. 88. [© 1952 M. C. Escher
Foundation, Baarn, Holland. All rights reserved.]
(See pp. 698 – 709.)

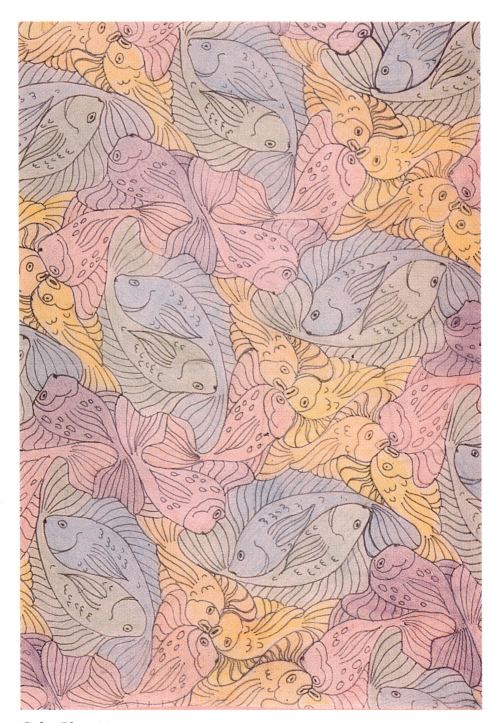

Color Plate 11
Fish, by Marjorie Rice, based on one or her unusual tilings by pentagons.
(See pp. 698 – 709.)

dollar then will buy only $1/(1 + i) = 1 - i/(1 + i)$ times as much. The quantity $d = -i/(1 + i)$ behaves like an interest rate, even though it is negative; the value of P dollars m years from now is given by $P(1 + d)^m$. Supposing 5% annual inflation from 1990 through 1995, what will be the value of a dollar in 1995 in constant 1990 dollars?

    c. Depreciation of the value of equipment is just like population growth, except that the rate of growth is negative. If you bought a car in 1990 for $10,000 and its value in current dollars depreciates steadily through 1995 at a rate of 15% per year, what will be its value in 1995 in 1995 dollars?

    d. If there is also 5% annual inflation from 1990 through 1995, what will be the value in 1995 of the car in part c in "inflation-adjusted" (constant 1990) dollars?

▲ 23. The *rule of 72* is a rule of thumb for finding how long it takes money at interest to double: If $100r\%$ is the annual rate of interest, then the doubling time is approximately $72/100r$ years.

    a. Calculate the balance at the end of the predicted doubling time for each $1000, with annual compounding, for the small growth rates of 3%, 4%, and 6%.

    b. As in part a, for the intermediate interest rates of 8% and 9%.

    c. As in part a, for the larger interest rates of 12%, 24%, and 36%.

    d. What do you conclude about the rule of 72?

▲ 24. (For this exercise you will need to use the EXP and LN buttons on your calculator.) More frequent compounding yields greater interest, but with diminishing returns as the frequency of compounding is increased. For small interest rates, there is little difference in yield for compounding annually, quarterly, monthly, daily, or continuously. Investigating doubling times with continuous compounding leads to understanding why the rule of 72 of Exercise 23 works. Recall that for continuous compounding at a (nominal) annual rate r, the balance A at the end of m years is Pe^{rm} for an initial principal of P. Let D be the number of years that it takes for the initial principal to double. Then we have $2P = A = Pe^{rD}$, so $e^{rD} = 2$. Taking the "natural logarithm" of both sides yields $rD = \ln 2$, where "ln" stands for the natural logarithm. The natural logarithm is represented on a calculator by a button marked either "ln" or "LN" (not "log," which stands for a different kind of logarithm). Using the button gives $\ln 2 = 0.693$. So we have $rD = 0.693$, from which we can determine D if we know r.

    a. Calculate the doubling times for continuous compounding at 3%, 6%, and 9%, and compare them with those predicted by the rule of 72. What do you conclude? Why do you think people prefer a rule of 72 rather than a rule of 69.3?

    b. Using the analysis for doubling as a model, devise a "rule of_____" for the time it takes money to triple. Use your rule to predict how long it will take $600 at 5% interest to triple.

▲ 25. In its estimates for doubling times for populations in the world, the Population Reference Bureau uses a rule of 70, similar to (but more accurate than) the rule of 72 used in banking. As noted in Exercise 24, a rule of 69.3 would be even more accurate; but the difference between that and the rule of 70 is only 1%. Apply the rule of 70 to estimate the doubling times for the following populations (figures are for mid-1990):

    a. China, 1.1 billion, 1.4%

    b. The world as a whole, 5.3 billion, 1.7%

▲ 26. As in Exercise 25, but for:

    a. Africa, 660 million, 2.9%

    b. United States, 250 million, 0.7%

▲ 27. Surprise! Just for fun, one of your friends wrote your name in on an Illinois State Lottery ticket, and you are the sole winner of $40 million! You discover, however, that you don't get the $40 million all at once; in fact, it is paid in 20 annual installments. All you get right away is the first installment of $2 million (minus 20% withheld against federal income tax due, and whatever you think your friend deserves for the favor). So, what is the prize really worth to you? That depends on the rate of inflation over the years. Assume a constant rate of inflation over the 19 years until your last payment and calculate the present value of your prize winnings by using the formula for present value combined with the formula for the sum of a geometric series. Do the calculation for rates of interest of:

    a. 3%

    b. 6%

    c. 9%

In fact, you will not be receiving your annual checks from the state of Illinois but from an insurance company from which Illinois has purchased an annuity (which is just a contract to pay a certain amount of money each year for a specified number of years). The price that Illinois pays for the annuity depends on current long-term interest rates.

▲ 28. In some cases we know the principal, the current balance, and the interval of time, and we want to learn the interest rate. For example, money market funds typically report earnings to investors each month, based on interest rates that vary from day to day. The investor may be interested in knowing some average rate of interest for the month, but usually no such figure is reported by the fund (nor are each of the daily rates given). For our purposes, we will simplify and assume that the same interest rate holds for each day of the month and that compounding is done daily. The key tool will be our general compounding formula, that the balance A is given by $P(1 + r)^n$, where P is the original principal, r is the interest rate per compounding period, and n is the number of compounding periods. For our purposes here, the compounding period is one day. We

▲   Advanced exercise.

have $A = P(1 + r)^n$, so $A/P = (1 + r)^n$. Taking the nth root of each side gives

$$1 + r = \left(\frac{A}{P}\right)^{1/n} \qquad \text{so} \qquad r = \left(\frac{A}{P}\right)^{1/n} - 1$$

We illustrate with an example. Suppose the monthly statement from the fund reports a beginning balance (P) of \$7373.93 and a closing balance (A) of \$7416.59 for 28 days (n). We thus have

$$r = \left(\frac{7416.59}{7373.93}\right)^{1/28} - 1 = (1.005785246)^{0.035714286} - 1 = 0.000206042$$

Thus the daily rate of interest is 0.0206042%. We obtain the annual effective yield by compounding at this rate 365 times: $(1 + 0.00206042)^{365} = 1.0780972$, for an effective annual yield of 7.81%.

Suppose that the preceding month the initial balance had been \$7331.35, and again the report had been for 28 days. Calculate the effective annual yield.

▲ 29. A 1990 advertisement reads, "If you had put \$100 per month in this fund starting in 1980, you'd have \$37,747 today." Assume that deposits were made on the first day of the month, starting on January 1, 1980, through December 1, 1989, and that interest is paid monthly on the last day of the month (120 months). How much money was deposited during this period? What annual rate of interest, compounded monthly, would lead to the result described in the advertisement? (Hint: This problem can be solved on a calculator by tediously adding deposits and crediting interest, or by writing a short computer program to do the same. But there is a more elegant and less tedious approach. Let m be the monthly rate of interest. The last deposit earns interest for only one month, so it contributes $\$100(1 + m)^1$ to the total. Similarly, the second last deposit contributes $\$100(1 + m)^2$, and so forth. The total of the contributions is

$$100(1 + m)^1 + 100(1 + m)^2 + \cdots + 100(1 + m)^{-120}$$

For simplicity, let $x = 1 + m$. Then the total is

$$100(x^1 + \cdots + x^{120}) = 100x(1 + x^1 + x^2 + \cdots + x^{119})$$

Apply the formula for the sum of a geometric series to the right-hand side and set it equal to 37,747. There is no formula to solve exactly the resulting equation, so you need to use a computer graphing program, a calculator equation solver, or successive guessing to find x. Then you can determine the monthly interest rate, m, the annual interest rate, and the effective annual yield.)

▲ 30. Some parents use savings accounts to accumulate funds for the college education of their children. Frequently the funds are deposited over a period of several years, and the funds together with accumulated interest are used to pay the costs of education.

Suppose that \$1000 is to be deposited each year into an account that pays interest at a rate of 8% per year compounded annually. If the first payment is made when a child is 2 years old and the last is made when the child is 17 (16 payments), how much is available when the child begins college at the age of 18? (You can make use of the formula for the sum of a geometric series, but be careful and think through the problem, as the formula itself will not give the final answer.)

For Exercises 31 and 32, refer to the following: radioactive elements emit particles and decrease in quantity at a predictable rate. For such elements, the length of time that it takes to decrease to half its original mass is called the *half-life* of the element. For instance, the radioactive isotope radium-226 has a half-life of approximately 1600 years. This means that if a substance contains 1000 grams of radium-226 now, then 1600 years from now, the substance will contain 500 grams of the isotope. Further, in 3200 years, the substance will contain 250 grams of the isotope, and in 4800 years, only 125 grams of the isotope will remain in the substance.

▲ 31. a. The isotope carbon-14 has a half-life of approximately 5730 years. If a substance contains 20,000 grams of the isotope now, in how many years will there be only 5000 grams of the isotope remaining in the substance?

   b. The isotope plutonium-239 (produced in breeder nuclear reactors and used in atomic bombs) has a half-life of approximately 24,400 years. If a substance contains 10,000 grams of the isotope now, in how many years will there be only 1250 grams of the isotope remaining in the substance?

▲ 32. [Inspired by John Oprea of Cleveland State University.]

   a. Radioactive iodine-125 caused great concern when it was released into the atmosphere across Europe in the nuclear reactor disaster at Chernobyl in 1986, since its radiation can cause thyroid cancer. Its half-life is 60 days. How long does it take for a quantity of iodine-125 to decay to 0.1% (one one-thousandth) of the original amount?

   b. The Nuclear Test-Ban Treaty of 1963 brought an end to atmospheric testing of nuclear weapons. Testing during the 1950s and 1960s released into the atmosphere the radioactive isotope strontium-90. It settled out of the air onto grass in fields, was eaten by cows, and wound up in children's milk. In the body, strontium-90 is absorbed into the bones, where its radiation can cause cancer; its half-life is 25 years. Of the strontium-90 absorbed into the bones of children in the 1950s, approximately how much will still remain 50 years later, during the first decade of the twenty-first century?

For Exercises 33 to 36, refer to the following: carbon-14 dating can be used to determine the age of materials that contain carbon. In a living body, for each gram of carbon, approximately 814 carbon-14 atoms decay each hour, each emit-

▲ Advanced exercise.

ting radiation. By measuring carefully the amount of carbon in a sample, and then counting the number of rays emitted over a period of time, we can determine the age of the sample. For example, if a sample is found to be giving off 26 rays per hour per gram of carbon sample, the estimated age of the sample can be determined by working backwards as follows:

Age of fossil (in years)	Rays emitted per gram per hour
0	814
5,730	$\frac{1}{2}(814) = 407$
11,460	$\frac{1}{2}(407) = (\frac{1}{2})(\frac{1}{2})(814) = (\frac{1}{2})^2(814) = 203.5$
17,190	$\frac{1}{2}(203.5) = (\frac{1}{2})^3(814) = 101.75$
22,920	$\frac{1}{2}(101.75) = (\frac{1}{2})^4(814) = 50.9$
28,650	$\frac{1}{2}(50.9) = (\frac{1}{2})^5(814) = 26$

Hence, the 814 rays would be decreased to approximately 26 rays in approximately 29,000 years, the estimated age of the fossil. (Note: An age of 0 for the fossil denotes the time of death of the living body.)

▲ 33. A carbon sample is emitting approximately 6.5 rays per gram per hour from carbon-14 decay. Determine the approximate age of the fossil.

34. A carbon sample is determined to be approximately 5100 years old. Approximately how many atoms of carbon-14 are breaking down per gram per hour?

▲ 35. The formula that relates Y, the age of the sample in years, and N, the number of carbon-14 atoms disintegrating per gram per hour, is

$$\left(\frac{1}{2}\right)^{Y/5730} = \frac{N}{814}$$

The formula can be solved simply for N, giving

$$N = 814 \times \left(\frac{1}{2}\right)^{Y/5730}$$

If you are familiar with natural logarithms, you can also solve for Y, getting the formula

$$Y = 55,403 - 8267 \ln N$$

(Despite the appearance of the formula, it can lead to a value of Y that is greater than 55,403 years. The quantity $\ln N$ can be negative if $N < 1$, that is, if there is less than one disintegration per gram per hour. Such rates are typical for small samples that are particularly old, and such samples must be observed for many hours to arrive at an accurate estimate of N.)

Use the formula to redo Exercise 33.

▲ 36. The "Ice Man" is the popular name for the body of a man that was found in 1991, having been preserved in a glacier in the Tyrolean Alps. At first researchers speculated that he had been a medieval messenger who perished in a storm. Carbon-14 dating of the Ice Man surprised everyone when it revealed that he had died about 5000 years ago. About how many atoms of carbon-14 are breaking down today per gram of carbon of his tissue per hour?

▲ 37. Suppose that a population of size P grows by an amount

$$Pk\left(1 - \frac{P}{100}\right)$$

between observations. We view k as an intrinsic growth rate (the rate of population growth without resource constraints) and the number 100 as a carrying capacity of the environment.

    a. For an initial population of 10 and $k = 0.8$, find the sizes of the population for the next 10 observations.

    b. As in part a, but for an initial population of 110. What differences do you observe between the results for these two initial populations?

    c. As in parts a and b, but with $k = 1.8$.

    d. On the basis of your analysis of the situations, what can you say about the dependence of the population growth on the parameter k?

▲ 38. Suppose that a population of size P grows by the amount

$$Pk\left(1 - \frac{P}{M}\right)$$

between observations, where k is the intrinsic growth rate and M is the carrying capacity of the environment. Suppose also that the carrying capacity is 100 for the first 5 observations and then drops to 70. (There is an environmental catastrophe at that time; for instance, a flood wipes out much of the food supply.)

    a. For an initial population of 20 and $k = 0.9$, find the population sizes for the first 10 observations.

    b. As in part a, but suppose also that the carrying capacity M is increasing steadily. This might be the case if, for example, the food supply was increasing steadily. Suppose that in the nth year the carrying capacity is $100 + 5n$. For an initial population of 10 and $k = 0.7$, find the population sizes for the first 10 observations.

    c. As in part b, but for an initial population of 10 and $k = 2$. Find the population sizes for the first 10 observations.

    d. What do you conclude? Describe what is happening in terms of the setting of the problem.

■ 39. In Exercise 22, we modeled inflation. When inflation proceeds at a constant rate, we can calculate its effect on prices by using the equation or exponential growth. However, in the real world, inflation does not remain constant. The U.S. Department of Labor keeps track of inflation and publishes a Consumer

▲ Advanced exercise.　　■ Discussion exercise.

Price Index for comparing the prices in different years. We reproduce that table below (Table 17.4). The table shows what a "market basket" of goods (including housing and transportation) that cost \$100 in 1982–1984 would have cost in other years.

TABLE 17.4   U.S. Consumer Price Index (1982–1984 = 100)

—	—	1941	14.7	1971	40.5
—	—	1942	16.3	1972	41.8
1913	9.9	1943	17.3	1973	44.4
1914	10.0	1944	17.6	1974	49.3
1915	10.1	1945	18.0	1975	53.8
1916	10.9	1946	19.5	1976	56.9
1917	12.8	1947	22.3	1977	60.6
1918	15.1	1948	24.1	1978	65.2
1919	17.3	1949	23.8	1979	72.6
1920	20.0	1950	24.1	1980	82.4
1921	17.9	1951	26.0	1981	90.9
1922	16.8	1952	26.6	1982	96.5
1923	17.1	1953	26.7	1983	99.6
1924	17.1	1954	26.9	1984	103.9
1925	17.5	1955	26.8	1985	107.6
1926	17.7	1956	27.2	1986	109.6
1927	17.4	1957	28.1	1987	113.6
1928	17.1	1958	28.9	1988	118.3
1929	17.1	1959	29.1	1989	124.0
1930	16.7	1960	29.6	1990	130.7
1931	15.2	1961	29.9	1991	136.2
1932	13.7	1962	30.2	1992	140.3
1933	13.0	1963	30.6		
1934	13.4	1964	31.0		
1935	13.7	1965	31.5		
1936	13.9	1966	32.4		
1937	14.4	1967	33.4		
1938	14.1	1968	34.8		
1939	13.9	1969	36.7		
1940	14.0	1970	38.8		

Note: This index covers all urban consumers, about 80% of the U.S. population. Each figure is an average for the year. The basis for the index is the period 1982–1984, for which the index was set equal to 100.

a. I bought my first LP record in 1965, at list price, for $4.98. How much would that be in 1992 dollars? How does that compare with the list price of a CD today?

b. My father bought a Royal portable typewriter in 1940 for $40 (I have the sales slip). What would be the equivalent price in 1992 dollars? How does that compare to the cost of a portable typewriter today?

c. My first-semester college mathematics book cost $10.75 in 1962. What would be the equivalent price in 1992 dollars? How does that compare to what you paid for this book? (My book did not have text in two colors, and it had no photographs, color or otherwise.)

d. In 1970, before the OPEC oil embargo, gasoline cost about 25 cents per gallon. In 1974, after the embargo, it cost about 70 cents per gallon. What would be the equivalent prices in 1992 dollars? How do they compare to the price of gasoline today?

■ 40. From the Consumer Price Index table, you can determine the rate of inflation from one year to the next. For example, you find the rate of inflation from 1991 to 1992 by subtracting the two index numbers and dividing by the earlier one: $(140.6 - 136.2)/136.2 = 0.032 = 3.2\%$. Similarly, knowing the rate of inflation, you can compute one index number from another. (Thus, from learning the rate of inflation from newspapers, you will be able to add entries to the Consumer Price Index table for years past 1992.)

a. What was the rate of inflation from 1980 to 1981?

b. For a 50% rate of inflation per year, what would the Consumer Price Index be for each of the years 1993 through 1997?

c. Gasoline cost about $1.10 per gallon in mid-1992, when Ross Perot proposed increasing the federal tax on gasoline by 10 cents each year for five years in order to pay for investments in U.S. infrastructure and to reduce dependence on oil imports. Assume, as in part b, that prices will go up 5% each year for the foreseeable future, including the price of gasoline. In addition to inflation, the price of gasoline would go up 10 cents more in 1993, 10 more cents in 1994, and so forth, through a fifth 10-cent hike in 1997. What would the price of gasoline be in 1997?

d. Convert your answer in part c to 1992 dollars and compare it with your answers to Exercise 39d. Discuss possible conclusions from your calculations.

■ 41. The TV series *All in the Family* featured Archie Bunker, his wife Edith, his daughter Gloria, and her friend (and later husband) Michael (called "Meathead" by Archie). This family provides a way to visualize the changes over the past 20 years in the economic situation of a U.S. blue-collar family. The series began in 1973. Archie is a factory worker, about 50 years old, with at most a

■   Discussion exercise.

high school degree earning about $13,000 then. Gloria, living at home and employed half-time, earns about $2000. Michael earns about $10,000 at his factory job.

We consider a similar family in 1979, the Trapps, with people of the same ages, education, and social background as before. The 50-year-old factory worker Art earns about $12,600. To help support them, his wife Enid is employed, earning about $9000. Their daughter Gina is married to Martin ("Cheddarhead"), who earns about $15,000 at the cheese factory, while Gina is at home with children.

We move now to 1988 and yet another family with the same age structure, the Sands. Patriarch Arnie earns about $34,000 at his job, while wife Eve makes $12,000. Their daughter Gwen is married to Matt ("Muttonhead"), who earns $25,000 (when he isn't laid off from the meat-packing plant). Gwen works, too, earning $12,000; but day-care expenses for their son cost $5000 per year.

Use the Consumer Price Index (Table 17.4) to convert all of these figures to 1992 dollars, and then compare the relative situations of these similar families in 1973, 1979, and 1988.

(Thanks to Paul Solman of PBS's *MacNeil-Lehrer Newshour* for the idea and the data.)

■ 42. In mid-1990, the population of the world was 5.3 billion and increasing at the rate of 1.7% per year.

  a. Project the world's population to mid-2000, to mid-2020 (by which time you will likely have finished having whatever children you may have), and to mid-2040 (by which time you will likely have retired).
  b. What are the assumptions involved in your projections?
  c. You can make a more refined model, which will give more realistic answers, by dividing the countries of the world into three groups that have differing rates of increase:

Group	Population mid-1990 (billions)	Rate of growth (%)
More developed countries	1.21	0.6
Less developed countries (excluding China)	2.98	2.4
China	1.12	1.4

  To get a projection for the world's population, project each group separately and add the totals. Redo your projections for the years 2000, 2020, and 2040. Do you find the differences from your earlier projections to be significant?
  d. Will the world be able to support the numbers of people that you project? What problems will these greater numbers of people cause? What could be done to avert those problems? Do you think that anything will be done before there is some kind of worldwide crisis?

▶ WRITING PROJECT

1. Based on the calculations you did in Exercise 42 and the discussion you had with other members of the class, write a short essay in the form of a guest editorial for a newspaper. Describe your projections and how you arrived at them, how serious a problem you think population growth is, what problems it is likely to cause, what you think needs to be done, and what the implications are for your own life.

Chapter 18

Inaccessible Distances

Before 1600, people had nothing more than their eyes and sighting rods with which to see the universe. Nevertheless, they made precise measurements of the size of objects on the earth, the size of the earth itself, and the distances to the moon and sun. Today we use the techniques developed long ago to make all kinds of measurements, for navigation on the earth, in the air, and in space.

The mathematical concepts used to determine these distances are congruence and similarity of triangles. Two triangles are **congruent** if one is an exact copy of the other; one congruent triangle can be made to fit exactly onto another. Two triangles are **similar** if they have the same shape but not necessarily the same size. Triangles that are congruent are also similar, but triangles that are similar may be of different sizes and hence not be congruent.

The basic ideas of congruence and similarity were set down by Euclid in his *Elements*, written sometime around 300 B.C. and consisting of 13 chapter-long "books." Although the *Elements* organized all the geometrical and arithmetical knowledge accumulated by the Greeks, thus incorporating earlier knowledge from Babylonia and Egypt as well, its greatest achievement was to show a natural sequence by which one result could be derived logically from another.

Book I of the *Elements* contains most of the standard facts about congruence for triangles. These facts are followed by the introduction of the famous parallel postulate (discussed in Chapter 20), which proved to be very important in the history of ideas and paved the way for Einstein's theory of relativity and other modern theories about the structure of the universe. Book I culminates with the **Pythagorean theorem**, which gives a way to calculate distances along a diagonal when the corresponding horizontal and vertical distances are known. This theorem is the basis of the standard ways of determining the distance between two points on a straight line or on a curve, and hence for all of analytic geometry. It is in turn the necessary preliminary for calculus, the tool that Newton invented for understanding the motions of the planets and of falling apples.

▶ DISTANCES VIA THE PYTHAGOREAN THEOREM

The Pythagorean theorem states that

In a right triangle, the square of the length of the side opposite the right angle equals the sum of the squares of the lengths of the other two sides.

The fundamental role of the Pythagorean theorem is illustrated by the following problem (see Figure 18.1): In a city laid out on a grid of east-west and north-south streets, a house B is known to be four blocks south and three blocks east of another house, A. What is the straight-line distance AB and A to B?

Here is the answer: Because $AC = 4$ and $BC = 3$, by the Pythagorean theorem we know that $(AB)^2 = (AC)^2 + (BC)^2 = 4^2 + 3^2 = 16 + 9 = 25$. Since $(AB)^2 = 25$, we find that $AB = 5$.

In other words, the Pythagorean theorem says that for any right triangle, the sum of the squares of the two short sides, or **legs**, is equal to the square of the long side, the **hypotenuse**.

In the formulation of the Pythagorean theorem in Euclid's *Elements*, the squares of the lengths of the sides of a right triangle were envi-sioned as areas of geometric squares constructed on those sides (Figure 18.2).

The Pythagorean theorem is a remarkable result. How was it discovered in the first place? How were the ancients sure that it was true? The presentation in Euclid is a highly systematic and organized presentation that hides the original roots. So, we don't know how the Pythagorean theorem was discovered. For the particular case of the 3-4-5 right triangle, we have a figure (but not a written-out proof) in a manuscript dating to about 2000 years ago in China. Like Euclid's *Elements*, this manuscript compiles mathematical facts learned over the previous several hundred years.

Figure 18.3 shows the proof. There is a 3-4-5 triangle colored blue in the upper right corner. You can identify a 3-by-3 square shaded in the upper left corner, a 4-by-4 square shaded in the lower right corner, and a 5-by-5 square askew in the middle. The proof proceeds by adding up the area of the overall 7-by-7 square in two different ways.

The first way: The 7-by-7 square is made up of the shaded 3-by-3 square, the shaded 4-by-4 square, a 3-by-4 rectangle in the upper right, and a 4-by-3 rectangle in the lower left. So

Area of 7-by-7 square
 = area of 3-by-3 square
  + area of 4-by-4 square
  + area of two 3-by-4 blocks

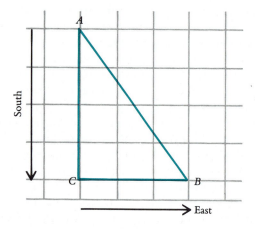

Figure 18.1 *Problem:* What is the distance, as the crow flies, from A to B?

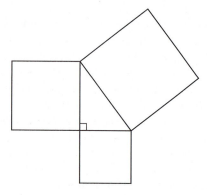

Figure 18.2 The Pythagorean theorem as visualized by the ancient Greeks, in terms of the areas of squares built on the sides of a right triangle.

The second way: The 7-by-7 square is made up of the 5-by-5 askew square in the middle, plus, surrounding it, four triangles that have been formed by splitting 3-by-4 blocks down their diagonals. So

Area of 7-by-7 square
  = area of 5-by-5 square
    + area of four halves of 3-by-4 blocks

Hence we have

Area of 3-by-3 square
  + area of 4-by-4 square
  + area of two 3-by-4 blocks
  = area of 5-by-5 square
    + area of four halves of
    3-by-4 blocks

The area of four halves of 3-by-4 blocks is the same as the area of two 3-by-4 blocks; subtracting this amount of area from each side of the equation gives us

Area of 3-by-3 square
  + area of 4-by-4 square
  = area of 5-by-5 square,

which is what was to be proved.

A slightly different idea leads to a general proof that $a^2 + b^2 = c^2$ for a right triangle with hypotenuse c and legs a and b.

Later books of the *Elements* go on to treat the standard theory of similarity, which is the basis for all map-making and for most other methods of calculating and representing inaccessible distances, including the tools of trigonometry. In the remainder of this chapter we show how these simple ancient tools enable us to measure other distances that are relevant to us today.

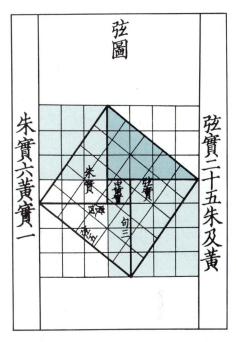

Figure 18.3 Illustration of the Pythagorean theorem for a 3-4-5 right triangle from the Chinese work *The Arithmetical Classic of the Gnomon and the Circular Paths of Heaven* (*Chou-pei suan-ching,* or *Zhōubì suànjīng* in the Pinyin system), the earliest Chinese writing on mathematics to survive to today. [From Frank J. Swetz and T. I. Kao, *Was Pythagoras Chinese?*, Pennsylvania State University Press, University Park, Pa., 1977.]

▶ ESTIMATING INACCESSIBLE DISTANCES

Our story concerns four men: Thales, Euclid, Aristarchus, and Eratosthenes, each of whom developed new techniques for measuring ever more distant objects. The last three lived about 300 B.C. and were probably born in the order named. Some 300 years earlier, Thales is supposed to have made two difficult measurements: (1) the distance of ships at sea, using *congruence* of triangles, and (2) the height of the Great Pyramid in Egypt, using *similarity* of triangles.

Using Congruent Triangles

Suppose you want to find the distance of a ship at position *B* straight out at sea from your position *A* on the shore, using the method attributed to Thales (see Figures 18.4 and 18.5). Here is Thales' solution: Starting from *A*, walk along the shore any distance in a direction perpendicular to *AB*. Put a marker at *S*, making sure it is tall enough to see from a distance. Then walk the same distance to point *C*. Now turn at a right angle and walk away from the shore until you reach a point *E* from which your marker *S* is exactly lined up with the ship *B*. The distance *CE* you walked away from the shore is exactly the same as the distance *AB* of the ship from shore.

Figure 18.5 shows why this works. The simple geometry that Thales used, long before Euclid, tells us that the two angles marked at *S* (known as **vertical angles**) are congruent. The angles at *A* and *C* are equal because both are right angles. The sides *AS* and *SC* are equal because they were paced off to be equal. By one of the congruence theorems for triangles, we know that triangle *ABS* is congruent to triangle *CES*, and hence that the corresponding lengths *AB* and *CE* are equal, as we claimed. The congruence theorem that we use here says that two triangles are congruent if two angles and the included side of one are congruent to two angles and the included side of the other.

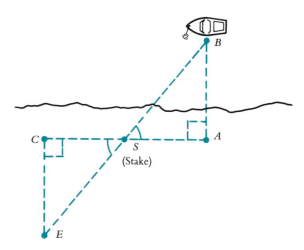

Figure 18.5 *Solution:* Thales found the distance of the boat from shore by putting a stake at *S* and then showing that triangles *SAB* and *SCE* are congruent.

Using Similar Triangles

If a triangle is enlarged or reduced by a photocopy machine, then the resulting triangle is similar to the original triangle. Thus, corresponding angles of similar triangles have the same size, and the lengths of corresponding sides are in the same ratio. For example, a photograph and its enlargement are similar to each other and to the original (we looked at enlargements, similarity, and scaling factors further in Chapter 16). Thales may have known that right triangles with corresponding angles are similar and hence have proportional sides. With this knowledge he could calculate the height of the Great Pyramid.

To take the height of the Great Pyramid (or any other vertical object such as a tall tree), hold an upright stick on the ground at the site of the object and measure its length and the length of its shadow. The right triangle whose legs are the stick and its shadow is similar to a right triangle whose legs are any other vertical object and its shadow (at the same place and time). Hence the ratio of these

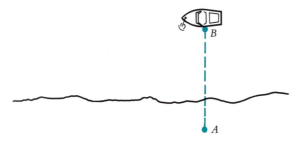

Figure 18.4 *Problem:* Find the distance of the boat from shore, using elementary geometry.

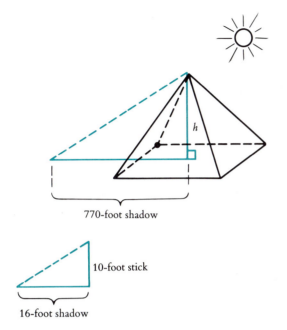

770-foot shadow

10-foot stick

16-foot shadow

Figure 18.6 Thales used similar triangles to determine the height of the Great Pyramid. Since the shadow of the Great Pyramid is 48.1 times the shadow of the stick, then its height must be 48.1 times the 10-foot height of the stick, 481 feet.

two lengths, stick and shadow, is the same as the ratio of the length (height) of the Great Pyramid to the length of its shadow (see Figure 18.6).

EXAMPLE: Finding the Height of the Great Pyramid

Suppose, to be specific, that in this case the length of the stick was 10 feet, it cast a 16-foot shadow, and Thales' measurement of the shadow of the Great Pyramid was 770 feet. From the equality of these ratios, $10/16 = h/770$, he could calculate the height h of the Great Pyramid to be $10/16 \times 770$, or 481 feet. ▲

Notice that ratios (such as the 10/16 of the height of the stick to the length of its shadow) are the key to finding all sorts of inaccessible heights. In fact, if the ratio of stick to shadow is a/b and the length of shadow of any object is s, then we can find the height of that object simply by multiplying the two: $h = (a/b) \times s$.

In later times extensive tables of such ratios showed, for any given angle of inclination of the sun, what the ratio of the length of the stick to its shadow would be (see Table 18.1).

With such a table, the heights of inaccessible objects can be calculated without the use of the stick if a suitable instrument, such as a sextant, is available for measuring the angle of the sun above the horizon. Table 18.1 immediately tells us, for each angle of the sun, the ratio for

$$\frac{\text{Length of stick}}{\text{Length of its shadow}}$$

To summarize, the procedure for measuring height is

1. Measure the length s of the shadow of the object whose height you want to determine.

2. Measure the angle of the sun above the horizon, and look up the corresponding ratio r in the table.

3. Multiply $r \times s$ to find the height of the object.

This procedure was the beginning of trigonometry (triangle measurement) as we know it. The ratio that we have tabulated is now called the **tangent** of the angle (see Figure 18.7).

EXAMPLE: Measuring the Height of a Flag Pole

The Department of Physical Plant at Beloit College needed to know the height of the college flag pole in order to buy rope of the right length to raise the flag. The shadow of the flag

TABLE 18.1   Table of Tangents

Angle of the sun above the horizon	Tangent of the angle = length of stick / length of its shadow
5° (Sun nearly on the horizon)	0.08749
10° (Sun somewhat higher)	0.17633
20°	0.36397
30°	0.57735
40°	0.83910
45° (Sun exactly halfway between horizon and directly overhead)	1.00000
50°	1.19175
60°	1.73205
70°	2.74748
80°	5.67128
87°	19.08114
89° (Sun almost directly overhead; higher than it ever gets in the United States)	57.28996

pole was measured to be $s = 25.5$ feet. A Brunton compass from the Geology Department was used to measure the angle of the sun above the horizon, which was 70°. From Table 18.1 (or from using the "tan" button on a calculator), we find the tangent of 70° is $r = 2.75$. So the height of the flag pole is $r \times s = 2.75 \times 25.5$ feet $= 70$ feet.   ▲

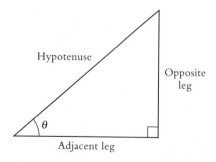

Figure 18.7 The tangent ratio for the angle θ is the length of the leg opposite θ divided by the length of the leg adjacent to θ.

We can verify the entries in the tangent table by using a pocket scientific calculator—one with buttons for the "trig" functions sin (short for "sine"), cos (short for "cosine"), and tan (short for "tangent"). Try this on your calculator: press the "tan" button for various angles and check that you get the same values given in Table 18.1. On some calculators there may be slight roundoff errors, so you may get tan 45° = 0.99999 instead of tan 45° = 1.00000. (Be sure that the calculator is in *degree* mode. If you get 1.62 instead of 1.00, you are in *radian* mode instead. Radians are another unit for measuring the size of angles.) Essentially, however, you should be able to use your calculator to reproduce the table and find the tangent of any angle.

EXAMPLE: How Low Are the Clouds?

One class of private pilot license allows a pilot to fly only when the cloud ceiling is at least 1000 feet and ground visibility is at least

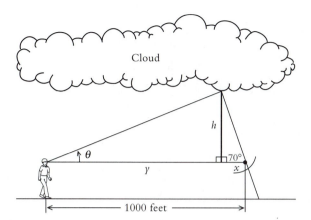

Figure 18.8 The geometry of determining the height of a cloud at night. The observer is at left and the light source at right. (The figure is not to scale.)

3 miles. With training, both of these factors can be determined easily in daylight. At night, ground visibility can be gauged from nearby lights whose distances are known; but what about cloud height?

Determining cloud height at night requires a special technique and some trigonometry. Light is directed toward the clouds, with the light kept in a narrow beam with the help of a parabolic reflector (how it works is a topic in Chapter 19). The light is directed upward at an angle of 70°, and an observer 1000 feet away sights toward the reflection of the light from the cloud and measures the angle of elevation θ. We show the situation in Figure 18.8, where we also introduce some notation for the various distances involved.

Our experience with right triangles tells us that

$$\tan \theta = \frac{h}{y} \qquad \tan 70° = \frac{h}{x}$$

which we can convert into

$$y = \frac{h}{\tan \theta} \qquad x = \frac{h}{\tan 70°}$$

We realize that $x + y = 1000$, so we have

$$1000 = \frac{h}{\tan \theta} + \frac{h}{\tan 70°} = h\left(\frac{1}{\tan \theta} + \frac{1}{\tan 70°}\right)$$

What we want is h, so we solve to get

$$h = \frac{1000}{\frac{1}{\tan \theta} + \frac{1}{\tan 70°}}$$

What's the angle that corresponds to a cloud height of 1000 feet, which is the dividing line between flying and not flying? We put in $h = 1000$ and find we have to solve the equation

$$\frac{1}{\tan \theta} + \frac{1}{\tan 70°} = 1$$

From our tangent table (or from your calculator) we find $\tan 70° = 2.75$, so

$$\frac{1}{\tan \theta} = 1 - \frac{1}{2.75} = 0.636$$

so $\tan \theta = 1.57$ and (from a tangent table or calculator) $\theta = 58°$. If the angle that the observer sights is larger than 58°, then the cloud ceiling is above 1000 feet. ▲

▶ DIGGING STRAIGHT TUNNELS

The famous Greek historian Herodotus, who lived some 100 years after Thales, described three engineering achievements on the Greek island of Samos. One was a tunnel that brought water through Mount Castro to the capital city, Samos (see Figure 18.9).

Nearly 2500 years later, in 1882, archeologists rediscovered the tunnel, exactly as Herodotus had described it. It was 1 kilometer (about 0.6 mile) in length and more than 2 meters (about 6 feet) high and wide. A deep ditch in its floor contained pipes,

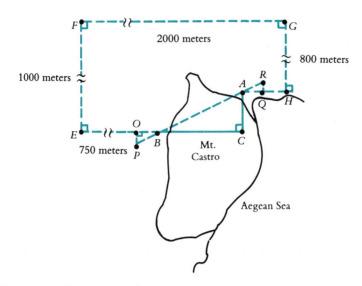

Figure 18.9 The plan for digging a tunnel through Mount Castro on the island of Samos. Using this plan, two digging teams starting at opposite ends A and B met at the center with only a small error.

and the tunnel had vertical vents for changing the air and cleaning away rubble, and niches where workers placed their lamps. The ditch had a depth of some 2 meters at the upper end and 8 meters at the lower end, and was probably dug because the drop that had originally been planned turned out to be too small.

The remarkable thing about this tunnel was that the digging teams, proceeding from each end, met at the center with an error of only 10 meters (33 feet) horizontally and 3 meters (10 feet) vertically. We know this because at the center of the tunnel there is a jog of that size to make the two ends meet.

King Hezekiah of Judea was less successful. When he had a similar aqueduct constructed through the rocks near Jerusalem around 700 B.C., his workers had to check the direction of digging in a very primitive way, by means of vertical shafts from the top. The result was a zigzag tunnel twice as long as the distance between its ends.

How was the Samos tunnel dug without the benefit of guiding shafts? We do not know for

sure, but a later writer, Heron, described a likely method, which we modify slightly here to bring out the essentials. In his view, the method used similar triangles in a considerably more complicated way than Thales had used them. We describe this method in detail, following Figure 18.9.

Suppose the tunnel entrances are to be at A and B, on opposite sides of Mount Castro in Figure 18.9. Begin by marking off some convenient distance BE on any line at B. Following the figure, make a right turn at E and go to F. At F turn again and go to G, then turn again and go to H, which is chosen so that a right turn takes you straight to A. Suppose, to be specific, that the distances in this detour around the mountain are $BE = 750$ meters, $EF = 1000$ meters, $FG = 2000$ meters, $GH = 800$ meters, and $HA = 250$ meters.

Now FG is parallel to EC, and these two lines are 1000 meters apart. Since it is 800 meters from G down to H, it must be 200 meters more to go from A down to C, so $AC = 200$ meters. Similarly, since EF and HG are parallel and 2000 meters apart, you can deduce that $BC = 1000$ meters.

So, in the triangle *ABC*, we have leg *BC* = 1000 meters and leg *AC* = 200 meters, for a 5-to-1 ratio. In order to decide in what direction to dig at *B*, we construct another 5-to-1 right triangle *OBP*, with, say, *OB* = 50 meters and *OP* = 10 meters. Because the proportions of the legs will be the same, this triangle will be similar to triangle *ABC*. So the corresponding angles *ABC* and *OBP* will be equal, and *PBA* will be a straight line. We will sight from *P* to *B* and dig in that direction, which the geometry tells us will be toward *A*.

We do the same at *A*, constructing a right triangle *RAQ*, with, say, *AQ* = 50 meters and *RQ* = 10 meters, in order to determine the direction to dig from *A*.

Thus, the clever use of similar triangles over 2600 years ago helped to solve a major problem of civil engineering.

Congruent Triangles or Similar Triangles?

In a particular problem, how do you decide which to use: congruent triangles or similar triangles? Recall, congruent triangles are the same *size*; similar triangles are just the same *shape*. In some special cases, like finding the distance out to a boat, it is easy and convenient to use congruent triangles, though we could have used similar triangles instead.

In general, though, similar triangles are more versatile than congruent triangles in practical applications. For example, we could have used congruent triangles in the previous tunnel example, instead of similar triangles, making triangles OBP and *RAQ* congruent to (the same shape and size as) triangle *ABC* instead of just similar to (the same shape as) it. Doing so would have provided longer sight lines from *P* to *B* and from *R* to *A*, thus allowing more accurate determination of the directions to dig. For even greater accuracy, we could even have used similar triangles *OBP* and *RAQ* that are *larger* than triangle *ABC*. At some point, though, the terrain and the need for clear

visibility along the sight lines would limit the size of the triangles that we could use, and congruent triangles might not be feasible.

▶ MEASURING THE EARTH

The modern use of similar triangles in engineering projects also depends on similarity principles. Our next example makes a big jump in the gradually increasing scale of distances: here we learn how to find the size of the earth itself. One of the truly spectacular achievements of ancient mathematical science was the determination, by a very simple method, of the **circumference** of the earth, or the distance around the earth measured along a circle passing through the poles. The most accurate of these calculations was that of Eratosthenes in about 200 B.C. His method is illustrated in Figure 18.10.

It was known that at a certain time the sun was directly overhead at Syene (now Aswan), point *S*, in Egypt. At exactly the same time in Alexandria, lying straight north of Syene at point *A*, the posi-

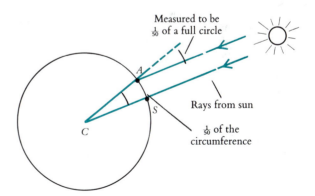

Figure 18.10 Eratosthenes noticed that the sun is about one-fiftieth of a circle south of the zenith at Alexandria when it is directly overhead at Syene. Thus the distance between Alexandria and Syene must be about one-fiftieth of the earth's circumference.

tion of the sun was measured to be $\frac{1}{50}$ of a full circle (that is, 7.2°) away from directly overhead. Because the sun is so far away from the earth, the two arrows in the figure that point to the sun are essentially parallel lines. Hence, the angle at C, the center of the earth, is also $\frac{1}{50}$ of a full circle because it is the **corresponding angle** when the two parallel lines are cut by the line AC (AC is traditionally called a **transversal** of the parallel lines).

Then, because the angle at C is $\frac{1}{50}$ of 360°, the full circle, the distance AS, from Alexandria to Syene, is also $\frac{1}{50}$ of the complete circumference of the earth. It is only necessary to measure the distance from Alexandria to Syene (not a triviality in those days!) to have all the information needed. When the distance from Alexandria to Syene was found to be 5000 *stadia* (a Greek unit of measure; singular, *stadium*), this yielded $5000 \times 50 = 250,000$ stadia for the circumference of the earth.

Although we are not sure how large the stadium unit was, one estimate from Pliny is that a stadium was 157.5 meters. Using this value, we get $157.5 \times 250,000$ meters $= 157.5 \times 250$ kilometers $\approx 39,000$ kilometers (24,500 miles) for the circumference of the earth. The symbol $\approx$ means "is approximately equal to." The kilometer was originally defined as $\frac{1}{10,000}$ of the distance from the North Pole to the equator — one-quarter of the earth's circumference — so that the earth's total circumference is 40,000 kilometers (24,900 miles). We see that Eratosthenes' result is very close to the true measurement. We can be excused for thinking that some of Eratosthenes' numbers seem to be rounded off, and hence only accidentally accurate. Even so, we must admire his achievement. After all, in later years there was even some doubt that the earth was round!

▶ MEASURING ASTRONOMICAL DISTANCES

Knowing the circumference of the earth (and hence its radius), the astronomer Aristarchus could consider even greater distances. His mea-

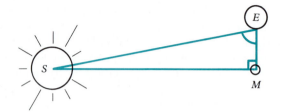

Figure 18.11 Aristarchus's method for estimating the ratio of the distances of the sun and moon from the earth.

surements of the distances of the moon and the sun from the earth were not as accurate as Eratosthenes' determination of the size of the earth, but his ingenious method is worth looking at. It shows how even a very simple understanding of triangle and circle geometry yielded information that completely revised his contemporaries' picture of the universe; they had imagined celestial distances to be much smaller than he showed them to be.

Using simple geometry, Aristarchus first determined how many times as far the sun is from the earth than the moon is from the earth (note that all our distances will be taken from the center of one object to the center of the other). He noticed that when the moon is exactly half full, that is, when we see exactly half the moon in shadow and half in the sun's light, the triangle MES formed by the moon, earth, and sun is a right triangle, with its 90° angle at the moon M (see Figure 18.11). (The two days in each month when the moon is exactly half full are marked on many modern calendars.)

We can learn the ratio MS/EM from the angle E — it is just the tangent of E, which we can look up in Table 18.1 once we have measured E. (We actually want the ratio ES/EM, but that will be close to MS/EM provided EM is small in comparison with ES, which is true.)

Aristarchus estimated angle E as 3° less than a right angle, that is, $E = 87°$. From the table we see that $\tan 87° \approx 19$; therefore, according to Aristarchus the sun is 19 times as far from the earth as the moon is.

Neither Thales nor Aristarchus actually had a table of tangents, but they were able to find these strictly geometric ratios in other, more complicated ways. Moreover, it is likely that Aristarchus realized that he had inaccurately estimated the crucial angle E, which could not easily be measured. A glance at the table shows that for large angles E, a *small* error makes a very *large* difference in the resulting ratio. The true value of E differs from $90°$ by less than one-sixth of a degree, so that E is more than $89\frac{5}{6}°$, and tan E is about 390, the true ratio of the distance of the sun to the distance of the moon from the earth. Even though Aristarchus's calculation was off by a factor of 20, his *method* was sound and his results revised upward the Greek estimates of the size of the universe substantially.

You will have noticed that Aristarchus's simple method gives only a ratio. To determine the actual distance to the sun, he needed to know the actual distance to the moon. By observing the time it takes for the shadow of the earth to cross the moon during a total eclipse of the moon, we can estimate this distance very accurately. Although Hipparchus, some 100 years after Aristarchus, used this method to come within 1% of the value we know today, the rougher estimates already available to Aristarchus were sufficiently accurate for his purposes.

Before describing this method, we need to remind ourselves of two simple facts from Euclidean geometry. In Euclid's treatise, facts about congruence appear at the very beginning. It is only after Euclid introduces properties of parallel lines that he is able to prove the fundamental result that the sum of the angles of a triangle is $180°$ (half of a complete circle). (This property of triangles — that all of them have the same angle sum, which is $180°$ — is one that dramatically distinguishes Euclid's geometry from *non-Euclidean geometries*, which are discussed in Chapter 20.)

A simple property of circles is also used in the derivation, namely, that the arc of a circle **subtended** (included) by an angle at its center is proportional to the radius of the circle. In Figure

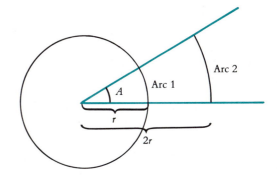

Figure 18.12   Arc 2 is twice as long as arc 1 because its circle has twice the radius of the other circle.

18.12, the arc subtended by angle A on the large circle is twice as long as the arc on the small circle because the radius of the large circle is twice that of the small one. When we measure the angle in degrees, the factor we have to multiply by to get the arc length is $\pi/180$ (approximately $\frac{3}{180}$, or $\frac{1}{60}$). Thus, for example, if a circle has a 10-foot radius, an angle of $90°$ at its center subtends (includes) an arc of $(\pi/180) \times 90 \times 10$, or approximately 16 feet.

Let's return to our main story: Aristarchus's measurement of the distance of the sun and the moon from the earth. Using the facts that the angle sum in a triangle is equal to a straight angle and that the arc subtended by a central angle of a circle is $(\pi/180) \times$ radius $\times$ angle, we can understand the following calculation of the distance of the moon from the earth. We here combine Aristarchus's own method, a similar method used by Hipparchus a century later, and modern notation.

By observing the amount of time it takes the earth's shadow to cross the moon, Aristarchus knew that the diameter of this shadow was about two times the diameter of the moon, as shown in Figure 18.13.

Because the moon's and sun's discs are both about the same size in the sky and because both are about $\frac{1}{720}$ of the whole circumference of the circle

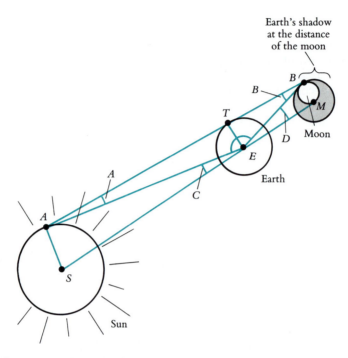

Earth's shadow
at the distance
of the moon

Figure 18.13 Aristarchus devised an ingenious method for determining the ratio between the moon's distance and the radius of the earth from the average duration of a lunar eclipse and the length of the month. His method resulted in an estimate for the moon's distance of 80 earth radii.

that they trace through the sky, the angles at C and D are easily found: $C = \frac{1}{1440}$ and $D = \frac{1}{720}$ of a complete circle. Hence $C + D = \frac{1}{1440} + \frac{1}{720} = \frac{1}{480}$ of a complete circle $= 0.75°$. Now, $A + B + E = C + D + E$ because $A + B + E$ is the angle sum in triangle AEB, and hence is equal to the straight angle $C + D + E$. Because $A + B$ is equal to $C + D$, we know $A + B$ is also $0.75°$. Moreover, angle A is very small compared to B (since the sun is much farther away than the moon), so angle B itself is approximately $0.75°$.

For angles as small as B, line segment TE is essentially equal to the arc TE of the circle with center at B and radius BT. Because the arc TE corresponding to angle B is proportional to the radius BT, we can write

$$TE = \frac{\pi}{180} \times BT \times B$$

Solving for angle B, we have

$$B = \frac{180}{\pi} \times \frac{TE}{BT}$$

Substituting $B \approx 0.75°$ and rearranging, we get

$$BT \approx \frac{180}{\pi} \times \frac{TE}{0.75} \approx 80 \times TE$$

That is, the distance from earth to moon, by this simple calculation, is about 80 earth radii. In fact, the distance is about 60 earth radii, a result that we could get from only slightly more refined versions of this same calculation.

Once the distance to the moon has been calculated, the ratio of the sun's distance to the moon's distance (which Aristarchus thought to be about 19, but which is in fact nearly 400) yields the distance to the sun in earth radii. Combined with the still earlier calculation of the earth's radius, this gives the distance to the sun, a value which we now know to be 92,956,000 miles. Using the same elementary ideas, we can calculate the distances to and between other planets in the solar system.

From Here to the Stars

An ancient argument against the heliocentric theory (that the earth orbits around the sun) was as follows. If the earth moves around the sun, then the stars should *appear* to move relative to one another as the earth travels through its orbit. (We observe a similar phenomenon today when we ride on a merry-go-round. If you look past the center pole toward bystanders and objects not on the merry-go-round, objects far away appear to move relative to closer ones, even though we know none actually move.) Since the ancients could not detect any such apparent motion, the stars appeared to them as fixed and unmoving. Copernicus, who argued for the heliocentric theory, surmised that the stars are so incomparably far away, compared to the distance from the earth to the sun, that any apparent motion is too slight to observe. This apparent relative motion, called **stellar parallax** (illustrated in Figure 18.14), was not confirmed until the nineteenth century.

(Today we know in addition that the stars not only appear to move because of parallax, but in fact they actually do move as the universe expands. But this actual motion, as viewed by us, is even tinier to us than the very small apparent motion due to parallax, and we will not need to consider it.)

Stellar parallax in fact is the key to determining how distant a star is. The more distant it is, the less it will appear to move. Put the other way round, the nearer a star is, the more it will appear to move, relative to more distant (and apparently fixed) stars. Figure 18.14 illustrates the principle (though the figure is certainly not to scale!).

You notice in the figure that the sun, the earth, and the nearer star form a right triangle (with the distances involved, it is a *very* skinny triangle). The short leg is the distance from the earth to the

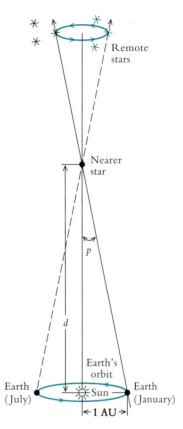

Figure 18.14 The geometry of stellar parallax (not to scale!). Remote stars *appear* in different positions, relative to the nearer star, as the earth moves around its orbit. The angle p is the parallax of the nearer star; the smaller the parallax, the larger the distance d to the nearer star.

sun, about 93 million miles, which astronomers take as one **astronomical unit,** written 1 AU. The tiny angle in the figure is the parallax of the star, which is measured in seconds of arc, that is, one second of a degree (1 degree = 60 minutes = 3600 seconds). Astronomers have devised another unit, the **parsec,** for measuring the very great distances to the stars. A star is one parsec away (from the earth or the sun) when its parallax is one second of arc. Put another way, in our right triangle, one astronomical unit subtends an angle of one second when the star is one parsec away.

A parsec is pretty large: 1 parsec = 3.26 light-years, the distance light would travel in 3.26 years, at 186,000 miles per second. Other conversions:

$$1 \text{ parsec} = 206,265 \text{ au} = 1.924 \times 10^{13} \text{ mi}$$
$$= 3.084 \times 10^{13} \text{ km}$$

A star that is two parsecs away will correspond to an even skinnier triangle, one whose angle is half as large. In terms of an equation, we have

$$d = 1/p$$

where d is the distance to the star (measured in parsecs) and p is the parallax of the star (measured in seconds of arc).

EXAMPLE: How Far Is the Nearest Star?

The nearest star is Proxima Centauri, and it has a parallax of 0.78 second. How far away is it? It is $1/0.78 = 1.3$ parsecs away, or 4.2 light-years. ▲

Because of blurring due to the atmosphere, from earth we can't measure parallaxes any smaller than about 0.01 second of arc, corresponding to stars that are within 100 parsecs (approximately 300 light-years) from the earth. The Hubble Space Telescope, which allows measurements to as small as 0.001 second of arc, greatly expands the number of stars whose distances we can measure.

▶ REVIEW VOCABULARY

Astronomical unit (AU)   The average distance from the earth to the sun, 93 million miles.
Circumference   The distance around a circle; for a sphere, the distance around a circle that passes through the poles.

Congruent   Two geometric figures are congruent if they have the same shape and the same size. In effect, they are the same figure in different positions.
Corresponding angle   When two parallel lines are cut by a third line (sometimes called a *transversal*), the angles formed match up; ones on the same side of the third line that are equal are called corresponding angles.
Elements   Euclid's compilation and organization of the geometric and arithmetic knowledge of his time. Most high school geometry texts are strongly influenced by this work.
Hypotenuse of a right triangle   The side opposite the right angle.
Legs of a right triangle   The two sides that include the right angle.
Parsec   A unit of distance, approximately 3.26 light-years. A star is one parsec away when its parallax is one second of arc.
Pythagorean theorem   "The sum of the squares on the two short sides of a right triangle is equal to the square on the long side." A fundamental tool for calculating distances, as in surveying.
Similar   Included; opposite to and delimited by.
Stellar parallax   Stars relatively close to the earth appear to move compared to more distant stars; how much a star seems to move is its parallax, which is measured in seconds of arc.
Subtended   An angle at the center of a circle is said to subtend the arc of the circle that it includes, and the arc is said to be subtended by the angle.
Tangent of an angle   The tangent of an angle in a right triangle is the ratio of the length of the side opposite the angle over the length of the side adjacent to the angle. (Two right triangles have to be similar if they have another pair of corresponding angles equal; so the tangent of an angle doesn't depend on what right triangle we pick to calculate it.)
Transversal   A third line cutting across two parallel lines.

Vertical angles   Two line segments that cross form two pairs of opposite angles, each pair of which is a pair of vertical angles.

▶ SUGGESTED READINGS

HEATH, T. L.: Introduction to Books I and II, in *The Thirteen Books of Euclid's Elements*, vol. 1, Dover Publications, New York, 1956.

JACOBS, HAROLD: *Geometry,* 2nd ed., Freeman, New York, 1987, pp. 364–374, 390–398, 404–408. Similar triangles, the Pythagorean theorem, and the tangent ratio.

KNILL, GEORGE: "Cloud Height at Night," *Mathematics Teacher* (October 1980). Reprinted in *Applications of Secondary School Mathematics: Readings from the MATHEMATICS TEACHER,* Joe Dan Austin (ed.), National Council of Teachers of Mathematics, 1991, pp. 203–205.

LAYZER, DAVID: *Constructing the Universe*, Scientific American Library, Freeman, New York, 1984.

PENNA, MICHAEL: *Surveying Outer Space.* UMAP Modules in Undergraduate Mathematics and Its Applications. Module 580. COMAP, Inc., Lexington, Mass., 1982. 2nd ed., 1993. Reprinted in *UMAP Modules: Tools for Teaching 1992,* edited by Paul J. Campbell, COMAP, Lexington, Mass., 1993.

VAN DER WAERDEN, B. L.: *Science Awakening,* Science Editions, Wiley, New York, 1963.

▶ EXERCISES

1. Triangles are *congruent* if they have the same shape *and* size. The corresponding parts (sides and angles) have the same measures. An *equiangular* triangle is one in which all three angles are the same size. Are all equiangular triangles congruent? Give an appropriate reason for your answer.

2. Does a diagonal of a square separate the square into two congruent triangles? Give an appropriate reason for your answer.

3. A right triangle has legs of lengths 6 centimeters and 8 centimeters. Determine the length of its hypotenuse.

4. A right triangle has a leg of length 12 feet and a hypotenuse with length 13 feet. Determine the length of the other leg.

5. A 26-foot ladder is placed against a building, with the foot of the ladder 10 feet from the base of the building. How far above the ground does the top of the ladder lean against the building?

6. The Great Pyramid is no longer as high as it was in Thales' time. A modern measurement using the same 10-foot stick with a 16-foot shadow would find a pyramid shadow of only 720 feet. How high is the pyramid now?

7. Two smaller pyramids near the Great Pyramid were, in Thales' time, 471 feet and 215 feet tall, respectively. What lengths of shadows would they have cast when Thales' 10-foot stick was casting its 16-foot shadow?

8. A 6-foot stick casts a 10-foot shadow at the same time that a tree casts a 120-foot shadow. How tall is the tree? What is the distance from the top of the tree to the tip of its shadow?

9. Suppose a 4-foot stick casts a 5-foot shadow at the same time that a pine tree casts a 50-foot shadow. How tall is the pine tree?

10. A man wishes to determine the height of the tree pictured in the figure below. He stands a yardstick vertically on the ground at D, 33 feet from A. From the ground, he then determines the point B on the ground such that the points C, E, and B are collinear. He next determines that the measure of the line segment DB is 6 feet. Determine the height of the tree.

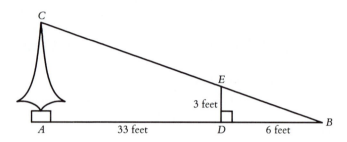

11. Suppose a woman notices that when her shadow is exactly the same length she is, the shadow cast by a neighboring building is 100 feet long. How tall is the building? (The earliest commentators say that this was the problem actually solved by Thales, not the slightly more complicated case requiring ratios.)

12. In the accompanying figure, $\triangle MNP$ is similar to $\triangle MRS$. The lengths of some line segments are indicated. Determine the value of x.

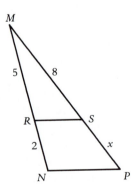

13. Consider $\triangle ABC$ in the accompanying figure. D is the midpoint of side AB. A line segment is drawn through D parallel to side AC and intersecting side AC at the point E. Is E the midpoint of AC? Give an appropriate reason for your answer. (Use the properties of similar figures that corresponding angles have the same measures and that the measures of corresponding sides are proportional.)

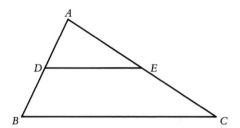

14. Two angles of $\triangle ABC$ have measures of 71° and 46°. Two angles of $\triangle DEF$ have measures of 46° and 63°. Determine if the two triangles are similar.

15. From Table 18.1, what is the angle of inclination of the sun (to the nearest 10°) when a 10-foot stick has a 16-foot shadow? Use a calculator to refine your answer, by calculating several nearby values, and find the angle of inclination to the nearest 1°.

16. An earlier calculation of the circumference of the earth was made using the same technique Eratosthenes used. The only difference was that the angle was measured at Lysimachia [now near Gelibolu (formerly Gallipoli), Turkey] instead of at Alexandria. This angle was found to be $\frac{1}{15}$ of a complete circle, or 24°. It was thought that Lysimachia was 20,000 stadia straight north of Syene. Using these figures, what would be the circumference of the earth, in stadia? In miles?

17. The calculation in Exercise 16 leads to a crudely accurate estimate of the circumference of the earth, off by less than 20%. But there was a serious error: in fact, the two cities are only 1180 miles apart. Using 25,000 miles as the circumference of the earth, and assuming that Lysimachia was directly north of Syene, what should the angle have been instead of 24°?

18. Use tangent ratios to show that two right triangles whose corresponding angles are equal have proportional sides.

19. *The American Heritage Dictionary*, Second College Edition (Houghton Mifflin, Boston, 1982), gives a different estimate for the length of a Greek stadium: 185 meters. (The dictionary actually says "185 kilometers, or 607 feet"; as noted in an exercise in Chapter 16, "kilometers" should be "meters.") Based on this length for a stadium, what was Eratosthenes' estimate for the circumference of the earth? What was the percentage error compared to the true circumference?

20. The angle subtended by the sun is 0.5°. Using the fact that the sun is 92,956,000 miles from the earth, find the radius of the sun in miles.

Modern Measurements of Astronomical Distances

Quantity	Symbol	Measurement (mi)
Radius of sun	s	432,000
Radius of earth	e	3,963.5
Radius of moon	l	1,080
Earth to sun (center-to-center average distance)	S	92,956,000
Earth to moon (center-to-center average distance)	L	238,857

21. The following figure shows a simple and practical method for finding the radius RM of the moon once the distance to the moon ER is known: assume that $ER = 238,857$ miles. The angle subtended by the whole moon, as seen by an observer at E on the earth, is 0.5°; hence, the angle E subtended by half the moon is 0.25°. Given that the tangent table says tan 0.25° = 0.00436, find RM.

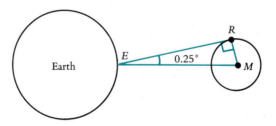

22. The sizes of the disks of the sun and moon as viewed from earth are nearly the same—a marvelous coincidence. The sun is 92,956,000 miles from earth and 432,000 miles in radius. The moon is (on average) 238,857 miles from earth; what is its radius?

23. An observer sights toward the reflection of light from a cloud and measures an angle of elevation of 70°. How high is the cloud?

24. For the technique described for measuring the height of clouds at night, what would be the angle of elevation corresponding to a cloud at 10,000 feet? Would the technique be useful for measuring the height of clouds that high or higher? Why or why not?

25. The brightest star in earth's sky (apart from the sun) is Sirius, which has a parallax of 0.375 ± 0.004 second of arc. How far away is this star, in light-years? Give an error bound, based on the error bound of the observation of parallax.

26. Another of the top ten brightest stars in earth's sky is Rigel, which has a parallax of approximately 0.004 second of arc. (It is so far away that the parallax is difficult to measure accurately.) How far away is this star, in light-years?

For Exercises 27 to 30, refer to the following: two triangles are congruent if:

▶ The three sides of one triangle are congruent, respectively, to the three sides of the other triangle (*SSS*).
▶ Two sides and the included angle of one triangle are congruent, respectively, to two sides and the included angle of the other triangle (*SAS*).
▶ Two angles and the included side of one triangle are congruent, respectively, to two angles and the included side of the other triangle (*ASA*).

▲ 27. In the figure to the left below, ∠1 is congruent to ∠2, and line segment *AB* is congruent to line segment *AC*. Is $\triangle ABD$ congruent to $\triangle ACD$? Give an appropriate reason for your answer.

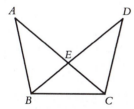 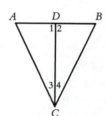

▲ 28. In the figure to the right above, ∠1 is congruent to ∠2, and ∠3 is congruent to ∠4. Is $\triangle ACD$ congruent to $\triangle BCD$? Give an appropriate reason for your answer.

▲ 29. In the figure at the left below, $\triangle ABC$ is congruent to $\triangle DCB$. List all of the pairs of corresponding parts that are congruent.

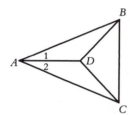 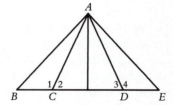

▲ 30. In the figure to the right above, ∠1 is congruent to ∠4, ∠2 is congruent to ∠3, and side *AB* is congruent to side *AE*. Is $\triangle ABC$ congruent to $\triangle AED$? Give an appropriate reason for your answer.

▲ Advanced exercise.

▲ 31. We saw earlier how Eratosthenes was able to compute the circumference and hence the radius of the earth. From this fact, we can easily determine the distance from the earth to the moon:

   a. In the following figure, A and B denote two points that measure 500 miles apart on the earth's surface. Compute the measure of angle AOB.
   b. Compute the measure of angles OAB and OBA.
   c. AH and BH represent the earth's horizontals at A and B; they are therefore tangent to the earth at these points. Use this fact to explain how to calculate the measures of angles ABH and BAH.
   d. Simultaneous observations of the moon are made from points A and B so that angles MAH and MBH are determined. Given these angles, you can easily determine the measures of angles MAB and MBA. Now, explain how to calculate the measure of angle AMB.
   e. Finally, explain how to calculate the distances from points A and B to the moon.

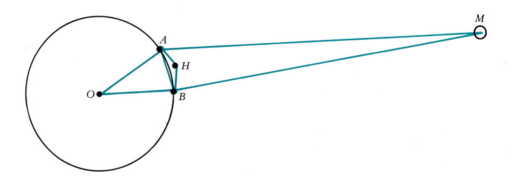

▲ 32. Assuming that the necessary measurements could actually be carried out, the figure at the top of the next page suggests a conceptually simple way of finding the distance of the moon from the earth. An observer at P sees the moon M directly overhead at exactly the same time as an observer at Q sees the moon right on the horizon. The two observers will be nearly a quarter of the way around the earth from each other. In fact, the central angle E is $89.07°$. With a calculator you can find that $\tan 89.07° = 61.60295$. Question: Assuming that the earth's radius is 3963.5 miles, what is the distance PM (from the earth to moon)

▲   Advanced exercise.

in miles? (Note that $PM = EM - 4000$ and that EM can be assumed to be equal to QM.)

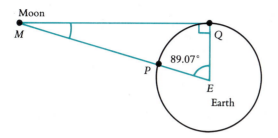

▲ 33. If we want to find the distance of the sun from Venus by the methods used to find the distance from the earth to the moon or the earth to the sun, we would have to be on Venus. Because this is not possible, we can use a more subtle and practical method. We can think of earth and Venus as points E and V moving in circular orbits around the sun. Notice in the following figure that the angle at E varies as the two planets travel around the sun and that it reaches its maximum value when the angle at V is a right angle.

Suppose we observe this angular separation of Venus and the sun throughout the year and find that the maximum value of angle E is $47°$. In right triangle EVS, we know $\tan 47° = VS/EV$. However, because we don't know either VS or EV, the tangent table is of no help. But another table, the *sine table*, would give the ratio VS/ES for various values of E. Using the value $\sin 47° = VS/ES = 0.73135$, find the distance of Venus from the sun to the nearest million miles.

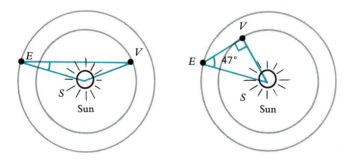

▲ 34. Hipparchus, who lived after Aristarchus, used an eclipse of the sun to estimate the distance to the moon. In this exercise we use a simpler calculation with his data to estimate the *radius* of the moon instead. As related later by Pappus, "Hipparchus starts from this observation: there was an eclipse of the sun which was exactly total in the region about the Hellespont [at Gelibolu in the

Dardanelles, the strait that connects the Aegean Sea with the Sea of Marmara], no portion of the sun being seen, whereas at Alexandria in Egypt about four-fifths only of its diameter was obscured." The figure below shows the situation. Following the ancients, we let L be the distance from the earth to the moon, S the distance from the earth to the sun, and l and s the radii of the moon and sun, respectively. Note that angles A and P are right angles and that triangles AQH and KQP have equal vertical angles at Q, so these two triangles are similar. Write down an equation for proportional sides, and approximate $S - L$ by S; use the fact that the disc of the moon exactly covers the disc of the sun to replace L/S by its equal, l/s; and arrive at an equation for l in terms of AH. Finally, use the fact that the north-south distance AH between A and H is 640 miles to calculate l.

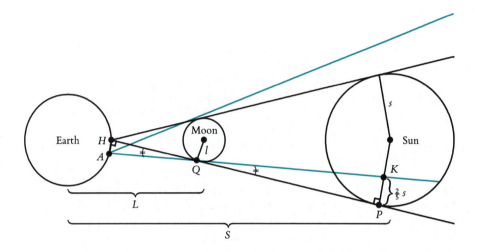

▲ 35. The situation of Exercise 34 required two observers a great distance apart. But if you by yourself observe a total eclipse of the sun, there is an easy way to estimate the distance to the moon. What you need to do is to *time* one of the next total eclipses that comes your way, from the first moment that the moon begins to move across the sun until the last instant before it is all the way across. In the United States, the eclipse of May 10, 1994 (not total) was visible in a band across the continental United States. (The next one visible in that area won't be until after year 2000.) Since you don't want to have to wait for an eclipse before you do this exercise, let us tell you that the elapsed time of an eclipse is about 125 minutes. The figure at the top of the next page shows the situation. Although the earth is moving during the eclipse (in fact, it moves more than 125,000 miles!), the earth and sun stay in the same relative positions and the

▲ Advanced exercise.

moon is carried along. In addition, the moon moves in its own orbit. How far? It moves two moon diameters in distance and about 1° of the 360° of its orbit.

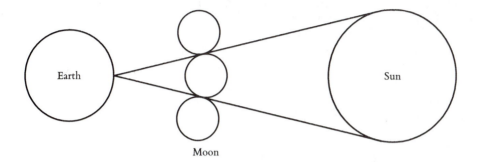

a. Write down two expressions for the circumference of the (assumed circular) orbit of the moon, one in terms of the distance from the earth to the moon and the other in terms of the radius of the moon. Use 1080 miles as the radius of the moon to arrive at an estimate of the distance to the moon.

b. You will notice that in part a we did not yet use any information from the eclipse! All we used is that the distance of two moon diameters is about 1° of the moon's full orbit, and it is this rough approximation that is the weak point in the calculation in part a and the explanation for the result being too large. We can use our timing of the eclipse to make a more refined estimate. We need to know one other easily measured quantity, the time it takes for the moon to complete one orbit (as seen from the earth); the ancients knew this quantity quite accurately. Modern measurements give 29 days, 12 hours, 44 minutes, and 2.8 seconds as the observed time from one new moon to the next. Convert this measurement to minutes, calculate what fraction of it 125 minutes is, arrive at a new estimate of the angular measure of the diameter of the moon, and make a new estimate of the distance to the moon.

▲ 36. A timing of a total eclipse of the moon also can be used to make a rough calculation of the distance of the moon. Lunar eclipses, which occur about once or twice a year, are not as common as solar eclipses; but since a lunar eclipse extends over a much wider area, it is likely to be seen by more people. Perhaps you observed the eclipse of November 29, 1993 (centered over Mexico City) or the one of May 25, 1994 (which was not total, even at its center over southern Brazil). Total eclipses of the moon at least partially visible in the Western Hemisphere between now and 2000 will take place on April 4, 1996 (centered over the west coast of Africa, lasting 216 minutes) and September 27, 1996 (French Guiana, 212 minutes).

The figure below shows an idealized situation: the observer is at O, on the side of the earth away from the sun, the moon is directly overhead, and the sun is on the other side of the earth.

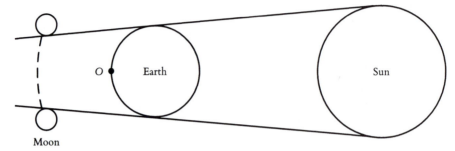

a. During the course of the eclipse, the moon moves approximately one earth diameter plus one moon diameter. Use this fact, the data in Exercise 35, and the fact that the radius of the earth is 3963.53 miles (as measured at the equator) to give an estimate of the distance to the moon.

b. In fact, the estimate in part a is way too large. The main reason is that the moon moves substantially *less* than one earth diameter plus one moon diameter, as the figure in fact suggests. For the sake of improving the model and method of part a, let's assume we know (by some other means) that the average distance from the sun to the earth is 92,956,000 miles and the sun's radius is 432,000 miles. Then the slant line through the top of the sun and the top of the earth dips 428,000 miles over a distance of 92.9 million miles, so it is dipping 1 mile for every 217 miles across. As it proceeds L miles farther to the moon, it dips another $L \times \frac{1}{217}$ miles. So in fact the moon during eclipse will travel one earth diameter ($2e$) plus one moon diameter ($2l$) minus $2(L \times \frac{1}{217})$. Use this new estimate of how far the moon travels to get a new estimate of the distance from the earth to the moon.

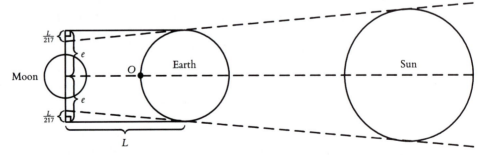

c. Even the new model gives a result that is too large. Can you think of defects in the new model?

d. If the moon were far enough away from the earth, it could never be completely eclipsed. The moon would have to be a distance M far enough away that the disc of the earth just barely fails to shade the moon, so that a ray from the top of the sun would graze the earth and pass on to graze the moon. That ray and the line through the centers of the bodies intersect at a point *z* in space, forming several similar triangles. Use the idea from part b to estimate how far the moon would have to be.

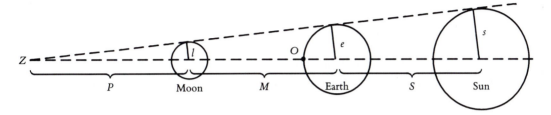

■ 37. What would be some practical difficulties involved in measuring the height of the Great Pyramid in the way that Thales did (short of getting a passport and raising money to travel to Egypt)?

■ 38. Eratosthenes was incorrect in assuming that Alexandria is 500 miles directly north of Syene (Aswan). Alexandria in fact is 490 miles north of Aswan (pretty close, Eratosthenes), but Aswan is 190 miles farther east and the cities are actually 520 miles apart. Regarding the situation of Exercises 16 and 17, Syene (Aswan) is 1130 miles south of Gelibolu but 390 miles farther east. How much does it matter that these cities are not in a straight north-south line? Please refer to the map below (redrawn from *Surveying Outer Space* by Michael Penna, UMAP Module 580).

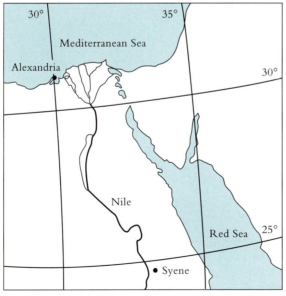

■   Discussion exercise.

■ 39. What are the observational difficulties of timing an eclipse? Which is easier to time accurately, a solar eclipse or a lunar one?

▶ WRITING PROJECT

1. Select a height or distance on your campus or in your area — perhaps the height of the tallest building, or the distance across a river. Use one or more techniques from this chapter to determine the height or distance involved. Write a report that describes the method that you used and gives the measurements and calculations that you made. Be sure to include a measure of precision (an estimate of possible error) for your result.

■   Discussion exercise.

19

Reflecting the Universe

Modern science arose in the seventeenth century with the work of Galileo Galilei (1564–1642), Johannes Kepler (1571–1630), and Isaac Newton (1642–1727). The first distinguishing characteristic of the new science was its *experimental method*. The second was its *quantitative character*. The physics of Aristotle (384–322 B.C.), which still held sway in the intellectual world of Galileo's time, gave only qualitative explanations for physical phenomena; for example, the old science tried to explain *why* an apple falls downward from a tree. In contrast, Galileo was more interested in *how*; he dismissed bare qualitative explanations as "fantasies" that are "not really worthwhile." He sought instead a mathematical description of such events as the motion of a freely falling body.

The third and crowning characteristic of modern science was its striving for a *mathematical theory*, which was highlighted in the work of Isaac Newton. A mathematical theory enables us to make predictions and—often incidentally—explains a wide variety of related phenomena. For example, Newton's work coordinated Kepler's observed laws of planetary motion with the laws of mechanics that appeared to govern terrestrial phe-

nomena. His mechanics explained much about gravitation and ocean tides as well as planetary motion.

▶ FROM GREECE TO GALILEO

Galileo began his assault on astronomy in 1609, after he learned that a Dutch lens maker had discovered how to achieve great magnification by arranging two lenses in a special way in a long tube. This, of course, was the invention of the telescope. Galileo then proceeded to build his own telescopes. He first achieved a threefold magnification. Then, after mastering the problems of grinding and polishing lenses and experimenting with the arrangement of the lenses in the tube, he was able to construct a telescope that magnified approximately 33 times (Figure 19.1). These instruments, although modest by today's standards, revealed some astonishing astronomical sights to Galileo.

Turning the telescope to the moon, he saw immediately that the surface of the moon had mountains and valleys and was not the "perfect"

(a)

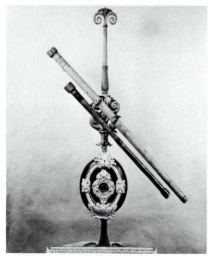

(b)

Figure 19.1 (a) After inventing the telescope, Galileo made some astonishing astronomical discoveries. (b) Two of Galileo's telescopes and the lens from another. (Photo (b): Scala/Art Resources.)

omer and geographer Claudius Ptolemy (second century A.D.). Briefly, the Aristotelian and Ptolemaic view maintained that the earth is the immovable center of the universe, which is a large celestial sphere that rotates about the earth and on which all the stars are fixed. Referred to as a *geocentric,* or earth-centered, theory, this view had prevailed for well over a thousand years; it had the support of almost all academicians and the official support of the Catholic Church, Martin Luther, and Jewish leaders.

An alternative to the geocentric theory had already been proposed by the ancient Greek astronomer Aristarchus of Samos in the third century B.C., but his work was largely ignored. Aristarchus held to a *heliocentric* theory, placing the sun at the center of the universe. This theory was revived in a modified form approximately 1800 years later by a young Polish student, Nicolaus Copernicus (1473–1543). Copernicus argued that all the planets, including the earth, moved in concentric spheres, with only slight modification, about the sun.

Perhaps most devastating to proponents of the geocentric theory was Galileo's discovery of four moons revolving about the planet Jupiter (Figure 19.2). If Jupiter, a planet, possesses moons, then the earth, too, might also be a planet. Moreover, these newly discovered moons of Jupiter were not circling the earth, the presumed center of the universe around which all bodies should revolve.

▶ IMPROVING THE TELESCOPE

A half-century after Galileo built his first telescope, Sir Isaac Newton turned his genius to improving the instrument. Galileo's was a *refractor telescope,* one that bent light rays by means of lenses. Such instruments have two shortcomings: (1) the glass used for the lenses must be of high quality and free of flaws in order to minimize distortions, and (2) the bending of the light rays separates the colors contained in white light, introduc-

sphere of accepted Aristotelian theory. Later, Galileo discovered sunspots, showing that the sun, too, was not "perfect." These observations shocked his contemporaries by contradicting long-held beliefs about the nature of the universe.

The prevailing conception of the universe in Galileo's time derived primarily from the Greek philosopher Aristotle and the Alexandrian astron-

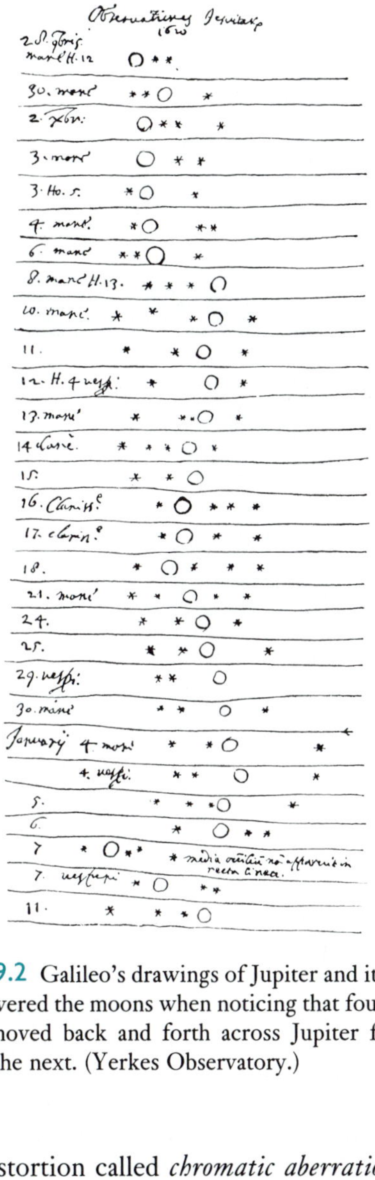

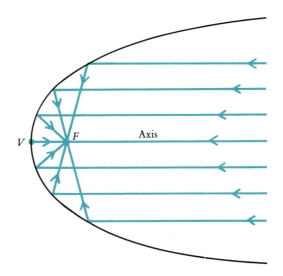

Figure 19.2 Galileo's drawings of Jupiter and its moons. He discovered the moons when noticing that four shining objects moved back and forth across Jupiter from one night to the next. (Yerkes Observatory.)

Figure 19.3 Parallel light rays reflect off a parabola and meet at its focus *F*.

ing a distortion called *chromatic aberration.* The first of these problems is eliminated, and the second reduced, by using a mirror instead of a lens for the light-gathering work of the telescope. It was this idea that Newton exploited when he constructed the first *reflector telescope,* using a mirror to replace the light-gathering lens of the refractor telescope.

As a student and great admirer of Greek geometry, and himself one of the most profound geometers in all history, Newton knew that the best shape for a light-gathering mirror would be a **parabola,** a shape related to the well-known curve discovered 2000 years earlier. Parabolas possess a remarkable **reflection property,** illustrated in Figure 19.3, that makes them especially suitable. At the point *V*, where the parabola crosses its axis of symmetry, the curve is "sharpest." Point *V* is called the **vertex of the parabola,** and the axis of symmetry through *V* and *F* is the **axis of the parabola.** Lines parallel to the axis that come in from afar meet the parabola (on its concave side) at some acute angle and then "bounce off" the curve at the same angle. The parabola's remarkable feature is that all the bouncing-off lines pass through a single point *F* called the **focus of the parabola.** If the parabola is a reflecting surface, then the lines parallel to its axis can be regarded as light rays radiating from a distant heavenly body; these light rays reflect off the surface and accumulate, or focus, at the focus of the parabola (Figure 19.3).

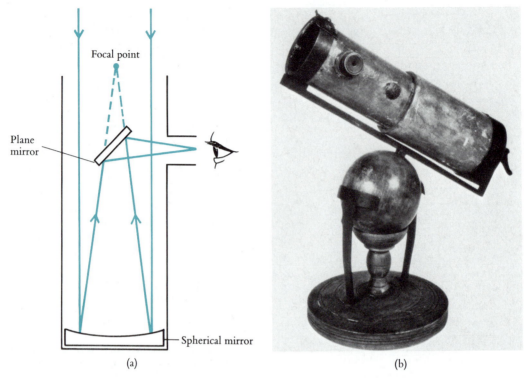

Focal point

Plane mirror

Spherical mirror

(a)

(b)

Figure 19.4 (a) The design of the Newtonian reflecting telescope. (b) A replica of Newton's reflecting telescope. (The Granger Collection.)

Because of the difficulties of grinding a parabolic mirror, Newton compromised and constructed a spherical mirror instead. Such a mirror, whose surface is a portion of a sphere, gathers light from afar and tends to accumulate it at the center, or focus, of the sphere. But such a light-gathering mirror presented another problem: the observer would have to be placed at the center of the sphere—directly in front of the mirror—thus blocking all the incoming light.

To make his telescope, Newton placed the spherical mirror at the bottom of a cylindrical tube so that the mirror would reflect the incoming rays of light onto one image point, the focus. To view the image, or focused light, from outside the tube, Newton placed a small plane mirror close to the focus in order to reflect the image to the side of the

telescope, where he made a small hole in the cylindrical housing (see Figure 19.4).

Less than four years after Newton built his first telescope, a report came to the French Academy that someone else, Guillaume Cassegrain, had invented still another reflecting telescope. Cassegrain had succeeded in grinding and polishing a large concave parabolic mirror to gather light, together with a smaller convex hyperbolic mirror to focus it (see Figure 19.5). Using both convex and concave mirrors makes distortions cancel. The main advantage of the Cassegrain design is the short physical length of the telescope compared to its long focal length: a long focal length is needed to focus distant objects accurately, while the small size of the telescope is crucial in some applications. For example, the NASA Hubble Space

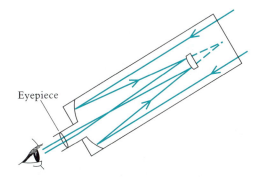

Figure 19.5 Diagram of the Cassegrain telescope, which uses both a parabolic mirror and a hyperbolic mirror to collect and focus light rays. The parabolic mirror focuses the image, and the hyperbolic mirror transmits the focused image to the viewer at the eyepiece.

Telescope launched into orbit in 1990 uses the Cassegrain system, with a parabolic mirror 94.5 inches in diameter. (See Spotlight 19.1, p. 602.)

▶ OTHER APPLICATIONS OF THE PARABOLA

Although the focal property of parabolas had been described by Apollonius (ca. 260–190 B.C.), the reflecting telescope appears to be its first technological application. The key ideas upon which the usefulness of the parabolic mirror rests are the focal property of the parabola and the fact that when light rays reflect off a smooth surface, the angle of incidence equals the angle of reflection (see Figure 19.6).

The conjunction of these two ideas has found a number of applications, including flashlights, searchlights, and the automobile headlight. These all reverse the job of the telescope. Instead of gathering incoming light and bringing it into focus at a point, they have the light source (a bulb) at the focus of a **paraboloid of revolution,** the surface made by rotating a parabola around its axis. The light is reflected outward along rays that are parallel to the axis (see the blue rays in Figure 19.7a). Headlights with both high and low beams make double use of the parabola. The light source for the high beam is at the focus of the parabola, but the light source for the low beam is to the left of and above the focus. While the high beam reflects rays directly ahead, the low beam sends rays down and to the right, away from oncoming traffic (see the black rays in Figure 19.7a). The amount of light reflected forward from the mirrored paraboloidal surface at the back of the headlight is about 6000 times as much as the light directed forward by the bulb alone.

Some fluorescent lamp tubes have housings above them with cross sections that are parabolas. The tube forms a line of foci for the different parabolas and the light is reflected straight down

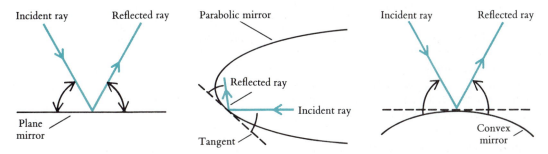

Figure 19.6 When light rays bounce off a smooth surface, the angle of incidence equals the angle of reflection.

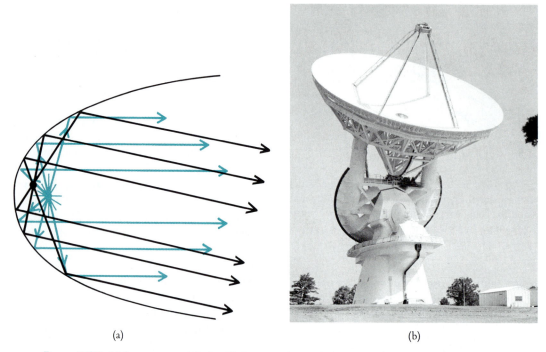

(a)   (b)

Figure 19.7 (a) In an automobile headlight, a parabola directs rays of light outward in parallel lines: straight out, for the high beam source located at the focus, and down and to the right, for the low beam source located above and to the right of the focus. (b) A 140-foot-wide radio telescope at Green Bank, West Virginia. (The National Radio Astronomy Observatory, operated by Associated Universities, Inc., under contract with the National Science Foundation.)

all across the length of the lamp (see Figure 19.8). The shape of the surface of the housing is called a **parabolic cylinder.** Another invention that uses a parabolic cylinder but focuses light for a different purpose is one kind of solar cooker.

Another major application of the parabola's focusing property is the dishlike antenna used for satellite television, radio telescopes, radar, microwave towers, and surveillance systems. Like the faint light rays caught and focused by the telescope, faint signals are caught by the antenna dish and reflected to its focus, where they are gathered and amplified into a strong signal (see Figure 19.7b).

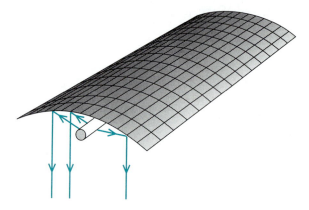

Figure 19.8 A parabolic lamp housing reflects all light from the fluorescent tube directly downward.

Figure 19.9 The main cables of the Golden Gate Bridge approximate a parabola.

In addition to its importance because of its focusing properties, the parabola plays a role in architecture. The main cables of a suspension bridge (such as the Golden Gate Bridge in Figure 19.9) approximate a parabola, and many other bridges are supported from below by approximately parabolic arches. The key feature is that the parabolic shape allows support of a uniform horizontal load to be spread out in such a way that there is uniform stress. Did you ever wonder why the largest dinosaurs, such as *Stegosaurus*, *Apatosaurus* (formerly *Brontosaurus*), and others, all seem to have big, humpy backs? They supported their enormous body weight with a spine that was approximately parabolic. Not all curves that look like parabolas are parabolas, however; see Spotlight 19.2, p. 603.

EXAMPLE: Finding the Focus of a Parabola

Given a parabola on paper, you can find the location of its focus approximately by drawing approximate tangents to the parabola and using the reflection property (more details follow in one of the exercises).

If you have the algebraic description of the parabola, in terms of its equation, you can easily find the focus exactly. With the appropriate choice of x and y axes so that the axis of the parabola lies along the y axis, the parabola will be described by the equation $y = ax^2 + bx + c$,

where a, b, and c are constants. You can locate the focus on the axis at a distance $1/(4a)$ from the vertex, on the inside side of the parabola [if a is negative, the distance is $-1/(4a)$]. So, for example, the sample parabola $y = x^2$ has its vertex at the point $(0, 1/4)$, one-fourth unit above the vertex at $(0, 0)$ (see Figure 19.10).   ▲

▶ PARABOLAS AND MOTION

Parabolas have been observed and used by humans from time immemorial. Projectiles, such as baseballs and bullets, follow approximately parabolic paths (modified somewhat by air resistance), as does the water from a garden hose or a fire hose. When you cup your hands in front of your mouth to project your voice, or around your ears to help your hearing, the ideal shape is a parabola.

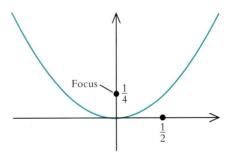

Figure 19.10 The parabola $y = x^2$.

In April 1990, the Hubble Space Telescope was was launched from the Space Shuttle *Discovery*. The Space Telescope was first proposed in 1946 by Lyman Spitzer, Jr., of Princeton University. Its name honors Edwin Hubble, an American astronomer who discovered in 1930 that the universe is expanding.

Dan Schroeder, Professor of Physics and Astronomy at Beloit College, Wisconsin, helped oversee its design and construction. He comments on the new potential for "exploring the universe through telescopes" that the Space Telescope represents:

The great advantages of a telescope in orbit are that it is above the atmosphere, with its air that attenuates light and the air currents that distort it, and away from the light pollution from human activity.

With the Space Telescope we should be able to see known objects much more clearly than before, and see many more objects than are now visible, including far-distant objects that may give us better clues to the age of the universe. To give you an idea of the telescope's resolution: it should enable us to read the inscriptions on a coin at a distance of ten miles.

The instruments on board the craft include cameras, spectrographs (to gather data on the chemical composition, temperature, pressure, and density of objects), and a photometer (to measure the brightness of objects).

Like other telescopes of the Cassegrain design (see Figure 19.5), the Space Telescope contains a hyperbolic mirror and a larger parabolic mirror (94.5 inches across). Early images from the Space Telescope revealed that one of the mirrors suffers from *spherical aberration*, a dis-

The Hubble Space Telescope deployed in space. (NASA.)

tortion that causes blurring. The distortion is caused by the mirror surface not being exactly hyperbolic or parabolic in shape. A true hyperbolic or parabolic shape would reflect all parallel light rays to the exact same point.

The defective mirror is only four millionths of an inch off from true at the edge, but the result is that the telescope cannot focus as sharply as scientists would like. One solution will be to replace a camera in the telescope with another one whose internal mirrors compensate precisely for the mirror imperfection, while other compensating mirrors will be installed between the telescope and the other instruments. The repairs were scheduled for December 1993.

SP TLIGHT 19.2   The Parabola's Rival

Many common shapes that may appear to be parabolas in fact are *not* exact true parabolas. Some, such as telephone and power lines, empty clotheslines, the string going up to a kite, the cross section of a parafoil or sail filled with wind, and Gateway Arch in St. Louis, have the shapes of *catenaries*.

A catenary is the curve formed by a freely hanging cable or rope that is supported at the ends. If a clothesline is loaded with clothes, with the same average weight of clothing per foot of line, the shape of the line deforms to a parabola.

The ideal design for a roof to support only itself would have a cross section that is a catenary (e.g., the stone roof of the church at Mission Carmel in California); the ideal cross section for a roof to support a uniform load of snow is a parabola. (Most roofs, of course, are either flat or have triangular cross sections, because those kinds of roof are easier to build; steep A-frame roofs avoid the snow-loading problem by being so steep that buildups of snow cascade off.)

Can you tell by eye the difference between a catenary and a parabola? The surprising but disappointing answer is no. The figure shows parts of a catenary and a parabola that "hang" from (pass through) the same two points (at the upper "corners") and have the same arc length. Near the vertex of the parabola, the parabola lies below the catenary; elsewhere, it lies above the catenary. The difference is very slight!

If a clothesline were completely inelastic (did not stretch) when clothes are added, its shape would change from parabola to catenary, but we wouldn't notice any difference. For a real clothesline, however, adding clothes also increases the length of the line, because the line stretches under tension. Thus, you see the line sag.

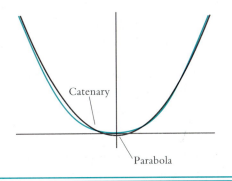

The Gateway Arch in St. Louis, Missouri, is in the shape of a catenary. (Tony Stone Worldwide — Click/Chicago Ltd.)

Catenary

Parabola

EXAMPLE: Galileo and the Leaning Tower of Pisa

Heavy objects fall with the same speed and acceleration as light ones (except for the effect of air resistance). A fanciful legend tells us that Galileo established this fact by dropping objects from the top of the Leaning Tower of Pisa. How are parabolas involved? An object dropped straight down follows a straight line to the earth; but the *graph* of distance fallen versus time forms half of a parabola (see Figure 19.11). The height y above the ground of an object t seconds after it is dropped is given by the formula

$$y = -\frac{1}{2} gt^2 \, h_0$$

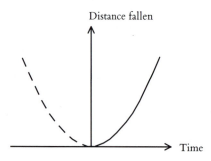

Distance fallen

Time

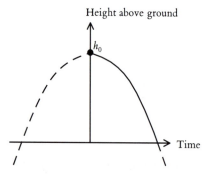

Height above ground

h_0

Time

Figure 19.11 Parabolic graphs for an object falling from height h_0.

where h_0 is the height from which the object is dropped, the minus sign indicates that the object gets lower with time (it falls down, not up), and g is a constant that stands for the acceleration due to gravity. In the U.S. Customary System of units, $g = 32$ feet per second per second; in the metric system, $g = 9.8$ meters per second per second. We take $t = 0$ to be the instant when the object is dropped. At the end of the first second, the formula gives $y = -16 + h_0$ feet, indicating that the object has fallen 16 feet from height h_0 since the drop. At the end of the second second, $y = -64 + h_0$, and the object is 64 feet below its original location. It has fallen 48 feet during the second second, compared to 16 feet during the first second; it is accelerating.

We can use the formula to determine how long it takes objects to fall from the Leaning Tower of Pisa. The Tower is 179 feet high, so $h_0 = 179$. The object hits the ground when $y = 0$, so that the time it takes to reach the ground is the value of t that solves $0 = -\frac{1}{2}(32)t^2 + 179$. Solving for t, we get $t = \sqrt{179/16} \approx 3.3$ seconds. Actually, we get $t = \pm\sqrt{179/16} \approx \pm 3.3$. The value $t = -3.3$ seconds corresponds mathematically to a point on the other half of the parabola and physically to throwing the object up from the ground (launched with the same speed as it later hits the ground) 3.3 seconds before the moment of dropping. ▲

EXAMPLE: Son of Galileo

According to an even less reliable legend, Galileo's son Claribel, who earned a mediocre living playing in a heavy-metal rock band, admired his father and wanted to pay him a tribute. (Claribel also needed more notoriety to help his own career along.) So for Galileo's birthday, Claribel booked a gig at the Leaning Tower and punctuated the act by throwing — not dropping, *throwing* — water balloons, guitars, and bathtubs from the top of the Tower. The anxious fans below were reassured once Galileo explained that all of the objects followed quite predictable parabolic paths, and that as long as Claribel threw the items in a somewhat upward direction, the fans had more than 3.3 seconds to get out of the way.

How much longer than 3.3 seconds? you ask. It's a good thing you asked. . . . The answer depends on the velocity and direction with which Claribel launched the object. We can analyze the problem by resolving the initial velocity into a horizontal speed component s_0 and a vertical speed component v_0. Then the motion of the object is described by

$$ y = -\frac{1}{2}gt^2 + v_0 t + h_0, \qquad x = s_0 t $$

where, as before, y is the height above the ground, and x is the distance away from the Tower.

To make these equations concrete, let's concentrate on the water balloons, which Claribel was launching at a modest $v_0 = 24$ feet per second (about 16 mph) and $s_0 = 30$ feet per second (about 20 mph) (the two combine according to the Pythagorean theorem, so that the total launch speed was $\sqrt{24^2 + 30^2} = 38$ feet per second $= 26$ mph). We can describe the motion of a balloon by $y = -16t^2 + 24t + 179$, $x = 30t$. To find out how long it takes to hit the ground, we have to find the t for which $y = 0$, that is, solve $0 = -16t^2 + 24t + 179$. Approximate solutions can be found by the quadratic formula (which Claribel didn't remember) or by trial and error; they are $t \approx 4.2$ and $t \approx -2.7$. So the balloons took 4.2 seconds to reach the ground, by which time they were $x = 30(4.2) = 126$ feet from the Tower.

By solving the second equation for t, getting $t = x/s_0$, and substituting this expression into the first equation, we get

$$ y = -\frac{1}{2}\left(\frac{g}{s_0}\right)x^2 + \left(\frac{v_0}{s_0}\right)x + h_0 $$

which in our case specializes to

$$ y = -\frac{16}{900}x^2 + \frac{24}{30}x + 179 $$

the equation of the parabola that the balloons followed (see Figure 19.12). ▲

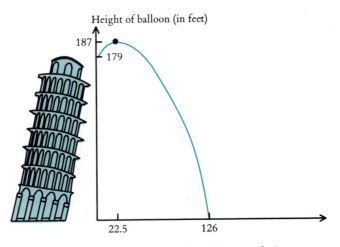

Height of balloon (in feet)

187

179

22.5   126

Distance from Tower (in feet)

Figure 19.12 Graph of the path of Claribel's water balloon.

EXAMPLE: Hang Time for Jordan

Professional basketball player Michael Jordan is renowned for his flights through the air on the way to the basket. Sports commentators often refer to his "hang time," during which he appears to be suspended at his maximum jump height while floating toward the basket. Is there really such a phenomenon as "hang time"? What is behind Michael's jump?

During a jump, Jordan's center of gravity (located at about the height of his navel) goes up and down. There are two parabolas involved, one that you can see and one that you can't (see Figure 19.13). The parabola that you can see take shape (Figure 19.13a) is Michael's flight toward the basket, which can be graphed as vertical height against horizontal distance traveled. How broad this parabola is—how far Jordan travels horizontally while in the air—depends on how fast Jordan is traveling horizontally when he takes off into the air. He continues to travel horizontally at the same speed throughout his flight. The parabola that you can't see dur-

ing the jump is the shape of the graph of Jordan's height against time (Figure 19.13b).

When Jordan is standing, his center of gravity is about four feet above the ground. In a jump, he can raise his center of gravity about 48 inches, or four more feet. Surprisingly, just knowing how high Jordan can jump is enough for us to determine how long he is in the air and what his initial upward speed is.

The height y in feet of his center of gravity is described by the quadratic equation

$$y(t) = -\frac{1}{2} gt^2 + v_0 t + 4$$

where t is the time (in seconds) elapsed since the start of the jump, v_0 is the initial upward velocity of his jump, and $g = 32$ ft/sec² is the acceleration due to gravity. The sign in front of $\frac{1}{2}gt^2$ is negative because the acceleration of gravity is downward toward earth, while the sign in front of v_0 is positive because Jordan's initial velocity is upward.

We arbitrarily decide that the jump ends when his center of gravity descends to 4 ft

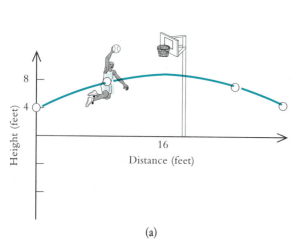

(a)

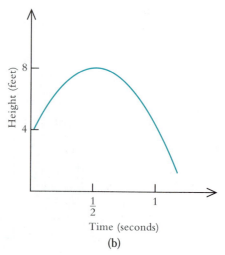

(b)

Figure 19.13 (a) Graph of Michael Jordan's flight toward the basket. (b) Height of Jordan's center of gravity as a function of time, including bending his knees as he lands.

again, even though he may follow through by bending his knees and letting his center of gravity sink further. We know that $y = 4$ at the start of the jump ($t_{start} = 0$) and also at the end of it (t_{final}). Setting $y = 4$ gives $4 = -16t^2 + v_0 t + 4$, or $-16t^2 + v_0 t = t(-16t + v_0) = 0$. The two solutions to this equation are $t_{start} = 0$ and $t_{final} = v_0/16$. Jordan reaches his maximum height at the vertex of the parabola, which occurs at $t_{max} = \frac{1}{2}(v_0/16) = v_0/32$. This result doesn't seem too informative until we plug it back into the equation for y,

$$y\left(\frac{v_0}{32}\right) = -\frac{1}{2} \cdot 32 \cdot \left(\frac{v_0}{32}\right)^2$$
$$+ v_0 \cdot \left(\frac{v_0}{32}\right) + 4 = \frac{v_0^2}{64} + 4$$

and realize that this quantity, the maximum height of Jordan's center of gravity, has to be 8 ft (the 4 ft above the ground when he is standing plus the 4 ft of jump). We can then solve to find $v_0 = 16$ ft/sec, from which we also find $t_{max} = 1/2$ sec and $t_{final} = 1$ sec. So the total duration of Jordan's flight is 1 sec.

What contributes to the impression that Jordan hangs in the air longer than other players? First, he can jump higher (48 inches) than most players; and a player with a higher jump will be in the air longer. However, nothing that Jordan can do (except jump higher) can alter the fact that he will be in the air just 1 sec, the same amount of time as any other player with a 48-inch jump. At the peak of a jump, a player is momentarily motionless in the vertical direction—the vertical speed is zero. With a little algebra, it can be shown that Jordan spends more than a third of the time in the air with his center of gravity between 7.5 and 8 feet above the floor (see Figure 19.13b). During that time he is moving up or down very slowly in comparison to the rest of the jump. That fact, plus his own unique motions while in the air—moving his arms (and the ball) up and

spreading his legs—contribute to the illusion that he is simply "hanging" in the air.   ▲

▶ CONIC SECTIONS

A parabola is one of several important curves that can be formed by cutting a (mathematical) **cone**. Imagine a circle drawn on a flat surface, such as a tabletop, with a line through the center of the circle perpendicular to the surface. Choose a point V on this line, above the table. The surface consisting of all the lines that simultaneously pass through both V and the circle is called a **right circular cone** with vertex V. A mathematical cone differs from a common cone in having two parts: the vertex separates the surface into two **nappes** (see Figure 19.14).

If we slice (intersect) a mathematical cone with a plane, we get a curve called a **conic section**. By changing the angle of the slice, we can see the variety of possibilities for conic sections. The plane we started with—the tabletop—intersects

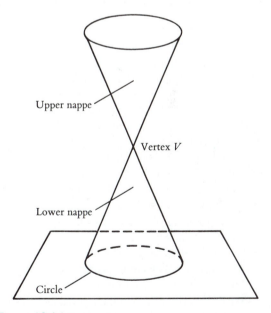

Figure 19.14 A cone.

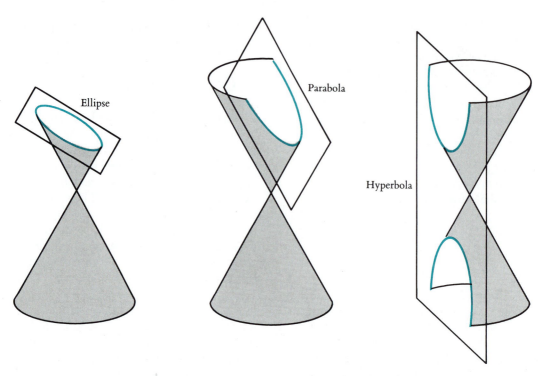

Figure 19.15 The conic sections. The circle is a special case of the ellipse.

the cone in a circle. When we tilt the plane a bit, the intersection becomes an **ellipse**. As we continue tilting the plane, the intersection remains an ellipse as long as the plane cuts the one nappe of the cone in a closed curve. However, there comes a point when the cutting plane, while still intersecting only one nappe, no longer intersects the nappe in a closed curve. If the slice is parallel to the side of the cone, the intersection is a parabola. Finally, by tilting the plane still further, it will cut both nappes of the cone, and the intersection curve is the twin branches of a **hyperbola** (see Figure 19.15). These four curves, the circle, ellipse, parabola, and hyperbola, are the conic sections. As we will see, they *all* have focusing properties.

The ellipse may have first been discovered by the Greeks in connection with sundials, since the tip of the shadow of the *gnomon* (peg) of a sundial follows an elliptic path as the sun moves. The

conic sections were probably investigated first by the Greek geometer Menaechmus in the fourth century B.C., and they were apparently studied by other Greek mathematicians, particularly Apollonius, whose studies were remarkably complete.

▶ JOHANNES KEPLER

Galileo had discovered the importance of the parabola in making telescopes. But conic sections play an even more fundamental role in the operation of the solar system: planets indeed travel around the sun, but not in the circular orbits hypothesized by Galileo. The man who established that they in fact travel in *ellipses* was Johannes Kepler (1571–1630), a brilliant mathematician

Figure 19.16 Johannes Kepler.

with a keen interest in geometry (see Figure 19.16).

Originally, however, Kepler had devised an elaborate mystical theory of the solar system, in which the six known planets were related to the five Platonic solids (see Spotlight 19.3, p. 611). In attempting to establish his mystical theory of celestial harmony, he had to use the ambiguous astronomical data available at the time. He realized that the construction of any theory would require more precise data. Those data, he knew, were in the possession of the Danish astronomer Tycho Brahe (1546–1601), who had spent 20 years making extremely accurate recordings of the planetary positions and the positions of 1000 stars.

Kepler became Brahe's mathematical assistant in 1600 and was assigned a specific problem: to calculate an orbit that would describe the position of Mars at any time to within the accuracy allowed by observation, which at that time was 4 seconds of arc, or $\frac{1}{900}$ of a degree. Kepler boasted that he would have the solution *in eight days*.

Both the Copernican and the Ptolemaic theories held that the orbit should be circular, perhaps with slight modifications. Thus, Kepler sought the appropriate circular orbits for earth and Mars. (The orbit for earth, from which all the observations were made, had to be determined before one could satisfactorily use the data for the positions of the planets.) *After four years*, Kepler found a solution that seemed to fit Brahe's observations. However, on checking his orbits — by predicting the position of Mars and comparing it with more of Brahe's data — he found that one of his predictions was off by at least 8 minutes (= 480 seconds) of arc!

This shocking failure led to *two more years* of struggle, in which Kepler finally took the revolutionary step of discarding the long-held conviction that all heavenly bodies move in circular paths (or circular paths modified in some way by the imposition of smaller circles). This decision permitted him to find an accurate solution to the Mars problem and to put forth a new theory of planetary motion. The results of Kepler's six years of research were published in 1609 in his *Astronomia Nova*, in which he announced two of his three remarkable laws (see Figure 19.17 and Table 19.1):

1. *Law of elliptic paths.* The orbit of each planet is an ellipse with the sun at one focus.

2. *Law of areas.* During each time interval, the line segment joining the sun and planet sweeps out an equal area anywhere on its elliptic orbit. (A brief version is: equal areas are swept out in equal times.)

Kepler's third law was published later and helped Isaac Newton formulate his law of gravity:

3. *Law of times.* The square of the time of revolution of a planet about the sun is proportional to the cube of that planet's average distance from the sun (where "average distance" means one-half the length of the ellipse's longer axis).

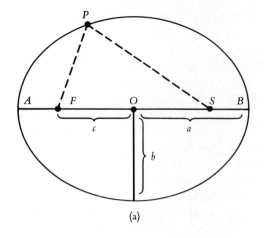

(a)

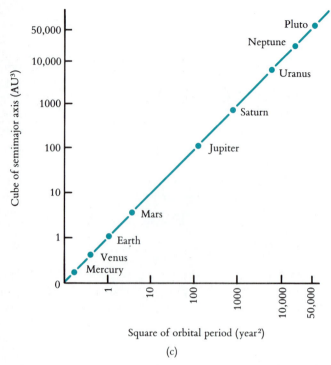

(c)

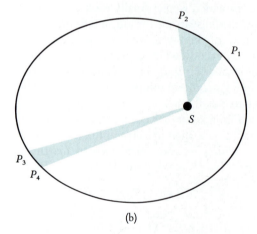

(b)

Figure 19.17 (a) The law of elliptic paths. The orbit of each planet is an ellipse with the sun at one focus. The sum $(PF + PS)$ of the distances from any point P of an ellipse to the two foci F, S is equal to the major diameter AB. (b) The law of areas. The shaded parts are of equal area; thus, the planet takes the same amount of time to move from P_1 to P_2 as to move from P_3 to P_4. (c) The law of times. The points in the graph fall along a straight line, verifying Kepler's discovery that the square of the orbital period equals the cube of the planet's average distance from the sun. (1 AU = 93,000,000 miles.) ((c) is from William J. Kaufmann III, *Universe*, W. H. Freeman, 1985.)

TABLE 19.1 A Demonstration of Kepler's Third Law

Planet	Orbital period P (in years)	Average distance from sun (in AU)	P^2	a^3
Mercury	0.24	0.39	0.06	0.06
Venus	0.61	0.72	0.37	0.37
Earth	1.00	1.00	1.00	1.00
Mars	1.88	1.52	3.53	3.51
Jupiter	11.86	5.20	140.7	140.6
Saturn	29.46	9.54	867.9	868.3

SP TLIGHT 19.3   # Kepler's Model of the Solar System

Tetrahedron	Cube	Octahedron	Dodecahedron	Icosahedron

The five Platonic solids.

Kepler, in the *Mysterium Cosmographicum (The Cosmographic Mystery),* published in 1596 a fantastic cosmological interpretation of the Platonic solids (here translated by Koyré, *The Astronomical Revolution,* p. 146):

The Earth [the sphere of the Earth] is the measure for all the other spheres. Circumscribe a Dodecahedron about it, then the surrounding sphere will be that of Mars; circumscribe a Tetrahedron about the sphere of Mars, then the surrounding sphere will be that of Jupiter; circumscribe a Cube about the sphere of Jupiter, then the surrounding sphere will be that of Saturn. Now place an Icosahedron within the sphere of the Earth, then the sphere which is inscribed is that of Venus; place an Octahedron within the sphere of Venus, and the sphere which is inscribed is that of Mercury.

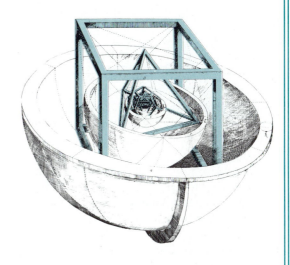

Once again, conic sections played a crucial role in the development of science. We now recognize, more than 2000 years later, the incredible genius of the ancient Greeks in identifying and exploring this and other fundamental areas of knowledge. Their work with conic sections developed a subject that is now known to be fundamental to the study of physics, astronomy, architecture, and engineering.

EXAMPLE: The Orbits of the Planets

Kepler's third law allows us to calculate how far from the sun other planets are. Take, for example, Jupiter, which takes 4332.4 earth days to complete its orbit. Let D_J and D_e be the average distances of Jupiter and the earth from the sun, and let T_J and T_e be the times it takes them to complete one orbit. Then the law says that $T_J^2 = KD_J^3$ and $T_e^2 = KD_e^3$, for the same K. In other terms,

$$\frac{T_J^2}{D_J^3} = K = \frac{T_e^2}{D_e^3}$$

Then substituting $T_J = 4332.4$ days, $T_e = 365.25636$ days, and $D_e = 92,956,000$ miles, we find $D_J^3 = 1.1300 \times 10^{26}$ and $D_J = 483.46$ million miles.

Alternatively, we could do the calculation in terms of earth years and astronomical units, where 1 **astronomical unit (AU)** equals the average distance of the earth from the sun, 92,956,000 miles. Jupiter completes its orbit in 11.861 earth years. The square of this time is $(11.861)^2 = 140.68$. By Kepler's law, the cube root of this number, $\sqrt[3]{140.68} = (140.68)^{1/3} = 5.20$, gives the average distance of Jupiter from the sun, in astronomical units. So Jupiter is (on average) 5.20 times as far from the sun as the earth is, and we can determine the distance in miles by multiplying by 92,956,000. ▲

▶ FURTHER APPLICATIONS OF THE ELLIPSE

We remarked earlier that neglecting air resistance, tossed objects follow approximately parabolic paths. To be precise, their paths are actually arcs of ellipses that have the center of gravity of the earth at one focus. Thus, just like planets around the sun or satellites around the earth, tossed objects follow Kepler's law of elliptic paths.

The ellipse has many applications beyond the magnificent ones in Kepler's work. We observe an ellipse whenever we look at a circle that is tilted or deformed, such as a round plate, the top of a round glass, or the surface of liquid in a glass.

An unusual application of the idea that an ellipse can result from the flattening of a circle is in paleozoology. In a bed of fossils, the squeezing from the layers above can result in flattened fossils that mistakenly appear to be distinct and separate species; an example is *Ellipsolithe,* which turned out to be just an elliptic distortion of the circular chambered nautilus of Figure 16.9. Similarly, the roofs of the underground stations of the DC Metro subway system also appear like flatten circles; they too have elliptical cross sections.

The ellipse has applications in mechanics. Elliptic gears in machinery can provide a quick-return mechanism and a slow power stroke (e.g., for heavy cutting) (see Figure 19.18a). Using the same idea, some racing bicycles have circular gears for the rear wheels (like ordinary 10-speed bikes) but an elliptic gear for one of the front gears, thereby allowing the gearing to match the natural cycle of available power in the rider's legs (Figure 19.18b). Ellipses even played a small role in the Battle of Britain in World War II. The British Spitfire fighter's excellent maneuverability and acceleration were due in part to the elliptic profile of its wings and tail (Figure 19.18c).

The most important geometric quality of an ellipse is its **reflection property:** A light ray passing through one focus of an elliptic mirror will

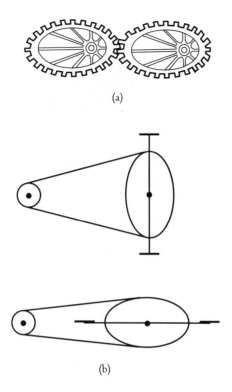

(a)

(b)

(c)

Figure 19.18 Applications of the ellipse. (a) elliptic gears; (b) bicycle with one elliptic gear; (c) British Spitfire airplane.

reflect off the mirror and pass through the other focus.

A familiar application of this reflection property is the lamp that a dentist uses in examining a patient's teeth and mouth. The inside surface of the lamp housing is a mirror-like part of a three-dimensional generalization of an ellipse, called an *elliptical cylinder*, whose cross sections are ellipses (see Figure 19.19). The bulb of the lamp runs through one focus of each ellipse, while the other is at the patient's mouth. A baffle over the bulb protects the patient's eyes from any glare, while the light reflects around the baffle to illuminate the patient's teeth, just a few inches away from the eyes.

A visual illustration of the reflection property of the ellipse is provided by an elliptic pool table with a single pocket at one focus (see Figure 19.20): any shot without spin that passes over the other focus will bounce off the cushion directly into the pocket. The reflection property of the ellipse has been used by acoustical engineers in designing whispering galleries, such as those in the Mormon Tabernacle in Salt Lake City and the Capitol building in Washington, D.C. If the shape of the cupola of a gallery or auditorium is elliptic, a weak whisper at one focus may be barely audible — even inaudible — in most of the room, except at the other focus, where the reflections of the whisper are brought together again.

A significant medical application of the reflection property of an ellipse is the use of the *lithotripter* in a noninvasive therapy to break up kidney stones or gallstones. The patient lies in an elliptic tub of water, placed so that the stone is located at one focus of the ellipse. A high-energy

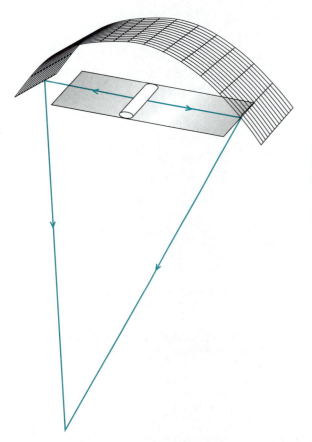

Figure 19.19 A dentist's examining lamp uses a mirror with elliptical cross section to reflect light to the patient's teeth.

Figure 19.20 An elliptic pool table. (From *Inventing, Discovery, and Creativity*, by A. D. Moore. ©1969 Doubleday & Company, Inc. Reproduced by permission of the publisher.)

shock is delivered at the other focus. The shock waves are reflected off the sides of the tub: at all other points, the waves cancel each other out (and the patient feels no pain or ill effects), while the waves concentrate at the other focus to blast the stone. It may take as many as 2000 shocks over a half-hour period to break a kidney stone into sand-sized particles that can pass through the urinary system; but the patient recovers in days, as opposed to up to six weeks if surgery is used instead.

EXAMPLE: Measuring How Skinny an Ellipse Is

As in the case of the parabola, you can use the reflection property and approximate constructions to locate approximately the foci of an ellipse. But it is easy to be exact. An ellipse is customarily described in terms of a, half of the length of its longer axis, and b, half of the length of its shorter axis (see Figure 19.17a). If you now place one end of a compass where the short axis and the ellipse intersect and mark off an arc of length a, the arc will intersect the long axis at the two foci. The distance of the foci from the center of the ellipse is customarily denoted by c, and the Pythagorean theorem gives $b^2 + c^2 = a^2$, or $c = \sqrt{a^2 - b^2}$.

If we think of the ellipse as the orbit of a planet around the sun at one focus (or a moon or satellite around a planet at one focus), we see that the maximum distance of the planet from the sun is $a + c$ and the minimum distance is $a - c$. The average of the two distances is just a. For example, consider the moon in orbit around the earth. The maximum distance of the moon from the earth (center to center) is 238,857 miles and the minimum distance is 221,463 miles. So we have $a = \frac{1}{2}(238,857 + 221,463) = 230,160$ miles, with $c = 8697$ miles, which locates the foci as being about one earth radius above the surface of the earth.

The length c is also involved in measuring how skinny the ellipse is. The technical term for skinniness is **eccentricity,** denoted by e, which is defined as $e = c/a$. Since c is always less than a and greater than or equal to zero, we have $0 \le e < 1$. For the moon we have an eccentricity of $8697/230,160 = 0.038$.

For the special case of an ellipse that is a circle, we have $a = b$, $c = 0$, and $e = 0$. A definition of eccentricity in more general terms allows us to make it meaningful for parabolas (for all of which we have $e = 1$) and hyperbolas (for which $e > 1$).   ▲

▶ THE HYPERBOLA AND ITS APPLICATIONS

It is likely that you most commonly see a hyperbola in the shadow cast by a shaded lamp, as in Figure 19.21a. But hyperbolas find their most significant applications in navigation, particularly in the LORAN (LOng RAnge Navigation) and OMEGA navigational systems. LORAN systems use stations with known locations that broadcast a signal simultaneously. A ship observes the time interval between receiving signals from one pair of stations and determines its location as lying on one hyperbola, which is the curve of constant time difference. The ship then does the same with another pair of stations, thereby placing itself on a different hyperbola. Where the hyperbolas cross is the location of the ship (Figure 19.21b). (Some small correction is needed to account for the earth not being flat.)

Rotating a hyperbola around the axis between its two branches produces what mathematicians

call a *hyperboloid of one sheet.* The hyperboloid is used as the design for cooling towers of nuclear power plants because it can be built from interlocking families of ordinary straight beams along the slant of the surface (Figure 19.21c).

Analogous to elliptic gears, there are hyperbolic gears, which can transmit rotation around one axis to rotation around another, as in the transmission of a car or truck, where rotation of the wheels is transmitted to rotation of the drive shaft (Figure 19.21d).

Like its other conic section cousins, the hyperbola has a **reflection property.** A light ray proceeding in a direction toward (or away from) the focus of a hyperbolic mirror will reflect off the mirror in a direction toward (or away from) the focus of the other branch of the hyperbola (Figure 19.22). It is this property that Cassegrain telescopes use (see Figure 19.5).

▶ NEWTON'S GREAT UNIFICATION

About 50 years after the death of Galileo, Sir Isaac Newton turned his attention to some of the same problems that had engaged Galileo and Kepler, particularly to the problems of terrestrial and celestial mechanics. In his famous *Principia,* whose full title is *Philosophiae Naturalis Principia Mathematica* (Mathematical Principles of Natural Philosophy), he unified terrestrial and celestial mechanics into one deductive mathematical science.

Writing in the spirit of Euclid, Newton began his *Principia* with definitions of terms such as *mass, force, inertia,* and *momentum.* He then presented three *laws of motion,* assumptions that constituted the starting point for his deductive system:

1. A body continues in a state of rest or in a state of constant unaccelerated motion in a straight line unless it is acted upon by an external force.

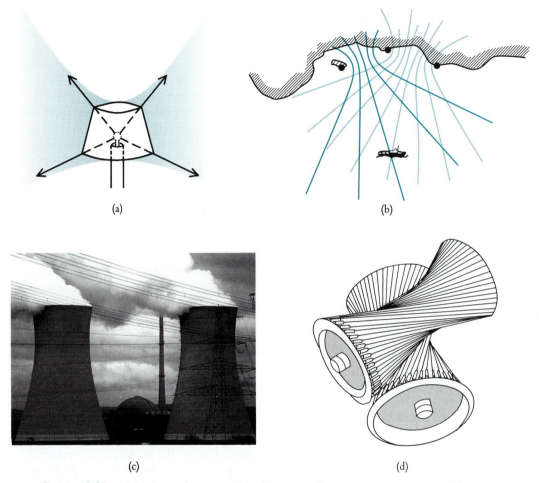

(a)

(b)

(c)

(d)

Figure 19.21 Applications of the hyperbola: (a) hyperbolic shadows cast by a lamp; (b) navigation by LORAN; (c) nuclear cooling tower in the shape of a hyperboloid; (d) hyperbolic gears.

2. At any instant of time, the force acting on a body is equal to the product of its mass and acceleration.

3. To every action there is always opposed an equal reaction.

Using Kepler's third law (the law of times), Newton was led to the formulation of his *universal law of gravitation:* between any two bodies is a gravitational force of attraction that is proportional to the mass of each and inversely proportional to the square of the distance between them.

The phrasing in words is not as easy to interpret as when the law is written as a formula:

$$F = \frac{Gm_1m_2}{r^2}$$

where F is the force of attraction, m_1 and m_2 are the masses of the bodies, r is the distance between them, and G is the universal gravitational constant.

Firmly convinced of the validity of the universal law of gravitation, Newton used it as an

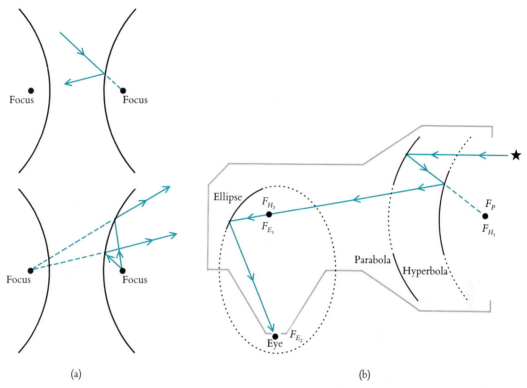

Figure 19.22 (a) Reflection property of the hyperbola; (b) its use in design of a telescope that uses the reflection properties of all three conics.

assumption. Together with his three laws of motion, the law of gravitation enabled Newton to erect a masterpiece of mathematics in which he deduced the dynamics of Galileo, the statics that was developed by Archimedes and Galileo, the planetary laws of Kepler, and much more. Imagine — all this in one mathematical system that simultaneously vindicated the "heresies" of Copernicus, Kepler, and Galileo!

Newton's friend Sir Edmund Halley (1656–1742) persuaded him to publish his discoveries and financed the publication of the *Principia*. It contained a wealth of mathematical and physical discoveries even beyond those already mentioned: It explained the perturbations in the path of the moon, the motion of comets, the flattened shape of planets, and the phenomenon of tides. The *Principia* was published in 1687, but Newton had discovered many of its great ideas at a much earlier date. In fact, his law of gravitation must certainly have been known to him a decade earlier, for in 1679 he verified the law by calculations based on a new measurement of the earth's radius, together with observations of the moon's position.

Halley was inspired by Newton's work to check historical records of comets. Halley concluded that the comet he observed in 1682 was a returning comet in an elliptic orbit and predicted it would appear again at 76-year intervals; we know it as *Halley's comet*. Also on the basis of Newton's work, later astronomers were able to predict the existence and the positions of the planets Neptune and Pluto (the discoverer of Pluto in 1930, Clyde Tombaugh, was still alive on the anniversary of the discovery in 1990!).

▶ REVIEW VOCABULARY

Astronomical unit (AU)   The average distance from the earth to the sun, 92,950,000 miles.

Axis of a parabola   The line dividing a parabola into two identical portions.

Cone   The set of all lines, each of which passes through some point of a given circle and all of which pass through the same point not in the plane of the circle.

Conic section   A curve formed when a plane intersects a cone.

Eccentricity of an ellipse   A measure of the deviation of an ellipse from circularity.

Ellipse   A conic section formed when a plane intersects one nappe of a cone in a closed curve.

Focus (of a parabola)   The single point at which rays parallel to the axis of a parabola come together after "bouncing off" a parabola. (An analogous definition can be given for the foci of ellipses and hyperbolas.)

Hyperbola   A conic section formed when a cutting plane intersects both nappes of a cone.

Nappes   The two surfaces of a cone separated by its vertex.

Parabola   A conic section formed when a cutting plane is parallel to a generating line of the cone, thus cutting only one nappe to result in a curve that is not closed.

Parabolic cylinder   A surface formed by translating a parabola in a direction perpendicular to the plane it lies in ("lifting it straight out of the plane").

Paraboloid of revolution   A surface formed by rotating a parabola around its axis.

Reflection property of an ellipse   A light ray passing through one focus of an elliptic mirror will reflect off the mirror and pass through the other focus.

Reflection property of a hyperbola   A light ray proceeding in a direction toward (or away from) the focus of a hyperbolic mirror will reflect off the mirror in a direction toward (or away from) the focus of the other branch of the hyperbola.

Reflection property of a parabola   A light ray entering a parabolic mirror parallel to its axis will reflect off the mirror and pass through the vertex; and vice versa, a light ray passing through the vertex will reflect off the mirror and leave the parabola on a line parallel to the axis.

Right circular cone   A cone in which the line joining the vertex V to the center of the circle C is perpendicular to the plane P.

Vertex of a parabola   The point where a parabola crosses its axis of symmetry.

▶ SUGGESTED READINGS

COHEN, I. BERNARD: *The Birth of a New Physics*, revised and updated, Norton, New York, 1985.

———: *Revolution in Science*, Belknap Press, Cambridge, England, 1985.

DRUCKER, DANIEL: "Reflection properties of curves and surfaces," *Mathematics Magazine*, 65(3): 147–157 (June 1992).

GARDNER, MARTIN: "The ellipse," in *Martin Gardner's New Mathematical Diversions from Scientific American*, Chapter 15, Simon & Schuster, New York, 1966, pp. 173–183, 252.

———: "Hyperbolas," in *Penrose Tiles to Trapdoor Ciphers*, Chapter 15, Freeman, New York, 1989, pp. 205–218.

———: "Piet Hein's superellipse," in *Mathematical Carnival*, Chapter 18, Knopf, New York, 1975, pp. 240–254, 274. Updated and revised edition, Mathematical Association of America, Washington, D.C., 1989, pp. 240–254, 277–279, 296.

MAESUMI, MOHSEN: "Parabolic mirrors, elliptic and hyperbolic lenses," *American Mathematical Monthly*, 99: 558–560 (June–July 1992).

THOMPSON, DICK: "Big gamble in space," *Time*, 62–63 (March 23, 1993). About "getting Hubble out of trouble."

VEST, FLOYD, AND VIRGINIA RAWLINS: "For earthlings, eccentricity is 0.01673," *Consortium* (Consortium for Mathematics and Its Applications, Lexington, Mass.) (Summer 1991) Pull-Out Section 1-6. Calculates eccentricity of the elliptical orbits of the earth and other planets.

WHITT, LEE: "The standup conic presents: The parabola and applications; The ellipse and applications; The hyperbola and applications," *The UMAP Journal*, 3(3): 285–313 (1982); 4(2): 157–183 (1983); 5(1): 9–21 (1984). Source for many of the applications of conics mentioned in this chapter; includes references to works with more details.

▶ EXERCISES

1. A stone is thrown upward with motion described by the equation

$$y = -16t^2 + 48t + 64$$

where y is the height (in feet) of the stone above the ground at time t (seconds).
   a. The person throwing the stone at time $t = 0$ is on top of a building. How high above the ground is her hand when she lets go of the stone?
   b. When will the stone hit the ground?
   c. How high does the stone go? (Hint: What kind of curve is the graph of the equation of motion, and what kind of symmetry does it have?)

2. As in Exercise 1, but for $y = -16t^2 - 56t + 32$. What is the significance of the minus sign in front of the 56?

3. Here is an alternative definition of parabola in terms of solely geometric considerations: a parabola is the set of points in a plane equidistant from a fixed point (called the *focus*) and a fixed line (called the *directrix*) in that plane. Carry out the following steps in drawing a parabola according to this new definition.
   a. On an ordinary-size sheet of paper, draw a line d and mark a point F 1 or 2 inches away from d.
   b. Locate a point V that is halfway between F and d. Why is V a point of the parabola?
   c. Using compasses or by trial and error, locate a point P (different from V) whose distance from F is equal to the perpendicular distance from P to d.
   d. Repeat part c several times, locating five or six such points P on the parabola.
   e. How does symmetry help you locate still more points on the parabola?
   f. Now connect the points of the parabola to obtain a smooth curve.

4. The point V constructed in the previous exercise is the vertex of the parabola, and line VF is the axis of the parabola.

    a. Explain why the axis of the parabola is the axis of symmetry of the parabola.

    b. On a fairly accurate drawing of a parabola (e.g., traced from a figure in this chapter), draw its axis.

    c. Draw FP, where P is a point of the parabola, and then draw a line through P that is parallel to the axis.

    d. Draw a tangent (by the eyeball method) to the parabola at P, and measure the angles that this tangent makes with FP and with the line through P parallel to the axis. Does this verify the focal property of the parabola?

    e. Repeat parts c and d with another point of the parabola that is not the reflection of P in the axis.

    f. What are the measures of the angles of incidence and reflection if $P = V$?

5. In this exercise we derive the equation for a parabola from the definition in Exercise 3. To implement the definition of Exercise 3, we need a formula for the distance between two points. If the two points are $P_1(x_1, y_1)$ and $P_2(x_2, y_2)$, then the distance between them is

$$d(P_1, P_2) = \sqrt{(x_1 - x_2)^2 + (y_1 - y_2)^2}$$

So, for example, the distance between the points $(2, 4)$ and $(5, -3)$ is

$$\sqrt{(2 - 5)^2 + [4 - (-3)]^2} = \sqrt{9 + 49} = \sqrt{58} \approx 7.6$$

One of the pleasures of coordinate geometry is that we can place the coordinate axes to suit our purposes. So, we will assume that our parabola is nicely centered with the y axis as its axis. Suppose the parabola has its focus at the point $(0, p)$ on the y axis and has as directrix the line $y = -p$, the horizontal line p units below the x axis (see figure below). What we do now is consider an arbitrary

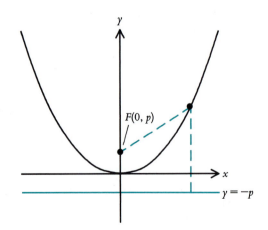

point $P(x, y)$ and see what must be true of it if it is to be on our parabola. The parabola consists of all points that are equally far from the focus point and the directrix line. From our distance formula, the first distance is

$$\sqrt{(x-0)^2 + (y-p)^2} = \sqrt{x^2 + (y-p)^2}$$

The distance to the directrix is along the vertical line from $P(x, y)$ to the directrix; this vertical line intersects the directrix at $(x, -p)$. So the distance of point $P(x, y)$ to the directrix is

$$\sqrt{(x-x)^2 + [y-(-p)]^2} = \sqrt{(y+p)^2}$$

Equating the two distances, we have that the point $P(x, y)$ will be on the parabola exactly when the coordinates x and y satisfy

$$\sqrt{x^2 + (y-p)^2} = \sqrt{(y+p)^2}$$

After we square both sides and multiply out, we have

$$x^2 + y^2 - 2py + p^2 = y^2 + 2py + p^2$$

which simplifies to

$$y = \frac{1}{4p} x^2$$

We've drawn the figure assuming that $p > 0$, in which case the parabola opens upward; if instead, $p < 0$, then it opens downward.

For the parabola described by the equation $y = \frac{1}{6}x^2$:
   a. Determine the coordinates of its vertex.
   b. Determine the coordinates of its focus.
   c. Determine the equation of its axis.
   d. If the point $(c, 6)$ lies on the parabola, what is the value of c?
   e. If the point $(-2, b)$ lies on the parabola, what is the value of b?

6. Consider the parabola described by the equation $y = x^2$.
   a. Starting with the graph of that parabola, indicate how to get the graph of $y = x^2 + 3$.
   b. Again starting with the given parabola, indicate how to get the graph of $y = -(x-2)^2 + 3$.

7. Here's a geometric definition of an ellipse: an ellipse is the set of all points P having the property that the sum of the distances from P to two fixed points F_1 and F_2 (the foci) is constant.
   a. Draw a horizontal line on a sheet of paper and mark two points F_1 and F_2 approximately 4 inches apart. Construct the perpendicular bisector of segment F_1F_2.
   b. On the perpendicular bisector you constructed in part a, mark a point Q approximately 2 inches from the line F_1F_2. Off to the side, draw a segment whose length is the sum of the lengths of segments QF_1 and

QF_2. Point Q will be a point of the ellipse you are constructing, so the length $QF_1 + QF_2$ will serve as the constant referred to in the definition.

c. Using compasses, determine several other points X, such that $XF_1 + XF_2 = QF_1 + QF_2$, thus giving you more points on the ellipse.

d. Explain why, in general, determining one point of the ellipse gives you three others almost immediately.

e. Now connect the points of the ellipse that you've constructed to form a smooth curve.

8. Pick a general point P on an ellipse and draw a tangent at P. Construct line segments from the two foci to P and measure the angles of incidence and reflection in a manner analogous to the one you used with the parabola. Do you have an analogous result?

9. A circle is a special kind of ellipse in which the two foci coincide; it is the set of points in a plane that are a given distance from a fixed point. The fixed point is the *center* and the given distance is the *radius*. In terms of coordinates, a circle in the xy-plane with center at the point (h, k) and radius equal to r is described by the equation $(x - h)^2 + (y - k)^2 = r^2$.

Determine the equation of a circle whose center is at the point $(-4, 3)$ and whose radius is 3.

10. A *diameter* of a circle is a line segment through the center of the circle with its endpoints on the circle. Determine the equation of a circle that has a diameter whose endpoints are $(2, 3)$ and $(-6, 5)$.

11. As in Exercise 9, but for the circle whose center is at the point $(5, -6)$ and which passes through the point $(1, 2)$.

12. Newton's law of gravitation tells us that the gravitational attraction of a body diminishes with the square of the distance. Use Newton's formula for the force of gravitational attraction to locate a point between the earth and the moon where the attractions of the two are exactly equal and opposite. (The mass of the earth is 5.979×10^{24} kg, the mass of the moon is 7.35×10^{22} kg, and the two are on average 383,403 km apart.)

13. As in Exercise 12, but for the earth and the sun. The sun has 330,000 times the mass of the earth, and it is on average 92.956 million miles away.

▲ 14. In this and the next few exercises, we use the definition of an ellipse in Exercise 7 to derive the equations of ellipses, together with the distance formula from Exercise 5. We choose our coordinate system so as to locate the two foci on the x axis at $(c, 0)$ and $(-c, 0)$. An ellipse is the set of points $P(x, y)$ with the property that the sum of the distances from P to the foci is constant. For reasons that will become clear shortly, let that constant be denoted by $2a$. So a point $P(x, y)$ will be on the ellipse if x and y satisfy the equation

$$\sqrt{[x - (-c)]^2 + (y - 0)^2} + \sqrt{(x - c)^2 + (y - 0)^2} = 2a$$

▲ Advanced exercise.

or

$$\sqrt{(x+c)^2 + y^2} + \sqrt{(x-c)^2 + y^2} = 2a$$

To simplify this equation, we subtract the second square root from both sides and then square both sides; simplifying that produces

$$a - \frac{c}{a}x = \sqrt{(x-c)^2 + y^2}$$

Well, there's nothing to do but square again and simplify some more, which produces the result of

$$\frac{x^2}{a^2} + \frac{y^2}{a^2 - c^2} = 1$$

Usually we make the definition $b = \sqrt{a^2 - c^2}$, so that the final equation is

$$\frac{x^2}{a^2} + \frac{y^2}{b^2} = 1$$

Ellipses are sometimes referred to as "flattened" circles. The equation $x^2 + y^2 = 1$ represents a circle of radius 1 centered at the origin. Solving for x gives $x = \pm\sqrt{1 - y^2}$.

    a. Let's stretch the circle in the horizontal direction by multiplying the x coordinates of its points by 2, so that $x = \pm 2\sqrt{1 - y^2}$. Show that the result is an ellipse, and identify the coordinates of its foci.

    b. As in part a, but stretch the circle in the vertical direction instead. Caution: In this situation, a and b change roles.

15. A point moves in the xy-plane such that the sum of its distances from $(3, 0)$ and $(-3, 0)$ is 10. What is the equation of the resulting ellipse?

16. As Kepler realized, the earth follows an elliptical orbit with the sun at one focus. Books of astronomical data usually give the *mean* distance of the earth from the sun—by which they mean one-half of the longer axis—as 92,956,000 miles. The eccentricity of the earth's orbit is 0.0167. Put these facts together to determine a, b, and c, and write an equation that describes the earth's orbit.

17. As in Exercise 16, but for the orbit of the moon around the earth, using the facts that the moon's greatest distance from the earth is 252,710 miles and its least distance is 221,643 miles.

18. The planet Neptune has an orbit with a mean distance (see Exercise 16) from the sun of 2.793 billion miles and an eccentricity of 0.0082. The planet Pluto has an orbit with average distance from the sun of 3.666 billion miles and an eccentricity of 0.2481. Show that Pluto is sometimes closer to the sun than Neptune is (such is the case for the years 1969–2009).

19. Kepler's third law, the law of times, applies not only to the planets of the sun but also to the moons and artificial satellites of any planet (but with a differ-

ent constant K). In this exercise we use what we know about the moon to calculate the height of an artificial satellite from its period of rotation; for simplicity, we assume circular orbits. The moon, which has as its true period of revolution 27 days, 7 hours, 43 minutes, and 11.5 seconds (the time from one new moon to the next is a couple days longer because of the motion of the earth), is at an average distance of 238,857 miles from the center of the earth. The first artificial satellite, *Sputnik I,* launched in 1957, took only 88 minutes to orbit the earth. How high was it? (Hint: Don't forget to take into account the earth's radius, 3963.5 miles.)

20. As in Exercise 19, apply Kepler's third law again, this time to calculate the period of rotation of the Hubble Space Telescope, launched in April 1990 into an orbit 381 miles above the surface of the earth.

21. We can apply the ideas of Exercise 19 to an application we take for granted today. If a satellite is at just the right height, it will orbit the earth in exactly 24 hours, so that it is always over the same spot on earth. Communications satellites, including those for television, are at this height. How high are they?

22. The orbits of the great majority of comets are parabolic, which means that the comet appears only once in our solar system and then departs forever. A few comets have hyperbolic orbits. Most of the rest, including Halley's comet, have elliptic orbits with very high eccentricity. From the period of Halley's comet, 76.1 years, and assuming an eccentricity of 0.999, calculate how far away from the sun it goes.

23. Here is a geometric definition of a hyperbola: the set of all points P having the property that the difference of the distances from P to two fixed points F_1 and F_2 (called the foci) is constant. Using this definition, construct a hyperbola in the same way you constructed the parabola and ellipse.

24. A hyperbola (with two branches) divides the plane into three mutually exclusive regions: one containing focus F_1, one containing the other focus F_2, and the other containing no focus.

    a. Let P be a general point on that branch of the hyperbola that isolates F_1. Draw a tangent to the hyperbola at this point. Now, draw a segment starting at F_1 and meeting the hyperbola at P. If you imagine F_1 to be a source of light and the hyperbola to be a mirror, then you have an angle of incidence. Finally, draw the line r that represents the ray of light reflected off the hyperbolic mirror at P, according to the usual law: the angle of incidence equals the angle of reflection.

    b. Extend line r "backward" and see how close it comes to passing through F_2. If your construction of the hyperbola is fairly accurate and if you have guessed right in constructing the tangent line at P, then your line r should pass very close — if not right through — F_2. (It can be proved, theoretically, that line r does pass through F_2.)

    c. Let Q be a point outside the region containing F_1. The line segment joining G to F_1 meets the hyperbola at a point that we call P. As before, draw a tangent at P and imagine the hyperbola to be mirrored on its convex side. A ray of light emanates from Q, hits the hyperbolic mirror at P, and is then reflected. Draw a line that represents the reflected ray and examine how close this line comes to passing through F_2.

  d. Summarize the discoveries you made in parts b and c to give a full and clear statement of the focal properties of the hyperbola.

25. Imitate the analysis of Exercise 14 to derive the equation that describes a hyperbola with foci at $(c, 0)$ and $(-c, 0)$:

$$\frac{x^2}{a^2} - \frac{y^2}{b^2} = 1$$

with $b = \sqrt{c^2 - a^2}$.

26. A point moves in the xy-plane such that the difference of its distances from the two points $(-4, 0)$ and $(4, 0)$ is 6. Use the result of Exercise 25 to find the equation that describes the point's motion.

27. Suppose a planet is moving at a constant velocity along a straight-line path. Show that the line from the sun to the planet sweeps out equal areas in equal time (see figure below).

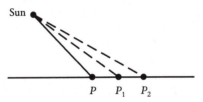

28. Since the earliest times, it has been known by observers in the Northern Hemisphere that the North Star is directly — or very close to directly — over the North Pole; hence, the name Polaris for the North Star. In the accompanying diagram, an observer at O is sighting Polaris in the direction of P. (Note the $23\frac{1}{2}°$ tilt of the earth as it spins on its north–south axis.)

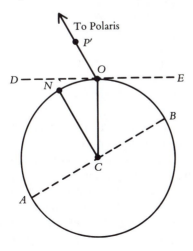

a. Which line in the diagram represents the equator and which line represents the horizon line for the observer?

b. The angle between the horizontal and the sighting of a star is called the *altitude* of the star. What angle is the altitude of Polaris for the observer at *O*?

c. What angle in the diagram represents the latitude of the observer?

d. What is the relationship between the latitude of the observer and the altitude of Polaris?

29. When Galileo observed the moon, he noted "lofty mountains and deep valleys." He proceeded to measure the shadows cast by the mountains in order to compute the approximate height of the mountains. Galileo concluded that the moon's mountains were 4 miles high (which he thought to be higher than any mountain on the earth).

a. Galileo determined that the ratio of the diameter of the earth to that of the moon was 7/2, and he believed that the earth's diameter was 7000 miles. (Today we know that the earth's diameter is closer to 7900 miles.) Using his data, compute the diameter of the moon, the circumference of the earth, and the circumference of the moon.

b. The following figure represents a view of the moon that shows one-quarter of its surface illuminated. Point *T* represents the top of a specific lunar mountain whose height Galileo was calculating. Line *TB*, which is tangent to the moon at *B*, represents a ray of sunlight. Thus, the surface of the moon immediately below segment *TB* was in the shadow of the mountain.

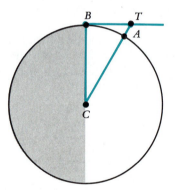

Draw segment *CT*, calling *A* the intersection of *CT* with the surface of the moon. Galileo measured the small arc *AB* to be $\frac{1}{20}$ of the diameter of the moon. He reasoned that this length was approximately equal to the length *TB*. What number did Galileo use for the length *TB*?

c. What theorem about circles enabled Galileo to conclude that triangle *BCT* was a right triangle? You can now do as he did and calculate the length *CT*, namely, the distance from the center of the moon to the top of the lunar mountain.

  d. Finally, you can determine the length of segment AT, the height of the lunar mountain. How close have you come to Galileo's calculation for the height of the mountain?

30. The surface of a solid figure whose faces are polygons is called a *polyhedron* (meaning "many planes"). The polygonal faces have vertices and edges that are the vertices and edges of the polyhedron. We want to explore the numerical relationship between the vertices, edges, and faces of an ordinary polyhedron (or solid) in which there is nothing peculiar, such as a hole.

  a. If V is the number of vertices, E the number of edges, and F the number of faces, find V, E, and F for the cube and the triangular prism.
  b. Notice that $V + F$ is larger than E. How much larger in the case of the cube? And how much larger in the case of the cube with a square pyramid pasted on a face? Write your discovery as a formula:
  $V - E + F =$ ___.
  c. Check this formula on all the polyhedrons that follow in Exercise 31.

31. A *regular polyhedron* (or solid) is one in which all the faces are alike and all the vertices are alike. To be more precise, we require that all faces be congruent regular polygons and that each vertex be surrounded by the same number of faces. Let p denote the number of edges on each face and q denote the number of faces that meet at a vertex. (A cube, for example, yields $p = 4$ and $q = 3$.) The climaxing theorem in Euclid's work on geometry is a proof that there are only five regular solids, now known as the Platonic solids (see Spotlight 19.3). Consider the polyhedrons in the figure below.

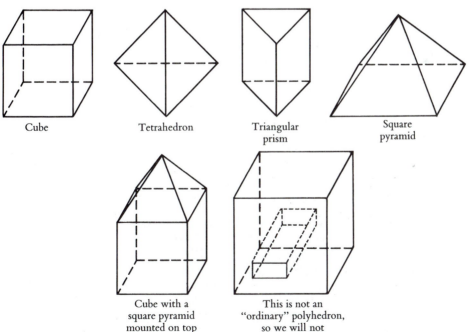

Cube   Tetrahedron   Triangular   Square
                                    prism   pyramid

Cube with a
square pyramid
mounted on top

This is not an
"ordinary" polyhedron,
so we will not
consider it.

a. Determine p and q for the tetrahedron and the octahedron.
b. Compare the values of pF and qV for the cube, tetrahedron, and octahedron.
c. What is the meaning of pF?
d. Show that $qV = 2E$.
e. The prefix *dodeca-* derives from the Greek word for "twelve"; hence, the regular solid with 12 faces is called a dodecahedron. Determine the number of vertices and edges of a regular dodecahedron.
f. The prefix *icosa-* derives from the Greek word for "twenty." Determine the number of vertices and the number of edges of a regular icosahedron.

■ 32. Conic sections that we haven't mentioned in this chapter occur when the plane that sections the cone passes through the vertex of the cone. These sections are called *degenerate*. Can you find three distinct types of degenerate conic sections?

▶ WRITING PROJECT

1. Examine the diagram of the Cassegrain telescope in Figure 19.22b and explain why it works.

■ Discussion exercise.

Chapter

20

▶▶▶▶▶ ▶▶▶

New Geometries for a New Universe

New kinds of geometry were conceived during the eighteenth century but were not to be born until the nineteenth century, when one of them became the basis for the next major revolution in physics and cosmology—the theory of relativity. We refer to **non-Euclidean geometries**—any of several sets of postulates, theorems, and corollaries that differ from Euclid's.

Ordinary geometry consists of statements, called *theorems* and *corollaries*, that are logical deductions from assumptions (other statements) called *postulates* or *axioms*. Euclid presented five postulates from which he developed a large body of theorems. We call this system **Euclidean geometry.**

From the fifth century B.C. until late in the nineteenth century, Euclidean geometry—including its extension to three dimensions—was thought to be the only science of space. Its theorems were thought to be statements of truth about the world we live in. No one could imagine a different geometry.

Euclid's five postulates, paraphrased somewhat, are as follows:

1. Two points determine a line.

2. A line segment can always be extended.

3. A circle can be drawn with any center and any radius.

4. All right angles are equal.

5. If *l* is any line and *P* any point not on *l*, then there exists exactly one line *m* through *P* that does not meet *l*.

These five statements were supposed to be absolute, self-evident truths. The first four are rather simple statements and are sufficiently unrelated that it can be shown fairly easily that they are **logically independent:** that is, none of them can be derived from the others by deduction. However, the postulate that *l* and *m* are parallel (i.e., they do not intersect) is another matter. Many early geometers thought this postulate was a logical consequence of the first four. In fact, Euclid himself may have thought his **parallel postulate** to be an unnecessary assumption for his geometry, for he derived nearly 30 theorems before using it.

The long history of these attempts to derive postulate 5 as the consequence of the first four postulates makes a fascinating story of failures; for whenever someone discovered a supposed proof, it was always found to be tacitly based on some assumption not contained in postulates 1 through 4. Attempts to prove the parallel postulate always failed because of an assumption that was **logically equivalent** to the parallel postulate; the reasoning was circular, hence invalid.

Among the hidden assumptions that were used in these purported proofs and that are logically equivalent to the parallel postulate are

1. The sum of the angles of a triangle equals 180°.

2. There is exactly one circle through any three points that are not on one line.

3. Parallel lines are equidistant.

▶ HYPERBOLIC GEOMETRY

Nikolai Ivanovich Lobachevsky (1792–1856) and János Bolyai (1802–1860), unlike many of their predecessors, did not try to derive Euclid's parallel postulate from the other postulates. Instead, working separately, they decided that the parallel postulate must be logically independent of those postulates. They realized that this means that those postulates really have nothing to say about the existence or nonexistence of parallels. Thus, including the parallel postulate with the other postulates will not lead to any inconsistencies. Lobachevsky and Bolyai reasoned that independence must mean even more: if they were to leave out Euclid's parallel postulate but include an *alternative* postulate about parallels—even one contradictory to Euclid's parallel postulate—they should once again get a consistent system of postulates. They both chose the same alternative postulate and proceeded to invent and explore the resulting "non-Euclidean" geometry, proving theorems from its postulates in the same style that Euclid had proved theorems from his (see Spotlight 20.1, p. 632).

The alternative postulate that Lobachevsky and Bolyai chose was:

Postulate H: If *l* is any line and *P* is any point not on the line, then there exists *more than one* line through *P* not meeting *l*.

The use of postulate H led to an entirely new system of theorems and corollaries, which we now call **hyperbolic geometry.** Some of the theorems in this system were exactly the same as those of the old, for those theorems that are derived only from postulates 1 through 4 must be valid in both systems. However, hyperbolic geometry provided some new and very surprising theorems.

Figure 20.1 shows a line *l* and a point *P* not on *l*. We drop a perpendicular from *P* to *l*, calling *A* the foot of the perpendicular. Now, consider the line *PE*, which is perpendicular to *PA*. The line *PE* is parallel to *l*, and according to postulate 5 it is the only parallel to *l* through *P*. However, if we are using postulate H, then there is *another* line *m* that passes through *P* and is parallel to *l*. Let us assume that *m* makes an acute angle with *PA*, as shown in Figure 20.1. Then there must be another line *n* that makes the same acute angle with *PA* on the other side of *PA* and is therefore also parallel to *l*.

From this construction, it is easy to see the first astonishing conclusion: through *P* there are *infinitely many* parallels to line *l*. This is clear as soon as one considers the set of all lines through *P*, which are separated into two classes by the lines *m* and *n*; one class contains *PA* and the other contains *PE*. The lines in the second class lie between *n* and *m*. All the lines in the second class are parallel to *l* (see Figure 20.2).

Using similar reasoning, Bolyai and Lobachevsky discovered many unusual theorems, of

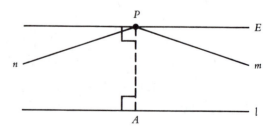

Figure 20.1 In hyperbolic geometry there is more than one parallel through a point *P* not on a given line *l*.

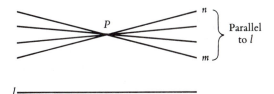

Figure 20.2 In hyperbolic geometry there are infinitely many parallels through a point P not on a given line l.

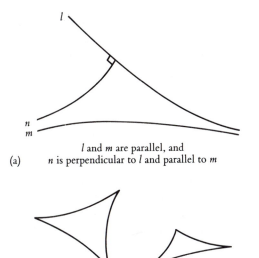

l and m are parallel, and
n is perpendicular to l and parallel to m

(a)

(b)

Typical triangles in hyperbolic geometry

Figure 20.3 In hyperbolic geometry, (a) given two parallel lines l and m, there exists a third line n perpendicular to one and parallel to the second; and (b) the sum of the angles of any triangle is less than $180°$.

which we will list only three (Figure 20.3 illustrates theorems 1 and 3):

1. The sum of the angles in any triangle is less than $180°$.

2. Similar triangles are congruent—that is, triangles having the same shape also have the same size.

3. Given two parallel lines, there exists a third line perpendicular to one and parallel to the second.

Similar results were obtained by Carl Friedrich Gauss (see Spotlight 20.2, p. 638), who investigated these unusual geometries before Lobachevsky and Bolyai, but did not publish his results.

▶ ELLIPTIC GEOMETRY

A generation after the discovery of hyperbolic geometry, G. F. Bernhard Riemann (1826–1866), a young German mathematician and disciple of Gauss, further scrutinized the basic assumptions of Euclidean geometry (see Figure 20.4). He analyzed postulate 2 and observed that "A line segment can always be extended" should be distinguished from "A line is infinite." That is, *unboundedness does not imply infinite extent.*

Think of the geometry on the surface of the earth. Going along what you would imagine was a line around the earth, you can travel another mile and another mile, and so on, and you would eventually return to your starting point. You have traveled on a finite path that is unbounded—that is, you can keep traveling on and on. When Riemann

Figure 20.4 Georg Riemann. (Deutsches Museum München.)

SPOTLIGHT 20.1

The Discovery of Non-Euclidean Geometry

▶ ▶ ▶ ▶ ▶ ▶ ▶ ▶ ▶ ▶ ▶ ▶ ▶

Nikolai Ivanovich Lobachevsky (1792–1856) and János Bolyai (1802–1860) independently discovered non-Euclidean geometry. Lobachevsky was the first to publish an account of non-Euclidean geometry (1829), which he first called "imaginary geometry" and later "pangeometry." His work attracted little attention when it appeared, largely because it was written in Russian and the Russians who read it were very critical.

Bolyai published his work on non-Euclidean geometry as a 26-page appendix to a book (the *Tentamen*, 1831) by his mathematician father Wolfgang, who proudly sent the work by his son to Carl Friedrich Gauss, the leading mathematician of his day. Gauss replied to Wolfgang that he himself had earlier discovered non-Euclidean geometry!

There is no direct mathematical connection between hyperbolic geometry and the hyperbola, or between elliptic geometry (the other type of non-Euclidean geometry that we examine) and the ellipse; but there *is* a significance in the common origins of their names. The name "hyperbolic" geometry was given to the geometry of Lobachevsky and Bolyai by Felix Klein, a famous geometer later in the nineteenth century. "Hyperbolic" comes from the Greek word *hyperbole*, meaning to be excessive (we get the slang word "hyper" from the same root): in hyperbolic geometry there are *too many* parallels. Klein at the same time christened another non-Euclidean geometry, that of Riemann, in which there are *no* parallels, "ellip-

Nikolai Ivanovich Lobachevsky. (Novasti Press Agency [A.P.N.].)

tic" geometry, from a Greek word meaning "to fall short." Ordinary Euclidean geometry, with Euclid's parallel postulate, fits into Klein's scheme as "parabolic" geometry, from a Greek word meaning "to compare." Apollonius, the Greek who discovered most of the properties of the conic sections, had named them in similar fashion according to their *eccentricity*, or departure from circularity: an ellipse has an eccentricity between 0 (for a circle) and up to but not including 1 (for very long and thin ellipses); a hyperbola has an eccentricity from just above 1 (for a hyperbola that is almost a pair of parallel lines) to infinity (for a very thin hyperbola). All parabolas come in at an eccentricity of exactly 1, in terms of the mathematical definition of this concept.

investigated the consequences of a line coming back on itself, he came to the conclusion that his rephrasing of Euclid's postulate 2 would also allow him to abandon Euclid's parallel postulate 5 in a strikingly novel way. Riemann replaced it with

Postulate E: *Every* two lines intersect.

Postulate E leads to **elliptic geometry.** We can see the plausibility of postulate E when we consider the geometry of the earth's surface, the **spherical geometry** used in navigation. What is a "straight line" in this geometry? The shortest distance between two points would be to tunnel directly through the earth, but to do so would be "out of bounds": we must stay on the surface of the sphere and consider our distances along it.

If we intersect the sphere in Figure 20.5 with a plane through A and B, the section is a circle passing through the given points; and the shortest arc, from A to B, of this circle is a candidate for the shortest path from A to B. Every plane section of the sphere is a circle, each with different curvature. *The larger the circle, the smaller the curvature; the smaller the curvature, the shorter the path between A and B.* Thus, the largest circle obtained as a section of the sphere gives rise to the shortest path. The largest circle is the **great circle**, obtained by having the plane cut through A and B and the center of the sphere (see Figure 20.5).

This result of spherical geometry is familiar to pilots who fly long distances. If an airplane is on the equator and the pilot wishes to fly to another point on the equator, the pilot would simply fly along the equator, a great circle. However, if the airplane were at 10° latitude north and the pilot wished to fly to a destination at the same latitude, then the shortest distance would require going farther north. An airplane flying from New York to Naples, both at the same latitude, would actually have to travel quite far north in the Atlantic Ocean, while the shortest flight path from Washington, D.C., to Ho Chi Minh City, Vietnam, is over the North Pole (see Figure 20.6).

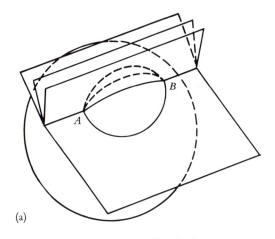

(a)

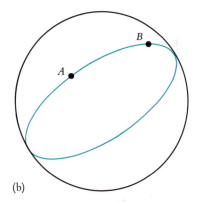

(b)

Figure 20.5 (a) The planes through A and B intersect the sphere in circles, the largest of which has the smallest curvature. (b) The great circle through A and B.

Returning to the idea of parallelism, we see that this concept simply doesn't exist in spherical geometry! Every pair of "lines"—that is, every pair of great circles—intersect. On the surface of the globe we note that triangles can have two or even three right angles: just put two vertices on the equator and one at a pole (see Figure 20.7). In fact, a general theorem of elliptic geometry states that *in an elliptic geometry, the sum of the angles of any triangle is greater than 180°.*

Ho Chi Minh City

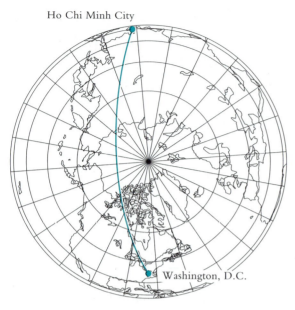

Washington, D.C.

Figure 20.6 The shortest path between two cities is an arc of a great circle.

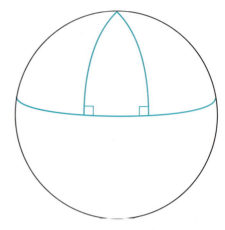

Figure 20.7 In elliptic geometry, a triangle can have two or more right angles.

Hence, in each of the three geometries we have investigated, we have a different sum for the angles in a triangle:

▶ $< 180°$ in hyperbolic geometry
▶ $= 180°$ in Euclidean geometry (sometimes called *parabolic geometry*)
▶ $> 180°$ in elliptic geometry

The property of ordinary planar Euclidean triangles — that all of them have the same angle sum, which is $180°$ — is one that dramatically distinguishes Euclid's geometry from spherical geometry and hyperbolic geometry. In spherical geometry, the angle sum in a triangle is always greater than $180°$, and it is not the same for all triangles. We find triangles whose angle sum is $190°$, as well as triangles whose angle sum is $250°$. However, any two triangles having the *same area* do have the same angle sum (see Spotlight 20.3, p. 640). In hyperbolic geometry (the geometry used by the artist M. C. Escher in his "Circle Limit"

prints), the angle sum in a triangle is always less than $180°$, but it is not the same for all triangles (see Spotlight 20.4, pp. 642–643).

As we have seen, any geometry that differs from Euclidean geometry is a non-Euclidean geometry. The first departures from Euclid's postulates involved denying, in some way, his parallel postulate. This led to the development of hyperbolic geometry and elliptic geometry, which are now considered the *classic non-Euclidean geometries*. However, we now have many more geometries that differ from Euclid's in a variety of ways.

▶ **MODELS FOR HYPERBOLIC GEOMETRY**

An example that satisfies a collection of axioms is called a *model* of the axioms. Showing that the example satisfies the axioms involves interpreting the language of the axioms for the particular example. For the postulates (axioms) of elliptic geometry, we have as a natural model the surface of the sphere. We interpret the "points" of the geometry to be the points on the sphere and the

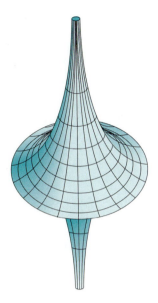

Figure 20.8 The pseudosphere, a model of hyperbolic geometry, with two kinds of "lines" shown.

"lines" to be its great circles. The postulates describe how "points" and "lines" should behave. Since the points and great circles on the sphere behave according to those postulates, our interpretation provides a model of the postulates.

The postulates of hyperbolic geometry have several intriguing models with differing interpretations of "points" and "lines":

▶ The *pseudosphere* of Figure 20.8 is formed by rotating a particular curve, called a *tractrix*, around the *y* axis. Imagine a girl with a dog on a leash that is always taut (see Figure 20.9). The dog starts at a point on the *x* axis. The girl starts at the origin and walks up the *y* axis, while the dog follows at leash length. The dog traces out a tractrix. The pseudosphere is formed by reflecting this curve in the *x* axis, then rotating the pair of curves around the *y* axis. The "points" of the geometry are the points on the surface of the pseudosphere. Like the great circles that are the "lines" on the sphere, the

"lines" in this model are the curves that give the shortest distances between points. Figure 20.8 shows examples of two different kinds of "lines," but there are also others, which spiral around the "horn."

To use the pseudosphere to gain much familiarity with hyperbolic geometry, you would have to make yourself a physical model. Even then, the varying kinds of "lines" make matters difficult. Fortunately, mathematicians have constructed other models of hyperbolic geometry that are more convenient to work with.

▶ In the *Klein-Beltrami disk model*, the "points" are the interior points of a fixed circle and the "lines" are chords of the circle with endpoints omitted. Figure 20.10 shows an instance of postulate H being fulfilled: for the line *l* and the point *P* not on the line *l*, lines are shown that go through *P* but do not meet *l*.

▶ In the *Poincaré disk model*, the "points" are again the interior points of a fixed circle.

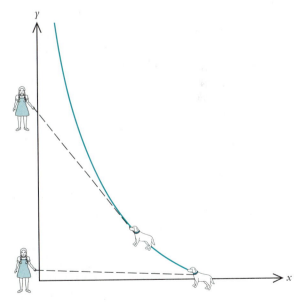

Figure 20.9 A tractrix traced out by a dog on a leash.

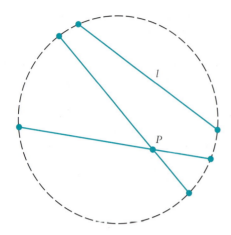

Figure 20.10 The Klein-Beltrami disk model of hyperbolic geometry.

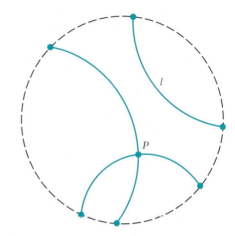

Figure 20.11 The Poincaré disk model of hyperbolic geometry.

The "lines" this time are circular arcs that meet the bounding circle at right angles. Figure 20.11 shows how postulate H is fulfilled in a particular instance, for a line *l* and a point *P* not on *l*.

▶ In the *Poincaré upper half-plane model,* the "points" are the points in a half-plane excluding the bounding line. There are two kinds of entities that we will consider as "lines": open semicircles with center on the bounding line, and open rays perpendicular to the bounding line. Figure 20.12 shows how postulate H is fulfilled in a particular instance.

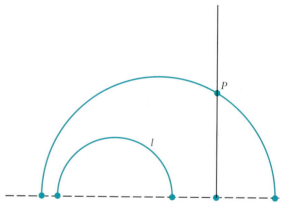

Figure 20.12 The Poincaré upper half-plane model of hyperbolic geometry.

▶ THE THEORY OF RELATIVITY

In 1905, Albert Einstein (Spotlight 20.6, p. 646) put forth his *special theory of relativity*, a complicated theory that constituted the first step in the greatest revolution in physics since Newton's *Principia*.

Einstein proposed a new way of thinking about events in the history of the universe. An event takes place in our three-dimensional space at a specific time in history. Thus, an event is located in *space-time* by four coordinates: three determine its position in space, and the fourth determines its position in time. Of course, these coordinates locate the event relative to a specific coordinate system. Einstein observed that the location of an event in space-time therefore depends on the position of the observer—that is, on the origin and orientation of the coordinate system being used. Different observers may obtain very different views of events, especially if one observer is traveling very fast with respect to the other.

Let's consider these ideas geometrically. The *distance* between two events, usually in relativity theory called an *interval*, is split into two parts: a *space-part* and a *time-part*. The space-part will be the part of the interval that comes from the position of the events in three-dimensional space, and the time-part will be the length of time that separates the events. This splitting up depends on the coordinate system and its orientation, so different results may be obtained by different observers (see Figure 20.13). However, the interval, being a line

segment joining the two events in four-dimensional space-time, is absolute—in the sense that it is the same for an observer at rest and for all other observers who are traveling at a constant velocity with respect to the one considered at rest.

For example, let's imagine that the eruption of Mount St. Helens in Washington in 1980 took place at the very same time that someone on Mount Palomar in California observed an astronomical phenomenon 100 light-years away. (A **light-year** is the *distance* that light travels in a year.) For those of us on earth (at rest relative to the earth), the eruption and the astronomical phenomenon took place one century apart on our time scale: the interval between the two events has a space-part of 100 light-years and a time-part of 100 years.

For observers traveling at constant velocity with respect to the earth, say, at 50 light-years

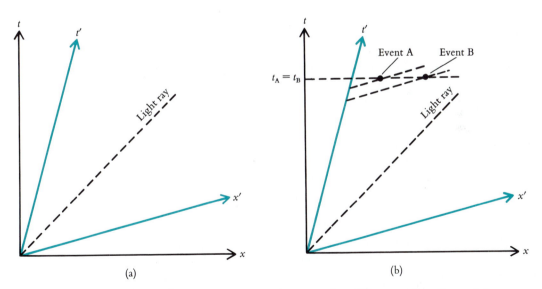

(a)   (b)

Figure 20.13 (a) A coordinate system representing space-time. The *t* axes show time and the *x* axes space. The black axes are a system at rest. Note that the blue, moving system tilts toward the 45° light ray line. (b) Observers in the system at rest (black axes) will say the events A and B occur at the same time. Observers in the moving blue system will say that event B occurs before event A.

SP TLIGHT 20.2   Carl Friedrich Gauss

Nicolai Ivanovich Lobachevsky and János Bolyai are justly given the credit for the invention of non-Euclidean geometry because they had the courage to publish their revolutionary work. However, Carl Friedrich Gauss (1777–1855) is also given credit as a coinventor. From correspondence and private papers that became available after his death, we know that Gauss, too, long believed that Euclid's parallel postulate could not be proved from the first four postulates.

Like Lobachevsky and Bolyai, Gauss derived many theorems based on postulate H and in fact produced some of the most beautiful proofs in hyperbolic geometry. The one most familiar to all students of the subject is Gauss's proof that triangles cannot be arbitrarily large; the maximum area that a triangle may possess

Carl Friedrich Gauss. (Deutsches Museum München.)

is the area contained by the trebly asymptotic triangle, the figure consisting of three lines, each of which is simultaneously parallel to the other two.

away from earth, the space-part and time-part of the interval would be very different. One observer might determine that the two events took place 200 years apart, while another might conclude that the two events took place simultaneously. Their splitting of the interval into space-parts and time-parts would be very different from ours. The geometry of space-time is indeed strange: in its four-dimensional space, the distance between two points—now the interval between two events—remains invariant (in the sense we have described), but its respective parts vary.

Three years after Einstein published his first paper on the subject, the mathematician Hermann Minkowski (1864–1909) gave Einstein's work a geometric interpretation that accepted Einstein's strange calculation of intervals and greatly simplified the theory. The geometry that was used, justifiably called *Minkowskian geometry*, is certainly non-Euclidean. Further, it makes use of one of Riemann's far-reaching ideas—that the nature of a mathematical space is determined by the way distance is measured; the distance formula therefore determines the nature of the geometry.

If the coordinate representations of two events are given by

$$(x_1, y_1, z_1, t_1) \quad \text{and} \quad (x_2, y_2, z_2, t_2)$$

then the interval I between them, in Minkowskian space, is calculated by the formula

$$I = \sqrt{\begin{aligned} & c^2 (t_2 - t_1)^2 - (x_2 - x_1)^2 \\ & - (y_2 - y_1)^2 - (z_2 - z_1)^2 \end{aligned}}$$

where c is the speed of light; whereas the distance d between them, if the points are in *Euclidean* four-dimensional space, would be computed by the formula

$$d = \sqrt{\begin{aligned} & c^2 (t_2 - t_1)^2 + (x_2 - x_1)^2 \\ & + (y_2 - y_1)^2 + (z_2 - z_1)^2 \end{aligned}}$$

The second formula is a direct generalization of the Pythagorean theorem from Euclidean plane geometry, while the first—with its minus signs—is not.

General Relativity

Little more than a decade after introducing his special theory of relativity, Einstein came forth with his *general theory of relativity*. This work, too, astonished the scientific world. Among other revolutionary ideas was his contention that space was "curved." By this he meant that light rays, which are considered to travel on paths of shortest distance, don't actually follow "straight lines" but bend to follow shortest distance paths in the curved space. Light rays even bend to different degrees, depending on where in the universe they are; if they pass through a strong gravitational field, then they bend considerably.

A test of this contention was made in 1919 during a total eclipse of the sun, when the light rays from a distant star passed close to the sun and could be studied. Einstein was right; the rays did bend

—and in an amount very close to his predictions. This observation showed that lines in the geometry of general relativity are not of the same character as Euclidean lines.

What sort of geometry was Einstein using? There are several answers to the question. First, the idea of "curved" space smacks of elliptic geometry, in the sense that a line through the universe comes back on itself. Second, Einstein used a variation on Minkowskian geometry in which the distance formula appropriate to the needs of physics varies from place to place in the universe, depending on the strength of the gravitational field. So, Einstein was using a form of Minkowskian geometry along with some considerably modified ideas of elliptic geometry. An appreciation of these non-Euclidean geometries very likely motivated his remark about the postulates of geometry in a famous 1921 lecture: "[They] are voluntary creations of the human mind. . . . To this interpretation of geometry I attach great importance, for should I not have been acquainted with it, I would never have been able to develop the theory of relativity."

Relativity and Length Contraction

The road that led Einstein to relativity is marked by one crucial experiment, conducted in 1887 by A. Michelson and E. C. Morley in Cleveland at what is now Case Western Reserve University. They were trying to determine whether there was a substance, an "ether," that served as the medium for the transmission of light and electromagnetic radiation. Sound waves do not travel in a vacuum but require a medium (e.g., air or water), relative to which we can measure the speed of sound. Nineteenth-century scientists reasoned that light waves, which do pass through the vacuum of space, must be carried through it in some medium that had not yet been detected.

The essence of their experiment was to send a lightbeam out to a mirror and back and measure the time that elapsed. They did this for two situations.

SPOTLIGHT 20.3 — Angle Sums in a Triangle

▶ ▶ ▶ ▶ ▶ ▶ ▶ ▶ ▶ ▶ ▶ ▶ ▶

The following simple proof shows that the angle sum in a triangle is equal to a straight angle. We will use triangle ABC as an example. Our goal will be to prove that $A + B + C = $ a straight angle $(= 180°)$.

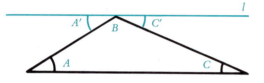

The Proof

At vertex B, draw a line l parallel to side AC. Then, by one of the first properties of parallels, which states that when parallel lines are cut by a transversal, the "alternate interior angles" are equal, angle $A = $ angle A'. Likewise, angle $C = $ angle C'. Now clearly $A' + B + C' = $ a straight angle. Hence, by substituting A for A' and C for C',

$$A + B + C = \text{a straight angle}$$

which is what we set out to prove.

We can't use this same proof on the sphere because it has no parallels at all. On the sphere every two straight-line paths, or great circle routes, eventually cross. In fact, on the sphere the sum of the angles of a spherical triangle is always *greater* than a straight angle.

On the other hand, in the plane of hyperbolic geometry there are "too many" parallels, and when the line l is drawn at B so that angle $A' = $ angle A, then angle C is always less than angle C'. Hence, the angle sum $A + B + C$ in triangle ABC is always less than the straight angle $A' + B + C'$.

In one, the lightbeam was aligned in the direction of the earth's rotation, so that on the way out the speed of the earth would add to the speed of light through the ether, and on the way back, it would subtract from the speed of the light.

In the other situation, they aligned the lightbeam perpendicular to the earth's rotation, so that the earth's motion would not affect the *speed* of the lightbeam in the ether. In this situation, however, the earth's rotation does affect *how far* the lightbeam has to travel, as we now demonstrate.

We will imagine an analogous situation. You are in a river that is a distance d wide and you

decide to conduct a swimming experiment. In place of bouncing lightbeams, you will swim back and forth. Your speed of swimming c corresponds to the speed of light, and the speed v of flow of the river corresponds to the speed of the rotation of the earth.

Corresponding to the first situation, you swim a fixed distance d downstream, then swim back again (see Figure 20.14a); let's say that it takes you time t_1 downstream and time t_2 upstream. On the trip downstream, you are swimming through the water at speed c but you are moving at a speed of $(c + v)$ relative to the bank. Similarly, on the trip

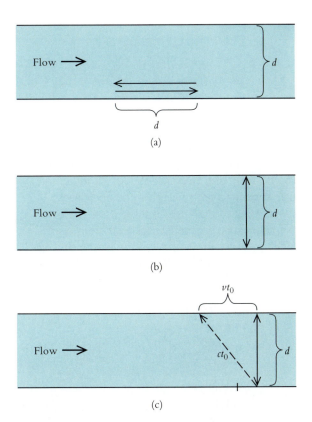

Figure 20.14 (a) Swimming with and against the current of a river. (b) Swimming across a river and back to the starting point. (c) To stay even with the starting point, the swimmer needs to head upstream to some degree.

upstream, you are really moving only at speed $(c - v)$.

Using the basic formula *distance = rate × time*, we have $d = (c + v)t_1$ and $d = (c - v)t_2$. Your total time for the round-trip is

$$t_{up/down} = t_1 + t_2 = \frac{d}{c + v} + \frac{d}{c - v}$$

$$= \frac{2dc}{(c + v)(c - v)} = \frac{2d/c}{1 - v^2/c^2}$$

You should check these calculations for a numerical example, for example, the distance d is 1/2

mile, you can swim at a speed c of 2 miles an hour, and the river flows at a speed v of 1 mile per hour. You should find that it takes you one-sixth of an hour downstream and one-half of an hour upstream, for a total time for the round-trip of two-thirds of an hour, or about 0.67 hr.

Corresponding to the second situation, you swim across the river and back to your starting point (see Figure 20.14b). As you swim, however, you have to fight the downstream current, so that you always stay even with your starting point. You not only have to swim the distance across the river, you also need to swim upstream the equivalent distance that the river is carrying you downstream; you need to always head part way upstream, so only part of your swimming effort is directed toward carrying you directly across the river. In effect, you are swimming the length of the hypotenuse of a right triangle, as shown in Figure 20.14c, and being carried downstream by the length of one leg.

Denote the time that it takes you to cross the river by t_0. The relationship between how far you travel and how long it takes is given by the Pythagorean theorem:

$$(ct_0)^2 = vt_0^2 + d^2$$

We solve the equation for the time:

$$c^2t_0^2 - v^2t_0^2 = d^2$$

$$t_0^2 = \frac{d^2}{c^2 - v^2}$$

$$t_0 = \frac{d}{c\sqrt{1 - v^2/c^2}}$$

The time for the return trip is the same, so the total time for the round-trip is

$$t_{back/forth} = \frac{2d/c}{\sqrt{1 - v^2/c^2}}$$

The Dutch artist M. C. Escher (1898–1972) was particularly interested in the metamorphosis of figures that change almost imperceptibly into other figures or into larger or smaller versions of themselves. He was able to discover a way to draw, inside a circle or square, figures that gradually get larger as they approach the outside of the enclosure. But it wasn't until he was shown a mathematician's representation of hyperbolic geometry (in which the sum of the angles of a triangle is always less than 180°) that he discovered how to make figures gradually get *smaller* toward the outside of a circle.

Douglas Dunham of the University of Minnesota, Duluth, has devised a computer program based on hyperbolic geometry that can produce an infinite variety of the type of drawings Escher so ingeniously drew. One of these, Dunham's "Circle Limit IV," is shown here. (For others, see Color Plates 5–6.)

The "lines" of this geometry are arcs of circles that are perpendicular to the outside circle. (The "lines" of spherical geometry are great-circle routes on the sphere.) This particular print is based on a regular tiling of the hyperbolic plane. The tiles shown here are regular quadrilaterals — their vertices are the points where the feet of three angels meet the feet of three devils. Six of these tiles meet at each vertex. Thus, the angle of each is 60° instead of 90°, which it would be for the corresponding tiling of the Euclidean plane, where four squares meet at each vertex.

The edges of some of the tiles (of the underlying tiling) have been drawn in so that they can be seen, and two are shaded. Although the tiles appear to get smaller (in the Euclidean sense) toward the edge of the outside circle, the hyperbolic geometry uses a different distance measure in which all of the tiles, including the two that are shaded, are congruent (the same size).

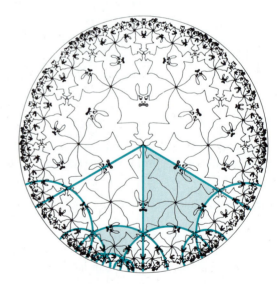

Dunham's Circle Limit IV plot. This computer-generated tiling in the hyperbolic plane creates an image very similar to that of M. C. Escher's *Angels and Devils*. Note how the positions of feet and heads are related by radii and arcs that define the underlying tiling pattern. (Courtesy of Douglas Dunham.)

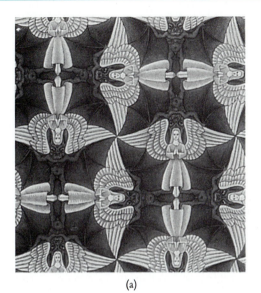

(a)

(b)

(c)

(a) M. C. Escher's *Heaven and Hell* (also known as *Angels and Devils*). Photographed from one of Escher's notebooks, this example demonstrates a repeating pattern of the Euclidean plane. Note the uniform size of the figures and the central meeting of the wingtips of four angels and four devils. (b) Escher's pattern of angels and devils carved on an ivory sphere by Masatoshi. This mapping of the pattern onto a sphere shows the different effects of a spherical geometry. Note that in this version, the wingtips of three angels and three devils meet. (c) M. C. Escher's *Circle Limit IV (Heaven and Hell)*. This example shows the repeating angels and devils pattern mapped onto a hyperbolic geometry. At the center, the feet of three angels and three devils meet. As one moves outward, the figures get smaller. Note that the wingtips of four angels and four devils meet. (a–c: © M. C. Escher Heirs, c/o Cordon Arts – Baarn – Holland.)

SP⬤TLIGHT 20.5   What's Hyperbolic about Hyperbolic Geometry?

The geometry of Lobachevsky is often called *hyperbolic* geometry and that of Riemann is often called *elliptic* geometry. No doubt these terms remind you of hyperbolas and ellipses — as we shall explain, rightfully so.

The word "hyperbola" comes from Greek, meaning "a throwing beyond," or excess. The word "ellipse" means "a falling short," or defect. The word "parabola" means "falling beside," or being parallel to. What all these terms are comparing are the angle at which the conic section (hyperbola, ellipse, or parabola) cuts the cone, compared to the angle of the cone itself (see Figure 19.15).

How do these terms come to be applied to whole geometries? You can think of hyperbolic geometry as having an "excess" of parallels, since given a line and a point not on the line, there is an infinite number of lines through the point that are parallel to the given line. Similarly, in an elliptic geometry, there is a "defect" of parallels, compared to Euclidean geometry, since there aren't any parallel lines.

In a deeper sense, the distance measure in a hyperbolic geometry has an algebraic form similar to the algebraic form of a hyperbola in analytic geometry coordinates ($y^2 - x^2 = 1$), while the distance measure in an elliptic geometry has a form similar to that of an ellipse ($x^2 + y^2 = 1$).

However, in the context of the sum of the interior angles of a triangle, there is a reversal of the meaning of the name of the geometry. The sum is greater than 180° in elliptic geometry, less than 180° in hyperbolic geometry, and exactly 180° in Euclidean geometry. Thus, hyperbolic geometry features triangles whose angle sums are "defective," while those in elliptic geometry are "excessive."

While Euclidean geometry corresponds to a space with no curvature, elliptic geometry describes a space of constant positive curvature, and hyperbolic geometry describes a space of constant negative curvature.

Finally, when it comes to trigonometry, the relevant formulas are different for each type of geometry. For example, the Pythagorean theorem for right triangles does not hold in spherical geometry nor in hyperbolic geometry, though in each case another formula does hold. For spherical geometry, the formulas for right triangles and trigonometry involve the familiar "circular" trigonometric functions (sine, cosine, tangent); in hyperbolic geometry, the related hyperbolic trigonometric functions (sinh, cosh, tanh — you might see keys for these on your calculator) apply.

Testing this algebra with the same numerical values as before, you should find that the round-trip across the river takes you $1/\sqrt{3} \approx 0.577$ hr.

You observe that the times for the two round-trips are different. The time for the trip along the direction of motion is longer, by a factor of $1/\sqrt{1 - v^2/c^2}$.

Michelson and Morley had a different experience: the times that they observed in their analogous experiment were *the same*. Because of their experience and reputation as experimenters, they and their colleagues were sure that this astonishing result was not the consequence of measurement or experimental error.

So what was wrong, then, with the theory? Scientists were baffled. One ingenious theory of how Michelson and Morley could observe equal times for the two alignments was that the speed of the rotation of the earth somehow *contracts lengths along the direction of motion*, in exactly the right proportion to cancel the difference in times. Thus, a planet moving at nearly the speed of light would flatten in the direction of motion, so as to turn into a pancake. The length l' along the direction of motion would have to be related to the length l in the direction perpendicular to the motion by

$$l' = l\sqrt{1 - \left(\frac{v}{c}\right)^2} = l/\gamma$$

where

$$\gamma = \frac{1}{\sqrt{1 - \left(\frac{v}{c}\right)^2}}$$

is known as the **Lorentz-Fitzgerald factor**.

Because the speed of light is so great—186,000 miles per second, or 3.0×10^{10} cm/sec—the value of the Lorentz-Fitzgerald factor is very nearly 1 until v reaches about 10% of the speed of light. See Table 20.1 for values of γ for various moving objects.

TABLE 20.1   **Gamma Corresponding to Various Speeds**

Moving object	v	v/c	Gamma (γ)
Automobile	100 km/hr	0.00000009	1.000000000
Concorde SST	2,000 km/hr	0.000002	1.000000000
Rifle bullet	1 km/sec	0.000003	1.000000000
Earth escape speed	11 km/sec	0.000037	1.000000001
Orbital speed of earth	30 km/sec	0.0001	1.000000005
10% of light's speed	30,000 km/sec	0.1	1.005
		0.9	2.294
		0.99	7.089
		0.999	22.37
Subatomic particles in an accelerator		0.9994	28.87
		0.9999	70.71
		0.999999	707.1
		0.99999999	7071

SP TLIGHT 20.6   Albert Einstein

▶ ▶ ▶ ▶ ▶ ▶ ▶ ▶ ▶ ▶ ▶ ▶ ▶ ▶ ▶

Einstein (1879–1955) was born in Ulm to a German Jewish family with liberal ideas. Although he showed early signs of brilliance, he did not do well in school. He especially disliked German teaching methods. In the mid-1890s he went to study in Switzerland, a country much more to his liking, where he went to work as a patent clerk. Einstein burst upon the scientific scene in 1905 with his theory of special relativity. In 1916 he published his theory of general relativity. General relativity was successfully tested in 1919, and his fame grew enormously. Nazism forced Einstein to leave Europe. He settled at the Institute for Advanced Study at Princeton, where he remained until his death at age 76.

Albert Einstein. (Yerkes Observatory.)

How come Michelson and Morley couldn't detect this shrinkage in the direction of motion? Because their measuring rods, when aligned along the direction of motion, shrank too. So the contraction theory could never be verified by direct measurements.

You would think that photography would help — for example, that you could see a ball moving at nearly the speed of light as the pancake shape that it must be. However, most surprisingly, even a camera can't see the contraction! An optical distortion compensates for the shrinkage, as we now explain.

You and the camera see by means of "particles" (photons) of light reflected from an object. Light from faraway objects can take a long time to reach us: light from the sun takes about eight min-utes, and the light that reaches us now from distant stars was emitted billions of years ago. Similarly, in the case of a moving object, we see at the same time images of close-up parts and (because of time delay from the fixed speed of light) *earlier* images of faraway parts of an object. Hence, the object appears (to us or on the film) stretched in the direction of motion. This stretching compensates for the contraction. Figure 20.15 shows a computer reconstruction of how an object moving at close to the speed of light would appear. Notice that it does not appear to shrink in the direction of motion (even though it actually does).

The Lorentz-Fitzgerald contraction theory was based on a complicated theory of matter interacting with the ether. Scientists eventually were forced to conclude that there is no medium in

(a)

(b)

Figure 20.15 (a) A teapot at rest. (© 1993 Pixar. All rights reserved.) (b) A view from the same angle of the teapot passing by at 99% of the speed of light. (© Ping-Kang Hsiung.)

which light waves move, no "ether" relative to which we can measure the speed of light. The appealing analogy between a light wave and a swimmer swimming through water is misleading.

Twenty-four years after the Michelson-Morley experiment, Einstein hypothesized that the speed of light is not affected by the motion of the source nor of the observer. The rotation of the earth cannot add to or subtract from the speed of light in the Michelson-Morley experiment. Einstein's theory predicts the same time, $2d/c$, for the round-trip in either alignment of their equipment.

Einstein's theory of relativity also predicts a contraction of length in the direction of relative motion, by exactly the Lorentz-Fitzgerald factor. The reason for this relativistic contraction, however, has nothing to do with ether or Lorentz's theory to explain the Michelson-Morley results on the basis of it. Einstein's theory eliminates the need to suppose that there is an ether. The underlying reason for the relativistic length contraction is relativity itself: the relative motion of object and observer. As viewed by you, the teapot is moving at close to the speed of light, and its length must contract in the direction that it appears to you to be traveling (even though you can't see the contraction, as we've explained). As viewed by an ant on the teapot, *you* appear to be moving at close to the speed of light and are thin as a pancake in the direction that you appear to be moving (even though the ant can't see you that way).

Another consequence of relativity is that *time, too, contracts with motion.* Consider two observers moving at a constant velocity v relative to each other. *Each* will observe that the other's clock runs slow compared to their own, by a factor of γ. This strange result is known as the *clock paradox.*

▶ **WHICH GEOMETRY IS TRUE?**

As far as measurement and travel on the surface of the earth go, we know that we live on a world with an elliptic geometry. When it comes to travel at velocities near the speed of light, the geometry that

SP   TLIGHT 20.7   The Implications of Non-Euclidean Geometry and Relativity

▶ ▶ ▶ ▶ ▶ ▶ ▶ ▶ ▶ ▶ ▶ ▶ ▶ ▶ ▶

The creation of non-Euclidean geometry affected scientific thought in two ways. First of all, the major facts of mathematics, that is, the axioms and theorems about triangles, squares, circles, and other common figures are used repeatedly in scientific work. Since these facts could no longer be regarded as truths, all conclusions of science that depended upon strictly mathematical theorems also ceased to be truths.

Second, the debacle in mathematics led scientists to question whether they could ever hope to find a true scientific theory.

Even on the level of engineering a serious question emerged. Since bridges, buildings, dams, and other works were based on Euclidean geometry, was there not some danger that these structures would collapse? Actually, there is no guarantee that they will not. But this thought did not alarm the scientists and engineers of the nineteenth century, who did not believe that the geometry of physical space could be other than Euclidean. However, the advent of the theory of relativity drove home the point that Euclidean geometry is not necessarily the best geometry for applications. For engineering involving motion with high velocities, such as modern accelerators of electrons or neutrons, the theory of relativity is used.

Past ages have sought absolute standards in law, ethics, government, economics, and other fields. They believed that by reasoning one could determine the perfect state, the perfect economic system, the ideals of human behavior, and the like. This belief in absolutes was based on the conviction that there were truths in the respective spheres. But in depriving mathematics of its claim to truth, the non-Euclidean geometries shattered the hope of ever attaining any truths.

Perhaps the greatest import of non-Euclidean geometry is the insight it offers into the workings of the human mind. No episode of history is more instructive. The view that mathematics is a body of truths was accepted at face value by every thinking being for 2000 years. This view, of course, proved to be wrong. We see, therefore, on the one hand, how powerless the mind is to recognize the assumptions it makes. It would be more appropriate to say of us that we are surest of what we believe, than to claim that we believe what is sure. Apparently we should constantly reexamine our firmest convictions, for these are most likely to be suspect. They mark our limitations rather than our positive accomplishments. On the other hand, non-Euclidean geometry also shows the heights to which the human mind can rise. In pursuing the concept of a new geometry, it defied intuition, common sense, experience, and the most firmly entrenched philosophical doctrines just to see what reasoning would produce.

Source: Adapted from Morris Kline, *Mathematics for the Nonmathematician,* Dover, New York, 1985, pp. 474–476.

applies to space-time is Minkowskian geometry, a kind of non-Euclidean geometry. But what about the universe of space beyond the earth's surface, without regard to time? Do we really live in a spatial universe that is Euclidean?

One way to tell would be to measure the angles in a large triangle, to see how their sum compares with 180°. Because all measurements have some imprecision, we cannot prove that the angles of a measured triangle add up to exactly 180°. Sufficiently precise measurements of very large triangles, however, could conceivably prove that space is not Euclidean.

Gauss was employed for a time by the government of Hanover in a geodetic survey, in the course of which he measured the angles in a triangle formed by three mountain peaks roughly 50 miles apart. The deviation from 180° was less than the error estimate for the measurement, so the sum could be equal to 180°, or greater, or less—the sum was consistent with all three hypotheses. In fact, if there is any difference from 180° for this triangle of mountain peaks, it was far too small for Gauss to detect—as he probably realized—and far too small even for us to detect today.

Lobachevsky considered even larger triangles and looked into the parallax of stars (the apparent relative motion, as the earth orbits the sun, of nearer stars compared to more distant ones—see Chapter 18, p. 581). But neither he nor others since have found a triangle whose angle sum is definitely different from 180°, despite the fact that in hyperbolic geometry, the larger the area of the triangle, the larger the defect must be.

If space does have a hyperbolic geometry, then there is a lower bound for the parallax of stars (though that would not mean that there is a limit to how far away stars can be). Although there is certainly a smallest observed parallax among the thousand or so stars whose parallax we know, there may be stars yet unmeasured whose parallax is even smaller. (The formula in Chapter 18 for deriving distance from parallax is not valid in hyperbolic geometry; a different formula applies, involving the *hyperbolic trigonometric functions*.)

Space could have an elliptic geometry, with triangles having angular excesses rather than defects. The universe could then be the three-dimensional analogue of the two-dimensional surface of a sphere. Just as the surface of a sphere has a finite area, the universe could have a finite volume, despite having no boundaries—just as the surface of a sphere has no boundaries. Such a space would have positive curvature.

Regardless of the true situation for actual three-dimensional space, we appear to *perceive* space visually as hyperbolic. Common visual illusions, classical experiments in perception, and the empirical truth of *Brentano's hypothesis* (that humans tend to overestimate small angles and underestimate large ones) all lead to the conclusion that "perceived space" is hyperbolic.

The question of which geometry is true would not have occurred to anyone before the nineteenth century. The discovery of non-Euclidean geometry and the theory of relativity have had profound intellectual implications in all fields (see Spotlight 20.7).

▶ REVIEW VOCABULARY

Elliptic geometry   A system of geometry in which there are no parallel lines.
Euclidean geometry   The "ordinary" system of geometry based on the five postulates Euclid used, including the parallel postulate.
Great circle   The set of points that is the intersection of a sphere and a plane containing its center.
Hyperbolic geometry   A system of geometry in which there exists more than one parallel line through a point P not on a given line l.
Light-year   The distance that light travels in a year.
Logically equivalent   Two statements are logically equivalent if each can be deduced from the other.
Logically independent   Refers to a set of statements, no one of which can be logically deduced from the others.

Lorentz-Fitzgerald factor   The factor by which length and time are contracted by motion.

Non-Euclidean geometry   Any geometry that differs from Euclidean geometry.

Parallel postulate   A basic assumption of geometry that states whether through a point P, not on a given line l, there exists none, one, or more than one line parallel to given line l.

Postulate E   *Every* two lines intersect.

Postulate H   If l is any line and P is any point not on the line, then there exists *more than one* line through P not meeting l.

Spherical geometry   The geometry of a sphere or the earth's surface.

▶ SUGGESTED READINGS

ABBOTT, EDWIN A.: *Flatland: A Romance of Many Dimensions*, Princeton University Press, Princeton, N.J., 1989.

BOLOTOVSKY, B. M.: "What's That You See? On the Perceived Shape of Rapidly Moving Objects," *Quantum* 3(4):5 – 8 (March/April 1993).

BOLTYANSKY, VLADIMIR: "Turning the Incredible into the Obvious: How Many Geometries Do You Know?" *Quantum* 3(1):19 – 23 (September/October 1992).

CARVER, MAXWELL: Brain Bogglers: "Chicken a la King," *Discover*, pp. 96, 92 (March 1988).

CROWE, DONALD W.: "Some exotic geometries," in Anatole Beck, Michael Bleicher, and Donald W. Crowe, *Excursions into Mathematics*, Chapter 4, Worth, New York, 1969, pp. 211 – 314. A comparison of spherical, Euclidean, and hyperbolic geometries, with sections on finite geometries.

DAVIS, A. S.: "The Relevance of Mathematics for the Remnant," University of Oklahoma Dept. of Mathematics Preprints No. 48, 1968.

DUBROVSKY, VLADIMIR: "Inversion: A Most Useful Kind of Transformation," *Quantum* 3(1):40 – 46 (September/October 1992); 80 – 81 (November/December 1992).

FABER, RICHARD L.: *Foundations of Euclidean and Non-Euclidean Geometry*, Dekker, New York, 1983.

GINDIKIN, SIMON: "The Wonderland of Poincaria," *Quantum* 3(2):21 – 26, 58 – 59 (November/December 1992).

GREENBERG, M. J.: *Euclidean and Non-Euclidean Geometries*, Freeman, New York, 1980.

HSIUNG, PING-KANG, ROBERT H. THIBADEAU, AND ROBERT H.P. DUNN: "Ray-Tracing Relativity," *Pixel* 1(1):10 – 18 (January/February 1990).

JACOBS, HAROLD R.: *Geometry*, 2nd ed., Freeman, New York, 1987. Chapters 6 and 16 treat parallel lines and non-Euclidean geometries.

MAURER, STEPHEN B.: "The King Chicken Theorems," *Mathematics Magazine* 53:67 – 80 (1980).

PARKER, GEORGE D.: "POINCARE: An Excursion into Hyperbolic Geometry." IBM-PC program for doing hyperbolic geometry in the Poincaré upper-half-plane model. Available from the author at 1702 West Taylor, Carbondale, IL 62901.

PETERSON, IVARS: "Space-Time Odyssey: Visualizing the Effects of Traveling Near the Speed of Light," *Science News* 137:232 – 233, 237 (14 April 1990).

PETIT, JEAN-PIERRE: *Everything Is Relative: The Adventures of Archibald Higgins*, William Kaufmann, Inc. Relativity as told in a cartoon tale, "for adults of any age."

SCHWINGER, JULIAN: *Einstein's Legacy: The Unity of Space and Time*, Scientific American Library, Freeman, New York, 1986.

SMART, JAMES R.: *Modern Geometries*, 3rd ed., Brooks/Cole, Pacific Grove, Calif., 1988.

SVED, MARTA: *Journey into Geometries*, Mathematical Association of America, Washington, D.C.,

1991. Introduces hyperbolic geometry in a delightfully informal style.

TAYLOR, E. F., AND J. A. WHEELER: *Spacetime Physics,* Freeman, New York, 1966.

WEEKS, JEFFREY R.: *The Shape of Space: How to Visualize Surfaces and Three-Dimensional Manifolds.*

Dekker, New York, 1985. Contains chapters on the hyperbolic plane (including how to make hyperbolic paper), spherical geometry, hyperbolic space, and three-dimensional spherical geometry.

WHEELER, JOHN ARCHIBALD: *A Journey into Gravity and Spacetime,* Scientific American Library, Freeman, New York, 1990.

▶EXERCISES

1. In $\triangle ABC$, the measure of $\angle A$ is twice the measure of $\angle B$, and the measure of $\angle C$ is three times the measure of $\angle B$. Determine all three measures.

2. By producing a specific example, show that there is a triangle in elliptic geometry in which all three angles are right angles, so that the sum of the angles of the triangle is 270°. (Hint: Spherical geometry is an elliptic geometry.)

For Exercises 3 and 4, refer to the following: The quadrilateral $ABCD$ in the following figure is drawn to suggest the situation in hyperbolic geometry. The angles at A and B are right angles, whereas the angles at C and D are acute.

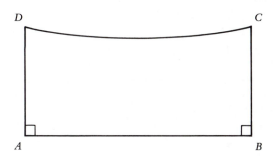

3. Prove that if $AD = BC$, then angles C and D are of equal measure. (Hint: Draw the diagonals. Then use the Euclidean theorems on the congruence of triangles, which do not depend on the parallel postulate.)

4. Prove that either the sum of the angles of triangle ABD is less than 180°, or the sum of the angles of triangle BCD is less than 180°, or both. (Hint: What can you say about the sum of the angles of the quadrilateral?) (In fact, *both* triangles have angle sum less than 180°, but that is harder to prove.)

For Exercises 5 to 12, refer to the following: A collection of trees is arranged in rows so that the following axioms are satisfied:

A There is at least one tree.
B Each row contains exactly two trees.
C Each tree belongs to at least one row.

D Any two trees have exactly one row in common.

E For any row, there is exactly one other row with no trees in common with the first row.

5. Show the following:
   a. There is at least one row.
   b. There are at least two rows.
   c. There are at least four trees.

6. Show the following:
   a. Every tree belongs to at least two rows.
   b. There are at least six rows.

7. Show the following:
   a. There are exactly four trees.
   b. There are exactly six rows.
   c. Each tree belongs to exactly three rows.

8. In this exercise, you create models of the axiom system by interpreting the terms used in the axioms:
   a. Interpreting "tree" as a point and "row" as a line, draw a model of the axiom system. Note that the lines that you draw in the plane have many other planar points that belong to them but which we don't consider, as they are not part of this particular axiom system. (This gives a *four-point geometry*.)
   b. Interpreting "tree" as a line and "row" as a point, draw a model of the axiom system. (This gives a *six-point geometry*.)

9. In this problem, you create further models of the axiom system:
   a. Interpreting "tree" as student and "row" as committee, construct a model of the axiom system. Name or label the students and list all of the members of each committee.
   b. Interpreting "tree" as committee and "row" as student, construct a model of the axiom system. Name or label the students and list all of the members of each committee.

10. Interpreting "tree" as a line and "row" as a point:
    a. How many triangles are there? (A triangle consists of three points and three lines joining them in pairs.)
    b. Are there two lines that are parallel? (Two lines are parallel if they do not have a point in common.)
    c. Does the parallel postulate hold?

11. Interpreting "tree" as a point and "row" as a line:
    a. How many triangles are there? (A triangle consists of three points and three lines joining them in pairs.)
    b. Are there two lines that are parallel? (Two lines are parallel if they do not have a point in common.)
    c. Does the parallel postulate hold?

12. (Adapted from Beck, Bleicher, and Crowe, *Excursions into Mathematics* [1969], p. 289.) Suppose that by "point" is meant the location of a store where

groceries are sold — say, Albany, Birmingham, Chicago, or Denver. Suppose that "line" means a particular line of groceries — say, apples, bananas, cheese, doughnuts, eels, or figs. Finally, suppose that "a tree belongs to a row" or "a row contains a tree" means that a particular line of goods is sold at a particular location. Give an explicit distribution of the six commodities in the four cities so that the axioms apply.

For Exercises 13 to 16, refer to the following: Suppose the following set of axioms concerns two classes of objects K and L, whose nature is left undetermined:

A Any two members of K are contained in exactly one member of L.
B No member of K is contained in more than two members of L.
C There is no member of L that contains all of the members of K.
D Any two members of L contain exactly one member of K in common.
E No member of L contains more than two members of K.

▲ 13. We are interested in possible *models* of these axioms, that is, examples that satisfy them. Check the axioms to verify that if neither K nor L has any members at all, the axioms still hold. We say that they are *vacuously true* for this "empty" model. For each of the following situations, is there a model? If so, verify that the axioms all hold; if not, show what axioms would have to be violated.
    a. K has no members and L has exactly one?
    b. K and L each have one member?

▲ 14. Is there a model in which:
    a. K has one member and L has two members?
    b. K has one member and L has three or more members?

▲ 15. Add the additional axiom

F K has at least two members.

    a. Show that K has at least three members.
    b. Show that K cannot have more than three members, so it has exactly three members. (This gives a *three-point geometry*.)
    c. How many members can L have?

▲ 16. As in Exercise 15:
    a. Using points and lines, give a model of the set of axioms.
    b. If axiom **C** is now omitted, are other models possible?

For Exercises 17 to 19, refer to the following: Consider the following system of axioms concerning two classes of objects M and N.

A N has at least one member.
B Every member of N contains exactly three members of M.

▲ Advanced exercise.

C Any two members of M are contained in just one member of N.
D There is no member of N that contains all of the members of M.
E Any two members of N have at least one member of M in common.

▲ 17. We explore how many members M and N may have:
   a. Show that each two members of N have exactly one member in common.
   b. Show that each of M and N has at least seven members. To help your intuition, you may want to try to construct a model of the axioms out of points and lines. Remember, though, that you must reason from the axioms, not from any picture that you draw.

▲ 18. We continue with determining the number of members of M and N.
   a. Suppose that M has an eighth member. Show that this supposition leads to a contradiction.
   b. Show that M and N must have exactly seven members each.

▲ 19. What are some models for this set of axioms?
   a. Interpret the members of M as students and the members of N as committees. Name or label the students and list all of the members of each committee.
   b. Interpret the members of M as points and the members of N as (not necessarily straight) lines, and draw an appropriate model. (This is *Fano's seven-point geometry*.)

▲ 20. For the model in terms of points and lines in Exercise 19(b):
   a. Are there two lines that are parallel? (Two lines are parallel if they do not have a point in common.)
   b. Does the parallel postulate hold?

For Exercises 20 to 25, refer to the following: Consider the set $\{0, 1, 2\}$ under a "clock" arithmetic, with 3 taking the role of the 12 in the usual clock arithmetic, so that $1 + 2 = 0$ and $2 + 2 = 1$. We can even introduce multiplication, with the usual results except that $2 \times 2 = 1$ (we would expect 4, which converts to 1 in our clock arithmetic). Using the set $\{0, 1, 2\}$ as the possible x- and y-coordinates, we can form the nine points in the figure below. These will be the points of a "miniature" analytic geometry. The lines will consist of points that satisfy linear equations, that is, equations of the form $ax + by = c$, with a, b, and c from the set $\{0, 1, 2\}$.

  (0,2)   (1,2)   (2,2)

  (0,1)   (1,1)   (2,1)

  (0,0)   (1,0)   (2,0)

▲ Advanced exercise.

▲ 21. List all of the points on the lines
   a. $x = 1$
   b. $y = 2$
   c. $x + y = 1$
   d. $x + 2y = 1$

▲ 22. Find the intersection of the lines $x + y = 1$ and $2x + y = 2$:
   a. By using algebra. (Hint: Subtract one equation from the other.)
   b. By listing all of the points on each line and comparing the lists.

▲ 23. We explore how many lines and points are in this geometry:
   a. How many different lines are there? (Hint: A line is either a vertical line, with equation $x = c$, or else it can be written in the form $y = mx + b$, with each of c, m, and b being either 0, 1, or 2.)
   b. How many points lie on each line?
   c. How many lines pass through each point?

▲ 24. This miniature analytic geometry is known as the *affine two-dimensional geometry over 3 elements*, an example of an *affine plane*. It satisfies the following three axioms:

  A For any pair of distinct points, there is exactly one line containing both of them.
  E Given a line and a point not on the line, there is exactly one line though the point that does not contain any points of the given line. (This is the parallel postulate, so this is a Euclidean geometry.)
  B There is a subset of at least four points, no three of which lie on the same straight line.

   a. Prove that axiom **A** is satisfied. (Hint: Use the two-point formula for the equation of a straight line, $(y - y_1)(x - x_1) = (x - x_1)(y_2 - y_1)$ to exhibit one such line. Then show that it would be a contradiction for two distinct lines to intersect in more than one point.)
   b. Prove that axiom **E** is satisfied. (Hint: Join the given point to every point on the given line. How many lines is that? Are there any lines left over, available to be "parallels"?)
   c. Prove that axiom **B** is satisfied.

▲ 25. Even in so miniature a geometry, we can do more than just play with points and lines. We can define circles, ellipses, hyperbolas, parabolas, and even tangents to circles! Here we'll just whet your appetite by introducing circles. A circle centered at (a, b) will consist of all of the points that satisfy an equation of the form $(x - a)^2 + (y - b)^2 = r$, where each of a, b, and r is one of 0, 1, or 2.
   a. Find the points on the circle $x^2 + y^2 = 0$.
   b. Find the points on the circle $x^2 + y^2 = 1$.
   c. Find the points on the circle $x^2 + y^2 = 2$.

▲ 26. (Adapted from Boltyansky [1992].) Take as the "points" of a geometry all points of the plane except a single point O. Take as "lines" all circles and straight lines that pass through the deleted point O.

a. Given two distinct "points" A and B, show how to construct a "line" that passes through both of them. Is this the only line with this property?

b. Given a "line" l and a "point" P not on l, show that there is a unique "line" through P that does not intersect l. In other words, the parallel postulate holds.

This geometry can be further furnished with angle measure, distance measure, "circles," triangles, and so forth, so that it satisfies all of the postulates of Euclidean geometry, and hence is a model of those postulates.

For Exercises 27 to 31, refer to the following (adapted from Maurer [1980]): Flocks of chickens, herds of horses, and groups of various other animals tend to be organized in a hierarchy of dominance. In the case of chickens, we call the hierarchy a *pecking order*, because one chicken displays dominance over another by pecking it on the head and neck. Although the flock could consist of one chicken that pecks all the others, a second that pecks all except the first chicken, and so forth, other social structures also occur. The major feature of a pecking order is that it determines the relationship between any pair of chickens, a feature that we set down as an axiom:

▶   For any pair of chickens, one of them pecks the other, but they don't both peck each other.

We will define as a *flock* any group of chickens for which this axiom is true. In a flock, is there necessarily some "top" chicken in the hierarchy? We will call a chicken an *empress* if she pecks every other chicken, and a *queen* if each other chicken is either pecked by her ("directly") or by another chicken who in turn is pecked by her ("indirectly").

▲ 27. We proceed to investigate what these notions involve.

a. Give an example of a flock in which it is *not* true that one chicken pecks all the others, a second pecks all but the first chicken, and so forth.

b. Does a flock have to have an empress? Either give an example of a flock without an empress, or else show that any flock has to have an empress.

c. Does a flock have to have a queen? (Hint: Your best candidate is a chicken that pecks the most other chickens.)

d. Is there a flock with only one queen but that queen is not an empress?

▲ 28. Can every chicken in a flock be a queen? Let's see:

a. Show that in a flock of three chickens, all three can be queens. Are there any other models of a flock with three chickens?

b. Show that in a flock with an odd number of chickens, all can be queens. (Hint: Start with a smaller flock in which all are queens, and

▲   Advanced exercise.

show how to add two more chickens so that all in the larger flock are queens.)

(Exercise 30 investigates flocks with an even number of chickens.)

▲ 29. Suppose a flock has a queen that pecks only one other chicken. Can the pecked chicken herself be a queen, or not?

▲ 30. What are the possible models for a flock with four chickens?
   a. In particular, show that not all four can be queens.
   b. Show that no flock (of whatever size) can have exactly two queens. (In fact, apart from this restriction and the fact that a flock with four chickens cannot have four queens, there are models with all other combinations of number of chickens in the flock and number of queens.)

▲ 31. Define a chicken to be a *slave* if it is pecked by all the other chickens in the flock, and to be a *serf* if every other chicken either pecks her directly or pecks another chicken that pecks her. These definitions are just the "opposites" of "empress" and "queen."
   a. What theorems can you assert about slaves and serfs?
   b. Here is a final theorem to prove: If every chicken in a flock is a queen, then every chicken is a serf.

For Exercises 32 to 35, refer to the following (adapted from Boltyansky [1992]): "In a certain town an association was founded that united philatelists, numismatists, collectors of the eyes of needles, and all other things worth collecting. It was decided that the association's administration must consist of an odd number of people (which is convenient for voting). So, it was decided that the administration should be accommodated in one building and every member of the administration should have a separate intercom unit. A switchboard was considered an unjustifiable luxury, but to keep the intercom system from becoming too complicated, each member was supposed to be connected directly to three others."

This proposed scheme can be described by the axiom system:

A The number of members is odd.
B Each line connects exactly two members.
C Each member is connected to exactly three members.

■ 32. Show that:
   a. There must be at least one member.
   b. There must be at least four members.
   c. There must be at least five members.
   d. Among any five members, there are always two who are not connected by a line.

■ 33. We can say that a line "leads" to a member. We make this precise by defining a *lead* as follows: If A is a member and a line l connects A to some other member, the pair (A, l) is a *lead*. Show that:
   a. Every member belongs to exactly three leads.
   b. The total number of leads is odd.

c. The total number of leads is even.

d. What do you conclude from parts b and c?

■ 34. Replace axiom **C** with

C′ Every member is connected to exactly four other members.

What can we prove about this new axiom system? How can we tell whether this system is consistent? If we can exhibit a model of the system, then it can't be inconsistent, since any inconsistency would have to show up in the model, too. Consider the numbers 1 through 7 as "members" and plot them on a circle, with 1 following 7. Define a "line" as a pair of numbers that are adjacent or separated by one other number. In the figure, draw little arcs between members that are connected by a line.

a. Verify that this interpretation provides a model of the axioms.

b. Can you find a model with 5 members?

c. Can you find a model with any given odd numbers of members?

■ 35. Replace axiom **C** with

C″ Every member is connected to exactly two other members.

Is this system consistent? What size models can it have?

▶ WRITING PROJECTS

For Projects 1 to 4, refer to the following (adapted from Davis [1968]): Folklore has it that everybody is (or should be) an enemy of their friends' enemies and a friend of their friends' friends, as well as a friend of their enemies' enemies and an enemy of their enemies' friends. If this is the case, what patterns of friendship are possible in a stable society? We investigate this question by converting the folklore into a axiom system and then seeing what conclusions can be drawn. We will use capital letters to denote people, together with the symbols =, $\heartsuit$, and #. Our interpretations of the symbols will be:

- $X = Y$ means X and Y are the same person,
- $X \heartsuit Y$ means that X is an immediate friend of Y,
- $X \# Y$ means that X is an immediate enemy of Y.

We also make a couple of definitions. We will say that X is *immediately involved with* Y if either $X \heartsuit Y$ or $X \# Y$. Also, we will define a chain of involvements from X to Y as a sequence of people $Z_0, Z_1, \ldots, Z_n$ (for some n) such that $Z_0 = X$, $Z_n = Y$, and each Z_{i-1} is immediately involved with Z_i, for

■   Discussion exercise.

$i = 1, 2, \ldots, n$. The Z_i don't have to be distinct. In fact, we will count $X\heartsuit X$ and $X\sharp X$ as chains of length one.

Our axioms will be:

A For all X and Y, if $X\heartsuit Y$, then $Y\heartsuit X$.
B For all X and Y, if $X\sharp Y$, then $Y\sharp X$.
C Every pair of people is connected by a chain of involvements.

Axiom **B** asserts that no person is ever completely isolated from another. We will call a chain of involvements *positive* if the number of immediate enmities in it is even, and *negative* if it is odd. We will also say that X is a *(distant) friend* of Y if there is a positive chain of involvements from X to Y, and is a *(distant) enemy* if there is a negative chain. Finally, we will say that X is *ambivalent* toward Y if X is both a friend and an enemy of Y.

1. Suppose that $X\heartsuit Z\sharp W\heartsuit Y$. Is X a friend of Y?

2. Show that if X is a friend of, an enemy of, or ambivalent toward Y, then Y is likewise related to X.

3. A *society* is a set of people for which each pair of (not necessarily distinct) people is connected by a chain of involvements consisting of members of that set.
   a. Show that the set of all people is a society.
   b. Can one person alone be a society?

4. A society is *stable* if there exist no ambivalences in it: no pair of people are friends and enemies both.
   a. Show that in an unstable society, everyone is everyone's friend and enemy both. In other words, everybody is ambivalent toward everybody, including themselves.
   b. Show that a society is stable if and only if it divides into two sets of people so that everyone is a friend of just the people in their own set and an enemy of exactly those in the other. (One of the sets may be empty, in which case everyone is a friend of everyone and an enemy of no one.)

Chapter

21

Symmetry and Patterns

"The senses delight in things duly proportional." So said the famous philosopher-theologian Thomas Aquinas more than 700 years ago, in noting human esthetic appreciation. In this chapter we examine some of the elements of that esthetic appreciation, particularly what we call *symmetry*.

Symmetry, like beauty, is very difficult to define. Dictionary definitions talk about "correspondence, equivalence, or identity among constituents of a system," "correspondence of form and arrangement of parts," and "beauty as a result of balance or harmonious arrangement" (*The American Heritage Dictionary*, 2nd ed.).

In the narrowest sense, symmetry refers to "mirror-image" correspondence between parts of an object. Crystals, in both their appearance and their atomic structure, provide examples of symmetry in this sense. Taken in a wider sense, though, symmetry includes notions of *balance, similarity*, and *repetition*.

It is our sense of symmetry that leads us to appreciate patterns. As we noted in the introduction to this part of the book, mathematics is the study of patterns, and we will see that mathematics gives important insights into symmetry.

Patterns abound in nature. The successive sections of the beautiful chambered nautilus grow according to a very strict and specific spiral pattern, a broader kind of symmetry. This spiral has the property that it has the same shape at any size: a photographic enlargement superimposed on it would fit exactly. This feature of self-similarity at a change in scale is a fundamental aspect of fractals and is no doubt associated with their esthetic appeal.

Botanists have long appreciated other spirals. In plant growth from a central stem, the shoots, leaves, and seeds often occur in a spiral pattern known as **phyllotaxis**. For instance, the seeds of a sunflower are arranged in spirals (Figure 21.2a), as are the scales on a pineapple or a pine cone (Figure 21.1), and the petals on a daisy. Like the chambers of the nautilus in Figure 21.2b, the spirals on these plants are geometrically similar to one another, and they are arranged in a regular way, with balance and "proportion." These plants have a kind of symmetry we would naturally call **rotational**.

Associated with the geometric symmetry of phyllotaxis, there is also a kind of numeric symmetry, with a "proportion" in the sense of a ratio

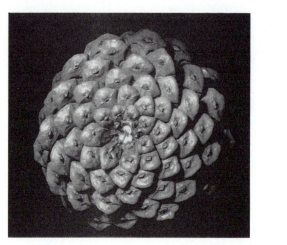

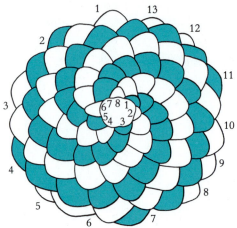

Figure 21.1 Spirals of scales on a pine cone: 8 right, 13 left. (From Verner E. Hoggatt, Jr., *Fibonacci and Lucas Numbers*, Houghton Mifflin, New York, 1969, p. 81.)

of numbers. Strangely, the number of spirals in these plants is never just any whole number but always comes from a particular sequence of numbers, called the **Fibonacci numbers:** 1, 1, 2, 3, 5, 8, 13, 21, 34, 55, 89, 144, 233, 377, ... This sequence begins with the numbers 1 and 1 again and each other number is obtained by adding the two preceding numbers.

Sometimes a sequence of numbers is specified by stating the value of the first term or first several terms and then giving an equation to calculate succeeding terms from preceding ones. This is called a *recursive* rule, and the sequence is said to be defined by **recursion**. Let's denote the nth Fibonacci number by F_n; then the Fibonacci sequence can be defined by

$$F_1 = 1, F_2 = 1, \text{ and}$$
$$F_{n+1} = F_n + F_{n-1} \text{ for } n \geq 2$$

(a)

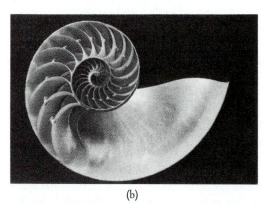

(b)

Figure 21.2 (a) This sunflower has 55 spirals in one direction and 89 spirals in the other direction. (b) A chambered nautilus shell. (Photo by Nancy Rodger.)

The recursive rule just expresses in algebraic form that the next Fibonacci number is the sum of the previous two.

Look at the sunflower in Figure 21.2a. You see a set of spirals running in a counterclockwise direction and another set in the clockwise direction. It is (just barely) possible to count the number of spirals in both directions; in the sunflower there are 55 in one and 89 in the other direction — two consecutive Fibonacci numbers. In the case of the pineapple, there are three sets of spirals, one each along the three directions through each hexagonally shaped scale. For the common grocery pineapple *(Ananas comosus)*, there are always 8 spirals to the right, 13 to the left, and 21 vertically — again, consecutive Fibonacci numbers.

Why are the numbers of spirals in plants the same numbers that appear next to each other in a purely mathematical sequence? The question has been the subject of extensive research, and there is no easy answer; there are several intricate theories about the dynamics involved in the plant's growth.

▶THE GOLDEN RATIO

During the last several centuries, a myth grew up that the ancient Greeks fastened on a specific numerical proportion as essential to their ideas of beauty and symmetry. Known variously by the modern names of **golden ratio, golden mean,** or **divine proportion,** the geometric aspects of this proportion had indeed been investigated by Euclid in Book II of his *Elements*. Though research in the past few years now reveals that there is no evidence connecting this proportion to Greek esthetics, we pursue the golden ratio briefly, because of its intimate connection to the Fibonacci sequence and because — whether the Greeks used it or not — it has some appeal as a standard for beautiful proportion.

The value of this ratio, which is usually denoted by the Greek letter phi (ϕ), is

$$\phi = \frac{1 + \sqrt{5}}{2} = 1.618034 \ldots$$

The basic esthetic claim is that a **golden rectangle** — one whose height and width are in the ratio of 1 to ϕ — is the most pleasing of all rectangles. The Greeks treated lengths geometrically, so for them it was important to construct lengths using straightedge and compass; in Spotlight 21.2 (p. 666) we show how to construct a golden rectangle that is 1 by ϕ.

What would make anyone think that this is such an attractive ratio? And where did it come from?

Given two line segments of different lengths, one method to find another length that will "strike a balance" between the smaller and the larger is to average the two. If l (the larger) and w (the smaller) are the original two lengths, their *arithmetic mean* (average) is $m = (l + w)/2$, and it satisfies

$$l - m = m - w$$

This equation reflects the fact that m strikes a certain balance between l and w, in terms of there being a common difference.

The Greeks instead sought a length s, the **geometric mean,** that would lead to a common ratio

$$l \div s = s \div w$$

Hence $lw = s^2$, so $s = \sqrt{lw}$. The equation $lw = s^2$ expresses the geometric fact that s is the side of a square that will have the same area as a rectangle that is l long and w wide (the Greeks thought in terms of geometry, not algebra).

Where ϕ arises is in the Greek problem of cutting a *single* line segment of length l into two lengths of s and $w = l - s$ so that the length s is the

SPOTLIGHT 21.1   Leonardo Pisano Bigollo ("Fibonacci")

▶ ▶ ▶ ▶ ▶ ▶ ▶ ▶ ▶ ▶ ▶ ▶ ▶

Leonardo Pisano Bigollo was born in Pisa in 1170. He has been known more popularly for the past century and a half as "Fibonacci." This nickname refers to his descent from an ancestor named Bonaccio, but the nickname is of modern origin, and there is no evidence he was ever known as Fibonacci in his own time.

Leonardo was the greatest mathematician of the Middle Ages. His stated purpose in his book *Liber abbaci* (1202) was to introduce Hindu-Arabic numerals and calculation with them into Italy, to replace the Roman numerals then in use. Other books of his treated topics in geometry, algebra, and number theory.

We know little of Leonardo's life apart from a short autobiographical sketch in the *Liber abbaci*:

I joined my father after his assignment by his homeland Pisa as an officer in the customhouse located at Bugia [Algeria] for the Pisan merchants who were often there. He had me marvelously instructed in the Arabic-Hindu numerals and calculation. I enjoyed so much the instruction that I later continued to study mathematics while on business trips to Egypt, Syria, Greece, Sicily, and Provence and there enjoyed discussions and disputations with the scholars of those places.

A "portrait" of Leonardo Pisano ("Fibonacci") of unlikely authenticity. (From Columbia University Library, D. E. Smith Collection.)

(Quoted from L. E. Sigler, *Leonardo Pisano Fibonacci, The Book of Squares: An Annotated Translation into Modern English,* Academic Press, New York, 1987.)

The *Liber abbaci* contains a famous problem about rabbits, whose solution is the sequence now called the Fibonacci sequence. Leonardo did not write further about it.

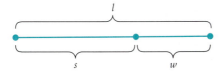

Figure 21.3 The line segment of length l is divided so that the length of s is the geometric mean between $w = l - s$; the dividing point divides the length l in the golden ratio.

mean proportional between l and w (see Figure 21.3). In this case, we have

$$\phi = \frac{l}{s} = \frac{s}{w}$$

Substituting $l = s + w$, we get

$$\phi = \frac{l}{s} = \frac{s + w}{s} = \frac{s}{s} + \frac{w}{s} = 1 + \frac{w}{s}$$

But w/s is just $1/\phi$, so we have

$$\phi = 1 + \frac{1}{\phi}$$

Multiplying through by ϕ gives

$$\phi^2 = \phi + 1$$

or

$$\phi^2 - \phi - 1 = 0$$

This is a quadratic equation of the form

$$ax^2 + bx + c = 0$$

with variable ϕ in place of x and $a = 1$, $b = -1$, and $c = -1$. To find the solutions of the equation, we apply the famous quadratic formula

$$x = \frac{-b \pm \sqrt{b^2 - 4ac}}{2a}$$

to get the two solutions

$$\phi = \frac{1 \pm \sqrt{5}}{2} = \frac{1 + \sqrt{5}}{2} = 1.618034 \ldots$$

and

$$\frac{1 - \sqrt{5}}{2} = -0.618034 \ldots$$

We discard this second, negative solution since it does not correspond to a length.

The first solution is the golden ratio. It occurs often in other contexts in geometry—for example, ϕ is the ratio of a diagonal to a side of a regular pentagon (see Figure 21.4).

Thanks to recent work of Roger Herz-Fischler (Wilfrid Laurier University) and George Markowsky (University of Maine), we now know that the term "golden ratio" was not used in antiquity and that there is *no evidence* that the Great Pyramid was designed to conform to ϕ, that the Greeks used ϕ in the proportions of the Parthenon, or that Leonardo da Vinci used ϕ. More-

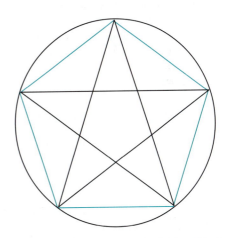

Figure 21.4 In a pentagon with equal sides, ϕ is the ratio of a diagonal to a side. The five-pointed star formed by the diagonals was the symbol of the followers of the ancient Greek mathematician Pythagoras.

SP TLIGHT 21.2   How the Greeks Constructed a Golden Rectangle

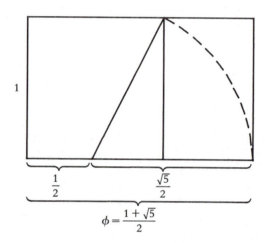

$$\frac{1}{2}$$

$$\frac{\sqrt{5}}{2}$$

$$\phi = \frac{1 + \sqrt{5}}{2}$$

First, we construct a 1-by-1 square: We start with a straight line with an interval of length 1 marked off on it. Placing one point of the compass at one end of the interval, we use the other end to mark off another interval of length 1. Using in turn each of the two opposite ends of these intervals as a center for the compass with an opening of length 2, we make arcs above and below the line. Their points of intersection determine a perpendicular bisector of the line, giving us the first right angle of a square of side 1. Continuing, we can go on to construct the remaining sides of the square.

Next we bisect the original segment to get a new point that divides it into two pieces of length one-half each. Using this new point and a compass opening equal to the distance from it to a far corner of the square we can cut off an interval of length ϕ.

over, experiments show that people's preferences for dimensions of rectangles cover a wide range, with golden rectangles not holding any special place.

The impressionists G. Caillebotte (1848–1894) and G. Seurat (1859–1891) may have used the golden ratio in the design of some of their paintings, but we do not have any historical evidence that they claimed or intended to do so.

It is true that human bodies exhibit ratios close to the golden ratio, as you can see by comparing your overall height to the height of your navel. The twentieth-century Swiss-born architect Le Corbusier (Charles-Édouard Jeanneret, 1887–

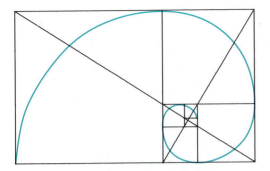

Figure 21.5 A logarithmic spiral determines a sequence of golden rectangles and corresponding squares.

1965) used the golden ratio (including a navel-height feature) as the basis for his "Modulor" scale of proportions.

There are intriguing connections between the spiral of the nautilus and the spirals of the sunflower and between the golden ratio and the Fibonacci sequence. The nautilus shape follows what is known as an *equiangular* or *logarithmic* spiral, which in its turning determines a sequence of golden rectangles (see Figure 21.5). The spirals of the sunflower are in fact approximations to a logarithmic spiral. The mathematical reason for this connection is that the ratios of consecutive Fibonacci numbers

$\frac{1}{1}$	1.0	$\frac{8}{5}$	1.6
$\frac{2}{1}$	2.0	$\frac{13}{8}$	1.625
$\frac{3}{2}$	1.5	$\frac{21}{13}$	1.615 . . .
$\frac{5}{3}$	1.666 . . .	$\vdots$	$\vdots$

provide alternately under- and overapproximations to $\phi = 1.618034. \ldots$

▶ BALANCE IN SYMMETRY

The spiral distribution of the seeds in a sunflower head and the spiraling of leaves around a plant stem are instances of *similarity* and *repetition*, two key aspects of symmetry.

Another ingredient of symmetry is *balance*, which refers to how the repetitions are arranged. A *single* spiral, although having a regularity, is not balanced, since there is no repetition; indeed, it may seem to be perpetually rotating, or just ready to rotate. Balance in symmetry is more common in such human-made objects as pottery, wallpaper, and buildings.

In considering patterns with repetition, we will distinguish the individual element or figure of the design (sometimes called the *motif*) from the pattern of the design — *how the copies of the motif are arranged.*

Rigid Motions

Mathematicians describe a variety of kinds of balance by using the geometric notion of **rigid motion**, also known as an **isometry** (which means "same size"). A rigid motion is a specific kind of variation on the original pattern: we pick it up and move it, perhaps rotate it, possibly flip it over — but we *don't change its size or shape.* (To connect this concept with the language of Chapter 16, the original figure and its image are *congruent*; but we won't need to use that terminology here. Fractals also possess symmetries that involve similarity, with figures whose shape remains the same but whose size is varied.)

Figure 21.6 shows the results of various motions applied to the rectangle in Figure 21.6a. Figure 21.6b shows the result of shrinking each side by 50%: not a rigid motion, because the size of the rectangle changes. For Figure 21.6c, we have imagined that the rectangle has rigid sides but hinges at the corner; like an unbraced bookshelf, it has sagged: again, this is not a rigid motion because the shape of the rectangle changes. In Figure 21.6d we have rotated the rectangle 90° (a quarter turn) clockwise around the center of the rectangle: this is a rigid motion. Similarly, in Figure 21.6e, rotating by 180° (a half turn) is a rigid motion; in fact,

Rotation by any angle, around any point as center, is a rigid motion.

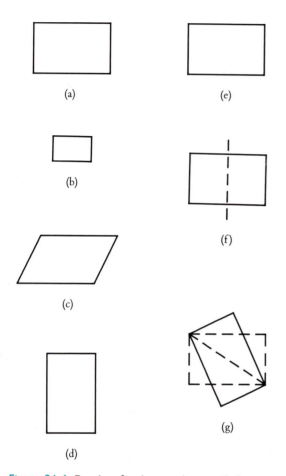

Figure 21.6 Results of various motions applied to a rectangle: (a) the original rectangle; (b) 50% reduction (not a rigid motion); (c) sagging (not a rigid motion); (d) quarter turn; (e) half turn; (f) reflection along the vertical line down the middle; (g) reflection along a diagonal line.

In Figure 21.6f we have reflected the rectangle along a vertical mirror down the middle: could you tell? The right and left halves have exchanged places.

A reflection across any line is a rigid motion.

Figure 21.6g shows the result of reflecting across a diagonal of the rectangle.

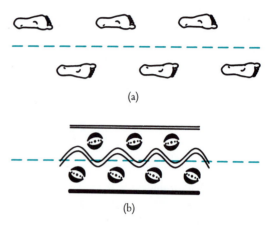

Figure 21.7 Glide reflection of (a) footprints; (b) design elements on a pot from San Ildefonso Pueblo. (From Dorothy K. Washburn and Donald W. Crowe, *Symmetries of Culture: Theory and Practice of Plane Pattern Analysis,* University of Washington Press, Seattle, 1988, p. 51.)

All reflections and all rotations are rigid motions. So are all **translations** — transformations that move every point in the plane a certain distance in the same direction.

A translation is a rigid motion.

The only remaining kind of rigid motion in the plane is a hybrid of reflection and translation. Known as a **glide reflection,** it is the kind of pattern your footprints make as you walk along: each successive element of the design (footprint) is a reflection of the previous one (see Figure 21.7). The motion combines *in an integral way* a translation ("glide") with a reflection across a line that is parallel to the direction of the translation.

A glide reflection along any axis is a rigid motion.

Performing one rigid motion after another results in a rigid motion that (surprisingly) must be one of the four types that we have just explored.

Any rigid motion of the plane must be one of:

1. *Reflection (across a line)*
2. *Rotation (around a point)*
3. *Translation (in a particular direction)*
4. *Glide reflection (across a line)*

▶ PRESERVING THE PATTERN

In terms of symmetry, we will be especially interested in rigid motions like those of Figure 21.6e and 21.6f that **preserve the pattern:** that is, ones for which the pattern looks exactly the same, *with all the parts appearing in the same places,* after the motion is applied.

You might enjoy thinking of applying these motions as a game, "The Pattern Game": *You turn your back, I apply a transformation, then you turn back and see if you can tell if anything is changed.*

The 90° rotation of Figure 21.6a into Figure 21.6d does not preserve the pattern. The moved rectangle doesn't fit exactly over the original rectangle.

On the other hand, the 180° rotation in Figure 21.6e does preserve the pattern. It's true that the top of the original rectangle is now on the bottom of the transformed version; but you can't tell that that has happened, because you can't distinguish the two. If you had turned your back while the motion was applied, you wouldn't be able to tell that anything had been done. A rotation by any multiple of 180° would also preserve the pattern.

Similarly, the mirror reflection along the vertical line in Figure 21.6f preserves the pattern, while the one in Figure 21.6g, where the mirror is along a diagonal, does not. (For an illustration of rotation and left-right reflection in calligraphy, see Spotlight 21.3, p. 670.)

The pattern of footsteps in Figure 21.7a is not preserved under just reflection along the direction of walking — there is not a left footprint directly across from a right footprint. The pattern is pre-

served under a glide reflection along the direction of walking, as well as by a translation of two steps, or one of four steps, etc. — but not by a translation of one step.

▶ ANALYZING PATTERNS

Given a pattern, we will analyze it by *determining which rigid motions preserve the pattern.* These are often referred to as the **symmetry operations (or symmetries) of the pattern.** We will then be able to classify the pattern by which rigid motions preserve it.

We may think of a pattern as a recipe for repeating a figure (motif) indefinitely. Of course, any pattern we see in nature or art has only finitely many copies of the figure; but if the recipe for repetition is clear, we may imagine that we are looking at just a part of a pattern that extends indefinitely.

Patterns in the plane can be divided into those that have indefinitely many repetitions in

▶ no direction — the **rosette patterns**
▶ exactly one direction (and its reverse) — the **strip patterns**
▶ more than one direction — the **wallpaper patterns**

A rosette pattern describes the possible symmetry operations for a flower. There is just one flower in the pattern; the repetition aspect of symmetry consists of the repetition of the petals around the stem. Translations and glide reflections do not come into play. The pattern is preserved under a rotation by certain angles, corresponding to the number of petals. There may or may not be reflections that preserve it, depending on whether the petal is symmetric. Most flowers have symmetric petals (Figure 21.8a), but some do not. An everyday example of the rosette pattern — a human-made one — that does not have reflec-

SPOTLIGHT 21.3   Symmetry in Modern Design: Scott Kim, an Artist in Symmetric Forms

▶ ▶ ▶ ▶ ▶ ▶ ▶ ▶ ▶ ▶ ▶ ▶ ▶

MAN

WOMAN

(Copyright Scott Kim, 1989.)

Scott Kim — whom author Isaac Asimov has called "the Escher-of-the-Alphabet" — created in the 1980s a new art form with words. His calligraphy is playful, surprising, elegant, and fun. He calls the results *inversions*: words that can be read right side up, upside down, and every which way. His inversions use symmetries, distorting letters a little here or there, reflecting them as in mirrors, or rotating them around central points. The two illustrations shown here, taken from his book *Inversions* (W. H. Freeman, 1989), illustrate rotation

(MAN) and left-right reflection (WOMAN). He comments:

> Deceptively simple constructions involving the letter *M*. Adding a single crossbar to a symmetric zigzag is enough to distinguish three asymmetrically placed letters: *M*, *A*, and *N*. A lowercase *a* is used in *WOMAN*. Notice that the two words, although closely related, have different symmetries: If you look at this design in a mirror, *WOMAN* will look the same but *MAN* will not. If you turn this design 180°, *MAN* will look the same but *WOMAN* will not.

tion symmetry is a pinwheel (Figure 21.8b). If there is no reflection symmetry, the motif of the pattern (the element that is repeated) is an entire petal; if there is reflection symmetry, the motif is just half a petal, because the entire pattern can be generated by rotation and reflection of a half petal. The fact that these are the only possibilities is

sometimes called *Leonardo's theorem*, after Leonardo da Vinci, who, in the course of planning the design of churches, needed to decide if chapels and niches could be added without destroying the symmetry of the central design.

Leonardo realized that there were two different classes of rosettes, the ones without reflection

Figure 21.8 (a) Flower with symmetric petals. (b) Pinwheel.

symmetry (*cyclic rosettes*) and the ones with reflection symmetry (*dihedral rosettes*) (see Figure 21.8). The respective notations for the patterns are *cn* and *dn*, where *n* is the number of times that the rosette coincides with its original position in one complete turn around the center. It coincides with itself for every rotation of $360°/n$. A cyclic pattern

has no lines of reflection (mirror) symmetry, while the dihedral pattern *dn* has *n* different lines of reflection symmetry. The flower of Figure 21.8a has a dihedral pattern, because each petal has reflection symmetry. The pinwheel of Figure 21.8b has pattern *c8*.

▶ STRIP PATTERNS

We will illustrate the different kinds of strip patterns, and their "ingredient" symmetry operations, with patterns in art of the Bakuba people of Zaire, who are noted for their fascination with pattern and symmetry (see Spotlight 21.4, p. 674).

All of the strip patterns offer repetition and translation symmetry along the direction of the strip. For our purposes, we will always position the pattern so that its repetition runs horizontally.

It may be that the pattern has no other rigid motions that preserve it apart from translation, as in Figure 21.9a.

The simplest other rigid motion to check for preservation of the pattern is reflection in a line, often called **bilateral symmetry** or *mirror symmetry:* the figure looks the same on both sides of a line, except that the two sides are mirror images of each other.

For a strip pattern, the center line of the strip may be a mirror line; if so, as in Figure 21.9b, we say that the pattern has symmetry across a horizontal line.

There may instead be mirror reflection across a *vertical* axis, such as the vertical lines through or between the V's in Figure 21.9c.

What kind of rotational symmetry can a strip pattern have? The only possibility for a strip pattern is a rotation by 180° (a half-turn), since any other angle won't even bring the strip back into itself. (We don't count rotations of 360° or integer multiples [full turns], since any pattern is preserved under these.) Figure 21.9d shows a strip pattern that is unchanged by a 180° rotation about any point at the center of the small crosshatched regions.

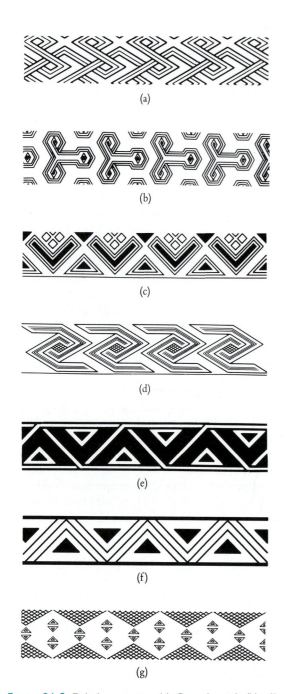

(a)

(b)

(c)

(d)

(e)

(f)

(g)

Figure 21.9 Bakuba patterns. (a) Carved stool; (b) pile cloth; (c) pile cloth; (d) embroidered cloth; (e) embroidered cloth; (f) carved back of wooden mask; (g) carved box.

What about glide reflections? A row of alternating p's and b's has glide reflection:

Glide	p p p p p p p p
Reflection	p p p p p p p p p b b b b b b b b b
Glide reflection	p b p b p b p b p

For glide reflection, a p is translated as far as the next b and is then reflected upside down. Figure 21.9e shows a Bakuba pattern whose only symmetry (except for translation) is glide reflection.

Having examined symmetry operations on strip patterns, we can ask: what *combinations* of the four are possible? It turns out that apart from the five kinds of patterns we have already seen, there are only two other possibilities: we can have vertical line reflection and half-turns, with either glide reflection but not horizontal line reflection (Figure 21.9f), or else with both glide reflection and horizontal line reflection (Figure 21.9g).

By analyzing cases, it is possible to show that these are the only seven possibilities; that is, mathematical analysis reveals that *there are only seven ways to repeat a pattern along a strip.* That this number is so small is quite surprising, since there is a myriad of different design elements (motifs). The key idea is that two designs may look entirely different yet share the same pattern of reproducing their design elements.

▶ NOTATION FOR PATTERNS

The standard notation of crystallographers for the strip patterns consists of four symbols; an example is *pma2*.

1. The first symbol is always a *p*, which indicates that the pattern repeats (is "periodic") in the horizontal direction.

2. The second symbol is *m* if there is a vertical line of reflection; *1* otherwise.

3. The third symbol is *m* (for "mirror reflection") if there is a horizontal line of reflection (in which case there is also glide reflection), *a* (for "alternating") if there is a glide reflection but no horizontal reflection, and *1* if there is no horizontal reflection or glide reflection.

4. The fourth symbol is *2* if there is half-turn rotational symmetry, and *1* otherwise.

A *1* always means that the pattern does not have the symmetry corresponding to that position.

So, for example, in the notation *p* __ __ __

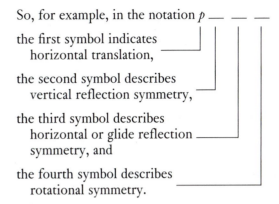

the first symbol indicates horizontal translation,

the second symbol describes vertical reflection symmetry,

the third symbol describes horizontal or glide reflection symmetry, and

the fourth symbol describes rotational symmetry.

Figure 21.10 gives a flowchart for identifying patterns, together with the notations for them.

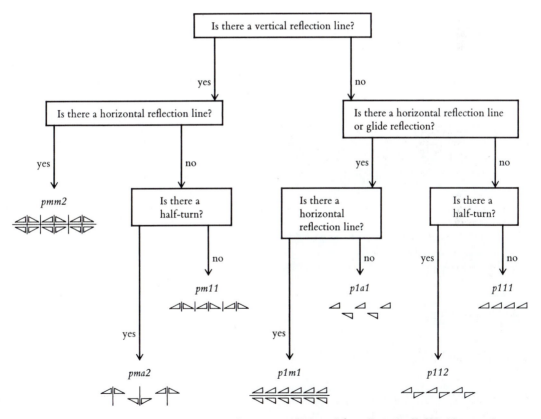

Figure 21.10 Flowchart for the seven strip patterns. (Adapted from Dorothy K. Washburn and Donald W. Crowe, *Symmetries of Culture: Theory and Practice of Plane Pattern Analysis*, University of Washington Press, Seattle, 1988, p. 83.)

Patterns Created by the Bakuba People

▶▶▶ ▶ ▶ ▶ ▶ ▶ ▶ ▶ ▶ ▶ ▶ ▶ ▶

Among the Bakuba people of Zaire (shaded area of map), it is considered an achievement to invent a new pattern, and every Bakuba king had to create a new pattern at the outset of his reign. The pattern was displayed on the king's drum throughout his reign and, in the case of some kings, on his dynastic statue.

When missionaries first showed a motorcycle to a Bakuba king in the 1920s, he showed little interest in it. But the king was so enthralled by the novel pattern the tire tracks made in the sand that he had it copied and gave it his name.

(Adapted from Jan Vansina, *The Children of Woot*, University of Wisconsin Press, Madison, 1978, p. 221.)

The pattern made by tire tracks fascinated the Bakuba people. (Travis Amos.)

Two women with raffia cloths from the Bakuba village of Mbelo, July 1985; Mpidi Muya with embroidered raffia cloth (left) and Muema Kenye with plush and embroidered raffia cloth (right). (Photo by Dorothy K. Washburn.)

EXAMPLES: Bakuba Patterns

We use the flowchart of Figure 21.10 to analyze some of the Bakuba patterns of Figure 21.9.

Figure 21.9a does not have a vertical reflection, so we branch right, and the pattern notation begins to take shape as *p1 _ _*. The figure does not have a horizontal reflection, nor a glide reflection, so we branch right again, filling in the third position in the notation, to get *p11 _*. A half-turn preserves part *but not all* of the pattern, so we conclude that we have a *p111* pattern.

Figure 21.9b does not have vertical reflection, so we branch right, to *p1 _ _*. The figure does have horizontal reflection, so we branch left and left, concluding that the pattern is *p1m1*.

Figure 21.9f has vertical reflection, so we branch left, to *pm _ _*. The figure does not have horizontal reflection, so we branch right but cannot yet fill in the third symbol. The figure does have a half-turn symmetry (and glide symmetry, too), with center on the middle of the three lines between any pair of closest triangles. So the pattern is *pma2*. ▲

▶ IMPERFECT PATTERNS

In applying these classification schemes to patterns on real objects, we need to take into account that the pattern itself may not be perfectly rendered. Also, patterns that are not on flat surfaces —for example, the pattern around the rim of a bowl, or around the body of a jar—require some latitude in our interpretation.

EXAMPLES: Patterns on Pueblo Pottery

The pitchers in Figure 21.11 are from a thousand-year-old Pueblo site at Starkweather Ruin near Reserve, New Mexico. We consider the

(a)   (b)   (c)

Figure 21.11 Reserve black-on-white pitchers from the Pueblo II horizon (900–1100 A.D.), excavated 1935–1936 from Starkweather Ruin by Prof. Paul H. Nesbitt and students from Beloit College. (Courtesy of Logan Museum of Anthropology, Beloit College. Photos by Paul J. Campbell.)

SP TLIGHT 21.5   Symmetry in Ancient Design

▶ ▶ ▶ ▶ ▶ ▶ ▶ ▶ ▶ ▶ ▶ ▶ ▶

Archeologist Dorothy Washburn comments on the significance of design as an indicator of a culture's history and how design can signal change in a culture:

> Human beings do things in a very consistent fashion; and over the years, within a given cultural group, their activities are nonrandom. In an archeological context we can see repetition of these behavior patterns in their material culture. But what we did not see before the mathematics of symmetry was that the structure of a culture's

decorative designs is consistent over time and through space within a given cultural group.

That is, although some people do produce random patterns, by far the largest number of decorative designs that cultural groups produce are based on symmetry. The point is that any time you repeat a motif in a systematic fashion, you're using one of the four rigid motions to make that repetition.

Early descriptions of design were largely idiosyncratic and dealt just with the individual types of material — types of textiles or types of pot-

patterns on the main bodies of the pitchers, which continue on the back sides. We suppose that they could be unwrapped and continued as strip patterns, and we consider them as such. We disregard the patterns on the spouts and handles.

We immediately come up against the question of the perfectness of the patterns. In Figure 21.11a the "teeth" on the left design element on the main body are "sharper" than those on the right. Is this lack of pattern, or just lack of perfection in executing one? For our analysis, we opt for the latter.

Similarly, what are we to make of the diagonal lines on the pitcher in Figure 21.11b? In the narrowest interpretation, these lines are part of the pattern and any rigid motion that is to qualify as a symmetry of the pattern must preserve them. More liberally, we may consider the lines as a kind of shading, a way to make the re-

gion appear gray; indeed, to an observer at a distance, that is the effect of the lines.

For the pattern on the body of the pitcher in Figure 21.11c, we notice that the jagged white line in the design element on the left has three "steps," while that in the one on the right has four. If we were really strict, we would decide that the two are different design elements. But we do detect a similarity of the two that we do not want to deny totally; we attribute the variations in the jagged lines to artistic license and for our purposes consider the two jagged lines to be the same.   ▲

▶   Figure 21.11a: Is there a vertical reflection? *No.* Is there a horizontal reflection or glide

tery, and the features that typified these types. But now we can study and compare the way the motifs in design are arranged in material from all different cultures throughout the world — regardless of whether they're contemporary cultures or past cultures — and see how these designs are put together and how they change through time and space.

One of the most interesting things we've found is that a given cultural group, a tribe or a band unit, will choose just a few symmetries to structure its designs. I've tested this observation through studies of California Indian baskets, with the work of Bakuba cloth weavers, and I've even found a consistency in the archeological record among the Anasazi, one of the prehistoric traditions of the American Southwest.

Let's take an example from material found on Crete. The site of Knossos had 3,000 years of uninterrupted prehistory. For 1,500 of those years, only two of the 7 one-dimensional symmetries were used. Then suddenly 5 more symmetries came into use. The design motifs were the same, but it was the rearrangement of these motifs into patterns based on different structures that suggested to us that something really interesting was happening.

That something interesting was the beginning of trade in the Aegean. We could see that simply by noting the introduction of new symmetry patterns from cultures outside that island. The increase in the number and variety of symmetries indicated that trade was coming in; the change in design structure was an incredibly sensitive marker of change.

reflection? *No.* Is there a half-turn? *No.* Hence the pattern is *p111*.

▶ Figure 21.11b (narrow interpretation of the diagonal lines): Is there a vertical reflection? *No.* Is there a horizontal reflection or glide reflection? *No.* Is there a half-turn? *Yes* (e.g., around the center of each cross). The pattern is *p112*.

▶ Figure 21.11b (liberal interpretation — diagonal lines as shading, their direction doesn't have to be preserved): Is there a vertical reflection? *Yes* (e.g., on a vertical line through the center of a cross). Is there a horizontal reflection? *Yes* (e.g., through the center of a cross). The pattern is *pmm2*.

▶ Figure 21.11c: Is there a vertical reflection? *No.* Is there a horizontal reflection or glide reflection? *No.* Is there a half-turn? *Yes* (e.g., around the center of each jagged white line). The pattern is *p112*. (This pitcher has the interesting feature that the patterns on the neck and the body are mirror images of each other.)

Women made the pots at Starkweather Ruin; they strongly preferred the symmetry of half-turns; very few of the pots have any reflection symmetry, either mirror or glide. The avoidance of mirror symmetry was a consistent feature of pottery of the indigenous peoples of the Western Hemisphere. Spotlight 21.5 discusses the significance of pattern classification for anthropologists.   ▲

▶FURTHER POSSIBILITIES

So far we have classified the patterns with no translation repetition (the rosette patterns) and those with repetition in one direction (the strip

There are exactly 17 wallpaper patterns. We give an example of each, together with a flowchart for identifying the patterns.

The International Crystallographic Union has established a standard notation for the wallpaper patterns. The full notation consists of four symbols:

▶ The first symbol is c (for "centered") if all rotation centers lie on reflection lines, and p (for "primitive") otherwise.

▶ The second symbol indicates rotational symmetry. It is either 1, 2, 3, 4, or 6, corresponding to 360° rotational symmetry (1), 180° symmetry (2), 120° symmetry (3), 90° symmetry (4), and 60° symmetry (6). The symbol is the largest

applicable number. For example, if 360°, 120°, and 60° symmetries are present, the symbol is 6.

▶ The third symbol is either m, g, or 1, corresponding to the presence of mirror, glide, or no reflection symmetry.

▶ The fourth symbol (m, g, or 1) is for describing symmetry relative to an axis at an angle to the symmetry axis of the third symbol.

(Note: The patterns p31m and p3m1 provide an exception to the notation.)

Below each pattern illustration we give both the standard abbreviation (on top) and the full notation (below).

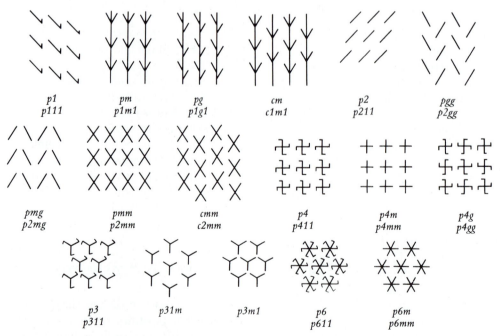

The 17 wallpaper patterns, with abbreviations and the full notation used by crystallographers.

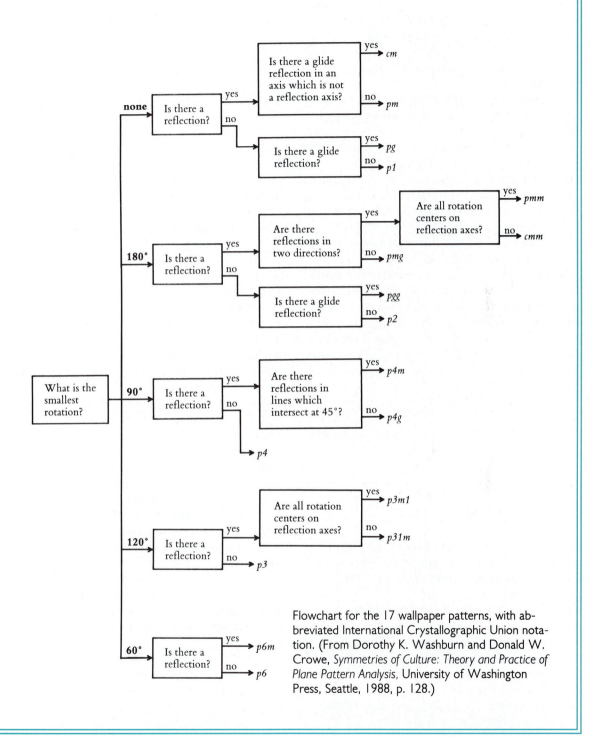

Flowchart for the 17 wallpaper patterns, with abbreviated International Crystallographic Union notation. (From Dorothy K. Washburn and Donald W. Crowe, *Symmetries of Culture: Theory and Practice of Plane Pattern Analysis,* University of Washington Press, Seattle, 1988, p. 128.)

patterns). What about those that have repetition in more than one direction — the wallpaper patterns? It turns out that there are exactly 17 of those. Illustrations, notation, and a flowchart are given in Spotlight 21.6 (pp. 678–679).

The method we have developed here for classifying patterns, by the combinations of symmetry elements present, was originally developed by crystallographers in the nineteenth century. They wanted to classify and recognize the three-dimensional patterns associated with crystal structure. They proved — after several years of different crystallographers coming up with different totals! — that there are exactly 230 *crystal patterns*.

We emphasize again that our analysis of patterns does not refer to the *design* in the pattern, but to how its repetition is structured across the plane. There is an infinite variety of possible designs that artists can devise. You should imagine that the artist has created one copy of the design and is then contemplating how to place equal-sized copies of it in other parts of the (infinite) plane, in a way that is symmetric. It is those strategies for placement of which there are very few.

A slightly more involved analysis allows mathematicians to refine the classification of patterns to take into account colors that are repeated in a symmetric way.

► REVIEW VOCABULARY

Bilateral symmetry   Ordinary mirror (that is, reflection) symmetry, as seen in the letter A.
Divine proportion   Another term for the **golden ratio**.
Fibonacci numbers   The numbers in the sequence 1, 1, 2, 3, 5, 8, 13, 21, 34, . . . (each number after the second is obtained by adding the two preceding numbers).
Geometric mean   The geometric mean of two numbers a and b is $\sqrt{ab}$.
Glide reflection   A combination of translation (=glide) and reflection in a line parallel to the

translation direction. Example:

$$. . . \, p \, b \, p \, b \, p \, b \, . . .$$

Golden ratio, golden mean   The number $\phi = (1 + \sqrt{5})/2 = 1.618. . . .$
Golden rectangle   A rectangle the lengths of whose sides are in the golden ratio.
Isometry   Another word for rigid motion. Angles and distances, and consequently shape and size, remain unchanged by a rigid motion. (For plane figures there are only four possible isometries: reflection, rotation, translation, and glide reflection.)
Phyllotaxis   The spiral pattern of shoots, leaves, or seeds around the stem of a plant.
Preserves the pattern   A transformation preserves a pattern if all parts of the pattern look exactly the same after the transformation has been performed.
Recursion   A method of defining a sequence of numbers, in which the next number is given in terms of previous ones.
Rigid motion   A motion that preserves the size and shape of figures; in particular, any pair of points is the same distance apart after the motion as before.
Rosette pattern   A pattern whose only symmetries are rotations about a single point and reflections through that point.
Rotational symmetry   A figure has rotational symmetry if a rotation about its "center" leaves it looking the same, like the letter S.
Strip pattern   A pattern that has indefinitely many repetitions in one direction.
Symmetry (operation) of the pattern   A transformation of a pattern is a symmetry operation (or symmetry) of the pattern if it preserves the pattern.
Translation   A rigid motion that moves everything a certain distance in one direction.
Translation symmetry   An infinite figure has translation symmetry if it can be translated (slid, without turning) along itself without appearing

to have changed. Example:

> . . . A A A A A A . . .

Wallpaper pattern A pattern in the plane that has indefinitely many repetitions in more than one direction.

▶ SUGGESTED READINGS

ASCHER, MARCIA: "Patterned strip decorations," in *Ethnomathematics: Mathematical Ideas in Other Cultures,* chapter 7, Wadsworth & Brooks/Cole, Pacific Grove, Calif., 1990.

BOLES, MARTHA, AND ROCHELLE NEWMAN: *The Golden Relationship: Art, Math & Nature,* Book 1: *Universal Patterns;* Book 2: *The Surface Plane,* Pythagorean Press, Bradford, Mass., 1992.

CROWE, DONALD W.: *Symmetry, Rigid Motions and Patterns,* High School Mathematics and Its Applications (HiMAP) Module 4, COMAP, Lexington, Mass., 1987. Reprinted in smaller format in *The UMAP Journal,* 8(3):207–236 (1987). Instructional module on rigid motions of the plane, strip patterns, and wallpaper patterns, with worksheets.

GALLIAN, JOSEPH A.: "Symmetry in logos and hubcaps," *American Mathematical Monthly* 97(3):235–238 (March 1990).

———: "Finite plane symmetry groups," *Journal of Chemical Education* 67(7):549–550 (July 1990). Hubcap examples.

HERZ-FISCHLER, ROGER: *A Mathematical History of Division in Extreme and Mean Ratio,* Wilfrid Laurier University Press, Waterloo, Ontario, Canada, 1987.

HOGGATT, VERNER E., JR.: *Fibonacci and Lucas Numbers,* Houghton Mifflin, New York, 1969.

HUNTLEY, H. E.: *The Divine Proportion,* Dover Publications, New York, 1970.

MARKOWSKY, GEORGE: "Misconceptions about the golden ratio," *College Mathematics Journal* 23(1):2–19 (January 1992).

O'DAFFER, PHARES G., AND STANLEY R. CLEMENS: *Geometry: An Investigative Approach,* Chapters 1–5, Addison-Wesley, Reading, Mass., 1976. A gentle introduction to the geometry of symmetry, with lots of examples and illustrations. Chapter 4 gives an elementary proof that there are only four kinds of rigid motions in the plane.

RUNION, GARTH E.: *The Golden Section and Related Curiosa,* Scott, Foresman, Glenview, Ill., 1972.

SIBLEY, THOMAS Q.: *Geometric Patterns: A Study in Symmetry,* Saint John's University, Collegeville, Minn., 1989.

WASHBURN, DOROTHY K., AND DONALD W. CROWE: *Symmetries of Culture: Theory and Practice of Plane Pattern Analysis,* University of Washington Press, Seattle, 1988. An introduction to the mathematics of symmetry, splendidly illustrated with photographs of patterns from cultures all over the world. Includes a complete analysis of patterns with two colors. Appendixes contain proofs of the facts that there are only four rigid motions in the plane and that there are exactly seven strip patterns.

▶ EXERCISES

1. Examine the "scales" on the surface of a pineapple, which are arranged in spirals (parastichies) around the fruit. Note that there are spirals in three distinct directions. For each direction, how many spirals are there?

2. There are two primitive models of natural increase of biological populations, similar to those Fibonacci hypothesized around the year 1200. A pair of newborn male and female rabbits is placed in an enclosure to breed.

a. Suppose that the rabbits start to bear young one month after their own birth. This may be unrealistic for rabbits, but we could substitute another species for which it is realistic; Fibonacci used rabbits. At the end of each month, they have another male–female pair, which in turn matures and starts to bear young one month later. Assuming that none of the rabbits dies, how many *pairs* of rabbits will there be at the end of six months from the start (just *before* any births for that month)? (Hint: Draw a month-by-month chart of the situation at the end of the month, just before any births.)

b. As in part a, but assume instead that the rabbits start to bear young exactly *two* months after their own birth.

3. Put the golden ratio $\phi = (1 + \sqrt{5})/2$ into the memory of your calculator.

a. Look at the value of ϕ. Now square it (either use the x^2 button or multiply it by itself). What do you observe?

b. Back to ϕ. Now take its reciprocal (either use the $1/x$ button or divide it into 1). What do you observe?

c. What formula explains what you saw in part a?

d. What formula explains what you saw in part b?

4. The golden ratio ϕ satisfies the equation $x^2 = x + 1$.

a. Show that $(1 - \phi)$ also satisfies the equation.

b. Use part a to show that $(1 - \phi) = (1 - \sqrt{5})/2$ is the other solution to $x^2 - x - 1 = 0$.

5. a. Find the geometric mean of 3 and 27.

b. Find the length of a side of a square that has the same area as a rectangle that is 4 by 64.

6. a. Find the geometric mean of 4 and 9.

b. You are to make a golden rectangle with 6 inches of string. How wide should it be, and how high?

7. Give the notation (e.g., *d4* or *c5*) for the symmetry patterns of the rosettes in hubcaps (a) through (c) in the figure, disregarding the logos in the centers. (Can you identify the make of car and year for each hubcap?)

(a)   (b)   (c)

8. As in Exercise 7, for hubcaps (d) through (f). (Photos by Joseph Gallian.)

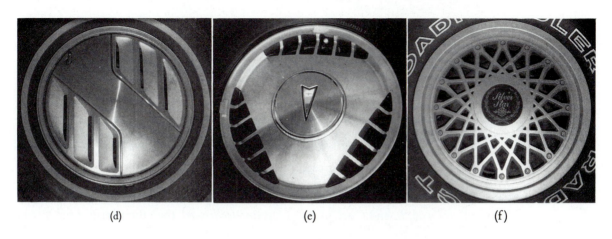

(d)   (e)   (f)

9. As in Exercise 7, for corporate logos (a) through (c). (Can you identify the corporations?)

(a)   (b)   (c)

10. As in Exercise 7, for corporate logos (d) through (f).

(d)   (e)   (f)

11. For each of the shapes in parts (a) through (e) of the accompanying figure, determine all lines of symmetry.

(a)   (b)   (c)   (d)   (e)

12. As in Exercise 11, but for the shapes in parts (f) through (j).

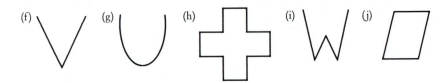

13. Determine whether each of the following statements is always true or sometimes false. (Drawing some sketches may be helpful.)
   a. A line reflection preserves *collinearity* of points. That is, if the points *A*, *B*, and *C* are in a straight line *(collinear)*, then their images reflected in some other line also lie in a straight line.
   b. A line reflection preserves betweenness. That is, if the collinear points *A*, *B*, and *C* (with *B* between *A* and *C*) are reflected about a line, then the image of *B* is between the images of *A* and *C*.
   c. The image of a line segment under a line reflection is a line segment of the same length.
   d. The image of an angle under a line reflection is an angle of the same measure.
   e. The image of a pair of parallel lines under a line reflection is a pair of parallel lines.

14. Determine whether each of the following statements is always true or sometimes false. (Drawing some sketches may be helpful.)
   a. The image of a pair of perpendicular lines under a line reflection is a pair of perpendicular lines.
   b. The image of a square under a line reflection is a square.
   c. Label the vertices of a square *A*, *B*, *C*, and *D* in a clockwise direction. Then their images *A′*, *B′*, *C′*, and *D′* under a line reflection also follow a clockwise direction.
   d. The perimeter of a geometric figure is equal to the perimeter of its image under a line reflection.
   e. The image of a vertical line under a line reflection is always a vertical line.

15. Which of the 26 capital letters of the alphabet have
   a. A horizontal line of reflection symmetry?
   b. A vertical line of reflection symmetry?
   c. Rotational symmetry?

(Assume that each letter is drawn in the most symmetric way. For example, the upper and lower loops of "B" should be the same size.)

16. As in Exercise 15, but for the lowercase letters.

17. In *The Complete Walker III*, 3rd ed., Knopf, 1984, p. 505, Colin Fletcher's answer to "What games should I take on a backpacking trip?" is the game he calls "Colinvert": "You strive to find words with meaningful mirror (or half-turn) images." Some of the words he found are

MOM   WOW   pod   MUd   bUM

    a. Which of his words reflect into themselves?
    b. Which of his words rotate into themselves?
    c. Find some more words or phrases of these various types — the longer, the better.

18. As in Exercise 17, but for words written vertically instead of horizontally.

19. For each of the following strip patterns, identify the rigid motions that preserve the pattern:
    a. A A A A A A A A A A   c. X X X X X X X X X X
    b. B B B B B B B B B B   d. F F F F F F F F F F

20. As in Exercise 19, but for
    a. N N N N N N N N N N   c. d b p q d b p q d b p q
    b. b d b d b d b d b d

21. Use the flowchart in Figure 21.10 to identify (by International Crystallographic Union notation) the types of the strip patterns from San Ildefonso Pueblo, New Mexico, shown in the accompanying illustration.

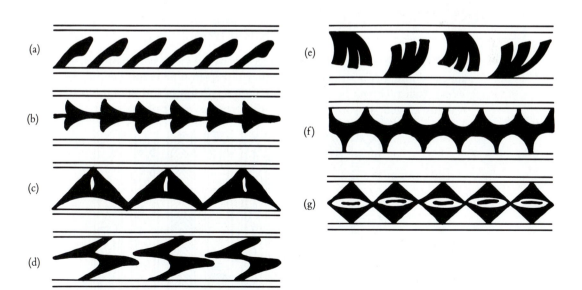

(a)

(b)

(c)

(d)

(e)

(f)

(g)

22. As in Exercise 21, for the accompanying patterns from Hungarian needlework.

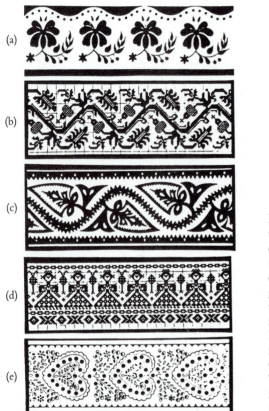

Hungarian needlework designs. (a) Edge decoration of table cover from Kalocsa, southern Hungary. (b) Pillow end decoration from Tolna County, southwest Hungary. (c) Decoration patched onto a long embroidered felt coat of Hungarian shepherds in Bihar County, eastern Hungary. (d) Embroidered edge decoration of bed sheet from the eighteenth century. (Note the deviations from symmetry in the lower stripes of the pattern.) (e) Shirt from Karád, southwest Hungary. (f) Pillow decoration pattern from Torockó (Rimetea), Transylvania, Romania. (g) Grape leaf pattern from the territory east of the river Tisza. (Courtesy of István Hargittai and Györgyi Lengyel, from the *Journal of Chemical Education* 61(12):1033–1034 (December 1984).)

23. As in Exercise 21, for the accompanying eight strip patterns, all of which appear on the brass straps for a single lamp from nineteenth-century Benin in West Africa. (From H. Ling Roth, *In Great Benin*.) Note that the patterns are roughly carved, so you will need to discern the *intent* of the artist.

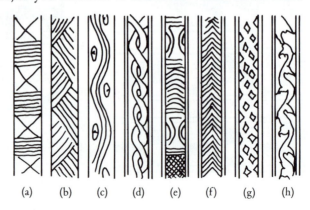

(a)   (b)   (c)   (d)   (e)   (f)   (g)   (h)

24. In each of the four accompanying examples, two adjacent triangles of an infinite strip are shown. For each example:

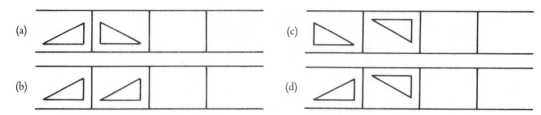

[Contributed by Margaret A. Owens, California State University, Chico.]

a. Determine a motion (translation, reflection, rotation, or glide reflection) that takes the first (=left) triangle to the second (=right) one.
b. Draw the next four triangles of the infinite strip that would result if the second triangle is moved to the next space by another motion of the same kind, and so on.
c. Identify (by notation) the resulting strip as one of the seven possible strip patterns.

▲ 25. Another sequence closely related to the Fibonacci sequence is the *Lucas sequence*, which is formed using the same recursive rule but different starting numbers. The nth Lucas number L_n is given by

$$L_1 = 1, L_2 = 3, \text{ and } L_{n+1} = L_n + L_{n-1} \qquad \text{for} \qquad n \geq 2$$

a. Calculate L_3 through L_{10}.
b. Calculate the ratio of successive terms of the Lucas sequence:

$$\frac{L_2}{L_1}, \frac{L_3}{L_2}, \ldots, \frac{L_{10}}{L_9}$$

What do you notice?

▲ 26. For a sequence specified by a recursive rule, finding an explicit expression for the nth term is not easy, nor is the form necessarily simple. An exact expression for the nth term of the Fibonacci sequence is given by the *Binet formula:*

$$F_n = \frac{1}{\sqrt{5}} \left(\frac{1 + \sqrt{5}}{2} \right)^n - \frac{1}{\sqrt{5}} \left(\frac{1 - \sqrt{5}}{2} \right)^n$$

a. Verify the formula for $n = 1$ and $n = 2$ (by multiplying out, not by using a calculator).
b. Use the Binet formula and your calculator to find F_5.
c. In fact, the second term on the right of the equation gets closer and closer to 0 as n gets large. Since we know that the Fibonacci numbers are integers, we can just round off the result of calculating the first

▲ Advanced exercise.

term. Find F_{13} by calculating the first term with your calculator and rounding.

▲ 27. Although the Fibonacci numbers get bigger and bigger, their units digit (the rightmost, or last, one) just keeps going through digits between 0 and 9. In fact, we can show that the sequence of units digits has to repeat.

Our idea is that if we ever come to two consecutive Fibonacci numbers that have the same units digits as two previous consecutive Fibonacci numbers, then the whole sequence of units digits from that previous point on has to repeat. The reason we need to look at two consecutive Fibonacci numbers is that their units digits completely determine the units digit of the next Fibonacci number: if one Fibonacci number ends in 5 and the next one in 9, then the following one has to end in 4. If we ever come to another 5 followed by a 9, the next has to be a 4 again, and so on.

So how do we know that we will ever get such a repetition? We make use of a simple but effective tool that mathematicians call the *pigeonhole principle:* if you have more pigeons than holes, then some hole has more than one pigeon. For our situation, the pigeons are the pairs of consecutive Fibonacci numbers (of which we may have as many as we please) and the pigeonholes are the possible pairs of units digits (of which there are only 100: 0, 0 through 9, 9). The units digits must have begun repeating by the time we get to the 101st Fibonacci number. In fact, the repetition starts a bit sooner, but not before

$$F_{49} = 7,778,742,049 \qquad F_{50} = 12,586,269,025$$

When does the repetition start? (Hint: Find just the units digits of F_{51}, F_{52}, etc.)

▲ 28. The Fibonacci sequence and the Lucas sequence of Exercise 25 are intimately related. Use your calculator to
   a. Make a table of values for $L_n - F_{n-1}$ for n from 2 through 10. What do you notice?
   b. Make a table of values for $F_n + L_n$ for n from 1 through 10. What do you notice?
   c. Make a table of values for $F_n L_n$ for n from 1 through 5. What do you notice?

29. In this exercise we construct geometrically the geometric mean between two segments of length a and b. We do it in three steps, constructing at each stage a right triangle. At each stage we employ the Pythagorean theorem (see Chapter 18). The text describes how to construct a right angle at a point (p. 666); the only fact we need is that a triangle inscribed in a semicircle is a right triangle.

▲ Advanced exercise.

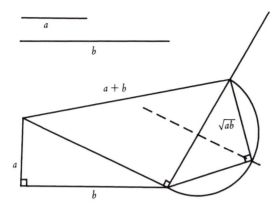

STEP 1.   Construct a right triangle with legs of lengths a and b.

STEP 2.   At one end of the hypotenuse of that first right triangle, construct a perpendicular. Center the compass at the other end of the hypotenuse and mark an arc of length $a + b$ that intersects the perpendicular. This forms a new right triangle.

STEP 3.   Draw the perpendicular bisector of the vertical leg of the new right triangle to find the center of the leg. Draw a semicircle using the center of the leg as center and half its length as radius. The two ends of the leg and the point where the semicircle intersects the perpendicular bisector form a triangle inscribed in a semicircle: our third right triangle.

Perform the construction and use the Pythagorean theorem (three times!) to show that a leg of the final right triangle has length $\sqrt{ab}$.

30. In this exercise we show that the length of a diagonal of a regular pentagon is the golden ratio times the length of a side. For simplicity we assume that the length of the side is 1, so we show that the length of the diagonal, which we will denote by x, is in fact ϕ. Consider the accompanying figure. From the facts that the measure of an interior angle of a regular pentagon is 108°, that the measures of the angles of a triangle sum to 180°, and that angles opposite equal sides of a triangle must have equal measure, conclude that triangles FAE and BDE are congruent, and that triangles FBD and BDE are similar. From the fact that similar triangles have proportional sides, arrive at an equation that you can solve to find $x = \phi$.

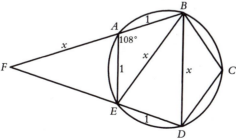

31. Consider a regular decagon (10-sided polygon) inscribed in a circle. Show that the ratio of the radius to the length of one of the sides is ϕ.

32. For each of the Bakuba cloths shown in the accompanying illustration, use the flowchart in Spotlight 21.6 to identify (by notation) the type of wallpaper pattern.

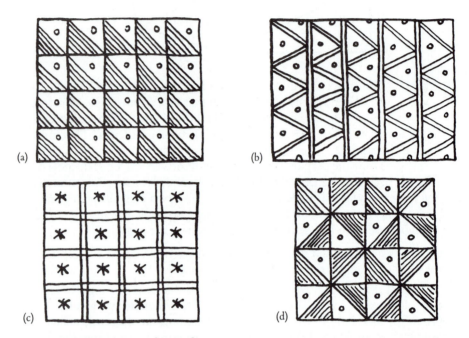

(a) (b) (c) (d)

Patterns on Yoruba (West Africa) *adire* cloth, made by starching a pattern onto white cloth, then dyeing the cloth blue before rinsing out the starch, so that the starched portion remains as a white design against a blue background. (Courtesy of Donald W. Crowe, in *Africa Counts: Number and Pattern in African Culture*, Claudia Zaslavsky, Prindle, Weber, & Schmidt, 1973, p. 195.)

33. The triangles in the grid at the top of the opposite page show beginning steps in forming instances of several of the wallpaper patterns, by putting together a vertical motion and a horizontal motion.
   a. Identify the horizontal motion.
   b. Identify the vertical motion.
   c. Fill in the remaining empty squares.
   d. Identify the wallpaper pattern.

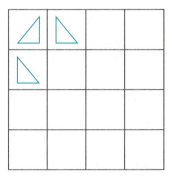

■ **34.** The following table shows comparative data about the frequency of occurrence of strip designs of various types on pottery (Mesa Verde, United States) and smoking pipes (Begho, Ghana, Africa) from two different continents.

Frequency of Strip Designs on Mesa Verde Pottery and Begho Smoking Pipes

Strip type	Mesa Verde		Begho	
	Number of examples	Percentage of total	Number of examples	Percentage of total
p111	7	4	4	2
p1m1	5	3	9	4
pm11	12	7	22	10
p112	93	53	19	8
p1a1	11	6	2	1
pma2	27	16	9	4
pmm2	19	11	165	72
Totals	174		230	

a. Which types of motions appear to be preferred for designs from each of the two localities?

b. What other conclusions do you draw from the data of this table?

c. On the evidence of the table alone, in which locality is each of the strip patterns in the accompanying figure most likely to have been found?

■ Discussion exercise.

■ 35. Which wallpaper patterns can be formed by the technique of Exercise 33?

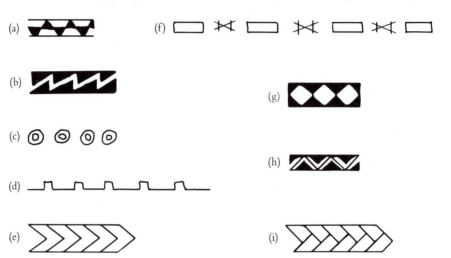

(a)

(b)

(c)

(d)

(e)

(f)

(g)

(h)

(i)

▶ WRITING PROJECTS

1. The Fibonacci Association is devoted to fostering interest in Fibonacci and related numbers. In November 1988, the society's journal, *The Fibonacci Quarterly*, published "Suppose More Rabbits Are Born" (pp. 306–311), by Shari Lynn Levine (a high school student when she wrote it). The article begins: "How would Fibonacci's age-old sequence be redefined if, instead of bearing one pair of baby rabbits per month, the mature rabbits bear two pairs of baby rabbits per month?" The article goes on to discuss properties of the resulting "Beta-nacci" sequence and the sequences that result from even greater rabbit fertility. Here we ask you to rediscover some of Shari's results about the Beta-nacci sequence:

    a. How many rabbits will there be each month for the first 12 months?
    b. What is the recursive rule for the nth Beta-nacci number B_n?
    c. For the terms of the sequence in part **a**, calculate the ratios B_{n+1}/B_n of successive terms. (Motivating hint: It's not the golden ratio this time.)
    d. Suppose that the ratio of successive terms approaches a number x. We show how to find x exactly. For very large n, $B_{n+1} \approx x B_n \approx x^2 B_{n-1}$. Substituting these values into the recursive rule for the sequence and dividing by B_{n-1} gives us the equation $x^2 = x + 2$. Solve this equation for x (you can use the quadratic formula). Make a table of values of $3B_n$ versus 2^n. From the evidence, can you suggest a formula for B_n?

2. Generalize Writing Project 1, parts a through d: to the case of each pair of rabbits having three pairs of rabbits (the "Gamma-nacci" sequence); to the case of each pair of rabbits having q pairs of rabbits.

Chapter
22
Tilings

When our ancestors used stones to cover the floors and walls of their houses, they selected shapes and colors to form pleasing designs. We can see the artistic impulse at work in mosaics, from Roman dwellings to Muslim religious buildings (see Figure 22.1). The same intricacy and complexity arise in other decorative arts — on carpets, fabrics, baskets, and even linoleum.

Such patterns have one feature in common: they use repeated shapes to cover a flat surface, without gaps or overlaps. If we think of the shapes as tiles, we can call the pattern a **tiling,** or *tessellation.* Even when efficiency is more important than aesthetics, designers value clever tiling patterns. In manufacturing, for example, stamping the components from a sheet of metal is most economical if the shapes of the components fit together without gaps — in other words, if the shapes form a tiling.

▶ REGULAR POLYGONS

The simplest tilings are those using only one size and shape of tile, known as **monohedral** tilings, and we begin our investigation with them.

In particular, we are interested especially in tiles that are **regular polygons,** figures all of whose sides are the same length and all of whose angles are equal. A square is a regular polygon with four sides and four equal interior angles; a triangle with all sides equal (an *equilateral* triangle) is also a regular polygon. A polygon with five sides is a pentagon, one with six sides is a hexagon, and one with n sides is an ***n*-gon.** Regular polygons are especially interesting because of their high degree of symmetry; each has the symmetry of a dihedral rosette pattern (see Chapter 21).

By a convention dating back to the ancient Babylonians, angles are measured in degrees, with an angle that goes all the way around a point measuring 360°. As we note in Chapter 19, the **interior angles** of a triangle add up to 180°; in an **equilateral triangle,** each angle equals 60°. Each of the interior angles of a square is one-fourth of 360°, or 90°.

Since we have deliberately referred to interior angles, you can be sure that there are also such things as exterior angles, and they too will come into play in our discussion. An **exterior angle** of a polygon is one formed by one side and the exten-

Figure 22.1 Mosaics from the Hakim Bey Mosque, Konya. (Bildarchiv Foto Marburg/Art Resource.)

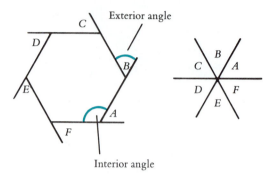

Figure 22.2 The exterior angles of a regular hexagon, like those of any regular polygon, add up to 360°. Each interior angle measures 60°.

sion of an adjacent side (Figure 22.2). Proceeding around the polygon in the same direction, we see that each interior angle is paired with an exterior angle. If we bring all the exterior angles together at a single point, they will add up to 360° (see Figure 22.2). If the polygon has n sides, then each exterior angle must measure $360/n$ degrees. For example, a square with $n = 4$ sides has 4 exterior angles, each measuring 90°; a pentagon with $n = 5$ sides has 5 exterior angles, each measuring 72°; while a regular hexagon with $n = 6$ sides has 6 exterior angles, each measuring 60°. Notice that each exterior angle plus its corresponding interior angle make up a straight line, or 180°. For a regular polygon with more than six sides, the interior angle is between 120° and 180°. This last consideration will prove crucial shortly.

▶ REGULAR TILINGS

A monohedral tiling whose tile is a regular polygon is called a **regular tiling**. A square tile is the simplest case. Apart from varying the size of the square, which would change the scale but not the pattern of the tiling, we can get different tilings by offsetting one row of squares some distance from the next.

However, there is only one tiling that is **edge-to-edge**, that is, the edge of a tile coincides entirely with the edge of a bordering tile (see Figure 22.3 for a tiling that is not edge-to-edge and another that is). For simplicity, from now on we will consider only edge-to-edge tilings. For edge-to-edge tilings (even ones with tiles of different shapes and sizes), edges of different tiles meet at points that are surrounded by tiles and their edges; the particular arrangement of polygons around a point is its **vertex figure**.

Any tiling by squares can be refined to one by triangles by drawing a diagonal of each square; but these triangles are not regular (equilateral). Equilateral triangles can be arranged in rows by alternately inverting triangles; as with squares, there is only one pattern of equilateral triangles that forms an edge-to-edge tiling.

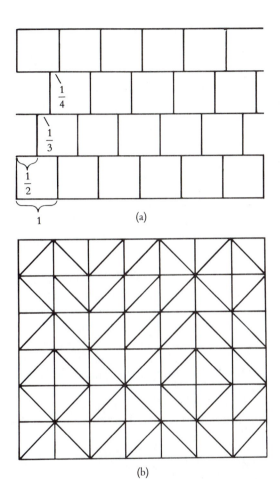

(a)

(b)

Figure 22.3 (a) A tiling that is not edge-to-edge; the horizontal edges of two adjoining squares do not exactly coincide. (b) A tiling by right triangles that is edge-to-edge.

What about tiles with more than four sides? An edge-to-edge tiling with regular hexagons is easy to construct (see the upper right pattern in Figure 22.5).

However, if we look for a tiling with regular pentagons, we won't be able to find one. How do we know whether we're just not being clever enough or there really isn't one to be found? This is the kind of question that mathematics is uniquely equipped to answer. In the other sciences, phe-

nomena may exist even though we have not observed them; such was the case for bacteria before the invention of the microscope. In the case of an edge-to-edge tiling with regular pentagons, we can conclude with certainty that there is no edge-to-edge tiling with regular pentagons.

The proof is very easy. As we calculated earlier, the interior angles of a pentagon are each 108°. At a point where several hexagons meet, how many can meet there? The total of all of the angles around a point must be 360°. Four pentagons at a point would be too many (they'd have to overlap), and three would be too few (some of the area wouldn't be covered). Since 108 does not evenly divide 360, *regular pentagons can't tile the plane*.

With this argument, we can do something that is a favorite with mathematicians: we can generalize it. Its main idea is a criterion for when a regular pentagon can tile the plane: when the size of its interior angles divides 360 evenly. We can apply this criterion to determine exactly which other regular polygons can tile the plane.

EXAMPLE: Identifying the Edge-to-Edge Regular Tilings

A regular hexagon has interior angles of 120°; 120 divides 360 evenly, and 3 regular hexagons fit together exactly around a point. A regular 7-gon — or any regular polygon with more than six sides — will have interior angles that are larger than 120° but smaller than 180°. Now 360 divided by 120 gives 3, and 360 divided by 180 gives 2 — and there aren't any other possibilities in between. Angles between 180° and 120° divided into 360° will give a result *between* 2 and 3, and consequently not an integer. So there are no edge-to-edge regular tilings of the plane with polygons of more than 6 sides.

The only edge-to-edge regular tilings are the ones with equilateral triangles, with squares, and with regular hexagons. ▲

The follow-up question, of course, is which *combinations* of regular polygons of different numbers of sides can tile the plane edge-to-edge? Recall that the particular arrangement of polygons around a vertex is called the *vertex figure*. A systematic tiling that uses a mix of regular polygons with different numbers of sides but in which *all vertex figures are alike*—the same polygons in the same order—is called a **semiregular tiling** (see Figure 22.4).

As before, our technique of adding up angles at a vertex (to be 360°) eliminates some impossible combinations, such as "square, hexagon, hexagon" (Figure 22.5). Having found the arrange-

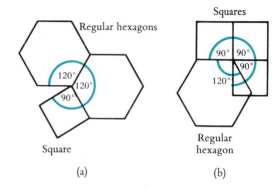

Figure 22.5 Polygons that come together at a vertex in a tiling must have interior angles that add up to 360°—no less, no more.

ments that are not numerically impossible, we must confirm the actual existence of each tiling by constructing it (i.e., show that it is geometrically possible). For example, even though a possible arrangement of regular polygons around a point is "triangle, square, square, hexagon," it is not possible to construct a tiling with that vertex figure at every vertex.

The result of this investigation is that in a semiregular tiling no polygon can have more than 12 sides. In fact, polygons with 5, 7, 9, 10, or 11 sides do not occur either. Figure 22.4 exhibits all of the semiregular tilings.

If we abandon any restriction about the vertex figures being the same at every vertex, then there are *infinitely many* systematic edge-to-edge tilings with regular polygons, even if we continue to insist that all polygons with the same number of sides have the same size.

▶ TILINGS WITH IRREGULAR POLYGONS

What about edge-to-edge tilings with irregular polygons, which may have some sides longer than others, or some interior angles larger than others? We will look just at monohedral tilings (in which all tiles have the same size and shape) and investi-

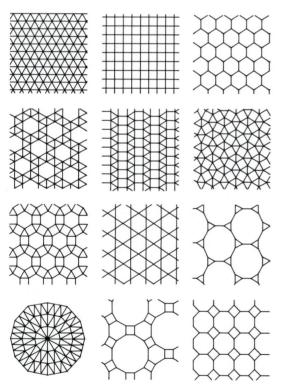

Figure 22.4 The three regular tilings and the eight semiregular tilings, plus one tiling that does not belong to either group. Can you identify it? (Courtesy of Darrah P. Chavey, Beloit College.)

gate in turn what triangles, **quadrilaterals** (four-sided polygons), hexagons, and so forth, can tile the plane.

The most general shape of triangle has all sides of different lengths and all interior angles of different sizes. Such a triangle is called a **scalene triangle,** from the Greek word for "uneven." We can always take two copies of a scalene triangle and fit them together to form a **parallelogram,** a quadrilateral whose opposite sides are parallel (Figure 22.6a). It's easy to see that we can then use such parallelograms to tile the plane, by making strips and then fitting layers of strips together edge-to-edge (Figure 22.6b). So:

Any triangle can tile the plane.

What about quadrilaterals? We have seen that squares tile the plane, and rectangles certainly will, too; and we have just noted that any parallelogram will tile. What about a quadrilateral (four-sided polygon) with its opposite sides not parallel, as in Figure 22.7a? The same technique as for triangles will work. We fit together two copies of the quadrilateral, forming a hexagon whose opposite sides

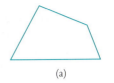

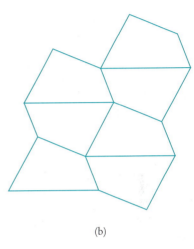

Figure 22.7 (a) A general quadrilateral. (b) Any quadrilateral tiles the plane.

are parallel. Such hexagons fit next to each other to form a tiling, as in Figure 22.7b.

A quadrilateral may be even more general in its shape, as in Figure 22.8a. A tile is **convex** if, when you take any two points on the tile (including the boundary), the line segment joining them lies entirely within the tile (again, including the boundary). The quadrilateral of Figure 22.8a is not convex, but the same approach works for using it to form a tiling (Figure 22.8b). So:

Any quadrilateral, even one that is not convex, can tile the plane.

We could hope that such success would extend to irregular polygons with any number of sides, but it doesn't. The situation for convex

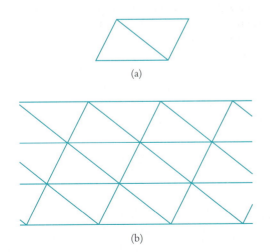

Figure 22.6 (a) A scalene triangle. (b) Every scalene triangle tiles the plane.

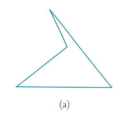

(a)

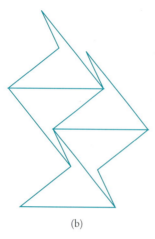

(b)

Figure 22.8 (a) A general nonconvex quadrilateral. (b) Any quadrilateral, convex or not, tiles the plane.

hexagons was determined by K. Reinhardt in his 1918 doctoral thesis. He showed that for a convex hexagon to tile, it must belong to one of three classes, and that every hexagon in those classes will tile. Examples of the three classes are shown in Figure 22.9, together with their characterizations. Notice that tilings with a hexagon of Type 2 (Figure 22.9b) use both ordinary and mirror-image versions of the hexagon.

Exactly three classes of convex hexagons can tile the plane.

Reinhardt also explored convex pentagons and found five classes that tile. For example, any

pentagon with two parallel sides will tile. Reinhardt did not complete the solution, as he did for hexagons, by proving conclusively that no other pentagons could tile; he claimed that it would be very tedious to finish the analysis. Still, he felt that he had found them all. In 1968, after 35 years of working on the problem on and off, R. B. Kershner, a physicist at Johns Hopkins University, discovered three more classes of pentagons that will tile. Kershner was sure that he had found all pentagons that tile, but again did not offer a complete proof, which "would require a rather large book."

When an account of the "complete" classification into eight types appeared in *Scientific American* (July 1975), the article provoked an amateur mathematician to discover a ninth type! A second amateur, Marjorie Rice, a housewife with no formal education in mathematics beyond high-school "general mathematics" 36 years earlier, devised her own mathematical notation and found *four more* types over the next two years (see Spotlight 22.2, pp. 704–705). A fourteenth type was found by a mathematics graduate student in 1985. Since then, no new types have been discovered, yet no one knows if the classification is complete.

With the situation so intricate for convex pentagons, you might think that it must be still worse for polygons with seven or even more sides. In fact, however, the situation is remarkably simple, as Reinhardt proved in 1927:

A convex polygon with seven or more sides cannot tile.

▶ M. C. ESCHER AND TILINGS

The Dutch artist M. C. Escher (1898–1972) was inspired by the great variety of decoration in tilings in the Alhambra, a fourteenth-century palace built during the last years of Islamic dominance in Spain. He devoted much of his career of making

TYPE 1

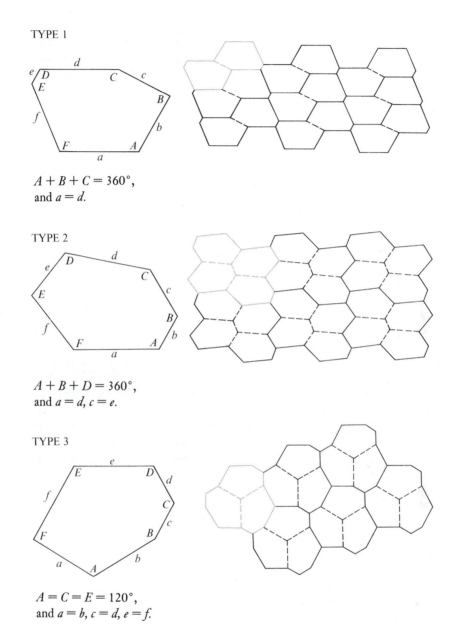

$A + B + C = 360°$,
and $a = d$.

TYPE 2

$A + B + D = 360°$,
and $a = d$, $c = e$.

TYPE 3

$A = C = E = 120°$,
and $a = b$, $c = d$, $e = f$.

Figure 22.9 The three types of convex hexagon tile. (From Martin Gardner, *Time Travel and Other Mathematical Bewilderments*, Freeman, New York, 1988, p. 168.)

SPOTLIGHT 22.1   Regular Polyhedra and Buckyballs

▶▶ ▶ ▶ ▶ ▶ ▶ ▶ ▶ ▶ ▶ ▶ ▶ ▶ ▶

The three-dimensional analogue of a regular polygon is a *regular polyhedron,* a convex solid whose faces are regular polygons all alike (same number of sides, same size), with each of the vertices of the polyhedron being surrounded by the same number of polygons. Although there are infinitely many regular polygons, there are only five regular polyhedra. Because they were known to the ancient Greeks, they are sometimes called the *Platonic solids;* in fact, Theaetetus (414–368 B.C.) proved that these five are the only ones possible. They are depicted in Spotlight 19.3 (p. 611). Of the five, only the cube and the tetrahedron can be used to tile (fill) space.

If the restriction about having the same number of polygons meet at each vertex is relaxed, five additional convex polyhedra are obtained, all of whose faces are equilateral triangles. If the restriction to having just one kind of regular polygon is relaxed, thirteen further convex polyhedra are obtained, known as the *semiregular polyhedra* or *Archimedean solids,* al-

A truncated icosahedron, which represents the structure of carbon atoms in a buckyball. The earliest drawing of this polyhedron appeared in a work by Piero della Francesca. This particular figure appeared as one of a series of illustrations by Leonardo da Vinci for the book *The Divine Proportion* by Luca Pacioli, published in 1509. (Courtesy of the Moffitt Library, University of California at Berkeley.)

prints to creating tilings with tiles in the shapes of living beings (a practice forbidden to Muslims). Those prints of interlocking animals and people have inspired awe and wonder among people all over the world. Color Plates 7–10 illustrate a few of his drawings and finished works. Like Marjorie Rice, he too developed his own mathematical notation for the different kinds of patterns for the tilings.

▶ TILING BY TRANSLATIONS

You may wonder just how much liberty can be taken in shaping a tile, and how you might be able to design an Escher-like tiling yourself.

The simplest case is when the tile is just *translated* in two directions, that is, copies are laid edge-to-edge in rows, as in Color Plate 7. Each tile must fit exactly into the ones next to it, including its

though there is no documented evidence that Archimedes studied them (but Kepler did catalogue them all). One of the Archimedean solids is the truncated icosahedron, whose faces are pentagons and hexagons, and when inflated is known throughout the world as a regulation soccer ball. Drawings of it appear in the work of Leonardo da Vinci.

The truncated icosahedron is also the structure of C_{60}, the new form of carbon known as buckminsterfullerene, and more familiarly, "buckyball." Sixty carbon atoms lie at the 60 vertices of this molecule, which was first discovered in 1985. It is named after R. Buckminster Fuller (1895 – 1983), the inventor and promoter of the geodesic dome. The molecule itself, with each carbon atom joined by bonds to three others, resembles one of Fuller's domes.

The buckyball is one of what is now known to be an entire family of carbon molecules, the fullerenes. In each of them, each carbon atom is joined to three others. This property, plus a famous equation due to Leonhard Euler (1707 – 1783), require that the polyhedra corresponding to fullerenes *all have exactly 12 pentagon faces,* no matter how many hexagon faces they have.

Euler's equation, which is true for any convex polyhedron, is very simple. It is just $v - e + f = 2$, where v is the number of vertices, e is the number of edges, and f is the number of faces of the polyhedron.

After discovering not only C_{60} but also C_{70}, C_{44}, and C_{540}, chemists wondered just whether the mathematics of polyhedra imposes any restrictions on the possibilities for the number of carbon atoms in fullerenes. But pure mathematics had already paved the way for applied mathematics. More than 30 years before the discovery of fullerenes, mathematicians had answered that question: a convex polyhedron in which every vertex has three edges must have 12 pentagon faces, but may have *any number* of hexagon faces, from 0 on up, except for 1.

neighbors above and below. We say that each tile is a **translation** of each other one, since we can move one to coincide with another without doing any rotation or reflection.

When is it possible for a tile to cover the plane in this manner? The boundary of the tile must be divisible into matching pairs of opposing parts that will fit together. Color Plates 7 and 8 illustrate two basic ways that this can happen. In the first, two opposite pairs of sides match; in the second, three opposite pairs of sides match.

A tile can tile the plane by translations *if either*
(a) there are four consecutive points A, B, C, D, E, and F on the boundary such that
(1) the boundary part from A to B is congruent by translation to the boundary part from D to C,

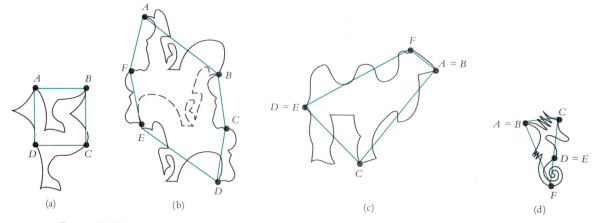

Figure 22.10 Individual tiles traced from the Escher prints of Color Plates 7–10, with points marked to show how they fulfill the criteria for tiling by translations or by translations and half-turns.

(2) the boundary part from A *to* C *is congruent by translation to the boundary part from* A *to* D *(see Figure 22.10a)*

or

(b) there are six consecutive points A, B, C, D, E, *and* F *on the boundary such that the boundary parts* AB, BC, *and* CD *are congruent by translation, respectively, to the boundary parts* ED, FE, *and* AF *(see Figure 22.10b).*

The tiles for each of the Color Plates 7 and 8 are shown in outline form in Figure 22.10, together with points marked to show how the tiles fulfill the criterion.

To create tilings, you can proceed exactly as Escher did. His notebooks show that he designed his patterns in just the way that we now describe.

EXAMPLE: Tiling the Plane Using a Parallelogram

For case (*a*) of the theorem, start from a parallelogram, make a change to the boundary on one side, then copy that change to the opposite side. Similarly, change one of the other two sides and copy that change on the side opposite it (Figure 22.11). Revise as necessary, always making the same change to opposite sides. You might find it useful (as Escher did) to make your designs on graph paper, or you can work by cutting and taping together pieces of heavy paper. ▲

EXAMPLE: Tiling the Plane Using a Hexagon

For case (*b*), start from a hexagon whose opposite sides are equal and parallel (a **par-hexagon**); this is one of the kinds of hexagons that tile the plane. Again, make a change on one boundary and copy the change to the opposite side, and do this for all three pairs of opposite sides (Figure 22.12). ▲

Of course, there is a real art to being able to make the resulting tile resemble an animal or human figure!

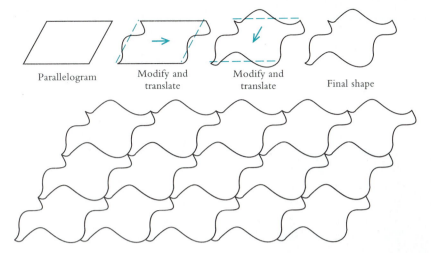

Figure 22.11 How to make an Escher-like tiling by translations, from a parallelogram base. (Adapted from Dale Seymour and Jill Britton, *Introduction to Tessellations*, Dale Seymour Publications, Palo Alto, Calif., 1989, p. 136.)

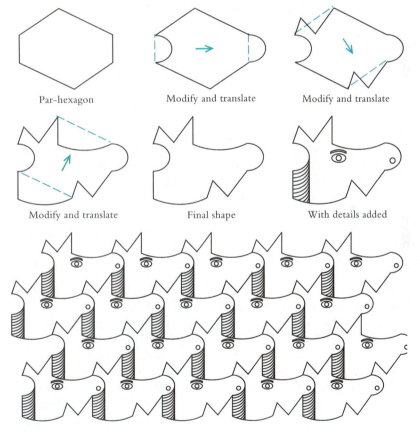

Figure 22.12 How to make an Escher-like tiling by translations, from a par-hexagon base. (Adapted from Dale Seymour and Jill Britton, *Introduction to Tessellations*, Dale Seymour Publications, Palo Alto, Calif., 1989, p. 137.)

Marjorie Rice.

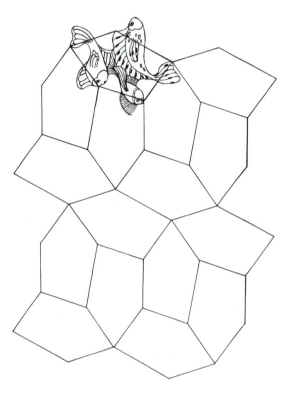

Underlying grid for Marjorie Rice's *Fish*, based on one of her unusual tilings by pentagons.

R. B. Kershner's claim to have found all of the types of convex pentagons that tile was reported by Martin Gardner in his column in *Scientific American,* which was read by many amateur puzzle enthusiasts. Among them was Richard James III, who found a tiling that Kershner had missed and wrote to Gardner with the news. Gardner reported James's discovery in a later column.

When that issue was delivered, another avid Gardner fan in California turned immediately to Gardner's "Mathematical Games" column. Marjorie Rice, a San Diego housewife and mother of five, was usually the first one in the household to read her son's magazine. She had been intrigued by the July article on tiling by pentagons and had "thought how wonderful it must have been [for Kershner] to discover the new types of pentagon tiles." Now, reading of James's newly discovered pentagon tile, her interest was strongly aroused and she set out to see if she might find still other new pentagons that tile. "I thought I would like to understand these fascinating patterns better and see if I could find still another type. It was like a delightful new puzzle to me and I considered how I could best go about this." Her search soon became a full-scale assault on the problem, extending over a period of two years.

Marjorie Rice had no formal education in mathematics beyond a general mathematics course required for graduation from high school in 1939. Thus, as she faced the challenge of finding new pentagonal tiles she not only worked out her own method of attack, but invented her own notation as well. Both her notation and her way of checking possibilities were far from the conventional ways that mathematicians use.

"This was the busy Christmas [1975] season which took much of my time but I got back to the problem whenever I could and began drawing little diagrams on my kitchen counter when no one was there, covering them up quickly if someone came by, for I didn't wish to have to explain what I was doing to anyone. Soon I realized that many interesting patterns were possible but did not pursue them further, for I was searching for a new type and a few weeks later, I found it." Over the next two years, she found three additional new tilings.

What makes a person pursue a problem so steadfastly as Marjorie Rice? She was not trained to do this, nor paid to do it, but obviously gained personal satisfaction in her patient and persistent search. No doubt her personal history is like that of many amateurs.

She was born in 1923 in St. Petersburg, Florida, a first child. At age 5, she began school in a one-room country school with eight grades and about two dozen pupils. "My mother wished me to have a good start and had taught me well at home so I was placed in the second grade.

"When I was in the 6th or 7th grade our teacher pointed out to us one day the Golden Section in the proportions of a picture frame. This immediately caught my imagination and though it was just a passing incident, I never forgot it. I've continued reading on a wide variety of subjects over the years and have been especially interested in architecture and the ideas of architects and planners such as Buckminster Fuller. I've come across the Golden Section again in my reading and considered its use in painting and design." She became especially interested in textile design and the works of M. C. Escher. As she pursued the problem of pentagons and their tilings, she produced some beautiful geometric designs and imaginative Escher-like patterns (see Color Plate 11 and the accompanying figure here).

After high school, Marjorie Rice worked until her marriage in 1945. She was drawn back into mathematics by her children, finding solutions to their homework problems "by unorthodox means, since I did not know the correct procedures."

"I enjoy puzzles of all kinds, crosswords, jigsaw puzzles, mathematical puzzles and games, and have purchased books of mathematical puzzles over the years. Those of a geometric nature are a special delight."

The mind and spirit are the forte of all such amateurs — the intense spirit of inquiry and the keen perception of all they encounter. No formal education provides these gifts. Mere lack of a mathematical degree separates these "amateurs" from the "professionals." Yet their curiosity and ingenious methods make them true mathematicians.

(Adapted from "In Praise of Amateurs," by Doris Schattschneider, in *The Mathematical Gardner*, edited by David A. Klarner, pp. 140–166 plus Plates I–III, Wadsworth, Belmont, Calif., 1981.)

▶ Tiling by Translations and Half-Turns

If the tiling is to allow half-turns, so that some of the figures are "upside down," the part of the boundary of a right-side-up figure has to match the corresponding part of itself in an upside-down position. For that to happen, that part of the boundary must be **centrosymmetric,** that is, symmetric about (unaltered by) a 180° rotation around its midpoint. The key to some of Escher's more sophisticated monohedral designs, and the fundamental principle behind some further easy recipes for making Escher-like tilings, is the **Conway criterion:**

A tile can tile the plane by translations and half-turns *if there are six consecutive points on the boundary (some of which may coincide, but at least three of which are distinct)—call them* A, B, C, D, E, *and* F—*such that*

(1) *the boundary part from* A *to* B *is congruent by translation to the boundary part from* E *to* D, *and*

(2) *each of the boundary parts* BC, CD, EF, *and* FA *is centrosymmetric.*

The first condition means that we can match up the two boundary parts exactly, curve for curve, angle for angle. The second condition means that each of the remaining boundary parts is brought back into itself by a half-turn around its center. Either condition is automatically fulfilled if the boundary part in question is a straight-line segment.

The tiles for each of the Color Plates 9 and 10 are shown in outline form in Figure 22.10, together with points marked to show how the tiles fulfill the Conway criterion.

Once again, you can make Escher-like tilings by starting from simple geometric shapes that tile. This time, the starting geometric tile can be any triangle or any quadrilateral.

Example: Tiling the Plane Using a Triangle

For a triangle, modify half of one side, then rotate that side around its center point to extend the modification to the rest of the side, thereby making the new side centrosymmetric. Then you may do the same to the second and third sides (Figure 22.13). ▲

Example: Tiling the Plane Using a Quadrilateral

For the quadrilateral, do the same, modifying each of the four sides, or as many as you wish (Figure 22.14). ▲

The same approach will work with some of the sides of some pentagons and hexagons that tile. Because not all sides can be modified, there is less freedom for designing tiles, so it is more difficult to make the resulting tiles resemble intended figures. Color Plate 11 shows the beautiful results achieved by Marjorie Rice, using one of the unusual tilings by pentagons that she discovered.

The sketches in Escher's notebook in Color Plates 7–10 indicate how he designed the prints whose tiles you see in Figure 22.10. For Figure 22.10a, he modified the two pairs of sides of a square. For Figure 22.10b, he modified the pairs of sides of a par-hexagon that became a tile made up of a pair of dark and light knights. This figure also has a reflection symmetry, taking a leftward-facing light knight to a rightward-facing dark knight. However, we have not discussed criteria for when you can start with a tile (e.g., a single knight) and produce a tiling with this symmetry. In Figure

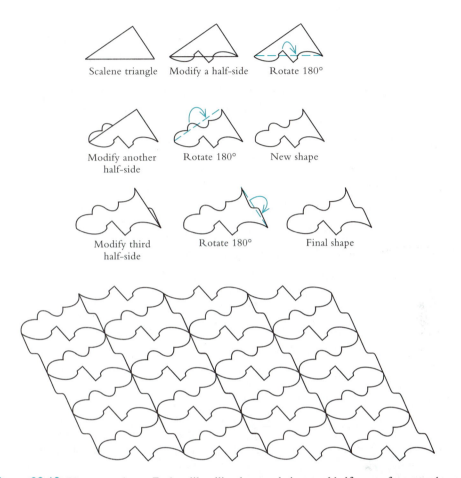

Scalene triangle   Modify a half-side   Rotate 180°

Modify another half-side   Rotate 180°   New shape

Modify third half-side   Rotate 180°   Final shape

Figure 22.13 How to make an Escher-like tiling by translations and half-turns, from a scalene triangle base. (Adapted from Dale Seymour and Jill Britton, *Introduction to Tessellations*, Dale Seymour Publications, Palo Alto, Calif., 1989, pp. 138–139.)

22.10c, the blue overlay shows how the tile could be made by modifying half of every side of a general quadrilateral, though Color Plate 9 shows that Escher actually designed the tiling from a parallelogram base. Regarding Figure 22.10d, Color Plate 10 shows that Escher used a triangle base. He did not use the procedure that we noted earlier, in which half of every side is modified. Instead, he treated the triangle as a quadrilateral, in which two adjacent sides (*CD* and *DF*) happen to continue on in a straight line.

▶ FURTHER CONSIDERATIONS

Some of the most impressive of Escher's prints use two or more interlocking tile shapes. Such is the case with *Heaven and Hell* ("Angels and Devils") in Spotlight 20.4 (pp. 642–643). All of his prints of tilings of the plane have underlying symmetries that are the wallpaper patterns, whose classification is discussed in Spotlight 21.5 (pp. 676–677). Often those symmetries are enhanced further by the use of color.

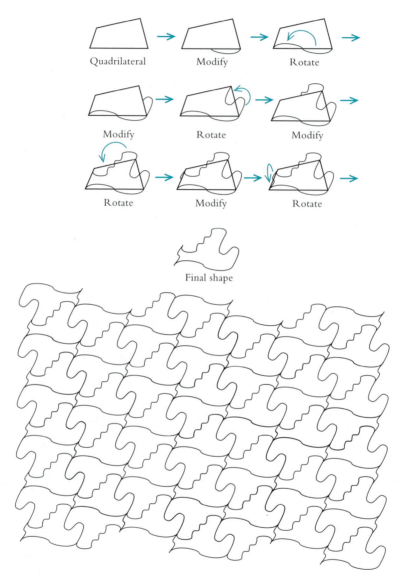

Quadrilateral → Modify → Rotate →

Modify → Rotate → Modify →

Rotate → Modify → Rotate →

Final shape

Figure 22.14 How to make an Escher-like tiling by translations and half-turns, from a quadrilateral base. (Adapted from Dale Seymour and Jill Britton, *Introduction to Tessellations*, Dale Seymour Publications, Palo Alto, Calif., 1989, p. 140.)

Several of Escher's more remarkable prints are tilings not of the Euclidean plane but of the Poincaré disk model of hyperbolic geometry, discussed in Chapter 20. Two examples, one by Escher and one by Douglas Dunham, are shown in Spotlight 20.4.

All the patterns that we have exhibited and discussed so far have been **periodic tilings.** If we transfer a periodic tiling to a transparency, it is possible to slide the transparency a certain distance horizontally, without rotating it, until the transparency exactly matches the tiling everywhere. We can also achieve the same result by moving the transparency some second direction (possibly vertically) a certain (possibly different) distance.

In a periodic tiling you can identify a **fundamental region** — a tile, or a block of tiles — with which you can cover the plane by translations at regular intervals. For example, in Color Plate 7, a single bird forms a fundamental region. In Color Plate 9, two adjacent camels, one right-side up and one upside down, form a fundamental region. In Escher's *Heaven and Hell* of Spotlight 20.4, a pair of foot-to-foot angels and the pair of foot-to-foot devils between them form a fundamental region. In the terminology of Chapter 21, the periodic tilings are ones that are preserved under translations in more than two directions. (The wallpaper patterns of Spotlight 21.5 are sometimes called *periodic plane patterns*. In this chapter we are concerned with the design elements more than with the patterns, which were the main topic of Chapter 21.)

▶ NONPERIODIC TILINGS

A nonperiodic tiling is a tiling in which either there is no regular repetition of the pattern by translation, or else the repetition is not completely regular.

EXAMPLES

The lower left pattern in Figure 22.4, with its expanding rings of triangles, does not have any regular repetition by translation.

In Figure 22.3a, the second row from the bottom is offset one-half of a unit to the right from the bottom row, the third row from the bottom is offset one-third of a unit further, and so forth. Since the sum $1/2 + 1/3 + 1/4 + \cdots + 1/n$ never adds up to exactly a whole number, this tiling does not repeat in a regular way: there is no direction (horizontal, vertical, or diagonal) in which we can move the entire tiling and have it coincide exactly with itself.

For another example of a nonperiodic tiling, consider the usual edge-to-edge square tiling. For each square, flip a coin; depending on the result, divide the square into two right triangles by adding either a rising or a falling diagonal (see Figure 22.3b). Because what happens in each individual square is unconnected to what happens in the rest of the tiling, the tiling by right triangles that is produced by this procedure has no chance of being periodic.   ▲

▶ THE PENROSE TILES

For all known cases, if a single tile can be used to make a nonperiodic tiling, then it can also be used to make a periodic tiling. Although it is still an open question whether this property is true for every possible shape whatsoever, mathematicians are inclined to think that it is so.

For a long time they also tended to believe the more general assertion that if you can construct a nonperiodic tiling with a set of one *or more* tiles, you can construct a periodic tiling from the same tiles. But in 1964 a set of tiles was found that permits only nonperiodic tiling. It contains 20,000 different shapes! Over the next several years, smaller sets were discovered with the same property, with as few as 100 shapes. But it was still amazing when in 1975 Roger Penrose, a mathematical physicist at Oxford, announced a set that would tile only nonperiodically—consisting of just 2 tiles! (See Figure 22.15.)

Penrose called his tiles "darts" and "kites." It is easy to specify their construction, as one of each can be obtained from a single rhombus. (A **rhombus** is a quadrilateral with four equal sides and equal opposite interior angles.) The particular rhombus from which the Penrose tiles are constructed has interior angles of 72° and 108°. If we cut the longer diagonal in two pieces so that the longer piece is the golden ratio ($(1 + \sqrt{5})/2 \approx$ 1.618) times as long as the shorter (see Chapter 21), and connect the dividing point to the remaining corners, we split the rhombus into a dart and a kite (Figure 22.15).

Since the two Penrose pieces come from a rhombus, and that rhombus can be replicated to tile the plane periodically, you must have guessed that the rules for fitting the Penrose pieces together do not allow the rhombus arrangement. The actual plastic pieces (produced in a limited edition in the 1970s) have little nicks and corresponding bumps

that enforce this prohibition. Instead, we may label the front and back vertices of the dart with H (for head) and its two wing tips with T (for tail), and do the reverse for the kite. Then our rule is that only vertices with the same letter may meet: heads must go to heads, and tails to tails.

A prettier method of enforcing the rules, proposed by John Conway of Cambridge University, is to draw circular arcs of different colors on the pieces and require that adjacent edges must join arcs of the same color. The result is the pretty patterns of Color Plate 4. In fact, Conway thinks of the darts as children, each with two hands. The rule for fitting the pieces together is that children are forced to hold hands. Penrose patterns become dancing circles of children.

Color Plate 1 shows a tiling by a different pair of pieces, both rhombuses, that tile the plane only nonperiodically. Color Plate 2 shows a modification of the Penrose pieces into two bird shapes. Color Plate 3 shows a coloring of one particular tiling with the Penrose pieces so that no two adjacent pieces have the same color.

Although tilings with Penrose's pieces cannot be periodic, the tilings possess unexpected symmetry. As you recall, we have explored our intuitions of symmetry in terms of *balance, similarity,* and *repetition.* Patterns made with the Penrose pieces certainly involve repetition, but it is the balance in the arrangement that we seek. What balance can there be in a nonperiodic pattern? It turns out that some Penrose patterns have a single line of reflection. But most surprising of all, every Penrose pattern has arbitrarily large regions with fivefold and tenfold rotational symmetry!

EXAMPLE: Fivefold Symmetry

Consider, for example, any one of the 10 red pieces of Figure 22.16. If we rotate the pattern around its center through one-fifth of a turn, the region surrounded by the red pieces looks exactly the same as before (but parts of the pattern farther away may not exactly match). Note

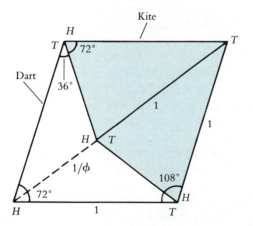

Figure 22.15 Construction of Penrose's "dart" and "kite" (colored area). The length $1/\phi \approx 0.618$; ϕ is the golden ratio.

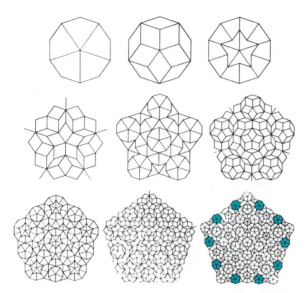

Figure 22.16 Successive deflation (that is, the systematic cutting up of large tiles into smaller ones) of patches of tiles of a Penrose nonperiodic tiling. (Courtesy of Roger Penrose.)

that rotation through one-tenth of a turn would not preserve all of that region. In Conway's metaphor, whenever a chain of children closes, the region inside has fivefold symmetry. ▲

There are, in fact, two—and only two—Penrose tilings for which the *entire* pattern has the fivefold rotation symmetry of a rosette. Figure 22.16 shows how to construct both of these, beginning with five kites meeting at a vertex. We cut the darts and kites up into smaller darts and kites and then enlarge the new ones to the same size as the old, that way covering more area each time (the figure doesn't show the enlargement). Conway calls this operation *deflation*. As we proceed with successive steps, we get partial tilings alternately by two different patterns, and each has a fivefold rotational center: one has five kites at the center, the other has five darts.

Where does this rotational symmetry come from? The original rhombus that we split up has

the angles shown in Figure 22.15. Except in the recess of the dart and matching part of the kite, all the internal angles of the kite and dart are either 72° or 36°. Now, 72° goes into 360° five times, and 36° goes 10 times. If we recall that it is the interior angles that matter in arranging polygons around a point, we see that fivefold or tenfold symmetry could conceivably result from using such tiles.

The reverse of Conway's deflation, *inflation*, is the key idea in a simple argument to show that a Penrose pattern must be nonperiodic. For the inflation process, cut each dart down its middle and put glue on the short edges of the resulting triangles (but not on the cut itself). The result is a pattern of larger kites and darts!

We show that a Penrose pattern is nonperiodic by proceeding *by contradiction*. Suppose (contrary to what we want to establish) that some Penrose pattern is periodic, that is, it has translation symmetry. Let d be the distance along the translation direction to the first repetition. Performing inflation does the same thing to each repetition, so the inflated pattern *must* still have translation symmetry and a distance d along the translation direction to the first repetition. Keep on performing inflation, time after time, until the darts and kites are so large that they are more than d across. The pattern, as we have just argued, must still have translation symmetry at a distance d; but it can't, because there's no repetition inside a single tile! We reach a contradiction. So what's wrong? Our initial supposition, that the pattern was periodic in the first place, must have been erroneous. We conclude that all Penrose tilings are nonperiodic.

Despite their being nonperiodic, all Penrose patterns are somewhat alike, in the following sense:

Any finite region in one Penrose pattern is contained somewhere inside every other Penrose pattern; in fact, it occurs infinitely many times in every Penrose pattern.

Penrose tilings have another feature that allows us to characterize them as "quasiperiodic." Robert Ammann introduced onto the two rhombic Penrose pieces used in Color Plate 1 lines that are now known as *Ammann bars*. In any Penrose tiling, these bars line up into five sets of parallel lines, each set rotated 72° from the next, forming a pentagonal grid (Figure 22.17). The distance between two adjacent parallel bars is one of only two values, either A or B. Do you want to guess what the ratio of the longer A is to the shorter B? You don't think it could possibly be anything but the golden ratio, do you? And so it is.

EXAMPLE: Musical Sequences

What about the order in which the A's and B's occur, as we move from left to right in Figure 22.17? Is there any pattern to that? From the limited part of the pattern we can observe, we see the sequence as

$$A B A A B A B A A B A B A \ldots$$

You might think from the figure that the pattern continues repeating the group

$$A B A A B \ldots$$

indefinitely; after all, there are five symbols in this group. But such is not the case. Known as a *musical sequence,* the sequence of intervals between Ammann bars is nonperiodic—it cannot be produced by repeating any finite group of symbols. We can think of it as a one-dimensional analogue of a Penrose tiling.

There is some regularity in musical sequences. Two B's can never be next to each other, nor can we have three A's in a row. Just as any finite part of any Penrose tiling occurs infinitely often in any other Penrose tiling, any finite part of any musical sequence appears infinitely often in any other one. The order of the symbols is neither periodic nor random, but between the two—quasiperiodic.   ▲

The ratio of darts to kites in an infinite Penrose tiling, or of A's to B's in a musical se-

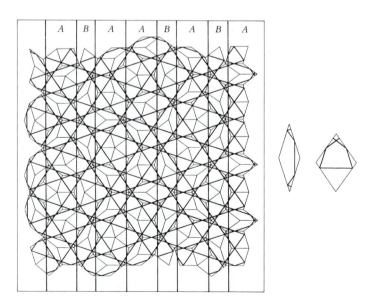

Figure 22.17 Penrose tilings with Ammann bars. Specially placed lines on the tiles produce five sets of parallel bars in different directions.

quence, is exactly the golden ratio, approximately 1.618. So if you are going to play with sets of Penrose pieces and see what kinds of patterns you can create, you will need about 1.6 times as many darts as kites.

As pointed out by geometers Marjorie Senechal (Smith College) and Jean Taylor (Rutgers University), Penrose tilings have three important properties:

1. They are constructed according to rules that force nonperiodicity.

2. They can be obtained from a substitution process (inflation and deflation) that features self-similarity (self-similarity at a change in scale is one of the hallmarks of fractals, which are discussed in the color photo essay on fractals.

3. They are quasiperiodic.

Research of the late 1980s indicates that these properties are somewhat independent, meaning that one or two may be true of a tiling without all three being true.

▶Shechtman's Crystals and Barlow's Law

Although Penrose's discovery was a big hit among geometers and in recreational mathematics circles in the mid-1970s, few people thought that his work might have practical significance. In the early 1980s some mathematicians even generalized Penrose tilings to three dimensions, using solid polyhedrons to fill space nonperiodically. Like the two-dimensional Penrose patterns, these have orderly fivefold symmetry but are nonperiodic.

Yet in 1982 scientists at the U.S. National Bureau of Standards discovered unexpected fivefold symmetry while looking for new ultrastrong alloys of aluminum (mixtures of aluminum with other metals).

Manganese doesn't ordinarily alloy with aluminum, but the experimenters were able to pro-

duce small crystals of alloy by cooling mixtures of the two metals at a rate of millions of degrees per second. Following routine procedures, chemist Daniel Shechtman began a series of tests to determine the atomic structure of the special crystals. But there was nothing routine about what he found: the atomic structures of the manganese-aluminum crystals were so startling that it took Shechtman 3 years to convince his colleagues they were real.

Why did he encounter such resistance? His patterns—and the crystals that produced them—defied one of the fundamental laws of crystallography. Like our discovery that the plane cannot be tiled by regular pentagons, **Barlow's law,** also called the **crystallographic restriction,** says that no crystal can have more than one center of fivefold symmetry.

Peter Barlow was a nineteenth-century British mathematician whose name survives today in the name of a book of mathematical tables. His argument was a very simple proof by contradiction, similar to Conway's proof in which we saw earlier that Penrose patterns are not periodic. Suppose (contrary to what we intend to show) that there is more than one fivefold rotation center. Let A and B be two of these that are closest together (see Figure 22.18). Rotate the pattern of Figure

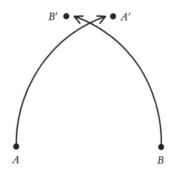

Figure 22.18 Barlow's proof that no pattern can have two centers of fivefold symmetry. (From Martin Gardner, *Penrose Tiles to Trapdoor Ciphers,* Freeman, New York, 1989, p. 27.)

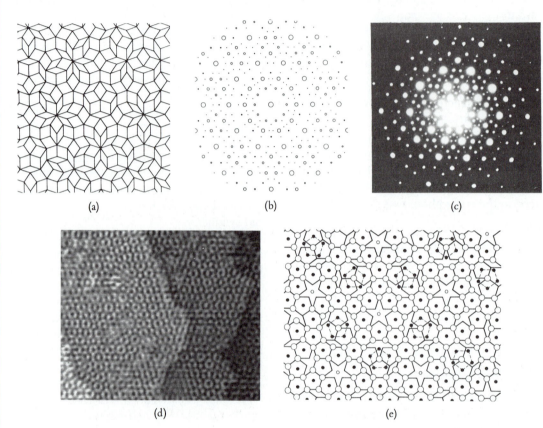

(a) A nonperiodic Penrose tiling. (b) A three-dimensional crystal-like structure based on this tiling. (c) The crystal pattern observed by chemist Daniel Shechtman in a special manganese – aluminum alloy. Note the similarity to the pattern in (a). (d) Scanning tunneling microscope image of a perfect Penrose tiling formed by four layers of atoms on the surface of an aluminum – cobalt – copper alloy. (e) The tiling corresponding to the image in (d); the aluminum atoms (open centers) tend to settle into pentagonal rings around copper or cobalt atoms (filled centers). ((a), (b), and (c) are from D. R. Nelson and B. I. Halperin, *Science*, 229:233 – 238 (1985); (d) and (e) courtesy of A. Refik Kortan, AT&T Bell Laboratories, and *Physical Review Letters* (8 January 1990).)

In 1984 Paul Steinhardt, a physicist at the University of Pennsylvania, and one of his graduate students, Don Levine, calculated the diffraction patterns that their three-dimensional Penrose patterns would produce if the building blocks were real atoms instead of imaginary tiles.

Diffraction patterns are the windows physicists use to peer inside materials. When a beam of electrons or x-rays passes through a solid material, they are diffracted, or scattered, by the atoms inside. The diffracted beams can be photographed head-on, and the images they form reflect the atomic architecture of the solid.

The most distinct diffraction patterns contain sharp, isolated dots. These patterns are the portraits of crystal, and they owe their clearly defined spots to the periodicity of the underlying structure. In a few preferred directions, depending on the arrangement of the atoms, the diffracted beams reinforce one another, producing bright spots on the film. A crystal is a little like an orchard planted in a rigid geometric grid. Most lines of sight are blocked by trees, but you can see right through to the other side in a few directions.

In another class of diffraction patterns the dots are either spread out into fuzzy rings or altogether absent. These are the images formed by glassy materials. Glasses, in contrast to crystals, are made up of atoms or molecules stuck together randomly; they're more like random forests than well-planned orchards. Because they offer no preferred directions for diffraction, the patterns they produce contain no sharp dots.

The computed diffraction pattern for Levine and Steinhardt's imaginary solid contained a surprise: unmistakable sharp points. Since the atomic arrangement of their solid was nonperiodic, it should have produced the fuzzy diffraction pattern characteristic of glassy substances. Since the dots in the pattern were arranged with fivefold symmetry, the solid wasn't a crystal either. Steinhardt decided to call it a *quasicrystal.*

In the fall of 1984 a colleague of Steinhardt's showed him a diffraction image made from a real substance, Shechtman's alloy of aluminum and manganese. The picture looked amazingly similar to Steinhardt and Levine's computer simulation.

In short order more than a hundred alloys with fivefold symmetry were discovered; and sevenfold, ninefold, and other symmetries proved to be possible. But no one could think of a mechanism by which millions upon millions of real atoms could arrange themselves spontaneously in those intricate patterns.

Anyone who tries to assemble Penrose pieces into tilings quickly realizes that it's not easy. You have to think ahead and keep the whole pattern in mind when adding a tile; otherwise, there is trouble. *Local rules,* or instructions for fitting a tile into a particular niche, don't seem sufficient to build the entire pattern without *global rules* that force you to plan ahead and check the configuration of tiles at distant points.

Local rules for adding tiles are analogous to forces that attract and hold new atoms to the surface of a growing quasicrystal. The atoms on a growing surface do not plan ahead. If quasiperiodic patterns could be constructed only with the help of global rules, they could not be assembled by real atoms in real alloys, and quasicrystals could not exist in nature.

In 1988 playfulness paid off once more. George Onoda, an IBM ceramics expert, started toying with about 200 Penrose tiles. Unconvinced that he wasn't supposed to be able to do it, he learned how to assemble flawless tilings of any size using only local rules.

For a complete theory of quasicrystals, the local rules will have to be generalized to three dimensions, and they must be shown to correspond to actual atomic forces. In the meantime, experimentalists continue to report bigger, more perfect quasicrystals.

(Adapted from Hans C. von Baeyer, "Impossible Crystals," in *Discover* 11(2):69–78, 84, February 1990.)

22.18 by one-fifth of a turn clockwise around B, which carries A to some point A'. Since the pattern has fivefold symmetry around B, the point A', which is the image of the fivefold center A, must itself be a fivefold center. Now we use A as a center and rotate the pattern by one-fifth of a turn counterclockwise, which carries B to some point B'; as we just argued for A', B' must also be a fivefold center. But A' and B' are closer together than A and B, which is a contradiction. Hence our original supposition must be false, and a pattern can have at most one fivefold rotation center (as the patterns in Figure 22.16 in fact do).

Barlow's law, as a mathematical theorem, shows that fivefold symmetry is impossible in a periodic tiling of the plane or of space. Chemists, for good theoretical and experimental reasons, believe that crystals are modeled well by three-dimensional tilings. An array of atoms with no symmetry whatever would not be considered a crystal. Yet until Penrose's discovery, no one realized that nonperiodic tilings — or arrays of atoms — can have the regularity of fivefold symmetry.

Chemists could simply say that Shechtman's alloys aren't crystals. In the classical sense they aren't, but in other respects they do resemble crystals. It is scientifically more fruitful to extend the concept of crystal to include them rather than rule them out; they are now known as *quasicrystals* (see Spotlight 22.3, p. 714–715).

Once again, as so often happens in history, pure mathematical research anticipated scientific applications. Penrose's discovery, once just a delightful piece of recreational mathematics, is now prompting a major reexamination of the theory of crystals.

▶ REVIEW VOCABULARY

Barlow's law, or the **crystallographic restriction**   A law of crystallography that states that a crystal may have only rotational symmetries that are twofold, threefold, fourfold, or sixfold.

Centrosymmetric   Symmetric by 180° rotation around its center.

Convex   A geometric figure is convex if for any two points on the figure (including its boundary), all the points on the line segment joining them also belong to the figure (including its boundary).

Conway criterion   A criterion for determining whether a shape can tile by means of translations and half-turns.

Edge-to-edge tiling   A tiling in which adjacent tiles meet only along full edges of each tile.

Equilateral triangle   A triangle with all three sides equal.

Exterior angle   The angle outside a polygon formed by one side and the extension of an adjacent side.

Fundamental region   A tile or group of adjacent tiles that can tile by translation.

Interior angle   The angle inside a polygon formed by two adjacent sides.

Monohedral tiling   A tiling with only one size and shape of tile (the tile is allowed to occur also in "turned-over," or mirror-image, form).

n-gon   A polygon with n sides.

Parallelogram   A convex quadrilateral whose opposite sides are equal and parallel.

Par-hexagon   A hexagon whose opposite sides are equal and parallel.

Periodic tiling   A tiling that repeats at fixed intervals in two different directions, possibly horizontal and vertical.

Quadrilateral   A polygon with four sides.

Regular polygon   A polygon all of whose sides and angles are equal.

Regular tiling   A tiling by regular polygons, all of which have the same number of sides and are the same size; also, at each vertex, the same kinds of polygons must meet in the same order.

Rhombus   A parallelogram all of whose sides are equal.

Scalene triangle   A triangle, no two sides of which are equal.

Semiregular tiling   A tiling by regular polygons; all polygons with the same number of sides must be the same size.

Tiling   A covering of the plane without gaps or overlaps.

Translation   A rigid motion that moves everything a certain distance in one direction.

Vertex figure   The pattern of polygons surrounding a vertex in a tiling.

▶ SUGGESTED READINGS

VON BAEYER, HANS C.: "Impossible crystals," *Discover* 11(2):69–78, 84 (February 1990). Tells how the playfulness of mathematicians and physicists led to the discovery of quasicrystals.

BARBER, FREDERICK, ET AL.: *Tiling the Plane*, Faculty Advancement in Mathematics Module. COMAP, Lexington, Mass., 1989.

BOLES, MARTHA, AND ROCHELLE NEWMAN: *The Golden Relationship: Art, Math & Nature. Book 2: The Surface Plane*, Pythagorean Press, Bradford, Mass., 1992.

CHOW, WILLIAM W.: "Automatic generation of interlocking shapes," *Computer Graphics and Image Processing* 9:333–353 (1979). Shows how to design a computer program to draw interlocking patterns.

———: "Interlocking shapes in art and engineering," *Computer Aided Design* 12:29–34 (1980). Discusses applications to sheet material manufacturing (e.g., fabrication of gloves, can openers, forks, key blanks, and bunk bed brackets).

CHUNG, FAN, AND SHLOMO STERNBERG: "Mathematics and the buckyball," *American Scientist* 81:56–71 (1993).

FLAHERTY, TERRY: *Escher-Sketch*, Intellimation Library for the Macintosh (P.O. Box 219, Santa Barbara, CA 93116). Apple Macintosh program that allows you to design Escher-like patterns, using any of the 17 wallpaper patterns discussed in Chapter 21.

GARDNER, MARTIN: "Mathematical games: Extraordinary nonperiodic tiling that enriches the theory of tiles," *Scientific American*, pp. 110–121, 132, and front cover (January 1977). Reprinted with additional material in *Penrose Tiles to Trapdoor Ciphers*, by Martin Gardner, Freeman, New York, 1989, pp. 1–29.

———: "Mathematical games: On tessellating the plane with convex polygon tiles," *Scientific American*, pp. 112–117, 132 (July 1975). Reprinted with additional material in *Time Travel and Other Mathematical Bewilderments*, by Martin Gardner, Freeman, New York, 1988, pp. 163–176.

———: "Mathematical games: More about tiling the plane: The possibilities of polyominoes, polyiamonds and polyhexes," *Scientific American*, pp. 112–115, 128 (August 1975). Reprinted with additional material in *Time Travel and Other Mathematical Bewilderments*, by Martin Gardner, Freeman, New York, 1988, pp. 177–187.

GRÜNBAUM, BRANKO, AND G. C. SHEPHARD: *Tilings and Patterns*, New York, Freeman, New York, 1987.

KERSHNER, R. B.: "On paving the plane," *APL Technical Digest* (Applied Physics Laboratory, Johns Hopkins University) 8(6):4–10 (July/August 1969). Gives an easy proof (depending only on Euler's formula) that no convex polygon with more than six sides can tile.

MARTIN, GEORGE: *Polyominoes: A Guide to Puzzles and Problems in Tilings*, Mathematical Association of America, Washington, D.C. 1991. Just as a domino is a tile made up of two adjacent squares, a polyomino is a tile made up of a number of squares.

This book explores which of the polyominoes can tile and how.

PENROSE, ROGER: *The Emperor's New Mind*, Oxford University Press, New York, 1989.

RANUCCI, ERNEST, AND JOSEPH TEETERS: *Creating Escher-Type Patterns*, Creative Publications, Oak Lawn, Ill.

SCHATTSCHNEIDER, DORIS: "Will it tile? Try the Conway criterion!" *Mathematics Magazine*, 53:224–233 (1980).

————: "In praise of amateurs," in David A. Klarner (ed.), *The Mathematical Gardner*, Wadsworth, Belmont, Calif., 1981, pp. 140–166 plus Plates I–III.

————: *Visions of Symmetry: Notebooks, Periodic Drawings, and Related Work of M. C. Escher*, Freeman, New York, 1990.

SENECHAL, MARJORIE, AND JEAN TAYLOR: "Quasicrystals: The view from Les Houches," *Mathematical Intelligencer* 12(2):54–64 (Spring 1990).

SEYMOUR, DALE, AND JILL BRITTON: *Introduction to Tessellations*, Dale Seymour Publications, Palo Alto, Calif. 1989. An excellent introduction to tessellations, including how to make Escher-like tessellations.

STEINHARDT, PAUL JOSEPH: "Quasicrystals," *American Scientist*, 74(6):586–597 plus cover (November/December 1986). Includes illustrations of nonperiodic tilings with sevenfold and ninefold symmetry.

TEETERS, JOSEPH L.: "How to draw tessellations of the Escher type," *Mathematics Teacher*, 67:307–310 (1974).

TESSELLATION WINNERS: *Escher-Like Original Student Art: The First Contest*, Dale Seymour Publications, Palo Alto, Calif., 1991.

▶ **EXERCISES**

Hint: For the exercises about determining whether a shape will tile the plane, you will want to make a number of copies of the shape and experiment with placing them. One easy way to make copies is to trace the shape onto a piece of paper, staple half a dozen other blank sheets behind that sheet, and use scissors to cut through all the sheets along the edges of the traced shape on the top sheet.

1. Determine the measure of an exterior angle and of an interior angle of a regular octagon (eight sides).

2. Determine the measure of an exterior angle and of an interior angle of a regular decagon (ten sides).

3. Discover a formula for the measure of an interior angle of a regular n-gon.

4. Using the formula from Exercise 3 and either your calculator or a short computer program, make a chart of the interior angle measures of regular polygons with 3, 4, . . . , 12 sides.

5. For each of the tiles below, show how it can be used to tile the plane.

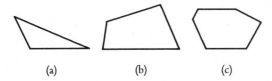

(a)   (b)   (c)

(From *Tiling the Plane*, by Frederick Barber et al., COMAP, Lexington, Mass., 1989, pp. 1, 8, 9.)

6. The lower left corner of Figure 22.5 shows a tiling by isosceles triangles.
   a. Use the center vertex to determine the measures of the angles of the isosceles triangle tile.
   b. Every vertex *except* the center vertex has the same vertex figure, in terms of the measures of the angles surrounding the vertex. What is that vertex figure?

Refer to tiles (a) through (g) below in doing Exercises 7 through 10.

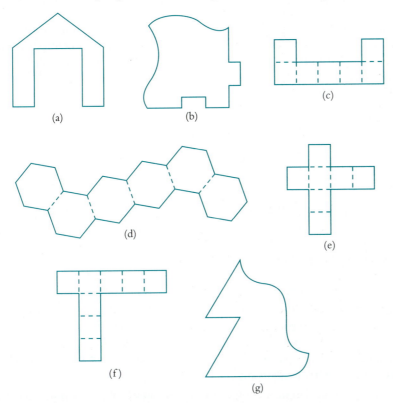

(a)   (b)   (c)

(d)   (e)

(f)   (g)

(Adapted from *Tilings and Patterns*, by Branko Grünbaum and G. C. Shephard, Freeman, New York, 1987, p. 25.)

7. For each of the tiles (a) through (c) on the previous page, determine if it can be used to tile the plane by translations.

8. As in Exercise 7, but for the tiles (d) through (g).

9. For each of the tiles (a) through (c) on the previous page, determine if it can be used to tile the plane by translations and half-turns.

10. As in Exercise 9, but for the tiles (d) through (g).

11. Show how an arbitrary pentagon with two parallel sides can tile the plane.

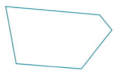

(Adapted from "In Praise of Amateurs," by Doris Schattschneider, in *The Mathematical Gardner*, edited by David A. Klarner, Wadsworth, Belmont, Calif., 1981, p. 142.)

12. Shown below is a pentagonal tile of type 13, discovered by Marjorie Rice. Show how it can tile the plane. (Hint: Carefully trace and cut out a dozen or so copies and try fitting them together.) The parts of this pentagon satisfy the following relations: $A = C = D = 120°$, $B = E = 90°$, $2A + D = 360°$, $2C + D = 360°$, $a = e$, and $a + e = d$.

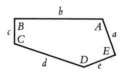

(Adapted from "In Praise of Amateurs," by Doris Schattschneider, in *The Mathematical Gardner*, edited by David A. Klarner, Wadsworth, Belmont, Calif., 1981, p. 162.)

13. Start from a parallelogram of your choice and modify it to tile the plane by translations. (You will probably find it useful to do your work on graph paper.) Can you draw a design on the tile so as to make an Escher-like pattern?

14. Start from a par-hexagon of your choice and modify it to tile the plane by translations. (You will probably find it useful to do your work on graph paper. If you choose a regular hexagon, there is special graph paper, ruled into regular hexagons, that would be particularly useful.) Can you draw a design on the tile so as to make an Escher-like pattern?

15. Start from a triangle of your choice and modify it to tile the plane by translations and half-turns. (You will probably find it useful to do your work on graph paper.) Can you draw a design on the tile so as to make an Escher-like pattern?

16. Start from a quadrilateral of your choice and modify it to tile the plane by translations and half-turns. (You will probably find it useful to do your work on graph paper.) Can you draw a design on the tile so as to make an Escher-like pattern?

▲ 17. Use the chart of interior angle measures from Exercise 4 to determine all of the possible vertex figures of regular polygons (with at most 12 sides) surrounding a point.

▲ 18. Which of the vertex figures of Exercise 17 do not occur in a semiregular tiling?

▲ 19. In addition to the vertex figures of Exercise 17, exactly five others are possible, each involving one polygon with more than 12 sides. None of these vertex figures leads to a semiregular tiling. The five many-sided polygons involved in these five vertex figures have 15, 18, 20, 24, and 42 sides. Determine the other polygons in these vertex figures.

▲ 20. In the text we have discussed criteria and methods for generating Escher-like patterns that involve just translations or translations and half-turns. A slight variation on one of those methods allows us to construct tilings that feature a tile and its mirror image.

Begin with a parallelogram made from two congruent isosceles triangles. Each of these triangles has two sides equal; be sure that the two triangles are arranged so that they have one of the equal sides in common, forming a diagonal of the parallelogram.

Start by modifying the two opposite sides of the parallelogram that are the third, unequal sides of their triangles. Make any modification to half of one side, then *mirror-reflect* that side across its center to extend the modification to the rest of the side. Take that entire modified side and reflect it across the original side. Let the mirror image be the pattern for modifying the opposite side.

Now modify in any way one of the other two remaining sides. Make the same change to the opposite side, without any rotation or reflection. The key step is to mirror-reflect one of these sides across its center and make this mirror-reflection change to the diagonal.

The result is a modified parallelogram that will tile by translation (though in a different way from the special cases that we used to introduce the Conway criterion) and that can be split into two pieces that are mirror images of each other. Escher used a similar technique, but starting from a par-hexagon made from two quadrilaterals, in his *Horseman* print, as shown in his sketch in Color Plate 8a.

Use this technique to produce a tiling of your own design. Can you draw a design on the tile so as to make an Escher-like pattern?

▲ 21. Show that the modified parallelogram in the previous exercise fulfills the Conway criterion, by identifying the six points of the criterion.

▲ Advanced exercise.

■ 22. The rabbit problem in Chapter 21 (Exercise 2) can lead us directly into nonperiodic patterns and musical sequences. Let A denote an adult pair of rabbits and B denote a baby pair. We will record the population at the end of each month, just before any births, in a particular systematic way—as a string of A's and B's. At the end of their second month of life, a rabbit pair will be considered to be adult. At the end of the first month our sequence is just A; and the same is true at the end of the second month. When an adult pair A has a baby pair B, we write the new B immediately to the right of the A. So at the end of the third month, our sequence is AB; at the end of the fourth, it is ABA, since the first baby pair is now adult; at the end of the fifth month we have $ABAAB$.

Mathematicians and computer scientists call this manner of generating a sequence a *replacement system*. At each stage we replace each A by AB and each B by A.

    a. What is the sequence at the end of the sixth month?

    b. Why can't we ever have two B's next to each other?

    c. Why can't we ever have three A's in a row?

    d. Show that from the fourth month on, the sequence for the current month consists of the sequence for last month followed by the sequence for two months ago.

▶ WRITING PROJECT

1. Just as for Penrose patterns, we will define inflation and deflation for any sequence of A's and B's. Both of these operations will preserve musicality: If we inflate or deflate a musical sequence, we get another musical sequence.

    a. Inflation can be used to generate musical sequences. Start at the first stage with just B. Inflation consists of replacing each A with AB and each B by A. Show that at the nth stage there are F_n symbols in the sequence.

    b. Deflation can be used to check whether a finite block of A's and B's can belong to a musical sequence or not. Each deflation stage has two parts: first replace each A by $(A/2)\, B\, (A/2)$ and each B by $(A/2)\, (A/2)$, then combine pairs of adjacent $(A/2)$'s into a single A so that no fractional A's remain. Another way to get the same result is to proceed from left to right, replacing B by A, AA by B, and deleting single A's. The deflated block will be shorter. If at any stage we have a block with two or more B's in a row, or three or more A's in a row, then the original block could not be part of a musical sequence; otherwise, the original block will eventually deflate to a single symbol, at which point we conclude that the original block is a part of a musical sequence (in fact, infinitely often, a part of every musical sequence). Check the two sequences $ABAABABAAB$ and $ABAABABABA$.

■   Discussion exercise.

Answers to Odd-Numbered Exercises

▶ ▶ ▶ ▶ ▶ ▶ ▶ ▶ ▶ ▶ ▶ ▶ ▶ ▶

CHAPTER 1

1. A:1; B:3; C:3; D:3; and E:0. The graph shows that geographically E is isolated, perhaps on an island or a different continent than the other cities.

3. Graph (b).

5.

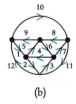

The valences are A:2; B:3; C:4; D:1; and E:2. The real-world consequences are that if you are at one city, you may or may not be able to travel to another particular city.

7. The supervisor is not satisfied because all of the edges are not traveled upon by the postal worker. The worker is unhappy because the end of the worker's route wasn't the same point as that where the worker began. The original job description is unrealistic because there is no Euler circuit in the graph.

9. Both graphs (b) and (c) have Euler circuits:

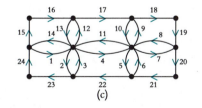

(b)   (c)

The valences of all of the vertices in (a) are odd, which makes it impossible to have an Euler circuit there.

11. h; g; a; b; e; d; h; e; c; m; l; h; i; e; f; k; j; h.

13.

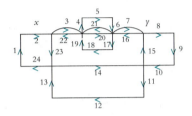

15. Remove the vertical edge in the middle.

17.

19.

21.

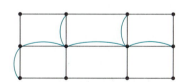

23.

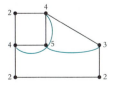

The graph is not connected.

A-1

25.

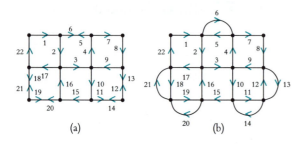

(a)   (b)

35.

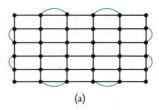

(a)

(b)

27.

29.

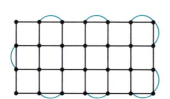

31.

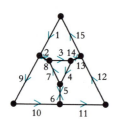

37.

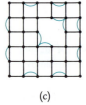

(a)

33.

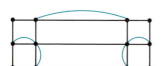

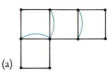

(b)   (c)

39.

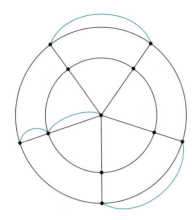

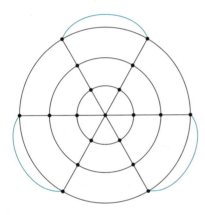

41. Yes, because every vertex has even valence and each street is represented by two edges.

43. Discussion. Answers will vary.

45. Discussion. Answers will vary.

CHAPTER 2

1. a. $X_1, X_6, X_5, X_2, X_3, X_4, X_1$.
  b. $X_1, X_6, X_7, X_8, X_9, X_{10}, X_{11}, X_{12}, X_5, X_4, X_3, X_2, X_1$.
  c. $X_1, X_4, X_5, X_8, X_9, X_6, X_7, X_2, X_3, X_1$.
  d. $X_1, X_2, X_5, X_8, X_3, X_4, X_7, X_6, X_1$.
  e. $X_1, X_9, X_8, X_7, X_6, X_5, X_4, X_3, X_2, X_1$.

3. Other Hamiltonian circuits include $ABIGDCEFHA$ and $ABDCEFGIHA$.

5. a. Add edge AB.   **b.** Add edge $X_1 X_7$.

7. a. There is no Hamiltonian circuit. Any Hamiltonian circuit would have to use edges $X_1 X_5$, $X_1 X_2$, and $X_1 X_4$. This would force X_1 to be revisited.
  b. There is no Hamiltonian circuit. Any Hamiltonian circuit would have to use edges $X_1 X_2$, $X_2 X_3$, $X_{10} X_{11}$, and $X_{11} X_{12}$. To visit X_6 and X_7 would require a revisit of X_2 or X_{11}.

9. a. For any $m \geq 2$ and $n \geq 1$, the graph has a Hamiltonian circuit.
  b. If either m or n is odd, the graph has a Hamiltonian circuit. If both m and n are even, the graph has no Hamiltonian circuit.
      A real-world application would be to design an efficient route to check that the traffic control equipment at each vertex was in proper working order.

11. a. No Euler circuit; Hamiltonian circuit.
  b. Euler circuit; Hamiltonian circuit.
  c. No Euler circuit; Hamiltonian circuit.
  d. Euler circuit; no Hamiltonian circuit.

13. Examples include inspection of traffic control devices at corners and placing new hour stickers on mailboxes located at street corners.

15. $9(9)(9)(9)(9) = 59{,}049$.

17. a. $26(26)(26)(10)(10)(10) - (26)(26)(26) = (26)^3(10^3 - 1)$.
  b. Answers vary.

19. $5 \times 10 \times 8 = 400$; $5 \times 10 \times 5 = 250$.

21. $5! = 120$, $6! = 720$, $7! = 5040$, $8! = 40{,}320$, $9! = 362{,}880$, $10! = 3{,}628{,}800$. The number of TSP tours in a 10-vertex complete graph is $181{,}440$.

23. She should follow the route FMCRF (or FRCMF), which takes 32 minutes.

25. a. $ACBDA$ (nearest neighbor). $ADBCA$ or $ACBDA$ (sorted edges).
  b. $ABCDA$ (nearest neighbor). $ABDCA$ (sorted edges).
  c. $ADBCEA$ (nearest neighbor). $ADCBEA$ (sorted edges).

27. A traveling salesman problem.

29. The complete graph shown has a different nearest-neighbor tour that starts at A ($AEDBCA$), a sorted-edges tour ($AEDCBA$), and a cheaper tour ($ADBECA$).

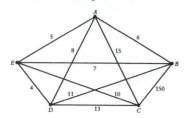

31. Graphs (b) and (d) are trees.

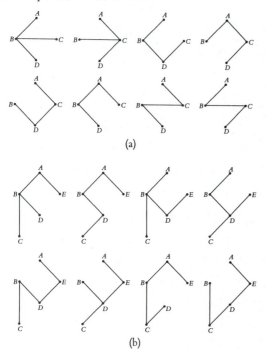

(a)

(b)

33.

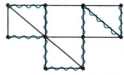

(c)

35. The wiggled edges in the figure constitute a spanning tree.

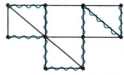

37. In drawing the graph model, do not include edges between two vertices where there is a hill of more than 800 feet between the towns represented by the vertices.

39. Yes. Change all the weights to negative numbers and apply Kruskal's algorithm. The resulting tree works, and the maximum cost is the negative of the answer you get. If the numbers on the edges represent subsidies for using the edges, one might be interested in finding a maximum-cost spanning tree.

41. A negative weight on an edge is conceivable, perhaps a subsidization payment. Kruskal's algorithm would still apply.

43. a. True. **b.** False. **c.** True. **d.** False.

45. Two different trees with the same cost are shown:

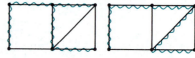

47.

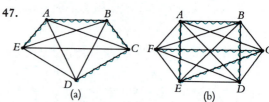

(a)   (b)

49. a. Critical path: $T_3 T_6 T_7$. Earliest completion time is 31.

b. There are two critical paths: $T_1 T_3 T_6 T_8$ and $T_1 T_3 T_5 T_7$. Earliest completion time is 35.

c. Critical path: $T_2 T_5 T_7$. The earliest completion time is 29.

51. The critical path is $T_1 T_5 T_7$. Hence, if T_1, T_5, or T_7 are shortened, the earliest completion time will decrease, while if T_2, T_3, T_4, or T_6 are shortened, the earliest completion time will not decrease. If T_5 is shortened to 7, the earliest completion time is 28 since $T_1 T_4 T_7$ is now the critical path.

53. One possibility is (times in minutes):

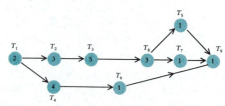

The earliest completion time is 16 minutes.

55. T_2, T_5, T_8.

57.

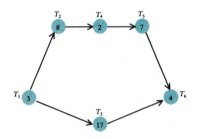

CHAPTER 3

1. Jacelyn must pack, get to the airport, make various connections, perhaps kennel her dog, etc. Processors include plane, bus, taxi (to get to the airport), and Jacelyn herself. Unless she can get a friend to help her pack or take her dog to the kennel, none of the tasks can be done simultaneously.

3. a. Operating room schedules, doctor schedules, emergency room staffing schedules, etc.
  b. Schedules for the trains or buses and their crews, etc.
  c. Scheduling runway use, reservation agents, food service for planes, etc.
  d. Schedules for each mechanic, radiator repair, etc.
  e. Schedules for drilling, welding, etc.
  f. Schedules for washing clothes, cleaning rooms, dusting, etc.
  g. Schedules for bus run, recess duty, etc.
  h. Day, night, afternoon shift schedules, etc.
  i. Firefighter shift schedules, schedules for checking if equipment on trucks is in repair, etc.

5. a. Processor 1: T_1, T_6, idle 13 to 15; T_5, T_7, T_{11}, idle 34 to 38; T_{10}. Processor 2: T_2, T_9, idle 21 to 27; T_8, idle 38 to 45. Processor 3: T_3, T_4, idle 15 to 45.
  b. Processor 1: $T_1, T_3, T_4, T_6, T_7, T_9, T_{11}$, idle 42 to 45. Processor 2: T_2, T_5, T_8, T_{10}.

7. a. The critical path, which has length 17, is $T_1 T_2 T_3$.
  b. $T_1, T_4, T_5, T_2, T_6, T_7, T_3$ is the list to be used. The one processor would have the tasks scheduled on it: $T_1, T_4, T_5, T_2, T_6, T_7, T_3$.

  c. $T_6, T_1, T_7, T_2, T_4, T_3, T_5$ would be the list. The resulting schedule on one processor would be: T_1, $T_7, T_4, T_2, T_5, T_6, T_3$.
  d. No idle time. Their completion times are the same.
  e. Earlier completion of tasks giving rise to cash payments.
  f. The required schedule is: Processor 1: T_1, T_2, T_7; Processor 2: T_4, T_5, T_6, T_3, idle 19 to 21.
  g. The completion time does not halve. As the number of processors goes up, the completion time may decrease, but at some point the length of the critical path will govern the completion time rather than the number of processors.

9. Such criteria include: decreasing length of the times of the tasks, order of size of financial gains when each task is finished, and increasing length of the times of the tasks.

11. a. Identical machines, typists who type the same number of words per minute, etc.
  b. Runways at an airport in different directions, humans with different levels of skills, etc.

13. a. Task times: $T_1 = 3, T_2 = 3, T_3 = 2, T_4 = 3, T_5 = 3, T_6 = 4, T_7 = 5, T_8 = 3, T_9 = 2, T_{10} = 1, T_{11} = 1$, and $T_{12} = 3$. This schedule would be produced from the list: $T_1, T_3, T_2, T_5, T_4, T_6, T_7, T_8, T_{11}$, T_{12}, T_9, T_{10}.
  b. Task times: $T_1 = 3, T_2 = 3, T_3 = 3, T_4 = 2, T_5 = 2, T_6 = 4, T_7 = 3, T_8 = 5, T_9 = 8, T_{10} = 4, T_{11} = 7, T_{12} = 9, T_{13} = 3$. This schedule would be produced from the list: $T_1, T_5, T_7, T_4, T_3, T_6, T_{11}, T_8$, $T_{12}, T_9, T_2, T_{10}, T_{13}$.

15. a. One reasonable possibility is (time in min):

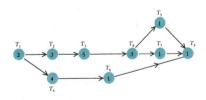

  The earliest completion time is 16.
  b. The decreasing-time list is: $T_3, T_4, T_2, T_8, T_1, T_5$, T_6, T_7, T_9. The schedule is: Processor 1: T_1, T_4, T_3, T_8, T_5, T_9; Processor 2: idle 0 to 2, T_2, T_3, idle 10 to 13, T_7, idle 14 to 15.

17. The task times total to 34. Since $(34/3)$ rounded up is 12, the earliest completion time is 12.

19. a. One such rule, admittedly artificial, could be: if by keeping a machine voluntarily idle, there is a longer task that becomes ready one time unit later, keep the machine idle.

b. If a longer task becomes ready at a certain time than the remaining time on a task currently scheduled, schedule the longer task and reschedule the interrupted task later. An assumption must be made whether or not an interrupted task must be resumed on the machine it was originally scheduled on or can be rescheduled on any machine that becomes free.

21. a. The tasks are scheduled on the machines as follows: Processor 1: 12, 13, 45, 34, 63, 43, 16, idle 226 to 298; Processor 2: 23, 24, 23, 53, 25, 74, 76; Processor 3: 32, 23, 14, 21, 18, 47, 23, 43, 16, idle 237 to 298.

b. The tasks are scheduled on the machines as follows: Processor 1: 12, 24, 14, 34, 25, 23, 16, 16, 76, idle 183 to 240; Processor 2: 23, 23, 21, 63, 43, idle 173 to 240; Processor 3: 32, 23, 53, 74, idle 182 to 240; Processor 4: 13, 45, 18, 47, 43, idle 166 to 240.

c. The decreasing-time list is: 76, 74, 63, 53, 47, 45, 43, 43, 34, 32, 25, 24, 23, 23, 23, 23, 21, 18, 16, 16, 14, 13, 12.

The tasks are scheduled on three machines as follows: Processor 1: 76, 45, 43, 24, 23, 18, 16, 13; Processor 2: 74, 47, 34, 32, 23, 21, 14, 12, idle 257 to 258; Processor 3: 63, 53, 43, 25, 23, 23, 16, idle 246 to 258.

The tasks are scheduled on four machines as follows: Processor 1: 76, 43, 24, 23, 16, idle 182 to 194; Processor 2: 74, 43, 25, 23, 16, 13; Processor 3: 63, 45, 32, 23, 18, 12, idle 193 to 194; Processor 4: 53, 47, 34, 23, 21, 14, idle 192 to 194.

d. The new decreasing-time list is: 84, 82, 71, 61, 55, 45, 43, 43, 34, 32, 25, 24, 23, 23, 23, 23, 21, 18, 16, 16, 14, 13, 12.

The tasks are scheduled as follows: Processor 1: 84, 45, 43, 25, 23, 23, 16, 12; Processor 2: 82, 55, 34, 32, 23, 18, 14, 13; Processor 3: 71, 61, 43, 24, 23, 21, 16, idle 259 to 271.

23. Examples include times to insert different chips into a circuit board, scheduling classes at a college, and putting dust jackets onto books of different sizes as part of a book manufacturing process.

25. Examples include jobs in a video tape copying shop, data entry tasks in a computer system, scheduling non-emergency operations in an operating room. These situations may have tasks with different priorities, but there is no physical reason for the tasks not to be independent, as would be the case with putting on a roof before a house had walls erected.

27. a. Each task heads a path of length equal to the time to do that task.

b. (1) The worst finish time is $(2 - \frac{1}{3})(450) = 750$.

(2) The worst finish time, if the decreasing-time list is used, is $[\frac{4}{3} - 1/(3)(3)](450) = 550$.

29. The times to photocopy the manuscripts, in decreasing order, are: 120, 96, 96, 88, 80, 76, 64, 64, 60, 60, 56, 48, 40, 32. Packing these in bins of size 120 yields: Bin 1: 120; Bin 2: 96; Bin 3: 96; Bin 4: 88, 32; Bin 5: 80, 40; Bin 6: 76; Bin 7: 64, 56; Bin 8: 64, 48; Bin 9: 60, 60. Nine photocopy machines are needed to finish within 2 minutes. The number of bins would not change, but the placement of the items in the bins would differ.

31. a. (1) The schedule with four secretaries is as follows: Processor 1: 25, 36, 15, 15, 19, 15, 27; Processor 2: 18, 32, 18, 31, 30, 18; Processor 3: 13, 30, 17, 12, 18, 16, 16, 16, 14; Processor 4: 19, 12, 25, 26, 18, 12, 24, 9.

The schedule with five secretaries is as follows: Processor 1: 25, 25, 31, 12, 16, 14; Processor 2: 18, 12, 17, 12, 15, 30, 9; Processor 3: 13, 32, 26, 16, 15, 18; Processor 4: 19, 36, 18, 19, 24; Processor 5: 30, 18, 15, 18, 16, 27.

(2) The decreasing-time list is: 36, 32, 31, 30, 30, 27, 26, 25, 25, 24, 19, 19, 18, 18, 18, 18, 18, 17, 16, 16, 16, 15, 15, 15, 14, 13, 12, 12, 12, 9.

The schedule using this list on four processors would be: Processor 1: 36, 25, 19, 18, 17, 16, 13, 9; Processor 2: 32, 26, 25, 18, 16, 15, 12; Processor 3: 31, 27, 24, 18, 16, 15, 12, 12; Processor 4: 30, 30, 19, 18, 18, 15, 14.

The schedule using this list on five processors would be: Processor 1: 36, 24, 18, 16, 14, 12; Processor 2: 32, 25, 18, 18, 15, 12; Processor 3: 31, 25, 19, 18, 15, 9; Processor 4: 30, 27, 18, 17, 15, 13; Processor 5: 30, 26, 19, 16, 16, 12.

(3) The five-processor decreasing-time schedule

is optimal (time 120), but the four decreasing-time schedule is not. One can see this since when the task of length 17 scheduled on processor 1 and the task of length 18 on processor 3 are interchanged, the completion time is reduced to 154 from 155 for the four-processor decreasing-time schedule.

b. As a bin-packing problem, each bin will have a capacity of 60. Using the decreasing list we obtain the following packings:

(1) (First-fit decreasing): Bin 1: 36, 24; Bin 2: 32, 27; Bin 3: 31, 26; Bin 4: 30, 30; Bin 5: 25, 25, 9; Bin 6: 19, 19, 18; Bin 7: 18, 18, 18; Bin 8: 18, 17, 16; Bin 9: 16, 16, 15, 13; Bin 10: 15, 15, 14, 12; Bin 11: 12, 12.

(2) (Next-fit decreasing): Bin 1: 36; Bin 2: 32; Bin 3: 31; Bin 4: 30, 30; Bin 5: 27, 26; Bin 6: 25, 25; Bin 7: 24, 19; Bin 8: 19, 18, 18; Bin 9: 18, 18, 18; Bin 10: 17, 16, 16; Bin 11: 16, 15, 15; Bin 12: 15, 14, 13, 12; Bin 13: 12, 12, 9.

(Best-fit decreasing): Bin 1: 36, 24; Bin 2: 32, 26; Bin 3: 31, 27; Bin 4: 30, 30; Bin 5: 25, 25; Bin 6: 19, 19, 18; Bin 7: 18, 18, 18; Bin 8: 18, 17, 16; Bin 9: 16, 16, 15, 12; Bin 10: 15, 15, 14, 13; Bin 11: 12, 12, 9.

(3) Since the total weight of all the objects is 596, a minimum of 10 bins is required. However, since there are no small weights, there is no way to achieve 10 bins, and in fact, 11 bins is optimal.

33. a. Using the next-fit algorithm, the bins are filled as follows: Bin 1: 12, 15; Bin 2: 16, 12; Bin 3: 9, 11, 15; Bin 4: 17, 12; Bin 5: 14, 17; Bin 6: 18; Bin 7: 19; Bin 8: 21; Bin 9: 31; Bin 10: 7, 21; Bin 11: 9, 23; Bin 12: 24; Bin 13: 15, 16; Bin 14: 12, 9, 8; Bin 15: 27; Bin 16: 22; Bin 17: 18.

b. The decreasing list is: 31, 27, 24, 23, 22, 21, 21, 19, 18, 18, 17, 17, 16, 16, 15, 15, 15, 14, 12, 12, 12, 12, 11, 9, 9, 9, 8, 7.

The next-fit decreasing schedule is: Bin 1: 31; Bin 2: 27; Bin 3: 24; Bin 4: 23; Bin 5: 22; Bin 6: 21; Bin 7: 21; Bin 8: 19; Bin 9: 18, 18; Bin 10: 17, 17; Bin 11: 16, 16; Bin 12: 15, 15; Bin 13: 15, 14; Bin 14: 12, 12, 12; Bin 15: 12, 11, 9; Bin 16: 9, 9, 8, 7.

c. The best-fit schedule using the original list is: Bin 1: 12, 15, 9; Bin 2: 16, 12; Bin 3: 11, 15; Bin 4: 17, 12; Bin 5: 14, 17; Bin 6: 18, 7; Bin 7: 19, 9; Bin 8:

21, 15; Bin 9: 31, Bin 10: 21, 9; Bin 11: 23, 8; Bin 12: 24; Bin 13: 16, 12; Bin 14: 27; Bin 15: 22; Bin 16: 18.

d. The best-fit decreasing schedule would be: Bin 1: 31; Bin 2: 27, 9; Bin 3: 24, 12; Bin 4: 23, 12; Bin 5: 22, 12; Bin 6: 21, 15; Bin 7: 21, 15; Bin 8: 19, 17; Bin 9: 18, 18; Bin 10: 17, 16; Bin 11: 16, 12, 8; Bin 12: 15, 14, 7; Bin 13: 11, 9, 9.

35. The bins have a capacity of 120. (First-fit): Bin 1: 63, 32, 11; Bin 2: 19, 24, 64; Bin 3: 87, 27; Bin 4: 36, 42; Bin 5: 63. This schedule would take five station breaks, however, the total time for the breaks is under 8 minutes.

The decreasing list is: 87, 64, 63, 63, 42, 36, 32, 27, 24, 19, 11. (First-fit decreasing): Bin 1: 87, 32; Bin 2: 64, 42, 11; Bin 3: 63, 36, 19; Bin 4: 63, 27, 24. This solution uses only four station breaks.

37. Discussion. Answers will vary.

39. The sum of the integers from 1 to n is $n(n + 1)/2$. Hence, the sum of the numbers from 1 to 20 is 210. Since each weight occurs twice, the weights total 420. Hence, at least 17 bins are required since the capacity of each bin is 25. For each weight occurring three times, at least 26 bins are required.

41. Such a heuristic will fill many bins to capacity, but the computation to find numbers summing to exactly W may be very time-consuming.

43. a. Packing boxes of the same height into crates; packing want ads into a newspaper page.

b. We assume, without loss of generality, $p \geq q$. One heuristic, similar to first-fit, orders the rectangles $p \times q$ as in a dictionary (i.e., $p \times q$ listed prior to $r \times s$ if $p > r$ or $p = r$ and $q \geq s$). It then puts the rectangles in place in layers in a first-fit manner; that is, do not put a rectangle into a second layer until all positions on the first layer are filled. However, extra room in the first layer is "wasted."

c. The problem of packing rectangles of width 1 in an $m \times 1$ rectangle is a special case of the two-dimensional problem, equivalent to the bin-packing problem we have discussed.

d. Two 1×10 rectangles cannot be packed into a 5×4 rectangle, even though there would be an area of 20 in this rectangle.

45. There is an example of a bin-packing problem for which a given list takes a certain number of bins, and when an item is deleted from the list, more bins are required. In this example, the deleted item is not first in the list.

CHAPTER 4

1.

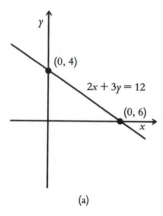

(a)

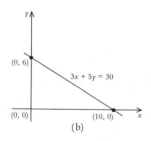

(b)

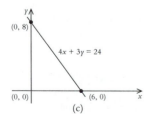

(c)

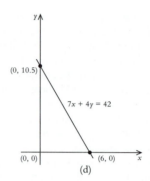

(d)

3.

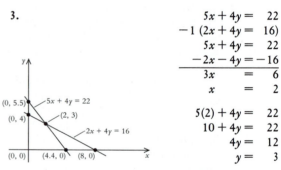

$$5x + 4y = 22$$
$$-1\,(2x + 4y = 16)$$
$$5x + 4y = 22$$
$$-2x - 4y = -16$$
$$\overline{3x = 6}$$
$$x = 2$$

$$5(2) + 4y = 22$$
$$10 + 4y = 22$$
$$4y = 12$$
$$y = 3$$

(2, 3) is point of in-
tersection.

5.

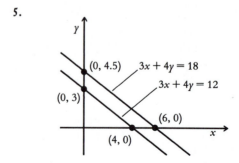

There is no point of intersection because these lines are parallel. The coefficients of x and y are the same in both equations, 3 for x and 4 for y.

7.

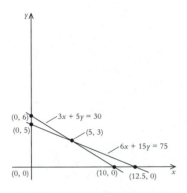

$$2(3x + 5y) = 30 \quad \text{gives} \quad 6x + 10y = 60$$
$$-1(6x + 15y) = 75 \quad \text{gives} \quad -6x - 15y = -75$$
$$\text{adding:} \quad -5y = -15$$
$$\text{so:} \quad y = 3$$

$$3x + 5(3) = 30$$
$$3x = 15$$
$$x = 5$$

Point of intersection is $(5, 3)$.

9.

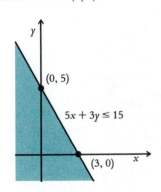

11.

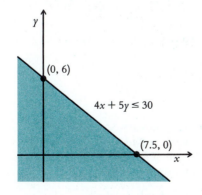

13.

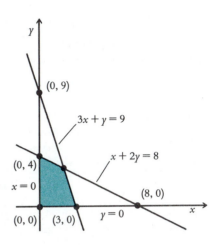

15.

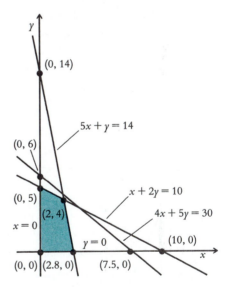

17. a. No. **b.** Yes. **c.** Yes. **d.** No. **e.** No.

19. a. No. **b.** Yes. **c.** Yes. **d.** No. **e.** Yes.

21. $3(20) + 4(25) = 60 + 100 = 160;$ $\quad 3(0) + 4(50) =$ $0 + 200 = 200; 3(30) + 4(10) = 90 + 40 = 130.$

23. $4x + 2y \leq 28.$

25. $6x + 4y \leq 240.$

27. a. $(0, 50) = 0$ gallons cranapple and 50 gallons apple-berry.

$(50, 25) = 50$ gallons cranapple and 25 gallons appleberry.

$(200/3, 0) = 200/3$ gallons cranapple and 0 gallons appleberry.

$(0, 0) = 0$ gallons of each.

b. $(50, 25)$; $(0, 50)$; $(200/3, 0)$.

29. a.

Mixture chart	High octane (500 gal)	Low octane (600 gal)	Profit
Premium gas (x gal)	.5 gal	.5 gal	$0.40/ gal
Regular gas (y gal)	.25 gal	.75 gal	$0.30/ gal

b.   Constraint inequalities

$$.5x + .25y \leq 500$$
$$.5x + .75y \leq 600$$
$$x \geq 0 \text{ and } y \geq 0$$

Profit formula
$$0.40x + 0.30y$$

c. Feasible region

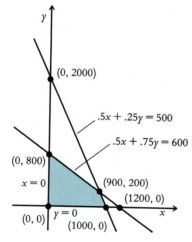

d. Profit

At $(0, 0)$, profit $= \$0$.
At $(0, 800)$, profit $= \$240$.
At $(900, 200)$, profit $= \$420$.
At $(1000, 0)$, profit $= \$400$.

Make 900 premium and 200 regular.

31. a.

Mixture chart	Oven (12)	Preparation/ decoration (16)	Profit
Bread, x	1.5	1	$0.50
Cake, y	1	2	$2.50

b.   Constraint inequalities

Oven: $1.5x + 1y \leq 12$
Preparation: $1x + 2y \leq 16$
Mins: $x \geq 0, y \geq 0$

Profit formula
$$0.50x + 2.50y$$

c. Feasible Region

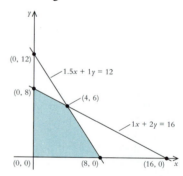

$$
\begin{array}{llll}
(2)(1.5x + 1y = 12) & \text{gives} & 3x + 2y = & 24 \\
(-1)(1x + 2y = 16) & \text{gives} & -1x - 2y = & \sim 16 \\
& \text{adding} & 2x & = \quad 8 \\
& \text{so} & x & = \quad 4
\end{array}
$$

Substitute: $4 + 2y = 16$ so $y = 6$, the point is $(4, 6)$.

d. Profit

At $(0, 0)$, profit $= \$0$.
At $(0, 8)$, profit is $\$20$.
At $(4, 6)$, profit is $\$17$.
At $(8, 0)$, profit is $\$4$.

Make 0 breads and 8 cakes.

33. a.

Mixture chart	Beef (500 lb)	Pork (300 lb)	Grain filler (400 lb)	Profit
Beef hot dogs (x packages)	1 lb	0 lb	0 lb	$0.80/ package
Regular hot dogs (y packages)	.25 lb	0.5 lb	.25 lb	$0.70/ package

b.   **Constraint inequalities**

$$x + .25y \le 500$$
$$0.5y \le 300$$
$$.25y \le 400$$
$$x \ge 0 \text{ and } y \ge 0$$

Profit formula
$$0.80x + 0.70y$$

c. Feasible Region

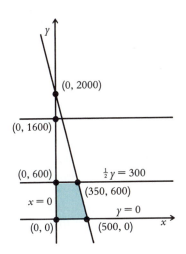

d. Profit

At (0, 0), profit = $0.
At (0, 600), profit = $420.
At (350, 600), profit = $700.
At (500, 0), profit = $400.

Make 350 beef and 600 regular.

35. a.

Mixture chart	Nurse (6,250 min)	Doctor (11,000 min)	Laboratory (5,000 min)	Profit
Routine visit (x times)	10 min	5 min	5 min	$30/ visit
Comprehensive visit (y times)	5 min	25 min	10 min	$50/ visit

b.   **Constraint inequalities**

$$10x + 5y \le 6{,}250$$
$$5x + 25y \le 11{,}000$$
$$5x + 10y \le 5{,}000$$
$$x \ge 0 \text{ and } y \ge 0$$

Profit formula
$$30x + 50y$$

c. Feasible Region

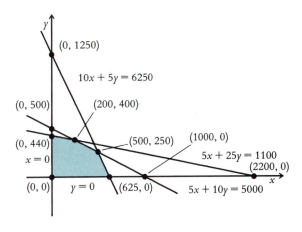

d. Profit

At (0, 0), profit = $0.
At (0, 440), profit = $22,000.
At (200, 400), profit = $26,000.
At (500, 250), profit = $27,500.
At (625, 0), profit = $18,750.

Schedule 500 routine and 250 comprehensive.

37. a.

Mixture chart	Bird count (100)	Cost $ (2,400)	Profit
Pheasants, x	1	20	$14
Partridges, y	1	30	$16

b.

Constraint inequalities

$$\text{count:}\quad 1x + 1y \le 100$$
$$\text{\$:}\quad 20x + 30y \le 2400$$
$$\text{mins:}\quad x \ge 0, y \ge 0$$

Profit formula
$$14x + 16y$$

c. Feasible Region

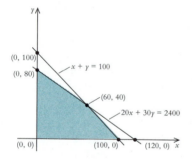

$$(20)(1x + 1y = 100) \text{ gives} \quad 20x + 20y = 2000$$
$$(-1)(20x + 30y = 2400) \text{ gives} -20x - 30y = -2400$$
$$\text{adding} \quad -10y = -400$$
$$\text{so} \quad y = 40$$

Substituting $1x + 40 = 100$ so $x = 60$, point is $(60, 40)$.

d. Profit

At $(0, 0)$, profit $= \$0$.
At $(0, 80)$, profit is $\$1280$.
At $(60, 40)$, profit is $\$1480$.
At $(100, 0)$, profit is $\$1400$.

Raise 60 pheasants and 40 partridges.

39. a.

Mixture chart	Shaper (50)	Smoother (40)	Painter (60)	Profit
Toy A, x	1	2	1	$4
Toy B, y	2	1	3	$5
Toy C, z	3	2	1	$9

b.

Constraint inequalities

$$\text{Shaper:}\quad 1x + 2y + 3z \le 50$$
$$\text{Smoother:}\quad 2x + 1y + 2z \le 40$$
$$\text{Painter:}\quad 1x + 3y + 1z \le 60$$
$$\text{mins:}\quad x \ge 0, y \ge 0, z \ge 0$$

Profit formula
$$4x + 5y + 9z$$

c. Optimal product policy

Make 5 of toy A, 0 of toy B, 15 of toy C for a profit of 155.

41. a.

	Lathe (50)	Grinder (36)	Polisher (81)	Profit
#1, w	10	6	4.5	$9
#2, x	5	6	18	$7
#6, y	2	2	1.5	$6
#8, z	1	2	6	$4

b.

Constraint inequalities

$$\text{Lathe:}\quad 10w + 5x + 2y + 1z \le 50$$
$$\text{Grinder:}\quad 6w + 6x + 2y + 2z \le 36$$
$$\text{Polisher:}\quad 4.5w + 18x + 1.5y + 6z \le 81$$
$$\text{mins:}\quad w \ge 0, x \ge 0, y \ge 0, z \ge 0$$

Profit formula
$$9w + 7x + 6y + 4z$$

c. Optimal production policy

Make 18 units of #6 and nothing else, for a profit of 108.

43. There would be four variables and six resource constraints.

45. A feasible region has infinitely many points; the corner point principle tells us we only need to evaluate the profit formula at a few of those points (the corner points), not all of them.

47. Answers vary.

49. Answers vary.

51. Answers vary.

CHAPTER 5

1. *Population:* All registered voters in the Second Congressional District. *Sample:* The 800 voters interviewed.

3. *Population:* All chips of this type made by the supplier, including future production. *Sample:* The 400,000 chips inspected.

5. Probably lower, because only households with phones were surveyed. Black households are poorer on the average than white households and so are more likely to lack a telephone. (More households have television sets than have telephones.)

7. a. RDD omits households without telephones (over 7% of U.S. households). **b.** Voter lists omit persons not registered to vote. Young adults and poor people are less likely to be registered than older or richer people.

9. If labels 00 to 29 are assigned to the 30 students in alphabetical order, the sample consists of 21 = Pirelli, 01 = Aspin, and 19 = Olds.

11. a. The answers obtained will depend on the parts of the table used. In the long run, we expect an average of 2 of the 5 tickets to go to women. **b.** A rough empirical answer is based on how many of your 20 samples included cases in which no women received tickets. (In fact, the probability that no tickets go to women is about .056. It is somewhat unlikely that no women will receive tickets.)

13. a. All 380 faculty members. **b.** Label the faculty members with three-digit labels, such as 000 to 379. Then look at three-digit groups in Table 5.1 to choose the sample. **c.** With labels 000 to 379, the first five members of the sample are the faculty members labeled 157, 274, 296, 001, and 338.

15. Ann Landers's poll relies on voluntary response. It attracts readers with strong feelings, especially those with

negative feelings toward their children. The random sample gives everyone the same chance, so is much more trustworthy. The voluntary response poll gives the result 70% "No" when the truth about the population is close to 90% "Yes." Such polls give *no* useful information about anyone except the actual respondents.

17. Forty-six percent of the sample believe in life on other planets. (A recent opinion poll found this result.) We can be confident that between 43% and 49% of all adults believe in extraterrestrial life.

19. The effect (if any) of the tea is confounded with the effect of visits and conversation with college students. The visits alone might make the residents more cheerful.

21. The unemployment rate is most strongly influenced by general economic conditions; it might have been still higher without the training program. The effect (if any) of the program on unemployment is confounded with the stronger effect of economic conditions.

23. No. The subjects had already been processed by the housing authority. They were not assigned to housing as part of the study. (Note that the effect of public housing on family stability is confounded with the effects of all factors that influenced the housing authority to accept some and reject others.)

25. Subjects who do not receive the drug should receive a placebo to avoid confounding the placebo effect with the effect of the drug. In the absence of reason to do otherwise, it is best to assign equal numbers of subjects to each treatment. The design is:

27. Because the diagnosis of mild heart attacks is somewhat subjective, it is best if the diagnosing physicians are blind in both experiments. The study of Exercise 19 should be double-blind; in Exercise 20c the subjects know whether or not they are in an exercise program, so they cannot be blind.

29. Randomly assign the 20 students into two groups of 10. One group uses the software that does trend analysis, the other uses software without this capability. Compare the average profits of the two groups. The design resembles Figure 5.5. To randomize, label the 20 students 00 to 19 (or 01 to 20) and use Table 5.1. The results will depend on which line of the table you use.

31. **a.** A direct comparison will eliminate the possibility that unusual nerve responses are due to something else in the diet or environment of the rats. **b.** The design is similar to Figure 5.4, with 10 rats in each group and diets with and without DDT as the treatments. If the rats are labeled 00 to 19, the DDT group contains rats 07, 10, 05, 00, 15, and so on. The randomization is tedious because half the rats must be chosen. The remaining 10 form the control group.

33. The outline is similar to Figure 5.4, with 10 subjects in each group and surgery and placebo as the treatments. Labeling the subjects 00 to 19, Group 1 consists of 03, 18, 07, 10, 04, 13, 08, 09, 00. The remaining subjects form Group 2.

35. There is variation among subjects in their response to the treatments and random variation in the results of the assignment to groups. Averaging this variation over a large number of subjects assures that there is very little systematic difference between the two groups, whereas two groups of 10 might by bad luck be quite different.

37. The average earnings of men exceeded those of women by so much that it is very unlikely that the chance selection of a sample would produce so large a difference if there were not a difference in the entire student population. But the black–white difference was small enough that it might be due to the accident by which students were chosen for the sample.

39. Discussion. Answers will vary.

41. **a.** The factors, or experimental variables, are type of corn (normal or floury-2) and protein level (12%, 16%, or 20%). **b.** The experimental units are 60 chicks; these are divided at random into six groups of ten chicks each; each group is fed one of the six diets; weight gains after 21 days are measured and compared. The outline is similar to Figure 5.5, but with six groups rather than three.

43. Here is a 3 × 3 Latin square, formed by starting with ABC and sliding each row one character to the left:

$$
\begin{array}{ccc}
A & B & C \\
B & C & A \\
C & A & B
\end{array}
$$

Here is a 4 × 4 Latin square formed in the same way:

$$
\begin{array}{cccc}
A & B & C & D \\
B & C & D & A \\
C & D & A & B \\
D & A & B & C
\end{array}
$$

CHAPTER 6

1. There are no outliers or other unusual features.

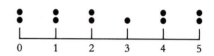

3. The distribution is roughly symmetric, with center near noon (12 hours from midnight). There are no outliers or gaps.

5. **a.** Your histogram will depend a bit on your choice of cells. Here is one choice.

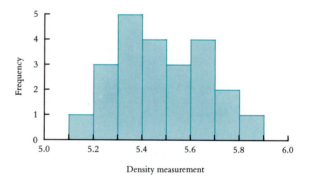

Density measurement

b. The distribution is roughly symmetric, with no outliers or other unusual features. (With only 23 observations, we cannot insist on a close approach to exact symmetry.)

7. **a.** $\bar{x} = 2.45$, $M = 2$. **b.** $Q_1 = 1$, $Q_3 = 4$.

9. **a.** The 1968 result, 43.4%, is a low outlier. In 1968 a third-party candidate drew over 13% of the vote.

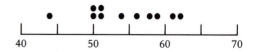

b. $M = 53.9\%$. **c.** $Q_3 = 58.8\%$ so that the 1964 (Johnson defeats Goldwater), 1972 (Nixon defeats McGovern), and 1984 (Reagan defeats Mondale) elections were landslides.

11. a. There are no outliers or other unusual features.

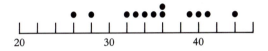

b. $\bar{x} = 35.3$, $M = 35.5$. **c.** $Q_1 = 32.5$, $Q_3 = 39.5$.

13. $M = 5.46$, $Q_1 = 5.34$, and $Q_3 = 5.63$; the five-number summary is 5.10, 5.34, 5.46, 5.63, 5.85. In a symmetric distribution, the two quartiles will fall about the same distance from the median, as will the two extremes. In this case, the quartiles are 0.17 and 0.12 from the median, and the extreme observations are 0.39 and 0.36 from the median.

15. a. Here is a histogram, using classes of width 2%.

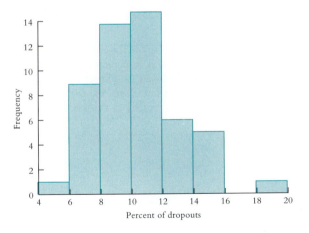

b. The distribution is very roughly symmetric, or at least not clearly skewed. There is an outlier: the District of Columbia, which is a city rather than a state. Dropout rates are higher in urban schools.

17. The median is $M = 10.2$ and the quartiles are $Q_1 = 8.7$ and $Q_3 = 11.9$. So $IQR = 3.2$. Both 4.3 and 19.1 are suspected outliers by the $1.5 \times IQR$ criterion.

19. Here, as an example, is the per capita consumption of fresh water per day for the 50 states (in alphabetical order across the rows). Sparsely populated western states that use large amounts of water for irrigation form the right tail of this skewed distribution, topped by Idaho's 22,200 gallons per person per day. The data are from the 1989 *Statistical Abstract of the United States*, Table 340.

2140	727	1960	2500	1420	4190	375
222	554	899	1100	22,200	1250	1470
960	2310	1130	2210	733	321	1070
1270	676	885	1210	10,500	6250	3860
688	307	2320	508	1260	1690	1180
386	2450	1210	152	2040	956	1770
1230	2540	235	853	1600	2810	1400
12,200						

21. Because a few very expensive houses pull up the mean but not the median, the mean is the larger of the two numbers.

23. A single high outlier is enough. For the data 1, 1, 2, 3, 3, 4, 28, the third quartile is 4 and the mean is $42/7 = 6$.

25. $s^2 = 15^2 = 225$.

27. The five-number summary of these data is 4.3%, 8.7%, 10.2%, 11.9%, 19.1%. Here is the boxplot.

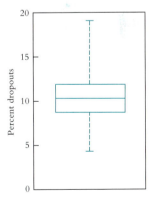

29. Choose the south to consist of: AL, AR, DC, FL, GA, KY, LA, MS, NC, SC, TN, and VA. (Other choices are possible). Then the five-number summary for these states is 10.4, 11.8, 12.8, 13.85, 19.1 and the five-number summary for the remaining 39 states is 4.3, 7.9, 9.6, 11, 14.9. As the boxplot shows, dropouts are much more common in the south.

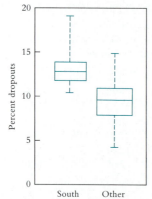

31. The five-number summary for males is 46.9, 47.4, 51.9, 62.0, 62.9; for females, it is 33.1, 38.2, 42.0, 49.55, 54.6. The generally higher lean body mass of males is apparent in the boxplots. The extremes in the male distribution do not extend far beyond the quartiles, but this is not surprising in a boxplot of only $n = 7$ observations.

33. Because both variables have the same units, it is best to use the same scale on both axes. U.S. prices, the explanatory variable, are plotted horizontally. There is a generally linear pattern, with many items costing much more in Japan. Cantaloupe falls outside the linear pattern; it costs much more in Japan than the overall pattern would suggest. Note that movie prices do fall in the overall linear pattern, although they are among the highest of those given in both countries.

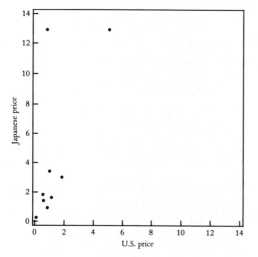

35. For $x = 3300$, $y = 4557$ pounds per square inch.

37. a. Higher. **b.** For $x = 422$, the prediction is $y = 461.7$. **c.** Hawaii's math score is higher (or its verbal score lower) than the general relationship suggests. Over 60% of Hawaii's population consists of Asians and Pacific islanders, so it is possible that less use of English may lower verbal scores.

39. a. There is a clear linear pattern. There are no pronounced outliers, although the first observation ($x = 7.2$, $y = 1.56$) lies a bit above the overall pattern. **b.** Use the "up and over" graphical method as in Figure 6.8b; the y obtained depends on the line drawn.

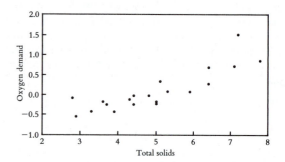

41. Use the "up and over" graphical method as in Figure 6.8b. The value predicted will depend on the line drawn in Exercise 40. Because the given x is almost three times as large as any of the x values in the data set, it refers to very different conditions about which our data say little.

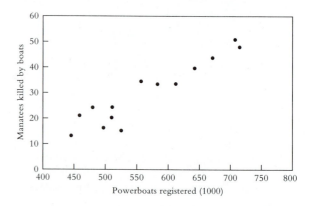

43. The basic sums are $\Sigma x = 1328$, $\Sigma y = 2317$, $\Sigma x^2 = 241{,}196$, and $\Sigma xy = 392{,}217$. The least squares regression line is $y = 58.588 + 1.304x$.

45. The basic sums are $\Sigma x = 99.3$, $\Sigma y = 2.4$, $\Sigma x^2 = 533.17$, and $\Sigma xy = 24.15$. The least squares regression line is $y = -1.393 + 0.3047x$. For $x = 4$, the prediction is $y = -0.174$.

CHAPTER 7

1, 3, 5. No answers provided for these exercises.

7. S can be taken to be all positive whole numbers, or you can choose a reasonable maximum, for example, S contains the whole numbers from 1 to 30.

9. $S = \{0, 1, 2, 3, \ldots\}$ (all nonnegative whole numbers) is the simplest choice because we do not have to decide a largest possible amount.

11. Models A and B are legitimate, because all probabilities are between 0 and 1 and their sum is 1. Model C has negative entries, so is not legitimate. Model D has entries with sum 1.5, so is not legitimate.

13. $P(AB) = 0.04$.

15. In a shuffled deck, all 13 possible outcomes are equally likely, so each has probability $\frac{1}{13}$. That is Model A. Model B is legitimate and would be correct if all the face cards were queens. Model C is not legitimate because the sum of the probabilities is greater than 1.

Outcome	Model A	Model B	Model C
Ace	$\frac{1}{13}$	$\frac{1}{13}$	$\frac{1}{10}$
King	$\frac{1}{13}$	0	$\frac{1}{10}$
Queen	$\frac{1}{13}$	$\frac{3}{13}$	$\frac{1}{10}$
Jack	$\frac{1}{13}$	0	$\frac{1}{10}$
Ten	$\frac{1}{13}$	$\frac{1}{13}$	$\frac{1}{10}$
Nine	$\frac{1}{13}$	$\frac{1}{13}$	$\frac{1}{10}$
Eight	$\frac{1}{13}$	$\frac{1}{13}$	$\frac{1}{10}$
Seven	$\frac{1}{13}$	$\frac{1}{13}$	$\frac{1}{10}$
Six	$\frac{1}{13}$	$\frac{1}{13}$	$\frac{1}{10}$
Five	$\frac{1}{13}$	$\frac{1}{13}$	$\frac{1}{10}$
Four	$\frac{1}{13}$	$\frac{1}{13}$	$\frac{1}{10}$
Three	$\frac{1}{13}$	$\frac{1}{13}$	$\frac{1}{10}$
Two	$\frac{1}{13}$	$\frac{1}{13}$	$\frac{1}{10}$

17. The probability that either A or B occurs is found by adding the probabilities of all outcomes that are either in A or in B. Because no outcome is in both A and B, there is no double counting if we first add the probabilities of outcomes in A and then separately add the probabilities of outcomes in B. So adding $P(A)$ and $P(B)$ is the same as adding the probabilities of all outcomes in either A or B.

19. The model is

Sum	2	3	4	5	6	7	8	9	10	11	12
Probability	$\frac{1}{36}$	$\frac{2}{36}$	$\frac{3}{36}$	$\frac{4}{36}$	$\frac{5}{36}$	$\frac{6}{36}$	$\frac{5}{36}$	$\frac{4}{36}$	$\frac{3}{36}$	$\frac{2}{36}$	$\frac{1}{36}$

a. $\frac{8}{36}$. **b.** $\frac{21}{36}$.

21. Repeats allowed: $(20 \times 20 \times 20)/(26 \times 26 \times 26) = 0.455$. No repeats allowed: $(20 \times 19 \times 18)/(26 \times 25 \times 24) = 0.438$.

23. No x: $(35 \times 35 \times 35)/(36 \times 36 \times 36) = 0.919$. No digits: $(26 \times 26 \times 26)/(36 \times 36 \times 36) = 0.377$.

25. There are $10^4 = 10,000$ possible PINs. Of these, $9^4 = 6561$ have no 0s. So the probability of at least one 0 is $3439/10,000 = 0.3439$.

27. The possibilities are *ags, asg, gas, gsa, sag, sga*, of which "gas" and "sag" are English words. The probability is $\frac{2}{6} = 0.33$.

29. $\frac{21}{6} = 3.5$.

31. $\frac{12}{8} = 1.5$.

33. a. The probabilities are all between 0 and 1 and have sum 1. **b.** $A = \{9, 10, 11, 12\}$, $P(A) = 0.931$. **c.** 11.251.

35. a. $\frac{12}{38} = 0.316$. **b.** $-\frac{2}{38} = -\$0.053$.

37. Group exercise; no answer provided.

39. Draw a normal curve, then mark the axis so that the center (mean) is at 69 and the change-of-curvature points are at 66.5 and 71.5. The curve should reach the horizontal axis at about 61.5 and 76.5.

41. The first quartile is 67.325 inches and the third quartile is 70.675 inches.

43. a. The median (same as the mean) is 10%. **b.** 9.6% to 10.4%. **c.** 9.866% to 10.134%.

45. a. 50%. **b.** 2.5% (half of 5%).

47. a. 234 and 298 days. **b.** Shorter than 234 days.

49. 245.5 days or shorter.

51. About 4.9% ($7/\sqrt{2}$).

53. 200 students (four times as many). Larger samples are more repeatable, that is, repeated samples would give closer to the same result than is the case for a smaller sample.

CHAPTER 8

1. 64.5 is a statistic, 63 a parameter.

3. Both are parameters.

5. Mean = 35%, standard deviation $\sigma_{\hat{p}} = 3.37\%$.

7. Mean = 3.0, standard deviation $\sigma_{\bar{x}} = 0.23$.

9. 1.18%, 1.26%, 1.29%, 1.26%, 1.18%.

11. If we repeated Gallup's sampling process we might get a different answer. But 95% of all samples will give an answer that is within ± 3% of the percent of all adults who jog. Gallup announces the margin of error to allow for this variation and indicate how accurate his result will usually be.

13. 59.3% ± 8.02%. Because the interval falls entirely above 50%, we are 95% confident that more than half of all visitors are in favor.

15. a. 2.53%. **b.** Gallup does not use a simple random sample, but rather a complex multistage sampling design. The margin of error for 95% confidence is somewhat larger for this sampling design than for a simple random sample of the same size.

17. Only **c.** The margin of error includes only the random sampling error described by the sampling distribution of the statistic.

19. The margin of error would be greater than ± 3 points. Smaller sample sizes result in wider intervals for the same level of confidence. (Because the poll did not use a simple random sample, we cannot calculate the margin of error.)

21. a. By the 68 – 95 – 99.7 rule, 68% of the observations in any normal distribution fall within ± 1 standard devia-

tion of the mean. For large samples, the sample proportion $\hat{p}$ follows approximately a normal distribution with mean p and standard deviation $\sqrt{p(100 - p)/n}$. So the unknown p is within this standard deviation of the observed $\hat{p}$ in 68% of all samples in the long run. Replacing the unknown p by $\hat{p}$ in the standard deviation changes its numerical value very little. So the unknown p falls in the interval given in 68% of all samples. That's what 68% confidence means. **b.** A 99.7% confidence interval is

$$\hat{p} \pm 3 \sqrt{\frac{\hat{p}(100 - \hat{p})}{n}}$$

23. 59.3% ± 4.01%. The 68% confidence interval is half as wide as the 95% confidence interval (one standard deviation rather than two) because a smaller margin of error is sufficient if we allow lower confidence in the result.

25. a. 18 ± 12 points. **b.** 18 ± 2.4 points.

27. 3.4137 ± 0.00115 grams.

29. 36.9 ± 1.74 points.

31. 0.7505 ± 0.00032 inch.

33. a. $\bar{x} \pm 1.28\sigma/\sqrt{n}$. **b.** 36.9 ± 1.11 points.

35. Center line = 5 grams, control limits = 5 ± 0.00173 grams.

37. The control charts in Exercises 37 to 39 have center line (drawn solid) at 101.5 and control limits (drawn dashed) at 101.2 and 101.8. Here there are no points outside the limits and no run of 8 or more on the same side of the center line.

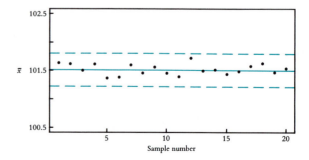

39. Samples 15, 18, and 19 fall above the upper control limit. The last six points lie above the center line. There is a clear upward drift in the plot.

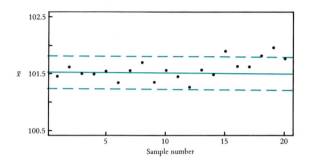

41. The percents of students who smoke for each parent condition are both parents smoke, 22.5%; one parent smokes, 18.6%; neither parent smokes, 13.9%. There is a clear association between parent smoking and student smoking, with students of smoking parents being more likely to smoke themselves.

43. a. The two-way table is

	Bill	Will
Hits	120	130
Outs	380	370
At bats	500	500

Bill gets a hit .240 of his times at bat, while Will's average is .260, so Will has the higher batting average. **b.** Bill hits better against right-handers (.400 vs. .300) and also against left-handers (.200 vs. .100). **c.** Both players hit much better against right-handers; Will bats against right-handers much more often than does Bill. So even though Bill does better against both types of pitchers than Will, Will has a higher average overall. We should choose Bill for our team.

Chapter 9

1. a. 51593-2067. **b.** 50347-0055.

3. If a double error in a block results in a new block that does not contain exactly two long bars, we know this block has been misread. If a double error in a block of five results

in a new block with exactly two long bars, the new block gives a different digit than the original one; if no other digit is in error, the check digit catches the error, since the sum of the ten digits will not end in 0. So, in every case an error has been detected. Errors of the first type can be corrected just as in the case of a single error. When a double error results in a legitimate code number, there is no way to determine which digit is incorrect.

5. No. The remainder upon dividing 2775042116 is 8.

7. 4.

9. March 29, female; September 17, male.

11. X.

13. 2.

15. +.

17. No. The computer only needs to know which digit is the check digit.

19. S000, S200, L550, L300, E663, O451.

21. M-263-4554-2258.

23. a. 7. **b.** 4. **c.** 7.

25. If you replace each short bar in Figure 9.7 by an a and each long bar in Figure 9.7 by a b, the resulting strings are listed in alphabetical order.

27. All Code 39 bar codes begin and end with the bar code for the character *. This character has the binary code 010010100. If the scanner picks up the code in this order, it is scanning left-to-right. If the scanner picks up the code in reverse order (that is, 001010010), it is scanning right-to-left.

29. 913*.

31. 0-669-09325-4.

33. Substitution of b for a where $|b - a| = 5$ in positions 1, 5, 7, 9, and 11 is undetected; all errors in position 3 are undetected; substitution of b for a where $b - a$ is even in position 8 is undetected; the transposition $ab \rightarrow ba$ increases the check digit by $(b - a)$ if $b < a$ and by $b - a + 10$ if $b < a$.

35. The check digit would be the same.

37. Z05428925245913640.

39. The bars signify that the envelopes are not postage-paid but do have the Postnet bar code.

41. They are the same.

43. Twins; sons named after their fathers (such as John L. Smith, Jr.); common names such as John Smith and William Johnson; states that do not include year of birth in the code.

47. Some examples are: libraries, hospitals, schools, large companies, and government agencies.

CHAPTER 10

1. Refer to pp. 304–305 if you have doubts about your answer.

3. a. 6. **b.** 3.

5. 1001101.

7. 000000, 100011, 010101, 001110, 110110, 101101, 011011, 111000.

9. 0000000, 1000001, 0100111, 0010101, 0001110, 1100110, 1010100, 1001111, 0110010, 0101001, 0011011, 1110011, 1101000, 1011010, 0111100, 1111101. No, since 1000001 has weight 2.

11. 000000, 100101, 010110, 001011, 110011, 101110, 011101, 111000. 001001 is decoded as 001101; 011000 is decoded as 111000; 000110 is decoded as 100110.

13. $2^5 = 32$.

15. a and b.

17. 00000000, 00010111, 00101110, 01001011, 10001101, 11000110, 10100011, 10011010, 01100101, 01011100, 00111001, 11101000, 11010001, 10110100, 01110010, 11111110. The code will detect any 3 errors or correct any single error.

19. 23, 49, 16.

21. 13.

23. The integer x is encrypted as $(x + 3) \bmod 26$.

25. 0000, 1012, 2021, 0111, 0222, 1120, 2210, 2102, 1201.

27. $3^4 = 81$, $3^6 = 729$.

29. 01111001110100010110110; ABAACBAAED.

31. t, m, and n; e.

33. In the Morse code a space is needed to determine where each code word ends. In a fixed-length code of length k, a word ends after each k digits.

CHAPTER 11

1. a. $3! = 6$. **b.** $4! = 24$. **c.** $n!$.

3. a. First (Not a) (60%) beats a (40%), and then b (75%) beats c (25%). **b.** First b (75%) beats c (25%), and then b (60%) beats a (40%). **c.** b wins with 3 (20%) + 3 (15%) + 2 (40%) + 2 (25%) = 2.35 points over 2.00 for a to 1.65 for c. **d.** b (60%) beats a (40%) and b (75%) beats c (25%). **e.** There are no possibilities for strategic voting in cases **a** and **c** (for secret or simultaneous voting). In case **b**, A may vote for c on the first ballot in an attempt to eliminate b, but this will fail if C votes for b on the first ballot (as in this case when described in the text). In case **d**, A could create a cycle (tie) in voting for c in the pairwise comparison between b and c.

5. a. First E (7 votes) beats H (3), and then D (7) wins over E (3). **b.** First D (7 votes) beats E (3), and then H (6) wins over D (4). **c.** Yes, in case **a**. They vote for H on the first (and second) ballots and their second choice H wins. No, in case **b**. **d.** With sincere voting, D wins (21 points) over E (20) and H (19). **e.** No.

7. Lolich, Munson, and Staub.

9. a. A.
  b. A (68 votes) beats B (32) in the runoff.
  c. In case **a**, A. In case **b**, C (55) beats A (45) in the runoff.
  d. If 6 (to 14) of these 45 voted B over A over C on the first ballot, then A beats B in the runoff.

11. a. W (242 points) beats M (238) beats A (120).
  b. M (238 points) beats W (204) beats A (158).

c. Not if they all ranked A second and M third. However, W wins if only 35 (to 43) of its group rank A second and M third.

13. a. D(27 points), B(22), C(14), A(12).
   b. D ties B (18 points), A(11), C(8).
   c. (1) D(27), B(21), C(15), A(12).
      (2) $D(18\frac{1}{3})$, $B(16\frac{2}{3})$, A(11), $C(9\frac{1}{3})$
   d. (1) D(27), B(17), C(16), A(15).
      (2) D(19), B(13), C(11), A(12).

15. a. D.
   b. A, B, D, F.
   c. A, B, D.
   d. A, B, D, F.

17. M should be selected so that two of the other three members prefer it most and the third prefers it least of all. Then M may beat M by 3 to 1 and 0 beats M by 2 to 2.

19. a. A wins when A breaks the 1-to-1-to-1 tie.
   b. C wins because B would vote insincerely for her second choice C rather than have her third choice A win.
   c. The chairman has additional power, but gets his least preferred outcome, C.

CHAPTER 12

1. a.

Voter	Combinations
A	Y Y Y Y Y Y Y Y Y N N N N N N N N
B	Y Y Y Y N N N N Y Y Y Y N N N N
C	Y Y N N Y Y N N Y Y N N Y Y N N
D	Y N Y N Y N Y N Y N Y N Y N Y N

   b. $\{A, B, C, D\}, \{A, B, C\}, \{A, B, D\}, \{A, B\}, \{A, C, D\}, \{A, C\}, \{A, D\}, \{A\}, \{B, C, D\}, \{B, C\}, \{B, D\}, \{B\}, \{C, D\}, \{C\}, \{D\}, \{\}$.
   c. Answers vary.
   d. (i) 1; (ii) 4; (iii) 6.

3. a. (1) $\{A\}$; (2) $\{A\}, \{A, B\}$; (3) $\{A\}$; (4) $\{A\}, \{A, B\}$; (5) None; (6) B.

b. (1) $\{A, B\}, \{A, C\}, \{B, C\}$; (2) $\{A, B\}, \{A, C\}, \{A, B, C\}$; (3) $\{A, B\}, \{A, C\}, \{B, C\}$; (4) $\{A, B\}, \{A, C\}, \{A, B, C\}$; (5) $\{A\}$; (6) None.
   c. (1) $\{A, B\}, \{A, C\}, \{B, C\}$; (2) $\{A, B\}, \{A, C\}, \{A, B, C\}$; (3) $\{A, B\}, \{A, C\}, \{B, C\}$; (4) $\{A, B\}, \{A, C\}, \{A, B, C\}$; (5) $\{A\}$; (6) None.
   d. (1) $\{A, B\}, \{A, C\}$; (2) $\{A, B\}, \{A, C\}, \{A, B, C\}$; (3) $\{A\}$; (4) $\{A, B\}, \{A, C\}, \{A, B, C\}$; (5) $\{A\}$; (6) None.
   e. (1) $\{A, B\}, \{A, C\}, \{B, C\}$; (2) $\{A, B\}, \{A, C\}, \{A, B, C\}$; (3) $\{A, B\}, \{A, C\}, \{B, C\}$; (4) $\{A, B\}, \{A, C\}, \{A, B, C\}$; (5) $\{A\}, \{A, D\}$; (6) D.
   f. (1) $\{A, B\}, \{A, C\}, \{B, C\}$; (2) $\{A, B\}, \{A, C\}, \{A, B, C\}$; (3) $\{A, B\}, \{A, C\}, \{B, C\}$; (4) $\{A, B\}, \{A, C\}, \{A, B, C\}$; (5) $\{A\}, \{A, D\}$; (6) D.
   g. (1) $\{A, B\}, \{A, C\}, \{B, C\}$; (2) $\{A, B\}, \{A, C\}, \{A, B, C\}$; (3) $\{A, B\}, \{A, C\}, \{B, C\}$; (4) $\{A, B\}, \{A, C\}, \{A, B, C\}$; (5) $\{A\}, \{A, D\}$; (6) D.
   h. (1) All coalitions with three voters; (2) all coalitions including A and at least two others; (3) all coalitions with at least three voters; (4) all coalitions including A and at least two others; (5) $\{A\}$, and all coalitions of A and at least one other voter; (6) None.

5. $\{A, X\}$, where $X = B, C$, or D; and $\{B, C, D\}$.

7. a. $BACD, BADC, CABD, CADB, DABC, DACB, BCAD, BDAC, CBAD, CDAB, DBAC, DCAB$.
   b. $ABCD, ABDC, CDBA, DCBA$.

9. $(\frac{1}{2}, \frac{1}{6}, \frac{1}{6}, \frac{1}{6})$.

11. a. (12, 4, 4, 4).
   b. (10, 6, 6, 2).

13. a. (24,8,8,0,0) (The Glen Cove and Long Beach supervisors are dummy voters.)
   b. (26,22,18,18,2,6).
   c. Answers vary.

15. a. In the following answer, a and b will always represent city officials (M, C, or P), d will represent a president of one of the larger boroughs (K or H), and y and z will represent presidents of the smaller boroughs (Q, X, or S). The minimal winning coalitions are of the forms $\{M, C, P\}, \{a, b, K, H\}, \{a, b, d, y\}, \{a, b, Q, X, S\}$, or $\{a, K, H, y, z\}$.
   b. No. In the notation used for the answer to part (i), the minimal blocking coalitions are of the forms $\{M, C, P\}, \{a, b, d\}, \{a, b, y, z\}, \{a, K, H, y\}$, or $\{a, d, Q, X, S\}$.

c. The city officials each have indices of 110, and K and H each have indices of 66, and Q, X, and S each have indices of 30.

17. a. $\{A, B\}, \{A, C\}, \{A, D\}$, and $\{B, C, D\}$.
b. A can sell 4 shares to B or C. A can only sell 1 share to E. However, A can sell 19 shares to D.
c. D can either sell 4 shares to B or C, or D can sell 1 share to A or E.
d. D can sell 2 shares to A, or 5 shares to B or C, or 4 shares to E.
e. 20 shares.

19. a. $[q:w(C), w(M_1), \ldots, w(M_8)] = [8:6, 1, 1, 1, 1, 1, 1, 1, 1]$.
b. $(492,16,16,16,16,16,16,16,16)$.
c. $(\frac{2}{3}, \frac{1}{24}, \ldots, \frac{1}{24})$.

21. (e).

23. a. Answers vary.
b. Each permanent member has a Banzhaf index of $1696 = 2 \times (C_4^{10} + C_5^{10} + C_6^{10} + C_7^{10} + C_8^{10} + C_9^{10} + C_{10}^{10})$, while an ordinary member has an index of $168 = 2 \times C_3^9$. By this index, a permanent member is about 10 times as powerful as an ordinary member.
c. An ordinary member A can only be a pivot in a permutation in which it is in the ninth position, preceded by all five permanent members and three ordinary members. There are C_6^9 ways to choose the six ordinary members to follow A in the permutation, and 6! ways to put them in order after choosing them. There are 8! ways to put the members preceding A in the permutation in order. By the multiplication principle, there are $C_6^9 \times 6! \times 8!$ permutations in which A is the pivot. Thus, A's Shapley Shubik index is

$$\frac{C_6^9 \times 6! \times 8!}{15!} = \frac{9! \times 8!}{3!15!} = \frac{4}{2145}.$$

Each ordinary member has the same index. Let x denote the index of a permanent member. Then

$$5x + 10 \times \frac{4}{2145} = 1,$$

so $x = \frac{421}{2145}$. Thus, by this index, a permanent member is over 100 times as powerful as an ordinary member.
d. Answers vary.

25. Each ordinary member gets one vote. The chairperson is a dummy voter, with 0 votes. The quota to pass a measure is 3.

27. If $n = 1$, the probability is 1. When $n = 2, 3, 4, 5, 6$, and 7, the probabilities are $\frac{1}{2}, \frac{1}{2}, \frac{3}{8}, \frac{3}{8}, \frac{5}{16}$, and $\frac{5}{16}$, respectively. If there are $2m$ voters, there are $2C_m^{2m-1}$ swings.

29. Answers vary.

31. Answers vary.

33. a. 16.
b. 6.
c. 10.
d. A has the same probability of being a swinger in either case. However, the probability that C will be a swinger is increased. This apparent paradox is resolved by noting that the quarrel makes the grand coalition in which the Senate votes unanimously impossible, and it also eliminates all but one of the four-member coalitions — in these coalitions there are no swingers. This increases the chance that there will be swingers, and the senators who were not involved in the quarrel benefit from that.

35. a. The Banzhaf index for the system (12.6), where E, F, and G are not allied, is $(16, 16, 8, 8, 8)$. The total Banzhaf index for the three senators is greater than the index of $\{A, B\}$ or $\{C, D\}$. If the senators form an alliance, that alliance will be just as powerful as $\{A, B\}$ of $\{C, D\}$.
b. For the system (12.7) the index is $(\frac{9}{30}, \frac{9}{30}, \frac{2}{15}, \frac{2}{15}, \frac{2}{15})$, so the total share of power held by E, F, and G, unallied, is $3 \times \frac{2}{15} = 40\%$. With the alliance, they have one-third of the power.

37.

n	Probability
40	0.021
100	0.00041
400	0.0000000000092

CHAPTER 13

1. Possible fair allocations for each table:

a.

1	1	(1)
(1)	0	0
0	(1)	0

b.

1	1	(1)
(1)	0	0
0	(1)	1

1	(1)	1
(1)	0	0
0	1	(1)

c.

1	(1)	1
(1)	0	0
1	1	(1)

1	1	(1)
(1)	0	0
1	(1)	1

d.

1	1	(1)
(1)	1	0
1	(1)	0

1	1	(1)
1	(1)	0
(1)	1	0

e.

1	(1)	1
(1)	1	0
1	0	(1)

1	1	(1)
1	(1)	0
(1)	0	1

(1)	1	1
1	(1)	0
1	0	(1)

f.

1	1	(1)
(1)	1	1
1	(1)	0

(1)	1	1
1	1	(1)
1	(1)	0

1	1	(1)
1	(1)	1
(1)	1	0

1	(1)	1
1	1	(1)
(1)	1	0

3. a.

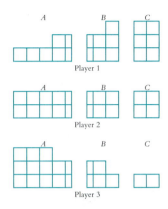

b. Player 2 finds *A* acceptable (9 square units), but not *B* (5 square units) or *C* (4 square units).

Player 3 finds *A* acceptable (12 square units), but not *B* (4 square units) or *C* (2 square units).

c. Players 2 and 3 both find *B* and *C* unacceptable. (*C* is on the right.)

d. (i) Assume *C* is given to Player 1. If Player 2 cuts the rest, he will make each piece 7 square units. Player 3 will choose the leftmost piece, which he thinks is 10 square units. Thus, Player 1 gets a piece he thinks is 6 square units, Player 2 gets a piece he thinks is 7 square units, and Player 3 gets a piece he thinks is 10 square units. (ii) If Player 3 cuts the rest, she will make each piece 8 square units. (This requires a vertical cut two-thirds of the way across the third triple of squares.) Player 2 will choose the leftmost piece, which she thinks is $8\frac{2}{3}$ square units. Thus, Player 1 gets a piece she thinks is 6 square units, Player 2 gets a piece she thinks is $8\frac{2}{3}$ square units, and Player 3 gets a piece she thinks is 8 square units.

5. a.

b. Player 2 will view A as being 9 square units, and will thus diminish it to yield A' as follows:

A'

c. Player 3 will view A' as being 9 square units, and will thus diminish it to yield A″ as follows:

A''

d. Player 3 receives the piece cut off the cake (A'') because she was the last to diminish it. She thinks it is 6 square units.

e. If Player 1 halves what remains, she will think each piece is 8 square units. Player 2 will choose the leftmost piece, which she thinks is $8\frac{2}{3}$ square units.

f. If Player 2 halves what remains, she will think each piece is 7 square units. Player 2 will choose the rightmost piece, which she thinks is 10 square units.

g. Player 1 will cut off 6 square units. Player 2, thinking 7 square units have been cut off, will trim 1 more square unit off it and receive this piece. Player 1 will receive what is left, which she thinks is 11 square units.

7. a.

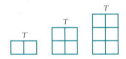

b.

X Y Z

$\frac{2}{3}$ square wide

c. Player 3 chooses any one of the three; he thinks they are all 2 square units.

Player 1 chooses either of the remaining two; he thinks both are $\frac{2}{3}$ square unit.

Player 2 receives the remaining piece; he thinks it is $\frac{4}{3}$ square units.

9. Player 4 chooses first, so he certainly envies no one. Player 3 created a two-way tie for largest, and at least one of these two pieces is still available after player 4 chooses. Player 2 created a three-way tie for largest. If all of these pieces were unavailable, then it is because two of the pieces were already chosen and the third destroyed by Player 3's trimming. But because of the proviso in step (6), the one trimmed piece would definitely have been one of those chosen. Hence, at least one of the three pieces is available to Player 2. Finally, Player 1 created a five-way tie for largest. At most two of these five pieces have been trimmed or chosen by Players 3 and 4. Moreover, two may have been trimmed by Player 2, and one chosen by Player 2. This appears to represent a potential of five pieces trimmed or chosen, but the proviso in step (5) guarantees that the set of two pieces trimmed by Player 2 cannot be disjoint from the set of pieces chosen. Hence, one of the original five pieces is untrimmed and unchosen. (In general, the total number of pieces chosen and trimmed by players n, $n - 1$, $n - 2$, $n - 3$, . . . is at most 1, 2, 4, 8, ...)

11. The total numbers of pieces trimmed and chosen by the different players are:

Player 5	1 piece (0 trimmed and 1 chosen)
Player 4	2 pieces (1 trimmed and 1 chosen)
Player 3	4 pieces (3 trimmed and 1 chosen)
Player 2	8 pieces (7 trimmed and 1 chosen)

Thus, Player 5 gets a piece he considers at least tied for largest, because he goes first. Player 4 also gets a tied piece, because he had a 2-way tie for largest and only one piece was chosen by Player 5 (and none was trimmed). Player 3 also gets a tied piece, because the number of pieces chosen and/or trimmed by Players 5 and 4 is at most $1 + 2 = 3$, and he had a 4-way tie for largest. Similarly, the number of pieces chosen and/or trimmed by Players 5, 4, and 3 is at most $1 + 2 + 4 = 7$, and Player 2 had an 8-way tie for largest. Finally, the number of pieces chosen and/or trimmed by Players 5, 4, 3, and 2 is at most $1 + 2 + 4 + 8 = 15$, and Player 1 had a 16-way tie for largest.

13. Player 1 will think the leftover from the eleventh stage is at most $(4/5)^{11}$ of the whole cake, which is approximately 0.086. Hence, this is less than one-tenth of the cake.

15. Sheila receives the Frisbee and gives Jean 69 cents.

17. Mary receives the house and the car and gives John $43,831.25.

19. a. A receives the farm plus $7,333.33; B receives $132,333.33; and C receives the house and sculpture and pays $139,666.66.   **b.** A receives the farm plus $80,000; B receives $120,000; and C receives the house and sculpture and pays $200,000.   **c.** A receives the farm plus $78,500; B receives $99,250; and C receives house and sculpture and pays $177,750.

21.   a. Heir I can divide the estate into 5 parts that she considers to be of 20 points value each as follows (where "$B + 5$" indicates a part consisting of the boat and 5 units of money):

$$B + 5 \quad C + 10 \quad F + 10 \quad P + 3 \quad A + D + 12$$

b. Heir II values the parts handed to her by Heir I as follows:

$$15 \quad 25 \quad 20 \quad 13 \quad 27$$

c. Heir II will trim the second and fifth part down to 20 units of value. Placing the trimmings in parentheses below the part they were trimmed from yields:

$$B + 5 \quad C + 5 \quad F + 10 \quad P + 3 \quad A + D + 5$$
$$(5) \qquad\qquad\qquad\qquad\qquad (7)$$

d. Heir III values the parts handed to her by Heir II as follows:

$$25 \quad 15 \quad 20 \quad 13 \quad 15$$

e. Heir III will trim the first part down to 20 units of value. Keeping all the trimmings in parentheses below, as before, yields:

$$B \quad C + 5 \quad F + 10 \quad P + 3 \quad A + D + 5$$
$$(5) \quad (5) \qquad\qquad\qquad\qquad (7)$$

f. Heir IV will now choose $F + 10$, which she thinks is 24 units of value.

g. Heir III will now choose B, which is one of the two tied for largest at a value to her of 20. Since she had trimmed this part and it was available, the proviso requires her to take it.

h. Heir II will now choose either $C + 5$ or $A + D + 5$ at a value to her of 20. Since Heir II really does not want the dog, she will choose $C + 5$.

i. Heir I will now choose $P + 3$ at a value to her of 20.

j. We have not yet distributed the fifth part $(A + D + 5)$ or the trimmings (a total of 17 units of money). Thus, we have left to distribute: the dog, the art, and 22 units of money. Selling the art for 10 units of money leaves the dog and 32 units of money.

k. Heir I can divide the remainder of the estate into 5 parts that she considers to be of 7 points value each as follows:

$$D + 4 \quad 7 \quad 7 \quad 7 \quad 7$$

l. Heir II already thinks there is at least a 3-way tie for largest (in fact, 4-way), as does Heir III. Hence, no trimming will be done.

m. Heir IV will take $D + 4$. The others will each take a part consisting of 7 units of money.

n. The only thing left is 7 units of money, and it can be divided equally among the 4 heirs.

23. Mary receives the house and car, and gives John $45,600.

25. a. A receives the farm plus $7,333.33; B receives $142,333.33; and C receives the house and the sculpture and pays $149,666.66.   **b.** A receives the farm plus $78,500; B receives $128,100; and C receives the house and sculpture and pays $206,600.   **c.** A receives the farm plus $78,500; B receives $106,750; and C receives the house and sculpture and pays $185,250.

27. Answers vary.

CHAPTER 14

1. The Hamilton apportionment is

$$14 + 33 + 27 + 14 + 12 = 100.$$

3.

Party	Vote	Quota	Lower Quota	Appor-tionment
Social Democrats	323829	23.0369	23	23
Democratic Socialists	880702	62.6524	62	63
Christian Democrats	5572614	396.4309	396	396
Greens	1222498	86.9674	86	87
Communists	111224	7.9124	7	8
Totals	8110867	577.0000	574	577

5. Discussion. Answers will vary.

7. The Alabama paradox occurs when the apportionment for the smallest state decreases from 2 to 1 as the house size increases from 83 to 84, and it occurs again as the house size increases from 89 to 90.

State	Population	Apportionments					
A	5,576,330	25	26	26	27	28	28
B	1,387,342	6	6	6	7	7	7
C	3,334,241	15	15	16	16	17	17
D	7,512,860	34	34	35	37	37	38
E	310,968	2	2	1	2	1	1
Total	18,121,741	82	83	84	89	90	91

9.

College	Hamilton	Jefferson	Webster	Adams
Arts	2	3	2	2
Science	2	1	2	2
Business	1	1	1	1

11.

State	Population	Hamilton	Jefferson	Webster	Adams
A	27774	10	11	10	10
B	25178	9	9	9	9
C	19947	7	7	8	7
D	14614	5	5	5	5
E	9225	4	3	3	3
F	3292	1	1	1	2
Total	100030	36	36	36	36
Divisor		—	2518	2659	3085

Divisors are not unique; answers in the following ranges are valid:

Jefferson	2517.9 – 2524.9
Webster	2651.1 – 2659.5
Adams	3075 – 3085.9

13. 10.8%.

15. a. Pennsylvania. b. 4. 97%.

17. a. Oregon's district size, 527,702.5, is the largest. South Dakota's district size is smallest, at 336,623.5. b. The absolute difference when Oregon has 4 seats and South Dakota has 2 is 191,079. If Oregon had 5 seats, and South Dakota had 1, the difference would be more: 251,085. c. The relative difference when Oregon has 4 seats and South Dakota has 2 is 57. Oregon had 5 seats, and South Dakota had 1, the relative difference would be more: 59%. d. South Dakotans have the largest share: 0.0000029707 seat. Oregonians have the smallest share: 0.0000018950 seat. e. The absolute difference when Oregon has 4 seats and South Dakota has 2 is 0.0000010757. If Oregon had 5 seats, and South Dakota had 1, the difference would be less: 0.0000008834. f. The relative difference when Oregon has 4 seats and South Dakota has 2 is 57%. If Oregon had 5 seats, and South Dakota had 1, the relative difference would be more: 59%.

19. a.

State	Population	Hamilton	Jefferson	Webster	Condorcet	Adams
National	92.15%	92	95	90	90	90
Splinter #1	1.59%	2	1	2	2	2
Splinter #2	1.58%	2	1	2	2	2
Splinter #3	1.57%	2	1	2	2	2
Splinter #4	1.56%	1	1	2	2	2
Splinter #5	1.55%	1	1	2	2	2
Total	100%	100	100	100	100	100
Divisor		—	0.97	1.02	1.02	1.03

b. Divisors are not unique; answers in the following ranges are valid:

Jefferson	0.960–0.970
Webster	1.019–1.029
Condorcet	1.020–1.030
Adams	1.024–1.035

c. All methods except the Hamilton method violate the quota condition in this apportionment problem.

d. With the Webster apportionment, each National Party legislator represents 1.024% of the voters; the splinter party legislators represent 0.795%, 0.79%, 0.785%, 0.78%, and 0.775%, respectively. The greatest difference in representation is 0.249%, favoring Splinter #5 over National. If one of the Splinter #5 seats were transferred to National, each National legislator would represent 1.013% of the voters, and the remaining Splinter #5 legislator would represent 1.55%. Now National would be favored by 0.537%, so the inequity would be increased by the transfer. The relative inequity before the transfer was 32%; after the transfer it would increase to 53%.

21. (1) 49; (2) 49; (3) Liberals, 48, Tories, 50, with a tie for the 99th seat.

23.

		Apportionments			
Class	Enrolled	Hamilton	Jefferson	Webster	Adams
Geometry	43	2	3	2	2
Algebra	42	2	2	2	2
Calculus	12	1	0	1	1
Total	97	5	5	5	5
Divisor	—		14.3	24	41

Divisors are not unique; answers in the following ranges are valid:

Jefferson	14.1–14.3
Webster	17.3–24
Adams	21.5–41.9

25.

		Apportionments			
Class	Enrolled	Hamilton	Jefferson	Webster	Adams
Geometry	67	4	4	4	3
Algebra	23	1	1	1	1
Calculus	5	0	0	0	1
Total	95	5	5	5	5
Divisor	—		16.75	19.14	33

Divisors are not unique; answers in the following ranges are valid:

Jefferson	13.5–16.7
Webster	15.4–19.1
Adams	23–33.4

27. Discussion. Answers will vary.

29. Discussion. Answers will vary.

31. The Hamilton method does minimize T. Suppose that $a_1, a_2, \ldots a_n$ differs from the Hamilton apportionment. Let $a_2', \ldots a_2'$ be the Hamilton apportionment, and assume that $a_1 > a_1'$. To maintain the house size, some other state must receive less than the Hamilton apportionment; we can assume that $a_2 < a_2'$.

We can then show that T will be reduced by transferring a seat from state 1 to state 2, and leaving the other apportionments unchanged. This will prove that the Hamilton apportionment minimizes T, by showing that T can always be decreased by transferral of a seat when the Hamilton apportionment is not used.

It is not hard to see that $a_1 > q_1$ and $a_2 < q_2$. If $a_1 \geq q_1 + 1$, then when a seat is taken from state 1, the quantity $|a_1 - q_1|$ decreases by 1. On the other hand, giving the seat to state 2 could conceivably *increase* the quantity $|a_2 - q_2|$. However, this increase would only occur if the original value of a_2 were equal to $\lfloor q_2 \rfloor$; and then the amount of increase would be

$$|(1 + \lfloor q_2 \rfloor) - q_2| - |\lfloor q_2 \rfloor - q_2| = 1 - 2(q_2 - \lfloor q_2 \rfloor)$$

which is *less than* 1. Therefore the total change in

$$\lfloor a_1 - q_1 \rfloor + \lfloor a_2 - q_2 \rfloor$$

is negative, and T is reduced.

A similar argument shows the transfer reduces T when $a_2 \leq q_2 - 1$. The remaining case to consider is when $a_1 - \lceil q_1 \rceil$ and $a_2 - \lfloor q_2 \rfloor$, while $a_1' = \lfloor q_1 \rfloor$ and $a_2' - \lceil q_2 \rceil$. Let t_1 and t_2 denote the fractional parts of q_1 and q_2, respectively. Then $t_2 > t_2$, since the Hamilton method gives state 1 its upper quota and state 2 its lower quota. Hence, before the transfer,

$$|a_1 - q_1| + |a_2 - q_2| = (1 - t_1) + t_2 > 1$$

It follows that the transfer decreases the value of T.

33. Discussion. Answers will vary.

35. a. 12,606. **b.** 8994.

37.

Enrollments			Sections		
Geom-etry	Alge-bra	Cal-culus	Geom-etry	Alge-bra	Cal-culus
43	42	12	2	2	1
55	25	20	3	1	1
67	23	5	3	1	1
76	19	20	3	1	1

39.

College	Dean	Hill
Arts	2	2
Science	1	1
Business	2	2

CHAPTER 15

1. a. and **b.** Saddle point at row 1, column 2, and value 5. **c.** Bad strategies: row 2, column 1.

3. a. No saddle point. **c.** No bad strategies.

5. a. No saddle point. **c.** None.

7. a. Saddle point at row 3, column 3, and value -20. **c.** Rows 1 and 2 and columns 1 and 2.

9. Batter's optimal mixed strategy is $(\frac{3}{4}, \frac{1}{4})$, pitcher's is $(\frac{1}{2}, \frac{1}{2})$, and value is .250.

11. Saddle point at knuckleball and .250.

13. Offense $(\frac{5}{8}, \frac{3}{8})$, defense $(\frac{3}{4}, \frac{1}{4})$, and value 0.575.

15. a.

		Officer does not patrol	Patrols
You park in	Street	0	-40
	Lot	-32	-16

b. You $(\frac{2}{7}, \frac{5}{7})$, officer $(\frac{3}{7}, \frac{4}{7})$, and value $-\$22.86$.
c. It is unlikely that the officer's payoffs are the (proportionally) opposite of yours. **d.** Use some available random device. Discuss.

17. Answers vary.

19. $\frac{1}{2}\begin{pmatrix} .4 \\ .3 \\ 0 \end{pmatrix} + \frac{1}{2}\begin{pmatrix} 0 \\ .2 \\ .4 \end{pmatrix} = \begin{pmatrix} .20 \\ .25 \\ .20 \end{pmatrix} \leq \begin{pmatrix} .2 \\ .4 \\ .3 \end{pmatrix}$

21. MD is best in Exercise 19.

23. From Exercise 20, $.160 < .240$.

25. Player II plays T and wins $\frac{1}{2}$ on average.

27. Player I plays $(0, \frac{1}{3}, \frac{2}{3})$, player II plays $(\frac{4}{5}, \frac{1}{5})$, and the value is $\frac{12}{5}$.

29. Saddle point at row 1, column 1, and value 6.

31. $(\frac{2}{3}, 0, \frac{1}{3})$, $(\frac{2}{3}, \frac{1}{3})$, $-(\frac{1}{3})$.

33. $(0, \frac{2}{3}, 0, 0, \frac{1}{3})$, $(\frac{1}{3}, \frac{2}{3})$, $\frac{5}{3}$.

35. Player I plays $(0, 1)$, Player II plays $(1 - p, p)$ where $\frac{1}{2} \le p \le 1$, and the value is 2.

37. $(0, \frac{1}{2}, \frac{1}{2})$ or $(\frac{1}{3}, 0, \frac{2}{3})$ or anything between these two, $(\frac{1}{2}, \frac{1}{2})$, and 0.

39. $(\frac{4}{5}, \frac{1}{5})$ and value \$420,000. See parts **a**, **b**, and **d** in Exercise 40 for alternate approaches.

41. Row 1, column 1 gives the equilibrium point $(5, 5)$, which is the overall best payoff. The point $(2, 2)$ is an inferior equilibrium point.

43. $(4,0)$, $(0,0)$, and $(0,4)$ are all in equilibrium. [It would be better if they could flip a coin to decide between $(4, 0)$ and $(0, 4)$.]

45. Answers vary.

47. **a.** Player 2 avoids "call" because "fold" dominates it. **b.** $(\frac{1}{3}, \frac{2}{3}, 0)$, $(\frac{2}{3}, 0, \frac{1}{3})$, and $-(\frac{1}{12})$. **c.** Player 2 since the value is nearest. **d.** Yes; Player I bets first while holding L with probability $\frac{2}{3}$. Player II raises while holding L with probability $\frac{1}{3}$.

CHAPTER 16

1. **a.** 1. **b.** 3; 9 times as large. **c.** 4; 24 sq in. **d.** 3.24. **e.** The 4-by-6 prints are almost twice as expensive per square inch of paper. **f.** 79 cents; \$1.46.

3. **a.** $\frac{1}{87} \approx 0.0115$. **b.** The volume of the real boxcar is $87^3 \approx 660,000$ times as large as the volume of the model. **c.** $\frac{1}{48}$.

5. **a.** Always. **b.** Sometimes. **c.** Always. **d.** Sometimes. **e.** Always. **f.** Sometimes. **g.** Sometimes (when the rectangle is also a square). **h.** Always. **i.** Never. **j.** Sometimes.

7. **a.** The new altar would have a volume 8 times as large—not "8 times greater than" or "8 times larger than," and definitely not twice as large, as the old altar. **b.** $\sqrt[3]{2} \approx 1.26$.

9. The writer of the ad meant that the volume was 2.5 times as much before packaging. So one finished bag has the same amount as 2.5 bags before compression. Since 2.5 bags have been compressed to one bag, the new volume is $1/2.5 = 0.4$ "times as much as" before. We could also correctly say that the peat moss has been compressed "to 40% of its original volume," that it has been "compressed by 60%," or that the compressed volume is "60% less than" the original volume.

11. 36 mpg.

13. **a.** 0.00013. **b.** We assume that all parts of the scale model are made of the same materials as the real locomotive. **c.** 0.27 lb. **d.** 0.12. **e.** 0.00012.

15. U.S. \$1.92/gal.

17. 185 meters, or 607 feet.

19. **a.** 900 lb per cu ft. **b.** Almost twice as dense. **c.** Since 230 lb of compost is supposed to add about 5%, the original should be about 230 lb divided by 0.05, or 4600 lb. So the revised quotation should say that the mineral soil weighs about 4500 lb.

23. 950,000 lb, or almost 500 tons.

25. $\sqrt{12} \times 20$ mph $= 69$ mph.

27. 9 ft 3 in to 11 ft 9 in (in modern times there have been men over 9 ft tall); 282 cm to 358 cm.

29. It has disproportionately large wings compared to geometric scaling up of a bird, hence lower wing loading and lower minimum flying speed. Also, in part it glides rather than flies.

31. It is the outside of the tree branches that the lights are strung around, so that in effect you are covering the outside "area" of the tree (thought of as a cone) with strings of lights. Hence, the number of strings needed grows in proportion to the square of the height: 30-ft tree will need $5^2 = 25$ times as many strings as a 6-ft tree. However, you could also argue that a 30-ft tree is meant to be viewed from farther away, so that the strings of lights would produce the same effect as on the shorter tree if they were strung farther apart, so you wouldn't need quite so many.

33. On log-log plot with calories/kilogram on the y axis and weight on the x axis, the middle four points lie fairly close to a straight line of slope -0.192 (found by fitting the least squares line). The guinea pig uses more calories per kilogram than that line would predict, and the whale uses fewer. The straight line on the log-log plot translates into the power relationship calories/kg $= 70(\text{weight})^{-0.192}$.

WP1. A human grows from a height (length) of between 1 and 2 ft to a height usually between 5 and 6 ft, hence by a scaling factor of between 2.5 and 6. Under proportional scaling, its weight would have to go up by the cube of the scaling factor, hence by a factor of between $2.5^3 = 15.6$ and $6^3 = 216$; so it would have a weight between $15.6 \times 10 = 156$ and $216 \times 5 = 1080$ pounds. But the vast majority of human adults weigh between 100 and 200 pounds.

WP3. a. Both the width and the height of the buildings are proportional to the cost, so that Warren, with a cost of about 1.5 times as much as South Mountain, has a building whose area on the page is about $(1.5)^2 = 2.25$ times as large, and whose implied volume is $(1.5)^3 = 3.4$ times as large. **b.** Simply monstrous! The line for 27.5 mpg, which is about $1\frac{1}{2}$ times as much as 18 mpg, is about 9 times as long as the line for 18 mpg. **c.** The picture shows the dollar bill shrinking in both length and width, even though the value shrinks only once. To use area to reflect the purchasing power of the dollar, the 1978 dollar should have about twice the area shown (and the other depictions also adjusted accordingly).

CHAPTER 17

1. a. $1080.00; 8.000%. **b.** $1080.00; 8.000%. **c.** $1082.43; 8.243%. **d.** $1083.29; 8.329%.

3. a. 2, 2.59, 2.705, 2.7169, 2.718280469. **b.** 3, 6.19, 7.245, 7.3743, 7.389041321. **c.** $e = 2.718281828 \ldots$; $e^2 = 7.389056098 \ldots$ Your calculator may give slightly different answers, because of its limited precision.

5. a. 698 days. **b.** After 24 months.

7. a. $95.24. **b.** $90.70.

9. 3.80 billion.

11. 3.72 billion.

13. The estimated population of Africa in 2025 is about 1800 million, which is 1300 more than Europe. Since $1300/500 = 2.60 = 260\%$, the correct answer would be "260% greater."

15. Equilibrium population size 25, maximum sustainable yield 7 for an initial population of 10.

17. a. 42 yr. **b.** 32 yr. **c.** Would tend to increase the indexes: more oil may be discovered; a low price for oil may increase the growth rate of consumption. Would tend to decrease the indexes: oil-exporting companies may reduce production, thereby raising prices and reducing consumption; societies may move more to other sources of energy.

19. $6806.56

21. $5712.40, assuming daily compounding.

23. a. $2032.79; $2025.82; $2012.20. **b.** $1999.00; $1992.56. **c.** $1973.82; $1906.62; $1849.60. **d.** For small and intermediate interest rates, the rule of 72 gives good approximations to the doubling time.

25. a. 50 yr. **b.** 41 yr.

27. a. $2,000,000 \times (1 + x + \cdots + x^{19})$, with $x = 1/1.03$, giving $30.6 million. **b.** $24.3 million. **c.** $19.9 million.

29. There were 120 payments of $100 each, for a total of $12,000. The equation resulting from the hint is $100x(x^{120} - 1)/(x - 1) = 37,747$. The solution is $x = 1.01650105$, for an annual interest rate of $12(0.01650105) = 0.1980 = 19.80\%$. The effective annual yield is $(1.01650105)^{12} - 1 = 0.2170$, or 21.70%.

31. a. 11,400 yr. **b.** 73,000 yr.

33. The emission rate is one-fourth of the rate for 28,650 years, so two more half-lives of 5730 years each must have elapsed, for a total of 40,000 years (rounded because 6.5 has only two digits of precision).

35. About 40,000 yr.

37. a. 17.2, 28.6, 44.9, 64.7, 83.0, 94.3, 98.6, 99.7, 99.9, 100.0. **b.** 101.2, 100.2, 100.0, 100.0 . . . In part **a** the population increases toward the carrying capacity; in part **b** it decreases toward the carrying capacity.   **c.** 26.2, 61.0, 103.8, 96.7, 102.5, 97.9, 101.6, 98.7, 101.0, 99.2, 90.2, 106.1, 94.4, 103.9, 96.6, 102.5, 97.9, 101.6, 98.7, 101.0. In both cases the population tends toward the carrying capacity but oscillates above and below it.   **d.** The bigger k is, the bigger the growth in the population; the closer the population to the carrying capacity, the smaller the growth. In fact, analysis using algebra can show that for $k \le 1$ and initial population P less than the carrying capacity, the population grows steadily toward the carrying capacity; for $k \le 1$ and initial population P greater than the carrying capacity, the population declines steadily toward the carrying capacity; for $k > 1$, the population oscillates (for fun, try $P = 50$ and $k = 3$) or worse (see what happens for $P = 50$ and $k = 4$).

39. a. $22.23.   **b.** $400.   **c.** $50.05.   **d.** 1970: $0.91; 1974: $2.00.

41. 1973: $41,200, $6,300, $31,700. 1979: $24,400, $17,400, $29,000. 1988: $40,400, $14,300, $29,700, $14,300, $5,900.

WP1. Answers will vary.

CHAPTER 18

1. No. Equiangular triangles have the same shape but not necessarily the same size.

3. 10 cm.

5. 24 ft.

7. 754 ft; 344 ft.

9. 40 ft.

11. 100 ft.

13. Yes. The triangles $\triangle ADE$ and $\triangle ABC$ are similar by ASA; note that $AE = EC$.

15. 32°.

17. 17°.

19. 46,250 km, for an error of $(46,250 - 39,375) = 6875$ km, which is 17.5% too much.

21. $\tan 0.25° = RM/ER$, so $RM = ER \tan 0.25° = 240,000 \text{ mi} \times 0.00436 = 1050$ mi.

23. 1400 ft.

25. 8.69 ± 0.09 light years.

27. Yes, SAS.

29. $\angle A$ and $\angle D$, $\angle ABC$ and $\angle DCB$, $\angle EBC$ and $\angle BCE$, $\angle BAC$ and $\angle CDB$, AB and DC, BC and BC, and AC and DB.

31. a. 500 miles is the same fraction of the earth's circumference of 25,000 miles — one-fiftieth — that $\angle AOB$ is of 360°, so the angle measures 7.2°.   **b.** 86.4°.   **c.** Because the tangents HA and HB make right angles with OA and OB, $\angle ABH$ and $\angle BAH$ each measure $90° - 86.4° = 3.6°$.   **d.** We add the measures of $\angle ABH$ and $\angle BAH$ to the measures of $\angle MAH$ and $\angle MBH$ to find the measures of $\angle MAB$ and $\angle MBA$, then subtract those values from 180° to get the measure of $\angle AMB$.   **e.** Think of the moon as the center of a circle of radius $MA = MB$. Then the distance from A to B along the arc of this circle is approximately the same as the distance along the circumference of the earth, or 500 mi. We have that 500 mi is to the circumference of this circle as the measure of $\angle AMB$ is to 360°, from which we can determine the circumference of this circle and hence its radius — the distance from the earth to the moon.

33. $VS = ES \times \sin 47° = 93,000,000$ mi $\times 0.73135 = 68$ million mi.

35. a. Let L be the distance to the moon and l be its radius. Then the circumference of the moon's orbit is $2\pi L = 360(4l) = 360(4)(1080)$, so $L = 248,000$ mi.   **b.** 42,524.05 min, of which 125 min is 1/340.19; 360 times this gives 1.06° for the angular measure of the diameter of the moon. Our new estimate of the distance L to the moon comes from $2\pi L = 340.19(4l) = 340.19(4)(1080)$, which gives 234,000 mi. (The distance of the moon varies between 221,463 and 238,857 mi; during the eclipse of January 4, 1992, the moon was near the farther distance, and the eclipse was not total but *annular*, meaning that the moon's disk did not completely cover the sun's.)

37. The length of the shadow must be measured from the center of the pyramid's base, which is inaccessible. Also, during the time it would take to measure such a long distance, the length of the shadow will have changed slightly. The Egyptians probably used rope to make measurements;

a problem suffered by nineteenth-century investigators who used metal tape measures or chains to try to make very accurate measurements is that their measurements were affected by expansion of the metal in the heat.

39. You need good observing conditions (weather, time of day) to be able to tell just when the eclipse begins and ends. (Also, to avoid damaging your eyes, you must not look directly toward the sun or a mirror reflection of it during a solar eclipse.) Because of the earth's atmosphere, the earth's shadow on the moon is not sharp, so it is difficult to tell exactly when a lunar eclipse starts or ends. Since the moon does not have an atmosphere, it is easier to see when the moon begins to block the sun during a solar eclipse.

CHAPTER 19

1. a. 32 ft. **b.** The stone hits the ground when $y = 0$, or $-16t^2 + 48t + 64 = -16(t^2 - 3t - 4) = -16(t - 4)(t + 1) = 0$. The only possibilities are $t = 4$ and $t = -1$. (If it had not been so easy to come up with the factors $t - 4$ and $t + 1$ by guesswork, we would have had to use the quadratic formula, a computer graphing program, or else trial-and-error approximation.) The stone hits the ground 4 seconds after it is thrown upward. **c.** The solution $t = -1$ in part **b** tells us that the stone would have the same trajectory (from the building height on) if it had been launched from ground level at $t = -1$. The symmetrical time between $t = -1$ and $t = 4$ is $t = 1.5$. Putting this value into the equation gives $y = 100$ ft.

3. Drawings will vary.

5. a. $(0, 0)$. **b.** $(0, \frac{1}{2})$. **c.** $x = 0$. **d.** -6 or 6. **e.** $\frac{2}{3}$.

9. $(x + 4)^2 + (y - 3)^2 = 9$.

11. $(x - 5)^2 + (y + 6)^2 = 80$.

13. The square root of the ratio of the masses, $\sqrt{330,000} = 574$, gives the ratio of the distances. So the object must be $\frac{1}{574}(92,956,000) = 162,000$ miles from the center of the sun (a location inside the sun).

15. $\dfrac{x^2}{5^2} + \dfrac{y^2}{4^2} = 1$.

17. We find $a + c = 252,710$ and $a - c = 221,643$, so $a = 237,177$, $c = 15,554$, and $b = 236,666$. The equation is $(x^2/237,177^2) + (y^2/236,666^2) = 1$.

19. The moon's orbit takes 39,343.2 minutes, while Sputnik's took 88, so Sputnik (which has long since reentered the atmosphere and burned up) took 447 times as long. Taking this ratio to the two-thirds power gives the ratio of their average distances from the earth: 58.5. So Sputnik was $\frac{1}{58.5}(238,857) = 4083$ miles from the center of the earth, or $4083 - 3963.5 = 120$ miles above the surface of the earth.

21. The moon takes 27.322 times as long to orbit. Taking this to the two-thirds power gives that the moon must be 9.0714 times as far away, so that the distance of the satellite is $\frac{238,857}{9.0714} = 26,331$ miles. Subtracting the radius of the earth, the satellite must be $26,331 - 3963.5 = 22,367$ miles above the surface of the earth.

27. Two equal times correspond to two line segments P_1P_2 and Q_1Q_2 of equal length. The triangles $\triangle SP_1P_2$ and $\triangle SQ_1Q_2$ have equal bases and equal altitudes, hence equal areas.

29. a. Diameter of the moon: 2000 mi. Circumference of the earth: 22,000 mi. Circumference of the moon: 6240 mi. **b.** 100 mi. **c.** Here BT is perpendicular to (forms a 90° angle with) CB, because a tangent to a circle is perpendicular to the radius of the circle through the point of tangency. **d.** Let r be the radius of the moon. Then $AT = CT - CA = CT - r = \sqrt{(CB)^2 + (BT)^2} - r = \sqrt{r^2 + (2r^2/20)} - r = r\sqrt{404/400} - r = 0.0050r = 5.0$ mi.

31. a. $p = 3$, $q = 3$. **b.** Cube: $pf = qV = 24$. Tetrahedron: $pf = qV = 12$. Octahedron: $pf = qV = 24$. **c.** Twice the number of edges. **d.** $V = 20$, $E = 30$. **e.** $V = 12$, $E = 30$.

CHAPTER 20

1. $m(\angle A) = 60°$, $m(\angle B) = 30°$, $m(\angle C) = 90°$.

3. $\triangle BAD$ is congruent to $\triangle ABC$ by SAS. Hence $BD = CA$. Then $\triangle ACD$ is congruent to $\triangle BDC$ by SSS. Hence $m(\angle C) = m(\angle D)$.

5. a. By **A**, there is at least one tree, and by **C**, that tree must belong to at least one row; so there is at least one row. **b.** By part **a**, there is at least one row; applying **E** to this row tells us that there is another row. **c.** By part **b**, there are at least two rows with no trees in common. By **B**, each row contains exactly two trees. So there are at least four trees.

7. a. By Exercise **5c**, there are at least four trees; call them T_1, T_2, T_3, and T_4. Suppose that there is a fifth tree, T_5. By **D**, T_5 and T_4 have a row in common. But neither the row containing T_1 and T_2 nor the row containing T_1 and T_3 has any tree in common with the row containing T_4 and T_5, which fact contradicts **E**. So there can be no fifth tree, and there must be exactly four trees. **b.** By **B**, each row contains exactly two trees. We can make exactly six pairs of trees from the four trees, so there are exactly six rows. **c.** Consider one of the four trees. By **D**, it has rows in common with each of the other three points; by **B**, these must be three different rows. If there were a fourth row through our given tree, by **B** it would have to contain one of the three other trees. But then, by **D**, it would be the same as one of the three rows we have already described. So each tree belongs to exactly three rows.

9. a. Let the students be A, B, C, and D. Then the committees are AB, AC, AD, BC, BD, and CD. **b.** Let the students be a, b, c, d, e, and f. One set of possible committees is abc, ade, bdf, and cef.

11. a. 4. **b.** Yes. **c.** Yes; with this interpretation, axiom **E** is the parallel postulate.

13. a. No; any member of L would (vacuously) contain all of the members of K, contradicting **C**. **b.** Yes, provided the member of L does not contain the member of K, so as not to contradict **C**.

15. a. By **F**, there are at least two members of K. By **A**, these two members of K are contained in a member of L. To avoid contradicting **C**, there must be yet another member in K. **b.** By **a**, K has at least three members; suppose there are four or more. Take two pairs of these members, with no member common to both pairs. By **A**, each pair of members of K is contained in a member of L. The two members of L corresponding to our two pairs have no member of K in common, contradicting **D**. (A variety of other proofs are possible, appealing to contradictions of other axioms.) **c.** 3.

17. a. By **E**, each two members of N have at least one member in common. If two members of N had two or more members of M in common, then **C** would be violated. **b.** By **A** and **B**, there are at least three members of M; call them a, b, and c. By **D**, there must be an additional member of M, call it d. By **C**, there must be three more distinct members of N containing d and a (by **B**, together with e, a new member of M), d and b (and f), and d and c (and g). By **C** and **E**, there must be three more members of N containing a, f, and g; c, f, and e; and e, b, and g.

19. a. Let the students be a, b, c, d, e, f, and g. One set of possible committees is abc, ade, afg, bdf, beg, cdg, and cef; other answers are possible.
b.

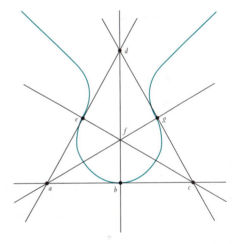

21. a. $(1, 0)$, $(1, 1)$, $(1, 2)$. **b.** $(0, 2)$, $(1, 2)$, $(2, 2)$. **c.** $(1, 0)$, $(0, 1)$, $(2, 2)$. **d.** $(1, 0)$, $(2, 1)$, $(0, 2)$.

23. a. 12. **b.** 3. **c.** 4.

25. a. $(0, 0)$. **b.** $(1, 0)$, $(0, 1)$, $(2, 0)$, and $(0, 2)$. **c.** $(1, 1)$, $(1, 2)$, $(2, 1)$, and $(2, 2)$.

27. a. a pecks b, who pecks c, who pecks a. **b.** The example in part **a** shows that a flock need not have an empress. **c.** Yes. Let a be a chicken who pecks the most other chickens (it's OK if a is tied with another chicken). Suppose a is not a queen. Then some other chicken b is not pecked by a and is not pecked by any chicken whom a pecks. So b must peck a as well as all of the chickens that a pecks — which is more chickens than a pecks, a contradiction to a pecking the most chickens. So a must be a queen. **d.** If a queen a is not an empress, then she is pecked by some other chicken. Consider all of the chickens who peck *a*; they are the chickens whom *a* does not peck directly but pecks indirectly. They form a flock in their own right, a subflock of the original flock. By part **c**, this subflock has its own queen, *b*. In fact, *b* is a queen of the original flock, as we now show: *b* pecks every chicken in the subflock, either directly or indirectly. The only other chickens in the original flock are the ones whom *a* pecks directly; since *b* pecks *a*, *b* indirectly pecks all of them, too. So *b* pecks, directly or indirectly, all chickens in the original flock. Hence, if there is a queen who is not an empress, there must be at least one other queen.

29. Yes. An example is the three-chicken flock of Exercise 27**a**, in which every chicken is a queen.

31. a. We can just turn every theorem about queens into one about serfs by changing the word "queen" to "serf" and "pecks" to "is pecked by." For example, every flock has a serf. The same applies to converting theorems about empresses into theorems about slaves. **b.** For any two chickens *a* and *b*, since *b* is a queen, either *b* pecks *a* directly or *b* pecks *a* indirectly. So *a* is a serf.

33. a. This is just a translation of **C**. **b.** By **A**, the number of members is odd, and by part **a**, the number of leads per member is 3. The total number of leads is just three times the number of members, which is an odd number. **c.** By **B**, each line produces two leads, so the total number of leads is twice the number of lines, hence even. **d.** We have proved two contradictory theorems! Our axiom system must be self-contradictory.

35. This theory has models of any odd size: Interpret the members as vertices of a polygon that has an odd number of vertices, and interpret the lines as the sides of the polygon.

WP1. By **A** and **B**, we can reverse any chain from *X* to *Y* to get one from *Y* to *X*, and the new chain will be negative or positive as the original is.

WP3. a. Suppose *a* and *b* are ambivalent toward each other. In any chain of involvements involving *a*, we can replace one occurrence of *a* by $a \heartsuit b \# a$, thereby producing a chain that is positive if the original was negative, and vice versa. **b.** Suppose we have a stable society. Let *a* be a person in the society, let $\mathcal{F}$ be the set of friends of *a* and $\mathcal{E}$ the set of enemies of *a*. Because the society is stable, $\mathcal{F}$ and $\mathcal{E}$ do not overlap. By definition of a society, everyone is related to *a* in some way, so everyone is in either $\mathcal{F}$ or $\mathcal{E}$. We need to show that the people in each of these sets are

all friends of each other and enemies of the people in the other set. If *b* and *c* are both in $\mathcal{F}$, then there is a positive chain from *b* to *a* and a positive chain from *a* to *c*. Joining the two chains gives a positive chain from *b* to *c*. The same idea can be used to show that any two people in $\mathcal{E}$ are friends and that any person in $\mathcal{E}$ is an enemy of any person in $\mathcal{F}$. To show the other half, that if the society divides as described, then it must be stable, we proceed by contradiction. Suppose that the society is not stable; then there are two people, *a* and *b*, who are ambivalent toward each other. Because they are friends, the two must both be in the same set; but because they are enemies, they must be in different sets, which is a contradiction, since the sets do not overlap. So the society must be stable.

CHAPTER 21

1. 5, 8, and 13.

3. c. $\phi^2 = \phi + 1$. **d.** $1/\phi = \phi - 1$.

5. a. 9. **b.** 16.

7. a. *c*5. **b.** *d*7. **c.** *c*22.

9. a. *c*6. **b.** *d*2. **c.** *c*16.

11. a. Vertical. **b.** Vertical and every multiple of 45°. **c.** Vertical and every multiple of 72°. **d.** Vertical and horizontal. **e.** None.

13. All are true.

15. a. B, C, D, E, H, I, K, O, X. **b.** A, H, I, M, O, T, U, V, W, X, Y. **c.** H, I, N, O, S, X, Z.

17. a. MOM, WOW (both either horizontally or vertically); MUd and bUM reflect into each other. **b.** pod rotates into itself; MOM and WOW rotate into each other. **c.** Here are some possibilities: NOW NO; SWIMS; ON MON; CHECK BOOK BOX; OX HIDE.

19. For all parts, translations. **a.** Reflection in vertical lines through the centers of the As or between them. **b.** Reflection in the horizontal midline. **c.** Reflection in the horizontal midline; reflections in vertical lines through the centers of the Xs or between them; 180° rotation around the centers of the Xs or the midpoints between them; glide reflections. **d.** None other than translations.

21. $p111, p1m1, pm11, p112, p1a1, pma2, pmm2$.

23. $pmm2, p1a1, pma2, p112, pmm2$ (perhaps), $p1m1$, $pma2, p111$.

25. a. 4, 7, 11, 18, 29, 47, 76, 123. **b.** 3, 1.333, 1.75, 1.571, 1.636, 1.611, 1.621, 1.617, 1.618. The ratios approach ϕ.

27. F_{61} and F_{62} both end in 1.

33. a. Reflection in a vertical line. **b.** Glide reflection. **c.**

d. cm.

35. Discussion. Answers will vary.

WP1. a. 1, 1, 3, 5, 11, 21, 43, 85, 171, 341, 683, 1,365. **b.** $B_n = B_{n-1} + 2B_{n-2}$. **c.** 1, 3, 1.667, 2.2, 1.909, 2.048, 1.977, 2.012, 1.994, 2.003, 1.999. **d.** $x = 2, -1$; we discard the -1 root.

e.
$$B_n = \frac{2^n - (-1)^n}{3}.$$

WP3. a. B_5 and B_6 both end in 1. **b.** $G_2 5$ and $G_2 6$ both end in 1. **c.** B_{22} and B_{23} end in 01 and 03, as B_1 and B_3 do.

CHAPTER 22

1. Exterior: $135°$. Interior: $45°$.

3. $180° - (360°/n)$.

5.

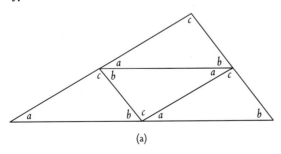

(a)

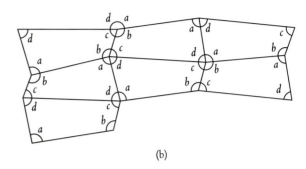

(b)

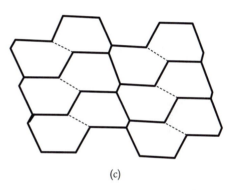

(c)

[From *Tiling the Plane*, by Frederick Barber et al.]

7. a. No. **b.** No. **c.** No.

9. a. Yes. **b.** No. **c.** No.

11.

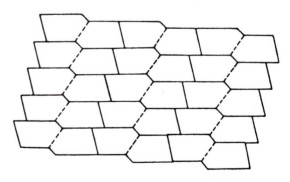

[From "In Praise of Amateurs," by Doris Schattschneider, in *The Mathematical Gardner*, edited by David A. Klarner.]

13. Answers vary.

15. Answers vary.

17. The usual notation for a vertex figure is to denote a regular n-gon by n, separate the sizes of polygons by periods, and list the polygons in clockwise order starting from the smallest number of sides, so that, e.g., 3.3.3.3.3.3 denotes six equilateral triangles meeting at a vertex. The possible vertex figures are 3.3.3.3.3.3, 3.3.3.3.6, 3.3.3.4.4, 3.3.4.3.4, 3.3.4.12, 3.4.3.12, 3.3.6.6, 3.6.3.6, 3.4.4.6, 3.4.6.4, 3.12.12, 4.4.4.4, 4.6.12, 4.8.8, 5.5.10, and 6.6.6.

19. 3.7.42, 3.9.18, 3.8.24, 3.10.15, and 4.5.20.

21. Answers will vary.

WP1. Let S_n, A_n, and B_n be the total number of symbols, the number of As, and the number of B's at the nth stage. We note that the only Bs at the nth stage must have come from As in the previous stage, so $B_n = A_{n-1}$. Similarly, the As at the nth stage come from both As and Bs in the previous stage, so $A_n = A_{n-1} + B_{n-1}$. Using both of these facts together, we have $A_n = A_{n-1} + A_{n-2}$. We note that $A_1 = 0, A_2 = 1, A_3 = 1, A_4 = 2, \ldots$. The A_n sequence obeys the same recurrence rule as the Fibonacci sequence and starts with the same values one step later; in fact, it is always just one step behind the Fibonacci sequence: $A_n = F_{n-1}$. Consequently, $B_n = A_{n-1} = F_{n-2}$, and $S_n = A_n + B_n = F_{n-1} + F_{n-2} = F_n$.

Index

▶▶▶▶▶▶▶▶▶▶▶▶▶